Controlled Atmosphere Storage of Fruit and Vegetables, 3rd Edition

Controlled Atmosphere Storage of Fruit and Vegetables, 3rd Edition

A. Keith Thompson, Robert K. Prange, Roger D. Bancroft
and Tongchai Puttongsiri

CABI is a trading name of CAB International

CABI
Nosworthy Way
Wallingford
Oxfordshire OX10 8DE
UK

Tel: +44 (0)1491 832111
Fax: +44 (0)1491 833508
E-mail: info@cabi.org
Website: www.cabi.org

CABI
745 Atlantic Avenue
8th Floor
Boston, MA 02111
USA

Tel: +1 (617)682-9015
E-mail: cabi-nao@cabi.org

A catalogue record for this book is available from the British Library, London, UK.

Library of Congress Cataloging-in-Publication Data

Names: Thompson, A. K. (A. Keith), author.
Title: Controlled atmosphere storage of fruit and vegetables / A.K. Thompson, R.K. Prange, R.D. Bancroft, Tongchai Puttongsiri.
Description: Third edition. | Boston, MA : CABI, 2018. | Includes bibliographical references and index.
Identifiers: LCCN 2018034923 (print) | LCCN 2018039620 (ebook) | ISBN 9781786393746 (ePub) | ISBN 9781786393753 (ePDF) | ISBN 9781786393739 (hardback : alk. paper)
Subjects: LCSH: Fruit--Storage. | Vegetables--Storage. | Protective atmospheres.
Classification: LCC SB360.5 (ebook) | LCC SB360.5 .T48 2018 (print) | DDC 634/.04--dc23
LC record available at https://lccn.loc.gov/2018034923

ISBN-13: 9781786393739 (Hardback)
9781786393746 (ePub)
9781786393753 (ePDF)

Commissioning Editors: Rachael Russell and Rebecca Stubbs
Editorial Assistants: Tabitha Jay and Alexandra Lainsbury
Production editor: Shankari Wilford

Typeset by SPi, Pondicherry, India
Printed and bound in the UK by Severn, Gloucester

Contents

About the Authors

A. Keith Thompson, PhD, is Visiting Professor at King Mongkut's Institute of Technology Ladkrabang in Thailand and was formerly: Professor of Plant Science, University of Asmara, Eritrea; Professor of Postharvest Technology and Head of Department, Cranfield University, UK; Team Leader, EU project at the Windward Islands Banana Development and Exporting Company; Principal Scientific Officer, Tropical Products Institute, London; Postharvest Expert for the UN in the Sudan, Yemen and Korea for the Food and Agriculture Organization, Ghana and Sri Lanka for the International Trade Centre and Gambia for the World Bank; Advisor to the British, Jamaican and Colombian Governments in postharvest technology of fruit and vegetables; Research Fellow in Crop Science, University of the West Indies, Trinidad; Demonstrator in Biometrics at University of Leeds; as well as Consultant for various commercial and Government organizations throughout the world and has published over 100 scientific articles and many scientific textbooks.

Robert K. Prange, PhD, is an Adjunct Professor in the Faculty of Graduate Studies and Research, Dalhousie University, Halifax, Nova Scotia, Canada and was formerly: Senior Research Scientist, Postharvest Biology and Technology, Agriculture and Agri-Food Canada; and Professor and Head, Department of Plant Science, Faculty of Agriculture, Dalhousie University. He has published over 100 scientific articles, primarily in the field of postharvest biology and technology with a particular interest in Controlled Atmosphere applications in fruits and vegetables. In recognition of his research, and numerous innovations adopted by the postharvest industry around the world, he has received a number of awards including Fellow of the American Society for Horticultural Science, Federal Partners in Technology Transfer Award and the Queen Elizabeth II Diamond Jubilee Medal for contributions to Canada.

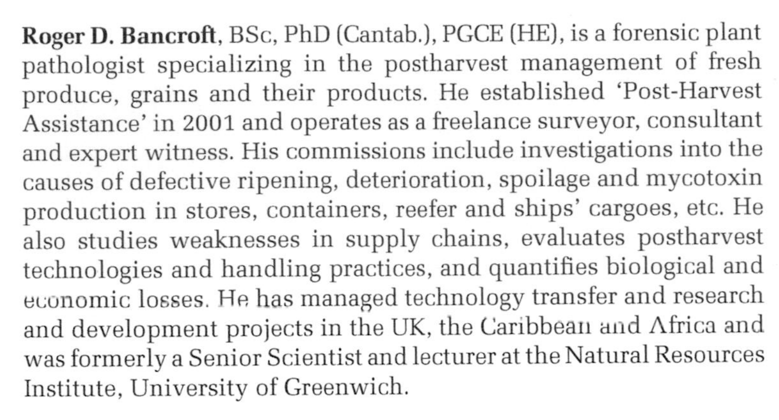

Roger D. Bancroft, BSc, PhD (Cantab.), PGCE (HE), is a forensic plant pathologist specializing in the postharvest management of fresh produce, grains and their products. He established 'Post-Harvest Assistance' in 2001 and operates as a freelance surveyor, consultant and expert witness. His commissions include investigations into the causes of defective ripening, deterioration, spoilage and mycotoxin production in stores, containers, reefer and ships' cargoes, etc. He also studies weaknesses in supply chains, evaluates postharvest technologies and handling practices, and quantifies biological and economic losses. He has managed technology transfer and research and development projects in the UK, the Caribbean and Africa and was formerly a Senior Scientist and lecturer at the Natural Resources Institute, University of Greenwich.

Tongchai Puttongsiri, PhD, is Assistant Professor and Associate Dean in the Faculty of Agro-Industry at King Mongkut's Institute of Technology Ladkrabang in Thailand. He teaches quality control of food products, food processing, design of experiments and model building. His research interests include postharvest and food preservation technology with chitosan as an edible coating to increase shelf life and quality and control of diseases of tangerines and mangoes.

Preface

The name 'controlled atmosphere storage' replaced the archaic term 'gas storage' and is synonymous with the term 'modified atmosphere storage', though the latter is little used and usually only specifically in modified atmosphere packaging. Modified atmosphere packaging refers to using packaging material (usually plastic film) that is partially permeable to gases and that uses the metabolism of the product it contains to change the atmosphere to one that is more beneficial for storage. Another way of changing the atmosphere is to coat the individual fruit or vegetable in such a way that gas exchange is restricted, resulting in a more beneficial atmosphere in the cells and intercellular spaces.

For almost a century an enormous volume of literature has been published on the subject of controlled atmosphere storage of fresh fruits and vegetables. It would be the work of a lifetime to begin to do those results justice in presenting a comprehensive and focused view, interpretation and digest for its application in commercial practice. Such a review is in demand to enable those engaged in the commerce of fruits and vegetables to be able to utilize this technology and reap its benefits in terms of the reduction of postharvest losses and maintenance of their nutritive value and organoleptic characteristics. The potential use of controlled atmosphere storage as an alternative to the application of preservation and pesticide chemicals is of continuing interest.

In order to facilitate the task of reviewing the literature we have had to rely on a combination of reviewing original publications as well as consulting reviews and technical books. The latter are not always entirely satisfactory, since they may not give their source of information and we may have inadvertently quoted the same work more than once. Much reliance has been made on conference proceedings, especially the International Controlled Atmosphere Research Conference held every few years, the European Co-operation in the Field of Scientific and Technical Research (COST 94) which held postharvest meetings throughout Europe between 1992 and 1995, the International Society for Horticultural Science's regular international conferences and the CAB Abstracts.

Different views exist on the usefulness of controlled atmosphere storage. Blythman (1996) described controlled atmosphere storage as a system that 'amounts to deception' from the consumer's point of view. The reason behind this assertion seems to be that the consumer thinks that the fruits and vegetables that they purchase are fresh and that controlled atmosphere storage technology 'bestows a counterfeit freshness'. Also she claimed that storage may change produce in a detrimental way and cited changes in texture of apples, 'potatoes that seem watery and fall apart when cooked and bananas that have no flavour'. Some of these contentions are true and need addressing, but others are oversimplifications of the facts. Another view was expressed by David Sainsbury in 1995 and reported in the press as: 'These techniques [controlled atmosphere storage], could halve the cost of fruit to the customer. It also extends the season of availability, making good eating-quality fruit available for extended periods at reasonable costs.'

The purpose of this book is primarily to help the fresh produce industry in storage and transport of fruit and vegetables, but it also provides an easily accessible reference source for those studying agriculture, horticulture, food science and technology and food marketing. In addition it will be useful to researchers in this area, giving an overview of our present knowledge of controlled atmosphere storage that will indicate areas where there is a need for further research.

Acknowledgements

Dr Graeme Hobson for an excellent review and correction of the first edition and Pam Cooke and Lou Ellis for help in typing the first edition. Dr John Stow, Alex van Schaik, Dr John Faragher, Dr Errol Reid and David Johnson for useful references, advice and comments on parts of the manuscript. Dr Rob Veltman, Managing Director, Fotein and Van Amerongen BV, for allowing the use of his unpublished paper on DCA storage and supplying photographs. Eric van der Zwet of Besseling Group BV for supplying text and photographs of equipment. Allan Hilton, Dr R.O. Sharples, Dr Nettra Somboonkaew, Dr Devon Zagory, the late Tim Bach of Cronos Controlled Atmosphere Systems, UK, David Bishop of Storage Control Systems Ltd, UK, Dave Rodden of Advanced Ripening Technologies Ltd, UK, Ramzi Amari of Van Amerongen CA Technology BV, The Netherlands, Greg Akins of Catalytic Generators Inc., Norfolk, Virginia, USA, and Its Fresh of Cranfield, UK, for supplying and giving permission to use photographs.

Glossary

Active packaging usually involves the inclusion of a desiccant or O_2 absorber within the package or as part of the packaging material.

Controlled atmosphere (CA) storage has been defined (Bishop, 1996) as 'A low O_2 and/or high CO_2 atmosphere created by natural respiration or artificial means ... controlled by a sequence of measurements and corrections throughout the storage period'.

Dynamic controlled atmosphere (DCA) storage has been defined as where the gas mixture will constantly change due to metabolic activity of the respiring fruit or vegetables in the store and leakage of gases through doors and walls.

Equilibrium modified atmosphere (EMA) packaging uses films that are highly permeable for O_2 and CO_2 to achieve an atmosphere inside the packages that remains constant during the marketing chain (Jacxsens *et al.*, 2002).

Humidity is referred to as percentage relative humidity (RH) (humidity relative to that which is saturated) but can also be referred to as **vapour pressure deficit**, which refers to the gaseous water in the atmosphere in relation to its maximum capacity at a given temperature.

Modified atmosphere (MA) packaging is where the fruit or vegetable is enclosed within sealed plastic film that is slowly permeable to the respiratory gases. The gases will change within the package, thus producing lower concentrations of O_2 and higher concentrations of CO_2 than exist in fresh air.

Permselectivity, used in modified atmosphere packaging, is the ratio of CO_2:O_2 permeation coefficients.

Scrubbing is the selective removal of gases from the atmosphere by adsorption or absorption. In some cases this is referred to as product generated controlled atmosphere or injected controlled atmosphere.

Shrink-films are plastic films usually made from polyolefin (PO), but polyethylene (PE) and polyvinyl chloride (PVC) are also used. The film should have a low permeability to water vapour and O_2 but high permeability to CO_2. It should be thin (about 17–19 µm thick) but with a high tensile strength and should be transparent. The film is placed over the product, which is then passed through a tunnel where heated air is blown over it, causing the film to shrink (i.e. the product is shrink-wrapped). The temperature and residence time in the tunnel are developed by experience, but should not be more than 180–210°C for individual wraps and 140–145°C for tray wraps. The residence time is usually 15–30 seconds.

Static CA is where levels of O_2 and CO_2 are set, constantly monitored and adjusted. It has been described as 'static CA storage' and 'flushed CA storage', to define the two most commonly used systems (D.S. Johnson, personal communication). 'Static' is where the product generates the atmosphere; 'flushed' is where the atmosphere is supplied from a flowing gas stream that purges the store continuously. Systems may be designed that utilize flushing initially to reduce the O_2 content and then either inject CO_2 or allow it to build up through respiration, after which the maintenance of this atmosphere is by ventilation and scrubbing. The gases are measured periodically and adjusted to the predetermined level by the introduction of fresh air or N_2 or by passing the store atmosphere through a chemical to remove CO_2.

Ultra-low oxygen (ULO) storage was first defined as O_2 levels of 2 kPa or less. It is now more commonly referred to as O_2 levels of less than 1 kPa.

Vacuum packing uses a range of low-permeable or non-permeable films (barrier films) or containers into which the fresh fruit or vegetable is placed and the air is sucked out.

Units of Measurement

Pressure

In **hypobaric and hyperbaric storage** of fruit and vegetables, different publications refer to the pressure data using different units. It would be cumbersome to give SI equivalents to all the examples given in this book, as would be normal practice. For guidance, these units might include millimetres of mercury (**mm Hg**) where atmospheric pressure at sea level is 760 mm Hg, but this is not an SI unit. Pressure can also be referred to in terms of a standard atmosphere (**atm**) and reduced pressures can be given as a fraction of atmosphere or of normal pressure. The most commonly used SI unit in hypobaric storage and in CA storage is the **pascal** (Pa), usually as kilopascals (**kPa**, i.e. 1000 Pa) and sometimes in hyperbaric storage as megapascals (**MPa**, i.e. 1000 kPa). Another SI unit is the **Newton** (**N**), a unit of force which, acting on 1 kg of mass, increases its velocity by 1 metre per second squared (1 m/s^2). By definition, 1 Pa = 1 N/m^2. Very occasionally measurements are given in pounds per square inch (**psi**), where 1 psi = 6.895 kPa or 6895 N/m^2. Note: 1 mm Hg = 0.133 kPa; 1 atm = 101.325 kPa = 760 mm HG or 14.7 psi.

In **controlled atmosphere storage**, the level of a gas in the store is measured as partial pressure in kPa or as a **percentage** (%), the latter being the proportion of the atmosphere that contains the gas. In this edition, the gas measurements are usually in kPa; 1% is approximately the equivalent of 1 kPa.

Volume

Volumes in this edition are measured on the basis of litres (**l**), including millilitres (**ml**) and microlitres (**µl**). There are also a few examples of parts per million (**ppm**), based on mass for mass, or volume for volume.

Some American data are given in **bushels** (e.g. in Chapter 1). By volume, 1 US bushel = 35.24 litres. By weight, the bushel varies according to the commodity being measured; for example, for oats 1 US bushel = 14.52 kg. For fruit and vegetables the equivalents also vary from State to State, though for apples it is typically 40 lb (18.15 kg) per bushel. In general, weights are given in metric tonnes (**t**), i.e. 1000 kg.

Abbreviations

1-MCP	1-methylcyclopropene
AA	ascorbic acid
ACC	1-aminocyclopropane 1-carboxylic acid
	ACC oxidase = an enzyme involved in biosynthesis of ethylene
	ACC synthase = an enzyme involved in biosynthesis of ethylene
	ACC synthesis = the reaction catalysed by ACC synthase
ACP	anaerobic compensation point
ACPd	granular activated carbon impregnated with palladium
ACR	advanced control of respiration
atm	atmosphere
ATP	adenosine triphosphate
AVG	aminoethoxyvinylglycine
BBD	Braeburn browning disorder
benomyl	methyl-1-[butylcarbamoyl] benzimidazol-1-yl carbamate = Benlate
BOP	bi-axially oriented polyester
BOPP	bi-oriented polypropylene
CA	controlled atmosphere
CA	(plastic film) = cellulose acetate
CF	chlorophyll fluorescence
CFC	chlorofluorocarbon
CFU	colony-forming units
CMC	carboxymethyl cellulose
CO_2e	carbon dioxide equivalent
COPP	coextruded oriented polypropylene
CPP	cast polypropylene
DAT	delivery air temperature
DCA	dynamic controlled atmosphere
DCR	dynamic control of respiration
DCS	dynamic control system
DLOCA	dynamic low oxygen controlled atmosphere
DPA	diphenylamine
DPPH	1,1-diphenyl-2-picrylhydrazyl
EMA	equilibrium modified atmosphere

EMR	East Malling Research
EMRS	East Malling Research Station
EP	ethanol production
EU	European Union
EVA	ethylene-vinyl acetate
EVAL	ethylene-vinyl alcohol copolymer; also trade name for EVOH resin
EVOH	ethylene-vinyl alcohol copolymer
FCA	fluorescence controlled atmosphere
GA	gibberellins
GA_3	gibberellic acid
HDPE	high-density polyethylene
HOA	high-oxygen atmosphere
ILOS	initial low oxygen stress
Imazalil	1-[allyloxy-2,4-dichlorophenethyl] imidazole
ISHS	International Society for Horticultural Science
kPa	kilopascal
l	litre
LDPE	low-density polyethylene
LLDPE	linear low-density polyethylene
LOL	lower oxygen limit
LPE	lysophosphatidylethanolamine
MA	modified atmosphere
MAP	modified atmosphere packaging
MB	methyl bromide
MDA	malondialdehyde
MDPE	medium-density polyethylene
MENA	methyl ester of naphthalene acetic acid
MH	maleic hydrazide
MHO	6-methyl-5-hepten-2-one
MIP	modified interactive packaging
MJ	methyl jasmonate
mRNA	messenger ribonucleic acid
NAA	naphthaleneacetic acid
NO	nitric oxide
NRI	Natural Resources Institute
OPP	oriented polypropylene
PA	polyamide, also called nylon
PAL	phenylalanine ammonia-lyase
PB	polybutylene
PBS	polybutylene succinate
PCR-PET	post-consumer recycled PET
PE	polyethylene
PET	polyethylene terephthalate
PFM	pulse frequency modulated
PG	polygalacturonase
PLA	polylactic acid
PME	pectin methylesterase
PO	polyolefin
POD	pyrogallol peroxidase
PP	polypropylene
PPO	polyphenol oxidase
PQC	Produce Quality Centre

PR pathogenesis-related
PRSV papaya ring spot virus
PS polystyrene
PSA pressure swing adsorption
PU polyurethane
PVB polyvinyl butyral
PVC polyvinyl chloride
PVDC polyvinylidene chloride
QR quinone reductase
RA refrigerated air
RAT return air temperature
RFID radio frequency identification
RH relative humidity
RNA ribonucleic acid
RQ respiratory quotient
RQB respiratory quotient breakpoint
SAM *S*-adenosyl-methionine
SAT supply air temperature
SCA standard controlled atmosphere
SOD superoxide dismutase
TA total acidity
TDCA two-dimensional dynamic controlled atmosphere storage
TBZ thiabendazole = 2-thiazol-4-yl benzimidazole
TPI Tropical Products Institute
TSS total soluble solids
ULO ultra-low oxygen
USDA US Department of Agriculture
UV ultraviolet
VLDPE very low-density PE
WTO World Trade Organization
XLO extreme low oxygen

1
Introduction

The maintenance or improvement of the postharvest life and quality of fresh fruit and vegetables is becoming increasingly important. This has been partly as a response to a free market situation where the supply of good quality fruit and vegetables constantly exceed demand. Therefore to maintain or increase market share there is increasing emphasis on quality. Also consumer expectation in the supply of all types of fresh fruit and vegetables throughout the year is often taken for granted. This latter expectation is partly supplied by long-distance transport but also long-term storage of many crops. With growing awareness and concern for climate change, long-distance transport of fruit and vegetables is being questioned.

Controlled atmosphere (CA) storage has been shown to be a technology that can contribute to these consumer requirements in that in certain circumstances, with certain cultivars of crop and appropriate treatments, the marketable life can be greatly extended. An enormous amount of interest and research has been reported on CA storage and modified atmosphere (MA) packaging of fruit and vegetables to prolong their availability and retain their quality for longer. This book seeks to evaluate the history and current technology reported and used in CA storage and MA packaging and its applicability and restrictions for the use in a variety of crops in different situations. While it is not exhaustive in reviewing the enormous quantity of science and technology that has been developed and published on the subject, it will provide an access into CA and MA for those applying the technology in commercial situations. The book can also be used as a basis for determination of researchable issues in the whole area of CA storage and MA packaging.

The scientific basis for the application of CA technology to the storage of fresh fruit and vegetables has been the subject of considerable research, which seems to be progressively increasing. Some of the science on which it is based has been known for over 200 years but was refined and applied commercially for the first time in the first half of the 20th century.

History of the Effects of Gases on Crops

The effects of gases on harvested crops have been known for centuries. In eastern countries some types of fruit were taken to temples, where incense was burned, to improve ripening. Bishop (1996) indicated that there was evidence that Egyptians and Samarians used sealed limestone crypts for crop storage in the second century BCE. He also quoted from the Bible suggesting that the technology might have been used in Old Testament

Egypt when Joseph prevented famine by storing grain for 7 years. Dilley (1990) mentioned the storage of fresh fruit and vegetables in tombs and crypts. This was combined with the gas-tight construction of the inner vault so that the fruit and vegetables would consume the oxygen (O_2) and thus help to preserve the crops. An interpretation of this practice would indicate that knowledge of the respiration of fruit pre-dates the work described in the 19th century (Dalrymple, 1967). Wang (1990) quotes a Tang dynasty 8th century poem which described how litchis were shown to keep better during long-distance transport when they were sealed in the hollow centres of bamboo stems with some fresh leaves. Burying fruit and vegetables in the ground to preserve them is a centuries-old practice (Dilley, 1990). In Britain crops were stored in pits, which would have restricted ventilation and may have improved their storage life. Currently CA research and commercial CA storage is used in many countries, some of which are described in this chapter.

France

The earliest documented scientific study of CA storage was by Jacques Etienne Berard at the University of Montpellier in 1819 (Berard, 1821), who found that harvested fruit absorbed O_2 and gave out carbon dioxide (CO_2). He also showed that certain types of fruit stored in atmospheres containing no O_2 did not ripen, but if they were held for only a short period and then placed in air they continued to ripen. These experiments showed that storage in zero O_2 gave a shelf-life of about 1 month for peaches, prunes and apricots and about 3 months for apples and pears. Zero O_2 was achieved by placing a paste composed of water, lime and iron sulfate in a sealed jar which, as Dalrymple (1967) pointed out, would also have absorbed CO_2. Considerable CA research has been carried out over the intervening period in France. Berard's method of achieving an atmosphere that had no O_2 and likely had CO_2 absorbed as well is still in use today. There are commercial sources of 'oxygen absorbers' available on e-commerce sites that will absorb virtually all O_2 in a sealed container. The product consists of food-safe sachets containing iron powder which consumes O_2 to form iron oxides. At least one of the products contains activated carbon, which can serve to absorb CO_2 (though the supplier does not make this claim) (Robert Prange, personal communication).

USA

In 1856 Benjamin Nyce built a commercial cold store in Cleveland, Ohio, using ice to keep it below 34°F. In the 1860s he experimented with modifying the CO_2 and O_2 in the store by making it airtight. This was achieved by lining the store with casings made from iron sheets, thickly painting the edges of the metal and having tightly fitted doors. It was claimed that 4000 bushels (about 72.6 t; see Glossary) of apples were kept in good condition in the store for 11 months. However, he mentioned that some fruits were injured in a way that Dalrymple (1967) interpreted as possibly being CO_2 injury. The carbonic acid level was so high in the store (or the O_2 level was so low) that a flame would not burn. He also used calcium chloride to control the moisture level in the mistaken belief that low humidity was necessary.

Dalrymple (1967) stated that R.W. Thatcher and N.O. Booth working in Washington State University around 1903 studied fruit storage in jars containing different gases. They found that 'the apples which had been in CO_2 were firm of flesh, possessed the characteristic apple colour, although the gas in the jar had a slight odour of fermented apple juice, and were not noticeably injured in flavour'. The apples stored in hydrogen, nitrogen (N_2), O_2 and sulfur dioxide (SO_2) did not fare so well. They subsequently studied the effects of CO_2 on raspberries, blackberries and loganberries and 'found that berries which softened in three days in air would remain firm for from 7 to 10 days in CO_2.

Fulton (1907) observed that fruit could be damaged where large amounts of CO_2

were present in the store, but strawberries were 'damaged little, if any ... by the presence of a small amount of CO_2 in the air of the storage room'. Thatcher (1915) published a paper describing work in which he experimented with apples sealed in boxes containing different levels of gases and concluded that CO_2 greatly inhibited ripening.

G.R. Hill Jr reported work carried out at Cornell University in 1913 in which the firmness of peaches had been retained by storage in inert gases or CO_2. He also observed that the respiration rate of the fruit was reduced and did not return to normal for a few days when removed from storage in a CO_2 atmosphere to air. C. Brooks and J.S. Cooley, working for the US Department of Agriculture, stored apples in sealed containers in which the air was replaced three times each week with air plus 5 kPa CO_2. After 5 weeks storage they noted that the apples were green, firm and crisp, but were also slightly alcoholic and had 'a rigor or an inactive condition from which they do not entirely recover' (Brooks and Cooley, 1917). J.R. Magness and H.C. Diehl in 1924 described a relationship between apple softening and CO_2 concentration in that an atmosphere containing 5 kPa CO_2 slowed the rate of softening with a greater effect at higher concentrations, but at 20 kPa CO_2 the flavour was impaired. Work on CA storage that had been carried out at the University of California at Davis was reported by Overholser (1928). This work included a general review and some preliminary results on Fuerte avocados. In 1930 Overholser left the University and was replaced by F.W. Allen, who had been working on storage and transport of fresh fruit in artificial atmospheres. Allen began work on CA storage of 'Yellow Newtown' apples. 'Yellow Newtown', like 'Cox's Orange Pippin' and 'Bramley's Seedling' grown in England, was subject to low-temperature injury at temperatures higher than 0°C. These experiments (Allen and McKinnon, 1935) led to a successful commercial trial on Yellow Newtown apples in 1933 at the National Ice and Cold Storage Company in Watsonville. Thornton (1930) carried out trials where the concentration of CO_2 tolerated by selected fruit, vegetables and flowers was examined at six temperatures over the range of 0–25°C. To illustrate the commercial importance of this type of experiment the project was financed by the Dry Ice Corporation of America.

From 1935 Robert M. Smock worked in the University of California at Davis on apples, pears, plums and peaches (Allen and Smock, 1938). In 1936 and 1937 F.W. Allen spent some time with Franklin Kidd and Cyril West at the Ditton laboratory in England and then continued his work at Davis, while in 1937 Smock moved to Cornell University. Smock and his PhD student, Archie Van Doren, conducted CA storage research on apples, pears and stone fruit (Smock, 1938; Smock and Van-Doren, 1938, 1939). New England farmers in the USA were growing a number of apple cultivars, particularly 'McIntosh'. 'McIntosh' is subject to chilling injury and cannot be stored at or below 0°C. It was thought that if the respiration rate could be slowed, storage life could be extended, lengthening the marketing period, and CA storage was investigated to address this problem (Smock and Van-Doren, 1938). Sharples (1989a) credited Smock with the 'birth of CA storage technology to North America'. In fact it was apparently Smock who coined the term 'controlled atmosphere storage' as he felt it better described the technology than the term 'gas storage' which was used previously by Kidd and West. Smock and Van-Doren (1941) stated: 'There are a number of objections to the use of the term *gas storage* as the procedure is called by the English. The term *controlled atmosphere* has been substituted since control of the various constituents on the atmosphere is the predominant feature of this technique. A substitute term or synonym is *modified atmosphere.*' The term 'controlled atmosphere storage' was not adopted in Britain until 1960 (Fidler *et al.*, 1973). Smock also spent time with Kidd and West at the Ditton laboratory. The CA storage work at Cornell included strawberries and cherries (Van Doren *et al.*, 1941). A detailed report of the findings of the Cornell group was presented in a comprehensive bulletin (Smock and Van Doren, 1941) which gave the results of research on atmospheres, temperatures and varietal response of fruit as well as store construction and operation. In addition to the research efforts in Davis and Cornell in the 1930s, several other groups in the USA were carrying out research into CA storage. CA storage research for a variety of fruit and vegetables was described by Miller and Brooks (1932) and Miller and Dowd (1936). Work on apples was described by Fisher (1939) and Massachusetts Agricultural Experiment Station (1941), work on citrus fruit by Stahl and Cain (1937) and Samisch (1937) and work on cranberries by Massachusetts Agricultural Experiment Station (1941). Smock's research in New York State University was facilitated in 1953 with the completion of large new storage facilities designed specifically to accommodate studies on CA storage. During the intervening years the work at Davis has been prolific, with an enormous number of publications (many are referred to in subsequent chapters), and the campus has become a widely recognized world leader in CA science and technology.

Commercial CA storage of apples in the USA began in New York State with 'McIntosh'. The first three CA rooms with a total capacity of 24,000 bushels were put into operation in 1940 with Smock and Van Doren acting as consultants. This had been increased to 100,000 bushels by 1949, but the real expansion in the USA began in the early 1950s. In addition to a pronounced growth in commercial operation in New York State, CA stores were constructed in New England in 1951, in Michigan and New Jersey in 1956, in Washington, California and Oregon in 1958, and in Virginia in 1959 (Dalrymple, 1967). According to Love (1988) the delayed adoption in Washington State in the 1940s and early 1950s may have been due to the high cost of CA installation and management coupled with a certain scepticism among USDA postharvest researchers. But after 1957, the Washington apple industry increased CA technology at a rapid rate. Washington's initial reluctance to adopt CA technology changed with new economic developments and new USDA scientific leadership in postharvest horticulture. A CA store for 'Red Delicious' was set up in Washington State in the late 1950s in a Mylar tent where some 1000 bushels were stored with good results. By the 1955–1956 season the total CA storage holdings had grown to about 814,000 bushels, some 684,000 bushels in New York State and the rest in New England. In the spring of that year Dalrymple did a study of the industry in New York State and found that typically CA stores were owned by large and successful fruit farmers (Dalrymple, 1967). The total CA storage holdings per farm were large, averaging 31,200 bushels and ranging from 7500 to 65,000 bushels. An average storage room held some 10,800 bushels. A little over three-quarters of the capacity represented new construction while the other quarter was remodelled from refrigerated stores. About 68% of the capacity was rented out to other farmers or speculators. In 2004 it was reported that some 75% of cold stores in USA had CA facilities (DGCL, 2004). On 1 December 2016, the amount of the apple crop remaining in storage in the USA was about 155 million bushels, representing 62% of the total crop. About 79% of these holdings were in CA storage (AAFC, 2016). Washington (WA) is the largest apple-producing state in the USA with a crop of c. 7,300,000 bins (one WA bin holds about 400–450 kg) in 2017 (Robert Blakey, personal communication). In a survey of packinghouses that represents some 50% of the crop, about 87.5% of the crop was stored in CA. Many of the respondents indicated that they planned to increase their CA capacity in the next 2 years and, based on their planned construction, CA capacity will increase to 88.9%

UK

Sharples (1989a) in his review in *Classical Papers in Horticultural Science* stated that '[Franklin] Kidd and [Cyril] West can be described as the founders of modern CA storage'. Sharples described the background to their work and how it came about. Dalrymple (1967) in reviewing early work on the effects of gases on postharvest of fruit and vegetables stated: 'The real start of CA storage had to await the later work of two British scientists [Kidd and West], who started from quite a different vantage point.'

During the First World War the British Government was concerned about food shortages. It was decided that one of the methods of addressing the problem should be through research, and the Food Investigation Organisation was formed at Cambridge in 1917 under the direction of W.B. Hardy, who was later to be knighted and awarded the fellowship of the Royal Society (Sharples, 1989a). In 1918 the work being carried out at Cambridge was described as 'a study of the normal physiology, at low temperatures, of those parts of plants which are used as food. The influence of the surrounding atmosphere, of its content of O_2, CO_2 and water vapour was the obvious point to begin at, and such work has been taken up by Dr. F. Kidd. The composition of the air in fruit stores has been suspected of being of importance and this calls for thorough elucidation. Interesting results in stopping sprouting of potatoes have been obtained, and a number of data with various fruits proving the importance of the composition of the air' (Food Investigation Board, 1919).

One problem, which was identified by the Food Investigation Organisation, was the high levels of wastage that occurred during the storage of apples. Kidd and West were working at that time at the Botany School in the University of Cambridge on the effects of CO_2 levels in the atmosphere on seeds (Kidd and West, 1917a, b) and Kidd was also working on the effects of CO_2 and O_2 on sprouting of potatoes (Kidd, 1919). Kidd and West transferred to the Low Temperature Laboratory for Research in Biochemistry and Biophysics (later called the Low Temperature Research Station) at Cambridge in 1918 and conducted experiments on what they termed 'gas storage' of apples (Sharples, 1986). The Low Temperature Research Station had a wide remit, including research problems basic to the preservation, storage and transport of perishable foodstuffs. It went on to focus on meat, eggs and poultry. With the assistance of the Empire Marketing Board the station was later enlarged and supplemented by the Ditton Laboratory at East Malling, which was set up in 1928 and focused on fruit and vegetables. By 1920 Kidd and West were able to set up semi-commercial trials at a farm at Histon in Cambridgeshire to test their laboratory findings in small-scale commercial practice. In 1929 a commercial gas store for apples was built by a grower near Canterbury in Kent. From this work they published a series of papers on various aspects of storage of apples in mixtures of CO_2, O_2 and N_2 (Kidd and West, 1925, 1934, 1935a, b, 1938, 1939, 1949). They also worked on pears, plums and soft fruit (Kidd and West, 1930). In 1927 Kidd toured Australia, Canada, the USA, South Africa and New Zealand discussing gas storage. In 1934 Kidd became Superintendent of the Low Temperature Research Station. By 1938 there were over 200 commercial gas stores for apples in England.

The Food Investigation Organisation was subsequently renamed, since the title did not describe fully the many agents used in food preservation. It was renamed the Food Investigation Board with the following term of reference: 'To organise and control research into the preparation and preservation of food'. The first step towards the formation of the Food Investigation Board was 'taken by the Council of the Cold Storage and Ice Association' (Food Investigation Board, 1919). The committee given the task of setting up the Food Investigation Board consisted of Mr W.B. Hardy, Professor F.G. Hopkins, Professor J.B. Farmer FRS and Professor W.M. Bayliss to prepare a memorandum surveying the field of research in connection with cold storage. The establishment of a 'Cold Storage Research Board' was approved.

Work carried out by the Food Investigation Board at Cambridge under Dr F.F. Blackman FRS consisted of experiments on gas storage of strawberries at various temperatures by Kidd and West at the Botany School (Food Investigation Board, 1920). Results were summarized as follows.

> Strawberries picked ripe may be held in cold store (temperature 1°C to 2°C) in a good marketable condition for six to seven days. Unripe strawberries do not ripen normally in cold storage; neither do they ripen when transferred to normal temperatures after a period of cold storage. The employment of certain artificial atmospheres in the storage chambers has

> been found to extend the storage life of strawberries. For example, strawberries when picked ripe can be kept in excellent condition for the market for three to four weeks at 1°C to 2°C if maintained:
>
> 1. in atmospheres of O_2, soda lime being used to absorb the CO_2 given off in respiration;
> 2. in atmospheres containing reduced amounts of O_2 and moderate amounts of CO_2 obtained by keeping the berries in a closed vessel fitted with an adjustable diffusion leak.
>
> Under both these conditions of storage the growth of parasitic and saprophytic fungi is markedly inhibited, but in each case the calyces of the berries lose their green after two weeks.

The Food Investigation Board (1920) said that CA storage of plums, apples and pears at low temperature 'has been continuing' with large-scale gas storage tests on apples and pears. It was reported that storage of plums in total N_2 almost completely inhibited ripening. Plums can tolerate, for a considerable period, an almost complete absence of O_2 without being killed or developing an alcoholic or unpleasant flavour.

The Food Investigation Board (1920) described work by Kidd and West at the John Street store of the Port of London Authority on 'Worcester Pearmain' and 'Bramley's Seedling' apples at 1°C and 85% relative humidity (RH), 3°C and 85% RH and 5°C and approximately 60% RH. 'Sterling Castle' apples were stored in about 14 kPa CO_2 and 8 kPa O_2 from 17 September 1919 to 12 May 1920. Ten per cent of the fruit were considered unmarketable at the end of November for the controls, whereas the gas-stored fruit had the same level of wastage 3 months later (by the end of February). The Covent Garden Laboratory was set up as part of the Empire Marketing Board in 1925. It was situated in London close to the wholesale fruit market with R.G. Tomkins as Superintendent. The Food Investigation Board (1958) described some of their work on pineapples and bananas as well as pre-packaging work on tomatoes, grapes, carrots and rhubarb.

Besides defining the appropriate gas mixture required to extend the storage life of selected apple cultivars, Kidd and West were able to demonstrate an interaction. They showed that the effects of the gases in extending storage life varied with temperature in that at 10°C gas storage increased the storage life of fruit by 1.5–1.9 times longer than when stored in air, while at 15°C the storage life was about the same in both gas storage and in air. They also showed that apples were more susceptible to low temperature breakdown when stored in controlled atmospheres than in air (Kidd and West, 1927b).

In 1928 the Ditton Laboratory (Fig. 1.1) was established close to the East Malling Research Station in Kent by the Empire Marketing Board with J.K. Hardy as Superintendent. At that time it was an out-station of the Low Temperature Research Station at Cambridge. The research facilities were comprehensive and novel with part of the Station designed to simulate the refrigerated holds of ships in order to carry out experiments on sea freight transport of fruit. Cyril West was appointed Superintendent of the Ditton Laboratory in 1931. West retired in 1948 and R.G. Tomkins was appointed Superintendent (later the title was changed to Director) until his retirement in 1969. At that time the Ditton Laboratory was incorporated into the East Malling Research Station as the Fruit Storage Section (later the Storage Department) with J.C. Fidler as head. When the Low Temperature Research Station had to move out from its Downing Street laboratories in Cambridge in the mid 1960s, part of it was used to form the Food Research Institute in Norwich. Subsequently it was reorganized, in November 1986, as the Institute of Food Research. Most of the staff of the Ditton Laboratory were transferred elsewhere, mainly to the new Food Research Institute. The UK Government's Agricultural Research Council had decided in the mid to late 1960s that it should reorganize its research institutes on a crop basis, rather than by discipline. For example, Dr W.G. Burton, who worked on postharvest and CA storage of potatoes at the Low Temperature Research Station at Cambridge and subsequently at the Ditton Laboratory, was appointed

Fig. 1.1. The Ditton Laboratory at East Malling in Kent. The photograph was taken in June 1996 after it had been closed and the controlled atmosphere storage work had been transferred to the adjacent Horticulture Research International.

Deputy Director at Food Research Institute. Some of the Ditton Laboratory staff thought this government action was just a ploy to dismember the Laboratory as it had got out of control. Apparently at one time visitors from Agricultural Research Council's headquarters were neither met off the train nor offered refreshment (John Stow, personal communication).

B.G. Wilkinson, R.O. Sharples and D.S. Johnson were subsequent successors to the post of Head of the Storage Department at East Malling Research Station. The laboratory continued to function as a centre for CA storage research until 1992, when new facilities were constructed in the adjacent East Malling Research Station and the research activities were transferred to the Jim Mount Building. In 1990 East Malling Research Station (EMRS) became part of Horticulture Research International and subsequently it became 'privatized' to East Malling Research (EMR). In an interview on 24 July 2009 David Johnson indicated that the team of scientists and engineers he joined at East Malling in 1972 'has come down to me', and he was about to retire (Abbott, 2009). Since then the Produce Quality Centre has been set up under the direction of Dr Deborah Rees. This is a collaborative initiative between the Natural Resources Institute (NRI) and East Malling Research. The NRI, which was a successor to the Tropical Products Institute (TPI), became part of the University of Greenwich in 1996. The work of TPI in London 'developed from that of the Plant and Animal Products Department of the Imperial Institute, which from the 1890s aimed to furnish scientific and technical information necessary for the better production and marketing of plant and animal products, which British territories overseas produced or might become able to produce'. The name Imperial Institute was changed to Colonial Products Laboratory in 1953 then to TPI in 1957 and changed again to Tropical Development and Research Institute (with the amalgamation of two other institutes) and then to NRI. The NRI Produce Quality Centre focused on tropical and subtropical perishable commodities and EMR on UK fruit. As UK government funding for agricultural and horticultural research was limited, the two organizations decided to benefit from the increased efficiency of pooling expertise and sharing facilities. In 2017 EMR was taken over by the National Institute of Agricultural Botany (NIAB) to become NIAB EMR and the NRI took over the management of the Jim Mount Building (Fig. 1.2). Previously, in 2015, as interest increased in the use of dynamic controlled atmosphere (DCA) storage, David Bishop and Jim Shaeffer of Storage Control Systems Ltd worked with the Product Quality Centre (PQC) (primarily D. Rees, R. Colgan and K. Thurston) to set out the key areas of research to determine how to use respiratory characteristics to optimize CA protocols, thereby creating DCA standards for the UK's important apple and pear cultivars (Deborah Rees, personal communication).

Although Kidd and West collaborated on 46 papers during their lifetimes, they rarely met outside the laboratory. Kidd was an avid walker, a naturalist, gardener and beekeeper. He also wrote poetry and painted. West was interested in systematic botany and was honoured for his contribution to that field (Kupferman, 1989). West retained an office in the Ditton Laboratory until the 1970s from which he continued to pursue his interests in systematic botany (John Stow, personal communication).

Fig. 1.2. The Jim Mount Building at East Malling containing the CA laboratory of the Natural Resources Institute of the University of Greenwich.

Australia

Little *et al.* (2000) provided an excellent review of the history of CA in Australia. In 1926 G.B. Tindale was appointed by the state of Victoria to carry out research into postharvest science and technology of fruit. He collaborated with Kidd during his visit there in 1927. The Council for Scientific and Industrial Research Organization(CSIRO) was formed about the same time and F.E. Huelin and S. Trout from CSIRO worked with Tindale on gas storage of apples and pears. For example, in one study in 1940 they used 5kPa CO_2 + 16 kPa O_2 for storage of 'Jonathan' apples by controlled ventilation with air and no CO_2 scrubbing. Huelin and Tindale (1947) reported on gas storage research of apples and CA work was subsequently started on bananas (McGlasson and Wills, 1972) but the work was not applied commercially until the early 1990s. In Australia most commercial fruit storage until 1968 was in air and in 1972 CA generators were introduced. So the reality was that commercial CA storage was probably not used before 1968 and presumably not to any significant extent until after 1972. It seems the real problem of introducing CA storage technology was that the old cold stores were leaky (Little *et al.*, 2000; and John Faragher, personal communication) and therefore not easily adapted to CA. In 2015 a successful commercial trial was conducted at Valley Pack Storage Pty Ltd in the Goulbourn Valley on 'Granny Smith' apples and 'Packham's Triumph' pears which maintained quality and controlled superficial scald without chemicals using DCA-CF (chlorophyll fluorescence). Combined with support from Australian supermarkets that are promoting less postharvest chemical usage, there is now an increase in modern CA construction in Australia (Mark Novotny, personal communication).

Brazil

Commercial apple production started in Brazil in the mid 1970s and the first commercial CA storage of apples occurred in the late 1970s. Apple production has increased and Brazil now produces 1–1.2 million tonnes of apples each year (62% 'Gala', 33% 'Fuji' and < 5% other cultivars). About 278,000 t are stored in air and about 645,060 t are stored in CA. In Brazil, 750–1000 t of kiwifruit are stored in CA and approximately 2000 t in air each year (R. Prange personal communication, 2017).

Canada

The commercial development of CA storage of apples and pears in Canada can be largely attributed to the research of Charles Eaves of the Canada Department of Agriculture (the current name is Agriculture and Agri-Food Canada (AAFC)). Eaves culminated a distinguished research career at the AAFC Kentville Research and Development Centre, Kentville, Nova Scotia, where he made numerous significant and innovative contributions (Dilley, 2006; Hoehn *et al.*, 2009). Eaves worked on fruit and vegetable storage in Nova Scotia and initiated the construction of the first commercial CA store in 1939 in Canada at Port Williams. Eaves was born in England but studied and worked in Nova Scotia and then spent a year in England with Kidd and West at the Low Temperature Research Station in Cambridge in 1932–1933. On returning to Canada he was active in the introduction of CA technology (Eaves, 1934). The use of high calcium hydrated lime (calcium hydroxide, $Ca(OH)_2$) (also known as just 'lime' in the CA literature) for removing CO_2 in CA stores was also first developed by him in the 1950s (Eaves, 1959). Hydrated lime is still used today as a simple means to scrub CO_2 in CA stores, particularly for some cultivars of apples and pears that are very sensitive to CO_2, i.e. stored at 0.5 kPa CO_2 or less. He was involved, with others, in the development of a propane burner for rapidly reducing the O_2 concentration in CA stores (Eaves, 1963). He retired in 1972. CA is important in Canada especially for apples and in December 2016 there were 172,020 t of apples in CA storage and 37,811 t in refrigerated storage (AAFC, 2016).

China

It was reported that in 1986 the first CA store was established in Yingchengzi in China to contain some 1000 t of fruit (DGCL, 2004). Apple is the most produced fruit in China (Zhang Xuejie, personal communication). In 2016, China produced about 43 million tonnes of apples (the total output of fruits was about 0.27 billion tonnes), and over 50% of the apples were stored. About 14.7% of the apples were stored in CA (which reached over 35% in some advanced areas like Shandong province), 55.2% of apples were stored in cold storage, and 30.1% were stored in short-term non-cold storage. The CA recommendations for apple and some other fruits in China are shown below (Table 1.1).

Holland

A considerable amount of CA research was carried out by the Sprenger Institute, which was part of the Ministry of Agriculture and Fisheries and eventually was incorporated within Wageningen University. Although the effects of CA on flowers dates back to work in the USA in the late 1920s (Thornton, 1930), much of the research that has been applied commercially was done by the Sprenger Institute, which started in the mid 1950s (Staden, 1986). From 1955 to 1960, research was carried out at the Sprenger Institute on 'normal' CA storage of apples, which meant that only the CO_2 concentration was measured and regulated. Also at that time the first commercial CA stores were developed and used with good results. From 1960, both O_2 and CO_2 were measured and controlled and several types of CO_2 scrubbers were tested, including those using lime, molecular sieves, sodium hydroxide and potassium hydroxide. For practical situations lime scrubbers were mainly used and gave good results. From 1967 to 1975, active carbon scrubbers were used, which gave the opportunity to develop central scrubber systems for multiple rooms. From 1975 to 1980, the first pull-down equipment was used in practice, mainly using the system of ammonia cracking called 'Oxydrain', which produced N_2 and hydrogen (H_2) gases. The N_2 gas was used to displace storage air to lower O_2 levels. At this time most apple cultivars were stored in 3kPa CO_2 + 3 kPa O_2. From 1980 to 1990, an enormous development in CA storage occurred with much attention on low O_2 storage (1.2 kPa) sometimes in combination with low CO_2. In this system the quality of the stored apples improved significantly (Schouten *et al.*, 1997). There

Table 1.1. CA recommendations for selected fruits in China.

Fruit	°C	% RH	O_2 kPa	CO_2 kPa	Storage time
Apple	0–5	85–95	2–4	3–5	4–6 months
Waxberry (*Myrica rubra*)	0.5–1.5	95	3–5	5–7	40 days
Strawberry	0	95–100	10	5–10	4 weeks
Hami melon	3~4	80	3~5	1–1.5	6 months
Date	0~1	90	3	5	2–3 months
Hawthorn	–1	90	5	3	7 months
Grape	3	90–95	5	5~20	7 months
Persimmon	0~–1	90	2	4–7	6 months
Peach	0~1	85	1–3	5	10 weeks
Plum	0~1	90	1~3	5	10–12 weeks
Cherry	0~1	90	3~5	20~25	6–7 weeks
Banana	11–13	90–95	0.5–10	0.5–7	100 days
Pineapple	7	85	2	0	40 days
Mango	10~12	85	3~5	2.5~10	40 days
Lychee	1~5	90	5	5	2 months
Kiwifruit	0~1	90–95	2~4	5	3~6 months

Fig. 1.3. An experimental CA storage room in Holland. Reproduced with permission from Wageningen Food & Biobased Research.

was also improvement in active CO_2 scrubbers, gas-tight rooms, much better pull-down systems for O_2 including membrane systems, pressure swing adsorption and centralized measurement and controlling systems for CO_2 and O_2. From 1990 to 2000, there was further improvement of the different systems and the measurement of defrosting water. During 2000–2009, there was further development of the Dynamic Control System using the measurement of ethanol as the control method for the O_2 concentration. This enables the level to be lowered sometimes down to 0.4 kPa in commercial storage. A lot of sophisticated ultra-low oxygen (ULO) storage rooms now use this technique (Fig. 1.3). At the same time the use of 1-methylcyclopropene (1-MCP) was introduced in combination with ULO storage. A next step in the 1-MCP application, especially for 'Elstar' apples, was the application during and after ULO storage, to get a better shelf-life. From 2015, a new dynamic CA (DCA) technique was introduced which is based on the respiration coefficient of the apples. This advanced control of respiration (ACR) system regularly measures the CO_2 production and oxygen uptake of the entire CA cell (Fig. 1.3). The oxygen content in the cell is adjusted on the basis of the calculated RQ (Alex van Schaik, personal communication).

Italy

By the late 1950s, CA was being used in the apple and pear growing regions of Northern Italy. Since then, CA has been extensively adopted. As an example, in the Alto Adige (South Tyrol) region, in the 2016–2017 storage season 88.43% of a total 864,000 t of apples was stored in CA (Angelo Zanella, personal communication).

India

Originally established in 1905, the Indian Agricultural Research Institute initiated a coordinated project on postharvest technology of fruit and vegetables in 1970. Under the direction of Dr R.K. Pal the IARI conducted a considerable amount of research on several fruit species, including work on CA. Jog (2004) reported that there are a large number of cold storages in India and some of the old ones have been revamped, generators added and the availability of CA and MA facilities was increasing, but CA stores remain rare.

Mexico

In countries like Mexico crops are not widely stored, partly because of long harvesting seasons. However, CA storage is increasingly used, for example, in Chihuahua and Coahuila for apples. It was estimated by Yahia (1995) that there were about 50 CA storage rooms in Mexico storing about 33,000 t of apples each year.

New Zealand

Research work for the Department of Industrial and Scientific Research on gas storage of apples between 1937 and 1949 was summarized in a series of papers (for example, Mandeno and Padfield, 1953). The first experimental CA shipments of 'Cox's Orange Pippin' apples were carried out to Holland and UK in the early 1980s supervised by Stella McCloud. These experiments were in reefer containers at 3.5–4°C and a small bag of lime was placed in each box to absorb CO_2, though this presented a problem with the Dutch customs officers who needed to be convinced that the white powder was in fact lime. Oxygen was probably controlled by N_2 injection. Before that time, apples were shipped only in reefer containers or reefer ships and the use of CA was in response to the apples developing 'bitter pit', which was controlled by the increased CO_2 levels (John Stow, personal communication). Subsequently the Department of Scientific and Industrial Research developed predictive levels of fruit calcium needed for 'Cox's Orange Pippin' apples free of bitter pit.

Poland

In Poland in 2017, there was about 1,450,000 t of fruit storage capacity of which 950,000 t was CA. There was 1,720,000 t of vegetable storage capacity but only about 20 t in CA in 2017 (Franciczek Adamicki, personal communication).

South Africa

Kidd and West (1923) investigated the levels of CO_2 and O_2 in the holds of ships carrying stone fruit and citrus fruit to UK from South Africa. However, the first commercial CA storage facilities in South Africa were installed in Elgin in 1935 and 1936 (Dilley, 2006). Others were commissioned in 1978 and by 1989 the CA storage volume had increased to a total of 230,000 bulk bins catering for > 40% of the annual apple and pear crop (Eksteen *et al.*, 1989). The adoption of CA has increased in South Africa since 1989. CA capacity was 660,000 bins in 1997 and > 1.3million bins in 2017, with 75% of CA capacity being used for apples (341,000 t assuming an apple bin contains 350 kg) and 25% used for pears (136,000 t assuming a pear bin contains 420 kg) (Richard Hurndall, personal communication.). There are no CA stores for avocados in South Africa but avocado export shipment takes place in CA containers. Export volumes are 50,000–60,000 t per annum.

South-east Asia

Considerable work has been carried out in many South-east Asian countries on the postharvest effects of gases on fruit, but since long-term storage of fruit is rarely carried out because of constant availability most of the work has been related to MA packaging and coatings. In Thailand, there is no commercial CA storage (Ratiporn Haruenkit, personal communication), but some CA research is currently carried out at Kasetsart University in Bangkok, mostly on rambutan. In the Philippines, CA transport trials by the Central Luzon State University were reported for mango exports by Angelito T. Carpio, FreshPlaza on 12 September 2005, but results were not published.

Turkey

In Turkey, CA research started in Yalova Central Horticultural Research Institute in 1979, mainly by Dr Umit Ertan, Dr Sozar Ozelkok and Dr Kenan Kaynaş, initially on apples and subsequently at the Scientific and Technical Research Council of Turkey (TÜBİTAK) and various Turkish universities. Commercial CA storage also started near Yalova and one private company was reported to have some 5000 t CA capacity (Kenan Kaynaş, personal communication, 2008). The most significant and accelerated expansion of CA storage in Turkey began between 2000 and 2010 and the CA storage capacity in 2017 for fruit and vegetables (especially apple, pear and pomegranate) has reached 200,000 t, which is about 10% of the total cold-storage capacity in Turkey, and the amount of CA-stored pomegranate was about 4000–5000 t (Kenan Kaynaş and Mustafa Erkan, personal communication, 2017).

West Indies

In 1928 the Low Temperature Research Station was established in St Augustine in Trinidad at the Imperial College of Tropical Agriculture at a cost of £5800. The initial work was confined to:

> improving storage technique as applied to Gros Michel [bananas] ... for investigating the storage behaviour of other varieties and hybrids which might be used as substitutes for Gros Michel in the event of that variety being eliminated by the epidemic spread on Panama Disease.

Due to demand the work was extended to include 'tomatoes, limes, grapefruit, oranges, avocados, mangoes, pawpaws, eggplant fruit, cucurbits of several kinds and to the assortment of vegetables that can be grown in the tropics' (Wardlaw and Leonard, 1938). An extension to the building was completed in 1937 (Fig. 1.4) at a cost of £4625. 'Dr F. Kidd and Members of the Low Temperature Research Station at Cambridge gave assistance and advice in technical and scientific matters' and CA storage work was carried out using imported cylinders of mixtures of N_2, O_2 and CO_2. Professor C.W. Wardlaw was Officer-in-Charge and a considerable amount of work was carried out by him and his staff on fresh fruit and vegetables, including work on CA storage (Wardlaw, 1938). Wardlaw returned to Manchester University as Professor of Cryptogamic Botany under Professor S.C. Harland FRS and in 1961, after Harland's retirement, he was appointed Head of Department.

Fig. 1.4. Laboratories at the Low Temperature Research Station in St Augustine, Trinidad, in 1937. Photo Tucker Picture Production Ltd (Wardlaw and Leonard, 1938).

With the formation of the University of the West Indies in the 1960s, postharvest work in Trinidad continued mainly under the supervision of Professor L.A. Wilson and subsequently by Keith Thompson and Lynda Wickham on postharvest science and technology of fruit and vegetables. There has been considerable work on MA packaging but limited work on CA storage. Errol Reid and Keith Thompson carried out experiments in 1997 on CA transport of bananas in reefer containers from the Windward Islands to Britain and subsequently Reid supervised the use of CA reefer ships in 1998. CA reefer ships were being used for transport of bananas from the Windward Islands to the UK with some 50,000–60,000 t /year in 2010, often combined with bananas from the Dominican Republic to reduce the risk of dead-freight.

2

Harvest and Pre-harvest Factors

The conditions and environment in which a crop is grown can affect its postharvest life and therefore can result in variability in storage recommendations and variable responses to recommendations. Factors that might affect optimum storage conditions include soil, climate and weather as well as the cultural conditions. When and how they are harvested can also have a substantial influence. For example, Johnson (1994b) found that the ester content of 'Cox's Orange Pippin' apples was influenced by growing season and source of fruit. Delaying harvesting maximized flavour potential, but harvesting too late reduced storage life and had adverse effects on texture. Johnson (2008) also reported that the significant variability between orchards in internal ethylene concentration and quality of 'Bramley's Seedling' apples demonstrated the need to maximize storage potential through attention to pre-harvest factors. In Moldova, Tsiprush *et al.* (1974) found that the place where 'Jonathan' apples were grown affected the optimum storage conditions. They recommended 5 kPa CO_2 + 3 kPa O_2 for fruit from the Kaushanskii and Kriulyanskii regions and 5 kPa CO_2 + 7 kPa O_2 for fruit from the Kotovskii region, all at 4°C and 83–86% RH. Variability in response of fruit to 1-MCP treatment has been shown, not only to the concentration used and exposure time, but also to the location of the orchard where the apples were grown (Johnson, 2008). At one site, 'McIntosh' apples treated with 1-MCP had more internal browning but there was no effect at another site (Robinson *et al.*, 2006).

Being able to identify a characteristic of the crop at harvest that could be used to predict its postharvest behaviour would have advantages. For example, Hatoum *et al.* (2016) studied the postharvest development of internal browning in 'Braeburn' apples in order to identify the main metabolic changes associated with their pre-harvest and post-harvest factors. They stored the apples in CA conditions and observed internal browning as early as 14 days after harvest, but they were unable to identify any metabolite at harvest that could be used to reliably predict the early incidence of internal browning. Hyperspectral imaging and magnetic resonance imaging of crops have been tested for use for this purpose before storage (Suchanek *et al.*, 2017). Resonant frequency was tested (Darko, 1984) and is used commercially to predict the ripening speed of avocados (Fig. 2.1).

The way that the crop is handled can clearly have an impact on its postharvest life, particularly if it is damaged. For example, the effect of impact stress on the expression of genes related to respiratory by-products and ethylene biosynthesis of damaged cabbages

Fig. 2.1. Avocados, on sale in a British supermarket, that had been graded by resonant frequency to predict ripening time.

was studied by Thammawong *et al.* (2014). The cabbages were damaged by dropping them and within 1 h this damage increased the expression of the genes *BoAPX2*, *BoPAL*, *BoSAMS* and *BoACS2* in the wounded and surrounding areas. They proposed a mathematical model based on a modified Weibull distribution to describe the stress response characteristics of the cabbage, which could be used to predict their stress-responsive cellular metabolisms and quality changes and therefore have an impact on their postharvest life.

Organic Production

There is conflicting information on the effects of organic production of fruit and vegetables on their postharvest characteristics. Organic production has been shown to result in crops having higher levels of postharvest diseases. Massignan *et al.* (1999) grew 'Italia' grapes both conventionally and organically and after storage at 0°C and 90–95% RH for 30 days they found that organic grapes were more prone to storage decay than those grown conventionally. Gasser and von Arx (2015) compared the storage of apples grown under an organic system with those produced under 'integrated production' and found that organic fruit had higher levels of rot during DCA storage than those from integrated production. The anaerobic compensation point (ACP) during DCA storage was not influenced by the production method. In another case there was evidence that organic production reduced disease level. In samples from organically cultivated 'Bintje' and 'Ukama' potato tubers the gangrene disease (*Phoma foveata*) levels were lower compared with conventionally cultivated ones. However, there was no such difference in the cultivars 'King Edward' and 'Ulama' tested 4 months later. The dry rot (*Fusarium solani* var. *coeruleum*) levels were generally lower in organically cultivated potatoes compared with tubers grown conventionally (Povolny, 1995). Louarn *et al.* (2013) compared storage of organically and conventionally grown carrots and found no effect of the system on the quality of the carrots during 6 months of storage. Vermathen *et al.* (2017) compared organic, low-input and integrated production of apples on their metabolic changes during storage. Metabolic changes over time were similar for all three production systems: mainly decreasing lipid and sucrose and increasing fructose, glucose and acetaldehyde levels. However, organic apples had consistently higher levels of fructose and monomeric phenolic compounds but lower levels of condensed polyphenols than integrated and low-input produced apples.

Fertilizers

The chemical composition of fruit and vegetables is affected by the nutritional status of the soil in which they are grown, which in turn can affect their storage life. An adequate supply of potassium (K) fertilizer in tomato production was shown to prolong their postharvest life as well as improving fruit colour and reducing the incidence of yellow shoulder (Hartz *et al.*, 2005). In the UK, Sharples (1980) found that 'Cox's Orange Pippin' apples were commonly stored at 3.5–4°C in 2 kPa O_2 + < 1 kPa CO_2. In order to achieve this storage period he found that the composition of the fruit at harvest should be 50–70% N, a minimum of 11% phosphorus (P), 130–160% K, 5% magnesium (Mg) and 4.5 kPa Ca (on a dry matter basis). However, an increase in N fertilizer to greenhouse-grown tomatoes, beyond a threshold level, may reduce the sugar content of the fruits and shorten their postharvest life (Passam *et al.*, 2007). N fertilizers over about 250 kg/ha reduced the TSS of tomatoes (Senevirathna and Daundasekera, 2010). In Brazil the effects of soil K application on storage of 'Fuji' apples from a long-term trial was evaluated. Fruits were stored at 0°C with 1 kPa O_2 + < 0.3 kPa CO_2, 1 kPa O_2 + 2 kPa CO_2 or air for 8 months and then 7 days at 20°C in air. Fruit weight losses during storage, ground colour development and rot incidence were not affected by soil K application. There was a significant interaction between K application and storage atmospheres for internal breakdown. When fruits were stored in 1 kPa O_2 + < 0.3 kPa CO_2 no differences were detected between treatments, but storing apples in 1 kPa O_2 + 2 kPa CO_2 resulted in higher breakdown in fruits with lower K concentrations (Hunsche *et al.*, 2003). Streif *et al.* (2001) found no relationship between CA storage disorders and K, Ca, Mg and P levels in pears, but boron (B) was related to the occurrence of browning development during storage. In citrus there is some indication that higher N may reduce the development of stem-end rot (*Diplodia natalensis*) and green mould (*Penicillium digitatum*) during storage (Ritenour *et al.*, 2009). Ortiz *et al.* (2011) reported that apple softening was mediated by Ca loss from the middle lamella, and accordingly pre-harvest Ca sprays (7 weekly applications at 1.6%, w/v, 81–123 days after full bloom) were applied to 'Fuji Kiku-8' apples, which improved cell-to-cell adhesion and reduced softening during CA storage. Holb *et al.* (2012) also evaluated the effects of pre-harvest Ca applications on brown rot (*Monilinia fructigena*) on 'Idared' apples during storage at 1°C for 4 months in air, CA or ULO. Both CA and particularly ULO storage reduced rot incidence compared with air and the Ca treatment significantly reduced the number of infected fruit from 75% in non-treated to 45.5% in treated fruit. Passam *et al.* (2007) showed that pre-harvest Ca application to tomatoes had a positive effect on the prevention of some diseases and delayed softening during storage. Grape clusters were sprayed with calcium chloride either from fruit set to veraison or from veraison to harvest and compared with non-treated controls over a 2-year period (veraison is the onset of ripening). In both years, Ca applications to bunches was effective in maintaining flesh firmness and berry breaking force as well as reducing *Botrytis cinerea* rots during storage. The applications were particularly effective if carried out between fruit set and veraison (Ciccarese *et al.*, 2013).

Climate and Weather

Tudela *et al.* (2017) investigated, over 26 months, whether different weather conditions influenced the quality of iceberg and romaine lettuce. After being grown at different times of the year the lettuce was fresh-cut and stored for 11 days in active MA packaging and then transferred to air for 24 h at 7°C. They determined that the mean temperature during cultivation was the climatic variable that contributed most to quality loss of the lettuce. Lau (1997) and Watkins *et al.* (1997) found that 'Braeburn' apples were most susceptible to BBD and internal CO_2 injury during CA storage after cool growing seasons. 'Cambridge Favourite' strawberries, harvested early, were less susceptible to fungal spoilage during storage at 2°C or 15°C than fruit harvested from the same crops later during each of these seasons (Browne *et al.*, 1984). Lachapelle *et al.* (2013) reported an association of soft scald in 'Honeycrisp' apples with weather conditions, particularly cool and relatively wet weather. They developed a model to test the effects and found it could be used, prior to storage, to predicted soft scald development.

Fungicides

Diseases, particularly those caused by fungal infection, are a major limiting factor in postharvest of fruit and vegetables. Kiwifruit harvested from orchards infected with bacterial canker (*Pseudomonas syringae* pv. *actinidiae*) showed significantly lower total soluble solids (TSS) during CA storage and after 1-MCP treatment, compared with those not treated with 1-MCP or stored in air. The TSS of kiwifruits from healthy orchards was not influenced by the storage conditions. They also showed that in kiwifruit the presence of bacterial canker affected postharvest quality, shelf-life and susceptibility to postharvest rots (Prencipe *et al.*, 2016). Youssef *et al.* (2012) sprayed aqueous solutions of various salts at 2% w/v at 2000 l/ha on citrus trees. After harvest, fruits were stored at 4 ± 1°C and 95–98% RH, followed by 7 days of shelf-life at 20 ± 2°C. They reported that the pre-harvest sprays of sodium carbonate, potassium carbonate and sodium silicate completely inhibited the incidence of decay during storage as compared with the water control. When the salts were applied after harvest, the activity was in general less pronounced, with sodium carbonate and potassium carbonate being the most effective. Kobiler *et al.* (2011) reported that *Alternaria alternata* infects persimmon fruit in the orchard and remains quiescent until harvest. During storage at 0°C the pathogen slowly colonizes the fruit, causing black spot symptoms. Pre-harvest treatment with a single spray of the fungicide polyoxin B 14 days before harvest in combination with a postharvest chlorine dip reduced the black spot infected area by 60%, compared with the chlorine dip alone. Thompson (1987) reviewed pre-harvest sprays of mangoes to control anthracnose (*Colletotrichum gloeosporioides*). Recommendations were for field application of chemical fungicides often followed by postharvest hot-water treatment usually combined with a fungicide (Fig. 2.2). In Florida (McMillan, 1972) cupric hydroxide at 2.4 g/l or tribasic copper sulfate at 3.6 g/l plus the organic sticker Nu-Film 17 at 0.125% applied at monthly intervals at 57 l per tree from flowering to harvest gave good anthracnose control. Benomyl at 0.3 g/l plus Triton B1956 at 0.15 ml/l at 57 l per tree at monthly intervals from flowering to 30 days before harvesting was also shown to be very effective against anthracnose on mangoes (McMillan, 1973). Mancozeb, chlorothalonil and ferbam were shown to be equally effective as field sprays against mango anthracnose in Florida (Spalding, 1982). Trials carried out on 'Keitt' mangoes in South Africa showed that two pre-flowering applications of copper oxychloride then two applications of triadimenol

Fig. 2.2. Mango hot-water treatment tank being tested in Jamaica on a commercial shipment of fruit to be exported to UK.

during flowering followed by monthly applications of copper oxychloride from fruit set to just before harvest gave effective control of anthracnose (Lonsdale, 1992). Sprays were applied to run off at about 20 l per tree. In Australia mancozeb applied at 1.6 g/l (active ingredient) as a weekly spray during flowering then as a monthly spray until just before harvesting gave good control of mango anthracnose (Grattidge, 1980). In studies in the Philippines, field sprays with mancozeb or copper were effective in controlling mango anthracnose and superior to either captan or zineb (Quimio and Quimio, 1974). Postharvest water and/or sodium hydroxide washes in the orchard and a hot water spray applied over rollers, without brushes, in the packhouse were effective in reducing the incidence of *Alternaria* and stem-end rots (*Phomopsis* spp.) in mangoes (Feygenberg *et al.*, 2014). Some fungicides are not permitted in many countries and checks should be made before they are used.

Clearly microorganisms exist on plants that can affect pathogens. For example, Korsten *et al.* (1993) showed that *Bacillus* spp. isolated from leaves and fruit of avocados were more effective in controlling anthracnose (*C. gloeosporioides*) and stem-end rot (*Botryodiplodia theobromae*) of avocados when applied as a postharvest dip than the fungicide prochloraz applied in the same way. Therefore they recommended that consideration should be given to field control of diseases that do not reduce the effects of these 'beneficial' organisms.

Water Relation

Irrigation and rainfall can affect the storage life of fruit and vegetables, but no information could be found on water relations having a differential effect specifically in CA storage. Citrus fruit harvested early in the morning, during rainy periods, or from trees with poor canopy ventilation, had a higher risk of postharvest decay (sour rot, brown rot and green mould) than fruit from trees on well drained soils with good canopy ventilation. Conversely, fruit grown in climates that are more arid tend to develop less rot during postharvest handling, transportation and marketing. If fruit must be harvested from trees during rainy periods, it may be best to avoid fruit from lower branches that may be exposed to more pathogens.

Increased irrigation has been reported to decrease postharvest incidence of stem-end rot but increased the incidence of green mould (Ritenour *et al.*, 2009). In New York State, McKay and Van Eck (2006) found that redcurrant plants that lacked water in the first growth phase (May–June) and had excess water in the phase just before harvest may burst during storage. They also reported that any stress in the first growth phase can cause the fruit to turn yellow. High rainfall or heavy irrigation increased skin cracking in fruit including cherries, apples and tomatoes. Generally crops that have higher moisture content have poorer storage characteristics. For example, hybrid onion cultivars that tend to give high yield of bulbs with low dry matter content had a comparatively shorter storage life (Thompson *et al.*, 1972b; Thompson, 1985). Srikul and Turner (1995) reported that low irrigation reduced the green life of bananas, but if bananas were allowed to mature fully before harvest and harvesting was shortly after rainfall or irrigation the fruit could easily split during handling operations, allowing micro-organism infection and postharvest rotting (Thompson and Burden, 1995). If oranges are too turgid at harvest the oil glands in the skin can be ruptured, releasing phenolic compounds and causing oleocellosis (Fig. 2.3).

Too much rain or irrigation can result in the leaves of leaf vegetables becoming brittle (Wardlaw, 1938). Irrigating crops can have other effects on their postharvest life. In carrots, heavy irrigation during the first 90 days after drilling resulted in up to 20% growth splitting, while minimal irrigation for the first 120 days followed by heavy irrigation resulted in virtually split-free carrots with a better skin colour and finish and only a small reduction in yield (McGarry, 1993). Shibairo *et al.* (1998) grew carrots with different irrigation levels and found that pre-harvest water stress lowered membrane

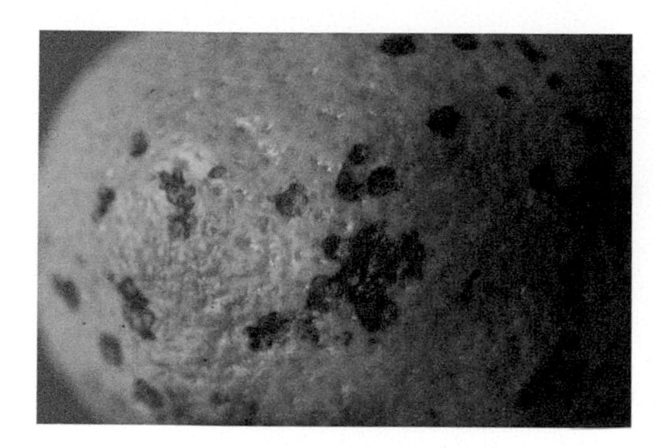

Fig. 2.3. Oleocellosis is a browning and pitting of the skin. It is due to the oil vesicles being ruptured, releasing phenolic compounds that damage the cells of the skin.

integrity of carrots, which may enhance moisture loss during storage.

The effects of water stress, applied for 30 or 45 days before flowering on 'Haden' mangoes, on their storage life was studied by Pina *et al.* (2000). They found that the 45-day fruits exhibited a higher incidence and severity of internal darkening, were firmer, more acid and had redder skins than 30-day fruits during storage at 13°C for 21 days.

In a study of the storage of onions grown in Tajikistan under various irrigation regimes, Pirov (2001) found that if the onions were to be used fairly quickly, maximum yields could be achieved by keeping the soil at 80–90% of field capacity. However, when they were to be stored for 7 months at 0–1°C and 75–80% RH, the best irrigation regime was 70% of field capacity throughout the growing season. Broccoli was cultivated under low soil water content (0.40 MPa) or field capacity (0.04 MPa). Low soil water content gave the best preservation of colour, antioxidant activity, L-ascorbic acid and 5-methyl-tetrahydrofolate contents (Cogo *et al.*, 2012).

Splitting during growth can affect postharvest losses. The incidence of damage in carrots was shown to be affected by the total amount of irrigation and the time when it was applied. Heavy irrigation during the first 90 days after sowing resulted in up to 20% growth splitting, while minimal irrigation for the first 120 days followed by heavy irrigation resulted in virtually no splitting, with a better skin colour and finish and only a small reduction in yield (McGarry, 1993). Carrots contain an antifreeze protein that improves storage performance and those grown in temperatures of less than 6°C accumulated higher levels of this protein and subsequently showed less electrolyte leakage from cells, slightly higher dry matter and less fungal infestation than carrots grown in warmer temperatures. However, levels also varied between the environments in which the carrots had been grown (Galindo *et al.*, 2004; Kidmose *et al.*, 2004).

Luna *et al.* (2012) studied the effects of five drip irrigation systems (excess 50%, excess 25%, control, deficit 25% and deficit 50%) on the postharvest quality of minimally processed iceberg lettuce. They found that visual quality was lower on those from the highest irrigation regime and they developed increased off-odours. Also the midrib tissue had a more than 17-fold increase in phenylalanine ammonia-lyase (PAL) activity from the highly irrigated lettuce. They concluded that the quality and shelf-life of the minimally processed lettuce was better preserved by reduced irrigation.

Harvest Maturity

The correct harvest maturity in terms of quality and postharvest changes of fruit and vegetables is crucial, but particularly so for long-term storage including CA. Many ways have been developed and used commercially to determine the stage of maturity or ripeness before or at harvest, including pressure testers, TSS, total acidity (TA) and starch content. These are all methods that rely on taking a representative sample, testing it and assuming the rest of the crop is similar within defined limits. Non-destructive methods of measuring fruit and vegetable maturity have been successfully tested, including surface colour and near-infrared hyperspectral imaging to measure TSS and juice content (Teerachaichayut and Ho, 2017) (Fig. 2.4).

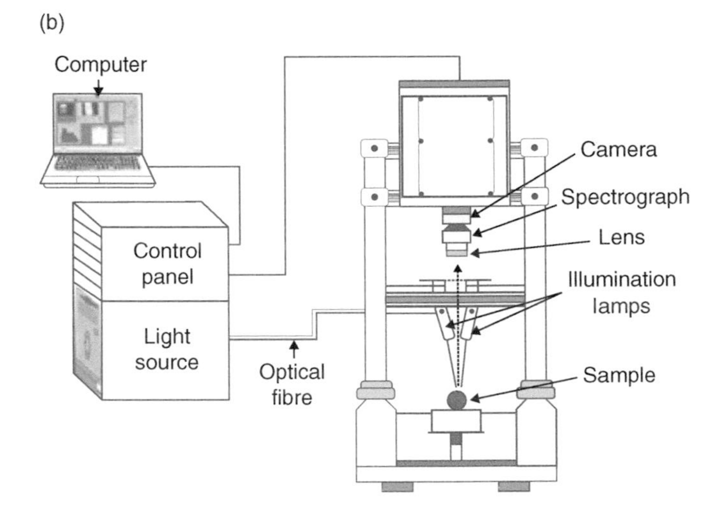

Fig. 2.4. Infra-red hyperspectral imaging being developed for non-destructive testing of fruit maturity at KMITL in Thailand. (a) Equipment being used by a student. (b) Schematic of the equipment. Reproduced with permission from Miss Huong HoThanh.

There can be interactions between optimum storage conditions and harvest maturity. For example, Trierweiler *et al.* (2004) indicated that 'Braeburn' apples harvested at advanced maturity had increased Braeburn browning disorder (BBD) during storage in 3.0 kPa CO_2 + 1.5 kPa O_2 compared with fruit harvested at a less mature stage. This confirmed the findings of Lau (1997) and Streif *et al.* (2001), who also found that late harvesting of pears increased the appearance and intensity of browning disorders. In 'Gala Must' apples, stored in air or CA at 0°C, flesh browning symptoms increased with delayed harvesting (Ben, 2001). Awad *et al.* (1993) found that for 'Jonagold' apples stored at 3 kPa O_2 + 1 kPa CO_2 the level of scald was 39% where the harvest was delayed compared with apples from a normal harvest date, which was 13%. However, in 'Fuji' apples Jobling *et al.* (1993) found that harvest maturity was not critical and all the maturity stages tested had a good level of consumer acceptability for up to 9 months in CA storage followed by 25 days at 20°C in air. Johnson (2001) recommended that in 'Gala' apples later harvests should be consigned to shorter-term storage and quoted previous work which suggested that 'Gala' stored well when harvested at above 7 kg firmness (measured with a Magness and Taylor penetrometer) and over a broad range of starch levels of 50–90% black. He also pointed out the difficulties in being sufficiently precise using firmness and starch levels in assessing the optimum maturity for CA storage of 'Gala'. Increased ethylene production was found to be the best physiological marker of ripening. Internal ethylene concentration of more than 100 ppb normally indicates that a fruit has commenced its ripening and fruits for storage are usually picked just prior to this. The internal ethylene concentration data for the six sites in Johnson's study indicated that a high proportion of fruit in five of the six orchards had already begun ripening when harvested. Whilst it is economically important that fruit achieves sufficient size and red coloration and can be harvested with stalks intact, these criteria should not be used as the primary indicators of maturity for storage. Johnson recommended that 'picking over' to ensure that fruit maturity matches the storage expectation appears to be the solution in orchards that are slow to develop the desired visual characteristics.

Harvest maturity has also been shown to affect postharvest life in other crops. Heltoft *et al.* (2016) stored 'Saturna' and 'Asterix' potatoes and found that immature potatoes had a lower dry matter content and higher weight losses and respiration rates during storage than more mature potatoes. In tomatoes, Batu (1995) showed that MA packaging was more effective in delaying ripening of fruit harvested at the mature green stage than for those harvested at a more advanced stage of maturity.

Chemical Sprays

Crops are sprayed during production for pest, disease and weed control. These sprays may affect their postharvest behaviour. Also sprays are applied, or have been investigated, specifically to improve maintenance of quality and reduce postharvest losses. Some of the sprays have also been tested as postharvest dips (see Chapter 7). Pre-storage sprays include abscisic acid, methyl jasmonate (MJ), daminozide, gibberellins (GA), aminoethoxyvinylglycine (AVG), lysophosphatidylethanolamine (LPE), Ethrel, maleic hydrazide (MH) and naphthaleneacetic acid (NAA).

Abscisic acid

Abscisic acid is common in plants and is involved in leaf retention and abscission and therefore is particularly important in the storage of leafy vegetables. Also Foukaraki *et al.* (2016) reported that abscisic acid and sucrose may play a role in sprout suppression when ethylene is exogenously applied to stored potatoes. Singh *et al.* (2014) sprayed litchis with abscisic acid at 150 or 300 mg/l at the colour-break stage. They found that this increased anthocyanins levels in the pericarp without adversely affecting the quality or stability of the fruit during 14 days of subsequent storage at 5°C.

Methyl jasmonate

MJ has been tested as a field spray as well as a postharvest treatment (see Chapter 7). Zapata *et al.* (2014) sprayed plum trees with MJ at 0, 0.5, 1.0 or 2.0 millimolar (mM) at 63, 77 and 98 days after full blossom. The fruit was harvested at the 'commercial ripening stage' and stored for 9 days at 20°C or for 50 days at 2°C then 1 day at 20°C. Treatment with 2.0 mM significantly accelerated ripening, whereas 0.5 mM delayed ripening. Ethylene production, respiration rate and softening were reduced significantly in both storage conditions for fruit treated with 0.5 mM. In these fruit, total phenolics, total antioxidant activity, pyrogallol peroxidase (POD), catalase and ascorbate peroxidase were higher in treated than non-treated plums during storage, which could account for the delay in ripening. Similarly Martínez-Esplá *et al.* (2014) found that the total phenolics, total antioxidant activity, firmness, colour, size and retention of weight of two plum cultivars were increased after MJ application to trees but there were interactions between cultivar and concentration. Application of 0.5 mM was the most effective for the cultivar 'Black Splendor' and 2.0 mM for 'Royal Rosa'.

Daminozide

N-dimethylamino succinamic acid, also called Alar, B9 and B995, was first marketed in the early 1960s mainly as a dwarfing agent (Brooks, 1964). It was also claimed to delay ripening of apples on the tree so that all the fruit on the same tree could be harvested at the same time. When applied to 'Cox's Orange Pippin' apples at 2500 μl/l in late June and mid-August, they developed a redder colour and were firmer than fruit from non-sprayed trees (Sharples, 1967). Sprayed apples were less susceptible to Gloeosporium rots but had more core flush during storage. There was some indication that sprayed fruits were slower to mature, since daminozide tended to retard the climacteric rise in respiration rate. In a comparison between pre-harvest and postharvest application of daminozide to 'Cox's Orange Pippin' apples, immersion of fruit in a solution containing 4.25 g/l for 5 min delayed the rise in ethylene production at 15°C by about 2 days, whereas orchard application of 0.85 g/l caused delays of about 3 days (Knee and Looney, 1990). Both modes of application depressed the maximum rate of ethylene production in ripe apples by about 30%. They also found that daminozide-treated fruit were less sensitive to exogenous ethylene during storage than non-treated fruit, but this response varied between cultivars. Daminozide has been used on other fruit; for example, Nagar (1994) found that treated litchis had increased shelf-life by up to 4 days. Daminozide has been withdrawn from the market in several countries.

Gibberellins

Pre-harvest sprays of gibberellic acid (GA_3) have been shown to have effects on the postharvest science and technology of fruit and vegetables. These include reduced yellowing and extended storage life of parsley (Lers *et al.*, 1998), reduced chilling injury in persimmon (Besada *et al.*, 2008), decreased fruit cracking in cherry (Yildirim and Koyuncu, 2010) and delayed ripening and reduced decay in prickly pear (Schirra *et al.*, 1999). Postharvest treatment with gibberellins has been shown to reduce degreening in lemons (Mizobutsi *et al.*, 2000) and limes (Sposito *et al.*, 2000) and improved retention of vitamin C also in limes (Keleg *et al.*, 2003). A negative effect of GA_3 was reported by Zoffoli *et al.* (2009) who found that it could induce shatter and predisposed grapes to grey mould caused by *B. cinerea*. Ding *et al.* (2015) found that storage of tomatoes in low temperatures inhibited increase of endogenous levels of GA_3. This effect was associated with lower expression of key GA metabolic genes (*GA3ox1*, *GA20ox1* and *GA2ox1*). GA_3 treatment reduced chilling injury and decreased electrolyte leakage and malondialdehyde (MDA) content, increased proline content, and improved antioxidant enzyme activities.

Aminoethoxyvinylglycine

AVG ([*S*]-trans-2-amino-4-(2-aminoethoxy)-3-butenoic acid hydrochloride) is an ethylene biosynthesis inhibitor, which is marketed commercially as ReTain® and is used in apple orchard sprays. Its mode of action is to inhibit the activity of 1-aminocyclopropane 1-carboxylic acid (ACC) synthase. Effects of AVG at both 100 and 200 mg/l applied to plums 2 weeks before harvesting decreased their firmness during subsequent storage compared with those not treated. Also with AVG treatment at 100 mg/l there was a decrease in total phenolics and total antioxidant activity (Ozturk *et al.*, 2012). They concluded that AVG generally had negative impacts on individual phenolic compounds. Mission muskmelon harvested from plots sprayed with AVG had lower rates of ethylene production, at harvest and after cold storage, than melons harvested from control plots. Melon ethylene production after storage was consistently lower when AVG was applied 7 days after full net formation, especially at the highest rate they tested (260 mg/l). AVG delayed initial development of an abscission zone and increased leaf chlorosis, but had no effects on flesh firmness, TSS or incidence of decay at harvest and after storage (Shellie, 1999). AVG at 1 g/l reduced ethylene production, respiration rate and softening of apricots (Muñoz-Robredo *et al.*, 2012). Lulai and Suttle (2004) found that treatment of potato tubers with AVG inhibited wound-induced ethylene production by about 90%, but did not affect wound-induced suberization. Cheema *et al.* (2013)

found that stored sweet potatoes could be exposed to ethylene to suppress sprouting, but it also resulted in increased respiration rate. However, AVG inhibited this increase in respiration rate and counteracted the decrease in monosaccharide concentrations. The effect of AVG was attributed to its possible inhibitory effect on protein synthesis.

AVG treatment delayed the onset of the climacteric in 'McIntosh' apples during storage but did not reduce internal ethylene concentration (Robinson *et al.*, 2006). Johnson and Colgan (2003) showed that 'Queen Cox' apples, which had been sprayed with AVG at 123.5g/ha before harvest, were firmer than those that had not been sprayed after 6 months of storage at 3.5°C under 1.2 kPa O_2 + 98.8 kPa N_2 combined with ethylene removal. The additive effect on firmness of AVG treatment and ethylene removal was negated by the development of core flush and after a simulated marketing period 57% of the fruit was affected after this combination of treatments. Drake *et al.* (2006) found that AVG reduced starch loss and ethylene production, retained firmness and reduced cracking in 'Gale Gala' apples, but reduced the sensory acceptance of apples and apple juice. AVG followed by ethylene treatment (Ethephon) reduced starch loss, ethylene production and cracking and maintained firmness. This combination also improved the sensory acceptance of apples but reduced sensory preference of apple juice. Harb *et al.* (2008b) showed that the biosynthesis of volatiles was highly reduced in apples after treatments with AVG, especially after extended storage periods in ULO. They found that after storage at 3.5°C in 1.2 kPa O_2 + < 1 kPa CO_2, with no removal of ethylene, until late March or early April, the AVG-treated fruit had a marked increase in the production of the corresponding volatiles. However, this effect was transitory in both AVG-treated fruits and those stored in ULO.

Application of 1-MCP has been investigated on fruits that had been treated with AVG. After 8 months in CA storage at 2°C, 'McIntosh' apples that had been treated with AVG had more internal browning disorders than non-treated controls but fruits with the combination of AVG + 1-MCP had less internal browning and were similar to the non-treated controls, but with better retention of firmness (Robinson *et al.*, 2006). Moran (2006) also reported that AVG + 1-MCP maintained firmness in 'McIntosh' apples more than 1-MCP alone after 120 or 200 days of CA storage. It was reported that AVG + 1-MCP could be used to maintain fruit firmness even when internal ethylene concentration at harvest was as high as 240 µl/l, but CA storage life was limited to 4 months. AVG was not effective in increasing the efficacy of 1-MCP on Cortland apples when internal ethylene concentration at harvest was not significantly different between AVG treated and non-treated fruit and internal ethylene concentration was less than 2 µl/l. AVG increased the efficacy of 1-MCP in the apples when internal ethylene concentration was 36 µl/l in non-treated fruits compared with undetectable in AVG treated fruits. Compared with non-treated control, AVG at 60 mg/l applied to 'Bartlett' pears 1 week before harvest suppressed ethylene production, respiration rate and softening and reduced senescence (Wang *et al.*, 2016). However, they found that the effectiveness of AVG was influenced by fruit maturity at the time of application.

Lysophosphatidylethanolamine

LPE is a naturally occurring lipid that has regulatory effects on senescence and ripening of fruit. In their review, Amaro and Almeida (2013) reported that the benefits of LPE include delayed leaf senescence, stimulation of maturation in grapes, acceleration of colour development and extension of shelf-life in cranberry and tomato. Özgen *et al.* (2015) also indicated that there is evidence that LPE can 'accelerate ripening of fruits and prolong shelf-life at the same time'. Hong *et al.* (2007) sprayed LPE (10 mg/l) on grape vines at either 4 or 6 weeks after fruit set and found that it increased berry size and TSS content. They also showed increases in anthocyanin levels, improved fruit quality and enhanced phytochemical characteristics of sweet cherries after pre-harvest application of LPE at 10 mg/l. Wan Zaliha and

Singh (2013) showed that spraying 'Cripps Pink' apples with LPE (125 mg/l or 250 mg/l) 2 and 4 weeks before 'commercial harvest' enhanced accumulation of total anthocyanins and polyphenolic compounds. LPE has also been shown to be effective as a postharvest treatment. For example, Ahmed and Palta (2016) dipped bananas at ripening stage 2 in a combination of LPE (200 mg/l) plus lecithin for 30 min and found delayed ripening during 10 days of storage at room temperature. LPE plus lecithin was better than LPE alone.

Ethylene

Ethylene is a common postharvest treatment to initiate ripening in climacteric fruit, de-greening of citrus and inhibition of sprouting in potatoes. Pre-harvest use of ethylene uses 2-chloroethyl-phosphonic acid, which is marketed as Ethrel or Ethephon. Ethrel C will release 74.4 l of ethylene gas from 1 l of Ethrel C. Ethrel has been used as a source of ethylene for decades. In order to reduce the difficulty of dislodging the fruit from the tree, Ethrel, abscissic acid and cycloheximide have all been shown to be effective but are not permitted for use in all countries. Ethrel is used to initiate flowering in pineapples. It has also been applied just before harvesting to accelerate de-greening and therefore the development of the orange colour in the skin. In Queensland, 'Smooth Cayenne' pineapples treated prior to harvest with Ethrel, at a concentration of 2.5 l in 1000 l of water, had superior eating quality, de-greened more evenly, but had a shorter shelf-life due to accelerated skin senescence than non-treated fruit 10 days after harvest. Treated fruit left on the plant for 23 days had inferior eating quality compared with the non-treated fruit (Smith, 1991). These effects are probably because of the ethylene speeding up the maturation of the fruit. Ferrer *et al.* (1996) showed that after the application of Ethrel (in the range of 0.1–1.0 mM) to lettuces, brown spots appeared in the leaves, but both soluble and cell wall-bound peroxidase activity in the leaves was reduced.

Maleic hydrazide

Chemicals may also be applied to certain crops in the field to prevent them sprouting during storage and thus to extend their storage period. An example of this is the application of MH (a plant growth regulator) to onions. MH is permitted in most countries, including the USA, Australia and most European countries. It has an acute mammalian oral LD_{50} of > 2000 mg/kg. Because it is necessary for the chemical to be translocated to the apex of the growing point towards the centre of the bulb for it to be effective, it must be applied to the leaves of the growing crop some 2 weeks before harvesting so that the chemical has time to be translocated into the middle of the bulb into the meristematic tissue. MH should be applied according to the manufacturer's instructions, but 2500 μl/l sprayed on the crop shortly before the leaves die down was recommended by Tindall (1983). Eshel *et al.* (2014) reported that onions grown in southern Israel were treated with MH before storage, and then stored for up to 8 months at 0°C with minimal losses. However, they recommended curing at 30°C and 98% RH for up to 9 days directly after harvest to reduced skin cracks.

Naphthalene acetic acids

In Trinidad pre-harvest spraying of sweet potatoes with MH or treating the tubers postharvest with methyl ester of naphthalene acetic acid (MENA) has been reported to inhibit sprouting. MENA was applied by interleaving it with layers of paper soaked in MENA in acetone at a ratio of 40 ml/100 kg of tubers (Kay, 1987). Paton and Scriven (1989) applied NAA either under reduced pressure as a 1 g/l solution containing a wetting agent or as a dust of 1, 10 or 100 mg/g of talc. During storage of at least 40 days, NAA applied as a solution reduced sprouting by more than 50% and when it was applied in talc the reduction in sprouting was 29%. However, in both cases there was an increase in water loss.

3

Effects and Interactions

CA storage is now used worldwide on a variety of fresh fruit and vegetables. The stimulation for the development of CA storage was arguably the requirement for extended availability of fruit and vegetables, especially certain cultivars of apple that were subject to chilling injury and therefore had reduced maximum storage periods. CA storage has been the subject of an enormous number of biochemical, physiological and technological studies and has been demonstrated to reduce the metabolism of fruit and vegetables in certain circumstances. However, for some crops in certain conditions high CO_2 or low O_2 or a combination of both can have either no effect or can even increase metabolism and therefore shorten storage life. The reasons for this variability are many. Interactions with temperature would mean that the metabolism of the crop could be changed so that it would be anaerobic and thus higher, especially where the O_2 content was low. High levels of CO_2 can actually injure the crop, which again could affect its rate of respiration. These effects on metabolism could also affect the eating quality of fruit and vegetables. Generally crops stored in CA have a longer storage life because the rates of the metabolic processes are slower. Particularly with climacteric fruit, this would slow ripening and deterioration so that when fruits have been stored for protracted periods they may well be less ripe than fruits stored in air. The actual effects that varying the levels of O_2 and CO_2 in the atmosphere have on crops varies with factors such as:

- species;
- cultivar;
- concentration of gases in store;
- temperature;
- stage of maturity of crop at harvest;
- degree of ripeness of climacteric fruit;
- growing conditions;
- ethylene in store; and
- pre-storage treatments.

There are also interactive effects of the two gases, so that the effect of CO_2 and O_2 in extending the storage life of a crop may be increased when they are combined. The effects of O_2 on postharvest responses of fruit, vegetables and flowers were reviewed and summarized by Thompson (1996) as follows:

- reduced respiration rate;
- reduced substrate oxidation;
- delayed ripening of climacteric fruit;
- prolonged storage life;
- delayed breakdown of chlorophyll;
- reduced rate of production of ethylene;
- changed fatty acid synthesis;
- reduced degradation rate of soluble pectin;
- formation of undesirable flavour and odours;

- altered texture; and
- development of physiological disorders.

Thompson (2003) also reviewed some of the effects of increased CO_2 levels on stored fruits and vegetables as follows:

- decreased synthetic reactions in climacteric fruit;
- delaying the initiation of ripening;
- inhibition of some enzymatic reactions;
- decreased production of some organic volatiles;
- modified metabolism of some organic acids;
- reducing the rate of breakdown of pectic substances;
- inhibition of chlorophyll breakdown;
- production of off-flavour;
- induction of physiological disorders;
- retarded fungal growth on the crop;
- inhibition of the effect of ethylene;
- changes in sugar content (potatoes);
- effects on sprouting (potatoes);
- inhibition of postharvest development;
- retention of tenderness; and
- decreased discoloration levels.

The recommendations for the optimum storage conditions have varied over time due mainly to improvements in the control technology over the levels of gases within the stores. Bishop (1994) showed the evolution of storage recommendations by illustration of the recommendation for the storage of the apple cultivar 'Cox's Orange Pippin' since 1920 (Table 3.1).

CA storage is still mainly applied to apples but studies of other fruit and vegetables have shown it has wide application and an increasing number of crops are being stored and transported under CA conditions. The technical benefits of CA storage have been amply demonstrated for a wide range of flowers, fruit and vegetables but the economic implications of using this comparatively expensive technology have often limited its commercial application. However, with technological developments, more precise control equipment and the reducing cost, CA storage is being used commercially for an increasing range of crops.

Table 3.1. Recommended storage conditions for Cox's Orange Pippin apples all at 3.5°C (modified from Bishop, 1994).

O_2 kPa	CO_2 kPa	Storage time in weeks	Approximate date of implementation
21	0	13	–
16	5	16	1920
3	5	21	1935
2	< 1	27	1965
1¼	< 1	31	1980
1	< 1	33	1986

The question of changes in quality of fruit after long-term storage is important. Johnson (1994a) found that the storage practices that retard changes in senescence in apples generally reduced the production of volatile aroma compounds. Reduced turnover of cell lipids under CA conditions is thought to result in lack of precursors (long-chain fatty acids) for ester synthesis. The lower O_2 levels required to increase storage duration and to maximize retention of the desired textural characteristics can further reduce aroma development. Using hydrated lime or activated carbon to reduce the level of CO_2 did not affect aromatic flavour development, though there may be concern over continuous flushing of N_2/air mixtures. The prospect of improving flavour in CA stored apples by raising O_2 levels prior to the opening of the store is limited by the need to retain textural quality.

Carbon Dioxide and Oxygen Damage

CO_2 and O_2 injury symptoms on some apple cultivars (Table 3.2) and selected fruit and vegetables (Table 3.3) have been reviewed. It should be noted that the exposure time to different gases will affect susceptibility to injury and there may be interactions with ripeness, harvest maturity or the storage temperature.

Fidler *et al.* (1973) gave a detailed description of injury caused to different cultivars of apples stored in atmospheres containing low O_2 or high CO_2 levels. Internal injury (Fig. 3.1) was described as often beginning

Table 3.2. Threshold level of O_2 or CO_2 required causing injury to apples and typical injury symptoms (Kader, 1989; Saltveit, 1989; Meheriuk, 1989b; Kader, 1993).

Cultivar	CO_2 injury level	CO_2 injury symptoms	O_2 injury level	O_2 injury symptoms
Boskoop	> 2 kPa	CO_2 cavitation	< 1.5 kPa	
Cox's Orange Pippin	> 1 kPa	Core browning	< 1 kPa	Alcoholic taste
Golden Delicious	> 6 kPa continuous > 15 kPa for 10 days or more	CO_2 injury	< 1 kPa	Alcoholic taint
Red Delicious	> 3 kPa	Internal browning	< 1 kPa	Alcoholic taste with late-picked fruit
Starking Delicious	> 3 kPa	Internal disorders	< 1 kPa	Alcoholic taste
Elstar	> 2 kPa	CO_2 injury	< 2 kPa	Coreflush
Empire	> 5 kPa	CO_2 injury	< 1.5 kPa	Flesh browning
Fuji	> 5 kPa	CO_2 injury	< 2 kPa	Alcoholic taint
Gala	> 1.5 kPa	CO_2 injury	< 1.5 kPa	Ribbon scald
Gloster	> 1 kPa	Core browning	< 1 kPa	
Granny Smith	> 1 kPa with O_2 at < 1.5 kPa > 3 kPa with O_2 < 2 kPa	Severe coreflush	< 1 kPa	Alcoholic taint, ribbon scald and core browning
Idared	> 3 kPa	CO_2 injury	< 1 kPa	Alcoholic taint
Jonagold	> 5 kPa	Unknown	< 1.5 kPa	Unknown
Jonathan	> 5 kPa	CO_2 injury, flesh browning	< 1 kPa	Alcoholic taint, core browning
Karmijn	> 3 kPa	Core browning and low temperature breakdown	< 1 kPa	Alcoholic taint
McIntosh	> 5 kPa continuous > 15 kPa for short periods	CO_2 injury, core flush	< 1.5 kPa	Corky browning, skin discoloration, flesh browning, alcoholic taint
Melrose	> 5 kPa	Unknown	< 2 kPa	Alcoholic taint
Mutsu	> 5 kPa with O_2 > 2.5 kPa	Unknown	< 1.5 kPa	Alcoholic taint
Rome	> 5 kPa	Unknown	< 1.5 kPa	Alcoholic taint
Spartan	> 3 kPa	Coreflush, CO_2 injury	< 1.5 kPa	Alcoholic taint
Stayman		Unknown	< 2 kPa	Alcoholic taint

in the vascular tissue and then increases to involve large areas of the cortex. At first the injury zones are firm, and have a 'rubbery' texture when a finger is drawn over the surface of the cut section of the fruit. Later, the damaged tissue loses water and typical cork-like cavities appear.

They also showed that the appearance of CO_2 injury symptoms is a function of concentration, exposure time and temperature. They described external CO_2 injury where 'initially the damaged area is markedly sunken, deep green in colour and with sharply defined edges. Later in storage the damaged tissue turns brown and finally almost black'.

Injury caused as a result of low O_2 levels is due to fermentation, also called anaerobic respiration, resulting in the accumulation of the toxic by-products alcohols and aldehydes. These can result in necrotic tissue, which tends to begin at the centre of the fruit (Fig. 3.2). It may be that the injury that is referred to as spongy tissue in 'Alphonso' mangoes and some other mango varieties may be related to CO_2 injury, since they display morphological symptom similar to those described for apples (Fig. 3.3).

The lower O_2 limit for apples was found to be cultivar dependent, ranging from a low of about 0.8 kPa for 'Northern Spy' and 'Law Rome' to a high of about 1.0 kPa for

Table 3.3. Threshold level of O_2 or CO_2 required causing injury to some fruits and vegetables and typical injury symptoms (Kader, 1989; Saltveit, 1989; Meheriuk, 1989a and b; Kader, 1993).

Crop	CO_2 injury level	CO_2 injury symptoms	O_2 injury level	O_2 injury symptoms
Apricot	> 5 kPa	Loss of flavour, flesh browning	< 1 kPa	Off-flavour development
Artichoke, Globe	> 3 kPa	Stimulates papus development	< 2 kPa	Blackening of inner bracts and receptacle
Asparagus	> 10 kPa at 3 to 6°C > 15 kPa at 0 to 3°C	Increased elongation, weight gain and sensitivity to chilling and pitting	< 10 kPa	Discoloration
Avocado	> 15 kPa	Skin browning, off-flavour	< 1 kPa	Internal flesh breakdown, off-flavour
Banana °	> 7 kPa	Green fruit softening, undesirable texture and flavour	< 1 kPa	Dull yellow or brown skin discoloration, failure to ripen, off-flavour
Beans, Green, Snap	> 7 kPa for more than 24 h	Off-flavour	< 5 kPa for more than 24 h	Off-flavour
Blackberry	> 25 kPa	Off-flavour	< 2 kPa	Off-flavour
Blueberry	> 25 kPa	Skin browning, off-flavour	< 2 kPa	Off-flavour
Broccoli	> 15 kPa	Persistent off-odours	< 0.5 kPa	Off-odours, can be lost upon aeration if slight
Brussels sprouts	> 10 kPa	–	< 1 kPa	Off-odours, internal discoloration
Cabbage	> 10 kPa	Discoloration of inner leaves	< 2 kPa	Off-flavour, increased sensitivity to freezing
Cauliflower	> 5 kPa	Off-flavour, aeration removes slight damage, curd must be cooked to show symptoms	< 2 kPa	Persistent off-flavour and odour after cooking
Celery	> 10 kPa	Off-flavour and odour, internal discoloration	< 2 kPa	Off-flavour and odour
Cherimoya	Not determined	Not known	<1 kPa	off-flavour
Cherry, sweet	> 30 kPa	Brown, discoloration of skin, off-flavour	< 1 kPa	Skin pitting, off-flavour
Cranberry	Not determined	Not determined	<1 kPa	Off-flavour
Cucumber	> 5 kPa at 8°C > 10 kPa at 5°C	Increased softening, increased chilling injury, surface discoloration and pitting	< 1 kPa	Off-odours, breakdown and increased chilling injury
Custard apple	15 kPa +	Flat taste, uneven ripening	< 1 kPa	Failure to ripen
Durian	> 20 kPa	Not known	< 2 kPa	Failure to ripen, grey discoloration of pulp
Fig	> 25 kPa (?)	Loss of flavour (?)	< 2 kPa (?)	Off-flavour (?)
Grape	> 5 kPa	Browning of berries and stems	< 1 kPa	Off-flavour
Grapefruit	> 10 kPa	Scald-like areas on the rind, off-flavour	< 3 kPa	Off-flavour due to increased ethanol and acetaldehyde contents
Kiwifruit	> 7 kPa	Internal breakdown of the flesh	< 1 kPa	Off-flavour
Lemon	> 10 kPa	Increased susceptibility to decay, decreased acidity	< 5 kPa	Off-flavour
Lettuce, Crisphead	> 2 kPa	Brown stain	< 1 kPa	Breakdown at centre

Lime	> 10 kPa	Increased susceptibility to decay	< 5 kPa	Scald-like injury, decreased juice content
Mango	> 10 kPa	Softening, off-flavour	< 2 kPa < 5 kPa	Skin discoloration, greyish flesh colour, off-flavour
Melon, Cantaloupe	> 20 kPa	Off flavour and odours, impaired ripening	< 1 kPa	Off-flavour and odours, impaired ripening
Mushroom	> 20 kPa	Surface pitting	Near 0 kPa	Off-flavour and odours, stimulation of cap opening and stipe elongation
Nectarine	> 10 kPa	Flesh browning, loss of flavour	< 1 kPa	Failure to ripen, skin browning, off-flavour
Olive	> 5 kPa	Increased severity of chilling injury at 7°C	< 2 kPa	Off-flavour
Onion, bulb	> 10 kPa for short term, > 1 kPa for long term	Accelerated softening, rots and putrid odour	< 1 kPa	Off-odours and breakdown
Orange	> 5 kPa	Off-flavour	< 5 kPa	Off-flavour
Papaya	> 8 kPa	May aggravate chilling injury at < 12°C, off-flavour	< 2 kPa	Failure to ripen, off-flavour
Peach, clingstone	> 5 kPa	Internal flesh browning severity increases with CO_2 kPa	< 1 kPa	Off-flavour in the canned product
Peach, freestone	> 10 kPa	Flesh browning, off-flavour	< 1 kPa	Failure to ripen, skin browning, off-flavour
Pepper, Bell	> 5 kPa	Calyx discoloration, internal browning, and increased softening	< 2 kPa	Off-odours and breakdown
Pepper, Chilli	> 20 kPa at 5°C, > 5 kPa at 10°C	Calyx discoloration, internal browning, and increased softening	< 2 kPa	Off odours and breakdown
Persimmon	> 10 kPa	Off-flavour	< 3 kPa	Failure to ripen, off-flavour
Pineapple	> 10 kPa	Off-flavour	< 2 kPa	Off-flavour
Plum	> 1 kPa	Flesh browning	< 1 kPa	Failure to ripen, off-flavour
Rambutan	> 20 kPa	Not known	< 1 kPa	Increased decay incidence
Raspberry	> 25 kPa	Off-flavour, brown discoloration	< 2 kPa	Off-colours
Strawberry	> 25 kPa	Off-flavour, brown discoloration of berries	< 2 kPa	Off-flavour
Sweetcorn	> 10 kPa	Off-flavour and odours,	< 2 kPa	Off-flavour and off-odours,
Tomato	> 2 kPa for mature-green, > 5 kPa for turning, also depends on length of exposure and temperature	Discoloration, softening and uneven ripening	< 2 kPa depending on length of exposure	Off flavour, softening and uneven ripening

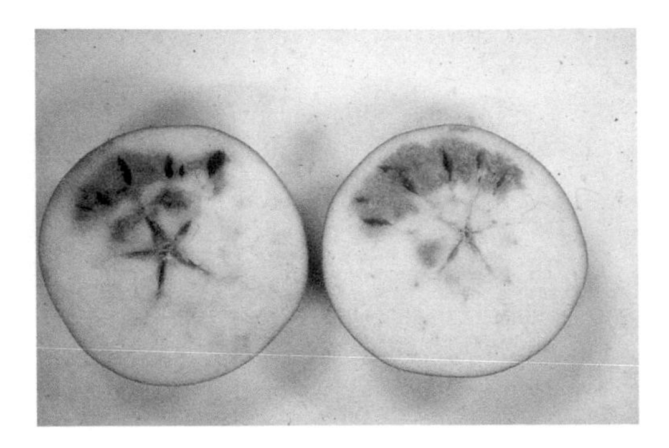

Fig. 3.1. CO_2 injury on apples from a CA store in UK. Photograph courtesy of Dr R.O. Sharples.

Fig. 3.2. O_2 injury on an apple taken from a CA store in UK. Photograph courtesy of Dr R.O. Sharples.

Fig. 3.3. Spongy tissue in mangoes.

'McIntosh' in cold storage. For blueberries, the lower O_2 limit increased with temperature and CO_2 level. Raising the temperature from 0°C to 25°C caused the lower O_2 limit to increase from about 1.8 kPa to approximately 4 kPa. Raising CO_2 levels from 5 kPa to 60 kPa increased the lower O_2 limit for blueberry fruits from approximately 4.5 kPa to > 16 kPa (Beaudry and Gran, 1993). 'Marshall McIntosh' apples held at 3°C in 2.5–3 kPa O_2 + 11–12 kPa CO_2 developed small desiccated cavities in the cortex associated with CO_2 injury (DeEll *et al.*, 1995). Mencarelli *et al.* (1989) showed that high CO_2 concentrations (5, 8 or 12 kPa) during storage resulted in CO_2 injury of aubergines, characterized by external browning without tissue softening. Wardlaw (1938), also working with aubergines, showed that high CO_2 can cause surface scald, browning, pitting and excessive decay, and these symptoms are similar to those caused by chilling injury. Mencarelli *et al.* (1989) showed that the spherical-shaped aubergine cultivars 'Black Beauty' and 'Sfumata Rosa' were more tolerant of high CO_2 concentrations than the long-fruited cultivar 'Violetta Lunga'. Burton (1974a) also found that the level of CO_2 that can cause damage varied between cultivars of the same crop and he put forward a hypothesis that these differences could be due to anatomical rather than biochemical differences. He stated that:

> Variability in plant material prevents precise control of intercellular atmosphere; recommendations can be designed only to avoid complete anaerobic conditions and a harmful level of CO_2 in the centre of the least permeable individual fruit or vegetable.

High CO_2 levels (10 kPa) have been shown to cause damage to stored onions, particularly resulting in internal browning (Gadalla, 1997). Adamicki *et al.* (1977) suggested that the decomposition of the cell walls was the result of the influence of hydrolytic enzymes of the pectinase group. Their preliminary comparative studies on cells of sound and physiologically disordered onions due to CO_2 injury indicated destructive changes in the ultra-structure of the mitochondria. The mitochondria displayed fragmentation, reduction in size and changes in shape from elliptical to spherical. These changes have also been observed in studies on the influence of CO_2 on the ultra-structure of pears (Frenkel and Patterson, 1974). Elevated CO_2 concentrations were shown to inhibit the activity of succinic dehydrogenase, resulting in accumulation of succinic acid, a toxicant to plant tissues (Hulme, 1956; Williams and Patterson, 1964; Frenkel and Patterson, 1974). Adamicki *et al.* (1977) discovered that the highest amino acid content was found in the physiologically disordered onion bulbs stored at 1°C, with lower values found at 5°C. They added that this may be due to greater enzyme activity, especially in physiologically disordered bulbs. Low O_2 levels (0.21 kPa and 3 kPa) on stored avocados from early-season harvests could lead to severe external browning (Allwood and Cutting, 1994).

There may be interactions between low O_2 and high CO_2 levels. Riad and Brecht (2003) reported that at 5°C sweetcorn tolerated 2 kPa O_2 or 25 kPa CO_2 alone for 2 weeks, but may be damaged by these two gas levels in combination.

High Oxygen Storage

Increased O_2 is also achieved by hyperbaric storage (see also Chapter 10). Kidd and West (1934) showed that storage of apples in pure O_2 can be detrimental and Yahia (1989, 1991) showed that exposure of 'McIntosh' and 'Cortland' apples to 100 kPa O_2 at 3.3°C for 4 weeks did not enhance the production of aroma volatiles. Pears kept in 100 kPa O_2 showed an increase in the rate of softening, chlorophyll degradation and ethylene evolution (Frenkel, 1975). Various publications have shown that high O_2 levels in citrus fruit stores can affect the fruit colour. Navel and 'Valencia' oranges were stored for 4 weeks at 15°C in a continuous flow of gases and after 2 weeks of storage in 40 or 80 kPa O_2, with the balance N_2, the endocarp of navel oranges, but not of 'Valencia', had turned a perceptibly darker orange than fruit stored in air or in air with ethylene. The orange colour intensified with increased storage time and persisted at least 4 weeks after the fruit was removed from high O_2. Juice from navel oranges, but not from 'Valencia', stored in 80 kPa O_2 for 4 weeks was slightly darker than juice from those stored in air (Houck *et al.*, 1978). Storage at 15°C for 4 weeks in 40 or 80 kPa O_2 deepened the red of flesh and juice of 'Ruby', 'Tarocco' and 'Sanguinello' blood oranges, but TSS and TA of the juice were not affected. 'Hamlin', 'Parson Brown' and 'Pineapple' oranges showed similar results with the respiration rate of 'Pineapple' being highest in 80 kPa O_2 and lowest in 40 kPa O_2 or air (Aharoni and Houck, 1980).

Ripe-green tomatoes were stored at either 19°C or 13°C followed by 19°C in air or in 100 kPa O_2 at normal or reduced atmospheric pressure until they were fully red in colour. Fruit softening was exacerbated in 100 kPa O_2 and reduced pressure (0.25 atmospheres) compared with a normal gaseous atmosphere and reduced pressure. Removal of ethylene in the atmosphere made little difference to the fruit but in 100 kPa O_2 ethylene accumulation was detrimental (Stenvers, 1977). Li *et al.* (1973) reported that ripening of tomatoes was accelerated at 12–13°C in 40–50 kPa O_2 compared with those ripened in air.

At 4°C the respiration rate of potato tubers was lower in 1 or 3 kPa O_2 compared with those stored in air or in 35 kPa O_2. Sprouting was inhibited and sugar content increased in 1 and 35 kPa O_2 compared with those in air, but those in 3 kPa O_2 showed no effects (Hartmans *et al.*, 1990). Storage atmospheres containing 40 kPa O_2 reduced mould infection compared with low O_2 levels, but increased sprouting and rooting of stored carrots (Abdel-Rahman and Isenberg, 1974). Blueberries in jars ventilated continuously with 60, 80 or 100 kPa O_2 had improved antioxidant capacity, total phenolic and anthocyanin contents and less decay compared with those in 0 or 40 kPa O_2 during 35 days of storage at 5°C (Zheng *et al.*, 2003). Dick and Marcellin (1985) showed that bananas held in 50 kPa O_2 during cooling periods of 12 h at 20°C had reduced high temperature damage during subsequent storage at 30–40°C. Belay *et al.* (2017) compared the storage of pomegranate arils at 5°C for 12 days in air, 5 kPa O_2 + 10 kPa CO_2, 10 kPa O_2 + 5 kPa CO_2 or 70 kPa O_2 + 10 kPa CO_2, all with balance N_2. They found that the highest relative composition of volatile organic compounds was in arils stored under 70 kPa O_2 + 10 kPa CO_2. Day (1996) indicated highly positive effects of storing minimally processed fruit and vegetables in 70 and 80 kPa O_2 in MA film packs.

He indicated that high O_2 inhibited undesirable fermentation reactions and delayed browning that was caused by damage during processing and that the O_2 levels of over 65 kPa inhibited both aerobic and anaerobic microbial growth. The technique used involved flushing the packs with the required gas mixture before they were sealed. The O_2 level within the package would then fall progressively as storage proceeded, due to the respiration of the fruit or vegetable contained in the pack. High O_2 levels have also been used in the marketing of other commodities. For example, Champion (1986) mentioned that O_2 levels above ambient were used to help preserve 'redness and eye appeal' of red meat.

Carbon Dioxide Shock Treatment

Treating fruit and vegetables with high levels of CO_2 prior to storage can have beneficial effects on their subsequent storage life. Hribar *et al.* (1994) described experiments where 'Golden Delicious' apples were held in 15 kPa CO_2 for 20 days prior to storage in 1.5 kPa O2 + 1.5 kPa CO_2 for 5 months. The CO_2 treated fruits retained a better green skin colour during storage, which was shown to be due to inhibition of carotenoid production. Other effects of the CO_2 shock treatment were that it increased fruit firmness during storage, ethanol production was greater due to fermentation, but TA was not affected. Pesis *et al.* (1993a) described experiments where apples were treated with 95 kPa CO_2 + 1 kPa O_2 + 4 kPa N_2 at 20°C for 24–48 h for 'Golden Delicious' and 24–96 h for 'Braeburn', then transferred to either 20°C or 5°C for 6 weeks in air followed by a shelf-life of 10 days at 20°C. High CO_2 pre-treatment increased respiration rate and induced ethylene, ethanol and ethyl acetate production in 'Golden Delicious' apples during storage at 20°C. The fruits became softer and more yellow but 'tastier' than non-treated fruits after 2 weeks at 20°C. For 'Braeburn', respiration rate and ethylene production were reduced and volatiles were increased by the high CO_2 pre-treatment during shelf-life at 20°C following 6 weeks of cold storage, and treated fruits remained firmer but more yellow than control fruits.

Two cultivars of table grapes were stored by Laszlo (1985) in CA: 'Waltham Cross' in 1–21 kPa O_2 + 0–5 kPa CO_2 and 'Barlinka' in 1–21 kPa O_2 + 5 kPa CO_2, both at –0.5°C for 4 weeks. Some of the grapes were exposed to 10 kPa CO_2 for 3 days before storage in air at 10°C for 1 week. With both cultivars the incidence of berry decay (mainly due to infections by *Botrytis cinerea*) was highest with grapes that had been subjected to the CO_2 treatment. The incidence of berry cracking was less than 1% in 'Waltham Cross' but in 'Barlinka' it was higher with fruits that had been subjected to the CO_2 treatment, high O_2 concentrations and controls without SO_2. In a trial with 'Passe Crassane' pears stored in air at 0°C for 187 days, internal browning was prevented by treatment with 30 kPa CO_2 for 3 days at intervals of 14–18 days. Also the fruits that had been subjected to CO_2 shock treatment had excellent quality after being ripened at 20°C (Marcellin *et al.*, 1979). They also reported that, with 'Comice' pears stored at 0°C for 169 days followed by ripening for 7 days, 15 kPa CO_2 shock treatment every 14 days greatly reduced the incidence of scald and internal browning. Park *et al.* (1970) showed that PE film packaging and/or CO_2 shock treatment markedly delayed ripening, preserved freshness and reduced spoilage and core browning of pears during storage.

'Fuerte' avocados were treated with 25 kPa CO_2 for 3 days, commencing 1 day after harvest, followed by storage for 28 days at 5.5°C in either 2 kPa O_2 + 10 kPa CO_2 or in air. Fruits from the CA and CO_2 shock treatments showed a lower incidence of physiological disorders (mesocarp discoloration, pulp spot and vascular browning) than fruits that had not had the CO_2 treatment. Total phenols tended to be lower in CO_2 shock fruits than in fruits from other treatments (Bower *et al.*, 1990). Wade (1979) showed that intermittent exposure of unripe avocados to 20 kPa CO_2 reduced chilling injury when they were stored at 4°C.

Pesis and Sass (1994) showed that exposure of feijoa fruits (*Acca sellowiana* cultivar 'Slor') to total N_2 or CO_2 for 24 h prior to storage induced the production of aroma volatiles including acetaldehyde, ethanol, ethyl acetate and ethyl butyrate. The enhancement of flavour was mainly due to the increase in volatiles and not to changes in TSS or the TSS:TA ratio. Eaks (1956) showed that high CO_2 in the storage atmosphere could have detrimental effects on cucumbers in that it appeared to increase their susceptibility when stored at low temperatures. Broccoli was treated with CO_2 at 5°C with 20, 30 and 40 kPa CO_2 for 3 and 6 days before storage. CO_2 treatment delayed yellowing and loss of both chlorophyll and ascorbic acid and retarded ethylene production, but 30 or 40 kPa CO_2 for 6 days resulted in the development of an offensive odour and flavour dissipated when the broccoli was transferred to air (Wang, 1979).

CO_2 behaves as a supercritical fluid above its critical temperature of 31.1°C and above its critical pressure of 7.39 MPa, and expands to fill the container like a gas but with a density like a liquid. Gui *et al.* (2006) exposed horseradish to supercritical CO_2 and found some reduction in enzyme activity which was reversed during subsequent storage in air at 4°C.

Total Nitrogen or High Nitrogen Storage

Since fresh fruit and vegetables are living organisms, they require O_2 for aerobic respiration. Where this is available to individual cells below a threshold level, the fruit or vegetable or any part of the organism can go into anaerobic respiration, usually called fermentation, the end products of which are organic compounds that can affect their flavour. It was reported by Food Investigation Board (1920) that storage of plums in total N_2 almost completely inhibited ripening. Plums were reported to be able to tolerate, for a considerable period, an almost complete absence of O_2 without being killed or developing an alcoholic or unpleasant flavour. Parsons *et al.* (1964) successfully stored several fruit and vegetables at 1.1°C in either total N_2 or 1 kPa O_2 + 99 kPa N_2. During 10 days storage of lettuce, both treatments reduced the physiological disorder russet spotting and butt discoloration compared with those stored in air without affecting their flavour. At 15.5°C, ripening of both tomatoes and bananas was retarded when they were stored in total N_2, but their flavour was poorer only if they were held in these conditions for longer than 4 days or for over 10 days in 99 kPa N_2 + 1 kPa O_2. In strawberries, 100 kPa and 99 kPa N_2 were shown to reduce mould growth during 10 days of storage at 1.1°C with little or no effect on flavour. Decay reduction was also observed on peaches stored in either 100 kPa or 99 kPa N_2 at 1.1°C; off-flavours were detected after 4 days in 100 kPa, but none in those stored in 99 kPa N_2. Klieber *et al.* (2002) found that storage of bananas in total N_2 at 22°C did not extend their storage life compared with those stored in air but resulted in brown discoloration. Storing potatoes in total N_2 prevented accumulation of sugars at low temperature, but it had undesirable side effects (Harkett, 1971). It was reported that the growth of fungi was inhibited by atmospheres of total N_2 but not in atmospheres containing 99 kPa N_2 + 1 kPa O_2 (Ryall, 1963). Initial low oxygen stress (ILOS) is a name that has been given to a pre-storage treatment where apples and pears are exposed to anaerobic conditions, commonly 7 days in 100% N_2, to inhibit superficial scald and increase their storage life (Eaves *et al.*, 1969a; Robatscher *et al.*, 2012) (see Chapter 9). Treatment of fruit in atmospheres of total N_2 prior to storage was shown to retard the ripening of tomatoes (Kelly and Saltveit, 1988) and avocados (Pesis *et al.*, 1993b). 'Fuerte' avocados were exposed to 97 kPa N_2 for 24 h or 40 h and then successfully stored at 17°C in air for 7 days by Dori *et al.* (1995) and Pesis *et al.* (1993b). They exposed 'Fuerte' avocados to 97 kPa N_2 for 24 h at 17°C then stored them at 2°C for 3 weeks followed by shelf-life at 17°C in air. Pre-treatment with N_2 reduced chilling injury symptoms and softening was also delayed and they had lower respiration rates and ethylene production during cold storage and shelf-life.

Ethylene

The physiological effects of ethylene on plant tissue have been known for several decades. Gane (1934), at the Low Temperature Research Station at Cambridge, showed that the volatile agent given off by ripening apples and pears is ethylene and at that time ethylene began to be considered as a ripening hormone and shipping companies stopped the transport of apples and bananas in adjacent holds. The concentration of ethylene causing half maximum inhibition of growth of etiolated pea seedlings was shown to be related to the CO_2 and O_2 levels in the surrounding atmosphere (Table 3.4) by Burg and Burg (1967). They suggested that high levels of CO_2 in stores can compete with ethylene for binding sites in fruit and the biological activity of 1 kPa ethylene was negated in the presence of 10 kPa CO_2. CO_2 levels in store affected ethylene biosynthesis and it was shown that increased levels of CO_2 in CA stores containing apples reduced ethylene levels (Tomkins and Meigh, 1968). Yang (1985) showed that CO_2 accumulation in the intercellular spaces of fruit acts as an ethylene antagonist. Woltering *et al.* (1994) therefore suggested that most of the beneficial effects of CA storage in climacteric fruit are due to suppression of ethylene action. However, Knee (1990) had pointed out earlier that ethylene was not normally removed in commercial CA apple stores, but CA technology was successful and so there must be an effect of reduced O_2 and or increased CO_2 apart from those on ethylene synthesis or ethylene action. Knee also mentioned that laboratory experiments on CA storage commonly used small containers which were constantly purged with the appropriate gas mixture. This meant that ethylene produced by the crop under such conditions was constantly being removed. In contrast, the concentration of gases in commercial CA stores was usually adjusted by scrubbing and limited ventilation, which could allow ethylene to accumulate to high levels in the stores. Knee also pointed out that ethylene concentrations of up to 1000 μl/l had been reported in CA stores containing apples and 100 μl/l in CA stores containing cabbage.

Table 3.4. Sensitivity at 20°C of pea seedlings to ethylene in different levels of O_2, CO_2 and N_2 (Burg and Burg, 1967).

O_2 kPa	CO_2 kPa	N_2 kPa	Sensitivity (μl ethylene/l)
0.7	0	99.3	0.6
2.2	0	97.8	0.3
18.0	0	82.0	0.14
18.0	1.8	80.2	0.3
18.0	7.1	74.9	0.6

Deng *et al.* (1997) found that as the O_2 level in the store reduced, the rate of ethylene production was also reduced. Wang (1990) also showed that ethylene production was suppressed in CA storage. He described experiments with sweet peppers stored at 13°C where exposure to combinations of 10–30 kPa CO_2 with either 3 or 21 kPa O_2 suppressed ethylene production down to less than 10 nl/100 g/h compared with fruits stored in air at 13°C which had ethylene production in the range of 40–75 nl/100 g/h. He also showed that the ethylene levels rapidly increased to a similar level to those stored in air when removed from exposure to CA storage for just 3 days. The concentrations of CO_2 and O_2 in such conditions should therefore inhibit ethylene biosynthesis. Plich (1987) found that the presence of a high concentration of CO_2 may strongly inhibit ACC synthesis in 'Spartan' apples and, consequently, the rate of ethylene evolution. The lowest ethylene evolution was found in fruit stored in 1 kPa O_2 + 2 kPa CO_2 or 1 kPa O_2 + 0 kPa CO_2, but in 3 kPa O_2 + 0 kPa CO_2 there was considerably more ethylene production whereas in 3 kPa O_2 + 5 kPa CO_2 ethylene production was markedly decreased.

Before the availability of gas chromatography it was generally thought that only climacteric fruit produced ethylene. Now it has been shown that all plant material can produce ethylene and Shinjiro *et al.* (2002) found that ethylene production in pineapples (a non-climacteric fruit) increased the longer the fruits were stored, but the maximum production rate was less than 1nl/g/h.

CA storage has also been shown to stimulate ethylene production. After 70 days in cold storage of celeriac in air the ethylene content did not exceed 2 ppm, but the ethylene level in CA stores of celeriac was nearly 25 ppm (Golias, 1987). Plich (1987) found that the lowest ethylene production was in apples stored in 1 kPa O_2 + 0 or 2 kPa CO_2, but there was considerably more ethylene production in 3 kPa O_2 + 0 kPa CO_2. Bangerth (1984) in studies on apples and bananas suggested that they became less sensitive to ethylene during prolonged storage.

The level of CO_2 and O_2 in the surrounding environment of climacteric fruit can affect the ripening rate by suppressing the synthesis of ethylene. The biosynthesis of ethylene in ripening fruit was shown in early work by Gane (1934) to cease in the absence of O_2. Wang (1990) reviewed the literature on the effects of CO_2 and O_2 on the activity of enzymes associated with fruit ripening and cited many examples of the activity of these enzymes being reduced in CA storage. This is presumably due, at least partly, to many of these enzymes requiring O_2 for their activity. Ethylene biosynthesis was studied in 'Jonathan' apples stored at 0°C in 0–20 kPa CO_2 + 3 kPa O_2 or 0–20 kPa CO_2 + 15 kPa O_2 for up to 7 months. Internal ethylene concentration, ACC levels and ACC-oxidise activity were determined in fruits immediately after removal from storage and after holding at 20°C for one week. Ethylene production by fruits was inhibited by increasing CO_2 concentration over the range of 0–20 kPa at both 3 and 15 kPa O_2. ACC levels were similarly reduced by increasing CO_2 concentrations even in 3 kPa O_2, but at 3 kPa O_2 ACC accumulation was enhanced but only in the absence of CO_2. ACC oxidase activity was stimulated by CO_2 up to 10 kPa but was inhibited by 20 kPa CO_2 at both O_2 concentrations. The inhibition of ethylene production by CO_2 may, therefore, be attributed to its inhibitory effect on ACC-synthase activity (Levin *et al.*, 1992). In other work, storing pre-climacteric apples of the cultivars 'Barnack Beauty' and 'Wagner' at 20°C for 5 days in 0 kPa O_2 + 1 kPa CO_2 + 99 kPa N_2 or in air containing 15 kPa CO_2 inhibited ethylene production and reduced ACC concentration and ACC-oxidase activity compared with storage in air (Lange *et al.*, 1993). Storage of kiwifruit in 2–5 kPa O_2 + 0–4 kPa CO_2 reduced ethylene production and ACC oxidase activity (Wang *et al.*, 1994). Ethylene production was lower in 'Mission' figs stored at 15–20 kPa CO_2 concentrations compared with those kept in air (Colelli *et al.*, 1991). Low O_2 and increased CO_2 concentrations were shown to decrease ethylene sensitivity of 'Elstar' apples and 'Scania' carnation flowers during CA storage (Woltering *et al.*, 1994).

Quazi and Freebairn (1970) showed that high CO_2 and low O_2 delayed the production of ethylene in pre-climacteric bananas, but the application of exogenous ethylene was shown to reverse this effect. Wade (1974) showed that bananas could be ripened in atmospheres of reduced O_2, even as low as 1 kPa, but the peel failed to de-green, which resulted in ripe bananas that were still green. Similar effects were shown at O_2 levels as high as 15 kPa. Since the de-greening process in 'Cavendish' bananas is entirely due to chlorophyll degradation (Seymour *et al.*, 1987a and b) the CA storage treatment was presumably due to suppression of this process. Hesselman and Freebairn (1969) showed that ripening of bananas, which had already been initiated to ripen by ethylene, was slowed in a low O_2 atmosphere. Goodenough and Thomas (1980, 1981) also showed suppression of de-greening of fruits during ripening; in this case it was with tomatoes ripened in 5 kPa O_2 + 5 kPa CO_2. Their work, however, showed that this was due to a combination of suppression of chlorophyll degradation and the suppression of the synthesis of carotenoids and lycopene. Jeffery *et al.* (1984) also showed that lycopene synthesis was suppressed in tomatoes stored in 6 kPa O_2 + 6 kPa CO_2.

Ethylene is used as a commercial sprouting inhibitor in potatoes, but ethylene is required to break dormancy, while continuous exposure inhibits sprout growth. In sweet potatoes Cheema *et al.* (2013) showed that continuous exposure to ethylene at 10 μl/l for 24 h, exposure to 1-MCP at 625 nl/l or dipping in AVG at 100 μl/l all inhibited sprout growth over 4 weeks of storage at 25°C. Both ethylene on its own and 1-MCP inhibited sprout growth, which they concluded indicated that while continuous exposure to exogenous ethylene leads to sprout growth inhibition, ethylene is also required for sprouting.

Carbon Monoxide

CO is a colourless, odourless gas which is flammable and explosive in air at concentrations between 12.5 and 74.2 kPa. It is extremely toxic and human exposure to 0.1 kPa for 1 h can cause unconsciousness and exposure for 4 h can cause death. Goffings and Herregods (1989) reported that CA storage of leeks could be improved by the inclusion of 5 kPa CO. If added to CA stores, with levels of 2–5 kPa O_2, CO can inhibit discoloration of lettuce on the cut butts or from mechanical damage on the leaves, but the effect was lost when the lettuce was removed to air at 10°C (Kader, 1992). Storage in 4 kPa O_2 + 2 kPa CO_2 + 5 kPa CO was shown to be optimum in delaying ripening and maintaining good quality of mature-green tomatoes stored at 12.8°C (Morris *et al.*, 1981). Reyes and Smith (1987) reported positive effects of CO during the storage of celery. CO has fungistatic properties, especially when combined with low O_2 (Kader, 1993). *Botrytis cinerea* on strawberries was reduced in CO levels of 5 kPa or higher in the presence of 5 kPa O_2 or lower (El-Goorani and Sommer, 1979). Decay in mature-green tomatoes stored at 12.8°C was reduced when 5 kPa CO was included in the storage atmosphere (Morris *et al.*, 1981). They also reported that in capsicums and tomatoes the level of chilling injury symptoms could be reduced, but not eliminated, when CO was added to the store.

Temperature

Kidd and West (1927a, b) were the first to show that the effects of the gases in extending storage life could vary with temperature. At 10°C, CA storage could increase the postharvest life of apples compared with those in air, while at 15°C their storage life was the same in both CA storage and in air. The opposite effect was shown by Ogata *et al.* (1975). They found that CA storage of okra at 1°C did not increase storage life compared with storage in air, but at 12°C CA storage increased postharvest life. Izumi *et al.* (1996a) also showed that the best CA conditions varied with temperature. They found that the best storage conditions for the broccoli cultivar 'Marathon' were 0.5 kPa O_2 + 10 kPa CO_2 at 0°C and 5°C, and 1 kPa O_2 + 10 kPa CO_2 at 10°C.

The relationship between CA storage and temperature has been shown to be complex. Respiration rates of apples were progressively reduced by lowering O_2 levels from 21 kPa to 1 kPa. Although lowering the temperature from 4°C to 2°C also reduced the respiration rate, fruits stored in 1 or 2 kPa O_2 were shown to be respiring faster after 100 days at 0°C than at 2°C or 4°C. After 192 days the fruit stored in air also showed an increase in respiration rate at 0°C. These higher respiration rates preceded the development of low temperature breakdown in fruit stored in air, 1 or 2 kPa O_2 at 0°C and in 1 kPa O_2 at 2°C. Progressively lowering the O_2 concentration reduced ethylene production but increased the retention of acidity, TSS, chlorophyll and firmness. In the absence of low temperature breakdown the effects of reduced temperature on fruit ripening were similar to those of lowered O_2 concentrations. The quality of apples stored at 4°C in 1 kPa O_2 was markedly better than in 2 kPa; the fruits were also free of core flush (brown core) and other physiological disorders (Johnson and Ertan, 1983).

Humidity

Kajiura (1973) stored *Citrus natsudaidai* at 4°C for 50 days in either 98–100 or 85–95% RH in air mixed with 0, 5, 10 or 20 kPa CO_2. It was found that, at the higher humidity, high CO_2 increased the water content of the peel and the ethanol content of the juice and produced abnormal flavour and reduced the internal O_2 content, causing watery breakdown. At the lower humidity, no injury occurred and CO_2 was beneficial, its optimum level being much higher. In another trial, *C. natsudaidai* fruits stored at

4°C in 5 kPa CO_2 + 7 kPa O_2 had abnormal flavour development at 98–100% RH but those in 85–95% RH did not. Bramlage *et al.* (1977) showed that treatment of 'McIntosh' apples with high CO_2 in a non-humidified room reduced CO_2 injury without reducing other benefits compared with those treated in a room with humidified air. In a comparison by Polderdijk *et al.* (1993) of 85, 90, 95 or 100% RH during CA storage of capsicums, they found that after removal from storage the incidence of decay increased as humidity increased but weight loss and softening increased as humidity decreased. The physiological disorder vascular streaking in cassava that develops rapidly during storage was almost completely inhibited at high humidity irrespective of the O_2 level (Aracena *et al.*, 1993).

Delayed CA Storage

Generally, placing fruit or vegetables in store as quickly as possible after harvest gives the longest storage period. Drake and Eisele (1994) found that immediate establishment of CA conditions of apples after harvest resulted in good-quality fruits after 9 months of storage. Reduced quality was evident when CA establishment was delayed by as little as 5 days, even though the interim period was in air at 1°C. Colgan *et al.* (1999) also showed that apple scald was controlled less effectively when establishment of CA conditions was delayed. In contrast, Streif and Saquet (2003) found that flesh browning in apples was reduced when the establishment of the CA conditions was delayed, with 30 days' delay being optimum. However, softening was quicker than in fruits where CA conditions were established directly after harvest. Landfald (1988) recommended prompt CA storage to reduce the incidence of lenticel rot in 'Aroma' apples. BBD appeared to develop during the first 2 weeks of storage and storage in air at 0°C prior to CA storage decreased incidence and severity of the disorder (Elgar *et al.*, 1998). Watkins *et al.* (1997) also showed that the sensitivity of 'Braeburn' to CO_2 induced injury was greatest soon after harvest, but declined if fruits were held in air storage before CO_2 application. The external skin disorders in 'Empire' apples could be reduced when fruits were held in air before CA was established (Watkins *et al.*, 1997). Streif and Saquet (2003) found that flesh browning in 'Elstar' apples was reduced when the establishment of the CA conditions was delayed for up to 40 days. A delay of 30 days was optimum, but firmness loss was much higher than in fruits where CA conditions were established directly after harvest. The storage conditions they compared at 1°C were 0.6 kPa CO_2 + 1.2 kPa O_2, 2.5 kPa CO_2 + 1.2 kPa O_2, 5 kPa CO_2 + 1.2 kPa O_2, 0.6 kPa CO_2 + 2.5 kPa O_2, 2.5 kPa CO_2 + 2.5 kPa O_2 and 5 kPa CO_2 + 2.5 kPa O_2. Wang *et al.* (2000c) also found that CO_2 linked disorders were reduced in 'Elstar' apples by storage at 3°C in air before CA storage, but excessive ripening and associated loss of flesh firmness occurred. Argenta *et al.* (2000) found that delaying establishment of CA conditions for 2–12 weeks significantly reduced the severity of brown heart in 'Fuji' apples, but resulted in lower firmness and TA compared with establishment of 1.5 kPa O_2 + 3 kPa CO_2. DeEll *et al.* (2016) showed that slow cooling of apples after harvest may exacerbate the development of external CO_2 injury during CA storage.

Storage of 'Jupiter' capsicums for 5 days at 20°C in 1.5 kPa O_2 resulted in post-storage respiratory rate suppression for about 55 h after transfer to air (Rahman *et al.*, 1995). Saquet *et al.* (2017) stored 'Rocha' pears for 257 days at –0.5°C under 0.6 kPa CO_2 combined with either 0.5 or 3.0 kPa O_2. The atmospheres were imposed either at the beginning of storage or after 46 days. They found that delaying CA for 46 days did not benefit the pears but development of internal disorders depended on O_2 level, in that when CA was imposed immediately under 0.5 kPa O_2 there were no internal disorders, but under 3.0 kPa O_2, 63.3% of fruit developed internal disorders.

In work on kiwifruit, Tonini and Tura (1997) showed that storage in 4.8 kPa CO_2 + 1.8 kPa O_2 reduced rots caused by infections with *Botrytis cinerea*. If the fruit was cooled to –0.5°C immediately after harvest then the effect was greatest, the quicker the CA storage conditions were established. With a delay of 30 days the CA storage conditions were ineffective in controlling the rots.

In pears, delaying CA conditions for 46 days did not benefit 'Rocha' pears during storage for 257 days. Pears tolerated immediate storage in 0.6 kPa CO_2 + 0.5 kPa O_2 without developing internal disorders during subsequent storage. However, 63.3% of fruit stored immediately in 0.6 kPa CO_2 + 3.0 kPa O_2 developed internal disorders, while delayed pull-down of O_2 for 46 days reduced their internal disorders to 35.5%, but increased the disorder incidence in fruit in 0.6 kPa CO_2 + 0.5 kPa O_2 to 27.3% (Saquet *et al.*, 2017). Both delayed storage of 'Flavourtop' nectarines held for 48 h at 20°C before storage, and CA (10 kPa CO_2 + 3 kPa O_2), alleviated or prevented chilling injury manifested as woolliness in nectarines stored for up to 6 weeks at 0°C (Zhou *et al.*, 2000).

Interrupted CA Storage

Where CA storage has been shown to have detrimental side effects on fruit and vegetables, the possibility of alternating CA storage with air storage has been studied. Results have been mixed with positive, negative and in some cases no effect. Neuwirth (1988) described storage of 'Golden Delicious' apples from January to May at 2°C in 1 kPa CO_2 + 3 kPa O_2, 3 kPa CO_2 + 3 kPa O_2 or 5 kPa CO_2 + 5 kPa O_2. These regimes were interrupted by a 3-week period of ventilation with air, beginning on 15 January, 26 January, 17 February or 11 March, after which the CA treatment was reinstated. Ventilation at the time of the climacteric rise in respiration in late February to early March produced large increases in respiration rate and volatile flavour substances, but ventilation at other times had little effect. After CA conditions were restored, respiration rate and the production of flavour substances declined again, sometimes to below the level of fruit stored in continuous controlled atmospheres.

Storage of bananas at high temperatures, as may happen in producing countries, can cause physiological disorders and unsatisfactory ripening. In trials with the variety 'Poyo' from Cameroon, storage at 30–40°C was interrupted by 1–3 periods at 20°C for 12 h either in air or in atmospheres containing 50 kPa O_2 or 5 kPa O_2. Damage was reduced, especially when fruits were stored at 30°C and received three cooling periods in 50 kPa O_2 (Dick and Marcellin, 1985). Parsons *et al.* (1974) interrupted CA storage of tomatoes at 3 kPa O_2 + 0, 3 or 5 kPa CO_2 each week by exposing them to air for 16 h at 13°C. This interrupted storage had no measurable effect on the storage life of the fruit, but increased the level of decay that developed on fruit when removed from storage to higher temperatures to simulate shelf-life. Intermittent exposure of 'Haas' avocados to 20 kPa CO_2 increased their storage life at 12°C and reduced chilling injury during storage at 4°C compared with those stored in air at the same temperatures (Marcellin and Chaves, 1983). Anderson (1982) described experiments where peaches and nectarines were stored at 0°C in 5 kPa CO_2 + 1 kPa O_2, which was interrupted every 2 days by removing the fruit to 18–20°C in air. When subsequently ripened, fruits in this treatment had little of the internal breakdown found in fruits stored continuously in air at 0°C.

Residual Effects of CA Storage

There is considerable evidence in the literature that storing fruit and vegetables in CA storage can affect their subsequent shelf-life or marketable life. Day (1996) indicated that minimally processed fruit and vegetables stored in 70 kPa and higher O_2 levels deteriorated more slowly on removal than those freshly prepared. Hill (1913) described experiments on peaches stored in increased levels of CO_2 and showed that their respiration rate was reduced, not only during exposure, but he also showed that respiration rate only returned to the normal level after a few days in air. Bell peppers exposed to 1.5 kPa O_2 for 1 day exhibited suppressed respiration rate for at least 24 h after transfer to air (Rahman *et al.*, 1993a). Berard (1985) showed that cabbages stored at 1°C and

92% RH in 2.5 kPa O_2 + 5 kPa CO_2 had reduced losses during long-term storage compared with those stored in air, but also the beneficial effects persisted after removal from CA. Burdon *et al.* (2008) showed that avocados that had been stored in CA had a longer shelf-life than those that had been stored in air for a similar period.

Goulart *et al.* (1990) showed that when 'Bristol' raspberries were stored at 5°C in 2.6, 5.4 or 8.3 kPa O_2 + 10.5 15.0 or 19.6 kPa CO_2 or in air the weight loss was greatest after 3 days for berries stored in air. When berries were removed from CA after 3 days and held for 4 days in air at 1°C, those that had been stored in 15 kPa CO_2 had less deterioration than any other treatment except for those stored in air. Deterioration was greatest in the berries that had previously been stored in the 2.6, 5.4 or 8.3 kPa O_2 + 10.5 kPa CO_2 treatments. Berries removed after 7 days and held for up to 12 days at 1°C showed least deterioration after the 15.0 kPa CO_2 storage.

The climacteric rise in respiration rate of cherimoyas was delayed by storage in 15 or 10 kPa O_2 and fruit kept in 5 kPa O_2 did not show a detectable climacteric rise and did not produce ethylene. All fruit ripened normally after being transferred to air storage at 20°C, but the time needed to reach an edible condition differed with O_2 level and was inversely proportional to O_2 concentration during storage. The actual data showed that following 30 days of storage in 5, 10 and 20 kPa O_2 fruit took 11, 6 and 3 days respectively to ripen at 20°C (Palma *et al.*, 1993).

Fruit firmness can be measured by inserting a metal probe into a fruit and measuring its resistance to the insertion. This is called a pressure test and generally the greater the resistance, the firmer and more immature is the fruit. The plum cultivars 'Santa Rosa' and 'Songold' were partially ripened to a firmness of approximately 4.5 kg pressure then kept at 0.5°C in 4 kPa O_2 + 5 kPa CO_2 for 7–14 days. This treatment kept the fruits in an excellent condition for an additional 4 weeks when they were removed to 7.5°C in air (Truter and Combrink, 1992). Mencarelli *et al.* (1983) showed that storage of courgettes in low O_2 protected them against chilling injury, but on removal to air storage at the same chilling temperature the protection disappeared within 2 days.

Khanbari and Thompson (1996) stored the potato cultivars 'Record', 'Saturna' and 'Hermes' in different CA conditions and found that there was almost complete sprout inhibition, low weight loss and maintenance of a healthy skin for all cultivars stored in 9.4 kPa CO_2 + 3.6 kPa O_2 at 5°C for 25 weeks. When tubers from this treatment were stored for a further 20 weeks in air at 5°C the skin remained healthy and they did not sprout while the tubers that had been previously stored in air or other CA combinations sprouted quickly. The fry colour of the crisps made from these potatoes was darker than the industry standard, but when they were reconditioned, tubers of 'Saturna' produced crisps of an acceptable fry colour while crisps from the other two cultivars remained too dark. This residual effect of CA storage could have major implications in that it presents an opportunity to replace chemical treatments in controlling sprouting in stored potatoes. The reverse was the case with bananas that had been initiated to ripen by exposure to exogenous ethylene and then immediately stored in 1 kPa O_2 at 14°C. They remained firm and green for 28 days, but then ripened almost immediately when transferred to air at 21°C (Liu, 1976a). Conversely, Wills *et al.* (1982) showed that pre-climacteric bananas exposed to low O_2 took longer to ripen when subsequently exposed to air than fruit kept in air for the whole period.

Hardenburg *et al.* (1977) showed that apples in CA storage for 6 months and then for 2 weeks at 21°C in air were firmer and more acid, and had a lower respiration rate, than those that had previously been stored in air. Storage of 'Jupiter' capsicums for 5 days at 20°C in 1.5 kPa O_2 resulted in post-storage suppression of the respiration rate for about 55 h after transfer to air and a marked reduction in the oxidative capacity of isolated mitochondria. Mitochondrial activity was suppressed for 10 h after transfer to air but within 24 h had recovered to values comparable to those of mitochondria from fruit stored continuously in air (Rahman *et al.*, 1993b, 1995).

4

Quality

Storage of crops *per se* can detrimentally affect eating quality through senescent breakdown or, in climacteric fruits, being overripe. There is also evidence in the literature that fruit ripened after storage does not taste as good as fruit ripened directly after harvest. This is not always the case and would depend on the length of storage and the type of crop. Generally the level of flavour volatile compounds is lower after storage than in freshly harvested fruit, but there is little evidence that CA storage affects flavour volatiles in any way that is different to storage in air. However, taking phenolics as an example, Tomás-Barberán and Gil (2008) showed that the method of preparation can affect their nutritional status. For example, steam-cooked spinach had 80 mg phenolic antioxidants per 100 g, and boiled, with the water removed, had only 33 mg/100 g. They also showed that the cultivar and parts that are consumed can have a very big effect on their nutritional value (Table 4.1).

Sugar levels clearly affect flavour. Since they are a primary source of energy for plant metabolism, generally the longer the storage period the lower are the sugar levels. However, in many fruits starch is the main storage carbohydrate, which is broken down to sugars during storage. In those cases there is a balance between sugars being produced and sugars being used in metabolism. The acidity of fruit and vegetables also has an effect on flavour. The data from storage trials show widely different effects on acidity, including levels of ascorbic acid (AA). Generally the acid levels of fruit and vegetables should be greater after CA storage in increased CO_2 atmospheres compared with air, because CO_2 in solution is acidic and there is lower acid metabolism associated with CA storage.

Flavour

CA storage can affect the flavour of fruit and vegetables and both positive and negative effects have been cited in the literature. CA-stored apples were shown to retain a good flavour longer than those stored in air (Reichel, 1974) and in many other crops CA storage has generally been shown to maintain better flavour than storage in air (e.g. Zhao and Murata, 1988; Wang, 1990). However, the stage of ripeness of fruit when they are put into storage has the major effect on their flavour, sweetness, acidity and texture. It is therefore often difficult to specify exactly what effect CA storage has, since the effect on flavour may well be confounded with the effect of maturity and ripening. An example of this is that the storage of tomatoes in low concentrations of O_2 had less effect on fruits, which were subsequently ripened,

Table 4.1. Amount of phenolic antioxidants contained in 150 g of selected fruit and vegetables (adapted from Tomás-Barberán and Gil, 2008).

Crop	Cultivar	Part	Phenolic antioxidants intake mg/150g
Peach	Snow King	With peel	110
	Flavour Crest	Without peel	15
Grape	Napoleon	With seeds and peel	150
	Napoleon	Without seeds and peel	5
Orange	Navel	With albedo	400
	Navel	Without albedo	100
Lettuce	Lollo Rosso	With external leaves	300
	Iceberg	White midribs	10

than the stage of maturity at which they were harvested (Kader *et al.*, 1978). Stoll (1976) showed that 'Louise Bonne' pears were still of good eating quality after 3-4 months storage in air but were of similar good eating quality after 5–6 months in CA storage. In contrast it was reported that the flavour of satsumas stored at low O_2 with high CO_2 was inferior to those that had been stored in air (Ito *et al.*, 1974). Truter *et al.* (1994) showed that after 6 weeks storage at –0.5°C under 1.5 kPa CO_2 + 1.5 kPa O_2 or 5 kPa CO_2 + 2 kPa O_2, apricots had an inferior flavour compared with those stored for the same period in air (Table 4.2).

Table 4.2. Effects of CA storage and time on the flavour of apricots which were assessed by a taste panel 3 months after being canned, where a score of 5 or more was acceptable (Truter *et al.*, 1994).

Storage period (weeks)	kPa CO_2	kPa O_2	Flavour score	
			Cultivar Bulida	Cultivar Peeka
0	–	–	7.5	4.6
4	1.5	1.5	6.3	6.3
5	1.5	1.5	5.0	5.0
6	1.5	1.5	1.3	1.3
4	5.0	2.0	6.3	5.0
5	5.0	2.0	5.5	5.4
6	5.0	2.0	2.9	5.4
4	0	21	6.7	5.8
5	0	21	4.2	5.4
6	0	21	2.1	4.6

Spalding and Reeder (1974b) showed that limes (*Citrus latifolia*) stored for 6 weeks at 10°C in 5 kPa O_2 + 7 kPa CO_2 or 21 kPa O_2 + 7 kPa CO_2 lost more acid and sugars than those stored in air, but the limes from all the treatments had acceptable flavour and juice content. 'Golden Delicious', 'Idared' and 'Gloster' apples were stored for 100 or 200 days at 2°C and 95% RH under 15–16 kPa O_2 + 5-6 kPa CO_2, then moved to 5°C, 10°C or 15°C (all at approximately 60% RH) for 16 days to determine their shelf-life. Flavour improvement occurred only in fruits that were removed from CA storage after 100 days and the apples removed from CA storage after 200 days showed a decline in flavour during subsequent storage at 5°C, 10°C or 15°C (Urban, 1995).

Blednykh *et al.* (1989) showed that Russian cultivars of cherry stored under 8–16 kPa CO_2 + 5–8 kPa O_2 still had good flavour after 3–3½ months. A 6-year experiment on the apple cultivars 'Golden Delicious', 'Auralia', 'Spartan' and 'Starkrimson' harvested at four different dates was carried out by Kluge and Meier (1979). These were stored at 3°C and 85–90% RH in air, 13 kPa O_2 + 8 kPa CO_2 or 3 kPa O_2 + 3 kPa CO_2. For prolonged storage, early harvesting and CA storage proved best in terms of flavour maintenance. There were interactions reported between storage conditions and cultivar on flavour. In a second experiment they stored 11 cultivars, including 'Golden Delicious', 'Jonathan', 'Gold Spur', 'Idared' and 'Boskoop'. For those stored in air only 'Boskoop was considered to have had a good flavour after 180 days but both 'Gloster 69' and 'Jonathan' fruit also had a good flavour after CA storage for 180 days. Storage of 'Abbe Fetel' pears at –0.5°C under 0.5 kPa O_2 resulted in losses in aroma and flavour and it was suggested that this may be overcome by raising the O_2 level to 1 kPa (Bertolini *et al.*, 1991). At 2°C sensory acceptance of peaches stored in air was lower than those stored in 3 kPa O_2 + 10 kPa CO_2 for up to 15 days when they were ripened after storage. Higher acceptance scores were associated mainly with the perception of juiciness, emission of volatiles and TSS (Ortiz *et al.*, 2009). Meheriuk (1989a) found that storage of pears in CA did not have a deleterious effect on fruit flavour compared with those stored in air and were generally considered better by a taste panel, although those in CA generally had significantly higher acid levels (Table 4.3). Litchis were packed in punnets and stored in air or over-wrapped with plastic film with 10 kPa CO_2 or vacuum packed. They were then stored for 28 days at 1°C followed by 3 days of shelf-life at 20°C. It was found that the taste and flavour of the fruit from both the plastic film packs were unacceptable, while they remained acceptable for the non-wrapped fruits (Ahrens and Milne, 1993).

Some of the effects on flavour of fruit and vegetables during storage are the result of fermentation. Off-flavour development is associated with increased production of ethanol and acetaldehyde. For example, Mateos *et al.* (1993) reported that off-flavour developed in intact lettuce exposed to 20 kPa CO_2 was associated with increased concentrations of ethanol and acetaldehyde. It was found by Magness and Diehl (1924) that the flavour of apples was impaired when CO_2 exceeded 20 kPa. Karaoulanis (1968) showed that oranges stored in 10–15 kPa CO_2 had increased alcohol content while those stored in 5 kPa CO_2 did not. He also showed that grapes stored in 12 kPa CO_2 + 21 kPa O_2 + 67 kPa N_2 for 30 days had only 17 mg alcohol/100 g fresh weight while those stored in 25 kPa CO_2 + 21 kPa O_2 + 54 kPa N_2 had 170 mg/100 g at the same time. Corrales-Garcia (1997) stored avocados at 2°C or 5°C for 30 days in air, 5 kPa CO_2 + 5 kPa O_2 or 15 kPa CO_2 + 2 kPa O_2. The fruits stored in air had a higher ethanol and acetaldehyde content than the fruits in CA. The alcohol content of strawberries increased with the length of storage and with higher concentrations of CO_2. Storage in 20 kPa CO_2 resulted in high levels after 30 days (Woodward and Topping, 1972). Colelli and Martelli (1995) stored 'Pajaro' strawberries in air, or air with 10 kPa, 20 kPa or 30 kPa CO_2 for 5 days at 5°C, followed by an additional 4 days in air at the same temperature. Ethanol and acetaldehyde accumulation was very slight, though sensory evaluation of the fruits showed that off-flavours were present at transfer from CA, but not after the subsequent storage in air. With storage at 5°C for 28 days, CO_2 concentrations of melons

Table 4.3. Effect of controlled atmosphere storage on the acidity and firmness of Spartlett pears stored at 0°C and their taste after 7 days of subsequent ripening at 20°C (adapted from Meheriuk, 1989a).

Storage atmosphere					
CO_2 kPa	O_2 kPa	Storage time (days)	Malic acid mg/100 ml	Taste (score 1 = like, 9 = dislike)	Firmness (Newtons)
0	21	60	450b	6.1	62a
5	3	60	513a	7.6	61ab
2	2	60	478ab	7.6	60b
0	21	120	382b	4.3	55b
5	3	120	488a	6.7	58a
2	2	120	489a	6.9	58a
0	21	150	342b	2.9	51b
5	3	150	451a	7.7	58a
2	2	150	455a	7.1	58a
0	21	180	318b	–	40b
5	3	180	475a	–	62a
2	2	180	468a	–	61a

Figures followed by the same letter were not significantly different (p = 0.05) by the Duncan's Multiple Range Test.

in sealed polyethylene (PE) and perforated PE packs were 1-4% and no off-flavours were found (Zhao and Murata, 1988). Storage of snow pea pods in either 2.5 kPa O_2 + 5 kPa CO_2 or 10 kPa CO_2 + 5 kPa O_2 resulted in the development of slight off-flavours, but this effect was reversible since it was partially alleviated after ventilation (Pariasca *et al.*, 2001). Delate and Brecht (1989) showed that exposure of sweet potatoes to 2 kPa O_2 + 60 kPa CO_2 resulted in less sweet potato flavour and more off-flavour.

Low O_2 levels have also been shown to result in fermentation and the production of off-flavours. Ke *et al.* (1991a) described experiments where 'Granny Smith' and 'Yellow Newtown' apples, '20th Century' pear and 'Angeleno' plums were kept in either 0.25 or 0.02 kPa O_2 with the consequent development of an alcoholic off-flavour. Ke *et al.* (1991b) also stored the peach cultivar 'Fairtime' in air or in 0.25 or 0.02 kPa O_2 at 0°C or 5°C for up to 40 days. They found that flavour was affected by ethanol and that acetaldehyde accumulated in 0.02 kPa O_2 at 0°C or 5°C or in 0.25 kPa O_2 at 5°C. The fruits kept in air or 0.25 kPa O_2 at 0°C for up to 40 days and those stored in 0.02 kPa O_2 at 0°C or in air, 0.25 or 0.02 kPa O_2 at 5°C for up to 14 days had good to excellent taste, but the flavour of the fruits stored at 5°C for 29 days was unacceptable. Mattheis *et al.* (1991) stored 'Delicious' apples in 0.05 kPa O_2 + 0.2 kPa CO_2 at 1°C for 30 days and found that they developed high concentrations of ethanol and acetaldehyde; these included ethyl propanoate, ethyl butyrate, ethyl 2-methylbutyrate, ethyl hexanoate, ethyl heptanoate and ethyl octanoate. The increase in the emission of these compounds was accompanied by a decrease in the amounts of other esters requiring the same carboxylic acid group for synthesis. In contrast, plums were shown to tolerate an almost complete absence of O_2 for a considerable period without developing an alcoholic or unpleasant flavour (Food Investigation Board, 1920). Ke *et al.* (1991a) reported that strawberries could be stored in 1.0, 0.5 or 0.25 kPa O_2 or air plus 20 kPa CO_2 at 0°C or 5°C for 10 days without detrimental effects on their eating quality. Fruit could also be stored in 0 kPa O_2 or 50 kPa or 80 kPa CO_2 for up to 6 days without affecting their appearance. The taste panel found slight off-flavour in the cultivar G3 kept in 0 or 0.25 kPa O_2 which correlated with ethanol, ethyl acetate and acetaldehyde in juice. Transfer of fruit to air at 0°C for several days after treatment reduced ethanol and acetaldehyde levels, leading to an improvement in final sensory quality.

Volatile Compounds

CA storage has also been shown to affect volatile compounds that are produced by fruit and give the characteristic flavour and aroma. There is some indication that CA storage can improve flavour development in grapes. Dourtoglou *et al.* (1994) found grapes exposed to 100 kPa CO_2 for 20 h had 114 volatile compounds after subsequent storage in air, compared with only 60 for those stored in air throughout. Forney *et al.* (2015) showed that the majority of fruit volatiles in raspberries were composed of C13 norisoprenoids (13-carbon butene cyclohexene degradation products formed by the cleavage of carotenoids), which totalled 65–93% of their total volatile content, and monoterpenes comprised 2.6–20.2% of total volatiles. They reported no effects of CA storage on these volatiles when compared with storage in air.

The onion cultivar 'Hysam' was stored at 0.5 ± 0.5°C and 80 ± 3% RH for 9 weeks in air, 2 kPa O_2 + 2 kPa CO_2 or 2 kPa O_2 + 8 kPa CO_2. Both the CA levels resulted in reduced pungency and flavour, compared with those stored in air, which was assumed to be due to the reduction both in flavour precursors and in enzyme activity (Uddin and MacTavish, 2003).

Willaert *et al.* (1983) isolated 24 aroma compounds from 'Golden Delicious' apples and showed that the relative amounts of 18 of these components declined considerably during CA storage. CA storage of apples in either 2 kPa O_2 + 98 kPa N_2 or 2 kPa O_2 + 5 kPa CO_2 + 93 kPa N_2 resulted in few organic volatile compounds being produced during storage (Hatfield and Patterson, 1974). Even when the fruits were removed from storage they did not synthesize normal amounts of esters during ripening. Yahia (1989) showed that 'McIntosh' and 'Cortland' apples stored in 100 kPa O_2 at 3.3°C for 4 weeks did not have enhanced production of aroma volatiles compared with those stored in air at the same temperature. He showed that most organic volatile compounds were produced at lower rates during ripening after CA storage than those produced from fruits ripened immediately after harvest. Hatfield (1975) also showed that 'Cox's Orange Pippin' apples, after storage in 2 kPa O_2 at 3°C for 3½ months and subsequent ripening at 20°C, produced lower amounts of volatile esters than they did when ripened directly after harvest or after storage in air. This was correlated with a marked loss of flavour. They showed that the inhibition of volatile production could be relieved considerably if the apples were kept in air at 5–15°C after storage and before being transferred to 20°C. Hansen *et al.* (1992) found that volatile ester production by 'Jonagold' apples was reduced after prolonged CA storage in low O_2. After removal from storage, large differences were seen in the production of esters. Esters with the alcohol 2-methylbut-2-enol were produced in negative correlation to increased O_2 concentration. Aroma volatiles in 'Golden Delicious' apples were suppressed during storage for up to 10 months at 1°C in CA compared with storage in air (Brackmann, 1989). 'Jonagold' apples were stored at 0°C in air or in 1.5 kPa O_2 + 1.5 kPa CO_2 for 6 months by Girard and Lau (1995). They found that CA storage decreased production of esters, alcohols and hydrocarbons by about half. The measurement of the organic volatile production of the apples was over a 10-day period at 20°C after removal from the cold store. In CA storage of 'Gala Must' apples, there was some slight reduction in volatile production compared with fruit stored in air. Similar results were obtained from sensory evaluation of aroma intensity after a further 1 week in air (Miszczak and Szymczak, 2000). Miszczak and Szymczak (2000) found that storage of 'Jonagold' at 0°C under 1.5 kPa O_2 + 1.5 kPa CO_2 for 6 months decreased production of volatile compounds by half. Girard and Lau (1995) also found that at 1°C in 1 kPa O_2 + 1.5 kPa CO_2 there was some slight reduction in volatile production compared with fruit stored in air. Harb *et al.* (1994) found that storage of 'Golden Delicious' apples in 3 kPa CO_2 + 21 kPa O_2, 3 kPa CO_2 + 3 kPa O_2, 3 kPa CO_2 + 1 kPa O_2, 1 kPa CO_2 + 1 kPa O_2 or 1 kPa CO_2 + 21 kPa O_2 suppressed volatile production but fruit stored in 1 kPa CO_2 + 3 kPa O_2 had higher volatile production. Belay *et al.* (2017) compared the storage of pomegranate arils at 5°C for 12 days in 5 kPa O_2 + 10 kPa CO_2 + 85 kPa N_2, 10 kPa O_2 + 5 kPa CO_2 + 85 kPa N_2, 70 kPa O_2 + 10 kPa CO_2 + 85 kPa N_2 or air. They found that the highest level of volatile organic compounds was in arils stored under 70 kPa O_2.

Harb *et al.* (2008b) found that the biosynthesis of volatiles in apples was greatly reduced following ULO storage, especially after an extended storage period. Treating fruit with volatile precursors (alcohols and aldehydes) stimulated the biosynthesis of the corresponding volatiles, mainly esters, but this effect was transitory with fruit that had been stored under ULO. Brackmann *et al.* (1993) found that the largest reduction in aroma production was in ULO at 1 kPa O_2 + 3 kPa CO_2. However, there was a partial recovery when they were subsequently stored for 14 days in air at 1°C.

Acidity

The acid levels in fruits and vegetables can obviously affect their flavour and acceptability. Knee and Sharples (1979) reported that acidity could fall by as much as 50% during storage of apples and that there was a good correlation between fruit acidity and sensory evaluation. Meheriuk (1989a) found that storage of pears in CA generally resulted in significantly higher acid levels than those stored in air. Girard and Lau (1995) also found that CA storage significantly reduced the loss of acidity in apples. Similarly Kollas (1964) found that the titratable acid of 'McIntosh' apples was much greater for fruits at 3.3°C in 5 kPa CO_2 + 3 kPa O_2 than those stored in air at 0°C and concluded that it was likely to be due to lower oxidation but that significant rates of 'CO_2 fixation' may have occurred. Kays (1997) described an experiment where 'Valencia' oranges lost less acid during storage at 3.5°C in 5 kPa CO_2 + 3 kPa O_2 than in air at 0°C.

Batu (1995) also confirmed changes in acidity in relation to CA storage and MA packaging in tomatoes. Those stored in sealed plastic film bags generally had a lower rate of loss in acidity than those stored without film bags and generally the fruit stored in the less permeable films had a lower rate of acid loss. The acidity levels of fruit sealed in polypropylene (PP) film, where the equilibrium atmosphere was 5–8 kPa O_2 + 11–13 kPa CO_2, were similar to fruit stored in air and both were lower than the fruit sealed in 30 or 50 μm PE film. He also showed that TA of tomatoes in CA storage generally increased during the first 20 days at both 13°C and 15°C (Figs 4.1 and 4.2). After 20 days, acidity levels tended to decrease until about 70 days. Although there was no correlation between the O_2 or CO_2 concentrations and acidity levels of fruits during CA storage, the acidity of tomatoes stored in 6.4 kPa CO_2 + 5.5 kPa O_2 were the highest among the treatments and the lowest was in 9.1 kPa CO_2 + 5.5 kPa O_2.

Nutrition

CA storage has been shown to have both positive and negative effects on the synthesis and retention of the phytonutrients required for human nutrition. The following are some examples reported of the different effects.

Ascorbic acid

Storage of 'Conference' pears in 2 kPa O_2 + 10 kPa CO_2 resulted in 60% loss in AA content (Veltman *et al.*, 1999). The AA content

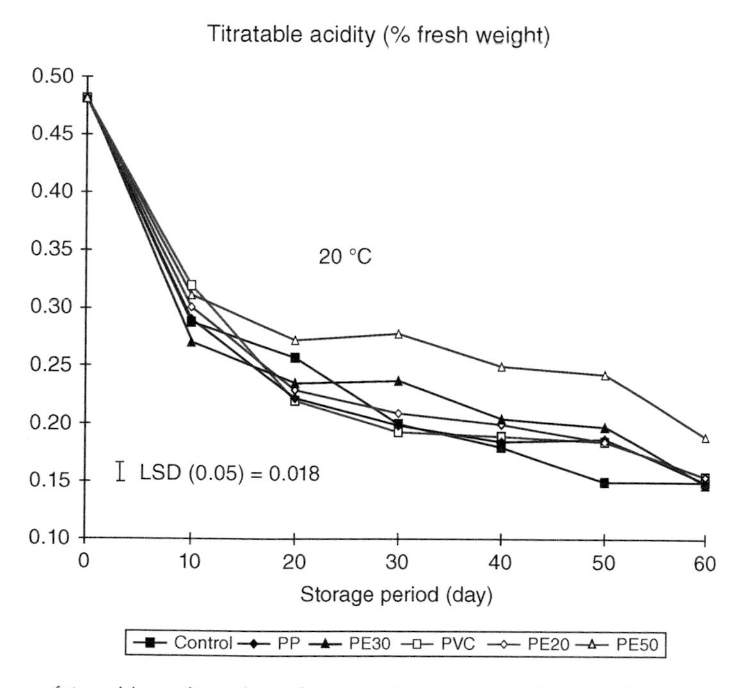

Fig. 4.1. Changes of titratable acidity values of tomatoes harvested at pink stage of maturity and sealed in various thickness of various packaging films and stored at 20°C (modified from Batu, 1995). *Control*: unwrapped; *PE20*: sealed with 20 μm polyethylene; *PE30*: sealed with 30 μm polyethylene; *PE50*: sealed with 50 μm polyethylene; *PVC:* sealed with 10 μm polyvinyl chloride; *PP*: sealed with 25 μm polypropylene.

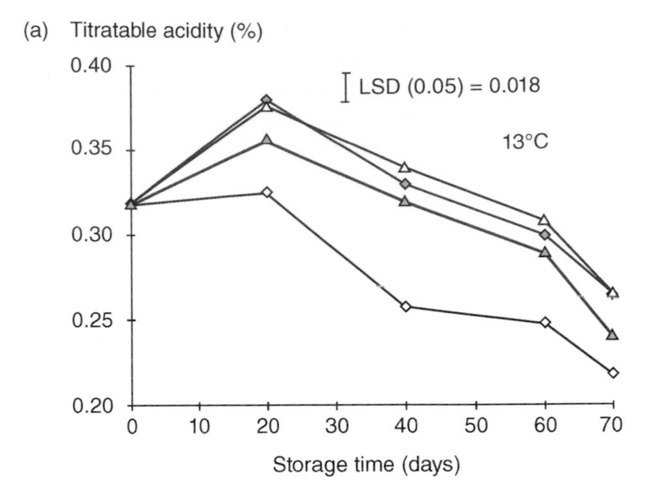

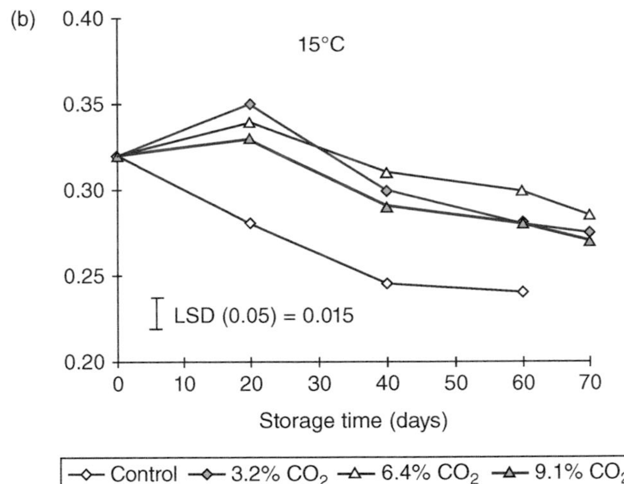

Fig. 4.2. Changes of titratable acidity values of tomatoes harvested at mature-green stage of maturity and kept in controlled atmosphere storage (all with 5.5 % O_2) for 60 days storage time at (a) 13°C and (b) 15°C, plus 10 days in air at 20°C (modified from Batu, 1995).

of the tomato cultivars 'Punjab Chuhara' and 'Punjab Kesri' decreased as the CO_2 concentration in the storage atmosphere increased and increased as the storage period was lengthened (Singh *et al.*, 1993). 'Elvira' strawberries and 'Thornfree' blackberries were stored at 0–1°C in up to 20 kPa CO_2 for strawberries and 30 kPa CO_2 for blackberries combined with either 1–3 kPa or > 14 kPa O_2. Loss of AA was highest in the higher CO_2 atmosphere and degradation after 20 days of storage was more rapid with the low O_2 concentrations. This degradation was even more pronounced during simulated shelf-life in ambient conditions following storage (Agar *et al.*, 1994a). Wang (1983) showed that AA levels of Chinese cabbage were not affected in storage in 10 or 20 kPa CO_2, but in 30 or 40 kPa CO_2 the rate of loss of AA was much higher than for cabbage stored in air. Ogata *et al.* (1975) stored okra at 1°C in air or 3 kPa O_2 combined with 3, 10 or 20 kPa CO_2 or at 12°C in air or 3 kPa O_2 combined with 3 kPa CO_2. At 1°C there was no effect of any of the CA storage combinations on AA, but at 12°C CA resulted in lower AA retention. Ito *et al.* (1974) found that storage of satsumas at 1–4°C in low O_2 and high CO_2 resulted in the AA content of the flesh and peel gradually declining but the dehydro-ascorbic acid content increasing. However, they found that such changes were smaller at high O_2 levels. Fruit held in CA storage lost more acid and sugars than those held in air. Haruenkit and Thompson (1996) showed that storage of pineapples in O_2 levels below 5.4% helped to retain AA levels but generally had little effect on TSS (Table 4.4). Kurki (1979) showed some greater loss in AA in CA storage compared with storage in air, but CA storage gave much better retention of AA than storage in air (Table 4.5). Trierweiler *et al.* (2004) found that the AA content was reduced during storage for 7 months in air or 3 kPa CO_2 + 1 kPa O_2 in 'Bohnapfel' apples but total antioxidant capacity remained constant.

In contrast, Vidigal *et al.* (1979) in studies of tomatoes found that the AA levels increased during CA storage at 10°C. Storage of peas in plastic film bags at 5°C had an equilibrium atmosphere of 5 kPa O_2 + 5 kPa CO_2, which resulted in better maintenance of AA, compared with those not stored in bags (Pariasca *et al.*, 2001). Serrano *et al.* (2006) found that broccoli in MA packaging film had half the loss of AA compared with those stored non-wrapped during 21 days. The AA content of fresh-cut kiwifruit at 0°C under 0.5, 2 or 4 kPa O_2 decreased by 7%, 12% or 18%, respectively, after 12 days while those in air, 5, 10, or 20 kPa CO_2 decreased by 14%, 22% or 34%, respectively. Generally, high CO_2 concentration in the storage atmosphere caused degradation of vitamin C in fresh-cut kiwifruit slices (Lee and Kader, 2000).

Table 4.4. Effects of CA storage on ascorbic acid, °brix and acidity (w/v) of Smooth Cayenne pineapples stored at 8°C for 3 weeks and then 5 days at 20°C (Haruenkit and Thompson, 1996).

Composition (kPa)			Ascorbic acid mg/100 ml
O_2	CO_2	N_2	
1.3	0	98.7	7.14
2.2	0	97.8	8.44
5.4	0	94.6	0.76
1.4	11.2	87.4	9.16
2.3	11.2	86.5	7.94
20.8	0	79.2	0.63
LSD (p = 0.05)			5.28

Chlorophyll

Ubiquitous in plants for photosynthesis, chlorophylls are also an important phytonutrient. For example, Ferruzzi *et al.* (2001) demonstrated the uptake of chlorophyll derivatives by human intestinal cells and their potential importance as health-promoting phytochemicals. Knee and Sharples (1979) found that chlorophyll content of 'Cox's Orange Pippin' apples stored in CA remained the same or only reduced slightly over a 6-month storage period. Storage of peas in plastic film bags at 5°C gave an equilibrium atmosphere of 5 kPa O_2 + 5 kPa CO_2 that resulted in better retention of chlorophyll compared with those not stored in bags (Pariasca *et al.*, 2001). Bangerth (1974) showed that hypobaric storage at 2–3°C improved retention of chlorophyll in stored parsley.

Lycopene

Lycopene consumption is important in human health and its antioxidant function has been shown to offer protection against lung, colorectal, breast, uterine and prostate cancers. Lycopene is a carotenoid in the same family as β-carotene and gives a red colour to fruit. It occurs in many fruit, including tomatoes, pink grapefruit, apricots, blood oranges, watermelon, rose hips and

Table 4.5. Effects of CA storage and storage in air on the retention of vitamins in leeks at 0°C and 95% RH (Kurki, 1979).

		After 4 months storage	
	Initial level before storage	Air	10 kPa CO_2 + 1 kPa O_2
Ascorbic acid (mg/100 g)	37.2	24.1	20.2
Vitamin A (IU/100g)	2525	62	1350

guavas. Plant breeders have developed high-pigment hybrid processing tomatoes that yield improved lycopene, which can constitute 80–90% of the total pigments present. Lycopene in fresh tomato fruits occurs essentially in the *all-trans* configuration (Shi and Le Maguer, 2000). Isomerization converts *all-trans* isomers to *cis*-isomers due to additional energy input and results in an unstable, energy-rich station. Lycopene in tomatoes was shown to increase during storage at 8°C or 16°C while at lower temperatures there was no change or a slight decrease. The optimum temperature for lycopene synthesis in tomatoes is 24°C. Lycopene is not formed at 30°C and above, and temperatures greater than 32.2°C during the growing season resulted in lower lycopene concentrations in the fruit (Garcia and Barrett, 2006). CA storage can also affect lycopene synthesis. Jeffery *et al.* (1984) showed that lycopene synthesis in tomatoes was suppressed during storage in 6 kPa CO_2 + 6 kPa O_2, while tomatoes stored in 5 kPa CO_2 + 5 kPa O_2 had suppressed synthesis of lycopene and xanthophylls, as well as suppressed chlorophyll degradation (Goodenough and Thomas, 1981). In an extensive study of Californian processing tomatoes over 3 years, Garcia and Barrett (2006) found that the range of lycopene content was 55–181 mg/kg, depending on year, cultivar and growing conditions. The mean lycopene concentrations were 106 mg/kg in 2000, 101 mg/kg in 1999 and 88 mg/kg in 2001. During maturation of red cultivars there was a 10–14-fold increase in the concentration of carotenoids, mainly lycopene (Fraser *et al.*, 1994). In general, dehydrated and powdered tomatoes have poor lycopene stability unless carefully processed and promptly placed in hermetically sealed containers in an inert atmosphere for storage. Frozen foods and heat-sterilized foods exhibit excellent lycopene stability throughout normal temperature storage shelf-life. Farneti *et al.* (2012) reported a survey among consumers that indicated that tomatoes were most often stored in refrigerators at well below 10°C. They tested two cultivars ('Cocktail' and 'Round Type') at the red-ripe stage during 20 days storage at 4°C, 8°C, 12°C and 16°C and found that temperatures below 12°C reduced their lycopene content. Application of ethylene to stored ripening tomatoes was shown to increase their lycopene content (Haard and Salunkhe, 1975). This appears to be additional to the increase that is normally associated with ripening.

Phenolics

Rogiers and Knowles (2000) stored four cultivars of amelanchier (*Amelanchier alnifolia*) at 0.5°C for 56 days in various CA and found that 5 kPa CO_2 + 21 O_2 or 5 kPa CO_2 + 10 kPa O_2 were most effective at minimizing losses in anthocyanins. In blueberries, CA storage had little or no effect on phenolic content (Schotsmans *et al.*, 2007). Matityahu *et al.* (2016) showed that storage of pomegranates at 7°C in 2 kPa O_2 + 5 kPa CO_2 was better than air in retaining their quality, but the anthocyanin levels and off-flavours in the CA-stored arils was higher than those in air at the same temperature. However, Zheng *et al.* (2003) found that total phenolics increased in blueberries during storage at 5°C in 60–100 kPa O_2 for 35 days to a greater extent than those stored in air or 40 kPa O_2. In grapes, anthocyanin levels were lower after storage in air at 0°C for those that had been pre-treated for 3 days in 20 kPa CO_2 + 20 kPa O_2 compared with those that had not been pre-treated (Romero *et al.*, 2008). Artés-Hernández *et al.* (2006) stored 'Superior Seedless' grapes in MA packaging and found no changes in the phenolics during storage at 0°C and 8°C, but a slight decrease during 2 days of subsequent shelf-life at 20°C. However, Sanz *et al.* (1999) found that 'Camarosa' strawberries in MA packaging had lower anthocyanin levels compared with those stored non-wrapped. Shin *et al.* (2008) found that the total flavonoid and phenolic concentrations and total antioxidant activity of strawberries harvested at the white-tip stage were greater than those harvested at the red-ripe stage and these differences were maintained during storage at 3°C or 10°C for 12 days. The latex that spurts from the broken pedicel of immature mango fruit is structurally similar to phenolic allergens found in other genera of Anacardiaceae. The main component of mango latex was identified as 5-[2(Z)-heptadecenyl] resorcinol (Bandyopadhyay *et al.*, 1985). Phenolic phytoalexins in carrots and parsnips accumulate in response to ethylene and give them a bitter flavour. In carrots, isocoumarin accumulation increased as respiration rate increased in response to ethylene and there was more accumulation of isocoumarin in immature carrots in response to ethylene than in more mature ones (Seljasen *et al.*, 2001).

Banana and plantain can contain high levels of phenolic compounds, especially in the peel (Von Loesecke, 1950). As fruit ripens the astringency becomes lower, which seems to be associated with a change in structure of the tannins, rather than a reduction in their levels, in that they form polymers. The enzyme polyphenol oxidase (PPO) is responsible for the oxidative browning reaction when the pulp of fruit is cut. In 'Carabao' mangoes, there was a progressive loss in total phenolic content during ripening which was associated with loss in astringency (Tirtosoekotjo, 1984). In guava, levels of polyphenols, particularly the high-molecular-weight polyphenols, were shown to decrease as the fruit matured (Itoo *et al.*, 1987). Young guava fruit contained 620 mg/100 g fresh weight, of which 65% were condensed tannins. In persimmon, the soluble tannin content in fruit varied between cultivars and in astringent cultivars it was particularly high in immature fruit (Itoo *et al.*, 1987).

5

Physiology, Ripening and Genetics

Respiration Rate

Within the physiological temperature range of the crop, the respiration rate increases exponentially with temperature, so that for every 10°C rise the increase in metabolism is in the order of two to three times. This is called the Q_{10} value and is predicated by the Van't Hoff rule. Respiration rate is also affected by the gaseous environment, especially O_2, CO_2 and ethylene. Respiratory quotient (RQ) is the ratio of moles (mol) CO_2 evolved to moles O_2 absorbed in plant cells. The value is 1 when the substrate is carbohydrate, about 0.7 for lipids and about 0.8 for proteins.

Generally fruit and vegetables have a lower respiration rate during CA storage compared with storage in air, but there are some exceptions. Most work on respiration rate has been reported on apples and Kidd and West (1927a, b) showed that the apple respiration rate in storage at 8°C in 12 kPa CO_2 + 9 kPa O_2 was 54–55% of that of fruit stored in air at the same temperature. Robinson *et al.* (1975) showed this lowering effect for a range of fresh fruit and vegetables during storage under 3% O_2 at various temperatures (Table 5.1). They studied the respiration rate of a variety of crops at different temperatures stored in air or in 3 kPa O_2 (Tables 5.2 and 5.3). Respiration rate was generally lower when crops were stored in 3 kPa O_2 than when they were stored in air, but there were some interactions between O_2 and temperature in that the effects of the reduced O_2 level were more marked at higher storage temperatures. Exceptions are also common; for example, Fonseca *et al.* (2004) found that 'Royal Sweet' watermelons had a higher respiration rate at 8 kPa O_2 compared with 14 kPa O_2. Respiratory heat at 'normal storage temperatures' in watts/tonne (W/t) was reported by Cambridge Refrigeration Technology (2001) as: avocado 183–465, asparagus 81–237, mango 133, banana 59–130, cauliflower 44, apple 9–18, grape 3.9–6.8, orange 18.9 and cherry 17–39.

Knee (1973) showed that CO_2 could inhibit succinate dehydrogenase in the tricarboxylic acid cycle which is part of the respiratory pathway. McGlasson and Wills (1972) also suggested that low O_2 limited the operation of the Krebs cycle between pyruvate and citrate and 2-oxoglutarate and succinate, but they apparently found no similar effect of high CO_2 in bananas. The tricarboxylic acid cycle is inhibited at very low levels of O_2 but the glycolytic pathway may not be affected. This results in less efficient energy production during respiration, since there is insufficient O_2 to metabolize stored carbohydrates to water and CO_2. Instead the blocking of the glycolytic pathway results in a build-up of acetaldehyde and

Table 5.1. Effects of temperature and reduced O_2 level on the respiration rate (CO_2 production[a], mg/kg/h[b]) and storage life of selected fruit and vegetables (Robinson *et al.*, 1975).

	In air					In 3 kPa O_2			
Temperature (°C)	0	5	10	15	20	0	10	20	Water loss[c]
Asparagus	28	44	63	105	127	25	45	75	3.6[d]
Beans, broad	35	52	87	120	145	40	55	80	(2.1)
Beans, runner	21	28	36	54	90	15	25	46	(1.8)
Beetroot, storing	4	7	11	17	19	6	7	10	1.6
Beetroot, bunching with leaves	11	14	22	25	40	7	14	32	(1.6)
Blackberries, Bedford Giant	22	33	62	75	155	15	50	125	0.5
Blackcurrants, Baldwin	16	27	39	90	130	12	30	74	–
Brussels sprouts	17	30	50	75	90	14	35	70	(2.8)
Cabbage, Primo	11	26	30	37	40	8	15	30	1.0
Cabbage, January King	6	13	26	33	57	6	18	28	–
Cabbage, Decema	3	7	8	13	20	2	6	12	0.1
Carrots, storing	13	17	19	24	33	7	11	25	1.9
Carrots, bunching with leaves	35	51	74	106	121	28	54	85	(2.8)
Calabrese	42	58	105	200	240	–	70	120	(2.4)
Cauliflower, April Glory	20	34	45	67	126	14	45	60	(1.9)
Celery, white	7	9	12	23	33	5	9	22	(1.8)
Cucumber	6	8	13	14	15	5	8	10	(0.4)
Gooseberries, Leveller	10	13	23	40	58	7	16	26	–
Leeks, Musselburgh	20	28	50	75	110	10	30	57	(0.9)
Lettuce, Unrivalled	18	22	26	50	85	15	20	55	(7.5)
Lettuce, Kordaat	9	11	17	26	37	7	12	25	–
Lettuce, Kloek	16	24	31	50	80	15	25	45	–
Onion, Bedfordshire Champion	3	5	7	7	8	2	4	4	0.02
Parsnip, Hollow Crown	7	11	26	33	49	6	12	30	(2.4)
Potato, maincrop (King Edward)	6[e]	3	4	5	6	5[g]	3	4	(0.05)
Potato, 'new' (immature)[f]	10	15	20	30	40	10	18	30	(0.5)
Pea (in pod) early, (Kelvedon Wonder)	40	61	130	180	255	29	84	160	(1.3) (cv. Onward)
Peas, main crop (Dark Green Perfection)	47	55	120	170	250	45	60	160	
Peppers, green	8	11	20	22	35	9	14	17	0.6
Raspberry (Malling Jewel)	24	55	92	135	200	22	56	130	2.5

Continued

Table 5.1. Continued.

	In air					In 3 kPa O_2			
Temperature (°C)	0	5	10	15	20	0	10	20	Water loss[c]
Rhubarb (forced)	14	21	35	44	54	11	20	42	2.3
Spinach (Prickly True)	50	70	80	120	150	51	87	137	(11.0) (glasshouse grown)
Sprouting broccoli	77	120	170	275	425	65	115	215	(7.5)
Strawberries, Cambridge Favourite	15	28	52	83	127	12	45	86	(0.7)
Sweetcorn	31	55	90	142	210	27	60	120	(1.4)
Tomato, Eurocross BB	6	9	15	23	30	4	6	12	0.1
Turnip, bunching with leaves	15	17	30	43	52	10	19	39	1.1 (without leaves)
Watercress	18	36	80	136	207	19	72	168	(35.0)

[a]These figures, which give the average rates of respiration of the samples, are a guide only. Other samples could differ, but the rates could be expected to be of the same order of magnitude, ± about 20 %.
[b]Heat production in Btu/t/h is given by multiplying CO_2 output in mg/kg/h by 10 (1 Btu approximates to 0.25 kcal and to 1.05 kJ).
[c]Values in parentheses were determined at 15°C and 6–9 mb water vapour pressure deficit (wvpd). Remainder at 10°C and 3–5 mb wvpd.
[d]Water loss as a % of the initial weight per day per millibar wvpd.
[e]After storage for a few weeks at 0°C, sufficient time having elapsed for low-temperature sweetening. The figures for potato are mid-season (December) values and could be 50 % greater in October and March.
[f]Typical rounded-off values, for tubers with an average weight of 60 g immediately after harvest. The rates are too labile for individual values to be meaningful. After a few weeks' storage they approximate to the values for maincrop potatoes.
[g]After storage for a few weeks at 0°C, sufficient time having elapsed for low-temperature sweetening. The figures for potatoes are mid-season (December values) and could be 50 % greater in October and March.

Table 5.2. Effects of temperature and reduced O_2 level on the respiration rate in CO_2 production in mg/kg/h of various fruit and vegetables (adapted from Robinson *et al.*, 1975).

Crop	Gaseous atmosphere	Storage temperature		
		0°C	10°C	20°C
Asparagus	Air	28	63	127
	3 kPa O_2	25	45	75
Beetroot without leaves	Air	4	11	19
	3 kPa O_2	6	7	10
Beetroot with leaves	Air	11	22	40
	3 kPa O_2	7	14	32
Blackberries	Air	22	62	155
	3 kPa O_2	15	50	125
Blackcurrants	Air	16	39	130
	3 kPa O_2	12	30	74
Broad beans	Air	35	87	145
	3 kPa O_2	40	55	80
Sprouting broccoli	Air	77	170	425
	3 kPa O_2	65	115	215
Brussels sprouts	Air	17	50	90
	3 kPa O_2	14	35	70
Green peppers	Air	8	20	35
	3 kPa O_2	9	14	17
Carrots	Air	13	19	33
	3 kPa O_2	7	11	25
Cauliflowers	Air	20	45	126
	3 kPa O_2	14	45	60
Cucumbers	Air	6	13	15
	3 kPa O_2	5	8	10
Gooseberry	Air	10	23	58
	3 kPa O_2	7	16	26
Onions	Air	3	7	8
	3 kPa O_2	2	4	4
Strawberry	Air	15	52	127
	3 kPa O_2	12	45	86
Sweetcorn	Air	31	90	210
	3 kPa O_2	27	60	120
Tomato	Air	6	15	30
	3 kPa O_2	4	6	12

ethanol, which can be toxic to the cells if allowed to accumulate. Compared with storage in air, storing broccoli in 50 kPa O_2 + 50 kPa CO_2 led to a decreased respiration rate, inhibition of the reduction of ATP levels, energy charge levels, succinic dehydrogenase and cytochrome oxidase and increases in glucose-6-phosphate dehydrogenase and 6-phosphogluconate dehydrogenas activity. Storage in 50 kPa O_2 + 50 kPa CO_2 also suppressed reduction of the tricarboxylic acid cycle and cytochrome oxidase pathway rate and increased the phosphopentose pathway, but had little impact on the Embden–Meyerhof–Parnas rate and phosphohexose isomerase activity (Li *et al.*, 2016c).

The effect of reduced O_2 and increased CO_2 on respiration rate can be affected by temperature. Kubo *et al.* (1989b) showed that the effects of high CO_2 on respiration rate varied according to the crop and its stage of development. The relationship between O_2 and CO_2 levels and respiration rate is not a simple one. It also varies between cultivars (Table 5.4) as in the comparison between 'Golden Delicious' and 'Cox's Orange Pippin' apples where the respiration rate

Table 5.3. Effects of temperature and reduced O_2 level on the respiration rate (CO_2 production in mg/kg/h) of potatoes (adapted from Robinson *et al.*, 1975).

	Air					3 kPa O_2		
Temperature (°C)	0	5	10	15	20	0	10	20
Main crop (King Edward)	6	3	4	5	6	5	3	4

Table 5.4. Rates of respiration rate of some cultivars of apples stored in different CA conditions; modified from Fidler *et al.* (1973).

	Storage conditions			Respiration rate (l/t/day)	
Cultivar	°C	CO_2 kPa	O_2 kPa	CO_2	O_2
Bramley	3.5	8–10	11–13	40–45	40
Cox's Orange Pippin	3.5	5	16	62	57
Cox's Orange Pippin	3.5	5	3	42	40
Cox's Orange Pippin	3.5	< 1	2.5	80	55
Golden Delicious	3.5	5	3	20	20
Delicious	0	5	3	18	–
Jonathan	3.5	7	13	33	38
McIntosh	3.5	5	3	35	–

was suppressed more in the former than the latter in CA storage (Fidler *et al.*, 1973). Olsen (1986) also showed an interaction between O_2 and CO_2 on the respiration rate of apples, with the strongest effect of O_2 at 0 kPa CO_2 and only a small effect of O_2 at 1 kPa CO_2. Andrich *et al.* (1994) showed some evidence that the respiration rate of 'Golden Delicious' apples was affected differentially by CO_2 concentration in storage at 21°C and 85% RH, depending on the O_2 concentration in the store. The effect was that in anaerobic conditions, or near-anaerobic conditions, respiration rates were more affected in increasingly high CO_2 levels than for the same fruit stored in aerobic conditions. When apples and melons were stored in 60 kPa CO_2 + 20 kPa O_2 + 20 kPa N_2 their respiration rate fell to about half the initial level. Ripening tomatoes and bananas also showed a reduction in respiration in response to high CO_2, but showed little response when tested before the climacteric (Kubo *et al.*, 1989a).

Where the O_2 level in plant cells is too low, the respiration rate becomes anaerobic. In some CA stores in England where they store apples in 1 kPa O_2, an alcohol detector is fitted which sounds an alarm if ethanol fumes are detected because of fermentation. This enables the store operator to increase the O_2 level and no damage should have been done to the fruit. The detector technology is based on that used by the police to detect alcohol fumes on the breath of motorists. The level of O_2 at which fermentation may occur in plant cells may be as low as 0.2 kPa, but the gradient of O_2 concentration between the store atmosphere and the cells of the crop requires the maintenance of a higher level around the crop (Kader, 1986). The gradient is affected by the cuticle or periderm of the crop, the ease of gas penetration through the cells and the utilization of O_2 by the crop.

Production of ethylene by guavas and tomatoes was substantially reduced by all levels of CO_2, but 30 kPa CO_2 accelerated ethylene production in bananas, carrots, cucumbers, onions and potatoes, possibly due to an injury response (Pal and Buescher, 1993). Green bananas were held in humidified gas streams comprising air, 5 kPa CO_2 + 20 kPa O_2 + 75 kPa N_2, 0 kPa CO_2 + 3 kPa O_2 + 97 kPa N_2 or 5 kPa CO_2 + 3 kPa O_2 + 92 kPa N_2. Ripening in all three CA combinations and the respiration rates were reduced over the period before the beginning of the climacteric compared with storage in air (McGlasson and Wills, 1972). Wade (1974) showed that the respiratory climacteric was induced in bananas by Ethephon (a source of ethylene) at O_2 concentrations of 3–21 kPa, but their respiration rate was not affected by Ethephon at O_2 concentrations of 1 kPa or less. Awad *et al.* (1975) showed that green bananas immersed for 2 min in Ethephon at 500 ppm had their climacteric advanced by 5 days, whereas fruits treated with GA at 100 ppm had their climacteric delayed by 2 days, compared with the control.

CA can also affect the respiration rate of vegetables and non-climacteric fruit. Kubo *et al.* (1989a) reported that exposure to high CO_2 had little or no effect in *Citrus natsudaidai*, lemons, potatoes, sweet potatoes or cabbage, but reduced the respiration rate in broccoli. High CO_2 stimulated the respiration rate of lettuce, aubergine and cucumber. Pal and Buescher (1993) found that short-term exposure to 20–30 kPa CO_2 reduced the respiration rate of pickling cucumbers but could increase the respiration rate of potatoes and carrots, and had no effect on respiration rate of oranges and onions. They found that changes in respiration rate seldom coincided with changes in ethylene production. The respiration rate of the strawberry cultivar 'Cambridge Favourite' held at 4.5°C in air or 1, 2 or 5 kPa O_2 fell to a minimum after 5 days. Thereafter the rate increased, more rapidly in air (in which rotting was more prevalent) than in 1 or 2 kPa O_2 (Woodward and Topping, 1972). Storage of strawberries in 10–30 kPa CO_2 or 0.5–2 kPa O_2 was shown to slow their respiration rate (Hardenburg *et al.*, 1990). Weichmann (1973) showed that the respiration rate of five cultivars of carrots stored at 1.5°C in atmospheres containing 0.03, 2.5, 5.0 or 7.5 kPa CO_2, but with no regulation of the O_2 level, increased with increasing CO_2 level. Weichmann (1981) found that horseradish stored in 7.5 kPa CO_2 had a higher respiration rate than those stored in air. Izumi *et al.* (1996b) showed that storage of freshly cut carrots in 10 kPa CO_2 + 0.5 kPa O_2 reduced their respiration rate by about 55% at 0°C, about 65% at 5°C and 75% at 10°C.

In summary, it has been shown that low O_2 levels in store can affect the crop in several ways (Richardson and Meheriuk, 1982), including:

- reduced respiration rate;
- reduced substrate oxidation;
- delayed ripening of climacteric fruit;
- prolonged storage life;
- delayed breakdown of chlorophyll;
- reduced rate of production of ethylene;
- changed fatty acid synthesis;
- reduced degradation rate of soluble pectins;
- formation of undesirable flavours and odours;
- altered texture; and
- development of physiological disorders.

The effects of CO_2 on postharvest responses of crops include:

- decreased synthetic reactions in climacteric fruit;
- delaying the initiation of ripening;
- inhibition of some enzyme reactions;
- decreased production of some organic volatiles;
- modified metabolism of some organic acids;
- reducing the rate of breakdown of pectic substances;
- inhibition of chlorophyll breakdown;
- production of 'off-flavours';
- induction of physiological disorders;
- retarded fungal growth on the crop;
- inhibition of the effect of ethylene;
- changes in sugar content (potatoes);
- effects on sprouting (potatoes);
- inhibition of postharvest development;
- retention of tenderness; and
- decreased discoloration levels.

Fruit Ripening

Fruit can be classified into two groups, climacteric and non-climacteric, though some fruits appear to be intermediate between the two. Climacteric can be defined as fruit that can be ripened after harvest and non-climacteric as fruit that do not ripen after harvest. The term climacteric was first applied to fruit ripening in the mid 1920s by Kidd and West

(1927b). They observed an increase in the respiration rate as measured by the production of CO_2 of Bramley's Seedling apples around the time of the normal commercial harvest. Biale and Barcus (1970) published measurements of the respiration rate of some selected fruits and classified them into climacteric, non-climacteric and indeterminate on the basis of their respiration rate. For example, they classified banana, biriba, breadfruit, mango, papaya and soursop as climacteric; cacao, cashew and guava as non-climacteric; and jackfruit, jambo, passionfruit and sapote as indeterminate. Other fruits that have been shown to display typical climacteric respiration patterns include annonas, apricot, avocado, blueberry, cantaloupe melon, durian, feijoa, kiwifruit, peach, persimmon, plum, pome fruits and watermelon. Other fruits that can be classified as non-climacteric include aubergine, blackberry, capsicum, cherry, citrus fruits, cucumber, grape, litchi, olive, pineapple, pomegranate, raspberry, strawberry and tamarillo. In some cases there is conflict of opinion on classification. For example, fig and melon were classified as climacteric by Wills *et al.* (1989), but Biale (1960) classified them both as non-climacteric and Pratt (1971) showed that both honeydew muskmelon and 'PMR-45' cantaloupes were climacteric. Biale and Barcus (1970) classified passionfruit as indeterminate while Akamine *et al.* (1957) classified them as climacteric. Grierson (1993), commenting on the classification of fruit into climacteric and non-climacteric, reported that parts of citrus fruits are climacteric and other parts of the same fruit are non-climacteric.

Initiation of ripening

In climacteric fruit, ripening is initiated when a threshold level of ethylene is reached in the cells. Ethylene biosynthesis involves a series of steps culminating in the synthesis of *S*-adenosyl-methionine (SAM) from the amino acid methionine. SAM is converted to ACC (1-aminocyclopropane-1-carboxylic acid) by the action of the enzyme ACC synthase and ACC is converted to ethylene by the action of ACC oxidase, which is sensitive to O_2. The synthesis of ACC was reported to be the rate-limiting step in the biosynthesis of ethylene (Kende, 1993). ACC synthase is controlled by a number of ACS genes (Li *et al.*, 2013). The biosynthesis of ethylene can be inhibited at temperatures above 35°C. High levels of CO_2 can compete with ethylene for binding sites in fruit and the biological activity of ethylene at 1 μl/l ceased in the presence of 10 kPa CO_2 (Burg and Burg, 1967). Yang (1985) showed that CO_2 accumulation in the intercellular spaces of fruit acts as an ethylene antagonist. The inhibition of ethylene production by CO_2 may be attributed to its inhibitory effect on ACC synthase activity, but ACC oxidase activity was stimulated by CO_2 up to 10 kPa but inhibited by 20 kPa CO_2 (Levin *et al.*, 1992). Koyakumaru *et al.* (1995) also found that ethylene production rates were suppressed at high CO_2 concentrations or at O_2 levels of less than 5 kPa in the fruit of *Prunus mume* (commonly known as Japanese apricot, or Chinese plum). In 3 kPa O_2, ethylene production rates peaked half or one day after the respiratory climacteric peak. Ethylene production in ripening fruit was shown to stop in the absence of O_2 (Gane, 1934). Treating fruit with chemicals, including 1-MCP and AVG, can inhibit ACC synthase activity. Han *et al.* (2016) identified three different ripening characteristics caused by ethylene that correlated well with ethylene production. Gene expression analysis showed a significant induction of *MaC2H2-1/2* transcripts during the ripening of bananas and they suggested that 'these genes are transcriptional repressors and may mediate a finely tuned regulation of ethylene production during ripening, possibly via transcriptional repression of ethylene biosynthetic genes'. Clendennen and May (1997) found that differential screening of complementary DNA libraries in banana pulp during ripening led to the isolation of 11 non-redundant groups of differentially expressed mRNA. Identification of these transcripts indicated that two of the mRNA encoded proteins were involved in carbohydrate metabolism, whereas the other nine were thought to be associated with pathogenesis, senescence or stress responses. Binding of ethylene requires a copper cofactor and proper receptor function relies on a copper transporter.

Changes in fruit during ripening

Ripening is associated with climacteric fruit, but it is sometimes used for non-climacteric fruit, especially as they develop on the plant. Perhaps a better term for non-climacteric would be maturation. The most obvious change in many fruits during ripening is their external colour, which is due to chlorophyll destruction and biosynthesis of pigments such as anthocyanins, carotenoids and lycopene. Textural changes are normally progressive softening during ripening. Softening is due to several factors; for example, Cordeiro *et al.* (2013) showed that softening of cherimoya was mainly attributed to depolymerization of pectin and lipid deterioration rather than hemicellulose degradation. During the developmental stage of climacteric fruit there is a general increase in starch content, which is hydrolysed to simple sugars during ripening. Acidity of fruit generally decreases during ripening. Acids help to form the desirable sugar-to-acid balance considered necessary for a pleasant taste. Organic acids present in fruit vary with different types of fruit, with malate and citrate being the most common. Some types of fruit can contain high levels of phenolic compounds – for example, in bananas and plantain. During ripening phenolics, such as tannins, are polymerized to insoluble compounds, resulting in a reduction of astringency in the ripe fruit. Phenolics are common in many fruit and are responsible for the oxidative browning reaction when the pulp is cut, especially if the fruit is immature. The enzyme PPO is responsible for this reaction. Besides the changes described above that affect the flavour of the fruit, the characteristic flavour and aroma of ripe fruit is due to the production of a complex mixture of individual volatile components that is characteristic to that particular fruit or cultivar. Toxins may exist in unripe fruit that reduce as the fruit ripens. Tomatoes at the green stage of maturity contain a toxic alkaloid called solanine that decreases during ripening (Laval Martin *et al.*, 1975) and ackee fruit contains the toxin hypoglycin in the arils that gradually reduces as the fruit matures (Larson *et al.*, 1994). Profuse ethylene biosynthesis occurs in climacteric fruit, which initiates the ripening process. This takes place on the plant or tree when the fruit reaches maturity. Avocados are an exception, since they will not ripen on the tree and are harvested and either placed in ripening rooms or allowed to ripen postharvest (Gazit and Blumenfeld, 1970). Immature fruit may also be harvested and exposed to exogenous ethylene, which initiates ripening. Also most pear cultivars, if allowed to become too mature or to ripen on the tree, develop a coarse, mealy texture and often have core breakdown.

Controlled atmosphere storage on ripening

In order to initiate climacteric fruit to ripen and control the ripening processes, the level of CO_2 and O_2 in the surrounding environment as well as ethylene and even moisture content have to be controlled. As indicated above, ethylene production and ripening can be suppressed at high CO_2 or low O_2 levels. Exposing fruit to sufficient stress can also initiate ripening. Stress during growing, caused by diseases or low water levels in the soil, can cause bananas to initiate to ripen before the fruit is fully grown or mature. Low humidity in the storage atmosphere can also result in stress and can initiate ripening. Maintaining high humidity around the fruit can help to keep fruit in the pre-climacteric stage (Thompson *et al.*, 1974d). Temperature is also critical; for example, Hatton *et al.* (1965) reported that ripening and softening rates of Florida mango cultivars increased as temperature increased over the range of 16–27°C and the optimum temperature range was 21–24°C. If mangoes were ripened at 27–30°C their skin became mottled and the fruit acquire a 'strong flavour', while ripening was retarded when they are held above 30°C (Paull and Chen,

2004). Exposure of Florida mango cultivars, picked at the mature-green stage, to ethylene at 20–100 μl/l for 24 h resulted in faster and more uniform ripening at 21°C and 92–95% RH (Barmore and Mitchell, 1977).

CA storage has been shown to suppress the biosynthesis of ethylene in ripening fruit and in early work by Gane (1934) biosynthesis of ethylene was shown to cease in the absence of O_2. Ethylene concentrations in the core of apples were generally progressively lower with reduced O_2 concentrations in the store over the range of 0.5–2 kPa O_2 (Stow, 1989). Wang (1990) reviewed the literature on the effects of CO_2 and O_2 on the activity of enzymes associated with fruit ripening and cited many examples of the activity of these enzymes being reduced in CA storage. This is presumably due, at least partly, to many of these enzymes requiring O_2 for their activity. Quazi and Freebairn (1970) showed that high CO_2 and low O_2 delayed the high production of ethylene in bananas, but the application of exogenous ethylene was shown to reverse this effect. Wade (1974) showed that bananas could be ripened in atmospheres of reduced O_2, even as low as 1 kPa, but the peel failed to de-green, which resulted in ripe fruits that were still green. Similar effects were shown at O_2 levels as high as 15 kPa. Since the de-greening process in 'Cavendish' bananas is entirely due to chlorophyll degradation (Seymour *et al.*, 1987a and b), the CA storage effect was presumably due to suppression of this process. Hesselman and Freebairn (1969) showed that ripening of bananas, which had already been initiated to ripen by ethylene, was slowed in low O_2 atmospheres. Goodenough and Thomas (1980, 1981) also showed suppression of de-greening of fruits during ripening; in this case it was with tomatoes ripened in 5 kPa CO_2 combined with 5 kPa O_2. Their work, however, showed that this was due to a combination of suppression of chlorophyll degradation and the suppression of the synthesis of carotenoids, lycopene and xanthophyll. Jeffery *et al.* (1984) also showed that lycopene synthesis was suppressed in tomatoes stored in 6 kPa CO_2 + 6 kPa O_2.

Ethylene biosynthesis was studied in 'Jonathan' apples stored at 0°C under CA storage conditions of 0–20 kPa CO_2 + 3 kPa or 15 kPa O_2. Fruits were removed from storage after 3, 5 and 7 months. Internal ethylene concentration, ACC levels and ethylene-forming enzyme activity were determined immediately after removal from storage and after holding at 20°C for 1 week. Increasing CO_2 concentration from 0 kPa to 20 kPa at both low and high O_2 concentrations inhibited ethylene production by the fruits. ACC levels were similarly reduced by increasing CO_2 concentrations even in low O_2; low O_2 enhanced ACC accumulation but only in the absence of CO_2. One of the metabolic pathways used for ethylene synthesis by some plants and microbes is via an ethylene-forming enzyme, which uses α-ketoglutarate and arginine as substrates. Ethylene-forming enzyme activity was stimulated by CO_2 up to 10 kPa, but was inhibited by 20 kPa CO_2 at both O_2 concentrations. The inhibition of ethylene production by CO_2 may, therefore, be attributed to its inhibitory effect on ACC synthase activity (Levin *et al.*, 1992). In other work, storing pre-climacteric fruit of apple cultivars 'Barnack Beauty' and 'Wagner' at 20°C for 5 days in 0 kPa O_2 with 1 kPa CO_2 (99 kPa N_2) or in air containing 15 kPa CO_2 inhibited ethylene production and reduced ACC concentration and ACC synthase activity compared with storage in normal atmospheres (Lange *et al.*, 1993).

Low O_2 and increased CO_2 concentrations were shown to have an effect on the decrease in ethylene sensitivity of 'Elstar' apples and 'Scania' carnation flowers during controlled atmosphere storage (Woltering *et al.*, 1994). Storage of kiwifruit in 2–5 kPa O_2 with 0–4 kPa CO_2 reduced ethylene production and ACC oxidase activity (Wang *et al.*, 1994). Ethylene production was lower in 'Mission' figs stored at 15–20 kPa CO_2 concentrations compared with those kept in air (Colelli *et al.*, 1991).

Genetics

The optimum storage conditions for fresh fruit and vegetables vary, not only between species (from very short in fruit such as strawberries to very long as in natural storage organs such as potato tubers), but also between varieties and cultivars of the same species. For example, in apples some cultivars will be damaged when exposed to low levels of CO_2 in storage while others can tolerate comparatively high levels. In pears, optimum CA conditions and maximum storage time have been shown to vary considerably between cultivars (Table 5.5), but these recommendations interact and can also vary between countries for the same cultivar (Table 5.6).

The effects of the various pre-harvest and postharvest conditions on the postharvest life of fresh fruit and vegetables are clear and are dealt with in detail in other parts of this book. However, these effects and variations in their postharvest life depend on not only the genetic make-up of the plants but also the expression of their genes. Gene expression can be influenced by various factors, including the environment in which the crop is grown, as well as the postharvest environment to which it is exposed. However, changes in gene expression of crops during storage, as well as their response to their environment, can affect their postharvest responses.

The genetic make-up of plants gives them their characteristics and their responses to postharvest conditions. Brizzolara *et al.* (2017) reported that the response of different cultivars of apple to low O_2 storage depended on their genetic make-up. They showed that, in ULO or DCA-CF storage, 'Granny Smith' apples had low levels of fermentation while fermentation was high under the same conditions in 'Red Delicious'. The expression of those genes may be different and can affect their responses to storage conditions. For example, Koch (1996) commented that 'Plant gene responses to changing carbohydrate status can vary markedly. Some genes are induced, some are repressed, and others are minimally affected.' Thammawong *et al.* (2015) found that changes in sugar concentration in cabbages during 2 days of storage seemed not to be influenced by CA and gene expression. They therefore interpreted it that the responses of crops to postharvest conditions can be a complex of genes, gene expression, the chemical changes occurring in the crop as well as the storage environment. They stored cabbages under different CA conditions and found that the changing pattern of each gene appeared to depend on the different percentages of CO_2 and O_2. Obvious elevated gene expression was detected in the *BoGAPDH* (glyceraldehyde 3-phosphate dehydrogenase [GAPDH]-encoding gene), *BoADH* (alcohol dehydrogenase [ADH]-encoding gene) and *BoCTS* (citrate synthase [CTS]-encoding gene) after 24 hours storage. At the same time the expression of *BoHXK1*

Table 5.5. Recommended CA storage conditions for pears stored at –0.5°C; adapted from Van der Merwe (1996).

Cultivar	O_2 kPa	CO_2 kPa	Storage duration (months)
Bon Chretien	1.0	0	4
Buerré Bosc	1.5	1.5	4
Forelle	1.5	0–1.5	7
Packham's Triumph	1.5	1.5	9

Table 5.6. Optimum CA conditions for selected pear cultivars; adapted from Kupferman (2001b).

Cultivar	Country	°C	O_2 kPa	CO_2 kPa	Duration (months)
Doyenne du Comice	Netherlands	–0.5	2.5	0.7	5
Doyenne du Comice	South Africa	–0.5	1.5	1.5	6
Doyenne du Comice	New Zealand	–0.5	2	< 1	3
Packham's Triumph	New Zealand	–0.5	2	< 1	5
Packham's Triumph	South Africa	–0.5	1.5	2.5	9
Williams Bon Chretien	South Africa	–0.5 to 0	1	0	4
Williams Bon Chretien	USA	–0.5 to 1	1.5	0.5	4

and *BoHXK2* (hexokinase [HXK]-encoding gene) and *BoGDH* (glutamate dehydrogenase [GDH]-encoding gene) was at a low level. Genes are expressed only under particular physiological conditions. Many genes might be involved, and interactions at different host–pathogen levels are continuously occurring during pathogen penetration and host reaction. Wild *et al.* (1997) found that failure to inhibit blue mould *Penicillium digitatum* in apples during the early stages of infection resulted in the subsequent development of the pathogen in tissue that is normally resistant. Li *et al.* (2018) reported that waxing pineapples reinforced their 'antioxidant system and enhanced the expression levels of genes related to defence, such as *PGIP*'.

The elements of the storage environment that have been shown to interact with genetic factors include temperature, ethylene, O_2 and CO_2. The postharvest changes that have been shown to be influenced by gene expression include ripening, textural and colour changes and susceptibility to diseases. Pre-storage treatments can affect the genetic response of the crop to storage conditions. For example, Sheng *et al.* (2018) found that several key genes involved in phenylpropanoid, flavonoid and stilbenoid pathways, including *PAL*, *CHS*, *F3H*, *LAR*, *ANS* and *STS*, were more expressed in response to pre-storage UV treatment, particularly with UV-C treatment.

Ethylene

A number of genes involved in ethylene signalling have been identified, mainly from the isolation of ethylene response mutants in *Arabidopsis*. Many have been cloned, defining a pathway from ethylene perception to changes in gene expression. Johnson and Ecker (1998) found a negative regulator of ethylene responses, *CTR1*, acting downstream of the ethylene receptors. This pathway suggests that the ethylene signal is propagated through a mitogen-activated protein kinase cascade. *EIN2*, *EIN5*, *EIN6* and *EIN7* function downstream of *CTR1* and appear to lead ultimately to the regulation of gene expression. During storage of Bartlett pears at 10°C, expression of ethylene biosynthesis genes (*ACS1a* and *ACO*) and ethylene receptor genes (*ETR2*, *ERS1a* and *ETR1a*) increased, while *CTR1* expression decreased (Nham *et al.*, 2017). Han *et al.* (2016) showed a significant induction of *MaC2H2-1/2* transcripts during the ripening of bananas with three different ripening characteristics caused by natural, ethylene-induced and 1-MCP delayed treatments, which correlated well with ethylene production. Also MaC2H2-1/2 bound to the promoters of the key ethylene biosynthetic genes *MaACS1* and *MaACO1* and repressed their activities. They concluded that MaC2H2-1/2 are transcriptional repressors and may mediate a finely tuned regulation of ethylene production during banana ripening, possibly via transcriptional repression of ethylene biosynthetic genes. Jian-fei Kuang *et al.* (2013) reported that EIN3 binding F-box protein (EBF) is an essential signalling component necessary for ethylene response. They isolated two EBF genes designated *MaEBF1* and *MaEBF2* from bananas and found that *MaEBF2* was enhanced by ethylene during fruit ripening, while *MaEBF1* changed only slightly. They concluded that *MaEBF* may be involved in banana ripening, at least partly via interaction with *MaEIL5*. Downstream, ethylene insensitive proteins (EIN2) are structurally novel containing an integral membrane domain. In the nucleus, the EIN3 family of DNA-binding proteins regulates transcription in response to ethylene, and an immediate target of EIN3 is a DNA-binding protein of the AP2/EREBP family (Alonso *et al.*, 1999).

Carbon dioxide

Exposure of fruit and vegetables to increased levels of CO_2 can have negative or positive effects. Increased levels are commonly used in CA storage (Table 5.7) and in other cases exposure to high levels has been used to control pests. Short exposures to high levels of CO_2 have been shown to be effective in controlling quarantine insect pests while maintaining fruit quality. It also enhances the storage life of fruit.

Table 5.7. Recommended CO_2 and O_2 levels for the apple cultivar Bramley's Seedling at 4–4.5°C; Johnson (1994b and personal communication 1997).

CO_2 kPa	O_2 kPa	Storage time (weeks)
8–10	11–13	39
6	2	39
5	1	44

Romero *et al.* (2008) stored grapes under 20 kPa CO_2 + 20 kPa O_2 + 60 kPa N_2 or air for 3 days, then air for 15 days. The grapes that had been exposed to CA for 3 days maintained their quality and activated the induction of transcription factors belonging to different families, such as ethylene response factors. Their results suggested that the beneficial effect of high CO_2 in maintaining grape quality appeared to be mediated by the regulation of ERFs and in particular *VviERF2-c* might play an important role by modulating the expression of PR genes. Ponce-Valadeza *et al.* (2009) reported that the effects of CO_2 on the metabolism of different fresh fruit and vegetables include changes in the activity of enzymes involved in glycolysis, fermentative metabolism, the tricarboxylic acid cycle and malate metabolism. Rothan *et al.* (1997) found that CO_2 induced the expression of certain stress-related genes and blocked the expression of both ethylene-dependent and ethylene-independent ripening associated genes. Responses of strawberries to high CO_2 concentrations have been shown to vary between cultivars (Pelayo *et al.*, 2003). For example, the cultivar 'Jewel' was shown to accumulate acetaldehyde and ethanol in storage in 20 kPa CO_2 while the cultivar 'Cavendish' did not. In 'Jewel', 168 gene sequences suggested differential expression, but in 'Cavendish' only 51 were differentially expressed (Ponce-Valadeza *et al.*, 2009).

Mature-green tomatoes were exposed to 20 kPa CO_2 for 3 days and then transferred to air. The high CO_2 effectively inhibited colour changes and ethylene production and reduced their protein content (Rothan *et al.*, 1997). They found that exposure of fruit to high CO_2 also resulted in the strong induction of two genes encoding stress-related proteins: a ripening-regulated heat-shock protein and glutamate decarboxylase. In addition, high CO_2 blocked the accumulation of mRNA for genes involved in the main ripening-related changes: ethylene synthesis (ACC and ACC oxidase), colour (phytoene synthase), firmness (polygalacturonase) and sugar accumulation (acid invertase). They postulated that the inhibition of tomato fruit ripening by high CO_2 was due, in part, to the suppression of the expression of ripening-associated genes, which was probably related to the stress effect exerted by high CO_2. Mathooko *et al.* (2004) found that exposure of peaches to 20 kPa CO_2 + 20 kPa O_2 + 60 kPa N_2 for 1 h, sealed in a jar, and then to injected ethylene, regulated ethylene biosynthesis during ripening, at least in part, by antagonizing ethylene action. They also found that exogenous application of ethylene stimulated ethylene production at concentrations of 100, 500 and 1000 ppm. ACC oxidase activity was stimulated in a concentration-dependent manner. *PP-ACO1* was slightly expressed and exogenous ethylene stimulated accumulation of its mRNA transcript in a concentration-dependent manner at 0.1, 1, 10 or 100 ppm, then the level remained constant at 100, 500 and 1000 ppm. CO_2 and 1-MCP inhibited the ethylene-stimulated ethylene production, ACC oxidase activity and accumulation of *PP-ACO1* transcripts by about 50%.

Oxygen

The molecular basis for the effects of low O_2 storage on fresh fruit and vegetables remains largely unknown. In climacteric fruit the final step in the biosynthesis of ethylene (ACC to ethylene) requires O_2 (Adams and Yang, 1979). Therefore a lower O_2 concentration might result in an inhibition of ethylene biosynthesis and slow the ripening process. Also all fresh fruit and vegetables synthesize O_2, many in very low levels, so the effects on ethylene biosynthesis may also be a factor. Loulakakis *et al.* (2006) concluded from their work on low O_2 storage of avocados that the:

> low oxygen, in addition to its inhibitory effect on ethylene biosynthesis and action, exerts its effect on prolonging the storage life of fruit by inducing a number of genes and proteins which possibly participate in the adaptation of fruit to low oxygen and by suppressing or by un-affecting others without the involvement of ethylene.

Pasentsis *et al.* (2007) showed that low O_2 regulated genes from *Citrus flavedo* responded differently in terms of their earliness, band intensity and their specificity to stress. They reported that this showed that some genes can be termed hypoxia-induced or anoxia-induced. Tonutti (2015) reported that in apples, in addition to the expression of genes involved in primary metabolism, ULO (about 1 kPa O_2) affects specific secondary metabolic pathways that appear to be selectively modulated by different low O_2 levels. Members of the ethylene responsive factor VII (transcription factors gene family) displayed differential expression, suggesting their involvement in the modulation or controlling mechanisms where there is O_2 deficiency in cells. Tomato fruit stored under low O_2 conditions showed that this induced *HSP17.7* and *HSP21* genes, indicating that their primary role is in regulating cellular actively so as to remain very nearly constant after this stress (Pegoraro *et al.*, 2012). Hypoxic conditions inhibited the accumulation of RNA, protein and DNA synthesis associated with wounding in potato tubers (Butler *et al.*, 1990). In a number of plant tissues, anoxia and hypoxia caused marked alterations in the profiles of proteins, stability of mRNA species and gene expression (Fukao and Bailey-Serres, 2004). Kanellis *et al.* (1989a, 1998b) found that storing ripening-initiated avocados to 2.5 kPa O_2 for 6 days suppressed cellulase activity immune-reactive protein and abundance of its mRNA and also produced an alteration in the profile of total proteins, which involved suppression, enhancement and induction of new polypeptides (Kanellis *et al.*, 1989a, 1993). In addition, it has been shown that O_2 levels of 2.5–5.5 kPa, which suppressed the appearance of ripening enzymes, at the protein and mRNA levels, was similar to those O_2 levels that induced the synthesis of new isoenzymes of alcohol dehydrogenase (Kanellis *et al.*, 1993).

Genetic engineering has been used to produce tomatoes that do not soften normally. These can be harvested green and ripened in an atmosphere containing ethylene. One genetically engineered cultivar of tomato called ‘Flavr Savr’ was marketed in the USA in 1993. It was also canned and marketed in the UK, but the market was restricted due to restrictions in marketing products of biotechnology in the European Union (EU) and other countries and consumer resistance. El-Sharkawy *et al.* (2016) showed that there is a direct involvement of growth substances in advancing fruit ripening independent of ethylene action through stimulating the transcription of several genes that encode cell wall metabolism-related proteins critical for determining fruit softening rate. Their results supported the hypothesis that the autonomous role played by growth substances is as important as that of ethylene in determining fruit ripening and in mediating other fruit quality traits, including their shelf-life.

Diseases

Controlling postharvest diseases and pests of fruit and vegetables has been shown to be effective using CA conditions as well as applying fungicidal chemicals. For example, storage of figs in high CO_2 atmospheres reduced mould growth without detrimentally affecting the flavour of the fruit (Wardlaw, 1938). Mitcham *et al.* (1997) showed that high CO_2 levels in the storage atmosphere could be used for controlling insect pests of grapes, using 45 kPa CO_2 + 11.5 kPa O_2 at 0°C, which gave complete control of *Platynota stultana*, *Tetranychus pacificus* and *Frankliniella occidentalis* without injury to the grapes. In papaya, low O_2 (1–5 kPa), with or without increased CO_2 (2–10 kPa), reduced decay and delayed ripening (Chen and Paull, 1986). However, new technologies have enabled the evaluation of gene expression for host defence enhancement for disease control; for example, increasing levels of phenolic compounds in the plant tissues. This approach to plant development meets the requirements of integrated disease management on sustainable use of pesticides that is included in the EU implementation through Directive 128/2009.

Fruit and vegetables have physical and chemical defence mechanisms, some of which are produced in response to an attack by a microorganism, to prevent or limit damage. The reaction of plant cells to a pathogenic attack leads to the induction of numerous genes encoding defence proteins. These include structural proteins that are incorporated into the cell walls and contribute to the confinement of the pathogen, enzymes involved in the biosynthesis of phytoalexins and pathogenesis-related (PR) proteins. Phytoalexins are antimicrobial compounds that are both synthesized by and accumulated in plants after an attack by microorganisms. Production of these chemical defences is associated with the activation of a large number of genes encoding various types of stress proteins, including PR proteins. Ding *et al.* (2002) suggested that the pre-treatment of tomato fruit with MJ or methyl salicylate induced the synthesis of some stress proteins, including PR proteins, which led to increased chilling tolerance and resistance to pathogens. The gene *MaPR1a* has a high level of identity with PR proteins and Kesari *et al.* (2007) also suggested that jasmonic acid and salicylic acid as well as ethylene could induce expression of *MaPR1a* in fruit tissue. Peng *et al.* (2014) reported that all calmodulin genes were upregulated by salicylic acid and MJ. Calmodulin is a multifunctional intermediate messenger protein expressed in plant cells. They performed expression studies on a family of six calmodulin genes (*SlCaMs*) in mature green tomatoes in response to mechanical injury and *Botrytis cinerea* infection. Both wounding and pathogen inoculation triggered expression of all those genes, with *SlCaM2* being the most responsive to both treatments. Nitric oxide acts as an important signal molecule in plants. Hu *et al.* (2014b) inoculated mangoes with a spore suspension of *Colletotrichum gloeosporioides*, which causes anthracnose disease. Nitric oxide treatment enhanced the activities of defence-related enzymes including PAL, cinnamate-hydroxylase, 4 coumarate: CoA ligase, peroxidase, β-1,3-glucanase and chitinase and also promoted the accumulation of total phenolics, flavonoids and lignin that might contribute to inhibition of the pathogen.

Several viruses attack papaya, including mosaic and bunchy top. Papaya ring spot virus (PRSV) causes chlorosis and if young plants are infected they remain stunted, while infection of adult plants can result in low yield of poor quality fruit. The virus is primarily spread by aphids and some control can be achieved by controlling the aphids. In 1985, a group from the US Pacific Basin Agricultural Research Center began work on a genetic modification of papaya (Gonzáles, 1998). They isolated and sequenced the gene that coded for the protein coat of PRSV and introduced it into callus cells from the papaya cultivar 'Sunset', which eventually resulted in a resistant plant line that was successfully field tested in 1991 (Tennant *et al.*, 2001). The result was the cultivar 'SunUp'. 'SunUp' was later crossed with the non-genetically modified 'Kapoho Solo' to create the 'Rainbow' cultivar. Both 'SunUp' and 'Rainbow' passed regulatory testing by the US government and were officially released to farmers in 1998; and in 2000 more than half papayas grown in Hawai'i were transgenic. 'Rainbow' subsequently passed Japanese regulatory testing in 2011. The team in Hawai'i has collaborated in producing PRSV-resistant transgenic papaya for other countries, including Brazil, Jamaica, Venezuela, Thailand, Bangladesh, Uganda and Tanzania (Ferreira *et al.*, 2002).

6

Pests, Diseases and Disorders

Pests and particularly diseases are among the main factors limiting the postharvest life of fresh fruit and vegetables. Clearly when fruit are infected in the field it can affect their postharvest characteristics. For example, bacterial canker (*Pseudomonas syringae* pv. *actinidiae*) in kiwifruit orchards affected the quality and storage life of the fruit (Prencipe *et al.*, 2016). Additionally the storage conditions, especially temperature and to a lesser extent humidity, can affect pests and diseases. CA storage has been shown to affect diseases and pest infestation but particularly physiological disorders. Changing the O_2 and CO_2 can also affect the metabolism of microorganisms and insect pests in storage and can therefore be a factor in controlling them.

Postharvest control of pests and diseases by CA storage can be used as a method of reducing the amount of chemicals used and thus chemical residues in the food we eat. Low O_2 but especially high CO_2 levels in storage have been generally shown to have a negative effect on the growth and development of disease-causing microorganisms. There is also some evidence that fruit develops less disease on removal from high CO_2 storage than after previously being stored in air. However, in certain cases the levels of CO_2 necessary to give effective disease control have detrimental effects on the quality of the fruit and vegetables. The mechanism for reduction of diseases appears to be a reaction of the fruit rather than the low O_2 or high CO_2 directly affecting the microorganism, though there is some evidence for the latter. The mode of action of changing the O_2 and CO_2 contents around the fruit or vegetable is related to modifying their physiology. Therefore the levels of the two gases or the balance between them can change the metabolism of fruit and vegetables in a way that is not beneficial resulting in physiological disorders. Physiological disorders of fruit can result from CA storage, but in other cases levels of disorders can be reduced. The mechanisms for reduction of disorders vary and are not always well understood. Exposing fruit and vegetables to either high levels of CO_2 or a combination of high CO_2 and low O_2 can control insects infecting fruit and vegetables. However, extended exposure to insecticidal levels of CO_2 may be phytotoxic. The CA treatment may be applied for just a few days at the beginning of storage, which may be sufficient to kill the insects without damaging the crop.

Physiological Disorders

There is a whole range of disorders that can occur in fresh produce during storage that are not primarily associated with infection with microorganisms. These are collectively referred to as physiological disorders, physiological diseases or physiological injury. As would be expected, CA storage has been shown to have positive, negative and no effect on physiological disorders of stored fresh fruit and vegetables. The incidence and extent of storage disorders are influenced not only by the concentration of CO_2 and O_2 and duration of exposure but also by the conditions in which they were grown, cultivar, harvest maturity, storage temperature and humidity. Cultivars and individual fruits vary in their susceptibility to injury because of biochemical and anatomical differences, including the size of intercellular spaces and the rate of gas diffusion through the cuticle, the epidermis and other cells. External injuries to the epidermis or internal disorders and cavities in the tissue usually become visible as brown spots as a result of oxidation of phenolic compounds. This is the last step in a reaction chain beginning with the impairment of the viability of cell membrane by fermentation metabolites, shortage of energy or possibly excess of free radicals (Streif *et al.*, 2003). Physiological disorders associated with high CO_2 are shown in Table 6.1.

A more comprehensive discussion of the subject can be found in Fidler *et al.* (1973) and Snowdon (1990,1992) but to indicate the symptoms and some of the causes of disorders mentioned in this book a few disorders are described in the following sections.

Bitter pit

The incidence and severity of bitter pit in apples is influenced by the dynamic balance of minerals in different parts of the fruit as well as the storage temperature and levels of O_2 and CO_2 in the store (Sharples and Johnson, 1987) and can be controlled by chemical treatment, especially with calcium. Jankovic and Drobnjak (1994) described experiments where 'Idared', 'Cacanska Pozna', 'Jonagold' and 'Melrose' apples were stored at 1°C and 85–90% RH, either in < 7 kPa CO_2 + 7 kPa O_2 or in air. Fruits stored in CA exhibited no physiological disorder, whereas bitter pit was observed in 'Melrose' stored in air. Bitter pit was observed in 'Red Delicious' apples stored in PE bags when CO_2 exceeded 5 kPa and O_2 fell to 15–16 kPa (Hewett and Thompson, 1988). In contrast, the incidence of bitter pit was reduced in 'Cox's Orange Pippin' apples in storage at 1°C in PE bags of 25 μm thickness in which there were 50–70 holes per bag to fit inside an 18.5 kg carton. However, it was not reduced to commercially acceptable

Table 6.1. Physiological disorders associated with high CO_2.

	Disorder	Source
Apples	Brown heart	Kidd and West, 1923
Apples	Core flush	Lau, 1983
Apples	Low temperature breakdown	Fidler, 1968
Broad bean	Pitting	Tomkins, 1965
Broccoli	Accelerated softening	Lipton and Harris, 1974
Broccoli	Off-flavours	Lipton and Harris, 1974
Cabbage	Internal browning	Isenberg and Sayles, 1969
Capsicums	Internal browning	Morris and Kader, 1977
Kiwifruit	Off-flavours	Harman and McDonald, 1983
Lettuce	Brown stain	Stewart and Uota, 1971
Mushrooms	Cap discoloration	Smith, 1965
Potato	Curing inhibition	Butchbaker *et al.*, 1967
Spinach	Off-flavours	McGill *et al.*, 1966
Strawberry	Off-flavours	Couey and Wells, 1970
Tomato	Uneven ripening	Morris and Kadar, 1977
Tomato	Surface blemishes	Tomkins, 1965

levels when the non-treated fruit had levels of bitter pit exceeding 20–30%. Stella McCloud (unpublished results), working in New Zealand, reported that CA, with increased CO_2 levels, reduced the development of bitter pit in 'Cox's Orange Pippin' apples (John Stow, personal communication). De Freitas *et al.* (2010) reported that the development of bitter pit was a complex process that involved the total input of calcium into the fruit, but also a proper calcium homeostasis at the cellular level. They found that calcium accumulation into storage organelles and calcium binding to the cell wall represented an important contribution to bitter pit development.

Cell wall destruction

Onions stored for 162 or 224 days by Adamicki *et al.* (1977) at 1°C in 10 kPa CO_2 + 3–5 kPa O_2 developed a physiological disorder where the cell walls of the fleshy scales of the epidermis and parenchyma were destroyed and the ultra-structure of mitochondria were altered.

Chilling injury

There is some evidence that CA storage can affect the development of chilling injury in stored fruit. Flesh browning and flesh breakdown of plums was shown to occur when they were stored in air at 0°C for 3–4 weeks (Sive and Resnizky, 1979), but when the same fruits were stored at 0°C in 2–8 kPa CO_2 + 3 kPa O_2 they could be stored for 2–3 months followed by 7 days in air without showing the symptoms. Wade (1981) showed that storage of the peach cultivar 'J.H. Hale' at 1°C resulted in flesh discoloration and the development of a soft texture after 37 days, but in atmospheres containing 20 kPa CO_2 fruit had only moderate symptoms even after 42 days. Conversely, Visai *et al.* (1994) showed a higher incidence of chilling injury in the form of internal browning in 'Passe Crassane' pears stored at 2°C in 5 kPa CO_2 + 2 kPa O_2 compared with fruit stored in air at 0°C. They accounted for this effect as being due to the stimulation of the production of free radicals in the fruit stored in CA.

Discoloration

Core flush has been described on several cultivars of apple where it develops during storage as a brown or pink discoloration of the core, while the flesh remains firm. It has been associated with CO_2 injury but may also be related to chilling injury and senescent breakdown. Wang (1990) reviewed the effects of CO_2 on brown core and concluded that it was due to exposure to high levels of CO_2 at low storage temperatures. Johnson and Ertan (1983) found that in storage at 4°C apples were free of core flush and other physiological disorders in 1 kPa O_2, which was markedly better than in 2 kPa O_2. Resnizky and Sive (1991) found that keeping 'Jonathan' apples in 0 kPa O_2 for the first 10 days of storage prevented core flush in early-picked apples from highly affected orchards. No damage due to fermentation was observed in any of the treatments.

Skin browning was reduced in litchis that were stored for 28 days at 1°C in plastic film bags with 10 kPa CO_2 or vacuum packed compared with non-wrapped fruit. However, their taste and flavour were unacceptable, while fruit stored non-wrapped remained acceptable (Ahrens and Milne, 1993). *Phaseolus vulgaris* beans, broken during harvesting and handling, developed a brown discoloration on the exposed surfaces; exposure to O_2 levels of 5 kPa or less controlled this browning, but resulted in off-flavours in the canned products (Henderson and Buescher, 1977). High CO_2 concentrations have also been reported to inhibit browning of beans at the sites of mechanical injury (Costa *et al.*, 1994) and were not injurious to quality as long as O_2 was maintained at 10 kPa or higher.

Wang *et al.* (2000c) reported a CO_2 linked disorder whose symptoms resembled superficial scald. The 'Empire' CO_2-linked disorder was effectively controlled by conditioning fruit at 3°C (but not at 0°C) for 3–4 weeks at 1.5–3% O_2 without CO_2 prior to CA storage at

1.5% O_2 + 3% CO_2. Reduction of the disorder was also achieved by storage at 3°C in air, but excessive ripening and associated loss of flesh firmness occurred during subsequent CA storage (Wang *et al.*, 2000c). Castro *et al.* (2007) reported flesh browning in 'Pink Lady' apples 2 months after harvest for those stored at 1°C in CA, but the browning did not increase after a longer storage time and was not observed during storage in air. Lipton and Mackey (1987) showed that Brussels sprouts stored in 0.5 kPa O_2 occasionally had a reddish-tan discoloration of the heart leaves and frequently an extremely bitter flavour in the non-green portion of the sprouts.

The apple cultivar 'Braeburn' is highly susceptible to internal browning (Braeburn browning disorder) (BBD), which has been associated with increased levels of CO_2 in the storage atmosphere (see Chapters 7, 8 and 12). Scott and Wills (1974) showed that all 'Williams' Bon Chretien' ('Bartlett') pears, which had been stored at −1°C for 18 weeks in about 5 kPa CO_2, were externally in excellent condition but were affected by brown heart. When the apples were placed in micro-perforated PE film bags after vacuum infiltration with up to 20% calcium chloride, they developed a brown heart-like disorder (Hewett and Thompson, 1989). Deuchande *et al.* (2017) stored 'Rocha' pears at −0.5°C in air and various atmospheres and found that only those in 2 kPa O_2 + 10 kPa CO_2 developed internal browning. They found faster depletion of ascorbate under high CO_2 and concluded that it was associated with the down-regulation of glutathione reductase, ascorbate peroxidase and monodehydroascorbate reductase. The genes involved were *PcGR*, *PcAPX* and *PcMDHAR*, respectively. They also reported a synergistic effect between high CO_2 and low O_2. When, after 60 days, the pears were transferred from 2 kPa O_2 + 10 kPa CO_2 to 1 kPa O_2 + 10 kPa CO_2 for 80 days, internal browning disorder incidence increased, which was associated with increases in fermentation.

Stewart and Uota (1971) showed that lettuce during cold storage had increasing levels of a brown stain on the leaves with increasing levels of CO_2 in the storage atmosphere.

Internal breakdown

Symptoms of internal breakdown are browning or darkening of the flesh, which eventually becomes soft and the fruit breaks down. It is associated with storage for too long, especially for fruit harvested too mature, and eventually all fruits that do not breakdown due to disease infection will succumb to internal breakdown. There was no internal breakdown of Cox's Orange Pippin apples after 4 months of storage at 2.5°C in 2.5 kPa O_2 + 3 kPa CO_2, but there was a perceptible but low incidence in air (Schulz, 1974).

Tonini *et al.* (1993) showed that storage of nectarine and plums for 40 days at 0°C in 2 kPa O_2 with either 5 or 10 kPa CO_2 reduced internal breakdown compared with those stored in air. When the nectarine cultivar 'Flamekist' was stored by Lurie *et al.* (1992) in CA for 6 or 8 weeks, the 10 kPa O_2 + 10 kPa CO_2 atmosphere prevented internal breakdown and reddening that occurred in fruits stored in air. Cooper *et al.* (1992) showed that CA storage with up to 20 kPa CO_2 reduced the incidence of internal browning in nectarine without any other adverse effect on fruit quality and good control was also shown in storage in 4 kPa O_2. Streif *et al.* (1994) found that exposure of nectarine to 25 kPa CO_2 prior to CA storage had little effect, but storage at high CO_2 levels, especially in combination with low O_2, significantly delayed ripening, retained fruit firmness and prevented both storage disorders. There was no deleterious effect on flavour with storage at high CO_2 levels.

Necrotic spots

Bohling and Hansen (1977) described the development of necrotic spots on the outer leaves of stored cabbage, which was largely prevented by low O_2 atmospheres in the store, but increased CO_2 had no effects.

Superficial scald

Scald is a physiological disorder that can develop in apples and pears during storage.

Fig. 6.1. Superficial scald on apples in New Zealand (photograph courtesy of Dr R.O. Sharples).

Kupferman (undated) described the typical symptoms of scald as a diffuse irregular browning of the skin with no crisp margins between affected and unaffected tissue and it is usually only skin deep (Fig. 6.1). It can occur on any part of the skin and can vary in colour from light to dark brown. In some cases the lenticels are not affected, leaving green spots.

Scald has been associated with ethylene levels in the store atmosphere and can be controlled by a pre-storage treatment with a suitable antioxidant. Ethoxyquin (1,2-dihydro-2,2,4-trimethylquinoline-6-yl ether) marketed as 'Stop-Scald' or DPA (diphenylamine) marketed as 'No Scald' or 'Coraza' can be effective when applied directly to the fruit within a week of harvesting (Hardenburg and Anderson, 1962; Knee and Bubb, 1975). Postharvest treatment with Ethoxyquin gave virtually complete control of scald and stem-end browning on fruit stored at 3.9°C or 5°C in 8–10 kPa CO_2 (Knee and Bubb, 1975). Ethoxyquin was initially registered as a pesticide in 1965 as an antioxidant used to prevent scald in pears and apples through a pre-harvest spray and postharvest dip or spray. The Codex Alimentarius Commission established maximum residue limits for Ethoxyquin residues in or on pears at 3 ppm under US Environment Protection Agency 738-R-04-011 November 2004. DPA was first registered as a pesticide in the USA in 1947. However, the Codex Alimentarius maximum residue limit for DPA on apples was set at 5 mg/kg, compared with the 10 ppm US tolerance for apples (*Guide to Codex Maximum Limits For Pesticides Residues* under EPA-738-F-97-010 April 1998). Residue levels of DPA in apples were found to vary depending on the application method and the position of the fruit in the pallet box. Villatoro *et al.* (2009) reported that there was an interaction between DPA and CA storage, in that 'Pink Lady' apples stored at 1°C in 2.5 kPa O_2 + 3 kPa CO_2 retained higher concentrations of DPA residues compared with those stored in 1 kPa O_2 + 2 kPa CO_2 or in air.

Very low levels of O_2 in CA storage have been used commercially to control scald in apples with 0.7 kPa O_2 in British Columbia and 1 kPa O_2 in Washington State, and 1 kPa for pears in Oregon State (Lau and Yastremski, 1993). Coquinot and Richard (1991) stored apples in 1.2 kPa O_2 + 1 kPa CO_2, with or without removal of ethylene, and found that in this atmosphere scald was controlled and ethylene removal was not necessary. Johnson *et al.* (1993) found that fruits respired normally for 150 days in storage in 0.4 O_2 + nominally 0 kPa CO_2 or 0.6 kPa O_2 + nominally 0 kPa CO_2 but ethanol accumulated thereafter. Retardation of scald development by 0.4 and 0.6 kPa O_2 was as effective as 5 kPa CO_2 + 1 kPa O_2 and ethylene removal from 9 kPa CO_2 + 12 kPa O_2 storage provided scald-free fruits for 216 days. However, rapid loss of firmness occurred in fruits stored in all low O_2 levels + nominally 0 kPa CO_2 after 100 days storage and was the major limitation to storage life.

It was recommended that scrubbed low O_2 storage, e.g. 5 kPa CO_2 + 1 kPa O_2, and ethylene removal from scrubbed or non-scrubbed CA stores should be considered as alternatives to chemical antioxidants for the control of scald. In 'Jonagold' apples, Awad *et al.* (1993) found that both O_2 and CO_2 affected the occurrence of scald. At 3 kPa O_2 + 1 kPa CO_2 the level of scald was 39%, which was significantly higher in early-harvested apples compared with 13% in apples from a normal harvest date. Scald in 'Granny Smith' apples was reported by Gallerani *et al.* (1994) to be successfully controlled by storing fruits in targeted low O_2 concentrations of 1 kPa O_2 + 2 kPa CO_2, and their findings enabled low O_2 application to be more precisely directed towards superficial scald control. Van der Merwe *et al.* (1997) showed that, for the cultivars 'Granny Smith' and 'Topred', exposure of fruit to initial low O_2 stress of 0.5 kPa O_2 for 10 days at either –0.5°C or 3°C (after 7 days in air at the same temperatures) could be used as an alternative to DPA for superficial scald control during storage at 1°C and 3 kPa CO_2 with 1 kPa O_2, since DPA had been banned in some countries. Kupferman (undated) reported that scald was reduced in apples that were well ventilated using a hollow fibre membrane air separator in a purge mode to supply air at 1.5 kPa O_2 + CO_2 below 3 kPa at 0°C or 3°C. The atmosphere was established within 7 days of loading the room and a continuous purging system was used to maintain the atmosphere. Fidler *et al.* (1973) also showed that scald development in storage could be related to the CO_2 and O_2 levels but that the relationship varied between cultivars (Table 6.2). 'Granny Smith' apples were stored at –0.5°C for 6 months in air or for 9 months in 1.5 kPa O_2 + 0 kPa CO_2 and then ripened at 20°C for 7 days. In air all the apples developed scald but in the CA only a few developed scald (Van Eeden *et al.*, 1992).

Kupferman (undated) also reported differences in susceptibility between apple cultivars and that 'Braeburn', 'Fuji', 'Gala', 'Golden Delicious', 'McIntosh' and 'Spartan' showed a low risk of scald, 'Rome Beauty' a moderate risk, 'Red Delicious' a moderate to high risk and 'Granny Smith' a high risk.

Table 6.2. Effects of CA storage conditions during storage at 3.5°C for 5–7 months on the development of superficial scald in three apple cultivars (Fidler *et al.*, 1973).

Storage atmosphere		Cultivar		
CO_2 kPa	O_2 kPa	Wagener	Bramley's Seedling	Edward VII
0	21	100	–	–
0	6	–	89	75
0	5	100	–	–
0	4	–	85	62
0	3	9	30	43
0	2½	–	17	43
0	2	0	–	–
8	13		24	3

Lau and Yastremski (1993) found that scald susceptibility in 'Delicious' apples was strain dependent. While storage in 0.7 kPa O_2 effectively reduced scald in 'Starking' and 'Harrold Red' picked over a wide range of maturity stages, it did not adequately reduce scald in 'Starkrimson' during 8 months of storage (Van der Merwe *et al.*, 2003). Yu and Wang (2017) claimed that the current commercial method of controlling superficial scald (ethoxyquin + CA storage) was not adequate for 'd'Anjou' pears produced in hot seasons. They found that a better combination was ethoxyquin at 2700 mg/l + 1-MCP at 0.3 μl/l + ethylene at 0.3 or 0.6 μl/l. This combination also extended storage quality, reduced decay and allowed recovery of ripening capacity after long-term CA storage.

Vascular streaking

This is a disorder of cassava where the vascular bundles in the tuberous roots turn dark blue to black during storage. The symptoms can develop within a day or so of harvesting and the disorder has been associated with the O_2 level in the atmosphere and other possible causes (Thompson and Arango, 1977). Aracena *et al.* (1993) found that waxed 'Valencia' cassava roots stored at 25°C and 54–56% RH for 3 days had 46% vascular streaking in air and 15% in 1 kPa O_2. However, at 25°C and 95–98% RH there was only 1.4% vascular streaking in air or 1 kPa O_2.

White core

Arpaia *et al.* (1985) stored the kiwifruit cultivar ‘Hayward’ for up to 24 weeks in 2 kPa O_2 + 0, 3, 5 or 7 kPa CO_2 at 0°C. The occurrence and severity of white core in CA plus ethylene was highest in 5 kPa CO_2. Two other physiological disorders were observed (translucency and graininess) and their severity was increased by the combination of high CO_2 and ethylene.

Fungal Diseases

Reports on the effects of CA storage on the development of diseases of fruit and vegetables have shown mixed results. Disease incidence on cabbages stored in 3 kPa O_2 + 5 kPa CO_2 or 2.5 kPa O_2 + 3 kPa CO_2 was lower than those stored in air (Prange and Lidster, 1991). *In vitro* studies of Chinese cabbage inoculated with *Phytophthora brassicae* and stored at 1.5°C in 0.5 kPa CO_2 + 1.5 kPa O_2 or 3.0 kPa CO_2 + 3.0 kPa CO_2 showed low levels of disease development compared with those stored in air. In contrast, *in vivo* studies showed that the infection caused by *P. brassicae* was significantly higher in the CA atmosphere than in air after 94–97 days of storage (Hermansen and Hoftun, 2005). In celery stored at 8°C, disease suppression was greatest in atmospheres with 7.5–30 kPa CO_2 + 1.5 kPa O_2, but there was only a slight reduction in 4–16 kPa CO_2 + 1.5 kPa O_2 or in 1.5–6 kPa O_2 + 0 kPa CO_2 (Reyes, 1988). A combination of 1 or 2 kPa O_2 + 2 or 4 kPa CO_2 prevented black stem disease development in celery during storage (Smith and Reyes, 1988). ‘Golden Delicious’ apple losses from rotting were lower in CA storage than at the same temperature in air (Reichel, 1974). ‘Cox’s Orange Pippin’ apples were stored at 2.5°C in air or in 2.5 kPa O_2 + 3 kPa CO_2 at 3.5°C by Schulz (1974). There was natural contamination of the fruits on the tree by *Pezicula* spp. and the disease was shown to be slightly retarded in the CA storage. There were no differences in the occurrence of Botrytis and Penicillium rots between CA and air storage. When both undamaged and damaged fruits were artificially inoculated at harvest, *Penicillium malicorticis* was shown to be more active on injured fruits in CA storage than in those stored in air. ‘Hayward’ kiwifruit were exposed to 60 kPa CO_2 + 20 kPa O_2 at 30 or 40°C for 1, 3 or 5 days by Cheah *et al.* (1994) for the control of *Botrytis cinerea* rot. Spore germination and growth of *B. cinerea* were completely inhibited *in vitro* by 60 kPa CO_2 at 40°C and partially suppressed at 30°C. Kiwifruits were inoculated with *B. cinerea* spores, exposed to 60 kPa CO_2 at 30°C or 40°C, and stored in air at 0°C for up to 12 weeks after treatment. Storage in 60 kPa CO_2 at 40°C reduced disease incidence from 85% in air at 20°C to about 50%, but exposure to CO_2 at 40°C for longer than 1 day adversely affected fruit ripening. ‘Flame Seedless’ and ‘Crimson Seedless’ grapes that were inoculated with the *B. cinerea* were exposed to 40 kPa CO_2 at 0°C for 48 h and then stored in 12 kPa O_2 + 12 kPa CO_2 also at 0°C. This combined treatment reduced *B. cinerea* incidence from 22% to 0.6% after 4 weeks and from 100% to 7.4% after 7 weeks of storage compared with fruit not treated and stored in air at 0°C (Teles *et al.*, 2014).

It was reported that the growth of the fungi *Rhizopus* spp., *Penicillium* spp., *Phomopsis* spp. and *Sclerotinia* spp. was inhibited by atmospheres of total N_2 but not in atmospheres containing 99 kPa N_2 + 1 kPa O_2 (Ryall, 1963). Kader (1997) indicated that levels of 15–20 kPa CO_2 could retard decay incidence on cherry, blackberry, blueberry, raspberry, strawberry, fig and grape. Storage of strawberries in 10–30 kPa CO_2 or 0.5–2 kPa O_2 are used to reduce disease levels, but 30 kPa CO_2 or less than 2 kPa O_2 was also reported to cause the development of off-flavour in some circumstances (Hardenburg *et al.*, 1990). Harris and Harvey (1973) stored strawberries in atmospheres with 0, 10, 20 or 30 kPa CO_2 + 21 kPa O_2 at 5°C for 3–5 days to simulate shipping conditions. The fruit were then held at 15.6°C for 1–2 days to simulate distribution. They found that fruit held in CO_2-enriched atmospheres had less softening and decay (mostly *B. cinerea*) than those held in air and that 20 or 30 kPa CO_2 was the most effective. However, fruit held in 30 kPa CO_2 developed off-flavours. Differences in decay in fruit held in air and at high CO_2 were greater after subsequent holding in air at 15.6°C than on removal from storage. They also showed that CA storage of strawberries not only reduced disease levels during storage, but also had an additional beneficial effect in disease reduction when it was removed for marketing (Table 6.3). Parsons *et al.* (1974) showed considerable reduction in disease levels on tomatoes in CA storage compared with storage in air. Most of the effect came from the low O_2 levels, with little additional effect from increased CO_2 levels (Table 6.4).

There is evidence of interactions between CA storage and temperature on disease development. After 24 weeks storage of currants (*Ribes* spp.) at –0.5°C and 20 or 25 kPa CO_2, the average incidence of fungal rots was lowest at about 5%. Rotting increased to about 35% at 1°C in 25 kPa CO_2 and to about 50% at 1°C and 20 kPa CO_2. With 0 kPa CO_2 at either temperature, the incidence of rots was about 95% (Roelofs, 1994).

Table 6.3. The effects of CO_2 concentration during storage for three days at 5°C on the percentage levels of decay in strawberries and the development of decay when removed to ambient conditions of approximately 15°C; modified from Harris and Harvey (1973).

Storage	0 kPa CO_2	10 kPa CO_2	20 kPa CO_2	30 kPa CO_2
3 days storage at 5°C	11	5	2	1
+ 1 day at 15°C in air	35	9	5	4
+ 2 days at 15°C in air	64	26	11	8

There remains the question of how CA storage actually reduces or controls diseases on fruit and vegetables. Parsons *et al.* (1974) showed that atmospheres containing 3 kPa O_2 reduced decay on stored tomatoes caused by *Rhizopus* or *Alternaria*. However, both genera of fungi grew well *in vitro* in 3 kPa O_2 or less. This led them to hypothesize that the reduction in decay was due to the CA storage conditions acting on the tomato fruit itself so that it developed resistance to the fungi, rather than acting only on the fungi. Probably the answer is a combination of both the reaction of the crop and the susceptibility of the microorganism.

Bacterial Diseases

Some *in vitro* and *in vivo* studies have indicated that CA may have a negative effect on bacteria. Brooks and McColloch (1938) reported that storage of lima beans in CO_2 concentrations of 25–35 kPa inhibited bacterial growth without adversely affecting the beans. Parsons and Spalding (1972) inoculated tomatoes with soft rot bacteria and held them for 6 days at 12.8°C in 3 kPa O_2 + 5 kPa CO_2 or in air. Lesions were smaller on fruits stored in 3 kPa O_2 + 5 kPa CO_2 than on those stored in air, but CA storage did not completely control decay. Amodio *et al.* (2003) stored slices of mushrooms at 0°C either in air or in 3 kPa O_2 + 20 kPa CO_2 for 24 days and found that there was a slight increase for mesophilic and psychrophilic

Table 6.4. Effects of CA storage conditions on the decay levels of tomatoes harvested at the green-mature stage; modified from Parsons *et al.* (1974).

Storage atmosphere	After removal from 6 weeks at 13°C	Plus 1 week at 15–21°C	Plus 2 weeks at 15–21°C
Air (control)	65.6 %	93.3 %	98.6 %
0 kPa CO_2 + 3 kPa O_2	2.2 %	4.4 %	16.7 %
3 kPa CO_2 + 3 kPa O_2	3.3 %	5.6 %	12.2 %
5 kPa CO_2 + 3 kPa O_2	5.0 %	9.4 %	13.9 %

bacteria in air but no increase or a slight decrease for those in 3 kPa O_2 + 20 kPa CO_2.

Potatoes are cured before long-term storage by exposing them to a higher temperature at the beginning of storage to heal any wounds. Weber (1988) showed that the defence reaction of potatoes to infection by *Erwinia carotovora* subsp. *atroseptica* during the curing period was inhibited by temperatures of less than 10°C, reduced O_2 levels of less than 5 kPa and CO_2 levels of over 20 kPa. CO_2 can retard bacterial growth by increasing the lag phase before they begin to develop. The degree of retardation was shown to increase with increasing concentrations of CO_2, but Daniels *et al.* (1985) reported that *Clostridium botulinum* may survive even at high CO_2 levels. Sobiczewski *et al.* (1999), in a study of the efficacy of antagonistic bacteria in protection of apples against fungal diseases, found that during storage at 2–3°C there was a tendency for the highest reduction in bacterial population in the atmosphere containing 1.5 kPa CO_2 and 1.5 kPa O_2 compared with air. Zhang *et al.* (2013) stored fresh-cut pineapple at 7°C and found that those packaged in air were not acceptable after 7 days, while those packaged in 50 kPa O_2 + 50 kPa CO_2 were still acceptable. Also 50 kPa O_2 + 50 kPa CO_2 retarded the growth of a lactic acid bacterium (*Leucono stoccitreum*) and the two spoilage yeasts tested (*Candida sake* and *C. argentea*).

Insects

CA storage has been used to control insects. It was suggested that the most promising application for fresh fruit, such as cherries, nectarines, plums, apples, pears, stone fruits, blueberries and strawberries, is to expose them to less than 1 kPa O_2 atmospheres for short periods to replace chemical treatments to meet quarantine requirements (Ke and Kader, 1989, 1992a). However, the levels that are necessary may be phytotoxic to the fruit. They reviewed the insecticidal effects of very low O_2 and/or very high CO_2 atmospheres at various temperatures and compared these levels with the responses and tolerance of fresh fruit and vegetables to similar CA conditions. They concluded that the time required for 100% mortality was shown to vary with insect species and their developmental stage, temperature, O_2 level, CO_2 level and humidity. Yahia and Kushwaha (1995) reported that 'Hass' avocados, 'Sunrise' papayas and 'Keitt' mangoes tolerated low O_2 (± 0.5 kPa) and/or very high CO_2 (± 50 kPa) for 1, 2 and 5 days, respectively, at 20°C. They suggested that these treatments could have some potential use as insecticidal atmospheres for quarantine insect control on the basis of fruit tolerance, insect mortality and costs. On stored vegetables, Cantwell *et al.* (1995) also showed that various combinations of O_2, CO_2 and temperature were effective in achieving complete insect mortality before the development of phytotoxic symptoms. They found that storage in 10–20 kPa CO_2 could control thrips (*Thysanoptera* spp.) in 7 days and the peach potato aphid (*Myzus persicae*) in 10–14 days. In 80–100 kPa CO_2 at 0°C, complete mortality was consistently achieved for both species within 12 h.

Ke *et al.* (1994a) showed that tolerances of peaches and nectarines to CA that were insecticidal were related to the time before occurrence of visual injury and/or off-flavour. The tolerances of 'John Henry' peaches, 'Fantasia' nectarines, 'Fire Red' peaches, 'O'Henry' peaches, 'Royal Giant' nectarine and 'Flamekist' nectarine to 0.25 kPa O_2 + 99.75 kPa N_2 at 20°C were 2.8, 4.0, 4.0, 4.4, 5.1 and 5.2 days, respectively. 'Fairtime' peaches tolerated 0.21 kPa O_2 with 99 kPa CO_2 at 20°C for 3.8 days, 0.21 kPa O_2 with 99 kPa CO_2 at 0°C for 5 days, 0.21 kPa O_2 at 20°C for 6 days, and 0.21 kPa O_2 at 0°C for 19 days. Comparison of fruit tolerance on the time to reach 100% mortality of some insect species suggested that 0.25 kPa O_2 at 20°C is probably not suitable for postharvest insect disinfestation, while 0.21 kPa O_2 with or without 99 kPa CO_2 at 0°C merited further investigation. Kerbel *et al.* (1989) showed that 'Fantasia' nectarines were quite tolerant to exposures to low O_2 and/or high CO_2 for short periods, but may be long enough for insect control. Yahia *et al.* (1992)

showed that insecticidal O_2 concentrations of less than 0.4 kPa O_2 with the balance being N_2 for less than 3 days at 20°C can be used as a quarantine insect control treatment in papaya without the risk of significant fruit injury. Scale insects (*Quadraspidiotius perniciosus*) on stored apples were completely eliminated by storing infested fruit in 2.6–3.0 kPa O_2 + 2.4–2.5 kPa CO_2 or 1.5–1.7 kPa O_2 + 1.0–1.1 kPa CO_2 at 1°C or 3°C for 31–34 weeks, plus an additional week at 20°C and 50–60% RH (Chu, 1992). Leon *et al.* (1997) disinfested 'Manila' mangoes by exposing them at 12°C to 1% O_2 combined with either 30% or 50% CO_2 for 3 days. After disinfestation, the mangoes were kept at 12°C for up to 27 days and no quality differences were found between CA-treated and non-treated mangoes. Respiratory activity of the mangoes held at 12°C for 18 days and then allowed to ripen at 25°C was higher than that of those kept at 25°C. Symptoms of chilling injury appeared by 18 days in mangoes that were treated with 50% CO_2, by 21 days in mangoes treated with 30% CO_2 and by 24 days in those not treated. It appeared that disinfestation by CA increased, to a limited degree, the susceptibility of 'Manila' mangoes to chilling injury.

Sweet potatoes were exposed to low O_2 and high CO_2 for 1 week during curing or subsequent storage to evaluate the effect on the weevil *Cylas formicarius elegantulus*. Exposure to levels required for insect control was found not to be feasible during curing, but cured sweet potatoes could tolerate CA combinations that have a potential as quarantine procedures. They tolerated 8 kPa O_2 during curing, but when exposed to 2 or 4 kPa O_2 or to 60 kPa CO_2 + 21 or 8 kPa O_2 they were unusable within 1 week after curing, mainly due to decay. Exposure of cured sweet potatoes to 2 or 4 kPa O_2 + 40 kPa CO_2, or 4 kPa O_2 + 60 kPa CO_2 for 1 week at 25°C had little effect on postharvest quality but exposure to 2 kPa O_2 + 60 kPa CO_2 resulted in increased decay, a reduction in flavour and more off-flavours (Delate and Brecht, 1989). Sweet potatoes infested with adult and immature stages of *C. f. elegantulus* were exposed to atmospheres containing low O_2 and increased concentrations of CO_2 for up to 10 days at 25°C and 30°C. Adults were killed within 4–8 days when exposed to 8 kPa O_2 + 40–60 kPa CO_2 at 30°C. At 25°C, exposure to 2 or 4 kPa + 40 or 60 kPa CO_2 at 25°C killed all the adult insects within 2–8 days but exposure to 8 kPa O_2 + 30–60 kPa CO_2 for 1 week at 30°C failed to kill all the weevils. However, no adult weevils emerged from infested roots treated with either 4 kPa O_2 + 60 kPa CO_2 or 2 kPa O_2 + 40 or 60 kPa CO_2 for 1 week at 25°C (Delate *et al.*, 1990).

Fumigation of 'Bing' cherries with methyl bromide (MB) to control codling moth (*Cydia pomonella*) negatively affected fruit and stem appearance (Retamales *et al.*, 2003). However, they reported that the cherries could be heated to 45°C for 41 min or 47°C for 27 min in an atmosphere of 1 kPa O_2 + 15 kPa CO_2 + 84 kPa N_2, which could provide quarantine security against codling moth and western cherry fruit fly (*Rhagoletis cingulata*). Treated cherries had similar incidence of pitting and decay and similar preference ratings after 14 days of storage at 1°C as non-treated or MB-fumigated fruit (Shellie *et al.*, 2001). They also showed that storage in CO_2 levels up to 30 kPa did not markedly influence the quality aspects compared with storage in air, with the exception of decay being significantly reduced. Chervin *et al.* (1999) suggested that exposure to 2 kPa O_2 for 3 days at 28°C approximately halved the time required in cold storage for effective control of late-instar light brown apple moth (*Epiphyas postvittana*) in apples. Preliminary observations suggested that there may not be substantial difference between the resistance of non-diapausing and pre-diapausing codling moth larvae to this treatment followed by cold storage. Consumer panels found that fruit were as acceptable as control fruit. In storage at 0°C the mortality responses of armoured scales, *Hemiberlasia* spp., on New Zealand 'Hayward' kiwifruit exposed to storage in air or 2 kPa O_2 + 5 kPa CO_2 were equally effective (Whiting, 2003). Mature scales were more tolerant of both storage treatments than immature scales. *Tuta absoluta* is a pest that affects tomatoes and other solanaceous crops in Europe and other Mediterranean areas.

As an alternative to fumigation with MB, Riudavets *et al.* (2016) found that exposure of *T. absoluta*-infected tomatoes to an atmosphere of 95 kPa CO_2 at 25°C for 48 h controlled all stages, but negatively affected fruit quality. However, they found that 40 kPa CO_2 for 72 h gave the same level of control with no negative effects on fruit quality. A cold storage treatment at 1°C for a total of 10 days was also effective in the control of the *T. absoluta* eggs.

Callosobruchus maculatus is the principal pest of the stored chickpea. The larvae of this insect feed on the grain, causing considerable damage that produces qualitative and quantitative losses. For control of this species, pesticides such as phosphine and MB have been used. Iturralde-García *et al.* (2016) tested CA storage (50, 70 or 90 kPa CO_2 + 10, 6 or 3 kPa O_2 for up to 120 h) on *C. maculates*. They found that exposure to 50 kPa CO_2 for 48 h resulted in 100% mortality of adult insects, but third-instar insects had higher resistance than adults, with only 18.9% mortality after 48 h. CA exposure had no significant effects on germination, water absorption, cooking time, texture and colour of the chickpeas. CA has been used in storage of dried cereal and grain legumes to control insect infestation. Tome *et al.* (2000) found that 0, 30, 40, 50 or 60 kPa CO_2 with the balance N_2 did not affect the moisture content, water absorption, cooking time or colour index of dried *Phaseolus vulgaris* beans during 20 days of storage. Brackmann *et al.* (2002b) found that storage of pinto beans (*P. vulgaris*) in 100% N_2 for 19 months had zero losses due to insects, compared with significant losses for those stored in air. The beans stored in total N_2 had a lighter tegument colour and a shorter cooking time than those stored in air. Darkening was a consequence of the oxidation of phenolic compounds. Pereira *et al.* (2007) used 50 ppm ozone to fumigate maize grain for 7 days against insect infestation and found no detrimental effects on the grain during 180 days of subsequent storage.

7

Pre-storage Treatments

The application of CA storage and MA packaging technologies to the fresh fruit and vegetable industry requires inputs of other technologies, including special treatments of the crop before it is put into store or before it is packaged. In most cases it is advisable to place fruit or vegetables into appropriate storage as quickly as possible after harvest. However, in certain cases it has been shown that some treatments can be beneficial prior to CA storage or MA packaging and these treatments are discussed in this chapter.

High Temperature

Root crops such as potatoes, yams and sweet potatoes have a cork layer over the surface of their tubers. These cork layers serve as a protection against microorganism infections and reduce water loss, but cork layers can be damaged during harvesting and handling operations. Exposing crops to high temperature and humidity for a few days directly after they have been loaded into the store can facilitate repair of the damaged tissue and reduce postharvest losses.

The term curing is also applied to some fruit, especially citrus fruits. The mechanism in citrus fruits is different to root crops but it effectively heals wounds and reduces disease levels. Drying is also carried out to aid preservation of onions and garlic, but this simply involves drying their outer layers to reduce microorganism infection and water loss. Various drying recommendations have been published, including exposure to about 30°C and 70% RH for 7–10 days before storage (Thompson *et al.*, 1972a).

Immersing fruit in hot water or brushing with hot water prior to storage is used to control microorganisms that may have infected fruit before storage. The optimum recommended conditions for disease control varies between crop and disease organism. For example, visible appearance of the fungal disease anthracnose (*Colletotrichum gloeosporioides*) during ripening of mangoes was effectively inhibited by hot-water treatments at 46°C for 75 min combined with storage for 2 weeks at 10°C in either 3 kPa O_2 + 97 kPa N_2 or 3 kPa O_2 + 10 kPa CO_2 (Kim *et al.*, 2007) (see Fig. 2.2). Hot-water brushing at 56°C for 20 s reduced decay development in grapefruit during storage (Porat *et al.*, 2004). In some cases exposure to temperatures slightly higher than those used in storage have been shown to have beneficial effects. For example, exposure of 'Granny Smith' apples to 0.5 kPa O_2 for 10 days at 1°C before CA storage at –0.5°C inhibited the development of superficial scald for fruit picked at pre-optimum maturity (Van der Merwe *et al.*, 2003). In apples, Gasser and von Arx (2015) found that hot-water treatment did not affect fruit quality or the incidence of physiological disorders during DCA storage. Hot-water treatment of jujube at 40°C increased storage life by slowing ripening and reducing the rate of ethylene evolution (Anuradha and Saleem, 1998).

Root crops

In root crops, curing involves suberin being deposited in the parenchymatous cells just below the damaged area of the tuber. In sweet potatoes, Stange *et al.* (2001) showed that the interior flesh tissues accumulate antifungal compounds (including 3,5-dicaffeoylquinic acid) under curing conditions of 30°C and 90–95% RH for 24 h, which helps resistance to infection. Sweet potato could be also cured at 32–36°C and 100% RH for 4 days (Thompson, 1972b) or 29–32°C and 85–90% RH for 7–15 days (Lutz, 1952). Curing potatoes has been standard pre-storage practice since the 1920s and can be achieved by exposure to 10°C for 30 days, 15°C for 15 days or 20°C for 10 days, all at high humidity (Burton, 1989). Curing involves suberin being deposited in the parenchymatous cells just below the damaged area of the tuber. Suberin refers to a polymer containing phenolics and long-chain decarboxylic fatty acids, hydroxy fatty acids and, perhaps, epoxy fatty acids, which provides initial protection to the tuber by forming a highly lipophilic coating within and on the surface of tissues. There are some indications that the fatty acids may also act as phytoalexins and the peroxides and epoxides may have toxic effects on infectious micro-organisms (Galliard, 1975). Subsequently, below the suberized cells, a meristematic layer of cells is formed which is the cork cambium. This produces new cells that contain lignin, which seal off the damaged area. Both these processes are temperature and humidity sensitive (Burton, 1989). Cassava could be successfully cured by exposure to 25–40°C and 80–85% RH (Booth, 1975) while cuts of 2 mm deep could be cured, but cuts of 8 mm or deeper showed no curing over a 10-day period (Akhimienho, 1999). In the 1980s a cassava root conditioning protocol was developed in Latin America by scientists of the Centro Internacional de Agricultura Tropical (Wheatley, 1989). It was found that the treatment of cassava roots with the fungicide thiabendazole and their subsequent exposure to conditions of high humidity at ambient temperatures in plastic bags permitted the storage of fresh cassava to be extended from 3 to 21 days without engendering significant postharvest losses or compromising culinary quality. In the 1990s this technology was transferred initially to Ghana and Tanzania by the NRI. On-farm trials and assessments in commercial market environments confirmed that the technology would extend the storage life of fresh cassava. However, due to the reliance on the use of a fungicide and plastic bags, Digges (1995) considered that the cost of the original protocol might militate against its widespread adoption.

A simplified methodology was therefore developed based on the careful out-grading of damaged roots and the temporary storage of water-dipped roots in locally available woven polypropylene sacks which were then held at high humidity under tarpaulins or plastic sheeting (Bancroft and Crentsil, 1995). This approach proved cheaper and was effective at maintaining cassava quality for a reasonable 7–10 days. Socio-economic surveys indicated a high level of satisfaction with this latter process, particularly amongst the snack-food vendors whose profit margins benefited from the extended shelf-life potential of fresh cassava and the reduction in the volume of waste (Bancroft *et al.*, 1998). In common with other root crops, the possibility of curing yams (*Dioscorea* spp.) after harvest in order to extend their storage life and reduce postharvest losses has attracted the attention of researchers for decades (Table 7.1).

In the early 2000s, over a series of consecutive storage seasons, rural farm trials were conducted in the yam-growing regions of Ghana to determine the effects of a range of different low-cost preconditioning and

Table 7.1. Optimum curing conditions for yams. Modified from Rees *et al.*, 2012.

Species	Temperature (°C)	Humidity (% RH)	Duration (days)	Reference
D. alata	29–32	90–95	4	González and Rivera (1972)
D. bulbifera	26–28	high	5–7	Martin (1974)
D. cayenensis	36–40	91–98	1	Thompson *et al.* (1977)
D. cayenensis	25–40	95–100	1–7	Been *et al.* (1976)
D. esculenta	26–28	high	5–7	Martin (1974)
D. esculenta	29	84	5	Thompson (1972a)
D. rotundata	25–30	55–82	5	Adesuyi (1973)
D. rotundata	25–40	95–100	1–7	Been *et al.* (1976)
D. rotundata	26	92	11–15	Nnodu and Nwankiti (1986)
D. rotundata	35–40	95–100	1	Thompson (1972a)

storage regimes on the postharvest quality of local varieties of *Dioscorea rotundata* (white yam). The study suggested that beneficial curing protocols could be deployed in rural communities (Bancroft *et al.*, 2005). For periods of 2–3 months traditional pit storage engendered less rot and moisture loss than storage above ground in barns. Eventually, however, pit-stored yams became prone to nematode infestation and sprouting. Curing in polyethylene bags was not found to be reliable. More successful was the use of temporary 'curing' clamps, rooms or enclosures. Under field conditions, curing was enhanced by exposure to temperatures of about 32°C for 5–6 days. Such conditioning helped to reduce the incidence of rots during subsequent storage but did not retard water loss. In contrast, super-optimal curing temperatures favoured the development of rots. Strict grading of the yam tubers prior to storage had a marked impact on the extent of postharvest losses. Although chronic internal infections and disorders can compromise the subsequent marketability of the yams, stored tubers are particularly susceptible to rots that originate at the site where the tuber is cut away from vine.

Fruit

For crops other than root crops, pre-storage exposure to high temperature is less well established and often still in an experimental stage. Sealing citrus fruits in plastic film bags and then exposing them to 34–36°C for 3 days resulted in the inhibition of *Penicillium digitatum* infection and reduced decay and blemishes through lignification and an increase in antifungal chemicals in the peel without deleterious effects on the fruit (Ben-Yehoshua *et al.*, 1989). Tariq (1999) found that placing damaged oranges in sealed PE film bags and exposing them to temperatures in the range of 25–35°C for 3–48 h (the lower the temperature, the longer was the exposure time required) completely eliminated the browning and fungal growth. The nature of curing appeared to be related only to temperature and humidity and not O_2 or CO_2, since where PE bags were used the thickness of the film (30 or 50 μm) had no effect on the process, although it changed the O_2 and CO_2 content around the fruit. Also, curing could be achieved without film packaging if the surrounding atmosphere was maintained at 95–98% RH.

'Granny Smith' apples were kept at 46°C for 12 h, 42°C for 24 h, or 38°C for 72 h or 96 h before storage at 0°C for 8 months in 2–3 kPa O_2 + 5 kPa CO_2. The heat-treated fruit were firmer at the end of storage and had a higher TSS:TA ratio and a lower incidence of superficial scald than fruits not heat treated. Pre-storage regimes with longer exposures to high temperatures (46°C for 24 h or 42°C for 48 h) resulted in fruit damage developing during storage (Klein and Lurie, 1992). Brigati and Caccioni (1995) recommended curing kiwifruits in well ventilated warehouses at 15–20°C for 48 h before refrigeration. 'Niitaka' Asian pears were cured at ambient temperatures of about 18°C for 5, 10 or 15 days followed by step-wise cooling down to 0.5°C, which reduced skin blackening without significantly altering other fruit quality characteristics during storage but affected the physical properties of the fruit's epidermis (Park and Kwon, 1999). Acorn squash tissue held at 28°C for 18 h developed considerable resistance to infection by *Penicillium italicum* (Stange and McDonald, 1999). Phillips and Armstrong (1967) and Purseglove (1968) also recommended curing at 23.9–29.4°C for 2 weeks to harden the shell, but other work claimed that curing was either ineffective or detrimental (Wardlaw, 1938; Lutz and Hardenburg, 1968).

Irradiation

UV irradiation has been demonstrated to increase the natural resistance of some commodities and has been studied as a direct method for decay control. In many countries irradiation of fresh fruits and vegetables has been approved at specified levels of exposure. UV-C has been shown to reduce postharvest decay on strawberries (Nigro *et al.*, 2000; Pombo *et al.*, 2011) and mangoes (González-Aguilar *et al.*, 2001) and bacterial counts on minimally processed baby spinach leaves (Escalona *et al.*, 2010), watermelon cubes (Artés-Hernández *et al.*, 2010) and white mushrooms (Guan *et al.*, 2012). Guan *et al.* (2012) showed that UV-C doses of 0.45–3.15 kJ/m^2 resulted in 0.67–1.13 log CFU/g reduction of *Escherichia coli* O157:H7 inoculated on mushroom cap surfaces. Sheng *et al.* (2018) exposed table grapes to UV-B or UV-C radiation at 3.6 kJ/m^2 and then stored them at 4°C for 28 days. They found that both UV treatments increased the phenolic compounds of grapes during storage. UV-C irradiation has also been reported to have other effects. Kaewsuksaeng *et al.* (2012) irradiated mature-green limes with UV-B doses at 19.0 kJ/m^2 and then stored them at 25°C in the dark. They found that the irradiation effectively suppressed chlorophyll degradation through the control of the action of the chlorophyll-degrading enzyme and slowed the loss of ascorbic acid. Jiang *et al.* (2010) reported that UV-C irradiation resulted in maintenance of firmness and enhanced antioxidant capacity in shiitake mushrooms. Kasim and Kasim (2012) found that it prevented yellowing in fresh-cut cress but increased electrolyte leakage due to tissue damage. UV-C irradiation extended postharvest shelf-life of banana by reducing ethylene production and respiration rate compared with those that had not been irradiated (Pongprasert *et al.*, 2011).

The mode of action seems to be related more to its effect on the crop rather than on microorganisms. Pongprasert *et al.* (2011) reported that UV-C could play a role in activation of plant defence mechanisms and antioxidant systems. Nigro *et al.* (2000) indicated that treatment with low UV-C doses produced a reduction in postharvest decay related to induced resistance mechanisms. UV treatment of carrot slices induced production of 6-methoxymellen, which is inhibitory to *Botrytis cinerea* and *Sclerotinia sclerotiorum* (Coates *et al.*, 1995). Sheng *et al.* (2018) found that several key genes involved in phenylpropanoid, flavonoid and stilbenoid pathways, including *PAL*, *CHS*, *F3H*, *LAR*, *ANS* and *STS*, were more expressed during storage of grapes at 4°C for 28 days in response to UV radiation at 3.6 kJ/m^2, particularly to UV-C.

Hydrogen

Fumigation of kiwifruit with water with a high hydrogen content (prepared by bubbling hydrogen gas through water until it was saturated) was shown to inhibit senescence of kiwifruit and postharvest production of hydrogen gas was shown to decrease during ripening (Hu *et al.*, 2014a). Hu *et al.* (2018) therefore tested whether fumigation of kiwifruit with hydrogen would affect ripening. They found that softening was delayed, incidence of decay reduced and ethylene production reduced with the downregulation of the corresponding gene transcripts after fumigation. Al Ubeed *et al.* (2018) reported that hydrogen sulfide can inhibit senescence of fruit and vegetables and reported that its mode of action included inhibiting the action of ethylene. They fumigated pak choi cabbages with hydrogen sulfide at 250 µl/l and stored them in an ethylene-free atmosphere. They found that the hydrogen sulfide fumigation was effective in inhibiting the loss of green colour and reduced the respiration rate, ion leakage and endogenous ethylene production. In a comparison with 1-MCP they found that both MCP and hydrogen sulfide were equally effective in the parameters they tested, but when ethylene was introduced to the containers 1-MCP was more effective than hydrogen sulfide.

Waxing and Polishing

Fruit and vegetables can be dipped or sprayed with a range of chemicals postharvest to delay deterioration, control diseases or improve their appearance. There are many treatments that have been tested; some are included in this chapter. Also many treatments have been used in combination with others and therefore may be included under a different heading in this chapter. Some coatings provide a partial barrier to gas exchange between the fruit and its environment that is generally beneficial. However, modification of the atmosphere in and around the fruit cells by the use of coating material can increase disorders associated with high CO_2 and low O_2 concentration, such as accumulation of ethanol and alcoholic off-flavours (Smith *et al.*, 1987).

Many fruits can benefit from the application of waxes and subsequent polishing. This is not simply to enhance the appearance, which is of course very important, but also to improve the storage quality of the crop. When the crop is in the field or even during harvesting, transport, washing or grading, it can undergo some scratching or abrasion that not only removes the natural layer of wax but can also scar the epidermis. Washing is essential in the removal of bird faeces, insect marks, chemical residues and general field dirt but can also remove much of the natural wax, especially if used with detergents. The natural layer of wax reduces moisture loss from the fruit. The appearance of crops such as citrus fruits may be improved by synthetic wax application and some producers may add dyes to the wax in order to modify the colour in addition to providing an enhanced shine. Commercial waxes, mainly used for citrus, are by-products of the petrochemical industry produced from proprietary formulae and include Fruitex, Britex and Seal Britex. There is public interest in development of edible natural biodegradable coatings to replace the synthetic waxes for maintaining postharvest quality of fruit. Natural waxes include sugarcane wax, carnauba wax, shellac and various resins. Carnauba is a wax extracted from the leaves of the palm *Copernicia prunifera*, which is a native of north-eastern Brazil and consists mostly of aliphatic esters, diesters of 4-hydroxycinnamic acid, hydroxycarboxylic acids and fatty alcohols. Lac is a resin secreted by the female lac bug (*Kerria lacca* or *Laccifer lacca*) found on the bark of many species of tree in the forests of India and Thailand. It can be refined and processed into shellac dry flakes or dissolved in ethanol to make liquid shellac. Rosins are obtained from the oleoresins of pine trees after the distillation of turpentine and are used for coating, especially for citrus, primarily to impart a gloss.

These coatings may be applied to the fruit in a foam or liquid bath, a liquid spray or by sponge or brush rollers. Application rates may be governed by legislation in the country where the coating is being applied, or in the country where the fruit is to be exported. It is essential that these regulations, as well as the manufacturers' recommendations, be strictly adhered to. Mechanized spray systems often apply an excess of fluid so that the unused surplus flows through the conveyor back into the holding vessel ready for reapplication. The fruit must be dried after waxing to give good adhesion and retention of the wax. Drying is usually done in a drying tunnel and may also incorporate soft brushes. A blast of warm air is often directed at the fruit as it passes along the conveyor under the fan, or else the heaters are under the fruit and the fan draws the air through it by suction (Clarke, 1996).

The yeast *Candida guilliermondii* was shown to be antagonistic towards *Penicillium* spp. (McGuire, 1993) and could be successfully incorporated into commercial citrus waxes where it was shown to survive for 2 months at 12°C. Kouassi *et al.* (2012) showed improved disease control when an essential oil was incorporated in shellac or carnauba wax. Coating cassava roots with paraffin wax was shown to keep them in good condition for 1–2 months at room temperature in Bogotá in Colombia (Young *et al.*, 1971) and this coating treatment is commonly applied to cassava roots exported from Costa Rica (Fig. 7.1). Another treatment is dipping the roots in a fungicide (benomyl or thiabendazole) and packing them in plastic film bags directly after harvest (Thompson and Arango, 1977), but there are restrictions on the use of benzinidazole fungicides.

Saberi *et al.* (2018) tested various coatings on storage of Valencia oranges for up to 4 weeks at 20°C or 5°C with an additional storage for 7 days at 20°C. Of the coatings they tested, pea starch and guar gum blended with a lipid mixture of shellac and oleic acid were the best in extending shelf-life and reducing respiration rate, ethylene production, weight loss, softening, peel pitting and decay. Li *et al.* (2018) waxed pineapples directly after harvest and found that waxing delayed their colour change, decreased respiration rate and ethylene production, and reduced internal browning.

Fig. 7.1. Cassava coated with paraffin wax in Costa Rica on arrival in UK.

Carboxymethyl Cellulose

Dan Cutts from Geest Industries and Peter Lowings from Cambridge University introduced a coating in 1982 that consisted of sodium carboxymethyl cellulose, sucrose fatty acid esters and mono-glycerides and di-glycerides. They claimed that it was non-phytotoxic, tasteless and odourless and that it increased the shelf-life of fresh fruit and vegetables. The product was called Tal-Prolong and Lowing supervised two PhD students, Nigel Banks and Roger Bancroft, to investigate the physiological response of different fruits to this preparation. Banks (1984) showed that Tal-Prolong could delay the ripening of bananas. He postulated that the effect was due to the restriction in gas exchange between the fruit and its surrounding atmosphere. This caused an increase of CO_2 and a depletion of O_2 in the cells, thus resulting in an effect similar to that achieved in CA storage. Bancroft (1989, 1995) confirmed these findings and also showed some reduction in levels of disease on top fruit coated with 1.5% Tal-Prolong. Bancroft observed that Tal-Prolong diminished the overall incidence of some common postharvest fungal infections of 'Conference' pears held in cold storage. At higher temperatures in both apples and pears, such coatings modified the spatial distribution of rots within the fruit and markedly reduced the rate of spread of lesions. In comparison with experimental controls, Tal-Prolong had a greater impact on rots caused by *Botrytis cinerea*, *Monilinia fructigena* and *Rhizopus nigricans* than those caused by *Alternaria alternata* and *Penicillium expansum* (Fig. 7.2). The efficiency of the treatments in reducing the rate of spread of rots was related to the concentration of Tal-Prolong, the type and cultivar of fruit, the timing of application and the temperature to which the fruit was subsequently exposed. Krishnamurthy (1989) ripened 'Alphonso' mangoes at 28–32°C and 30–60% RH coated with various concentrations of Tal-Prolong and found that 1% inhibited the development of yellow peel colour and delayed some of the other ripening changes for 4–5 days. At higher concentrations the fruit

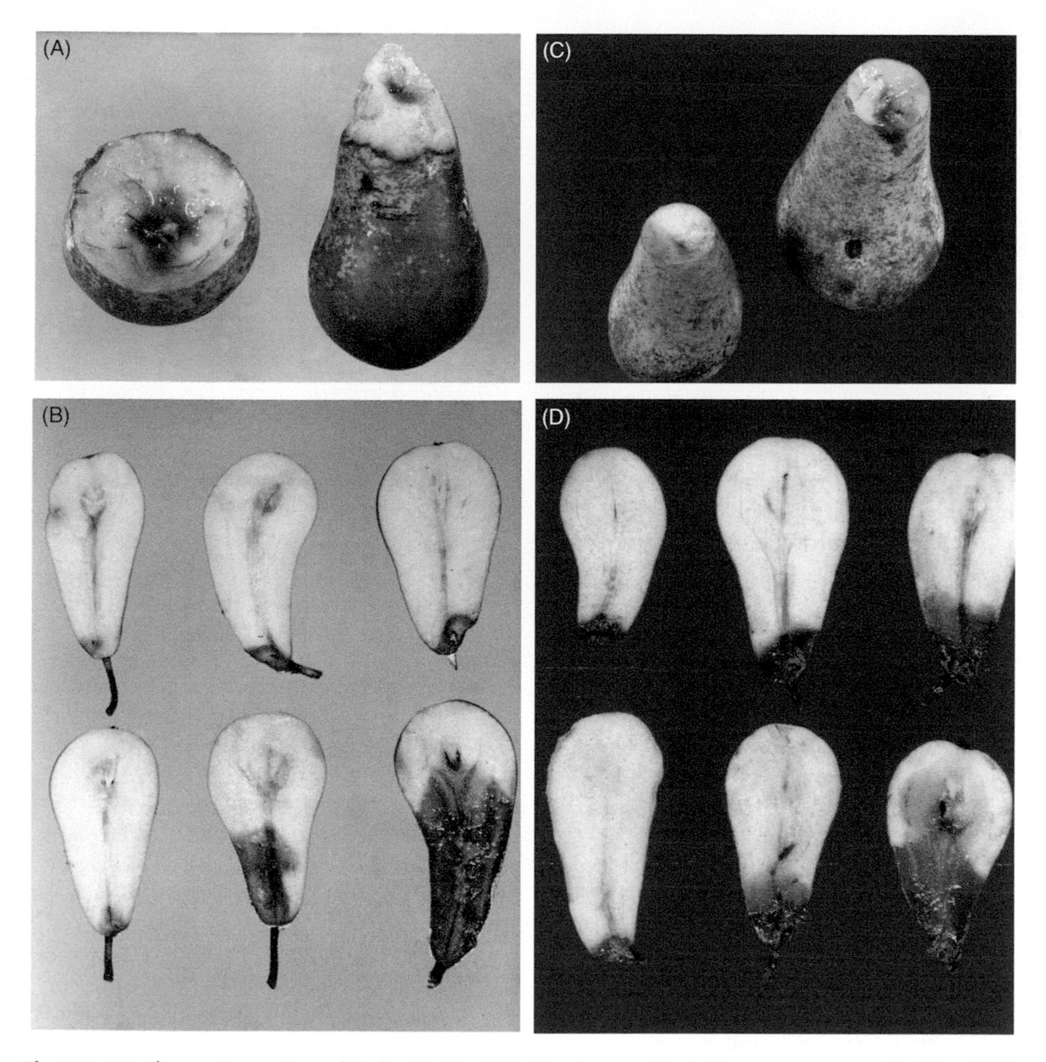

Fig. 7.2. 'Conference' pears inoculated with *Monilinia fructigena* or *Penicillium expansum* and held at 20°C in the presence and absence of a surface treatment with 1.5% Tal-Prolong.
(A) The configuration of residual 'healthy' tissues recovered from 'Conference' pears in the presence (right) and absence (left) of Tal-Prolong some 6 days after inoculation with *M. fructigena*.
(B) The extent of lesions caused by *M. fructigena* at 2, 4 and 6 days after inoculation in the presence (top row) and absence (bottom row) of Tal-Prolong.
(C) The configuration of residual 'healthy' tissues recovered from 'Conference' pears in the presence (left) and absence (right) of Tal-Prolong some 6 days after inoculation with *P. expansum*.
(D) The extent of lesions caused by *P. expansum* at 4, 7 and 10 days after inoculation in the presence (top row) and absence (bottom row) of Tal-Prolong.

remained green, but softened, with a high percentage of spoilage.

Tal-Prolong was subsequently reformulated and marketed as Semperfresh. The company that manufactured and marketed both Tal-Prolong and Semperfresh was called Surface Systems International. Al-Zaemey *et al.* (1993) found that banana crowns coated with Semperfresh showed delayed development of crown rot caused by infection with *Colletotrichum musae*. This effect could be enhanced by the inclusion of organic acids to Semperfresh. Kerbel *et al.* (1989) showed that for 'Granny Smith' apples coated with

Semperfresh (0.75% and 1.5%) and stored at 0°C for 4 or 5 months, ripening was delayed 'somewhat' compared with fruit with no coating. The treatments did not reduce levels of bitter pit or superficial scald. Carrillo-Lopez *et al.* (2000) observed ripening delays of several days in 'Haden' mangoes coated with Semperfresh. Dang *et al.* (2008) evaluated a combination of Semperfresh + *Aloe vera* gel coatings on 'Kensington Pride' mangoes and observed that coated fruit ripened more slowly; however, the coatings also reduced the development of aroma volatiles. Mohamed *et al.* (1992) found that guavas coated with Semperfresh and stored for 2 months at 10°C were in better condition than non-treated fruit, but coating fruit with palm oil was more effective, and cheaper, than Semperfresh. In a comparison between 'Granny Smith' apples that had been coated with either 1.5% Semperfresh or 1.5% Nutri-Save, Kader (1988) found that the internal CO_2 level was lower and the internal O_2 level was higher in the Semperfresh-coated fruits than those that had been coated with Nutri-Save. He also showed that coating with Nutri-Save reduced ethylene production and respiration rate to a larger degree than Semperfresh coating. From these results he concluded that Nutri-Save at 1.5% formed a surface barrier on the fruit that was less permeable to gases than 1.5% Semperfresh.

Arnon *et al.* (2014) coated mandarins, oranges and grapefruit with carboxymethyl cellulose (CMC) and chitosan and found that it was equally as effective as polyethylene wax in enhancing fruit gloss. It also slightly delayed softening, but was mostly not effective in reducing weight loss. There were only slight increases in internal CO_2 levels in fruit and ethanol accumulation in juice and flavour after storage was slightly impaired in mandarins but not in oranges and grapefruit. A coating similar to CMC was described by Deng *et al.* (2017). This was cellulose nanofibre-based emulsion (0.3% cellulose nanofibre + 1% oleic acid +1% sucrose ester fatty acid, w/w on a wet basis), which they found reduced respiration rate and ethylene production in bananas, thus delaying their ripening. The coated bananas ripened normally during 10 days of ambient storage; the coating simply delayed the ripening processes.

Starch

Starches that have been used for fruit coatings include oxidized banana starch (Torres *et al.*, 2008), cassava starch (Oriani *et al.*, 2014; Brandelero *et al.*, 2016), corn starch (Peano *et al.*, 2014), rice starch (Kerdchoechuen *et al.*, 2011) and phosphorylated *Swartzia burchelli* seed starch (Gomes *et al.*, 2016). Films for MA packaging have also been prepared from cereal starches, e.g. Mater-Bi®. Kerdchoechuen *et al.* (2011) coated minimally processed pummelos with rice and cassava starches that had been mixed with either 5% or 10% pummelo juice as an anti-browning agent and stored them at 4°C for 21 days. The coatings had no effect on their colour, but the coated pummelos had a lower weight loss, higher sugar content, but lower TA and vitamin C compared with the non-coated pummelos. Strawberries were coated with 2% cassava starch, 1% chitosan, 2% cassava starch + 1% chitosan or left not coated and stored at 10°C packed in plastic hinged boxes. Those coated with cassava starch + chitosan had the lowest weight loss, lowest counts of yeast and psychrophilic microorganisms and the best appearance according to the sensory evaluation (Campos *et al.*, 2011). Oriani *et al.* (2014) coated apple slices with cassava starch at 0%, 2% or 3% (w/v) with or without fennel or cinnamon bark essential oils at 0.05–0.3% (v/v). The coatings reduced weight loss and respiration rate and the coatings that contained 0.3% cinnamon bark oil resulted in increased antioxidant capacity and inhibition of the growth of *Staphylococcus aureus* and *Salmonella choleraesuis*, while coatings containing fennel oil inhibited only *S. aureus*. Gomes *et al.* (2016) tested starch extracted from the seeds of *S. burchelli* that had been phosphorylated with sodium tripolyphosphate. They coated cherry tomatoes and then stored them for up to 21 days at 10 ± 2°C and 80 ± 5% RH. Coated tomatoes retained better appearance with slower ripening and softening than non-treated fruit.

Carrageenan

Carrageenan is a complex mixture of polysaccharides derived, by solvent extraction, from red seaweed after which the solvent is evaporated, forming a three-dimensional network of polysaccharide double helices (Karbowiak *et al.*, 2007). Carrageenan-based coatings carrying antimicrobial or antioxidant compounds have been applied to a variety of foods to reduce microbial damage, oxidation, moisture loss and tissue disintegration (Vargas *et al.*, 2008). Hamzaha *et al.* (2013) tested various combinations of carrageenan films with or without glycerol on papayas. They found that 0.78% (w/v) carrageenan + 0.85% (w/v) glycerol delayed, but did not impede, ripening.

Chitosan

Chitin is a natural polysaccharide that occurs in many organisms as crystalline microfibrils forming structural components in the exoskeleton of arthropods, the cell walls of fungi and yeast and in other living organisms. Chitosan is produced by partial deacetylation of chitin under alkaline conditions or by enzymatic hydrolysis in the presence of a chitin deacetylase. Chitosan can be used on fruit and vegetables to give a coating that is selectively permeable to water vapour, CO_2, O_2 and ethylene (Sivakumar *et al.*, 2016). Chitosan coatings have been shown to have fungistatic properties and can modify the internal atmosphere in fruit (No *et al.*, 2007), but they must be dissolved in an acid to activate its antimicrobial and eliciting properties. Chitosan-based products are available commercially, e.g. Nutri-Save, which have shown similar effectiveness as the polymer dissolved in an acid. 'Golden Delicious' apples, coated with Nutri-Save, were kept in storage at 0°C for 6 months and were found to be of superior quality to those stored without coating (Davies *et al.*, 1988). Chitosan has been successfully used as a postharvest coating of peppers and cucumbers to reduce water loss and maintain their quality (El-Ghaouth *et al.*, 1991) and has been shown to reduce the growth of decay-causing fungi, induce resistance responses in host tissues and produce other effects contributing to reduced decay (Romanazzi *et al.*, 2012). It has been successfully used to control brown rot (*Monilinia fructicola*) on peaches (Li and Yu, 2001), *M. fructicola* infections on nectarines (Casals *et al.*, 2012), grey mould in inoculated grapes (Romanazzi *et al.*, 2006) and reduce the decay of blueberry during storage (Duan *et al.*, 2011). Chitosan has other properties and has been shown to improve the postharvest life of apples (Davies *et al.*, 1988; Kader, 1988), peppers and cucumbers (El-Ghaouth *et al.*, 1991), longans (Jiang and Li, 2001), oyster mushrooms (Cho and Ha, 1998), breadfruit (Worrell *et al.*, 2002) and broccoli (Ansorena *et al.*, 2011). Coating of mangoes with chitosan has been the subject of continuing research in many countries (Fig. 7.3). Zhu *et al.* (2008) coated mangoes with 2% chitosan and found that the coating delayed ripening and reduced disease development in inoculated fruit during subsequent storage. Nongtaodum and Jangchud (2009) found that chitosan coatings (0.5% and 0.8%) could maintain sensory attributes, reduce weight loss and slow changes in TSS, colour and growth of microorganisms. Conversely, chitosan was shown to have no positive effects on papaya (Maharaj and Sankat, 1990) and strawberries (Perdones *et al.*, 2012). Papayas coated in 1.5% Nutri-Save resulted in no significant difference in weight loss, abnormal browning and limited colour development compared with non-treated fruit during storage at 16°C (Maharaj and Sankat, 1990). All the fruits were of the cultivar Tainung Number 1 and were harvested at the colour break stage and placed in hot water at 48°C for 20 min and then benomyl (1.5 g/l) at 52°C for 2 min.

Essential Oils

Essential oils have been used both as fruit coatings and also successfully as volatiles inside MA packaging (see Chapter 14).

Fig. 7.3. Texture of mangoes being tested after coating with chitosan and storage at room temperature at KMITL in Thailand.

Essential oils extracted from the peel of oranges have fungitoxic properties (Sharma and Tripathi, 2006) and essential oils from aromatic plants such as oregano, fennel and lavender have also been reported to control, to a certain degree, postharvest citrus fruit pathogens (Soylu *et al.*, 2005). Kouassi *et al.* (2012) reported good disease control when *Cinnamomum zeylanicum* essential oils were incorporated in shellac or carnauba wax, but were less effective when they were incorporated into paraffin wax or polyethylene wax formulations. Disease control by essential oils appeared to depend on the volume that remained on the fruit skin and probably on the retention of essential oils components. Hashemi *et al.* (2017) coated fresh-cut apricots with basil seed gum containing 6% oregano essential oil and stored them at 4°C for 6 days. The addition of essential oil significantly decreased water vapour permeability of the coating. In comparison with the non-coated apricots, total soluble phenolic and antioxidant activity of oregano essential oil-coated fruit was significantly higher after storage. Taste panels could perceive rosemary and peppermint on grapes, which had been exposed to their essential oils for 24 h, after subsequent storage at 4°C or 20°C for 24 h, but after 48 h there was no perception (Servili *et al.*, 2017). Both essential oils showed good control of grey mould.

Alginate

Various forms of alginate gums are used as fruit coatings, including alginic acid, sodium alginate (extracted from the cell walls of brown algae) and potassium alginate (extracted from seaweed). Díaz-Mula *et al.* (2012) coated cherries with sodium alginate (1%, 3% or 5% w/v) and stored them for up to 16 days at 2°C plus 2 days at 20°C. Alginate-coated cherries could be stored, retaining their optimal quality and enhanced antioxidant activity, for 16 days while those that had not been coated could be stored for only 8 days. Valero *et al.* (2013) coated plums with 1% or 3% alginate before storage at 2°C for 7, 14, 21, 28 or 35 days followed by a 3-day period at 20°C. Both treatments, especially 3%, were effective in inhibiting ethylene production, reducing weight loss, improving acidity retention and slowing softening and colour change, but coated fruit had lower anthocyanin and carotenoid contents. Various concentrations of alginate were combined with citral to coat fresh strawberries and fresh-cut apples. The concentrations used were alginate 1%, alginate 2%, alginate 1% + citral 0.15%, alginate 1% + citral 0.3%, alginate 2% + citral 0.15% and alginate 2% + citral 0.3%, all on a w/v basis. The fruits were then stored for 28 days at 0.5°C (apples) and 14 days at 4°C (strawberries). Antioxidant activity was improved with 1% alginate followed by 1% alginate + 0.3% citral. The highest reductions in microbial spoilage were with 2% alginate with either 0.15% or 0.3% citral concentrations. For strawberries treated with alginate at 2% with citral 0.3%, colour and firmness were better preserved and microbial spoilage was reduced but the TSS, weight loss and antioxidant activity were not affected. Taste panels scored 0.3% citral-coated fruit lowest, therefore alginate 2% with 0.15% citral was the most appropriate among the combinations studied (Guerreiro *et al.*, 2015). Citral is an alkene-type terpene that occurs in the oil of lemon, orange and lemon grass and is used in perfumery and as a flavour. It has strong antimicrobial properties (Onawunmi, 1989; Lawless, 2014).

Salicylic Acid

Salicylic acid has been used for medical purposes (aspirin) for over a century. It is extracted from willow trees (*Salix* spp.) and also occurs in the fruit of *Ribes*, *Fragaria* and *Rubus* (Hulme, 1970). It has been investigated for its effects on plant material. For example, immersing peaches in salicylic acid at 0.1 g/l delayed their peak of ethylene production, reduced electrolyte leakage (chilling injury) and polyphenol oxidase (PPO) activity during storage compared with fruits that had not been treated (Han *et al.*, 2000). Luo *et al.* (2012) found that treating Japanese plums with salicylic acid suppressed chilling injury, reduced respiration rate and ethylene production and delayed the onset of the climacteric peak. Sayyari *et al.* (2009) also found that dipping fruit in salicylic acid, especially at 2 mM concentration, was highly effective in reducing chilling injury and electrolyte leakage in the peel of pomegranate, as well as ascorbic acid loss, compared with that observed in non-treated fruit during storage at 2°C for 3 months. Sayyari *et al.* (2012) subsequently reported that dipping fruit in 0.1, 0.5 or 1.0 mM acetyl salicylic acid immediately after harvest reduced chilling injury and helped to maintain quality during storage at 2°C. Fruit that had been treated with acetyl salicylic acid had improved retention of sugars, organic acids, total phenolics and anthocyanins. Pan *et al.* (2007) reported that treatment with hydrogen peroxide, salicylic acid or methyl jasmonate alleviated chilling injury symptoms during storage of bananas. They also found that all these treatments could retard the change of the fruit colour and increase PPO activity, relative electrical conductivity and respiration rate.

Xu and Tian (2008) found that 2 mM salicylic acid did not inhibit *Penicillium expansum* growth *in vitro* but it activated antioxidant defence responses of fruit of the sweet cherry cultivar 'Hongdeng', which may play a role in the resistance against *P. expansum*. However, Yao and Tian (2005) found that 2 mM salicylic acid was toxic to *Monilinia fructicola* on cherries and significantly inhibited mycelial growth and spore germination of the pathogen *in vitro*. Dessalegn *et al.* (2013) treated mangoes with either salicylic acid or potassium phosphonate at 1000 mg/l combined with dipping in a sodium bicarbonate solution for 3 min at 51.5°C. Both treatments significantly reduced disease development compared with the sodium bicarbonate hot water treatment alone or those not treated. These combinations kept anthracnose (*Colletotrichum gloeosporioides*) lesion development to below 5% and resulted in 93.3% marketable fruit compared with the mean disease severity on untreated fruit that reached 100% with zero marketability within 12 days of storage.

Trehalose

Trehalose is sugar that has been tested as a coating material for food preservation (Roser and Colaco, 1993; Abboudi, A.H., 1993, unpublished results). It is known as mycose or tremalose and is also referred to as mushroom sugar since it is prevalent in some mushrooms, including shiitake (*Lentinula edodes*), maitake (*Grifola fondosa*), nameko (*Pholiota nameko*) and Judas's ear (*Auricularia auricula-judae*), which can contain 1–17% of trehalose. It is a natural α-linked disaccharide formed by an α,α-1,1-glucoside bond between two α-glucose units. This makes trehalose a very stable reducing sugar that is chemically quite inert and biologically non-toxic. It is thought to form a gel phase as cells dehydrate. These properties make it a useful preservative. The use of trehalose in preservation of dehydrated material is covered by the European patent 0 297 887 B1 22 April 1992. Li and Tian (2007) reported that treatment with trehalose could help to retain the viability of the biocontrol yeast *Cryptococcus laurentii* in controlling postharvest diseases of apples. They also found that citric acid could induce accumulation of intracellular trehalose in *C. laurentii*.

Gum Arabic

Gum arabic is the dried, gummy exudates from the stems and branches of *Acacia* spp. and is used as an edible coating in the food industry. Coating papayas with 10% gum arabic significantly slowed the skin colour changes, retained their firmness and reduced weight loss compared with non-coated fruits (Ali *et al.*, 2010). Ali *et al.* (2010) coated tomatoes with aqueous solutions of 5%, 10%, 15% or 20% gum arabic and then stored them at 20°C and 80–90% RH for 20 days. Coated tomatoes had lower weight loss, were firmer and had higher TA, TSS, AA and slower decay and colour development compared with non-coated fruit, with 10% gum arabic proving optimum. Sensory evaluation tests showed the 10% gum arabic coating had no detrimental effects on quality during storage. Subsequently Ali *et al.* (2013) showed that coated tomatoes had a lower respiration rate and ethylene production and their total antioxidant capacity, lycopene content, total phenolics and total carotenoids were higher than fruit treated with 5% gum arabic or not coated. Mahfoudhi and Hamdi (2015) stored cherries at 2°C and 90–95% RH for 15 days coated with 10% gum arabic, 10% almond gum or not coated. Both coatings decreased respiration rate and ethylene production and delayed changes in weight, firmness, TA, TSS and colour changes compared with non-coated fruit. Green-mature papayas were coated with either 5% or 10% gum arabic in aqueous solutions. They were then stored at 13 ± 1°C by Addai *et al.* (2013). Fruit coated with 10% gum arabic and stored for 7 days had significantly ($p < 0.05$) higher antioxidant capacity (TPC, TFC, FRAP, DPPH and ABTS), physico-chemical properties (colour, pH, TA, moisture and TSS), sensory scores and lower microbial levels than 5% gum arabic and non-coated fruit.

Maqbool *et al.* (2011) dipped bananas in a composite coating of 10% gum arabic + 1% chitosan. They showed that, after 33 days of storage, coated fruit had 24% lower weight loss and 54% lower TSS than those not coated and were firmer, had higher total carbohydrates and reducing sugars than those not coated. Subsequently Maqbool *et al.* (2012) found that postharvest treatment with 0.05% lemongrass oil at 0.05% or cinnamon oil at 0.4% showed fungicidal effects against *C. gloeosporioides*, but the combination of 10% gum arabic with 0.4% cinnamon oil gave even better control.

Zein

Zein is an alcohol-soluble hydrophobic protein found in the endosperm of maize, where it comprises about 45–50% of its protein. It is biodegradable and used on confectionery products. It has good film-forming properties and is a good barrier to O_2. The water vapour permeability of zein films was reported to be about 800 times higher than a 'typical shrink wrapping film' (Aydt *et al.*, 1991). Permeability to CO_2, O_2 and water vapour was strongly dependent on the concentration of zein in the coating and the internal CO_2 of 'Gala' apples ranged from 4 to 11 kPa and O_2 6–19 kPa for zein-coated fruit, with an optimum formulation of 10% zein and 10% propylene glycol (Bai *et al.*, 2003). Coatings that are made from zein may swell and deform in prolonged contact with water, which may 'restrict their use in MA packaging of fruit and vegetables' (Yoshino *et al.*, 2002). Park *et al.* (1994) reported that application of zein coating on tomatoes delayed colour change, weight loss and softening during storage. Chávez-Murillo *et al.* (2015) produced a zein and ethyl cellulose-based coating and tested it on tomatoes. They also found that it decreased their weight loss and respiration rate and delayed changes in colour and softening compared with non-coated fruit. Rakotonirainy *et al.* (2008) reported that broccoli florets packaged in zein films (laminated or coated with tung oil) maintained their original firmness and colour during 6 days of storage. Bai *et al.* (2003) tested zein dissolved in aqueous alcohol with propylene glycol to give a high-gloss coating for apples. Increasing levels of both zein and propylene glycol resulted in increased gloss.

Methyl Jasmonate

MJ is a plant growth regulator involved in the antioxidant systems used to ameliorate oxidative reactions induced by biotic and abiotic stress. It is volatile and can diffuse through plant membranes and may have a function in mediating intra-plant and inter-plant communications and modulating plant defence responses (Wasternack and Hause, 2013). It appears to up-regulate genes that induce the synthesis of some stress proteins that affect chilling tolerance and resistance to pathogens (Ding *et al.*, 2002; Peng *et al.*, 2014). MJ significantly inhibited lignification of kiwifruit at later stages of storage at 0°C (Li *et al.*, 2017a). Exposure of Sunrise papayas to MJ vapours (10^{-5} or 10^{-4} M) for 16 hours at 20°C inhibited fungal decay, reduced chilling injury, retained higher levels of organic acids and delayed softening during storage at 10°C for 14–32 days followed by 4 days shelf-life at 20°C compared with non-treated fruit. MJ-treated fruit also retained higher organic acids (González-Aguilar *et al.*, 2003). It was concluded that the postharvest quality of papayas was significantly enhanced by combining MJ and MA packaging in low-density polyethylene (LDPE) film. Pan *et al.* (2010) reported that treatment of bananas with MJ could alleviate chilling injury symptoms, retard the change of the fruit colour and increase PPO activity and respiration rate. Development of chilling injury in loquats was associated with a reduction in unsaturated:saturated fatty acid ratio. Postharvest treatment with MJ or 1-MCP delayed the decrease in this ratio, thereby increasing chilling resistance (Cao *et al.*, 2009 a, b). MJ (10 μM) reduced hydrogen peroxide in cucumber during storage, suggesting that it alleviated chilling injury by inhibiting hydrogen peroxide generation (Liu *et al.*, 2016b).

Gibberellins

Gibberellins are a group of plant growth substances that were first isolated from the fungus *Gibberella fujikuroi*. They are identified by numbers, e.g. gibberellic acid = GA_3. Postharvest treatment with gibberellins has been shown to reduce de-greening in lemons (Mizobutsi *et al.*, 2000) and limes (Sposito *et al.*, 2000), improve retention of vitamin C in limes (Keleg *et al.*, 2003), reduce yellowing and extend storage life of parsley (Lers *et al.*, 1998), reduce chilling injury in persimmon (Besada *et al.*, 2008), decrease cracking in cherries (Yildirim and Koyuncu, 2010) and delay ripening and reduce decay in cactus pear (Schirra *et al.*, 1999). A negative effect of GA_3 was reported by Zoffoli *et al.* (2009), who found that it could induce shatter and predisposed grapes to grey mould (*Botrytis cinerea*).

Biton *et al.* (2014) tested the effects of spraying persimmon plants with a mixture of GA_4, GA_7 and benzyl adenine, starting 40 days after fruit set, applied once a month for three consecutive months, on subsequent storage at 0°C. This treatment combination delayed chlorophyll degradation and reduced fruit cuticle cracking and black spot disease (*Alternaria alternata*) susceptibility during the late stages of fruit growth and during storage.

Bacteria

Biological control of postharvest diseases is becoming of increasing importance. Some bacteria isolated from fruit can be used to control human pathogens that may contaminate minimally processed fruit. For example, the strain CPA-7 of *Pseudomonas graminis*, isolated from fruit, can inhibit the growth of human pathogens including *Salmonella enterica* subsp. *enterica*, *Listeria monocytogenes* and *Escherichia coli* O157:H7 (García-Martin *et al.*, 2018).

Alegre Vilas *et al.* (2013) confirmed previous work that showed *P. graminis* was effective in controlling bacteria and reported that concentrations of *P. graminis* CPA-7 of 105 and 107 cfu/ml extracted from apples avoided *Salmonella* growth at 10°C and lowered *L. monocytogenes* population increases. They also found no effect on fruit quality and the bacteria did not survive to simulated gastric stress. They concluded that it could be used during storage as part of a combination with disinfection, low storage temperature and modified atmosphere packaging (MAP). Collazo *et al.* (2017) confirmed that *P. graminis* CPA-7 treatment had no effect on the fruit quality parameters that were tested and their results suggested that *P. graminis* CPA-7 may trigger the activation of the fruit defence response, thereby mitigating its oxidative damage. Such activation may play a role in the control mechanism against pathogenic bacteria. Previously they had shown that on minimally processed pears *P. graminis* CPA-7 inhibited the growth of *L. monocytogenes* and *S. enterica* subsp. *enterica* by 5.5 and 3.1 $\log_{10}$, respectively, after 7 days of storage at 10°C.

Daminozide

N-dimethylaminosuccinamic acid is marketed as DPA, daminozide, Alar, B9 or B995 and is a plant growth regulator that has been used to retard tissue senescence and cell elongation. However, it has been withdrawn from the market in several countries because of suggestions that it might be carcinogenic. In a comparison between pre-harvest and postharvest application of daminozide to 'Cox's Orange Pippin' apples, immersion of fruit in a solution containing 4.25 g/l for 5 min delayed the rise in ethylene production at 15°C by about 2 days, whereas an orchard application of 0.85 g/l resulted in delays of about 3 days. Both modes of application depressed the maximum rate of ethylene production by about 30%. Daminozide-treated apples were shown to be less sensitive to the application of ethylene than non-treated fruit, but this response varied between cultivars (Knee and Looney, 1990). Graell and Recasens (1992) sprayed 'Starking Delicious' apple trees with daminozide (1000 mg/l) in mid-summer in 1987 and 1988. Controls were not sprayed. Fruit was harvested on 12 and 24 September 1987, and 7 and 17 September 1988, and placed in CA storage. Two CA storage rooms were used: a low-ethylene CA room, fitted with a continuous ethylene removal system, with a store ethylene concentration ranging from 4 to 15 μl/l; and a high-ethylene CA room, where ethylene was not scrubbed and store ethylene concentration was > 100 μl/l. In both seasons, after 8 months at 0–1°C in storage in 3 kPa O_2 + 4 kPa CO_2, with scrubbed ethylene, daminozide-sprayed and earlier-harvested fruits were firmer and more acid after storage than non-sprayed or later-harvested fruits stored in high ethylene CA. Generally, after a 7-day post-storage shelf-life period at 20°C, the differences between treatments were maintained. The TSS content of fruit samples was similar for all treatments.

1-Methylcyclopropene

Since the discovery of the properties of 1-MCP as a postharvest ethylene inhibitor, there has been an explosion of relevant publications. 1-MCP is applied to fruit, vegetables and flowers, but mainly apples, as a fumigant for a short period directly after harvest to preserve quality and reduce losses during subsequent storage. However, Warren *et al.* (2009) applied it as an aqueous dip and Jiang *et al.* (1999) and De Reuck *et al.* (2009) sealed 1-MCP in plastic bags with bananas or litchis. 1-MCP is available as commercial preparations, including SmartFresh™ and Ethylbloc™. It can be applied to the store on completion of loading or in containers or gas-tight tents prior to loading into the store, or it can be applied in gas-tight rooms. Gas-tight curtains are available that can be used temporally to partition part of a store, making it independently gas tight with access through a door closed with a gas-tight zip. This is useful where only a relatively small amount of fruit or vegetables needs to be fumigated. Pallet covers are also available that are sufficiently gas tight for fumigation.

Ma *et al.* (2009) reported that the mode of action of 1-MCP was by inhibiting the activities of ACC and delaying the peaks in the ACC synthase activity and ACC concentration and gene expression of these enzymes and of ethylene receptors at the transcript level. Fu *et al.* (2007) indicated that the beneficial effect of 1-MCP might be due to its ability to increase the antioxidant potential as well as to delay fruit ripening and senescence. They found that it retarded the activities of pectin methylesterase and polygalacturonase during ripening of pears and the activities of the antioxidant enzymes catalase, superoxide dismutase and peroxidase were all significantly higher in 1-MCP-treated pears than in the control fruit. Kesari *et al.* (2010) reported that ethylene exposure of unripe mature bananas induced the expression of the gene *MaPR1a*, which increased with ripening, and 1-MCP treatment prior to ethylene exposure inhibited expression. Liu *et al.* (2016a) suggested that the mode of action of 1-MCP was related to the expression of acid transport genes, including *MdVHA-A*, *MdVHP* and *Ma1*. This was because 1-MCP maintained fruit acidity by regulating the balance between malate biosynthesis and degradation. 1-MCP delayed the postharvest loss of malate and citrate.

De Wild (2001) suggested that short-term CA storage might be replaced by 1-MCP treatment. Watkins and Nock (2004) supported this view for apples, but, they stated, 'we do not advocate the use of 1-MCP as an alternative to CA storage for medium to long term periods. With increasing storage periods, the two technologies are always more effective when used in combination.' Rizzolo *et al.* (2005) also concluded that 1-MCP could not be a substitute for CA storage but could reinforce the CA effects. Johnson (2008) reported that for apples the greatest benefit from 1-MCP was derived when combined with CA storage, though worthwhile extensions in storage life were achieved in air storage of 1-MCP-treated fruit. 1-MCP is an ethylene inhibitor but in certain circumstances treatment can stimulate ethylene production (Jansasithorn and Kanlavanarat, 2006; McCollum and Maul, 2009). Shinjiro *et al.* (2002) found that treatment of immature, mature-green or ripe pineapples with 1-MCP applied several times during storage resulted in temporary stimulation of ethylene production immediately after each treatment. Most of the reported effects of 1-MCP treatment have been positive, but Ella *et al.* (2003) reported that the use of too low concentrations may lead to some degree of senescence acceleration in leafy vegetables, possibly because of relief from ethylene auto-inhibition. Also in certain circumstances 1-MCP treatment has been associated with decreased quality of fruit (Fabi *et al.*, 2007) and increased rotting (Hofman *et al.*, 2001; Jiang *et al.*, 2001; Bellincontro *et al.*, 2006; Baldwin *et al.*, 2007). Jiang *et al.* (2001) suggested that comparatively low levels of phenolics in strawberries treated with 1-MCP could account for the decreased disease resistance. With 'Cavendish' bananas harvested at the mature-green stage, treatment with 1-MCP delayed peel colour change and fruit softening, extended shelf-life and reduced respiration rate and ethylene production (Jiang *et al.*, 1999). Ripening was delayed when fruits were exposed to 1-MCP at 0.01–1.0 μl/l for 24 h and increasing concentrations of 1-MCP were generally more effective for longer periods of time. Similar results were obtained with fruits sealed in PE bags 0.03 mm thick containing 1-MCP at either 0.5 or 1.0 μl/l, but delays in ripening were longer at about 58 days.

In their review, Li *et al.* (2016b) reported that the main observed effects of 1-MCP on postharvest performance of non-climacteric fruits were that they inhibited:

- senescence, including as rachis browning in grapes, scale senescence in pitaya, and leaf senescence in mandarins marketed with attached leaves;
- the development of physiological disorders, including scald development in pomegranate, pericarp browning in litchi and internal browning in loquat; and
- de-greening and colour change in citrus fruits, strawberry, pitaya, prickly pear and olive.

Legislation

It may be surprising that permission has been received for the use of 1-MCP as a postharvest treatment when there is considerable public resistance to any chemical treatment of fruit and vegetables, especially postharvest. In the USA (US Federal Register, 2002) it was stated:

> This regulation establishes an exemption from the requirement of a tolerance for residues of 1-methylcyclopropene (1-MCP) in or on fruits and vegetables when used as a post harvest plant growth regulator, i.e., for the purpose of inhibiting the effects of ethylene. AgroFresh, Inc. (formerly BioTechologies for Horticulture) submitted a petition to EPA under the Federal Food, Drug, and Cosmetic Act, as amended by the Food Quality Protection Act of 1996, requesting an exemption from the requirement of a tolerance. This regulation eliminates the need to establish a maximum permissible level for residues of 1-MCP.

The overall conclusion from the evaluation reported on 23 September 2005 by European Commission Health & Consumer Protection Directorate-General (European Commission, 2005) was:

> that it may be expected that plant protection products containing 1-methylcyclopropene will fulfil the safety requirements laid down in Article 5(1) (a) and (b) of Directive 91/414/EEC.

As stated by the World Trade Organization (WTO, 2008):

> The World Trade Organization does not expect any human health concerns from exposure to residues of 1-MCP when applied or used as directed on the label and in accordance with good agricultural practices. The data submitted by applicant and reviewed by the Agency support the petition for an exemption from the requirement of a tolerance, for 1-MCP on pre-harvested fruits and vegetable, when the product is applied or used as directed on the label. This regulation is effective 9 April 2008.' (WTO, 2008).

Diseases

The effects of 1-MCP on postharvest disease development are variable. Li *et al.* (2017c) reported that 1-MCP (5 μl/l) suppressed postharvest blue mould (*Penicillium expansum*) of apples by inhibiting the growth of mycelium. Xu *et al.* (2017) showed that 1-MCP suppressed postharvest development of anthracnose disease (*Colletotrichum gloeosporioides*) in mangoes by directly inhibiting spore germination and mycelial growth. However, tomatoes were artificially inoculated with an *Alternaria alternata* conidial suspension then exposed to 1-MCP at 0.6 μl/l for 24 h and stored for 6 days at 10°C and one more week at 20°C. After storage, tomatoes treated with 1-MCP showed significantly more *A. alternata* disease (Estiarte *et al.*, 2016). Jiang *et al.* (2001) found that disease development was quicker in strawberries treated with 500 and 1000 nl/l. 1-MCP inhibited PAL activity and reduced increases in anthocyanin and phenolic content. They suggested that comparatively low levels of phenolics in strawberries treated with 1-MCP could account for their decreased disease resistance.

Apple

Watkins and Nock (2004), in discussing the beneficial effects of 1-MCP on apples, pointed out that the response can vary between cultivars as well as harvest maturity, handling practices, time between harvest and treatment and type and duration of storage. On the negative side they found that 1-MCP may increase susceptibility to chilling injury, CO_2 injury and superficial scald. Ethylene production, respiration rate and colour changes in 'Fuji' apples were all inhibited following 1-MCP treatment at 0.45 mmol/m^3 (Fan *et al.*, 1999). The threshold concentration of 1-MCP inhibiting *de novo* ethylene production and action was 1 μl/l. 1-MCP-treated 'McIntosh' and 'Delicious' apples were significantly firmer than non-treated apples following cold storage and post-storage exposure to 20°C for 7–14 days (Rupasinghe *et al.*, 2000). De Wild (2001) made a single application of 1-MCP directly after harvesting to 'Jonagold' and 'Golden Delicious' apples, followed by 6 months storage at 1°C in either 1.2 kPa O2 + 4 kPa CO2 or air, followed by shelf-life of 7 days at 18°C. No loss of firmness was found after the use of 1-MCP and yellow discoloration of 'Jonagold' decreased. Apples stored in air softened quicker than in CA, unless treated with 1-MCP. Treated apples did not show significant softening during subsequent shelf-life. Exposure to 1-MCP improved retention of flesh firmness and reduced ethylene production of 'Scarletspur Delicious' and 'Gale Gala' after storage in 0°C both in air and in 2 kPa O_2 + < 2 kPa CO_2 (Drake *et al.*, 2006). In 'McIntosh' apples, 1-MCP treatment reduced internal ethylene concentrations and maintained firmness of fruit during storage compared with the non-treated controls, but the location of the orchard where the apples were grown affected their response to 1-MCP. At one site 'McIntosh' treated with 1-MCP had more internal browning, but there was no effect at another (Robinson *et al.*, 2006). In 'Pink Lady' apples, 1-MCP treatment at 0.5 μl/l at 20°C for 24 h significantly inhibited respiration rate, ethylene production and ACC oxidase activity while it slowed down the decrease in flesh firmness and TA during storage at 0°C (Guo *et al.*, 2007). Internal ethylene concentration in 'Bramley's Seedling' also tended to be higher where treatment was delayed and fruit were softer. In general a lower dose combined with a shorter treatment time resulted in the highest internal ethylene concentration in 'Cox's Orange Pippin' and 'Bramley's Seedling' and a lower dose in combination with delayed treatment resulted in the highest internal ethylene concentration and lowest firmness in 'Cox's Orange Pippin' (Johnson, 2008). Moya-Leon *et al.* (2007) stored 'Royal Gala' at 0°C and 90–95% RH for 6 months followed by 7 days at ambient. They found that fruit stored in 2.0–2.5 kPa CO_2 + 1.8–2.0 kPa O_2 or those treated with 1-MCP at 625 nl/l and stored in air had reduced ethylene production, softening and acidity loss and were preferred by consumers because their preferred textural characteristics were said to be maintained. However, storage in air gave the highest levels of aromatic volatiles, which were depressed by both CA storage and 1-MCP treatment. 1-MCP at 42 mol/m^3 for 24 h at 20°C inhibited ethylene production regardless of storage temperature over the range of 0–20°C. Results indicated that 1-MCP is an effective means that can be used to delay ripening and to retain fruit quality during transport and retailing at 10°C or 20°C. Fruit stored at 20°C for longer than 40 days showed some shrivelling and decay, regardless of 1-MCP treatment (Argenta *et al.*, 2001).

Combining 1-MCP treatment with DCA has been shown to be effective. Storage of 'Braeburn' at 1°C in DCA at 0.3–0.4 kPa O_2 + 0.7–0.8 kPa CO_2 maintained firmness at levels comparable to 1.5 kPa O_2 + 1.0 kPa CO_2 plus 1-MCP applied at 625 ppb for 24 h at 3.5°C. Storage was for some 7 months, followed by 7 days of shelf-life at 20°C. DCA storage plus 1-MCP resulted in no softening during storage, but 1-MCP-treated fruit in both 1.5 kPa O_2 + 1.0 kPa CO_2 and DCA showed a higher incidence of BBD (on average + 53%) than non-treated fruit. DCA storage reduced BBD to 46% compared with fruit stored only in 1.5 kPa O_2 + 1.0 kPa CO_2. Fruit susceptibility to low-temperature breakdown and CO_2 injury was significantly higher in DCA storage than in 1.5 kPa O_2 + 1.0 kPa CO_2 or in air (Lafer, 2008).

The protocol recommended by Johnson (2008) for apples grown in the UK was

1-MCP at 625 nl/l (based on empty store volume) applied for 24 h with minimal delay between harvest and application. He reported that differences in firmness between air and CA stored 'Cox's Orange Pippin' increased with time in store due to the more rapid softening of fruit stored in air. Only storage in 5 kPa CO_2 + 1 kPa O_2 gave complete control of scald in 1-MCP-treated 'Bramley's Seedling' apples stored for long periods. Delaying establishment of CA conditions from 10 to 28 days to avoid CO_2 injury in 'Bramley's Seedling' did not affect internal ethylene concentration, scald incidence or firmness in 2002 but in 2003 there was an adverse effect of a 21-day delay on the firmness of fruit stored in 5 kPa CO_2 + 1 kPa O_2 for 284 days but no effects on scald development. 'Fiesta', 'Jonagold', 'Idared' and 'Braeburn' treated with 1-MCP failed to soften during 90 days of CA storage despite a reduced internal ethylene concentration. 'Meridian' apples treated with 1-MCP were not significantly firmer than non-treated and 'Spartan' did not respond to 1-MCP in terms of either internal ethylene concentration or firmness. Bulens *et al.* (2012) found that when 'Jonagold' were treated with 1-MCP, ethylene biosynthesis was almost completely suppressed throughout their whole postharvest life, with the exception of late-harvested apples that regained some ethylene-forming capacity after 2 weeks of shelf-life.

As reported above, the development of physiological storage disorders has also been associated with 1-MCP treatment. Watkins and Nock (2012b) found that its effects on storage disorders were inconsistent, though they detected slight increases in the risk of CO_2 injury. DeEll *et al.* (2005) reported that 1-MCP resulted in CO_2 injury in 'McIntosh' and 'Empire', and a higher incidence of core browning in 'Empire' and internal browning in 'Gala'. 'Gala' with no 1-MCP in storage at 0°C for 240 days with zero CO_2 resulted in a large reduction in firmness, but 'Gala' in storage in 1 kPa O_2 + 0 kPa CO_2 had similar respiration rates, ethylene and total volatiles production to those fruits stored in 2.5 kPa O_2 + 2 kPa CO_2. 'Empire' stored either in air at 0°C or in CA of 5.0% CO_2 + 3.0% O_2 at 2.5°C developed large incidences of core browning, which was worse in 1-MCP-treated fruits. Subsequently DeEll and Ehsani-Moghaddam (2012) suggested that delaying CA storage for 1 or 2 months at 1°C to reduce the development of certain storage disorders in 'Empire' could be utilized in combination with 1-MCP treatment to retard fruit softening during storage at 3°C with 2.5 kPa O_2 + 2.0 kPa CO_2 for 7–9 months.

Li *et al.* (2017c) showed that fumigation with 1-MCP (5 μl/l) inhibited postharvest blue mould (*Penicillium expansum*) of apples during subsequent storage. The mode of action of the 1-MCP was thought to be due to the enhanced oxidative damage to *P. expansum* and destruction of the integrity of plasma membrane of its spores

Apricot

Fan *et al.* (2000) treated the cultivar 'Perfection' with 1-MCP at 1 μl/l for 4 h at 20°C then stored them at either 0°C or 20°C. The 1-MCP treatment reduced ethylene production and respiration rate and resulted in slower colour change, less firmness loss and loss in TA during storage at both temperatures. There was also delayed production of alcohols and esters during storage at 20°C. They also found a relationship between 1-MCP and maturity of the fruit in that the effects of MCP were less pronounced at more advanced maturity.

Asian pear

Cold storage of the cultivars 'Gold Nijisseiki' and 'Hosui' plus 1-MCP treatment was effective for long-term storage (Itai and Tanahashi, 2008).

Avocado

'Ettinger', 'Hass', 'Reed' and 'Fuerte' were treated with various concentrations of 1-MCP for 24 h at 22°C and, after ventilation, were exposed to ethylene at 300 μl/l for 24 h also at 22°C. The fruits were then stored at 22°C in ethylene-free air for ripening assessment.

1-MCP at very low concentrations inhibited ripening. Treatment for 24 h with 1-MCP at 30–70 nl/l delayed ripening by 10–12 days, after which the fruit ripened normally (Feng *et al.*, 2000). Hofman *et al.* (2001) showed that 1-MCP at 25 µl/l for 14 h at 20°C increased the ripening time by 4.4 days (40% increase) for 'Hass' compared with non-treated fruit. 1-MCP treatment was associated with slightly higher rot severity compared with fruit not treated with 1-MCP. They also showed that 1-MCP treatment could negate the effects of exogenous ethylene and concluded that 1-MCP treatment has the potential to reduce the risk of premature fruit ripening due to accidental exposure to ethylene.

Banana

Williams *et al.* (2003) found that at 15°C bananas could be stored for 6 weeks in CA, but this was doubled by application of 1-MCP at 300 nl/l for 24 h at 22°C directly after ripening initiation with ethylene. However, 3 nl/l had no effect and 30 µl/l slowed ripening 'excessively'. Eating quality was not affected by 1-MCP at 300 nl/l (Klieber *et al.*, 2003). With 'Cavendish' bananas, treatment with 1-MCP delayed peel colour change and softening, extended shelf-life, reduced respiration rate and reduced ethylene production (Jiang *et al.*, 1999). Ripening was delayed when fruits were exposed to 1-MCP at 0.01–1.0 µl/l for 24 h and increasing concentrations of 1-MCP were generally more effective for longer periods of time. Similar results were obtained with fruits sealed in PE bags 0.03 mm thick containing 1-MCP at either 0.5 or 1.0 µl/l, but delays in ripening were longer at about 58 days. Jiang *et al.* (2004), working on the cultivar 'Zhonggang', found that fruits treated with 1-MCP for 24 h at 20°C had significantly delayed peaks of respiration rate and ethylene production, but the peak height of those exposed to ethylene at 50 ml/l was reduced. They also found a reduction in all the changes associated with ripening. The 1-MCP effects on all these changes were higher during low-temperature storage and lower in higher-temperature storage. Bananas treated with 1-MCP were shown to result in uneven de-greening of the peel. De Martino *et al.* (2007) initiated 'Williams' to ripen by exposure to ethylene at 200 µl/l for 24 h at 20°C. These bananas were then treated with 1-MCP at 200 nl/l at 20°C for 24 h, then stored in > 99.9 kPa N_2 + < 0.1 kPa O_2 in perforated plastic bags at 20°C. No differences in the accumulation of acetaldehyde and ethanol were detected during storage compared with fruits not treated with 1-MCP. Peel de-greening, the decrease in chlorophyll content and chlorophyll fluorescence were delayed after the 1-MCP treatment. There was some general browning throughout the 1-MCP-treated peel in both the green and yellow areas of the ripening peel. They concluded that it appears the 1-MCP-treated peel 24 h after the ethylene treatment may still undertake some normal senescence that occurs during banana ripening. Green-mature 'Apple' bananas were treated with 1-MCP at 0, 50, 100, 150 and 200 nl/l for 12 h and then stored at room temperature of 20 ± 1°C and 80 ± 5% RH. The best concentration in extending the shelf-life based on total sugars, soluble pectins, firmness and external appearance in the end of storage was 1-MCP at 50 nl/l. This concentration delayed the onset of peel colour change by 8 days, while with concentrations at 100, 150 and 200 nl/l it was by 10 days compared with non-treated fruits (Pinheiro *et al.*, 2005). 'Khai' bananas were treated with 1-MCP at 50, 100 or 250 ppb for 24 h and then stored at 20°C. 1-MCP delayed ripening and prolonged storage to some 20 days, with the higher concentration being more effective, but bananas treated with 250 ppb 1-MCP had the lowest respiration rate and ethylene production. However, there was no significant difference among treatments for colour changes (Jansasithorn and Kanlavanarat, 2006).

Blueberry

Changes in the marketable percentage of the cultivars 'Burlington' and 'Coville' after storage for up to 12 weeks at 0 ± 1°C were not affected by pre-treatment for 24 h at 20°C with 1-MCP at rates up to 400 nl/l (DeLong *et al.*, 2003).

Breadfruit

Treatment with 1-MCP at 100 nl/l for 12 h reduced respiration rate, but treatment at 10, 20 or 50 nl/l did not affect respiration rate or ethylene production. All the treatments tended to delay softening and improve shelf-life. However, at the later stage of ripening, some fruitlets were abnormally hard, causing uneven mushy softening with light brown streaks (Paull *et al.*, 2005). They also reported that fruits treated with 1-MCP at 25 or 50 nl/l followed by storage in perforated PE bags at 17°C had a shelf-life of 10–11 days compared with 4 days for non-treated fruits at the same temperature; and fruits treated with 1-MCP alone also had a 4-day shelf-life.

Broccoli

Ku and Wills (1999) showed that 1-MCP markedly extended the storage life of the cultivar 'Green Belt' through a delay in the onset of yellowing during storage at either 20°C or 5°C and in development of rotting at 5°C. The beneficial effects at both temperatures were dependent upon concentration of 1-MCP and treatment time. Subsequently Ma *et al.* (2009) also showed that 1-MCP treatment delayed the yellowing of broccoli florets and that exogenous ethylene exposure did not accelerate yellowing in the florets pre-treated with 1-MCP. Xu *et al.* (2016) showed that broccoli florets treated with 1-MCP at 2.5 µl/l decreased the loss of chlorophyll and maintained higher levels of sugars. The expression of genes encoding sucrose transporters (*BoSUC1* and *BoSUC2*) and carbohydrate metabolizing enzymes (*BoINV1*, *BoHK1* and *BoHK2*) was induced by the 1-MCP treatment. Therefore 1-MCP was shown to delay senescence of broccoli florets, which may be attributed to maintaining higher sugar content through regulation of sugar metabolism.

Capsicum

'Setubal' red peppers were treated with 900 ppb 1-MCP for 24 h at 20°C or left non-treated and packed in perforated PP bags by Fernández-Trujillo *et al.* (2009). They were stored at either 20°C or 8°C for 4½ days. then at 20°C for 3 days and finally in a domestic refrigerator at 5.6°C for 4½ days to simulate retail distribution. 1-MCP prevented the increase in skin colour and ethylene production but it may have increased susceptibility to shrivelling and weight loss and, to a greater extent, pitting and grey mould.

Carambola

Generally, submerging fruit in aqueous 1- MCP at 200 µg/l suppressed total volatile production in the cultivar Arkin at both ¼ and ½ yellow harvest maturity or when stored at 5°C for 14 days and transferred to 20°C or stored constantly at 20°C until turning ¼ orange. 1-MCP suppressed volatile production for fruit stored at 20°C where the fruits were harvested at the ¼ and ½ yellow stage and total volatiles were suppressed by 14% and 28%, respectively. Treatment with 1-MCP delayed the time for them to turn ¼ orange by 2–5 days at 20°C (Warren *et al.*, 2009).

Sudachi (*Citrus sudachi*)

Sudachis were treated with 1-MCP at 0, 0.01, 0.1, 0.5 or 1.0.µl/l for 6, 12 or 24 h at 20°C and then stored at 20°C by Isshiki *et al.* (2005). Treatment with 1-MCP at 0.1 µl/l was the most effective in delaying de-greening and it did not induce peel browning and internal quality degradation. However, at 0.5 and 1.0 µl/l, 1-MCP had little effect on delaying de-greening or inducing peel browning.

Cucumber

Lima *et al.* (2005) found that exposure of different cultivars to 1-MCP at 1.0 µl/l resulted in significant retention of firmness and surface colour in the cultivar 'Sweet Marketmore' but it had minimal effects on the cultivar Thunder. They found that those exposed to ethylene at 10 µl/l at 15°C showed rapid and acute cellular breakdown within 4 days; this effect was reduced, but not eliminated, by the 1-MCP treatment. Nilsson (2005) also found some delay in de-greening with 1-MCP, especially when cucumbers were exposed to ethylene. However, it was concluded that they probably showed little benefit from 1-MCP unless exogenous ethylene was present.

Custard apple

Hofman *et al.* (2001) found that exposure of the *Annona reticulata* cultivar 'Africa Pride' to 1-MCP at 25 µl/l for 14 h at 20°C increased the number of days to ripen by 58% compared with non-treated fruit. 1-MCP treatment was associated with slightly higher severity of external blemishes and slightly higher rot severity compared with non-treated fruit. Since the 1-MCP treatment countered the effects of exposure to exogenous ethylene, they concluded that it had the potential to reduce the risk of premature ripening due to accidental exposure to ethylene.

Durian

Pulp of the *Durio zibethinus* cultivar 'Mon Thong' was exposed to 0, 50, 100, 200 or 500 ppb of 1-MCP for 12 h at 20°C and then stored in 15 µm PVC film bags at 4°C by Sudto and Uthairatanakij (2007). After 8 days of storage at 20°C, treated fruits were firmer than non-treated ones. The most effective in delaying the accumulation of CO_2 inside the packages was 50 ppb, but none of the treatments affected ethylene production and 1-MCP treatment had no significant effect on firmness. In contrast, they found that whole fruits that had been exposed to 1-MCP and stored at 20°C were firmer than non-treated ones after 8 days. Treatment with 1-MCP resulted in a reduction of the starch content, an increase of TSS and increased lipoxygenase activity. Also working with the cultivar 'Mon Thong', Amornputti *et al.* (2016) showed that 1-MCP (500 nl/l at 25°C) inhibited ethylene biosynthesis during subsequent storage at 15°C mainly by preventing an increase in ACC oxidase activity in the peel.

Fig

Fruits of the cultivar 'Bardakci' were harvested at optimum harvest maturity and half were treated with 10 ppb 1-MCP at 20°C for 12 h and the other half left non-treated; both were stored at 0°C with 90–92% RH for 15 days by Gözlekçi *et al.* (2008). They found that the 1-MCP treatment slowed fruit softening but had no effect on TA and TSS.

Grape

Grapes of the cultivar 'Aleatico' harvested at 20.4–20.6°brix were left non-treated or were treated with ethylene at 500 mg/l or 1-MCP at 1 mg/l for 15 h at 20°C and 95–100% RH, then stored at 20°C and 85% RH for 13 days. 1-MCP-treated grapes did not show ethylene production for 2 days after treatment, whereas non-treated grapes produced constant amounts of 3–4 µl/kg/h but after 6 days. There were no significant differences in their respiration rates. 1-MCP-treated grapes showed a significant loss of terpenols and esters, though levels of alcohols were maintained, especially hexan-1-ol. Levels of grey mould (*Botrytis cinerea*) were also higher in 1-MCP treated fruit (Bellincontro *et al.*, 2006).

Grapefruit

Treatment with 1-MCP at concentrations equal to or greater than 75 nl/l inhibited ethylene-induced de-greening but the effect was transient, and increasing 1-MCP concentrations greater than 150 nl/l did not cause additional inhibition of de-greening. 1-MCP at both 15 and 75 nl/l resulted in a slight suppression of respiration rate while 150 or 300 nl/l resulted in higher respiration rates than non-treated fruit. 1-MCP treatment also resulted in an increase in the rate of ethylene production that was dose and time dependent. The effects of 1-MCP on respiration rate and ethylene production were significantly reduced if fruits were subsequently exposed to ethylene. Rates of ethylene evolution were some 200 nl/kg/h from control and ethylene-treated fruits compared with about 10,000 nl/kg/h from those treated with 1-MCP. Fruits treated with ethylene following 1-MCP treatment had ethylene production rates of about 400 nl/kg/h (McCollum and Maul, 2009).

Guava

In the cultivar 'Allahabad Safeda', 1-MCP treatment at 20 ± 1°C reduced ethylene production, respiration rate, changes in fruit firmness, TSS and TA during storage at both 10°C and 25–29°C, depending upon 1-MCP concentration and duration of exposure. Vitamin C content in 1-MCP-treated fruit was significantly higher and there was a significant reduction in the decay compared with non-treated fruit. Also the development of chilling injury symptoms was ameliorated to a greater extent in 1-MCP-treated fruit. It was reported that 1-MCP at 600 nl/l for 12 h and storage at 10°C seemed a promising way to extend the storage life while 1-MCP at 300 nl/l for 12 h and 24 h or 600 nl/l for 6 h could provide 4–5 days of extended marketability of fruit under ambient conditions of 25–29°C (Singh and Pal, 2008a).

Kiwifruit

Kiwifruit are particularly susceptible to postharvest exposure to ethylene. CA storage and 1-MCP treatment delayed kiwifruit softening compared with storage in air or with no 1-MCP treatment (Prencipe *et al.*, 2016). The application of 1-MCP at 1 μl/l for 20 h at room temperature immediately after harvest showed beneficial effects in retaining quality during 3 months of storage at 1°C. Fruits treated with 1-MCP were firmer and had higher TSS than non-treated fruits. AA was higher in the non-treated fruits but taste panellists did not find significant differences between treatments, except that 1-MCP-treated fruits had better appearance (Antunes *et al.*, 2008). Li *et al.* (2017a) also reported that treatment with 1-MCP maintained fruit firmness during cold storage, though with enhanced activity of PAL, cinnamyl-alcohol dehydrogenase and peroxidase as well as increased lignification. Lignification occurred after prolonged storage. They found that the expression levels of the lignin-related gene AcPOD1 were increased in fruit core tissue after treatment with 1-MCP, but treatment with MJ or methyl salicylate significantly inhibited lignification.

Lettuce

The cultivar 'Butterhead' treated with 1-MCP had reduced respiration rates. Ethylene production and russet spotting were both retarded. Retention of AA was much better than in non-treated lettuce during 12 days of storage at either 2°C or 6°C (Tay and Perera, 2004).

Lime

1-MCP conserved the colour of 'Tahiti' limes (*Citrus aurantifolia*) during 30 days of storage at 10°C (Jomori *et al.*, 2003). Win *et al.* (2006) showed that chlorophyllase and chlorophyll-degrading peroxidase activities in flavedo tissue were delayed in 1-MCP-treated fruit at concentrations of 250 and 500 nl/l. 1-MCP at 250 or 500 nl/l effectively suppressed endogenous ethylene production, but a concentration of 1000 nl/l accelerated yellowing within 9 days, while fruit treated at 750 nl/l turned completely yellow in 15 days. Ethylene production rate of fruits that had been treated with 1-MCP at 1000 nl/l was 1.6 times higher than that of non-treated fruits. Also AA content was reduced in fruits treated with 1-MCP at 1000 nl/l but not in fruits treated with 250, 500 or 750 nl/l.

Litchi

Sivakumar and Korsten (2007) investigated the possibility of replacing SO_2 fumigation as a postharvest treatment with 1-MCP on the cultivar McLean's Red. They simulated export marketing conditions by fumigation with a moderate concentration of 1-MCP for 3 h, then storage in 17 kPa O_2 + 6 kPa CO_2 for 21 days at 4°C and packaging in MA at 10°C to simulate marketing. They found that with these treatments fungal decay was controlled and the natural pinkish/red colour was retained. They also found that storage in 3 kPa O_2 + 7 kPa CO_2 retained a deep red colour, showed limited yeast decay and only slightly affected the TSS, TA and AA contents. De Reuck *et al.* (2009) packed the cultivars 'Mauritius' and 'McLean's Red' in bi-oriented polypropylene (BOPP) bags and exposed them to 1-MCP at 300, 500 or 1000 nl/l within the packaging, heat sealed them and stored them at 2°C for up to 21 days. This combination of treatments was effective in preventing browning and retention of the pinkish-red colour in both cultivars. 1-MCP significantly reduced the PPO and peroxidase activity, retained membrane integrity and anthocyanin content and prevented the decline of pericarp colour during storage. At concentrations higher than 300 nl/l, 1-MCP had negative effects on membrane integrity, pericarp browning, PPO and peroxidase activity in both cultivars, but at 1000 nl/l there was a significant suppression of respiration rate and better retention of TSS, TA and firmness.

Lotus

Li *et al.* (2017d) exposed lotus pods (*Nymphaeas lotus*) to 1-MCP at 0.5 μl/l or 3% lacquer wax and then stored them at 25 ± 1°C and 90–95% RH. Both 1-MCP and lacquer wax reduced browning and retained their quality, but lacquer wax had little influence on the respiration rate. 1-MCP-treated pods were firmer than non-treated pods for only 6 days. The combination of lacquer wax followed by 1-MCP gave equivalent effects on respiration rate and firmness to lacquer wax alone. However, treatment of lotus pods with 1-MCP first, followed by lacquer wax coating, gave better retention of firmness and total phenol concentration, and reduced respiration rate, accumulation of MDA and browning. This treatment also increased superoxide dismutase (SOD) and catalase activities and decreased PPO and POD activity when compared with non-treated and those in the other treatments.

Mango

Hofman *et al.* (2001) showed that 1-MCP at 25 μl/l for 14 h at 20°C increased the time to ripening by 5.1 days (37%) in the cultivar 'Kensington Pride' but at least doubled the severity of stem-end rot compared with non-treated fruit. However, they concluded that 1-MCP has the potential to reduce the risk of premature fruit ripening due to accidental exposure to ethylene. Mangoes of the cultivar 'Nam Dok Mai' were treated with 500 ppb 1-MCP and then stored at 13°C in 3 kPa O_2 + 5 kPa CO_2 by Kramchote *et al.* (2008). They reported that storage in CA was the most effective way to delay fruit yellowing, while 1-MCP was more effective in maintaining firmness and suppressing ethylene production. The production of aroma volatile compounds emitted after 28 days of storage showed the accumulation of high levels of ethanol for fruits in 3 kPa O_2 + 5 kPa CO_2 while 1-MCP mainly suppressed volatile emission. Wang *et al.* (2007a) harvested the cultivar 'Tainong' at the green-mature stage and treated them with 1-MCP at 1 μl/l for 24 h and stored them at 20°C for up to 16 days. 1-MCP maintained fruit firmness and AA but inhibited the activities of antioxidant enzymes including catalase, SOD and ascorbate peroxidase.

Melon

Softening, respiration rate and ethylene production were reduced in the cultivar 'Hy-Mark' by 1-MCP, with 300 nl/l being the optimum rate. Treated melons were acceptable for 27 days, whereas the non-treated fruits could be stored for no longer than 7 days. 1-MCP-treated galia melons retained their firmness for up to 30 days and the postharvest life of treated charentais melons was 15 days at room temperature (25 ± 3°C and 65 ± 5% RH). In the cultivar 'Orange Flesh', 1-MCP delayed ripening and controlled decay caused by *Fusarium pallidoroseum* (Alves *et al.*, 2005). Sa *et al.* (2008) stored cantaloupes in X-tend film bags for 14 days at 3 ± 2°C, then removed them from their bags and stored them for a further 8 days at room temperature (23 ± 2°C and 90 ± 2% RH). The fruits retained their quality in the bags, but pretreatment with 1-MCP or inclusion of potassium permanganate on vermiculite within the bags gave no additional effects.

Onion

During storage at 18°C, exogenous ethylene suppressed sprout growth in the cultivar 'Copra' of both dormant and already-sprouting bulbs by inhibiting leaf blade elongation, but treatment of dormant bulbs with 1-MCP resulted in premature sprouting. The dormancy-breaking effect of 1-MCP indicates a regulatory role of endogenous ethylene in onion bulb dormancy (Bufler, 2009). In contrast, Chope *et al.* (2007) reported that sprout growth was reduced in bulbs treated with 1-MCP at 1 μl/l for 24 h at 20°C and stored at 4°C or 12°C, but not when they were stored at 20°C. Higher concentrations of sucrose, glucose and fructose were found in 1-MCP-treated bulbs of the cultivar SS1 stored at 12°C and dry weight was maintained compared with non-treated bulbs.

Papaya

Hofman *et al.* (2001) showed that exposure to 1-MCP at 25 μl/l for 14 h at 20°C increased the number of days to ripening by 15.6 days (325%) in the cultivar 'Solo' compared with non-treated fruit. They concluded that 1-MCP treatment has the potential to reduce the risk of premature fruit ripening due to accidental exposure to ethylene. However, Fabi *et al.* (2007) compared non-treated, ethylene-treated or ethylene + 1-MCP-treated 'Golden' papayas and found that 1-MCP could decrease the quality of fruit and that even the use of ethylene for initiating ripening could result in lower quality when compared with that of fruit allowed to ripen 'naturally'. Hofman *et al.* (2001) also found that 1-MCP could have a negative effect, since treatment was associated with slightly higher severity of external blemishes and slightly higher severity of rots compared with those not treated. Zerpa-Catanho *et al.* (2017) reported that polygalacturonase and endoxylanase correlated negatively with firmness in the hybrid 'Pococí' and that 1-MCP treatment repressed and reduced the expression of genes that controlled the two enzymes, which indicates that treating papaya with 1-MCP could delay softening.

Parsley

Ella *et al.* (2003) reported that a single application of 1-MCP at 10 μl/l could be used to retard senescence of parsley leaves early in the storage life, but they also found that this treatment could increase ethylene biosynthesis.

Peach

Treatment with1-MCP at 1 μl/l at 20°C for 14 h reduced the rate of softening, loss of sugars and organic acids and inhibited increases in ethylene production and respiration rate during 16 days storage at 8°C in 90–95% RH for the cultivar 'Mibaekdo' and 5°C in 90–95% RH for the cultivar 'Hwangdo' (Park *et al.*, 2009). Treating the cultivar 'Jiubao' with 1-MCP at 0.2 μl/l at 22°C for 24 h slowed the decline in fruit firmness. The minimum concentration of 1-MCP that inhibited fruit softening was 0.6 μl/l and repeated treatment resulted in more effective inhibition of ripening. Changes in TSS, total sugars, TA, soluble pectin and ethylene production were also significantly reduced or delayed by 1-MCP. The activities of PAL, PPO and peroxidase in the inoculated fruit were also enhanced by treatment with 1-MCP (Liu *et al.*, 2005). In the cultivar

'Tardibelle', treatment with 1-MCP was effective in preserving firmness during subsequent cold storage for 21 days in air but had no effect on those in CA (Ortiz and Lara, 2008). Postharvest decay of the cultivar 'Jiubao' that had been inoculated with *Penicillium expansum* was reduced by treatment with 1-MCP and disease progress was reduced during cold storage for 21 days (Liu *et al.*, 2005). Yu *et al.* (2017) found that 1-MCP significantly inhibited chilling injury and reduced softening in the two cultivars they tested during storage at 5°C. Treatment with 1-MCP reduced the activity and expression of enzymes related to sucrose degradation and increased activity and expression of sucrose synthase synthesis, resulting in a lower rate of sucrose degradation and an increase of glucose and fructose content.

Pear

1-MCP prolonged or enhanced the effects of CA storage of the cultivar 'Conference' (Rizzolo *et al.*, 2005). However, they found that the effects of 1-MCP at 25 nl/l and 50 nl/l declined with duration of storage for up to 22 weeks at –0.5°C (followed by 7 days in air at 20°C) in both CA and air storage. Even re-treating the fruit after 7 and 14 weeks of storage had little additional effect on subsequent ripening. Ethylene production was lower and firmness was higher in fruits treated at 50 nl/l, while those treated at 25 nl/l were similar to non-treated fruit. Non-treated and 25 nl/l-treated fruits reached their best sensory quality after 14 weeks of storage, while 50 nl/l-treated fruits reached the same sensory quality later, retaining a fresh flavour when the quality of non-treated fruits declined and became 'watery or grainy'. The cultivars 'Williams', 'Bosc' and 'Packhams Triumph' were stored for approximately 300 days at –0.5°C in 2.5 kPa O_2 + 2.0 kPa CO_2 or in air by Lafer (2005). The fruits that had been pre-treated with 625 ppb 1-MCP had delayed softening and no loss of TA in all three cultivars, whereas non-treated fruits showed excessive softening and reduction in TA during subsequent shelf-life. The cultivar 'Yali' was exposed to 1-MCP at 0, 0.01, 0.1, 0.2 or 0.5 μl/l for 24 h or 0.2 μl/l for 0, 12, 24 or 48 h and then stored at 20°C and 85–95% RH by Fu *et al.* (2007). Those treated with 1-MCP retained their firmness and TSS content better than those not treated. After 100 days of storage the incidence and index of core browning was reduced by 91% and 97%, respectively, by 1-MCP treatment and the occurrence of black and withered stems was also reduced by 59% by 1-MCP after 32 days of storage. 1-MCP also reduced the ethylene production and respiration rate. Rizzolo *et al.* (2005) showed that the development of superficial scald was not prevented by 1-MCP, but the severity of the symptoms was less in the cultivar 'Conference' after treatment. Isidoro and Almeida (2006) reported that 1-MCP can be used to replace diphenylamine as a postharvest treatment to control scald in the cultivar 'Rocha', though ripening after storage was delayed by relatively high 1-MCP concentrations. Spotts *et al.* (2007) reported that 1-MCP reduced decay (bull's-eye rot, phacidiopycnis rot, stem-end rot and grey mould) in the cultivar 'd'Anjou'. They concluded that a combination of 1-MCP and hexanal at optimal rates may reduce storage decay caused by fungal infection and control superficial scald while allowing normal ripening. Lafer (2005) found that rotting caused by *Penicillium expansum* and *Botrytis cinerea* was the main problem during long-term storage and neither CA nor 1-MCP was effective in preventing fruit rotting. The effect of 1-MCP on fungal decay varied considerably among the cultivars tested and the stage of maturity. Yu and Wang (2017) claimed that the current commercial practice for controlling superficial scald in the cultivar 'd'Anjou' (ethoxyquin and CA storage) was not adequate for the fruit produced in hot seasons. They tested treatment with 1-MCP and found that the combination of ethoxyquin at 2700 mg/l plus 1-MCP at 0.3 μl/l + ethylene at 0.3 or 1.5 μl/l controlled scald, reduced decay and allowed recovery of ripening capacity of 'd'Anjou' after storage at –1.1°C in 1.5 kPa O_2 + < 0.5 kPa CO_2 for 8 months. Escribano *et al.* (2016) treated the cultivar 'Bartlett' after cold storage with 1-MCP at 0.6 μl/l at 0°C for 24 h, which slowed fruit ripening at 20°C, but also resulted in fully ripened pears with higher sweetness, juiciness and aroma compared with those treated with ethylene at 100 μl/l at 20°C for 24 h and non-treated pears. However, Escribano *et al.* (2017) reported that the response of 'Bartlett' to 1-MCP could be variable, due to competition with ethylene. They tested a liquid formulation of 1-MCP (immersion in 500 μg/l for 60 s), which provided a more consistently effective treatment than the fumigation (0.6 μl/l at 0°C for 24 h). The most effective treatment was 1000 μg/l for 60 s, but the extended time to ripen at 20°C after storage 'could be too long for some commercial applications'.

Persimmon

Kaynaş *et al.* (2009) reported that the cultivar 'Fuyu' treated with 1-MCP at either 625 or 1250 ppb for 24 h at 18–20°C then storage in sealed LDPE bags at 0–1°C and 85–90% RH could be kept for 120 days. Ben Arie *et al.* (2001) reported that there was very little fruit softening during storage at 1°C in an ethylene-free atmosphere and the rates of ethylene evolution and fruit softening declined with the increase in 1-MCP concentration from 0 to 600 ppb. The commercial shelf-life of the fruit was doubled or trebled following exposure to 1-MCP, the effect becoming stronger as the season advanced, and the rate of softening of non-treated fruit accelerated. No adverse effects of the treatment were observed. Black spot (*Alternaria alternata*) was the limiting factor in storage, but 1-MCP had no observable effects on infection levels. 'Fuyu' stored by Pinto *et al.* (2007) at 10°C, especially in 1 kPa O_2 + 0 kPa CO_2 and 2 kPa O_2 + 0 kPa CO_2 plus 1-MCP, had a higher percentage of firm fruits and lower decay incidence after 17 days. The cultivar 'Nathanzy' was harvested when the orange colour had developed on the peel (commercial maturity) by Ramm (2008) and then treated with 1-MCP at 0.5, 1 or 1.5 μl/l for 24 h at 20°C and stored at 20°C. The non-treated fruits softened within 15 days; those treated with 1-MCP at 1 or 1.5 μl/l remained firmer for 30 days after harvest and had lower respiration rates, but 0.5 μl/l had a limited inhibitory effect on softening. Changes in TSS, TA and peel colour occurred during storage, but all were significantly delayed by 1-MCP. It was concluded that 1-MCP was effective for quality maintenance and extension of shelf-life in persimmon and could allow harvesting at the orange stage of maturity, at which stage the most desirable organoleptic attributes had been developed on the tree. Luo (2007) treated the cultivar 'Qiandaowuhe' with 1-MCP at 3 μl/l for 6 h and stored them at 20°C which greatly extend their postharvest life compared with those not treated. 1-MCP treatment delayed the onset of ethylene production and the respiration rate and delayed the depolymerization of chelator-soluble pectic substances and alkali-soluble pectic substances. These are cell wall hydrolysis enzymes, and the increase in water-soluble pectic substances was reduced compared with non-treated fruit.

Pineapple

Shinjiro *et al.* (2002) found that the rate of colour change of immature, mature-green or ripe fruits was slower for those that had been treated with 1-MCP. The delaying effect was greater in the immature fruits. They also showed that 1-MCP treatment, several times during storage, actually resulted in temporary stimulation of ethylene production immediately after each treatment but, at the same time, 1-MCP decreased their respiration rate.

Plum

Manganaris *et al.* (2008) reported that both European (*Prunus domestica*) and Japanese (*P. salicina*) plum responded to 1-MCP by delayed ripening. Postharvest application of 1-MCP at 0, 0.5, 1.0 or 2.0 μl/l for 24 h at 20 ± 1°C to the Japanese cultivar 'Tegan Blue' was reported by Khan and Singh (2007) to delay ethylene production significantly and also suppress its magnitude. Also there was a reduction in the activity of ethylene biosynthesis enzymes and fruit-softening enzymes in the skin and pulp tissues. Subsequently Khan *et al.* (2009) showed similar effects, but additionally they reported reduced TSS, AA, total carotenoids and total antioxidants in 1-MCP-treated 'Tegan Blue' compared with those not treated. They found that 1-MCP at 1 μl/l was effective in extending the storage life for up to 6 weeks in cold storage with minimal loss of quality.

Potato

Ethylene is used as a sprout suppressant in potatoes. Foukaraki *et al.* (2016) stored potatoes at 6°C for 30 weeks with ethylene applied from harvest or from the first indication of sprouting. In both cases ethylene reduced sprout growth significantly but induced sugar accumulation. However, sugar accumulation was prevented by 1-MCP application prior to storage. Also a rise in AA content in tubers was observed in response to ethylene application while 1-MCP blocked the ethylene-induced AA rise. Prange *et al.* (2005a) concluded that 1-MCP at 1 μl/l for 48 h at 9°C could be used to control fry colour darkening induced by ethylene (4 μl/l) without affecting ethylene control of tuber sprouting. They found that the number of 1-MCP applications required varied between the two cultivars they tested, in that one application was sufficient in 'Russet Burbank' but not in 'Shepody', which started darkening 4 weeks after exposure in the single ethylene + 1-MCP treatment.

Sapodilla

In Brazil, Manilkara achras fruits treated with 1-MCP at 300 nl/l for 12 h and then stored in MA packaging at 25 ± 2°C and 70 ± 5% RH had delayed softening for 11 days longer than those not treated (Morais *et al.*, 2008). In Mexico, treatment with 1-MCP at 100 or 300 nl/l resulted in a storage life of 38 days at 14°C (Arevalo-Galarza *et al.*, 2007). 1-MCP treatment at 40 or 80 nl/l for 24 h at 20°C delayed the increases in respiration rate and ethylene production and resulted in increased PGA activity by 6 days during storage at 20°C. Decreases in AA, TSS, TA and chlorophyll content were also delayed (Zhong *et al.*, 2006). 1-MCP-treated fruit showed an inhibition of cell wall-degrading enzyme activity and less extensive solubilization of polyuronides, hemicellulose and free neutral sugars when compared with non-treated fruit. It was suggested that delayed softening of sapodilla is largely dependent on ethylene synthesis (Morais *et al.*, 2008).

Spinach

Grozeff *et al.* (2010) found that treatment with 1-MCP at 0.1 or 1.0 μl/l inhibited ethylene sensitivity, which could be successfully used to retain quality and chlorophyll content at 23°C for up to 6 days.

Strawberry

The cultivar 'Everest' was treated with 1-MCP at various concentrations from 0 to 1000 nl/l for 2 h at 20°C and then stored for 3 days at 20°C and 95–100% RH. 1-MCP treatment lowered ethylene production and tended to maintain fruit firmness and colour, but disease development was quicker in fruit treated with 500 and 1000 nl/l. 1-MCP inhibited PAL activity and reduced increases in anthocyanin and phenolic content (Jiang *et al.*, 2001).

Tomato

1-MCP was applied for 12 h at 22 ± 1°C and 80–85% RH. During subsequent storage at 20 ± 1°C and 85–95% RH, 1-MCP delayed ripening by 8–11 days for concentrations of 250 ml/l, 11–13 days for 500 ml/l and 15–17 days for 1000 ml/l compared with non-treated tomatoes. After 17 days, fruits treated with 1000 ml/l were firmer and greener, but total carotenoids were lower than in non-treated fruits (Moretti *et al.*, undated). Both CA storage at 2 kPa O_2 + 3 kPa CO_2 and 1 ppm 1-MCP for 24 h significantly delayed colour development and softening. The effect of 1-MCP at 1 ppm was greater than the effect of CA but 1-MCP at 0.5 ppm had no effect (Amodio *et al.*, 2005). Baldwin *et al.* (2007) harvested the cultivar 'Florida 47' at immature-green, green, breaker, turning and pink stages and treated half of them with 1-MCP and stored them all at 18°C. Those treated with 1-MCP had slightly lower levels of most aroma volatiles but ripened more slowly by 3–4 days compared with non-treated fruit, but at 13°C the delay in ripening of 1-MCP-treated fruit was only 2–3 days. They reported that 1-MCP seemed to 'synchronize' ripening of pink-stage tomatoes.

Watercress

After 5 days of storage in air there was 25% yellowing at 5°C and 100% at 15°C. 1-MCP applied at 500 or 1000 nl/l for 8 h at 10°C had no significant effect on subsequent storage life at 5°C or 15°C. It was concluded that watercress could be stored for a maximum of 4 days at 5°C (Bron *et al.*, 2005).

8

CA Technology

It is crucial for the success of CA storage technology that the precise levels of gas are achieved and maintained within the store. Where these are too high, in the case of CO_2, or too low, for O_2, then the fruit or vegetable may be irrevocably damaged. The range of manifestations of the symptoms of damage varies with the types of product and the intensity of the effect. The effects of CA storage are not simple. In many cases they are dependent on other environmental factors, including the effects of temperature where CA storage may be less or totally ineffective at certain temperatures on some crops. This effect could be related to crop metabolism and gas exchange at different temperatures. From the limited information available it would appear that the store humidity and CO_2 level in the store could interact, with high CO_2 being more toxic at high than low humidity. The physiological mechanism to explain this effect is difficult to find. CA can reduce or eliminate detrimental effects of ethylene accumulation, possibly by CO_2 competing for sites of ethylene action within the cells of the fruit.

The literature is ambivalent on the quality and rate of deterioration of fruit and vegetables after they have been removed from CA storage. Most of the available information indicates that their storage life is adequate for marketing, but, in certain cases, the marketable life is even better than freshly harvested produce. The physiological mechanism that could explain this effect has so far not been demonstrated.

The equipment and methods used in the control of the atmosphere inside stores is constantly being developed to provide more accurate control. Use of the physiological responses of the fruit and vegetables to CA storage is increasing, which may eventually replace the methods that were applied during the 20th century. A review of the design, construction, operation and safety considerations of CA stores is given by Bishop (1996). In fresh fruit and vegetable stores the CO_2 and O_2 levels will change naturally through their metabolism because of the reduced or zero gas exchange afforded by the store walls and door. Levels of the respiratory gases in non-CA stores have shown these changes. For example, Kidd and West (1923) showed that the levels of CO_2 and O_2 in the holds of ships from South Africa carrying stone fruit and citrus fruit arriving in UK at 3.9–5°C were 6 kPa CO_2 and 14 kPa O_2 and at 2–3.3°C the levels were 10.0 kPa CO_2 and 10.5 kPa O_2. The level of CO_2 in a banana ripening room can rise from near zero to 7 kPa in 24 h (Thompson and Seymour, 1984).

Much of the current commercial practice is basically the same as that described in the early part of the 20th century. The crop is loaded into an insulated storeroom whose walls and door have been made gas tight. The first CA stores were constructed in the same way as refrigerated stores from bricks or concrete blocks with a vapour-proof barrier and an insulation layer that was coated on the inside with plaster. In Australia, Little *et al.* (2000) pointed out the difficulty of converting refrigerated stores to CA stores in the early 1970s because the cork insulation made them so leaky. In order to ensure that the walls were gas tight for CA storage, they were lined with sheets of galvanized steel (Fig. 8.1). The bottoms and tops of the steel sheets were embedded in mastic (a kind of mortar composed of finely ground oolitic limestone, sand, litherage and linseed oil) and a coating of mastic was also applied where sheets abutted on the walls and ceiling (Fig. 8.2). The walls were insulated using wooden frames with cork, wood shavings or Rockwool insulation. They often had wooden floors, which could result in O_2 leakage into the stores at a faster rate than it was utilized by the fruit giving an equilibrium level of 5–8 kPa O_2.

Modern CA stores are made from metal-faced insulated panels (usually polyurethane foam and its fire-resistant form, called isophenic foam), which are fitted together with gas-tight patented locking devices (Fig. 8.3). The joints between panels are usually taped with gas-tight tape or painted

Fig. 8.1. The inside of a traditional controlled atmosphere store in England showing the internal walls covered with galvanized sheets to render them gas tight. Photograph courtesy of Dr R.O. Sharples.

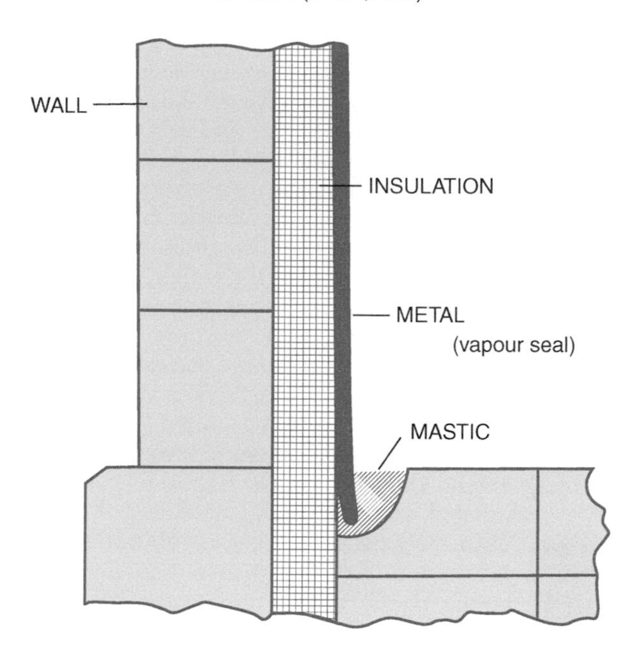

Fig. 8.2. Detail of internal wall of a traditional controlled atmosphere store in England covered with galvanized sheets and filled with mastic to render them gas tight. Photograph courtesy of Professor H.D. Tindall.

Fig. 8.3. Access corridor of modern CA stores showing the doors with the fruit ready to be loaded. Reproduced with permission from David Bishop, Storage Control Systems Ltd, UK.

with flexible plastic paint or resin sealants to ensure that they are gas tight. Cetiner (2009) described some of the properties of the insulated panels as rigid polyurethane foams having a closed cell structure and natural bonding to steel facers. In China, one company used a thermal insulating board, which is of light structure, high strength, good air-tightness, good temperature-insulation, convenient construction, substantial short construction period and less construction cost. They used ester urethane thermal insulating board and polystyrene heat board for constructing the framework.

Major areas of the store where leaks can occur are the doors. These are usually sealed by having rubber on the jamb or frame (Fig. 8.4) so that when the door is closed the two meet to seal the door. In addition a flexible soft rubber hose may be hammered into the inside between the door and its frame to give a double rubber seal to ensure gas tightness. In other designs screw jacks are spaced around the periphery of the door. Some modern CA stores and fruit ripening rooms have inflatable rubber door seals.

Fig. 8.4. Door of a traditional CA store in the UK showing rubber seals on jamb matched by rubber seals on door to ensure that they are gas tight. There is also a circular sampling port and small door to ensure that anyone trapped inside can escape. Photograph courtesy of Dr R.O. Sharples.

With the constant changes and adjustments of the store temperature and concentrations of the various gases inside the sealed store, variation in pressure can occur. Pressure difference between the store air and the outside air can result in difficulties in retaining the store in a completely gas-tight condition. Stores are therefore fitted with pressure release valves (Fig. 8.5), but these can make the maintenance of the precise gas level difficult, especially the O_2 level in ULO stores. An expansion bag may be fitted to the store to overcome the problem of pressure differences. The bags are gas tight and partially inflated and are placed outside the store, with the inlet to the bag inside the store. If the store air volume increases, this will automatically further inflate the bag; and when the pressure in the store is reduced, air will flow from the bag to the store. The inlet of the expansion bag should be situated before the cooling coils of the refrigeration unit in order to ensure that the air from the expansion bag is cooled before being returned to the store.

Testing the gas tightness of a store is crucial to the accurate maintenance of the required CO_2 and O_2 levels. One way of measuring this was described by MAFF

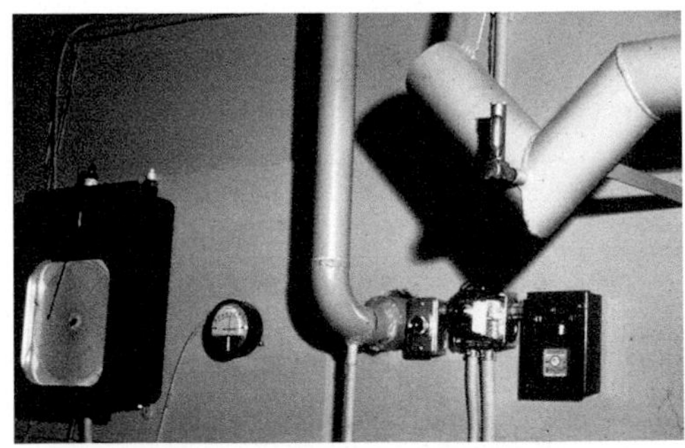

Fig. 8.5. Pressure release valve on CA store to ensure equalization of pressure between the store and externally. Photograph courtesy of Dr R.O. Sharples.

ADAS (1974). A blower or vacuum cleaner is connected to a store ventilator pipe and a manometer connected to another ventilator pipe. It is essential that the manometer is held tightly in the hole so that there are no leaks. A blower then blows air into the store which is maintained until a pressure of 200 Pa has been achieved and then the blower is switched off. The time for the pressure inside the store to fall is then monitored. They recommended that the time taken to fall from 187 to 125 Pa should not be less than 7 min in order to consider the store sufficiently gas tight. If the store does not reach this standard, efforts should be made to locate and seal the leaks.

An example of this CA technology was the largest commercial store in Europe in the late 20th century, which has now gone out of business (East Kent Packers, 1994, personal communication). CA was achieved by sealing the store directly after loading and reducing the temperature to below 10°C. Bags of hydrated lime were loaded into the store prior to sealing as a method of removing respiratory CO_2. The temperature was then reduced further and when it was below 5°C, N_2 was injected to reduce the O_2 level; for 'Cox's Orange Pippin' apples this was to about 3 kPa. The O_2 level was allowed to decline to about 2 kPa for 7 days through fruit respiration and then progressively to 1.2 kPa over the next 7 days. Headspace monitoring was used that set off an alarm if alcohol fumes were detected.

Temperature Control

The main way of preserving fruit and vegetables in storage or during long-distance transport is, of course, by refrigeration. CA storage and MA packaging are considered supplements to increase or enhance the effect of refrigeration. There is evidence that the inhabitants of the island of Crete were aware of the importance of temperature in the preservation of food, even as early as 2000 BCE (Koelet, 1992). Mechanical refrigeration was developed in 1755 by a Scotsman, William Cullen, who showed that evaporating ether under reduced pressure, caused by evacuation, resulted in the temperature of the water in the same vessel being reduced, thus forming ice. Cullen then patented a machine for refrigerating air by the evaporation of water in a vacuum. John Leslie, also a Scot, subsequently developed the technique in 1809 and invented a differential thermometer and a hygrometer, which led him to discover a process of artificial congelation. Leslie's discovery improved upon Cullen's equipment by adding sulfuric acid to absorb the water vapour and he developed the first ice-making machine. In 1834 an American, Jacob Perkins, was granted a British patent (Number 6662) for a vapour compression refrigeration machine. Carle Linde, in Germany, developed an ammonia compression machine. This type of mechanical refrigeration equipment was used in the first sea-freight shipment of chilled beef from Argentina in 1879 and was employed in many of the early experiments in fruit and vegetable stores.

Kidd and West (1927 a, b) produced evidence that CA storage was only successful when applied at low temperatures. Standard refrigeration units are therefore integral components of CA stores. Temperature control consists of pipes containing a refrigerant inside the store. Ammonia or chlorofluorocarbons (CFCs) R-12 and R-22 (the R-number denotes an industry standard specification and is applied to all refrigerants) are common refrigerants, but because of concern about depletion of the ozone layer in the 1980s new refrigerants have been developed that have potentially less detrimental effects on the environment, such as R-410A, which is 50:50 blend of R-32 and R-125 and is used as a substitute for R-22. The pipes containing the refrigerant pass out of the store; the liquid is cooled and passed into the store to reduce the air temperature of the store as it passes over the cooled pipes. A simple refrigeration unit consists of an evaporator, a compressor, a condenser and an expansion valve. The evaporator is the pipe that contains the refrigerant mostly as a liquid at low temperature and low pressure, and is the part of the system that is inside the store. Heat is absorbed by the evaporator, causing the refrigerant to vaporize. The vapour is drawn along the pipe through the compressor, which is a pump that compresses the gas into a hot high-pressure vapour. This is pumped to the condenser, where the gas is cooled by passing it through a radiator. The radiator is usually a network of pipes open to the outside air. The high-pressure liquid is passed through a series of small-bore pipes, which slows the flow of liquid so that a high pressure builds up. The liquid then passes through an expansion valve, which controls the flow of refrigerant and reduces its pressure. This reduction in pressure results in a reduction in temperature, causing some of the refrigerant to vaporize. This cooled mixture of vapour and liquid refrigerant passes into the evaporator, thus completing the refrigeration cycle. In most stores a fan passes the store air over the coiled pipes containing the refrigerant, which helps to cool the air quickly and distribute it evenly throughout the store.

In commercial practice for long-term CA storage of most crops, the store temperature is initially reduced to 0°C for a week or so, whatever the subsequent storage temperature will be. This would clearly not be applicable to fruit and vegetables that can suffer from chilling injury at relatively high temperatures (10–13°C). Also CA stores are normally designed to a capacity that can be filled in 1 day, so fruit is loaded directly into store and cooled the same day. In the UK the average CA store size was given as about 100 t with variations between 50 and 200 t; in continental Europe it was about 200 t and in North America about 600 t (Bishop, 1996). In the UK the smaller rooms were preferred, because they facilitated the speed of loading and unloading.

Humidity Control

Most fruit and vegetables require a high humidity when kept in storage. Generally the closer the atmosphere is to saturation the better, as long as moisture does not condense on the fruit or vegetables. The amount of heat absorbed by the cooling coils of the refrigeration unit is related to the temperature of the refrigerant they contain and the surface area of the coils. Water will condense on the evaporator if the refrigerant temperature is low compared with the store air temperature. Condensate will freeze if the refrigerant temperature is well below freezing, reducing the efficiency of the cooling system. Removal of moisture from the store air results in the stored crop losing moisture by evapotranspiration. In order to reduce crop desiccation, the refrigerant temperature should be kept close to the store air temperature. However, this must be balanced with the removal of the respiratory heat from the crop, temperature leakage through the store insulation and doors and heat generated by fans, otherwise the required crop temperature cannot be maintained. There remains the possibility of having a low temperature differential between the refrigerant and the store air by increasing the area of the cooling surface and this may be helped by adding devices such as fins to the cooling pipes or coiling them into spirals. In a study on evaporator coil refrigerant temperature, the room humidity, cooling rate and mass loss rates were compared in commercial 1200-bin apple storage rooms by Hellickson *et al.* (1995). They reported that rooms in which evaporator coil refrigerant temperatures were dictated by cooling demand required a significantly longer time to achieve desired humidity levels than rooms in which evaporator coil temperatures were controlled by a computer. The overall mass loss rate of apples stored in a room in which the cooling load was dictated by refrigerant temperature was higher than in a room in which refrigerant temperature was maintained at approximately 1°C during the initial cooling period.

A whole range of humidifying devices can also be used to replace the moisture in the air that has been condensed on the cooling coils of refrigeration units. These include spinning disc humidifiers where water is forced at high velocity on to a rapidly spinning disc. The water is broken down into tiny droplets, which are fed into the air circulation system of the store. Sonic humidifiers utilize energy to detach tiny water droplets from a water surface that are fed into the store's air circulation system. Dijkink *et al.* (2004) described a system that could maintain humidity very precisely in 500 l containers. They were able to maintain 90.5 ± 0.1% RH using a hollow fibre membrane contactor that allowed adequate transfer of water vapour between the air in the storage room and a liquid desiccant. The membrane was made of polyetherimide, coated on the inside with a thin non-porous silicone layer. The desiccant was a dilute aqueous glycerol solution, which was pumped through the hollow fibres at a low flow rate.

Another technique that retains high humidity within the store is secondary cooling so that the cooling coils do not come into direct contact with the store air. One such system is the 'jacketed store'. These stores have a metal inner wall inside the store's insulation, with the refrigeration pipes cooling the air space between the inner wall and the outer insulated wall. This means a low temperature can be maintained in the cooling pipes without causing crop desiccation and the whole wall of the store becomes the cooling surface. In a study of export of apples and pears from Australia and New Zealand to the UK, the use of jacketed stores gave a uniform temperature within 0.5°F (−17.5°C) in a bulk of a stack of 50,000 cases (Food Investigation Board, 1937). Ice bank cooling is also a method of secondary cooling, where the refrigerant pipes are immersed in a tank of water so that the water is frozen. The ice is then used to cool water and the water is converted to a fine mist, which is used to cool and humidify the store air (Neale *et al.*, 1981).

Where crops such as onions are stored under CA conditions, they require an optimum humidity of about 70% RH. This can be achieved by having a large differential between the refrigerant and air temperature of about 11–12°C with natural air circulation or 9–10°C where air is circulated by a fan.

Gas Control Equipment

Burton (1982) indicated that in the very early gas stores (CA stores) the levels of CO_2 and O_2 were controlled by making the storeroom gas tight with a controlled air leak. This resulted in the two gases being maintained at approximately equal levels of about 10%. Even in more recent times the CO_2 level in the store was controlled while the level of O_2 was not and was calculated as $O_2 + CO_2 = 21$ kPa (Fidler and Mann, 1972). Kidd and West (1927a, b) recognized that it was desirable to have more precise control over the level of gases in the store and to be able to control them independently. In subsequent work the levels of CO_2 and O_2 were monitored and adjusted by hand and considerable variation in the levels could occur. Kidd and West (1923) showed that the levels monitored in a store varied between 1 kPa and 23 kPa for CO_2 and between 4 kPa and 21 kPa for O_2 over a 6-month storage period.

In the 1940s R.E. Smock at Cornell University introduced the addition of a sodium hydroxide solution to the system for removal of CO_2. The use of hydrated lime for removing CO_2 was subsequently developed by Charles Eaves in Canada. The Whirlpool Corporation in the USA built 'Tectrol' units for apple CA stores. Tectrol and other types used the combustion of propane or natural gas to develop an atmosphere low in O_2. In the USA, Donald Dewey and George Mattus developed some automatic controls for CA storage (Mattus, 1963). N_2 generators using selective gas permeable membranes (Fig. 8.6) and N_2 cylinders are now more commonly used.

Generating equipment

In early work the method used to achieve the required O_2 and CO_2 levels in CA rooms was product generated. This obviously resulted in a delay in achieving the required condition which, in turn, could reduce the maximum storage life or increase losses during storage. One method of reducing O_2 rapidly was to introduce air into the store that had first been passed over a gas burner. The room was ventilated to prevent the increase in pressure; therefore the method was referred to as gas flushing. This gave a rapid reduction in O_2 but many of the products of burning gas, such as carbon monoxide, can act like ethylene and have a negative effect

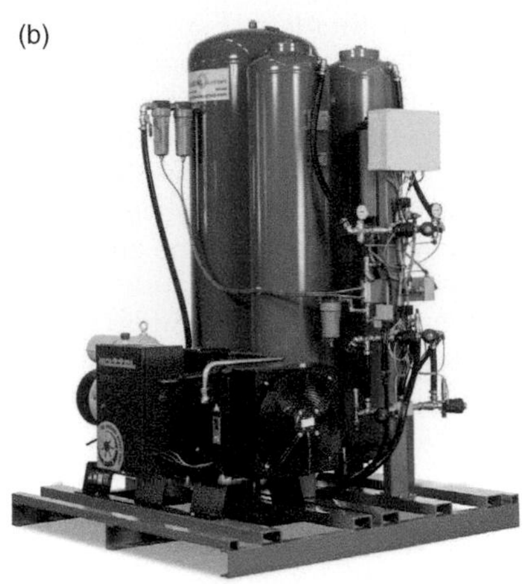

Fig. 8.6. Nitrogen generation. (a) Nitrogen generator in use at a commercial controlled atmosphere store in Kent in England in the 1990s. (b) Modern Pressure Swing Adsorption nitrogen generator. Reproduced with permission from Eric van der Zwet, Besseling Group BV, The Netherlands.

on maximum storage life of the stored crop. Natural or propane gas was combined with outside air and burned, which lowered the O_2 and increased CO_2. Most of the CO_2 was removed before the atmosphere was fed into the CA room. An Italian method used an 'Isolcell' unit where the effluent was passed through a catalyst to ensure complete oxidation of any partially oxidized compounds to CO_2. Activated carbon was used to scrub out most of the CO_2 before the generated atmosphere entered the store room. A different type of atmosphere generator was developed in The Netherlands in the early 1980s, called 'Oxydrain'. This utilized ammonia rather than propane. The ammonia was cracked, forming N_2 and H_2, and the H_2 was burned with the O_2 from the room so that no CO_2 was produced. The 'Arcat' system recirculated the air from the room through the natural gas or propane burner and there was a separate system to scrub out the CO_2 produced. The use of a carbon molecular sieve N_2 generator in CA storage was described by Zhou *et al.* (1992b).

Oxygen control

In traditional CA storage systems, when the O_2 had reached the level required for the particular crop it was maintained at that level by frequently introducing fresh air from outside the store. Sharples and Stow (1986) reported that tolerance limits were set at ± 0.15 kPa for O_2 levels below 2 kPa, and ± 0.3 kPa for O_2 levels of 2 kPa and above. This method is still in use. In early CA stores the samples were analysed for both O_2 and CO_2 levels using an Orsat gas analyser, which measured the levels of the two gases by absorbing them and noting the changes in volume. With continuing equipment development, the precision with which the set levels of CO_2 and O_2 can be maintained is constantly increasing. With recent developments in the control systems used in CA stores, it is possible to control O_2 levels close to the theoretical minimum. This is because modern systems can achieve a much lower fluctuation in gas levels and ULO storage (levels around 1 kPa or less) is now common. Modern stores use electrochemical cells (Fig. 8.7) or sometimes a paramagnetic analyser (Fig. 8.8), which are monitored and controlled by a computer. Instruments have been developed that combine analysis of both CO_2 and O_2 from the same sample. These are used for sampling gases in MA packages but can be adapted for CA storage (Fig. 8.9).

The benefits of O_2 levels as low as 1 kPa, or less, have been shown to be optimum in

Fig. 8.7. Electrochemical O_2 sensor with typical resolution down to 0.001 kPa O_2. Reproduced with permission from Storage Control Systems Ltd.

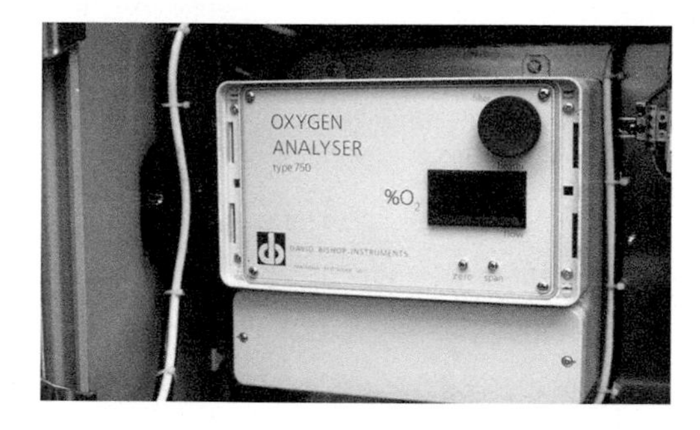

Fig. 8.8. Paramagnetic O_2 analyser installed at Cranfield University, UK, in the 1980s for CA storage research.

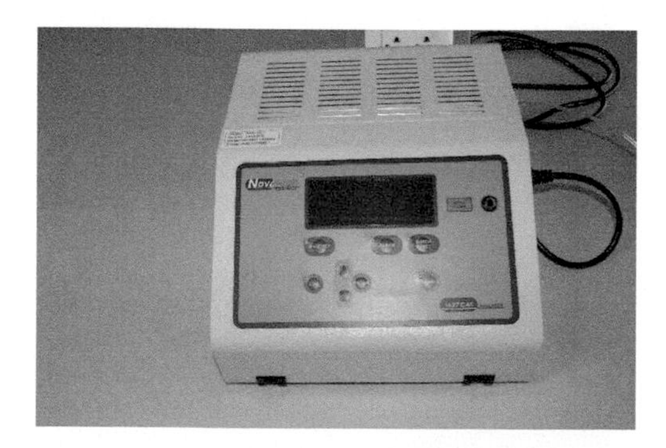

Fig. 8.9. Monitoring for gas levels. A small gas sample is taken from the package or the air and injected into the apparatus that measures and displays oxygen and CO_2 concentration. The analyser uses zirconia ceramic technology for O_2 and non-dispersive infrared spectroscopy for CO_2.

the extension of the storage of some fruit. These precise O_2 levels can also be used to control physiological disorders. For example, in apples O_2 levels down to 0.6 kPa were found to control BBD and with Bramley's Seedling O_2 levels down to 0.4 kPa were shown to maintained firmness and gave control of scab (Rees, 2014).

At these very low O_2 levels precise control is vital in order not to damage the fruit. Methods that have been developed are based on approaches to the physiology of the fruit. There are three main approaches: one based on respiratory quotient (RQ), one based on ethanol biosynthesis and one based on chlorophyll fluorescence (CF). These monitoring and control methods have been called dynamic controlled atmosphere (DCA) storage while monitoring the O_2 and CO_2 predetermined levels in the store air has been referred to as static controlled atmosphere storage.

Carbon dioxide control

There are many different types of scrubbers that can be used to remove CO_2 from CA stores, but they can basically be divided into two types. One, sometimes called 'passive scrubbing', uses a chemical that reacts with CO_2 and thus removes it from the store; and the other is a renewable system sometimes called 'active scrubbing'. Sharples and Stow (1986) stated that the CO_2 level in the store should be maintained at ± 0.5 kPa of the recommended level. Burdon *et al.* (2005) reported that the method of CO_2 control may affect the volatile composition of the room atmosphere, which in turn may affect the volatile synthesis of fruit. They compared activated carbon scrubbing, hydrated lime scrubbing, N_2 purging and storage in air on 'Hayward' kiwifruits at 0°C in 2 kPa O_2 + 5 kPaCO_2. After storage the fruits were allowed to ripen at 20°C and although volatile profiles differed between CA-stored and air-stored fruits, and also among fruits from the different CO_2 scrubbing systems, the different CO_2 scrubbing systems did result in measurable differences in ripe fruit volatile profiles. They also reported (Lallu *et al.*, 2005) that there was little effect of the CO_2 removal system on fruit softening and the incidence of rots, but N_2 flushing resulted in kiwifruit with the lowest incidence of physiological pitting. They concluded that different CO_2 removal systems altered room volatile profiles but did not consistently affect the quality of kiwifruit.

Dual-bed scrubbers are where one bed is removing CO_2 while the other is being renewed and the two beds are constantly switched to provide continuity of CO_2 removal. Hermann A. Zinnen, Anil R. Oroskar and Chin-Hsiung Chang invented CO_2 removal using aminated carbon molecular sieves, a method that was patented by Allied-Signal Inc. in 1989 (US4810266 A). Carbon molecular sieves are prepared from O_2-free precursors carbonized in the absence of O_2. The pores of the material are enlarged by high-temperature steaming and alcohol amines are used to impart an amine functionality to the sieves. The store air is passed through the sieve, with the absorbed CO_2 being released by heating to a moderate temperature.

Occasionally N_2 purging is used for flushing out CO_2. Measurement of CO_2 levels forms part of the measurement used in some modern DCA systems for calculating RQ. Some commercial stores still use bags of hydrated calcium hydroxide powder, which reacts with CO_2; this is used solely, or as a supplement to the activated carbon, especially when the fruit or vegetable is very sensitive to CO_2. Some of the scrubbing systems used commercially are reviewed below.

Passive scrubbing

The simplest method is to place bags or pallets of a CO_2-absorbing chemical (usually calcium hydroxide) inside the store, which can keep CO_2 levels low (usually about 1 kPa). This method is referred to as 'product generated', since the gas levels are produced by the crop's metabolism. Calcium hydroxide reacts irreversibly with CO_2 to produce calcium carbonate, water and heat. The disadvantage of this method is the space occupied by the lime and the inefficiency of CO_2 absorption, because less than 20% of the lime in the bags is used under normal stacking procedures (Olsen, 1986). In a store of 55 t capacity containing hydrated lime, the CO_2 concentration fell to less than 1 kPa while the lime was fresh, then rose gradually; the average CO_2 concentration during the 128-day storage period was 2.2 kPa. Blank (1973) found that the apple cultivars 'Cox's Orange Pippin', 'Ingrid Marie', 'Boskoop' and 'Finkenwerder' stored well under these conditions and the annual total cost was reduced to between one-half and one-quarter of that for stores with activated carbon scrubbers. This simple method of passive scrubbing is thought suitable for stores holding up to 100 t.

For greater control the bags or pallets of lime may be placed in a separate airtight room. When the CO_2 level in the store is above that which is required, a fan draws the store atmosphere through the room containing the bags of lime (Fig. 8.10) until the

Fig. 8.10. A simple lime scrubber that was commonly used in CA stores in UK in the mid to late 20th century. Photograph courtesy of Dr R.O. Sharples.

required level is reached. After scrubbing, the air should re-enter the store just before it passes over the cooling coils of the refrigeration unit. There may also be some dehydration of the air as it is passed through the lime and the air may need humidification before being reintroduced to the store. The amount of lime required depends on the type and variety of the crop and the storage temperature. For example, Koelet (1992) stated that for 1 t of apples, 7.5 kg of 'high calcium lime' is needed every 6–10 weeks for most cultivars. Bishop (1996) also calculated the amount of lime required to absorb CO_2 from CA stores on the theoretical basis. He calculated that 1 kg of lime will absorb 0.59 kg of CO_2, but for a practical capacity he estimated 0.4 kg of CO_2 per 1 kg of lime. On that basis the requirement of lime for 'Cox's Orange Pippin' apples stored in < 1 kPa CO_2 + 2 kPa O_2 was 50 kg of lime per tonne of fruit. For other apple cultivars, storage regimes and periods, Bishop indicated that probably less lime would be required. Thompson (1996) quoted a general requirement for apples of only 25 kg for 6 months. Olsen (1986) found that a total of 23 kg of lime per tonne of apples was required for a 128-day storage period given in two portions, with one change during storage.

With passive scrubbers the time taken for the levels of these two gases to reach the optimum (especially for the O_2 to fall from the 21 kPa in fresh air) can reduce the maximum storage life of the crop. It is common, therefore, to fill the store with the crop, seal it and inject N_2 gas until the O_2 has reached the required level and then maintain it in the way described above. The N_2 may be obtained from large liquid N_2 cylinders or from N_2 generators (see Fig. 8.6).

Active scrubbing

Active scrubbing consists of two containers of material that can absorb CO_2. Store air is passed through one of these containers when it is required to reduce the CO_2 in the store atmosphere while the CO_2 is being removed from the other one. When one is near-saturated with CO_2 the containers are reversed and the CO_2 is removed from the first while the second absorbs CO_2 from the store. Active scrubbers have the advantage of being compact and are particularly suitable for use in CA transport systems. Koelet (1992) described such a system where a ventilator sucks air from the store through a plastic tube to an active carbon filter scrubber. When the active carbon becomes saturated with CO_2 the tube is disconnected from the store and outside air is passed over it to release the CO_2. When this process is complete the active carbon scrubber is connected to a 'lung', which is a balloon of at least 5 m^3 capacity (depending on the size of the equipment) containing an atmosphere with low O_2. This process is to ensure that when the active carbon scrubber is reconnected to the store it does not contain high O_2 levels. Baumann (1989) described a simple active scrubbing system that could be used in stores to remove both CO_2 and ethylene using activated charcoal. A chart was also presented that showed the amount of activated charcoal required in relation to the CO_2 levels required and ethylene output of the fruit in the store.

Pressure swing adsorption (PSA) uses a compressor that passes air under pressure through a molecular sieve, made from alumino-silicate minerals called zeolites, which separates the O_2 and the N_2 in the air. PSA is a dual-circuit system so that when one circuit is providing N_2 for the store, the other circuit is being renewed.

Hollow fibre technology is similar to PSA but hollow fibre technology is also used to generate N_2 to inject into the store. The walls of these fibres are differentially permeable to O_2 and N_2. Compressed air is introduced into the fibres and by varying the pressure it is possible to regulate the purity of the N_2 coming out of the equipment and produce an output of almost pure N_2. The O_2 content at the output of the equipment will vary with the throughput. For a small machine with a throughput of 5 nm^3/h the O_2 content will be 0.1 kPa, at 10 nm^3/h it will be 1 kPa and at 13 nm^3/h it will be 2 kPa.

Molecular sieves and activated carbon can hold CO_2 and organic molecules such as ethylene. When fresh air is passed through these substances the molecules are released. This means that they can be used in a two-stage system where the store air is being passed through the substance to absorb the ethylene while the other stage is being cleared by the passage of fresh air. After an appropriate period the two stages are reversed. Hydrated aluminium silicate or aluminium calcium silicate are used. The regeneration of the molecular sieve beds can be achieved when they are warmed to 100°C to drive off the CO_2 and ethylene. This system of regeneration is referred to as 'temperature swing', where the gases are absorbed at low temperature and released at high temperature. Regeneration can also be achieved by reducing the pressure around the molecular sieve, which is called 'pressure swing'. During the regeneration cycle the trapped gases are usually ventilated to the outside, but they can be directed back into the container if this is required.

In China, a carbon molecular sieve N_2 generator was developed (Zhou and Yu, 1989) that was made from fine coal powder, which was refined and formed to provide apertures similar in size to N_2 (3.17 Å) but smaller than O_2 (3.7 Å). When air was passed through it under high pressure, the O_2 molecules were absorbed on to the fine coal powder and the air passing through had an enriched N_2 level.

A 40% solution of ethanolamine can be used in temperature swing scrubbers, which can be regenerated by heating the chemical to 110°C. However, ethanolamine is corrosive to metals and is not commonly used for CA stores.

Pressurized water scrubbers work by passing the store air through a water tower which is pumped outside the store, where it is ventilated with fresh air to remove the CO_2 dissolved in it inside the store. However, O_2 is dissolved in the water while it is being ventilated and will therefore provide a store atmosphere of about 5 kPa O_2 + 3 kPa CO_2. A pressurized water CO_2 scrubber was described by Vigneault and Raghavan (1991) which could reduce the time required to reduce the O_2 content in a CA store from 417 h to 140 h.

Sodium hydroxide and potassium hydroxide scrubbers work when the store air is bubbled through a saturated solution, at about 50 l/min, when there is excess CO_2 in the store. This method requires about 14 kg sodium hydroxide for each 9 m^3 of apples per week and, since the chemical reaction which absorbs CO_2 is not reversible, the method can be expensive.

Selective diffusion membrane scrubbers have been used commercially (Marcellin and LeTeinturier, 1966) but they require a large membrane surface area in order to maintain the appropriate gaseous content around the fruit.

Nitrogen cylinders

This is a simple system where cylinders or a tank containing N_2 are connected to a vaporizer and pipes leading to the CA store. There is a sampling pump and meters to monitor O_2 and CO_2 levels in the storeroom and an air pump to inject outside air. As the O_2 is depleted fresh air is introduced and as CO_2 accumulates N_2 is introduced. N_2 gas pressurized to about 40 pounds per square inch (psi) (about 276 kPa) from liquid N_2 tanks can be fed into the room through a vaporizer, which can establish the required O_2 levels in the store quickly with purge rates of up to 20,000 cubic feet (566 m^3) per hour. Care must be taken to allow adequate ventilation to avoid a potentially catastrophic over-pressurization of the room. Additionally, very high purge rates may let purge gas escape through the vent before it is completely mixed with the store air. Introducing N_2 gas behind the refrigeration coils and venting low on the return wall assures high mixing effectiveness (Olsen, 1986).

CO_2 addition

For crops that have only a very short marketable life, such as strawberries, the CO_2 level may also be increased to the required level by direct injection of CO_2 from a pressurized gas cylinder. This is commonly only used in transport of fruit in CA containers.

Ethylene control

Ethylene is synthesized by plant cells. In climacteric fruit its synthesis initiates ripening. Ethylene can also have negative effects on fresh fruit and vegetables, especially when accumulated in stores. For example, Wills *et al.* (1999) showed that the postharvest life of some non-climacteric fruit and vegetables could be extended by up to 60% when stored in an atmosphere containing less than 5 ppb ethylene compared with those stored in 100 ppb. There are various ways in which ethylene can be removed from stores, including absorption, reaction, ozone scrubbers and catalytic converters.

Absorption

Molecular sieves and activated carbon can hold organic molecules such as ethylene. When fresh air is passed through these substances the organic molecules are released. So two-stage systems have been developed where one stage is absorbing ethylene while the other is ventilating to fresh air. The system automatically switches when one side is judged to be saturated.

Reaction

Ethylene can be oxidized at room temperatures when it comes into contact with potassium permanganate. Various commercial systems are available, such as Ethysorb ®, Purafil ® Bi-On 4® and Sorbilen ®, which can be placed inside the CA store or even inside the actual package containing the crop.

Ozone scrubbers

Ozone is a powerful oxidizing agent and reacts with ethylene to produce CO_2 and water. It can be easily generated from O_2 by ultraviolet (UV) radiation or electrical discharge. Scott and Wills (1973) described a UV lamp giving out radiation at 184 and 254 nanometres which produced both ozone and atomic oxygen. The store atmosphere was passed through the reaction chamber with a fan and any ethylene it contained was rapidly oxidized. The outlet of the reaction chamber contained rusty steel wool, which would react with any excess ozone and prevent it entering the store.

Catalytic converters

Air from the store is passed through a device where it is heated to over 200°C (usually 230–260°C) in the presence of an appropriate catalyst, usually platinum, that oxidizes ethylene and any other unsaturated hydrocarbon to CO_2 and water. A catalytic converter was marketed by Tubamet AG of Vaduz in Liechtenstein in 1993 and called ‘Swingcat’. Another is produced by the French company Absoger (2016) who claim that it can maintain store ethylene levels below 20 parts per billion. Huang *et al.* (2002) described a scrubber that maintained ethylene concentration below 100 ppb in a sealed plastic tent containing cabbages. Sfakiotakis *et al.* (1993) reported that an ethylene scrubber retained ethylene levels below 300 ppb in an apple CA store. A two-column ethylene converter is available in which the store air passes through a catalyst bed where the ethylene is broken down and then to the second catalyst bed, where the remaining ethylene is broken down and the air cooled (Fig. 8.11a). It is claimed that this system can reduce the ethylene level to about 1 ppb and its heat recovery and temperature control mean that it consumes minimal energy (Eric van der Zwet, 2018, personal communication). Ethylene scrubbers can be incorporated in the gas control system or portable ethylene scrubbers are available that can be placed in a store when required (Fig. 8.11b).

Plastic Tents and Membranes

Xie *et al.* (1990) compared storage of apples for 117 days in small plastic bags and large plastic tents. After storage the proportion of fruits of good quality among those stored non-packed was 54%, or more than 98% for those in small plastic bags and 87–89% for those in the large plastic tents. Fruit firmness and vitamin C were highest and acidity

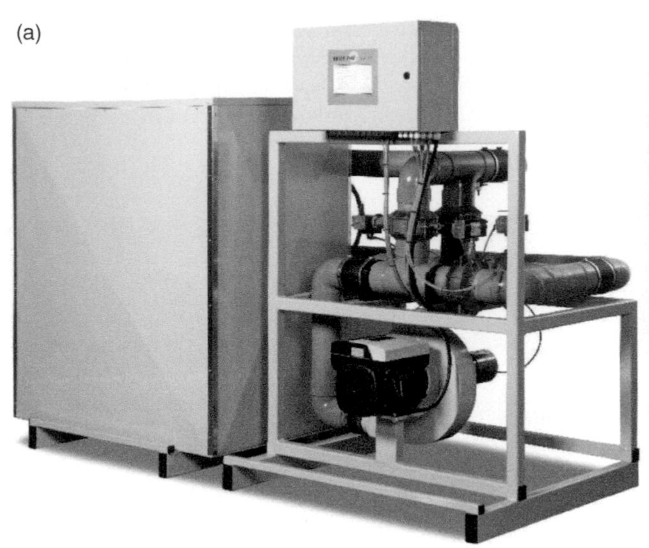

Fig. 8.11. (a) A modern Besseling ethylene converter (Reproduced with permission from Eric van der Zwet, Besseling Group BV, The Netherlands). (b) A portable Tubamet Swingcat ethylene scrubber in use in a wholesale packhouse in UK in the 1990s.

lowest in small plastic bags, followed by plastic tents and then non-packed. Leyte and Forney (1999) designed and constructed a tent for the CA storage of small quantities of fruit and vegetables suspended from pallet racking in a cold room. The tents could hold two standard pallets stacked 1.8 m high with fruit or vegetables and were sealed with two airtight zips and a small water trough, resulting in an airtight chamber that successfully maintained a CA storage environment. Huang *et al.* (2002) described studies using a sealed plastic tent inside a cold storage room. A 2 × 2 × 2 m tent made of LDPE was set up inside a cold room maintained at 0–1°C. The composition of the atmosphere within the tent was maintained at 2–6 kPa O_2 (mostly at 3 kPa) and 3–5 kPa CO_2 (mostly at 5 kPa). Ethylene concentration was controlled at below 0.1 ppm using an ethylene scrubber. The temperature inside the tent was always about 0.7°C higher than the temperature in the rest of the room. In 1997 a company developed a system that involved enclosing a large trolley in a six-sided insulating blanket (a material with the trade name of 'Tempro') that was closed with heavy-duty Velcro. This was shown to maintain the product temperature within 1°C, at a starting temperature of 0°C, during a 22 h period. The inner walls of the insulating blanket had pockets in which dry ice could be placed to help to maintain product temperature and presumably increase the CO_2 content. The Palliflex storage system was developed for CA storage by Van Amerongen CA Technology, USA, in about 2002. The system can be used to set individual CA conditions on individual pallets, which means that different types of fruit and vegetables requiring the same temperature but different atmospheres can be stored in the same room. The system consists of a single plastic pallet of fruit or vegetables with a transparent gas-tight plastic film cover made from polyamide/polyethylene (PA/PE) or polyurethane (PU) film (Fig. 8.12). The sealed pallet is connected to a regulating system that has built-in O_2 and CO_2 analysers and an air pump and is connected to CO_2 and N_2 cylinders and an O_2 scrubber (Fig. 8.12). The sealed pallets can be linked together so that one monitoring and control unit can be used for many pallets or just one. Doäÿan and Erkan (2014) described experiments with the Palliflex storage system that had been successfully completed in Turkey and it was found to be

Fig. 8.12. Palliflex storage system: (a) control unit; (b) pallet of fruit sealed in plastic film. Reproduced with permission from Ramzi Amari, Van Amerongen CA Technology BV, The Netherlands.

advantageous for use in storage and transport of horticultural commodities. Sahin *et al.* (2015) compared storage of 'Red Globe' grapes at 0°C with 95% RH for 90 days in different CA conditions using a Palliflex storage system and Selcuk and Erkan (2015) used the Palliflex storage system for experiments on CA storage of medlar fruit.

Besseling have also developed a similar system that they called the 'Pallet Fresh System' (Fig. 8.13). Each pallet is covered with a polyethylene film cover that can be sealed and can be used for individual pallets or groups of pallets. The covered pallets are placed in a cold store and the atmosphere inside the pallet cover can be monitored and controlled by an external CO_2 cylinder, O_2 air compressor and PSA N_2 generator (Eric van der Zwet, 2018, personal communication). Another Dutch company, Fotein BV, supplies a modified atmosphere system for pallets called 'Ship it Fresh!', which consists of a nine-layer foil comprising four different types of plastic providing tensile and flexible strength, impact resistance, gas tightness and ease of sealing. Covers are equipped with a gas exchange membrane so that the gas conditions in the cover remain virtually unchanged during transport. The unit, which is 1 m wide and 1.2 m deep, consists of a preformed top cover 1.80 m high and a bottom cover 30 cm high. The unit has a maximum height of 2.25 m (Fig. 8.14).

Browne *et al.* (1984) tested PE-covered pallets of 'Cambridge Favourite' strawberries sealed with a patented mechanism and injected with CO_2 and found after transport at 2°C that the pallets contained between 3 and 16 kPa CO_2, most containing < 10 kPa. In subsequent experiments they found that 10 kPa CO_2 did not consistently retard spoilage of fruit by *Botrytis cinerea* or *Mucor piriformis*, as they had hoped, and also caused persistent off-flavours. However, in spite of these results they recommended that PE covers be used on pre-cooled strawberries to 'help keep the fruit cool and reduce moisture loss during distribution'.

Marcellin and LeTeinturier (1966) reported a system of controlling the gases in store by a silicone rubber diffuser. In trials they reported that they were able to maintain an atmosphere of 3 kPa CO_2 + 3 kPa O_2 + 94 kPa N_2 with the simple operation of only gas analysis. In later work by Gariepy *et al.* (1984) a silicone membrane system was shown to maintain 3.5–5 kPa CO_2 + 1.5–3 kPa O_2. They found that, in this system, total mass loss was 14% after 198 days of storage of cabbage compared with 40% in air. On a commercial-scale experiment (472 t), an 'Atmolysair System' CA room was compared

Fig. 8.13. The 'Pallet Fresh System'. Reproduced with permission from Eric van der Zwet, Besseling Group BV, The Netherlands.

Fig. 8.14. A 'Ship it Fresh' modified atmosphere pallet: (a) being assembled; and (b) loaded with fruit for shipment. Reproduced with permission from Rob Veltman, Managing Director, Fotein (www.fotein.com).

with a conventional cold room for storing 'Winter Green' cabbages for 32 weeks at 1.3°C. The atmosphere in the Atmolysair System room was 5–6 kPa CO_2 + 2–3 kPa O_2 + 92 kPa N_2 and traces of other gases. This was compared with a cold room maintained at 0.3°C. The average trimming losses were less than 10% for the Atmolysair System room but exceeded 30% in the cold room (Raghavan *et al.*, 1984). Rukavishnikov *et al.* (1984) described a 100 t store with the crop sealed inside 150 or 300 μm PE films, with windows of membranes selective for O_2 and CO_2 permeability made of polyvinyl trimethyl silane or silicon-organic polymers. Apples and pears stored at 1–4°C under the covers had 93% and 94% sound fruit, respectively, after 6–7 months of storage, whereas the fruit storage in air at similar temperatures had less than 50% sound fruit after 5 months. A design procedure for a silicon-membrane CA store, based on parametric relationships, was suggested by Gariepy *et al.* (1988) for selecting the silicone membrane area for long-term CA storage of leeks and celery. A 'modified gaseous components system' for CA storage using a gas separation membrane was designed, constructed and tested by Kawagoe *et al.* (1991). The membrane was permeable to CO_2, O_2 and N_2 but to different degrees, thus the levels of the three gases varied. A distinctive feature of the system was the application of gas circuit selection to obtain the desired modified gaseous components. It was shown to be effective in decreasing CO_2 content and increasing O_2 content, but N_2 introduction to the chamber was not included in the study.

Ripening Rooms

Design

Ripening rooms are designed for climacteric fruit that are harvested before they are ripened and placed in temperature-controlled rooms under CA conditions in order to initiate and control ripening. Also some citrus, which are non-climacteric fruit, are placed in temperature and CA rooms simply to change the colour of the peel. These are called degreening rooms and ethylene is commonly used. During the 1990s there was an increasing demand for all the fruit being offered for sale in a supermarket to be of exactly the same stage of ripeness so that it has an acceptable and predictable shelf-life. This led to the development of a system called 'pressure ripening', which is used mainly for bananas but is applicable to any climacteric fruit. The system involves direction of the circulating air in the ripening room by channels through boxes of fruit so that exogenous ethylene gas, which initiates ripening, is in contact equally with all the fruit in the room. At the same time the CO_2, which can impede ripening initiation, is not allowed to concentrate around the fruit (Fig. 8.15).

The primary requirements for ripening rooms (Fig. 8.16) are that they should:

- have a good temperature control system;
- have good and effective air circulation;
- be gas tight; and
- have a good system for introducing fresh air.

Air circulation around the fruit is important to prevent local accumulations of the CO_2 given out by the fruit and to ensure good contact between the fruit and the ethylene gas applied to initiate ripening. In the case of bananas and some other types of fruit, the boxes are lined with polyethylene film and usually transported to ripening rooms stacked on pallets. Several systems have been used. One is to remove each box from the pallet, pull back the plastic film and re-stack them on pallets so that there is a space between boxes (Fig. 8.17). In other cases the pallets are stacked so that one handhold in each box is facing outwards. As the fruit are loaded into the rooms, the plastic just inside each handhold is torn to facilitate air exchange. This is especially important where fruit have been vacuum packed.

Air circulation systems are usually largely convectional. Air is blown through the cooler and then across the top of the store just below the ceiling. The cooled air falls by convection through the boxes of fruit and is taken up at floor level for recirculation. Many modern ripening rooms have air channels in the floor through which air is circulated at high pressure. This forces it up through the pallets of boxes and should give better air circulation. Special devices such as inflatable air bags placed between pallets are now used to ensure better air circulation and, therefore, more even fruit ripening.

Good ventilation to enable fresh air to be introduced is very important for successful fruit ripening. During the period of initiation of ripening, which is usually 24 h, no fresh

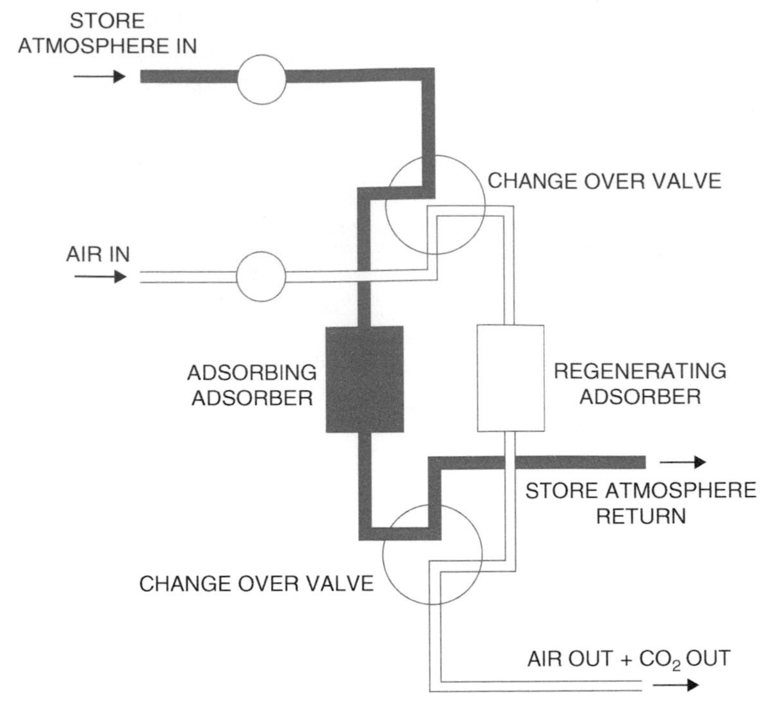

Fig. 8.15. Schematic of CO_2 scrubber.

Fig. 8.16. Banana ripening room in UK in the 1990s.

air is introduced into the rooms. This is the period when ethylene is introduced. Directly after this period the rooms must be thoroughly ventilated. Setting up a ripening system for bananas in Ecuador, Thompson and Seymour (1984) found that the CO_2 level of ripening rooms had gone up from about zero to 7 kPa during this initial 24 h initiation period. Even with a good fan extraction system it took 40 min of ventilation to bring the CO_2 levels to below 1 kPa. This ventilation with fresh air must be repeated every 24 h during subsequent ripening to prevent levels of CO_2 becoming too high. If rooms are not frequently ventilated, ripening can be delayed, or abnormal ripening can occur.

Ripening rooms need to be gas tight in order to ensure that threshold levels of ethylene are maintained around the fruit during the initiation period. As in CA stores, the most common place where leakage occurs is around the doors.

It is advisable to have high humidity of 90–95% RH in ripening rooms. To this end many rooms are fitted with some humidification device such as a spinning disc

Fig. 8.17. (a) Pressure ripening rooms. (b) Pallets of bananas loaded into a pressure ripening room. Reproduced with permission from Dave Rodden, Advanced Ripening Technologies Ltd, UK.

humidifier. However, if the rooms are full of fruit and the refrigerant in the cooling coils used to maintain the room temperature is regulated to within a few degrees of the required room temperature, this should be sufficient to keep the humidity high.

Ethylene generators

Ethylene can be applied in a liquid formulation as Ethrel. For example, recommended ripening conditions of Otaheite apple (*Spondias cytherea*) were 2–4 days at ambient conditions or 20–22°C and 90–95% RH with 500 ppm Ethrel for 1–1½ days followed by 7–8°C and 90–95% RH, since fully ripe fruit were found to be less sensitive to chilling injury (Mohammed *et al.*, 2011). In commercial ripening rooms in the Yemen, buckets of sodium hydroxide were placed throughout the ripening rooms. When all these were in place, measured amounts of Ethrel were added, which gave an instant release of ethylene gas into the room (Thompson, 1985). Steel cylinders containing ethylene are also used (Fig. 8.18), as are calcium carbide and other chemicals that release ethylene analogues.

However, ethylene generators are currently used almost exclusively. These are devices that are placed in ripening rooms. A liquid is poured into them and they are connected to an electrical power source and they produce ethylene over a protracted period (Fig. 8.19). The manufacturers of these generators do not provide information on exactly what the liquid is that they supply for use in the generators or the process by which the ethylene is generated. A possible way of generating ethylene would be to heat ethanol in a controlled way in the presence of a copper catalyst. Care should be exercised in doing this because of the inflammability of the alcohol.

The way that generators are used is to calculate the volume of the store and place the correct number of generators in the store to provide the required ethylene concentration. This method has the advantage of supplying ethylene to the store over 16 h rather than applying it in one dose from cylinders. This means that there is a better chance of achieving the desired ripening or de-greening effect where there are problems of the room being perfectly gas tight. Gull (1981) used this method for ripening tomatoes. Blankenship and Sisler (1991) reported that effluent produced by the catalytic generators had a distinctly different odour than that produced by ethylene from cylinders. When they compared the two methods for ripening tomatoes, they found inconclusive

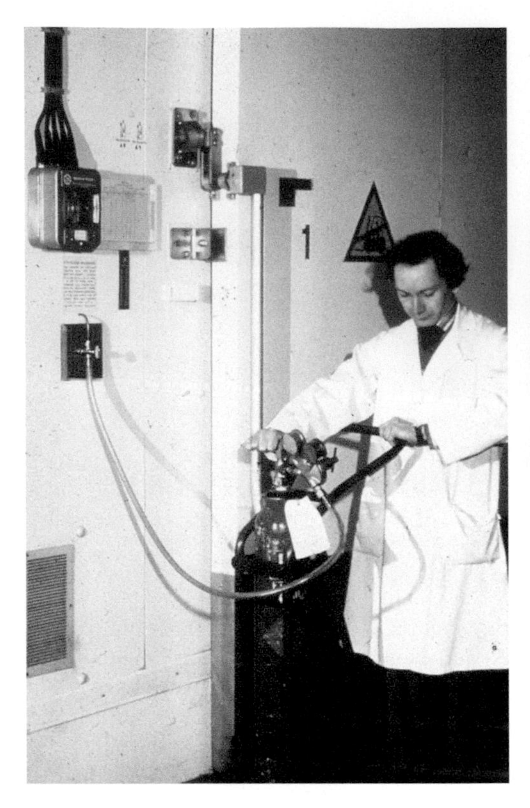

Fig. 8.18. Ethylene being injected into a banana ripening room at Geest, UK, in the 1970s.

Fig. 8.19. An an ethylene generator being filled and prepared to be used in a banana ripening room. Reproduced with permission from Greg Akins, Catalytic Generators Inc., Norfolk, Virginia, USA.

results in that taste panellists could detect a difference between the fruit ripened by the two methods, but they did not express a preference for either treatment. There tended to be less variation in colour between fruits ripened with ethylene from cylinders compared to those ripened from ethylene generators.

Bananas are sensitive to physiological levels of ethylene as low as 0.3–0.5 µl/l if the O_2 and CO_2 levels are similar to those found in outside fresh air (Peacock, 1972). The three main factors affecting response to external ethylene are: fruit maturity; time after harvest when ethylene exposure began; and the length of exposure to ethylene.

Modelling

Modelling has been used for CA storage and considerable work has been reported at the Catholic University of Leuven in Belgium led by Dr Bart Nicolä. However, in DCA storage Yearsley *et al.* (1996) experienced problems finding an acceptable mathematical method to estimate low O_2 using ethanol and had to resort to subjective visual estimation by a human panel. Lee *et al.* (1991, 2003a) developed a model that fitted a polynomial equation to the respiration rate of different fruits to determine optimum CA storage conditions, and successfully tested it on cherry tomatoes. Nahor *et al.* (2003) described a model for simulation of the dynamics of heat, moisture and gas exchange in CA stores consisting of three interacting systems: the cooled space, the refrigeration unit and the gas handling unit. They reported that model predictions were usually found to have a good agreement with the experimental results. Due to the inclusion of product respiration rate and product quality models, the advantage of the simulation model was that the optimization of CA plant performance with respect to the final quality of the stored fruit or vegetable could be carried out. Owing to the hierarchical and object-oriented approach, greater flexibility with respect to model modification and reusability was achieved. Nahor *et al.* (2005) later reported that the Michaelis–Menten type gas exchange models were appropriate for storage of apples and pears, as opposed to the empirical ones, due to their generic behaviour. They recommended that the kinetic parameters of the model should be expressed as a function of the physiological age of the crop to include the influence of age when the models were to be used to simulate gas exchange of bulk stored fruit for longer storage period. Gwanpua *et al.* (2012) proposed kinetic equations for a model to describe softening of 'Braeburn' apples in CA storage. They measured changes in pectin-degrading enzymes, pectin degradation and ethylene production and found that their model could explain most of the total variance of the measured data. Gwanpua *et al.* (2018) also used the Michaelis–Menten constant for a mathematical model to describe autocatalytic ethylene production on softening and colour changes during storage. They calibrated, and successfully validated, their model on avocados stored at 5°C, 10°C or 20°C in air with the addition of ethylene at 0, 1 or 10 μl/l. CO_2 evolution, O_2 consumption and ethanol production were fitted to a polynomial equation used in 'response surface methodology' and tested on cherry tomatoes in order to determine optimum CA storage conditions. The regression equations fitted well with the experimental data of the respiration characteristics having a coefficient of determination $R^2 > 0.8$ and the effective CA conditions for cherry tomatoes were estimated at 4–6 kPa O_2 + 3.5–10 kPa CO_2. Ho *et al.* (2016) evaluated gas exchange in pears during CA storage using a continuum diffusion respiration model. Simulation results showed that internal O_2 and CO_2 variation in the gas profiles in the pears were affected by variable diffusion, maximum respiration rate and the 'three dimensional morphology' of the fruit. They also modelled the onset of fermentation during low O_2 levels in ULO storage. Their model was used to show that harvest maturity and fruit size were important criteria to be taken into account prior to CA storage.

Peppelenbos *et al.* (1996) used a Michaelis–Menten kinetic model to describe the O_2 uptake and the CO_2 production depending on the O_2 partial pressure and validated the models by respiration measurements. Castellanos *et al.* (2016) successfully developed models to represent changes in the fruit firmness and colour during storage of feijoa fruits in MA packaging with R^2 values of 0.95 for firmness and 0.94–0.95 for colour. They were able to estimate, in advance, the equilibrium concentration of the gases in the packaging headspace for each MA package they tested, using a previously developed model. Pereira *et al.* (2017) developed a model for defining optimum packaging and storage conditions. They tested the model on minimally processed rocket leaves during 10 days of storage in MA packages at 0°C, 5°C or 10°C factorially combined with 2.5, 5, 10 and 20 kPa O_2. The RQ was found to range between 0.6 and 1.3 and the temperature contributed more than 80% of the Michaelis–Menten model variance, while O_2 pressure contributed only with 13%.

Safety

The use of CA storage has health and safety implications. One factor that should be taken into account is that the gases in the atmosphere could possibly have a stimulating effect on microorganisms. Also the levels of these gases could have detrimental effects on the workers operating the stores. In Britain the Health and Safety Executive (HSE, 1991) showed that work in confined spaces could be potentially dangerous and entry must be strictly controlled, preferably through some permit system. Also it is recommended that anyone entering such an area should have proper training and instruction in the precautions and emergency breathing apparatus. Stringent procedures need to be in place with a person on watch outside and the formulation of a rescue plan. MAFF ADAS (1974) also indicated that when a store is sealed anyone entering it must wear breathing apparatus. Warning notices should be placed at all entrances to CA stores and an alarm switch located near the door inside the chamber in the event that someone may be shut in. There should be a release mechanism so that the door can be opened from the inside. When the produce is to be unloaded from a store, the main doors should be opened and the circulating fans run at full speed for at least an hour before unloading is commenced. In the UK the areas around the store chamber must be kept free of impedimenta in compliance with the appropriate Agricultural Safety Regulations.

Ethylene

Ethylene is a colourless gas with a sweetish odour and taste, has asphyxiate and anaesthetic properties and is flammable. Its flammable limits in air are 3.1–32% volume for volume and its auto ignition temperature is 543°C. Care must be taken when the gas is used for fruit ripening to ensure that levels in the atmosphere do not reach 3.1%. As added precautions, all electrical fitting must be of a 'spark-free' type and warning notices relating to smoking and fire hazards must be displayed around the rooms. For the USA, Crisosto (1997) said that 'All electrical equipment, including lights, fan motors and switches, should comply with the National Electric Codes for Class 1, Group D equipment and installation.'

Oxygen

O_2 levels in the atmosphere of 12–16 kPa can affect human muscular coordination, increase respiration rate, and affect a person's ability to think clearly. At lower levels vomiting and further impediment to coordination and thinking can occur. At levels below 6 kPa human beings rapidly lose consciousness, and breathing and heart beating stop (Bishop, 1996). O_2 levels of over 21 kPa can create an explosive hazard in the store and great care needs to be taken when hyperbaric storage or high O_2 MA packaging is used.

Carbon dioxide

Being in a room with higher than ambient CO_2 levels can be hazardous to health. The limits for CO_2 levels in rooms for human occupation were quoted by Bishop (1996) from the Health and Safety Executive, publication EH40-95 *Occupational Exposure Limits*, as 0.5 kPa CO_2 for continuous exposure and 1.5 kPa CO_2 for 10 min exposure.

Microorganisms

Microorganisms, especially fungi, can grow on fruit and vegetables during storage and can produce toxic chemicals. One example is the mycotoxin patulin (4-hydroxy-4H-furo [3,2-c], pyran-2(6H)-one). The EU (European Commission, 2006) has established patulin limits of 50 µg/kg for apple juice, 25 µg/kg for solid apple products and 10 µg/kg for baby foods. Patulin can occur in apples and pears that have been contaminated with *Penicillium patulum*, *P. expansum*, *P. urticae*, *Aspergillus clavatus* or *Byssochlamys nivea* and stored for long periods. Morales *et al.* (2007) inoculated the apple cultivar 'Golden' with *P. expansum* and stored them at 1°C for up to 2½ months in either low O_2 or ultra-low O_2. Generally, bigger lesions were observed under ULO than under low O_2, but no patulin was detected in either condition. There was an increase in lesion size when fruits were subsequently stored at 20°C and patulin was detected, but there was no difference in patulin content between ULO and low O_2. Dos Santos *et al.* (2018) found that healthy fruits were not contaminated with patulin, even when stored together with infected decayed apples. Application of 1-MCP increased the percentage of fruit with decay and patulin contamination in 'Galaxy' apples. Neither CA nor DCA storage (DCA-CF or DCA-RQ 1.3) prevented patulin production in either 'Galaxy' or 'Fuji Kiku' apples.

9

Dynamic CA Storage

For optimum conditions for CA storage, it is essential that the precise levels of gases in the store are known and maintained. As has been described in Chapter 8, these levels have traditionally been determined by experimentation. Basically the lower the O_2 content in the CA store, the better are the storage conditions for the crop, down to a critical O_2 level, called the anaerobic compensation point (ACP), when the effects become negative. Dynamic controlled atmosphere (DCA) storage uses ultra-low oxygen (ULO) levels and monitors these by measuring physiological responses (respiration rate, chlorophyll fluorescence, fermentation) of the stored crop and controls the levels through a computer. DCA therefore uses the responses of the actual crop being stored as monitors of the store atmosphere, instead of an arbitrary setting of the O_2 and CO_2 levels in the store based on previous experience and experimentation. In other words, CA conditions are adjusted dynamically between and within seasons to optimize the product's response to CA. CA storage has therefore involved two phases. In the first phase, the dynamic nature of the CA conditions was solely determined from empirical data. In the second and current phase, bio-sensing of the product physiology is employed to reset the CA conditions to reflect the changing metabolic condition of the stored product.

History of DCA

Phase 1: DCA using empirical data

Alique and De la Plaza (1982) are believed to be the first to use the term 'dynamic controlled atmosphere'. Taking advantage of research showing that apples varied over storage time in their susceptibility to CO_2 damage, they proposed to increase the CO_2 in the store after 1 month from 0 to 2 kPa and after 2 months to 5 kPa, at constant 3 kPa O_2, successfully demonstrating its effect on 'Golden Delicious' apples (Alique and De la Plaza, 1982) and 'Blanquilla' pears (Alique *et al.*, 1984). Qi *et al.* (1989) introduced the term 'two-dimensional dynamic controlled atmosphere storage' (TDCA) in which the first dimension is a change in storage temperature from 10–15°C to 0°C and the second dimension is a change in storage CO_2 from an initial 10–15 kPa to about 6 kPa with a constant 3 kPa O_2. Mattheis *et al.* (1998) used the term 'dynamic atmosphere storage' in which temperature and CO_2 are kept constant and O_2 is varied from 1 kPa to ambient air for periods of 1, 2 or 3 days and then returned to 1 kPa O_2.

Phase 2: DCA using product physiology

DCA storage using product physiology, rather than static general recommendations, was described conceptually by Wollin *et al.* (1985) at the 4th National Controlled Atmosphere Research Conference. They speculated that the respiratory quotient (RQ) may be the method to determine the lowest O_2 level, but identified several constraints to its use outside carefully controlled laboratory conditions. The concept of Wollin *et al.* (1985) was used to devise an automated test system that sought out the optimum combination of temperature, O_2 and CO_2 to achieve minimum respiration (Wolfe *et al.*, 1993). Jameson (1993) described a similar research project in the UK but no data or evidence of success was provided by either group. The minimum point of respiration rate they sought was the ACP, which is the O_2 concentration that produces the lowest CO_2 production. The assumption was that this was the O_2 concentration where the respiration rate changes from primarily aerobic (above the ACP) to primarily anaerobic (below the ACP) and fresh fruit or vegetables held exactly at or just above the ACP would have the maximum storage life (Fig. 9.1).

Yearsley *et al.* (1996) attempted to integrate terminology by introducing the general term lower oxygen limit (LOL) and stated that the optimum storage atmosphere occurs just above the LOL at which aerobic respiration is at the lowest level that can be achieved without development of anaerobic metabolism. They stated that, from a physiological perspective, there are two approaches to determining the LOL: (i) the ACP that is the O_2 concentration at which CO_2 production is minimal; and

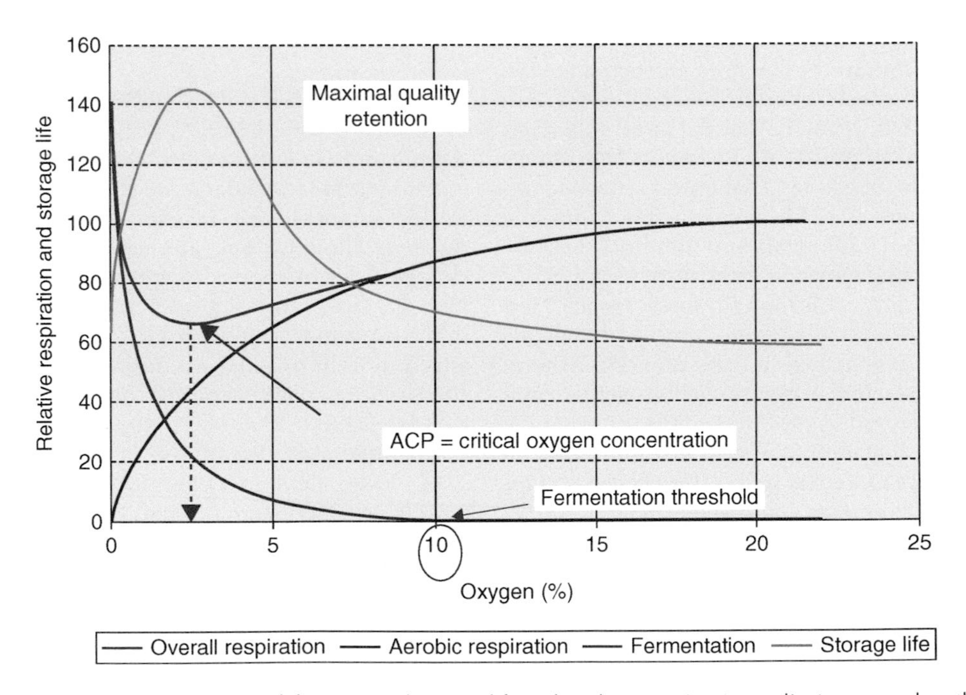

Fig. 9.1. It is generally accepted that maximal storage life and quality retention (green line) occurs when the product is stored at a storage O_2 concentration (% or kPa) exactly at or just above the Anaerobic Compensation Point (ACP) but not below the ACP. The blue line (aerobic respiration) is measured as O_2 consumed and the red line (overall respiration) is measured as total CO_2 produced. The ACP occurs at the lowest total CO_2 production. The Fermentation threshold can also be called the Extinction Point (EP). The ACP can also be called the lower oxygen limit (LOL).

(ii) measures of the onset of ethanol production or respiratory quotient breakpoint. Using experimental data on 'Cox's Orange Pippin' apples, they identified how these two approaches of determining the LOL produce different LOL values and how internal gas concentrations are more relevant than external gas (headspace) measurements. At that time of writing (1996) the use of chlorophyll fluorescence to determine LOL was not known.

Respiratory Quotient

RQ is the measure of moles CO_2 evolved to moles O_2 absorbed in plant cells. RQ = 1 when the substrate is carbohydrate, but it is lower for lipids and proteins. Burton (1952) measured RQ in potatoes stored at 10°C in 5–7 kPa CO_2 for up to 14 weeks. The increased CO_2 reduced both O_2 uptake and CO_2 output by about 25–30%, but the RQ was unaffected and remained close to 1. Wollin *et al.* (1985) discussed the possibility that RQ may be used to calculate the lowest O_2 level that can be tolerated in fruit during storage to be incorporated in an automated CA system. Yearsley *et al.* (1996) also discussed the possibility of using RQ in CA stores and considered that the fermentation threshold could best be represented by RQ. They reported that RQ gave the safest estimate of the true lower O_2 limit for optimizing storage atmospheres. The methods used for research in the Federal University of Santa Maria in Brazil were described by dos Santos *et al.* (2018) as:

> DCA-RQ conditions were measured twice a week during nine months of storage. For RQ determination, O_2 and CO_2 partial pressure were measured immediately after the chambers were closed. After 13 h, the RQ was calculated as the ratio between CO_2 produced and O_2 consumed within this period. The O_2 partial pressure was controlled by RQ variation. If the RQ calculated was higher than the pre-established value (RQ 1.3), O_2 levels were increased by ± 0.02 kPa to reduce anaerobic metabolism and RQ. If the calculated RQ was less than pre-established RQ, O_2 partial pressure was decreased by ± 0.02 kPa to increase anaerobic metabolism.

Van Schaik *et al.* (2015) stored 'Elstar' apples for 5 months at 1.8°C in 1.2 kPa O_2 + 2% CO_2, using the DCR system, and measured RQ and ethanol. In all the rooms the final O_2 levels were set by the measurement of RQ. Their results indicated that 'the DCR system is a promising interactive storage system' and that it also 'reduced respiratory heat production up to a factor of 2 compared to standard ULO, resulting in significant energy savings'. Bessemans *et al.* (2016) found that for 'Granny Smith' apples in storage containers, the DCA-RQ system controlled O_2 and CO_2 partial pressures 'in an autonomous way', with the RQ breakpoint varying between 0.25 kPa and 0.4 kPa O_2. They reported some difficulties in obtaining reliable RQ measurement at very low O_2 levels, due to the low respiration rate of the fruit. Another issue was reported to be the variation in the estimated level of the CO_2 production rate (Gasser *et al.*, 2003). The storage of 'Galaxy' apples under DCA-RQ 1.3 showed lower ethylene production, lower respiration rates, lower mealiness and higher flesh firmness in comparison with static CA-stored fruit (Thewes *et al.*, 2017c). As the storage temperature increased from 0.5°C to 1.0°C, the incidence of low O_2 injury and flesh breakdown was reduced in 'Royal Gala' apples stored in DCA-RQ (Weber *et al.*, 2015). The meaning of DCA-RQ 1.3, 1.5 and 2.0 is that the numbers refer to the RQ ratio that is set during storage. To achieve this, the O_2 partial pressures are changed during the entire storage period to obtain these RQ levels. Thus, the fruit stored in DCA-RQ 1.5 had a higher anaerobic metabolism (lower O_2 partial pressures) compared with the ones stored in DCA-RQ 1.3 (F.R. Thewes, 2017, personal communication). Bessemans *et al.* (2018) tested a method to correct real-time measurements of RQ for gas leakage in an empty storage container The model was successfully implemented in an automated DCA-RQ control system under 0.4 kPa O_2 and 1.3 kPa CO_2 using 'Braeburn' apples, which showed that the RQ estimates without a leakage correction were heavily biased when leakage of atmospheric air into the storage container occurred and would lead to erroneous control of the storage atmosphere composition. Gasser *et al.* (2003) assessed the possibility of adopting this strategy of using RQ by examining, in carefully controlled laboratory conditions, the effect of decreasing O_2 levels on the CO_2 production and ACP of 'Idared' apples. They also evaluated the possible influence of harvest date, storage duration and storage conditions for the sample fruit prior to the measurements. They detected an ACP in all the samples, though the CO_2 production below the ACP was dependent on rate of O_2 reduction, degree of maturity, storage duration of the sample fruit and the previous storage conditions. More specifically, as storage duration increased, apples stored in CA, compared with refrigerated air, showed less increase in CO_2 production below the ACP.

Safepod™

A commercial system, which measures the CO_2 given out and the O_2 taken in by fruit and calculates RQ, was developed by International Controlled Atmosphere Ltd in the UK and Storage Control Systems in the USA. They called the system Safepod™ and the patent for SafePod™ is held jointly by David Bishop and Jim Schaefer (US Patent No. 8739694, Canadian Patent No. CA2746152) (David Bishop, Storage Control Systems, 2018, personal communication). The Safepod™ container (Fig. 9.2) sits in the CA store and thus has the same temperature, humidity, pressure and atmosphere as the store. Periodically the valves are closed and the CO_2 and O_2 are measured and the RQ is calculated. Safepod™ is based on stress detection on a representative sample of fruit (some 70 kg) and automatic measurement of respiration rate and RQ, which can then be used to automatically adjust the store atmosphere to its optimum O_2 and CO_2 levels. The SafePod™ is a hermetically sealed enclosure with a stainless steel base and a clear moulded cover that sits in a water trough for sealing (Fig. 9.3). Motorized valves and a circulation fan are normally open to maintain the atmosphere within the SafePod™ at the same conditions as the surrounding CA room. DeEll and Lum (2017) used Safepod™ to store 'Empire' apples at 0.6 kPa O_2 + 0.5 kPa CO_2 for 8 months at 1.5°C and 3°C and found that they were firmer after storage than those stored under CA at 2.5 kPa O_2 + 2 kPa CO_2 or 1.5 kPa O_2 + 1.2 kPa CO_2, but their results 'were from a single season, and further research is needed to confirm these observations'. There are about 70 Safepod™

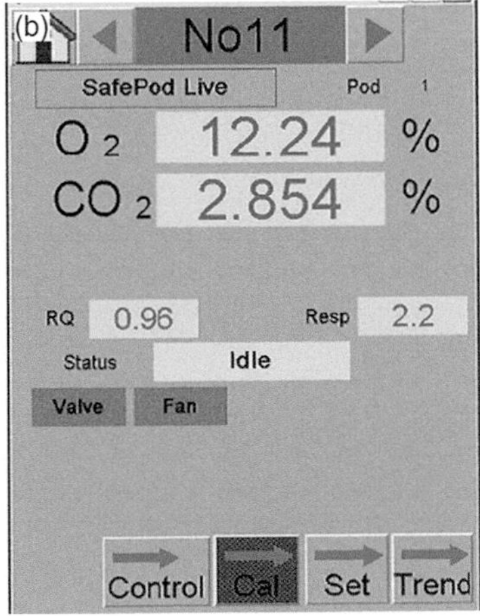

Fig. 9.2. (a) Safepod™ unit with (b) close-up of visual controls. Reproduced with permission from Storage Control Systems Ltd (www.storagecontrol.com).

Fig. 9.3. Safepod™ unit, containing apples, on a pallet ready for loading into DCA apple store. Reproduced with permission from Storage Control Systems Ltd (www.storagecontrol.com).

systems in the UK and about 150 in the USA (David Bishop, personal communication, 2018).

Advanced Control of Respiration

Another commercial DCA-RQ system is marketed as Advanced Control of Respiration (ACR) by Van Amerongen CA Technology BV (2018). The company has been producing equipment for CA storage of fruit and vegetables since 1969, with equipment installed in more than 50 countries throughout the world. At the time of writing, there are 70 rooms equipped with ACR technology in South Africa, though not all rooms are being used for low O_2 storage (R. Hurndall, H. Ackermann, personal communication), plus 550 worldwide (Rob Veltman, 2018, personal communication).

The following information is taken from an unpublished paper supplied by Rob Veltman of Van Amerongen. The development of ACR was initially guided by the need to control scald in some cultivars of pome fruit during long-term storage and to avoid fermentation. The DCA system was therefore designed to continuously check for fermentation and search for the lowest O_2 level possible throughout the entire storage period. It was considered vital to determine RQ values of the total cold store load, avoiding compartments with samples of fruit because these samples might not be representative of those in the cold store and there could be a microclimate effect in a container. The principle of ACR is based on 'ACR runs' during which the respiration rate of the fruit is determined by bringing the store under a slight over-pressure, and preventing any O_2 entering the store during the run. Then the partial pressure of O_2 and CO_2 is measured with an interval time of 30–60 min. Based on these measurements, the control system calculates the respiration rate in terms of O_2 uptake and CO_2 production, calculates the RQ and adjusts the O_2 set point automatically by using a cultivar specific algorithm (Figs 9.4 and 9.5).

Van Amerongen, in collaboration with with Wageningen University, showed that during Dynamic Control System (DCS) storage of 'Elstar' apples, O_2 partial pressures were adjusted based on the first traces of alcohol in the storage atmosphere. In containers with about 200 kg of apples this led to storage at O_2 partial pressures much lower

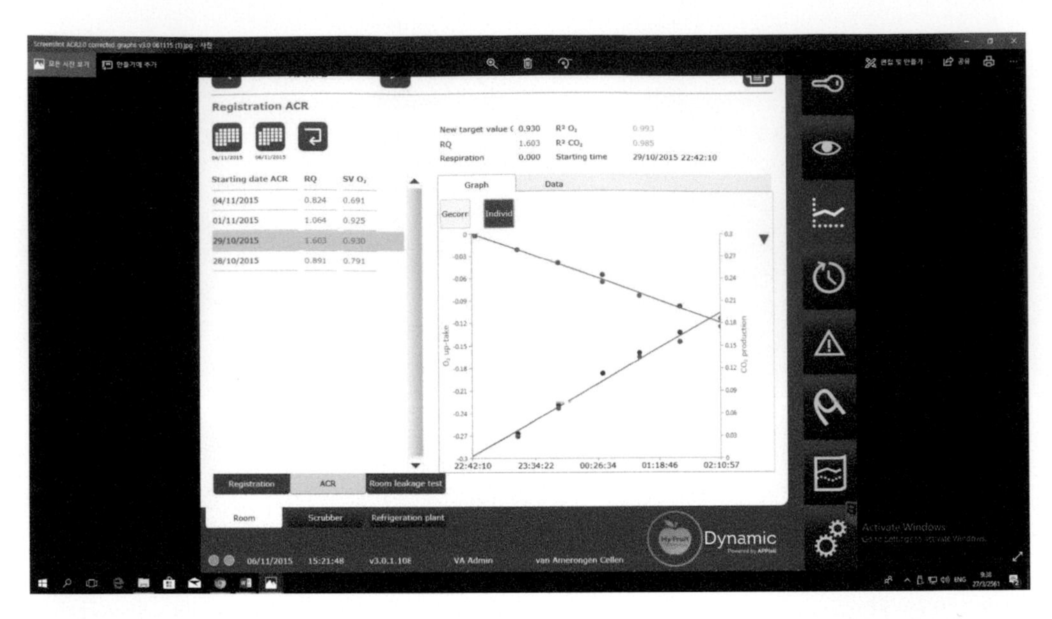

Fig. 9.4. Screen shot of ACR measuring CO_2 output, O_2 uptake and RQ calculation. Reproduced with permission from Rob Veltman, Managing Director, Van Amerongen BV (www.van-amerongen.com).

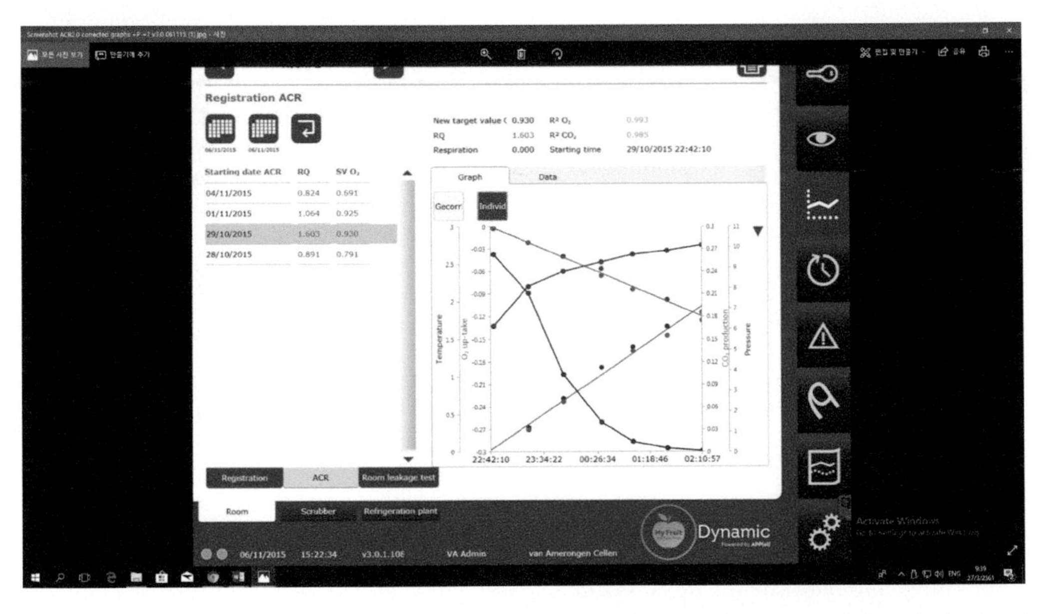

Fig. 9.5. Screen shot of ACR measuring CO_2 output, O_2 uptake, temperature pressure and RQ calculation. Reproduced with permission from Rob Veltman, Managing Director, Van Amerongen BV (www.van-amerongen.com).

than 1 kPa, even as low as 0.5 kPa. Scaling up DCS to cold stores of typically 150 t showed that ethanol in the storage air was not always a reliable marker for fermentation. Variation in, amongst others, growing location and harvest date appeared to cause re-metabolization of alcohol. Alcohol produced by fermenting apples seems to be consumed by apples that still respire normally. Re-metabolization is thought to cause a build-up of ethanol in a part of the apple population in a cold store, while ethanol in the storage air is zero or not detectable.

The next step in the trials was trying to avoid re-metabolization. In experiments a compartment was used in the storeroom with a selection of apples from the different sources. Ethanol measurements were carried out on this 20–40 kg sample. However, the compartment introduced several new issues. One of them was the question of which fruits should be selected for the measurements. Next, the compartment was warmer inside, which led to condensation and finally increased rotting. Rotting fruit also produce alcohol, and this made the interpretation of the measurements difficult. Finally, ethanol is not easy to detect. It requires special, expensive meters, but above all, ethanol sticks to tubing and it therefore may not be detected centrally as is the case for O_2 and CO_2. In 2006 there were seriously doubts about whether alcohol was the right practical indicator to detect product response. After successful experiments in several projects with Wageningen University (see Fig. 1.3) a new dynamic CA system was developed by Van Amerongen called ACR (Veltman, 2013). Up to 2015 the system was called Dynamic Control of Respiration (DCR) and the company was partnered, for a while, with Agrofresh. The name was changed to the preferred Advanced Control of Respiration (ACR). Some of the challenges that needed to be overcome in applying ACR were as follows.

- The cooling system had to be interrupted during an 'ACR run' (the period the respiration rate was determined).
- The same for the activity of all CA equipment, such as the CO_2 scrubber, the N_2 generator and aeration.
- A cold store should be gas-tight enough. To overcome slight leakages that would let in O_2 and disturb the measurement of O_2 uptake, the cold store was brought under a slight overpressure at the start of an ACR run by injecting N_2 (10 mm water column).
- The sensitivity of the O_2 and CO_2 meters could be an issue, and calibration of the meters needed to be reconsidered.

Experiments were successfully carried out at two local apple-packing houses. The lines representing the respiration rates in Fig. 9.6 had R^2 levels that were always higher than 97%. ACR has been patented (Veltman, 2013) and a newer version has been developed where ethylene production is also monitored.

Ethanol

Another approach to using product physiology to control the O_2 content in CA stores is the measurement of ethanol content in the store room, based on the assumption that the appearance of ethanol indicates the LOL and onset of anaerobic respiration (fermentation) in the stored crop. Measuring atmospheric ethanol is further supported by the observation that there is a stable ratio between the ethanol content in the tissue and the atmosphere around the crop (North and Cockburn, 1975). However, they warned that this stable ratio is suitable for stores equipped with lime CO_2 scrubbers but not for stores equipped with activated carbon scrubbers. North and Cockburn (1975) described detection for ethanol in the stores using indicator tubes called 'Alcotest' that contained crystals, which changed colour from yellow to green in the presence of alcohol. An approximate quantitative assessment of the alcohol level was achieved by measuring the quantity of store air required to achieve this colour change. For use in commercial CA stores, the alcohol detecting device could be connected to an alarm system so that when fermentation was just beginning the O_2 level could be slightly increased to stop fermentation levels, so that the cell tissues would return to aerobic respiration and the crop would not be damaged. For the apple cultivars studied by Fidler *et al.* (1973) the fruit could fully recover from alcohol contents of less than 40 mg/kg when subsequently aerated. The next report of O_2 control using ethanol was 20 years later by Schouten (1995), who concluded that a DCS for the O_2 content in CA storage rooms was possible, based on his results with 'Jonagold' apple and Brussels

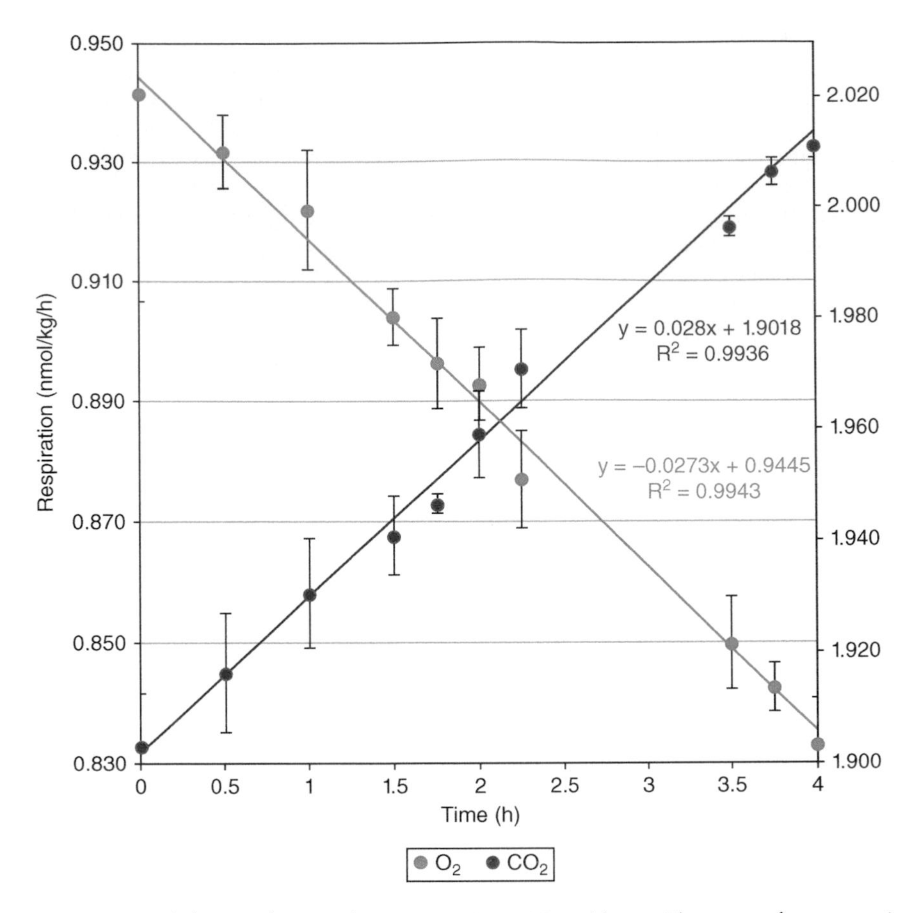

Fig. 9.6. Respiration of Elstar apples stored at 0.9% O_2 in a 125 t cold store. There were four measuring points that were averaged. There was minimal variation in the O_2 and CO_2 values at the four different measuring points. Reproduced with permission from Rob Veltman, Managing Director, Van Amerongen BV (www.van-amerongen.com).

sprouts. Schouten *et al.* (1997) confirmed that DCS storage of 'Elstar' apple, in which O_2 is adjusted to maintain ethanol below 1 ppm in the headspace (0.3–0.7 kPa O_2 + < 0.5 kPa CO_2), maintained fruit quality better than ULO-stored fruit (1.2 kPa O_2 + 2.5 kPa CO_2). See also the previous section, 'Advanced Control of Respiration', where some of the challenges in measuring ethanol levels in CA stores are discussed by Veltman.

A closely related approach, ILOS+, was introduced during the same period. Initial Low Oxygen Stress (ILOS) is a technique, first described by Eaves *et al.* (1969a, b), which uses short (up to 2 weeks) low O_2 stress treatments, i.e. typically but not always ≤ 0.5 kPa. ILOS+ is repeated applications of low O_2 stress at various times in the storage period with the addition of destructive ethanol tissue analysis used to determine when to stop treatment.

Dynamic Control System

DCS was developed and introduced by Food & Biobased Research at Wageningen Agricultural University in The Netherlands in 1997. It was first licensed to Van Amerongen (VA), which eventually abandoned the DCS method and switched to the ACR method. The DCS system carries out a pulp analysis for the presence of ethanol. When no ethanol is detected, the O_2 level in the CA room can be decreased, but when ethanol is detected the O_2 can be set to increase automatically. DCS uses small sample chambers integrated into the main store enabling low-level measurements of ethanol from the fruit or vegetable samples when the ACP is reached. Veltman *et al.* (2003) described DCS as 'the oxygen concentration is slowly lowered until the product emits a stress-signal. More specifically, DCS uses ethanol, the final product of fermentation, to establish the lowest possible oxygen concentration under which a product can be stored.' They described storage of 'Elstar' apples and showed that they were firmer after DCS storage, especially after shelf-life, with better colour retention and less skin spots compared with traditional CA storage. Paillart (2013) reported DCS being successfully used for long storage of white cabbage where the minimum O_2 concentration was adjusted by controlling the ethanol concentration within the storage facilities. He also claimed that the energy consumption of storage under DCS was lower than other ULO systems. DCS is marketed by Storex BV and is used in more than 200 storerooms (Storex, 2018), mainly for 'Elstar' and 'Jonagold' apples in The Netherlands (Prange *et al.*, 2013) plus at least 8000 t of stored apples (cultivars not specified) in other countries, e.g. Poland.

Initial Low Oxygen Stress

1-MCP, CA, DCA and ILOS have all been shown to have potential to inhibit superficial scald and increase the storage life of apples and pears (Robatscher *et al.*, 2012). ILOS is essentially a pre-storage treatment where fruit are exposed to anaerobic conditions; commonly 7 days in 100% N_2 (Eaves *et al.*, 1969a) or 0.5 kPa O_2 (Little *et al.*, 1982) who showed that the optimum for scald control in 'Granny Smith' apples was 0.5 kPa O_2 rather than 0 kPa O_2, for 9–14 days followed by CA storage. ILOS exposure of 'Granny Smith' to 0.5 kPa O_2 reduced scald compared with continuous storage at 1 kPa in Australia, but this did not positively affect 'Red Delicious' in the same trial. In another trial, ILOS (10 days at 0.5 kPa O_2) followed by CA (1 kPa O_2, 3 kPa CO_2 at 1°C) without DPA controlled scald on 'Granny Smith' and 'Red Delicious' apples in South Africa. ILOS also markedly reduced scald in 'Granny Smith', 'Rome' and 'Red Delicious' in Michigan trials (Kupferman 2001a). Sabban-Amin *et al.* (2011) also found that apples stored under ILO showed better post-storage quality, due to lower superficial scald, higher flesh firmness and greener skin colour. ILOS+ (ILOS with ethanol monitoring) was reported to be used on about 19,000 t of stored apples (cultivars not specified) in the South Tyrol region of Italy (Robert Prange, 2018, personal communication).

Chlorophyll Fluorescence

The above methods rely on measuring changes in ACP, RQ or ethanol in the atmosphere, or in tissue, that could signal that the product is at or near its LOL. The third system to determine the LOL is the use of chlorophyll fluorescence (CF), which does not need to measure ethanol or other fermentative products. Beginning in the 1980s, CF was being examined as method to detect various stresses (Prange, 1986; Prange *et al.*, 1990). Prange and research associates in his laboratory continued research into the 1990s on the effect of postharvest stresses on CF, culminating in the discovery in 1999–2000 of a sudden and reversible 'spike' in CF when O_2 is lowered below the product's apparent LOL. They concluded that it is a very sensitive, non-destructive method of dynamically controlling the O_2, and possibly the CO_2 environment, according to the unique requirements of each product.

Prange *et al.* (2003b) described the commercial version of this technology, which measures CF on a group of fruit or vegetables using a periodic low irradiance. It was first adopted commercially in the 2003–2004 storage season in Washington State, USA, and South Tyrol, Italy. This system is a pulse frequency modulated (PFM) proprietary technology that uses extrapolation to produce a theoretical estimate of the minimum fluorescence (F_0) parameter for which they coined a new term, F_α. With DCA, one can detect changes in the LOL and immediately alter the O_2 level in the storeroom. The LOL varies with the product, cultivar, season, growing regions and harvests, as well as time in storage (Table 9.1).

In the example in Table 9.1, three of the four apple cultivars had a drop in the LOL of about 0.4 kPa O_2 and in the fourth cultivar, 'Empire', there was no change in LOL. Subsequent research shows that this CF system can detect not only the LOL but other stresses experienced by stored product, e.g. CO_2, low temperature (chilling), 1-MCP application, and the presence of toxic ammonia, refrigeration gas and water loss (Prange *et al.*, 2010, 2012). These stresses, except for water loss, result in an immediate increase in F_α, similar to the increase caused by O_2 below the LOL.

When it was first introduced, the application of the CF method in DCA was described as dynamic low-O_2 controlled atmosphere (DLOCA) (DeLong *et al.*, 2004). Another term used by a few researchers was fluorescence CA (FCA) (Vanoli *et al.*, 2007). Other users have embraced neither of these terms. Although DCA did not initially refer solely to the DCA-CF method, many researchers and users referred to the two DCA-ethanol methodologies above as DCS and ILOS+, respectively, and shortened DCA-CF to just DCA (e.g. Zanella *et al.*, 2005, 2008; Raffo *et al.*, 2009; Gasser *et al.*, 2010; Streif *et al.*, 2010). Since 2012, in order to avoid confusion, especially with the RQ methods, Prange and Zanella began using DCA-CF to distinguish it from DCA-RQ.

Table 9.1. The effects of time in storage on the LOL (kPa O_2) detected by DCA-CF on four apple cultivars (Prange *et al.*, 2013).

Apple cultivar	10–19 October	1–4 December
Delicious	0.85	0.47
Golden Delicious	0.92	0.45
Honeycrisp	0.90	0.50
Empire	0.90	0.88

HarvestWatch™

The discovery of a sudden and reversible 'spike' in CF when O_2 was lowered below the product's apparent LOL resulted in a commercial system, using this discovery, being first presented at the ISHS CA symposium in Rotterdam, The Netherlands in 2001, and subsequently published (Prange *et al.*, 2002, 2003b). The patented DCA system is globally marketed as HarvestWatch™ and in Italy it is marketed by Isolcell S.p.A. as IsoStore® (Isolcell Italia, 2018) and used around the world, mainly in commercial apple (Prange *et al.*, 2010) and pear storage (Prange *et al.*, 2011) (Figs 9.7, 9.8 and 9.9). The current estimate of its usage at the end of 2017 was about 1979 DCA-CF rooms of > 20 cultivars of apples and pears in > 26 countries supported by research in > 19 research laboratories. The growth in DCA-CF was initially in Italy (Fig. 9.10) and after 2013 adoption outside Italy became more dominant, with a steady growth rate overall. Robert Prange (2018, personal communication) reported that DCA-CF accounts for about 85–90% of the DCA technology usage with the remaining 10–15% being ethanol and RQ-related technologies.

Fruit Observer™

A DCA-CF system was developed by Besseling Group BV and their partners in Germany and Brazil to monitor and control fruit during DCA storage. The system continuously measures CF, O_2, CO_2, temperature and relative humidity in CA rooms and shows a reaction of the chlorophyll when the LOL has been reached. The Fruit Observer™ software provides continuous information about the physiological condition of the observed fruit (Fig. 9.11). The initial screen of the software is split into two graphs showing: (1) actions – levels of O_2, CO_2, temperature and relative humidity; and (2) reactions – chlorophyll response. A description of the system was given in an article in a commercial fruit journal by Pullano (2017).

Fig. 9.7. A DCA-CF (HarvestWatch™) system consists of multiple FIRM sensors, each mounted within a kennel and connected using CAT 6/7 cables to a hub and then to a monitoring computer with HarvestWatch™ software.

Fig. 9.8. Typical arrangement of six DCA-CF (HarvestWatch™) FIRM kennels connected to one hub in a commercial CA room.

Comparison of Ethanol, RQ and CF Systems

Although it is not technically possible to compare concurrently the RQ, ethanol and CF commercial systems, one can simultaneously measure ethanol production, CO_2 production and CF and their accuracy in identifying the LOL. Hourly values of CF from 'Honeycrisp' and 'Delicious' apples, compared with CO_2 production (ACP), are presented in Fig. 9.12 and ethanol production in Fig. 9.13. They show that lowering O_2 had measurable effects on all three responses, i.e. ethanol production, CO_2 production and CF. The difference was mostly in the ease of interpretation. Chlorophyll response was easily interpreted, whereas the ethanol and CO_2 measurements were less precise, resulting in uncertainty in the LOL value, compared with CF. The CF LOL, which was the O_2 concentration at which the CF value did not return to the baseline, corresponds with the ACP (lowest CO_2 production) (Fig. 9.12).

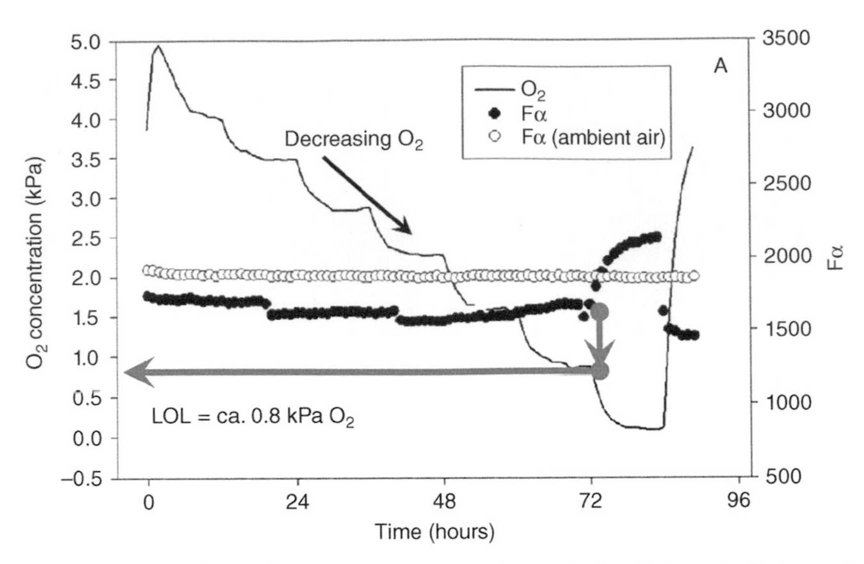

Fig. 9.9. Using HarvestWatch™ to determine LOL in an apple sample. The O_2 is gradually lowered and the chlorophyll fluorescence signal (Fα, solid dots) suddenly increases at ca. 72 h. The LOL is the O_2 value (0.8 kPa) when the increase (spike) begins. The O_2 is increased above the LOL at 84 h and the Fα signal immediately returns to its baseline.

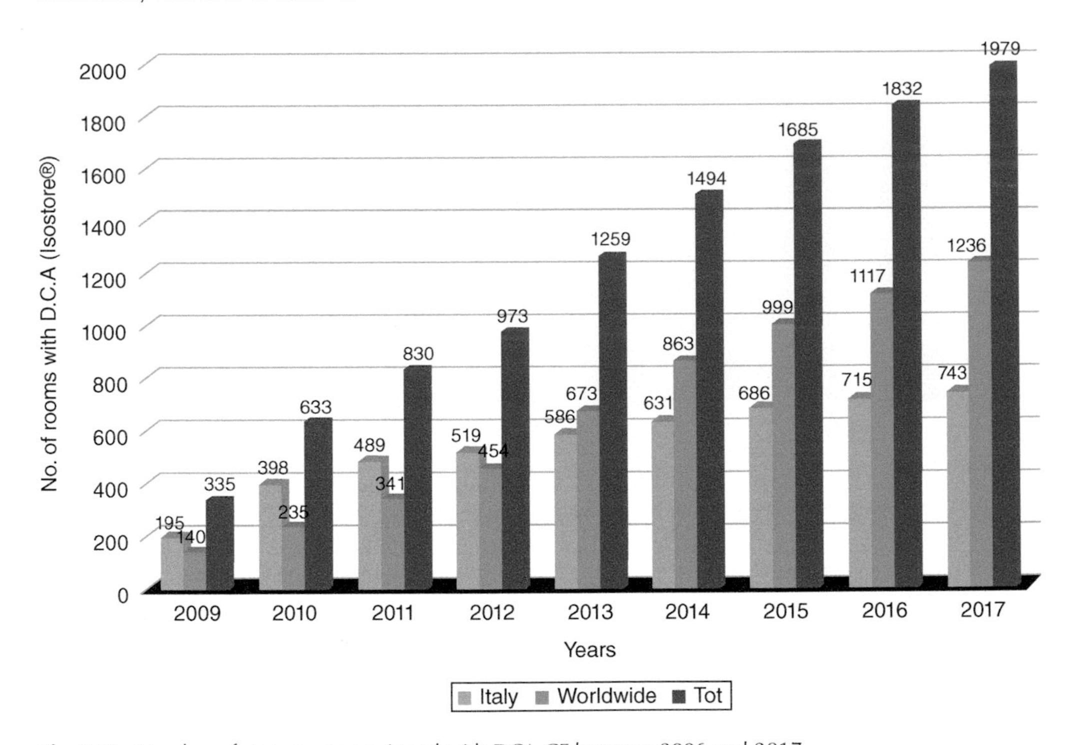

Fig. 9.10. Number of storerooms equipped with DCA-CF between 2006 and 2017.

Furthermore, the CF returns to the baseline after anoxia is terminated, whereas the CO_2 production does not return to its normoxia values and is highly variable. Ethanol production (Fig. 9.13) was even more difficult to interpret. 'Honeycrisp' ethanol production occurs in normoxia and after anoxia, ethanol production continues in normoxia.

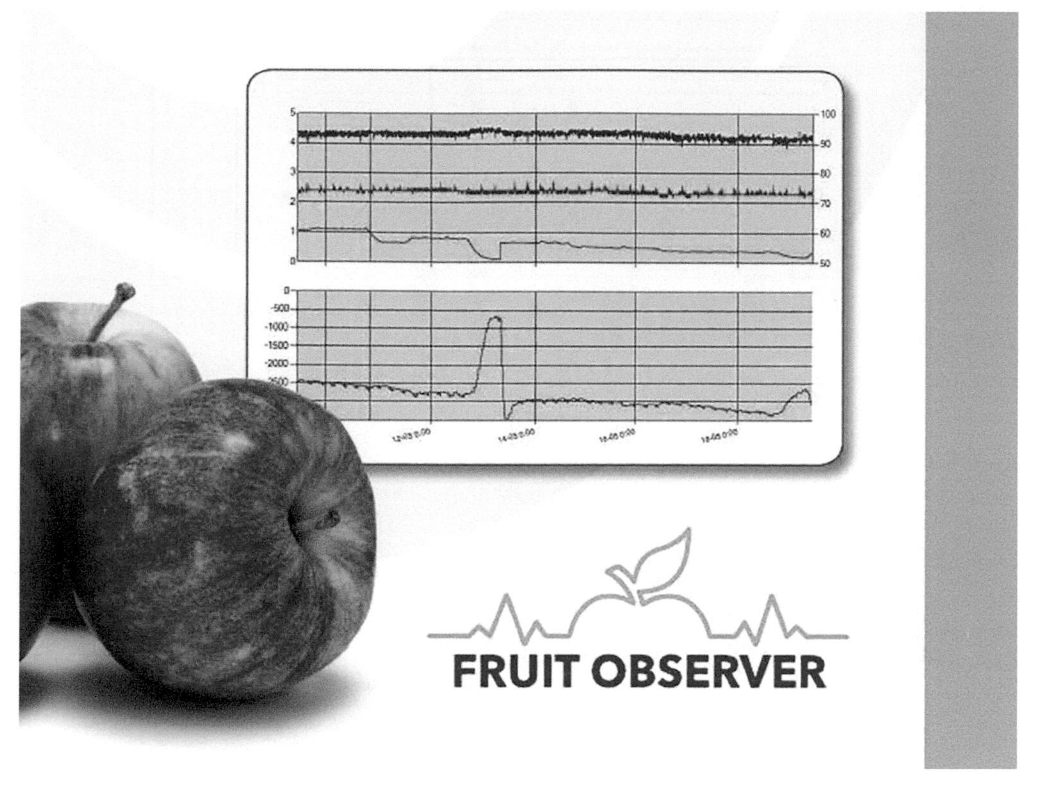

Fig. 9.11. Fruit Observer™. The upper graph shows which parameter triggered the reaction of chlorophyll, which is simultaneously displayed in the lower graph. The upper graph shows the actions: blue is humidity; red is temperature; and purple is O_2 level. The lower graph shows the reaction on the actions. The O_2 is too low, which caused a peak. The O_2 was then slowly lowered and again there was a peak at exactly the same LOL point. This shows that the chlorophyll recovers as long as the O_2 is set at the right level. Reproduced with permission from Eric van der Zwet of Besseling Group BV, The Netherlands.

Weber *et al.* (2015) compared static CA with DCA-RQ 2 and DCA-CF on 'Royal Gala' apples. Flesh firmness, ethylene production and respiration rates were lower in apples stored in DCA-RQ and DCA-CF than in CA. They concluded that storage in DCA-RQ 2 at 1°C or DCA-CF at 0.5°C maintained fruit quality during 8 months of storage. In a comparison between CA, DCA-CF and DCA-RQ on 'Fuji Suprema' apples, Thewes *et al.* (2017a) found that storage under both DCA methods had lower ethylene production and retained higher flesh firmness compared with those under more conventional CA. The fruit stored under CA had the highest butyl acetate, 2-methylbutyl acetate and hexyl acetate. DCA-RQ resulted in lower decay incidence when the atmosphere was established immediately, compared with a 30-day delay in establishment. However, the apples stored in DCA-RQ had the highest total ester concentration and the volatile compounds that are characteristic to 'Fuji' apples (ethyl 2-methyl butanoate, ethyl butanoate and ethyl hexanoate). The storage under DCA-CF resulted in the lowest production of volatile compounds. Thewes *et al.* (2017b) stored 'Galaxy' apples in CA at 1.2 kPa O_2 + 2.0 kPa CO_2, DCA-RQ 1.3 + 1.2 kPa CO_2 or DCA-RQ 1.5 + 1.2 kPa CO_2 for 9 months plus 7 days of shelf-life at 20°C. DCA-RQ suppressed internal ethylene concentration, ethylene production and respiration rate and the highest ester concentration occurred in overripe fruit stored at DCA-RQ 1.3 and DCA-RQ 1.5, while those stored in DCA-RQ 1.5 had higher concentrations of total esters and characteristic aroma volatile

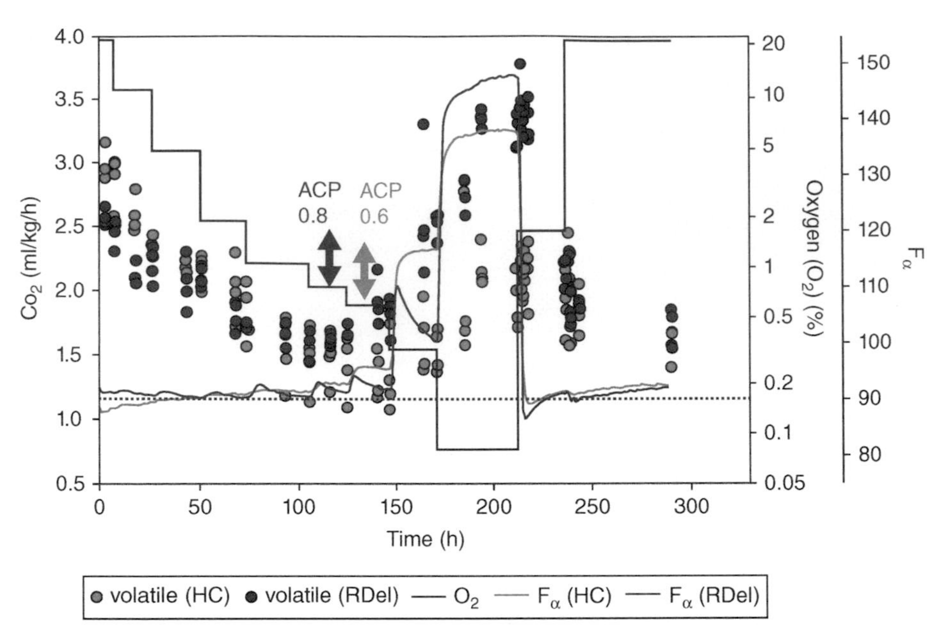

Fig. 9.12. Effect of O_2 reduction, prolonged anoxia and return to air on CO_2 production at 3.5°C in six 'Delicious' and six 'Honeycrisp' apples. RDel = 'Delicious' (Red); HC = 'Honeycrisp' (Green); ACP = anaerobic compensation point = LOL; dotted line = chlorophyll fluorescence baseline.

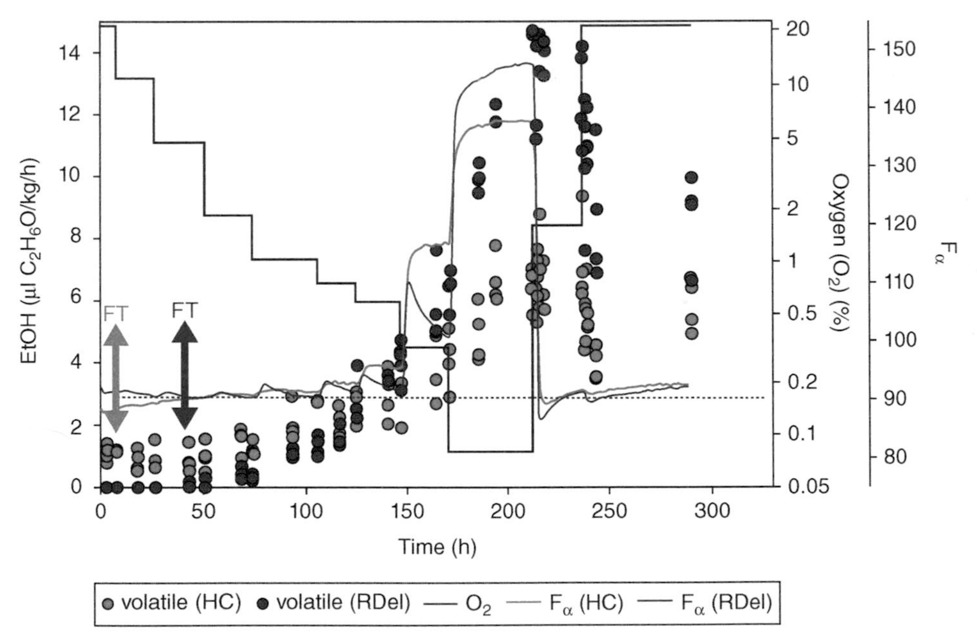

Fig. 9.13. Effect of O_2 reduction, prolonged anoxia and return to air on ethanol (EtOH) production at 3.5°C in six 'Delicious' and six 'Honeycrisp' apples. RDel = 'Delicious' (Red); HC = 'Honeycrisp' (Green); FT = Fermentation Threshold; dotted line = chlorophyll fluorescence baseline.

compounds. DCA storage reduced ethylene production and respiration rate of 'Royal Gala' apples stored for 9 months at 1.0°C plus 7 days at 20°C, resulting in higher flesh firmness, TA and less physiological disorders. Storage in static CA, DCA-RQ 1.5 and DCA-RQ 2.0 had higher levels of butyl acetate, 2-methylbutyl acetate and hexyl acetate compared with fruit stored in DCA-CF. Fruit stored in DCA-RQ 1.5 and DCA-RQ 2.0 had higher increments of ethanol and ethyl acetate, but far below the odour threshold (Both *et al.*, 2017). Patulin is a mycotoxin produced postharvest by some fungi infecting apples (see Chapter 8). Dos Santos *et al.* (2018) tested whether DCA storage could reduce patulin production in apples. They stored 'Galaxy' at 2.0 ± 0.1°C and 'Fuji Kiku' at 0.5 ± 0.1°C for 9 months plus 7 days shelf-life at 20°C in CA (1.0 kPa O_2 + 0.8 kPa CO_2), DCA-CF + 1.2 kPa CO_2 or DCA-RQ 1.3 + 1.2 kPa CO_2. They found that none of these storage conditions prevented patulin production.

Current Situation

Most of the published DCA research has been conducted on apples and pears (Table 9.2) but there are research and commercial trial reports on a variety of other fruit and vegetables. The first two regions to adopt commercial DCA-CF were South Tyrol (Italy) and Washington State (USA) in 2003, after a trial season. In the South Tyrol, 88% of the apple crop was stored in CA in 2016–2017 (Fig. 9.14). About 21.8% of this amount was stored in DCA: 19.4% in DCA-CF + 2.4% in ILOS+, a DCA-ethanol technology. In Washington State, Stemilt Growers first used DCA-CF commercially in 2003 on organic 'Delicious' apples and there has been a steady increase in DCA since then. In a survey in November 2017 that covered about 50% of the apple storage in Washington State (Table 9.3), 88% of the apples were in CA, which is similar to the South Tyrol, but only 3% were in DCA, compared with 21.8% in the South Tyrol. However, the planned construction of new storages will be 29% DCA.

Even though ethanol and RQ methods were known before 2001, significant commercial adoption of DCA only began with the availability of DCA-CF in 2001. The explanation is a combination of factors. Firstly, DCA-CF offered a package of technical features that were appealing (Prange *et al.*, 2013):

- accurate and fast detection of the internal flesh ACP (= LOL);
- durable and reliable (never need to calibrate or replace components);
- non-destructive repeat measurements on large surface areas;
- rapid and frequent measurement;
- non-chemical;
- real-time monitoring of produce allowing for on-site or remote monitoring and archiving of data for future reference; and
- detection of changes due to senescence, decay or incorrect storage conditions, i.e. temperature, unwanted toxic gases such as ammonia refrigerant.

Secondly, independent research and commercial trials around the world showed that using DCA-CF to hold O_2 to just above the ACP not only preserved quality in apples and pears, it also controlled several disorders, especially superficial scald. This bolstered the commercial adoption, especially in South Tyrol, and in 2010–2011 the South Tyrol industry decided to rely solely on DCA-CF and/or 1-MCP for superficial scald control rather than the postharvest chemical DPA (Zanella and Stürz, 2013). The third factor occurred in 2012 when the EU decided that DPA and ethoxyquin were no longer acceptable in the EU. This meant no apples or pears stored in the EU or imported into the EU could be treated with these two chemicals. The significant increase in demand for organic fruit has also been a major factor, especially in Washington State where organic apple production is approaching 20%. DCA is an acceptable postharvest technology for organic fruit, but 1-MCP, DPA and ethoxyquin are not acceptable.

Table 9.2. List of some publications providing information on DCA use in fruits and vegetables.

Commodity	Cultivar	Publications
Fruits		
Apple	Braeburn	Gasser *et al.*, 2008; Lafer, 2008; Lafer, 2009; Poldervaart, 2010a; Prange *et al.*, 2012; Withnall, 2008
	Cortland	DeLong *et al.*, 2004, 2007; Prange *et al.*, 2012; Wright *et al.*, 2012
	Cripp's Pink (Pink Lady)	Withnall, 2008; Prange *et al.*, 2012
	Delicious (Red)	DeLong *et al.*, 2004, 2007; Lafer, 2009; Prange *et al.*, 2012; Stephens and Tanner, 2005; Withnall, 2008; Wright *et al.*, 2012; Zanella and Stürz, 2013
	Elstar	Gasser *et al.*, 2008; Köpcke, 2009; Poldervaart, 2010a; Prange *et al.*, 2012
	Fuji	Withnall, 2008; Zanella *et al.*, 2008
	Gala	Withnall, 2008; Zanella *et al.*, 2008
	Golden Delicious	DeLong *et al.*, 2004; Gasser *et al.*, 2008; Poldervaart, 2010a; Prange *et al.*, 2012; Withnall, 2008; Zanella *et al.*, 2008
	Honeycrisp	DeLong *et al.*, 2004; Prange *et al.*, 2010; 2012; Wright *et al.*, 2008, 2010, 2011, 2012
	Granny Smith	Lafer, 2009; Prange *et al.*, 2012; Wright *et al.*, 2012; Withnall, 2008; Zanella *et al.*, 2005
	Idared	Poldervaart, 2010a; Prange *et al.*, 2012
	Jonagold	DeLong *et al.*, 2004; Prange *et al.*, 2012
	Maigold	Gasser *et al.*, 2008; Poldervaart, 2010a
	McIntosh	DeLong *et al.*, 2004; Prange *et al.*, 2002; 2003
	Northern Spy	Prange *et al.*, 2005
	Pinova (Piñata)	Raffo *et al.*, 2009
	Topaz	Lafer, 2009; Poldervaart, 2010a,b
Avocado	–	Prange *et al.*, 2003; Yearsley *et al.*, 2003
	Hass	Burdon, 2009; Burdon and Lallu, 2008; Burdon *et al.*, 2008; 2010; Prange *et al.*, 2002; Yearsley *et al.*, 2003
Banana	–	Prange *et al.*, 2002; 2003
Kiwifruit	–	Prange *et al.*, 2002; 2003
	Hayward	Lallu and Burdon, 2007
Mango	–	Prange *et al.*, 2002; 2003
Pear	–	Prange *et al.*, 2003
	Abbé Fétel (Abaté Fétel)	Rizzolo *et al.*, 2008; Vanoli *et al.*, 2007, 2010a,b; Zerbini and Grassi, 2010
	Bartlett (Williams Bon Chrétien)	Mattheis, 2007; Prange *et al.*, 2002, 2011; 2012
	Conference	Zerbini and Grassi, 2010
	d'Anjou	Mattheis, 2007; Mattheis and Ruddell, 2011; Prange *et al.*, 2011; 2012
	Forelle	Prange *et al.*, 2011
	Packham's Triumph	Prange *et al.*, 2011, 2012
	Spadona	Prange *et al.*, 2011
	Uta	Lafer, 2011
Vegetables		
Asparagus	–	Prange *et al.*, 2005
Broccoli	–	Prange *et al.*, 2005
Cabbage	–	Prange *et al.*, 2003
Pepper (green)	–	Prange *et al.*, 2003
Lettuce	Iceberg	Prange *et al.*, 2003, 2005
	Romaine	Prange *et al.*, 2003
Spinach	–	Prange *et al.*, 2012; Wright *et al.*, 2011, 2012

Future Trends

Two benefits of DCA that were not initially anticipated but are becoming more evident are energy conservation and flavour enhancement. DCA results in a substantial reduction in fruit respiration, with the result that higher storage temperatures are possible, thus saving energy (J. Streif, as quoted in Poldervaart, 2010a). In New Zealand, experiments have been conducted using a combination of DCA and high temperature storage to achieve energy savings during storage, and/or avoid chilling-related storage disorders in 'Royal Gala' and 'Pink Lady' apples (N. Lallu, personal communication). Preliminary results show that, at 5°C, storage savings of 35% were possible during cooling and 15% during storage. In addition, flesh browning in 'Pink Lady' could be avoided by storage at 3°C in DCA. Energy saving and quality benefits have also been reported by Kitteman *et al.* (2015a) and Neuwald *et al.* (2016) on 'Golden Delicious', 'Jonagold' and 'Pinova' apples in Germany. In yellow-fleshed kiwifruit, chilling injury was avoided by storage at 7°C in DCA whilst maintaining the same firmness as fruit stored at 0°C in air (N. Lallu, personal communication). Thus, with more research

Table 9.3. Current apple storage and plans for the future for apples in Washington State apple industry (adapted from R. Blakey, 2018, personal communication).

	Current (2017)		Plans for next 2 years	
Storage type	Bins	%	Bins	%
Refrigerated Air (RA)	450,000	12	0	0
Controlled Atmosphere (CA)	3.2 million	85	332,000	71
Dynamic CA (DCA)	101,000	3	133,000	29

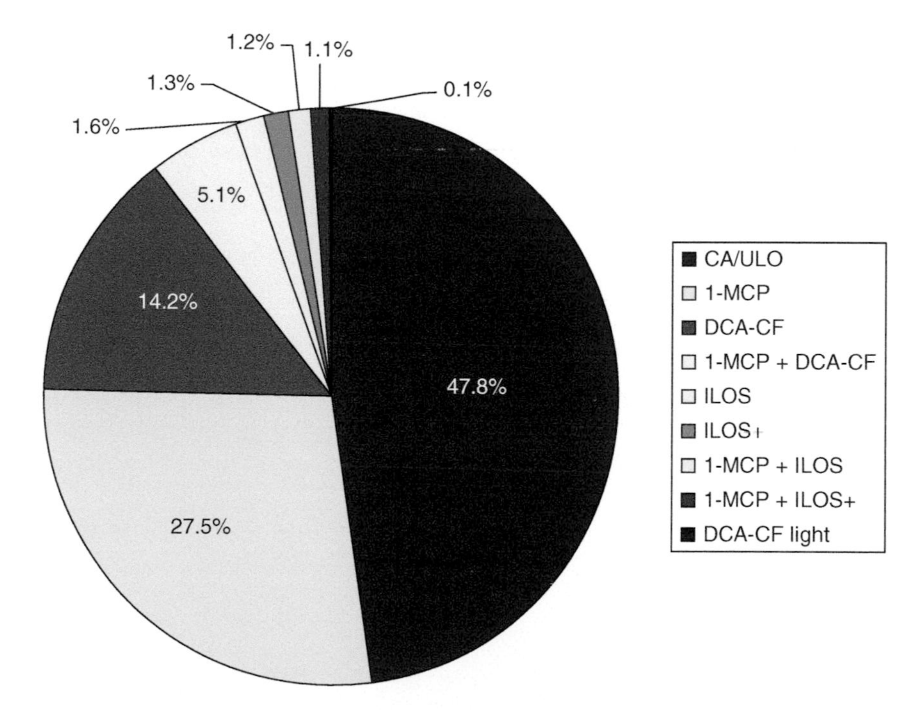

Fig. 9.14. CA storage technologies used in the apple industry in South Tyrol, Italy, in 2016–2017. Total CA storage = 764,315 t. An additional 100,000 t were stored in Regular Atmosphere (RA). Initial Low Oxygen Stress (ILOS) technology is not a DCA technology since it does not use fruit physiology to control the CA conditions. DCA-CF light uses fewer sensors per store and a higher O_2 safety margin. Total DCA-CF (DCA-CF, 1-MCP + DCA-CF, DCA-CF light) = 19.2%. Total ILOS+ (ILOS+, 1-MCP + ILOS+) = 2.4%. Adapted from Zanella (personal communication).

and commercial trials, it is likely that the use of DCA will allow the use of warmer storage temperatures.

After its introduction, commercial users of DCA have reported anecdotally that DCA enhances flavour of apple cultivars such as 'Pink Lady' and 'Granny Smith', compared with other storage technologies being used. These reports were confirmed by Zanella *et al.* (2005) and Raffo *et al.* (2009), who concluded that DCA storage technology, besides avoiding any chemical treatment, can preserve apple flavour/aroma compounds better than 1-MCP + ULO during long-term storage. Raffo *et al.* (2009) observed that DCA favoured the production of branched-chain esters over straight-chain esters. In the case of apples, the flavour enhancement is adding demand and value to the product in the market place. This is an unforeseen benefit of DCA that may extend beyond apple cultivars and suggests that DCA may be able to alter flavours and aromas in a predictable way and/or increase flavour-life of the product.

10

Hyperbaric and High-oxygen Storage

Hyperbaric storage was recently reviewed in more detail by Thompson (2015) and much of the information contained here has been updated from that book. Increasing the pressure (hyperbaric conditions) in a sealed room or cabinet can increase the partial pressure of O_2. Hyperbaric conditions have been used in therapy to control human medical conditions, usually in a specially designed pressure chamber. Hyperbaric O_2 therapy enables the patient to breathe pure O_2 while under increased atmospheric pressure. Its use has included the treatment of decompression sickness, gas gangrene and carbon monoxide poisoning as well as being tested for the control of cerebral palsy and multiple sclerosis (Mathieu, 2006).

High pressures are used in food processing and were perhaps first proposed by Royer in 1895 to kill bacteria (Naik *et al.*, 2013). Ahmed and Ramaswamy (2006) and Baba and Ikeda (2003) stated that hyperbaric treatment of fresh fruit and vegetables is different from high-pressure treatment used in processing foods where O_2 pressures of between 400 and 1200 MPa are used. Goyette *et al.* (2011) defined hyperbaric storage as 'exposing fruit or vegetable to compressed air in a range lower than 10 atmospheres'. Hite *et al.* (1914) reported that exposure of fruit or vegetables to 680 MPa for 10 min at room temperature gave a 5–6 log cycle reduction in microorganisms. The first reports of commercial hyperbaric processing were probably in the early 1990s in Japan for acid foods to be stored in chilled conditions. Such high pressures for fresh fruit and vegetables can cause irreversible damage to cell structure. Goyette *et al.* (2007) stated that it may be possible to use much lower hyperbaric pressures for fruit and vegetables than for processed products and pressures greater than 100 MPa may be above the threshold for irreversible tissue damage, thus causing substantial injuries (Goyette *et al.*, 2012). High-pressure inactivation of yeast and moulds has also been reported, for example in citrus juices that had been exposed to 400 MPa for 10 min at 40°C. Juice treated in this way did not spoil during storage for up to 3 months (Olsson, 1995). Hyperbaric conditions have also been used successfully in increasing the storage time of meat and eggs at room temperature for several days where compressed air, at 15 to 44 atmospheres (atm), was used. It was reported that bacteria, yeasts and moulds were killed in foods by this treatment (Bert, 1878 quoted by Kader and Ben-Yehoshua, 2000). Charm *et al.* (1977) reported that at high pressure the storage life of Atlantic cod fish was greatly extended compared with storage at atmospheric pressure. After 30 days the fish held at –3°C and 238 atm had significantly lower bacterial counts and a higher sensory evaluation than those stored −25°C under atmospheric pressure. Moreira *et al.* (2015) compared storage of soup at different pressures and temperatures. They concluded that 100 MPa compared with 150 MPa and 4 h exposure compared with 8 h resulted in more pronounced microbial growth inhibition and microbial inactivation. Aerobic mesophiles showed less susceptibility to hyperbaric storage compared with Enterobacteriaceae and yeast and moulds. Hyperbaric storage at 25°C or 30°C generally maintained the physicochemical parameters at levels similar to refrigeration at 4°C under normal atmospheric pressures. High pressure was reported to have potential as a treatment to control quarantine insect pests in fresh or minimally processed fruit and vegetables (Butz *et al.*, 2004).

A novel way of increasing pressure is storage in sealed containers underwater. Storage of cereals underwater at a depth of 30 m maintained a constant low temperature as well as increased pressure and successful experiments were carried out in Japan using this technique (Mitsuda *et al.*, 1972). Saraiva (2014) observed that food that had been recovered from a sunken submarine in the 1990s was still in a consumable condition. The submarine had been sunk 10 months earlier to a depth of 1540 m where the pressure was about 15 MPa and the temperature 3–4°C.

Hyperbaric conditions, even at variable room temperatures of up to 37°C, have been shown to preserve foods and thus achieve significant energy savings (Fernandes *et al.*, 2015). Hyperbaric storage, at room temperature, could be more energy efficient than refrigeration, since the only energy costs are during compression and no additional energy is required to subsequently maintain the product under pressure. Liplap *et al.* (2012) showed that hyperbaric storage of avocados in ambient conditions used some 3% of the energy required for commercial refrigerated storage at 5°C but could have similar effects in delaying ripening. The capital costs of high-pressure equipment are high (Saraiva, 2014), but Balasubramaniam *et al.* (2008) commented that the expansion of high-pressure food processing should help to reduce equipment costs by increasing the scale.

Technology

Hyperbaric tanks and chambers are used mainly in the medical and diving industries. As indicated above, they have also been used to improve the postharvest life of some fruit and vegetables. For example, Robitaille and Badenhop (1981) described a completely autonomous storage system with CO_2 removal and automatic O_2 replenishment for hyperbaric storage that was successfully used to store mushrooms. Goyette (2010) described the small stainless steel test chamber he used as consisting of an outor chamber and inner chamber. The purpose of the outer chamber was to resist pressures up to 15 atm. The inlet of the outer chamber was connected to a cylinder of compressed air with a manometer to regulate the pressure. The airtight vessel could be pressurized up to 7 atm and instrumented to automatically monitor gas concentration using computer controlled valves, flow meter and CO_2 gas analyser. Liplap *et al.* (2014c) developed a method to determine the metabolic respiration rate of fruit or vegetables during the transient period at the beginning of a hyperbaric treatment for the correction in the apparent respiration rate by considering the dilution effect of flushing the system and the error associated with gas solubilization as the gas partial pressure varied. They simulated the dilution process by using the general equation for exhaust ventilation, thus allowing for the elimination of the dilution effect during the calculation of the net respiration rate. See also Chapter 13 for a description of a hyperbaric container.

Effects of High Pressure

As indicated above, the effect of high-pressure processing is mainly to control microorganisms but there is little or no detrimental effect on flavour, as can happen in thermal processing (Hill, 1997). However, Hill quoted work where some fruit juices were affected by high-pressure processing; for example, in grapefruit juice, which had good retention of vitamin C and many consumers found the juice more acceptable after high-pressure processing because it was less bitter. Small molecules, which were reported to be the characteristics of flavouring, and nutritional components typically remained unchanged when fruit or vegetables were exposed to high pressure (Horie *et al.*, 1991). Ludikhuyze *et al.* (2002) and Yordanov and Angelova (2010) reviewed the effects of high pressures on fruit and vegetable processing and reported that vegetative cells were inactivated at about 300 MPa at ambient temperature, while spore inactivation required 600 MPa or more in combination with a temperature of 60–70°C. Previously, Larson *et al.* (1918, quoted by Ludikhuyze *et al.*, 2002) had observed that pressures up to 1800 MPa at room temperature were not sufficient to achieve commercial sterility of food products. A combination of pressure with temperatures of 60°C and higher was required for extensive inactivation of spores, and the lower the pressure applied, the higher was the temperature required to achieve inactivation (Sale *et al.*, 1970). At temperatures below 60°C in combination with a pressure of about 400 MPa, there was a maximal three log-cycle reduction of spores of *Bacillus coagulans* (Roberts and Hoover, 1996) and *Clostridium sporogenes* (Mills *et al.*, 1998). Generally, it was found that Gram-positive bacteria were more resistant to pressure than Gram-negative bacteria, while fungi, yeasts and *Enterococcus hirae* were more resistant to high pressure than *Listeria monocytogenes* and *Aeromonas hydrophila*. The most resistant to high pressure were bacterial spores (Fonberg-Broczek *et al.*, 2005).

In addition to its effect on bacteria, it was noted that high-pressure treatment helped to preserve colour in fruit and vegetable products, including tomato juice (Poretta *et al.*, 1995), orange juice (Ferrari and Di Matteo, 1996), guava purée (Yen and Lin, 1996) and avocado purée (Lopez-Malo *et al.*, 1998). The decrease in vitamin C content in strawberry purée and guava purée during storage after pressure treatment of 400–600 MPa for 15–30 min was much lower than loss of vitamin C content in non-pressure-treated purée (Sancho *et al.*, 1999). Luscher *et al.* (2005) used 250 MPa during the freeze processing at –27°C of potatoes and found that after thawing there was a considerable improvement in texture, colour and visual appearance compared with freezing at atmospheric pressure.

There are various effects of high O_2 storage on the chemical changes and enzyme activity of some fruit and vegetables, which would be expected to be reproduced by hyperbaric conditions. The effects of hyperbaric storage appear to be mainly on cell metabolism of the fruit or vegetables and on microorganisms that infect them. Under hyperbaric pressures, a large change in the respiration rate was observed immediately after the pressure was applied and its amplitude decreased during the initial period of the hyperbaric treatment, which was described as an unsteady or transient state (Goyette *et al.*, 2012). Eggleston and Tanner (2005) found that at pressure of 600 MPa the respiration rate of carrot sticks decreased and this effect was greater the longer they were exposed over periods of 2–10 min. Liplap *et al.* (2012, 2013c, 2014a) described the effects of hyperbaric conditions on the respiration rate of lettuce, maize and avocados. From these studies, it can be concluded that the method used to measure respiration rate was limited, as it only allowed the respiration rate to be determined after the system reached equilibrium, which might require several hours. It was also proposed that gas dilution, solubilization and/or de-solubilization processes could take place under hyperbaric pressure treatment. It is these processes that may be responsible for the apparent large changes in the respiration rate of the produce exposed to hyperbaric conditions. Liplap *et al.* (2013b) found that the trend in antioxidant activity observed from both O_2 radical absorbance capacity and trolox equivalent antioxidant capacity assays was generally similar. In storage of tomatoes at 13°C and 20°C there was no significant effect observed of hyperbaric exposure on lipophilic antioxidant and hydrophilic antioxidant compared with tomatoes under atmospheric pressure.

Liplap *et al.* (2014b) showed that bacterial growth was affected by hyperbaric pressure at 20°C under 100, 200, 400, 625 and 850 kPa. As hyperbaric pressure increased, the bacterial growth significantly decreased, but the effect varied with species. The maximum growth at 850 kPa was reduced by 56% for *Pseudomonas cichorii*, by 43% for *Pectobacterium carotovorum* and by 71% for *Pseudomonas marginalis*. In a review, San Martín *et al.* (2002) reported inhibitory effects of hyperbaric pressure on the growth of several microorganisms but very high pressure was required to kill or inactivate their growth. Liplap *et al.* (2014a) exposed lettuce at 20°C to pressures ranging from 100 to 850 kPa for 5 days and found that the development of decay was delayed under hyperbaric pressure, especially at 850 kPa, in comparison with those under normal atmospheric pressure. They considered the mode of action of hyperbaric pressure on the growth of microorganisms could be due to the direct impact of elevated pressure itself on the microorganisms, as had been demonstrated on infections in strawberry juice (Segovia-Bravo *et al.*, 2012). Another explanation could be that the elevated O_2 caused toxicity to bacteria, yeasts and moulds or the enhancement of the host pathogen defence compound synthesis, induced by mild stress. This latter effect of stress has been demonstrated for tomatoes by Lu *et al.* (2010). High CO_2 can also affect bacterial growth and Enfors and Molin (1980) found that the growth of *Bacillus cereus* was completely inhibited at 3 atm of CO_2 and *Streptococcus cremoris* at 11 atm of CO_2. Vigneault *et al.* (2012) commented that exposure to hyperbaric pressures could be an alternative to chemical treatment for preserving postharvest quality of fruit and vegetables.

In addition to the control of microorganisms and some reductions in susceptibility to pathogens (Baba *et al.*, 1999; Romanazzi *et al.*, 2008) the apparent effects of hyperbaric conditions on fresh fruit and vegetables, as well as decreased respiration rate, include decreased ethylene production, slowing of the ripening processes and the possible extension of the synthesis of certain chemicals (Baba and Ikeda, 2003; Eggleston and Tanner, 2005; Goyette *et al.*, 2012). Other effects of hyperbaric exposure described by Baba and Ikeda (2003) and Goyette *et al.* (2012) include reduced weight loss, maintenance of peel colour and TSS:TA ratio, improved lycopene synthesis and some evidence of reduction in chilling injury in tomatoes. There was some indication that hyperbaric storage could improve the retention of flavour in stored fruit compared with storage under atmospheric pressure. Yang *et al.* (2009) found that the composition of volatile compounds produced by peaches was higher after storage under hyperbaric pressure compared with CA storage or storage in air at atmospheric pressure.

Hendrickx *et al.* (1998) reported that exposure to pressures of 100 MPa or higher can induce structural rearrangements in enzymes, which can cause their activation or their partial or total inactivation in a reversible or irreversible manner. The specific effect of pressure depends on several factors, including the structure of the enzyme, its origin, the medium composition, pH, or the temperature and the pressure levels applied. For example, pectin methylesterase (PME) from peppers, tomatoes, white grapefruit, plums and carrots has been shown to be resistant to exposure to high pressure, since pressures higher than 700 MPa are usually required to induce short-term inactivation at room temperature (Segovia-Bravo *et al.*, 2012). With tomatoes, polygalacturonase (PG) is much more pressure labile than PME and almost complete PG inactivation was shown to occur in cherry tomatoes at 500 MPa in ambient temperature (Tangwongchai *et al.*, 2000). Verlent *et al.* (2004) found that the optimal temperature for tomato PME activity at atmospheric pressure was about 45°C at pH 8.0 and about 35°C at pH 4.4, but at both pH levels the optimal temperature increased as pressure was increased over the range of 0.1 to 600 MPa. Also at both pH levels the catalytic activity of tomato PME was higher at elevated pressure than at atmospheric pressure.

Oxidation

One effect of hyperbaric conditions is to increase the partial pressure of O_2 above that found in the air at normal atmospheric pressure. It has been shown by many workers that high O_2 levels in storage can affect various postharvest processes in whole fresh fruit and vegetables as well as those that have been minimally processed. As well as oxidative processes, the effects of the increased O_2 levels include respiration rate, ethylene production, volatile compounds, chlorophyll degradation, softening, pigments, nutrient content, sprouting, free radicals, diseases and physiological disorders. González-Roncero and Day (1998) reported that high O_2 MAP had beneficial effects on the degree of lipid oxidation. Shredded lettuce in MAP stored at 5°C for 10 days showed browning, but those in 80 kPa O_2 + 20 kPa CO_2 did not (Heimdal *et al.*, 1995). Sliced apples stored at 1°C for 2 weeks had less browning in 100 kPa O_2 than in air (Lu and Toivonen, 2000). Jiang (2013) coated button mushrooms (*Agaricus bisporus*) with 2% alginate and stored them in jars continuously ventilated with 100 kPa O_2 at 4°C for up to 16 days and found that this treatment delayed browning. Since the browning reaction is oxidative, it is counter-intuitive that levels should be lower in high O_2. Day (1996) reported that there were highly positive effects of storing minimally processed fruit and vegetables in 70 and 80 kPa O_2 in plastic film bags. He indicated that it inhibited undesirable fermentation reactions and delayed enzymic browning and that the O_2 levels of > 65 kPa inhibited both aerobic and anaerobic microbial growth. He also showed that the cut-surface browning of slices of apples was inhibited during storage in 100 kPa O_2 compared with those that had been stored in air. He explained this effect by suggesting that the high O_2 levels may cause substrate inhibition of PPO or that, alternatively, high levels of colourless quinones formed may cause feedback inhibition of PPO.

Respiration rate

Kidd and West (1934) found that storage of apples in 100 kPa O_2 accelerated the onset of the climacteric rise in respiration rate and Biale and Young (1947) reported a similar increase in respiration rate in lemons exposed to 34.1, 67.5 or 99.2 kPa O_2. With avocados, Biale (1946) found that there was only a small acceleration in the time of the onset of the climacteric rise in respiration rate when they were exposed to 50 or 100 kPa O_2. Exposure of potato tubers for 'some hours' to 100 kPa O_2 had little or no effect on their respiration rate, but prolonged exposure led, at first, to an increase in their respiration rate, over a period of some 2 weeks, but thereafter the effects of 'O_2 poisoning' became apparent and after 5–6 weeks the effect on respiration rate was negligible (Barker and Mapson, 1955). Cherries or apricots exposed to 30, 50, 75, and 100 kPa O_2 showed no effect on their respiration rate, but in plums the respiration rate was stimulated in proportion to the increasing O_2 concentration (Claypool and Allen, 1951). However, Escalona *et al.* (2006a and b) concluded that 80 kPa O_2 must be used in MAP of fresh-cut lettuce in combination with 10–20 kPa CO_2 to reduce their respiration rate and avoid fermentation. Zheng *et al.* (2008) found that zucchini squash exposed to 100 kPa O_2 had the lowest respiration rate compared with storage in air or 60 kPa O_2.

Ethylene

Effects on increased O_2 levels during storage have been reported to have various effects on ethylene biosynthesis. 'Russet Burbank' potatoes stored at 7°C had a higher ethylene production rate in 80 kPa O_2 + 12 kPa CO_2 compared with storage in air (Creech *et al.*, 1973) and 'Bartlett' pears in storage at 20°C had a higher ethylene production rate in 100 kPa O_2 than in air (Frenkel, 1975). Similar results were reported by Morris and Kader (1977) for mature-green and breaker tomatoes stored at 20°C when they were exposed to 30 or 50 kPa O_2 but exposure to 80 or 100 kPa O_2 reduced ethylene production rates and muskmelons stored in 100 kPa O_2 at 20°C had similar ethylene production levels as those stored in air (Altman and Corey, 1987). Zheng *et al.* (2008) found that zucchini squash exposed to 60 kPa O_2 had the lowest ethylene production compared with those stored in 100 kPa O_2 or in air.

Volatile compounds

It appears that exposure to high O_2 does not affect the volatile content of fruit. For example, Rosenfeld *et al.* (1999) found that blueberries in MAP stored at 4°C or 12°C for up to 17 days had similar sensory quality whether flushed with air or 40 kPa O_2. Yahia (1989) found that apples that had been stored at 3.3°C in 3 kPa O_2 + 3 kPa CO_2 for up to 9 months did not have increased volatile formation when subsequently exposed to 100 kPa O_2 at 3.3°C for up to 4 weeks.

Texture

Jiang (2013) coated button mushrooms with 2% alginate and stored them in jars continuously ventilated with 100 kPa O_2 at 4°C for up to 16 days and found that this treatment maintained their firmness and delayed cap opening. Additionally the treatments delayed changes in the TSS, total sugars and AA and inhibited the activity of PPO and POD throughout storage. 'Bartlett' pears kept at 20°C in 100 kPa O_2 had higher rates of softening than those kept in air (Frenkel, 1975). Day (1996) reported that softening of slices of apples was inhibited during storage in 100 kPa O_2 compared with those stored in air.

Pigments

'Bartlett' pears stored at 20°C in 100 kPa O_2 had higher rates of chlorophyll degradation than those stored in air (Frenkel, 1975). Biale and Young (1947) reported that exposure of lemon to 99.2 kPa O_2 accelerated de-greening and 'Hamlin' oranges stored in 50 kPa O_2 had an increased rate of de-greening (Jahn *et al.*, 1969). Aharoni and Houck (1980) exposed oranges for 4 weeks at 15°C to 40 or 80 kPa O_2, followed by 2 weeks in air. They found that fruits kept in 80 kPa O_2 had the palest coloured peel, but their endocarp and juice were the deepest orange. The response was intermediate for oranges kept in 40 kPa O_2. Aharoni and Houck (1982) reported that storage in 40 or 80 kPa O_2 increased anthocyanin synthesis of flesh and juice of blood oranges. Li *et al.* (1973) reported that ripening of tomatoes at 12–13°C was accelerated in 40–50 kPa O_2 compared with air; and Frenkel and Garrison (1976) found that lycopene synthesis in *rin* tomatoes was stimulated in storage in 60 or 100 kPa O_2 in the presence of ethylene at 10 ml/l. Brown spots on the skins of bananas is a normal stage of ripening that usually occurs when the skin has turned from green to yellow or fully yellow, depending on the variety. Maneenuam *et al.* (2007) compared the effect of storage in different O_2 partial pressures at 25°C and 90% RH on peel spotting in the variety 'Sucrier' (*Musa* AA). The fruits had first been initiated to ripen and were turning yellow. They were then transferred to atmospheres containing either 90 kPa O_2 or 18 kPa O_2 in gas-tight chambers. Peel spotting and a decrease in dopamine levels (a free phenolic compound) were quicker in fruit in 90 kPa O_2, indicating that dopamine might be a substrate for the browning reaction. The browning reaction is related to the activity of PAL that converts phenylalanine to free phenolic substances that form the substrate that is converted to quinones by PPO.

Nutrition

Barker and Mapson (1955) reported that AA content of potato tubers kept in 100 kPa O_2 was lower than in those stored in air. Day *et al.* (1998) reported that high O_2 MAP had beneficial effects on the retention of AA and degree of lipid oxidation. They also stated that high O_2 in MAP of minimally processed lettuce did not decrease antioxidant levels in comparison with low O_2 MAP.

Sprouting

Abdel-Rahman and Isenberg (1974) found that exposure of carrots to 40 kPa O_2 increased sprouting and rooting during storage at 0°C compared with those stored in air.

Free radicals

A free radical is an element or compound that remains unaltered during its ordinary chemical changes. Increased O_2 concentrations around and within fruit or vegetables were shown by Fridovich (1986) to result in higher levels of free radicals that can damage plant tissues.

Chilling injury

Zheng *et al.* (2008) reported that there was an indication of low chilling injury in zucchini squash exposed to high O_2 and that the O_2 radical absorbance capacity and total phenolic levels in the skin were both induced by cold storage and further enhanced by storage under 60 kPa O_2. The enhanced antioxidative enzyme activities and the O_2 radical absorbance capacity and phenolic levels appeared to correlate with the reduced chilling injury.

Decay

Day (1996) stated that high O_2 levels could influence both aerobic and anaerobic growth of microorganisms. Amanatidou *et al.* (1999) found that exposure to 80–90 kPa O_2 generally did not inhibit microbial growth strongly, but caused a significant reduction in the growth rate of some of the microorganisms they tested, including *Salmonella enteritidis*, *S. typhimurium* and *Candida guilliermondii* (a yeast used in biological control). Among the ten microbial species studied, growth of some was even stimulated by high O_2. The combined application of 80–90 kPa O_2 plus either 20 or 10 kPa CO_2 had an inhibitory effect on the growth of all the microorganisms they tested. They concluded that when high O_2 or high CO_2 was applied alone, the inhibitory effect on microbial growth was highly variable, but stronger and more consistent inhibition of microbial growth occurred when the two gases were used in combination. Wszelaki and Mitcham (1999, 2000) found that 80–100 kPa O_2 inhibited the *in vitro* growth of *Botrytis cinerea* on strawberries. However, only 100 kPa O_2 inhibited growth of *B. cinerea* more than 15 kPa CO_2 in air and then only after exposure for 14 days. No residual effect on *in vitro* fungal growth was observed upon transfer to air. In *in vivo* studies, they found that *B. cinerea* on strawberries was reduced during 15 days of storage at 5°C in 80–100 kPa O_2, but there was some fermentation in the fruit. However, storage in 40 kPa O_2 was effective and gave good control of rotting without detrimental effects on fruit quality during storage at 5°C for 7 days. Deng *et al.* (2005) also reported that high O_2 storage conditions significantly reduced fruit decay in strawberries. High O_2 has been combined with other treatments; for example, strawberries were treated with a solution of 1% chitosan, packaged in modified atmosphere film packages with either 80 kPa O_2 or 5 kPa O_2, with the balance N_2, and then stored at 4°C, 8°C, 12°C and 15°C. The coating inhibited the growth of microorganisms at all temperatures, especially when combined with high O_2 that also seemed to help in retaining their colour (Tamer and Çopur, 2010).

High O_2 levels have been tested on minimally processed fruit and vegetables. Day (1996) suggested that high O_2 atmospheres could be advantageous for MAP by directly inhibiting the decay- causing organisms, particularly fungi on soft fruits. Day *et al.* (1998) reported that 99 kPa O_2 alone did not prevent the growth of *Pseudomonas fragi*, *Aeromonas hydrophila*, *Yersinia enterocolitica* and *Listeria monocytogenes* but growth of *P. fragi* was reduced by 14% and *A. hydrophila* by 15%. The combination of 80 kPa O_2 + 20 kPa CO_2 was more effective in inhibiting growth of all the organisms tested at 8°C than either 80 kPa O_2 or 20 kPa CO_2 alone. Amanatidou *et al.* (2000) found that a combination of 50 kPa O_2 + 30 kPa CO_2 prolonged the shelf-life of sliced carrots by 2–3 days compared with storage in air. In *in vitro* studies, Caldwell (1965) reported that exposure to 10 atm pressure of pure O_2 completely suppressed growth in some fungi and two species of bacteria. However, unlike the tissues of higher plants, the fungi, when removed from the pressure and returned to air, recovered and began to grow apparently quite normally if the period of exposure was not too long. In these cases a period of some days normally elapsed before the growth of the colonies in air began again.

Physiological disorders

Kidd and West (1934) showed that storage of 'Bramley's Seedling' apples at 4°C for 4 months in 100 kPa O_2 resulted in disorders whose symptoms included mealy flesh and browning of skin and flesh. Solomos *et al.* (1997) reported that the apple cultivars 'Gala' and 'Granny Smith' exposed to 100 kPa O_2 developed extensive peel injury compared with that which occurred under lower O_2 atmospheres, especially in 1 kPa O_2. However, Lurie *et al.* (1991) found that storage of 'Granny Smith' apples at 0°C in 70 kPa O_2 for 1 month did not accelerate the severity of the disorder caused by sunscald. Production of α-farnesene and trienol, related to development of storage scald, increased in apples stored at 0°C in 100 kPa O_2 atmospheres for up to 3 months. 'Granny Smith' apples stored in 100 kPa O_2 were completely 'bronzed' after 3 months and contained high ethanol concentrations (Whitaker *et al.*, 1998). An atmosphere of 100 kPa O_2 potentiated the effect of ethylene at 0.5 ml/l on isocoumarin formation in carrots, resulting in a fivefold increase over that found in carrots treated with ethylene in air (Lafuente *et al.*, 1996). Super-atmospheric O_2 levels increased ethylene production and the incidence and severity of pink rib and ethylene-induced russet spotting on lettuce (Klaustermeyer and Morris, 1975).

Effects on Selected Crops

Avocado

Liplap *et al.* (2012) stored avocados for 7 days at ambient temperature using pressure levels of 1, 3, 5, 7 or 9 atm compared with those stored in commercial cold storage conditions of 5°C under atmospheric pressure. Hyperbaric pressure decreased ripening, resulting in an extension of the storage life, and the respiration rates were inversely proportional to the pressure applied. They concluded that hyperbaric storage had potential for extending the storage life and maintaining the quality of avocado while using little energy.

Cherry

Romanazzi *et al.* (2008) exposed 'Ferrovia' cherries for 24 h under approximately 1.5 atm or 1 atm for 4 h, then stored them at 0 ± 1°C for 14 days followed by 7 days at 20 ± 1°C under atmospheric pressure. Cherries that had been exposed to 1.5 atm had reduced incidence of brown rot, grey mould and blue mould, compared with those exposed to 1 atm. They concluded that induced resistance from the hyperbaric conditions was likely to be responsible for the decay reductions.

Grape

Romanazzi *et al.* (2008) artificially wounded 'Italia' grapes and inoculated the wounds with 20 µl of a *Botrytis cinerea* conidial suspension (5 × 10^4 spores/ml) and then exposed them for 24 h under either approximately 1.5 atm or 1 atm. They were then removed from the hypobaric chambers and stored at 2 ± 1°C for 3 days under atmospheric pressure. Hyperbaric storage resulted in a significant reduction of grey mould lesion diameter and percentage of *B. cinerea* infections on the fruit. As with cherries, they concluded that induced resistance was likely to be responsible for the decay reductions.

Lettuce

Liplap *et al.* (2013a) stored lettuces in a range of pressures from 100 to 850 kPa at 20°C and 100 kPa at 4°C. The hyperbaric storage at 20°C resulted in noticeable changes in sensory quality, but they were still considered marketable after 3 days, while those stored at atmospheric pressure and 4°C showed little degradation even during 7 days storage. After 5 days of storage the respiration rate of those under 625 and 850 kPa remained fairly stable, while the respiration rate began to increase in those in lower pressures, which they assumed was an indication of the initiation of decay. At 20°C those under 850 kPa showed better quality than those under atmospheric pressure. They concluded that 'overall, hyperbaric treatment has the potential of being used as an alternative technique for short-term storage of lettuce without refrigeration'.

Mango

Apelbaum and Barkai-Golan (1977) tested the effect of hyperbaric pressure storage on mangoes for 16 days and found that pressures from 250 to 700 kPa did not result in a shelf-life longer than commercial storage in air at atmospheric pressure.

Melon juice

Queirós *et al.* (2014) stored melon juice for 8 h at 25°C, 30°C or 37°C, under atmospheric pressure and under hyperbaric pressures within the range of 25–150 MPa. These were compared with storage at 4°C under atmospheric pressure. They found that juice under hyperbaric storage of 50 and 75 MPa had similar or lower microbial counts (total aerobic mesophiles, Enterobacteriaceae, and yeasts and moulds) compared to juice stored at 4°C, while at 100 and 150 MPa the counts were lower for all the tested temperatures, indicating an additional microbial inactivation effect. At 25 MPa no microbial inhibition was observed. Juice stored under hyperbaric conditions had similar pH, TA, TSS, browning and cloudiness levels to juice that had been in storage at 4°C.

Mume

At an average temperature of 22°C, the shelf-life of mume fruit (*Prunus mume*) was reported to be only 2 or 3 days (Miyazaki, 1983). Mume fruits were subjected to pressures of 0.5–5 MPa (500–5000 kPa) for 10 min and maintained at 0.5 MPa (500 kPa) for 5 days by Baba and Ikeda (2003). Treatment at 0.5 MPa decreased the respiration rate, ethylene production and weight loss during storage and showed reduced chilling injury symptoms. During storage at 0°C or 5°C, chilling injury occurred as surface pitting and/or peel browning (Goto *et al.*, 1988).

Mushroom

Robitaille and Badenhop (1981) stored mushrooms under 35 atm, which did not affect their respiration rate, but significantly reduced moisture loss and cap browning compared with storage at normal pressure. Neither pressurization nor gradual depressurization over 6 h injured the mushrooms.

Peach

The composition of volatile compounds emanating from peach fruit varied quantitatively and qualitatively during 4 weeks of storage. Yang *et al.* (2009) identified 21 compounds prior to storage and 59 after storage. Storage under hyperbaric pressures contributed most to the concentration of total volatile compounds compared with CA storage and storage in air.

Tomato

Romanazzi *et al.* (2008) observed that hyperbaric storage had been shown to have variable effects on the shelf-life of tomatoes. In their review, Ahmed and Ramaswamy (2006) reported that at 20°C hyperbaric exposure at ≥ 0.3 MPa resulted in respiration rates equal to or higher than those in fruit stored in normal atmospheric pressure. Liplap *et al.* (2013a) subjected early-breaker tomatoes to normal atmospheric pressure, 0.3, 0.5, 0.7 or 0.9 MPa at 20°C or 13°C for 4 days, followed by ripening at 20°C for up to 10 days at atmospheric pressure. Hyperbaric treatment initially inhibited lycopene synthesis but then enhanced lycopene accumulation during exposure and subsequent ripening. All antioxidants were found in lower concentrations in tomatoes subjected to atmospheric pressure at 13°C. They concluded that, overall, hyperbaric treatment at 20°C had potential to extend tomato shelf-life during short treatment durations without adverse impact on quality during ripening. They showed that the only consistent effect of hyperbaric treatment at 0.5, 0.7 and 0.9 MPa was a reduction in weight loss and enhanced firmness retention for up to 5 days of ripening after treatment. Hyperbaric storage at 0.5, 0.7 and 0.9 MPa significantly reduced weight loss and retained colour, firmness, TSS, TA and TSS:TA ratio at similar levels to the tomatoes treated at 13°C and 0.1 MPa. Firmness after treatment was highest for fruit from 0.1 MPa at 13°C and from 0.5, 0.7 and 0.9 MPa at 20°C. The higher firmness advantage declined during the 5 days of ripening after treatment, with higher firmness only being retained for fruit at 0.9 MPa at 20°C and at 0.1 MPa at 13°C. After 10 days of ripening, firmness was similar for all treatments. The lowest respiration rate was in those stored at 0.1 MPa at 13°C. They also exposed early breaker-stage tomatoes to pressures of 1, 3, 5, 7 or 9 atm for 5, 10 or 15 days at 13°C, followed by a storage at 20°C for 12 days (Goyette *et al.*, 2012). Based on firmness values, those that had been stored in normal atmospheric pressure were no longer acceptable for consumption after 12 days of subsequent storage at 20°C. Those that had been stored at 7 and 9 atm for 15 days had irreversible physiological damage, while those exposed to 3, 5 or 7 atmospheres for 10 days or 5 atm for 5 days maintained marketable firmness. Lycopene content was improved in all the fruit that had been stored under hyperbaric pressures followed by 12 days of maturation compared with those that had been under atmospheric pressure, with the highest lycopene content (28% more than in those from atmospheric pressure) from fruit that had been stored under 5 atm for 10 or 15 days. Goyette (2010) reported that hyperbaric exposure of tomatoes showed that their respiration rate was inversely proportional to the pressure applied. Respiration rate was reduced by 20% under 9 atm compared with those under 1 atm. At the onset of hypobaric storage the RQ was low and increased to reach a value of approximately 1 within 120 h. Low RQ values were caused by solubilization of CO_2 in the tomato cells at the beginning of the process. Liplap *et al.* (2013b) concluded that, overall, hyperbaric exposure at 20°C had potential to extend tomato shelf-life during short-duration treatment without adverse impact on quality during ripening.

Watermelon juice

Fidalgo *et al.* (2014) compared the preservation of watermelon juice at room temperature and 5°C at atmospheric pressure with preservation under 100 MPa at room temperature. After 8 h exposure to 100 MPa, the initial microbial loads of the watermelon juice were reduced by 1 log-unit, for total aerobic mesophiles, to levels of about 3 log-units and 1–2 log-units for Enterobacteriaceae, yeasts and moulds that were below the detection limit. These levels remained unchanged for up to 60 h. Similar results were obtained at 30°C under 100 MPa after 8 h. At atmospheric pressure for 24 h at room temperature and for 8 h at 30°C, microbial levels were above quantification limits and unacceptable for consumption. Storage at 5°C after the hyperbaric exposure gave an extended shelf-life.

11

Hypobaric and Low-oxygen Storage

Hypobaric storage has also been referred to as low-pressure storage and sub-atmospheric pressure storage. With hypobaric storage the control of the O_2 level in the store can be very accurately and easily measured and achieved by measuring the pressure inside the store with a vacuum gauge. Hypobaric storage was recently reviewed in more detail by Thompson (2015) and much of the information contained here has been modified from that book. The application of hypobaric conditions to CA storage is that the reduced pressure reduces the partial pressure of O_2 and thus O_2 availability to fruit or vegetables in the store. The reduction in the partial pressure of the O_2 is proportional to the reduction in pressure in the store. The humidity in the store atmosphere has to be taken into account when calculating the partial pressure of O_2 in the store, using a psychometric chart. Also as air pressure is reduced, so the boiling point of water is reduced, thus increasing water loss from the stored fruit or vegetable. Store humidity, therefore, must be kept high.

Burg (2004) reviewed the considerable literature on the effects of hypobaric conditions on the storage and transport of meat and horticultural produce as well as fruit, vegetables and flowers. An example of meat transport was given by Sharp (1985), who reported on the successful transport of lamb carcasses at 0.006 atmosphere (atm) (≈ 4.6 mm Hg; 1 atm = 101.325 kPa) in a hypobaric container with an inner wall temperature of about −1.5°C and found that the carcasses were still in good condition with a mean weight loss of only 2.5% during the 40 days of transport. Burg (1967) took out a patent in the USA for the application of hypobaric storage and he gave a detailed evaluation and justification of the award of the patent in Burg (2004). Stanley P. Burg was listed as the inventor and Grumman Allied Industries Incorporated as the original assignee. In support of the application several examples of the effects on horticultural products, meat and fish were given, including those in Table 11.1.

Burg (2004) refers to legal actions in which he was involved, related to 'LP patents' and tax implications of those patents, that were eventually resolved. Most of the research on hypobaric storage and its commercial application has been carried out in the USA, particularly by Stanley P. Burg and many other workers, including David Dilley, but research work has also been carried out in the UK (Langridge and Sharples, 1972; Ward, 1975; Hughes *et al.*, 1981), Germany (Bangerth, 1973, 1974, 1984; Bangerth and Streif, 1987), Spain (Alvarez, 1980), Canada (Lougheed *et al.*, 1974, 1977), Israel (Apelbaum and Barkai-Golan, 1977; Apelbaum *et al.*, 1977a, b; Aharoni *et al.*,1986), China (Chang, 2001; Cao, 2005) and Italy (Romanazzi *et al.*, 2001, 2008). Burg

Table 11.1. Effects of hypobaric storage on the postharvest life of some fruit and vegetables. Taken with modifications from Burg, S.P. 1977 Patent US4061483A, Low temperature hypobaric storage of metabolically active matter. Available at https://patents.google.com/patent/US4061483A/en?oq=Patent+US4061483A+ (accessed September 2018).

Crop	Temperature	Pressure	Storage time
Bartlett, Clapp and Comice pears	–1 to 1°C	Atmospheric	1½–3 months
		60 mm Hg	4–6 months
McIntosh, Red Delicious, Golden Delicious and Jonathan apples	–1 to 2°C	Atmospheric	2–4 months
		60 mm Hg	6 months
Waldin avocados	10°C	Atmospheric	12–16 days
		60–80 mm Hg	30 days
Lula avocados	8°C	Atmospheric	23–30 days
		40 and 80 mm Hg	75–100 days
Booth 8 avocados	8°C	Atmospheric	8–12 days
		40 and 80 mm Hg	About 45 days
Fresh green onions (scallions)	0–3°C	Atmospheric	2–3 days
		60–80 mm Hg	More than 3 weeks
Green peppers (capsicums)	8–13°C	Atmospheric	16–18 days
		60–80 mm Hg	46 days
Snap-beans	5–8°C	Atmospheric	7–10 days
		60 mm Hg	26 days
Cucumbers	10°C	Atmospheric	10–14 days
		80 mm Hg	49 days
Pole-beans	8°C	Atmospheric	10–13 days
		60 mm Hg	30 days
Tioga and Florida 90 strawberries	0–2°C	Atmospheric	5–7 days
		80–200 mm Hg	4–5 weeks
Blueberries	0–1°C	Atmospheric	4 weeks
		80–200 mm Hg	At least 6 weeks
Iceberg lettuce	0–4°C	Atmospheric	2 weeks
		150–200 mm Hg	About 4 weeks
Ruby Red grapefruit	6°C	Atmospheric	4–6 weeks
		80–150 mm Hg	90 days
Mature-green tomatoes	13°C	Atmospheric	2 weeks
		80 mm Hg	8 weeks

(2014) evaluated the reasons why hypobaric technology has had only limited impact on the food industry. His contentions included criticisms of the research techniques and also included:

- its high cost;
- experimental error caused by humidifying air at 1 atm pressure instead of at the low storage pressure;
- insufficient air changes so that the stored commodity consumed all available O_2 and suffered anaerobic damage;
- laboratory apparatus installed in cold-rooms that had a non-uniform air distribution pattern that produced a 'cold spot' on the vacuum chamber's surface (this created an evaporation/condensation cycle between the commodity and chamber cold spot); and
- the commodity being stored in a sealed system that accumulated an active ethylene concentration between occasional ventings.

Burg (2014) also commented that 'experimental errors by academics and other concerns have prevented hypobaric storage from achieving more widespread adoption'.

The literature reports several effects of hypobaric conditions on fresh fruit and vegetables that can lead to extension in post-harvest life. Workman *et al.* (1957) found that the respiration rate of tomatoes was reduced when they were stored at 20°C under 88 mm Hg compared with those stored at 20°C under atmospheric pressure (note that 1 mm Hg = 0.133 kPa). At the same time Stoddard and Hummel (1957) stored several types of fruit and vegetables in household refrigerators and found that those stored under 658–709 mm Hg had increased postharvest lives of 20–92%, depending on the crop, compared with those stored in the same refrigerators at atmospheric pressure. Burg and Burg (1966b) showed delays in ripening of bananas when they were stored in 125–360 mm Hg compared with those in atmospheric pressure. They also showed that limes stored at 15°C took about 10 days for 50% of them to turn from green to yellow in 760 mm Hg, while this took up to 56 days in storage in 152 mm Hg at the same temperature. Burg (1975) reported that 'Lula' avocados remained firm for 3½ months under hypobaric conditions and then ripened normally upon removal. This length of storage is almost twice that reported for Lula avocados held under CA storage at normal atmospheric pressure (Hatton and Reeder, 1965).

It has been well established that the reduced O_2 in CA storage is a major factor in extending postharvest life. However, in addition to the effects of reduced O_2 on respiration rate, hypobaric conditions have other effects. Since hypobaric chambers are constantly ventilated and air is removed, ethylene produced by the fruit or vegetable will be constantly removed. Ethylene has also been shown to be removed more quickly from the plant cells where it is being synthesized. This ethylene removal would therefore reduce its effect in the ripening process and also reduce the effects on the development of some physiological disorders associated with ethylene accumulation. Chen *et al.* (2013b) reported that fresh bamboo shoots stored at 2°C in 50 mm Hg for 35 days had reduced ethylene production, compared with those stored in atmospheric pressure, which delayed their softening. Burg (2004) reviewed several publications that reported inhibition of growth and sporulation of pathogenic fungi in hypobaric conditions that then resumed growth and sporulation when they were removed to atmospheric pressure, which may also be related to reduced O_2. There is also strong evidence that hypobaric conditions can help to control insect infestation in fruit. For example, Davenport *et al.* (2006) reported that all the eggs and larvae of Caribbean fruit fly (*Anastrepha suspensa*) in mangoes stored at 13°C in 15–20 mm Hg with ≥ 98% RH were killed within 11 days, while a substantial number of eggs survived for the 14-day storage period under atmospheric pressure. Hypobaric conditions may have other effects on fruit physiology. For example, Wang *et al.* (2015) concluded that the increased shelf-life of honey peach (*Prunus persica*) in hypobaric storage could be due to increased energy status, enhanced antioxidant ability and less membrane damage.

Technology

As indicated above, hypobaric storage has hardly been taken up commercially. Probably the only commercial application was the Grumman Corporation in the USA in the 1970s. The company developed and constructed a hypobaric container which it called 'Dormavac' (similar to the one in Fig. 11.1). It was operated commercially at 2.2–2.8°C and a pressure of 15 mm Hg, but they were unable to make it profitable, resulting in eventual losses of some $50 million (Grumman Park, undated). However, as indicated above, there has been considerable experimentation on hypobaric storage.

Since fresh fruit and vegetables in hypobaric store are constantly respiring, it is essential that the store atmosphere is constantly being changed in order to maintain the required partial pressure, especially the desired O_2 level. This is achieved by a vacuum pump evacuating the air from the store and constantly replenishing fresh air from the outside. The air inlet and the air evacuation from the store are balanced in such a way as to achieve the required reduced pressure within the store. There are two important considerations in developing and applying this technology to crop storage. The first is that the store needs to be designed

Fig. 11.1. Reefer container that has been tested for hypobaric storage.

to withstand low pressures without imploding. The second is that the reduced pressure inside the store can result in rapid water loss from the crops. To overcome the first, stores have to be strongly constructed; for example, with thick steel plate with a curved interior. For the second, the air being introduced into the store must be as close as possible to saturated (100% RH). If it is less than this, serious dehydration of the crop can occur.

Various systems have been developed for experimental work to test hypobaric storage, but most are similar to the one described by Dilley (1977). This type of system was also used by Hughes *et al.* (1981) where specially constructed 100 l steel barrels were used that had transparent Perspex lids 4 cm thick through which the produce could be observed during storage. Air was introduced through an inlet that had been bubbled through water in order to saturate it (though no measurement of the relative humidity was made). A vacuum pump and vacuum gauge were attached to the outlet and the flow rate was adjusted on the pump to give a flow rate of 5 l/h (Fig. 11.2).

The systems were kept in temperature-controlled cabinets that were adjusted to ± 1°C. Romanazzi *et al.* (2001) used a vacuum pump in 64 l gas-proof tanks in which the fruit was placed at atmospheric pressure as controls or a vacuum was applied with a vacuum pump and measured with an external vacuum gauge. Al-Qurashi *et al.* (2005) used quart-sized pressure cookers that were continuously evacuated by a belt-drive pump. The inlet air was humidified by bubbling it through water in 5-gallon (22.7 l) containers then through a filter to prevent water from getting into the pressure cookers from the ventilated air. The air filter was quarter-filled with water and cellulose pads were inserted to increase the humidity of the air flowing to the pressure cookers. The pressure cookers were also sealed at the lids

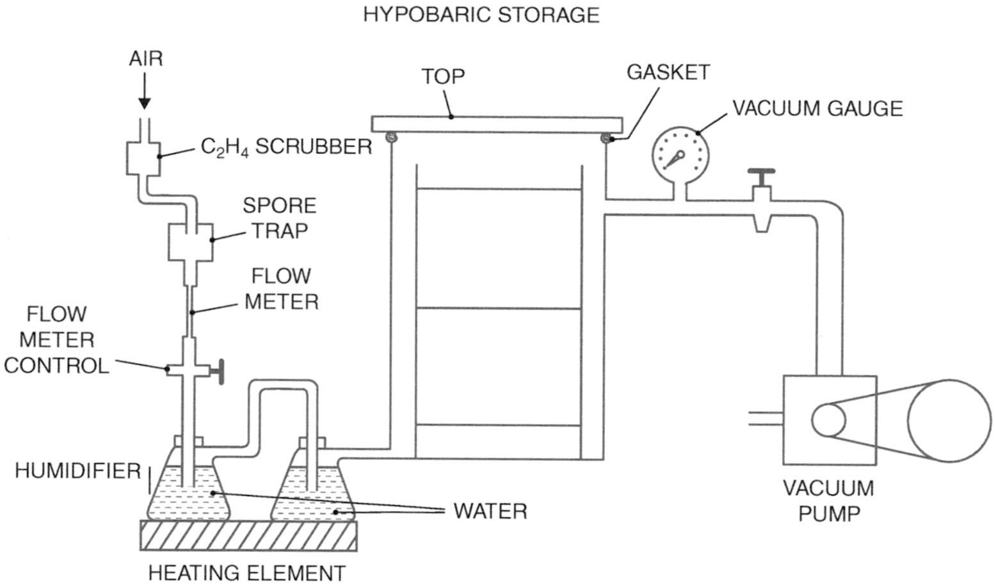

Fig. 11.2. Schematic diagram of the experimental hypobaric storage system set up at the Tropical Products Institute in London in the late 1970s.

with 'Play-Doh', allowing an air tight seal. Valve regulators, located between the filter and the pressure cookers, were used to maintain the desired pressures by admitting air at the proper rate. The pressure within the pressure cookers was monitored with pressure gauges placed at the top of the pressure cookers. Jiao *et al.* (2012) used aluminium chambers (0.61 × 0.43 × 0.58 m) with a two-stage rotary vacuum pump regulated by a compact proportional solenoid valve controlled by a proportional/integral/derivative computer control system. Chamber pressure was monitored with a digital pressure gauge. A rotameter was used to adjust the air exchange rate and the ingoing rarefied air was passed through a humidifier before entering the hypobaric chamber, in order to keep the humidity near saturation. The relative humidity was calculated by measuring wet-bulb and dry-bulb temperatures having relatively high accuracy (± 0.1%). They experimented with various modifications and showed that added foam covering the chambers maintained the temperature of the inside air to within ± 0.1°C. The regulating system kept pressure to within ± 1% of the set point and maintained humidity at > 98% RH under various air exchange rates and pressures (the measurements inside the cabinet varied between 98.40 and 99.35 RH) with a chamber leakage rate of 0.009 kPa/h and hypobaric system leakage rate of 0.48 kPa/h. Spalding and Reeder (1976) used pure CO_2 and O_2 from gas cylinders that were metered into a 150 ml glass mixing chamber and their flow rates regulated to supply the required mixture of O_2 and CO_2. Flow into the chamber was controlled at 110 ml/min (1 air change per hour). Chamber pressures were maintained at about 2 mm Hg and the humidity at 98–100% RH as determined by a humidity-sensing element. Jamieson (1980) used vacuum desiccators for the fruit or vegetables to be tested. A vacuum pump sucked air from the desiccators and air was allowed in by first passing it through a flask containing water that was constantly heated on an electric plate to maintain near-saturation humidity. Air flow and pressure inside the desiccators were measured and controlled with a flow meter, a vacuum gauge, a pressure regulator and a needle valve. Burg (2004) described equipment used in laboratory-scale hypobaric experiments with the fruit in desiccators supplied with air passed through water in a beaker with a vacuum pump attached to the outlet of the desiccators and a manometer for monitoring the pressure inside the desiccators.

Jiao *et al.* (2012) described a hypobaric chamber called 'VivaFresh' that had been made in 2007 by Atlas Technologies (Port Townsend, Washington, USA). The VivaFresh system comprised aluminium chambers (0.61 m long × 0.43 m wide × 0.58 m high) with a two-stage rotary vacuum pump regulated by a compact proportional solenoid valve controlled by a proportional/integral/derivative computer control system. Chamber pressures were monitored with a digital pressure gauge and a rotameter was used to adjust the air exchange rate. The ingoing air passed through a humidifier before entering the hypobaric chamber in order to keep the humidity near saturation. The temperature inside the chamber and the exterior chamber wall and humidity was calculated by measuring wet-bulb and dry-bulb temperatures using calibrated YSI 55000 Series GEM thermistors with a relatively high accuracy of ± 0.1%. Temperature variation of the chamber wall was controlled to within ± 0.2°C and the inside air to within ± 0.1°C. Humidity measured inside the chambers varied between 98.4% and 99.35% RH. The pressure was within 1% of the set point and O_2 concentration could be controlled at less than 0.6% when the pressure was less than 3.3 kPa. The leakage rate of the chamber was given as 0.01 kPa/h.

A different hypobaric system was described by Knee and Aggarwal (2000) who used plastic containers, capable of being evacuated to 380 mm Hg with a vacuum pump, for their experiments. They found that this required 18 strokes of the pump for a container with a nominal volume of 500 ml capacity and 24 strokes for a 750 ml capacity container. These were placed in a refrigerator run at either 4°C or 8°C, depending on the product to be stored. Overall the vacuum containers showed little advantage over conventional plastic containers for the types of produce tested. The multinational company VacuFresh also makes small plastic containers of 1.2–2.8 l capacity that are sold together with a hand pump. They are used for storing fruit and vegetables under hypobaric conditions in a domestic refrigerator. The company claims in its advertisements, 'Keeps food fresh up to 5 times longer. Vacuum locks in freshness and prevents premature spoilage and food decay.' In the mid-1970s Prodesarrollo in Colombia (A.K. Thompson, unpublished) investigated the possibility of storing vegetables (potatoes, carrots and cabbages) in a brick-built store room at a café/ski centre at an altitude of about 4800 m in Colombia that would have a barometric pressure of about 57 kPa (430 mm Hg), where the temperature was also low. The trial was on the slopes of the volcano Nevado Del Ruiz (near the town of Manizales in the Departamento de Caldas) whose peak is 5321 m above sea level. The trial was unsuccessful, due to rapid desiccation of the vegetables, and because of transport problems it was not repeated.

Adding Carbon Dioxide

Burg (2004) observed that hypobaric conditions, which can decrease ambient and intercellular CO_2, are an important advantage, providing benefits that cannot be duplicated by increasing CO_2 levels given in some controlled atmosphere storage recommendations. Spalding and Reeder (1976) concluded that high CO_2 was necessary for the successful storage of Waldin avocados, even in a hypobaric system. Yahia (2011) commented that CO_2 was considered essential in controlling decay and ameliorating chilling injury in avocados and CO_2 cannot be added in a low pressure system. Burg (2004) contended that the effects of hypobaric conditions on removing CO_2, produced by respiration, from cells and intercellular spaces may result in reduced bacterial and fungal growth, better AA retention, inactivation of the ethylene-forming enzyme and the prevention of succinate formation. In confirmation of the effect of CO_2 on bacterial growth, Wells (1974) reported that the postharvest pathogens *Erwinia carotovora*, *E. atroseptica* and *Pseudomonas fluorescens* were unable to multiply in the very low CO_2 levels, as well as the low O_2 levels, that are to be found in the cells of vegetables stored under hypobaric conditions. However, Enfors and Molin (1980) found that when *P. fragi* was grown at O_2 limitation (0.0025 atm O_2) and exposed to 0.99 atm CO_2, the inhibiting effect of the CO_2 was added to that of the O_2 limitation. They did not note any indications of a synergistic effect between CO_2 inhibition and O_2 limitation. Laurin *et al.* (2006) reported that cucumbers exposed to hypobaric conditions of 532 mm Hg for only 6 h could exhibit an indirect stress response that occurred only when they were transferred to atmospheric pressure, preventing closure of stomata. They explained that this residual effect may have been due to the possibility that hypobaric conditions enhanced outward diffusion of CO_2, reducing intercellular CO_2 concentration and causing stomata to open. When the cucumbers were transferred to atmospheric pressure, stomata could still have remained open to restore the CO_2 concentration.

Pre-storage Treatments

As with any other technique, storage under hypobaric conditions has been successfully used in combination with other treatments. For example, Romanazzi *et al.* (2003) found that the combination of spraying with chitosan at 0.1%, 0.5% or 1.0% 7 days before harvest and exposure to 380 mm Hg for 4 h directly after harvest effectively controlled fungal decay of sweet cherries during 14 days of storage at 0 ± 1°C, followed by a 7-day shelf-life. Fungi associated with rots included brown rot (*Monilinia* sp.), grey mould (*Botrytis cinerea*), blue mould (*Penicillium expansum*), alternaria rot (*Alternaria* sp.) and rhizopus rot (*Rhizopus* sp.). Kashimura *et al.* (2010) found that applying 1-MCP to 'Jonagold' and 'Fuji' apples and 'Shinsei' and 'Shinsui' Japanese pears under 152 mm Hg reduced the exposure time required to have the same effect as applying 1-MCP at atmospheric pressure. Dong *et al.* (2013) concluded that high efficacy of 1-MCP applied to tomatoes under hypobaric conditions was due to rapid ingress and accumulation of the gaseous 1-MCP. They had found that mid-climacteric tomatoes exposed to 1-MCP at 500 nl/l under 76 or 16 mm Hg for 1 h showed acute disturbance of ripening. Spalding and Reeder (1976) found that limes coated with wax containing 0.1% of either of the fungicides thiabendazole or benomyl remained green and suitable for marketing after 3–4 weeks under hypobaric storage of 170 mm Hg at 21.1°C.

Burg (2013, personal communication) described a method for generating hypochlorous acid vapour into a commodity packed for shipment in a hypobaric intermodal container. Within 1 h the vapour was 100% effective in killing bacteria, fungi and viruses both on the plant surfaces and within its interior, without injuring the commodity. The US Food and Drug Administration was reported to have given clearance for use of hypochlorous acid vapour in a hypobaric system and Burg has filed a new patent on hypobaric storage.

Effects on Physiology, Quality and Deterioration

Volatiles

Hypobaric storage has the advantage of constantly removing ethylene from the store. This constant removal would have the effect of reducing the detrimental effects of ethylene on postharvest life. Burg and Kosson (1983) reported that hypobaric storage lowers the internal equilibrium content of volatiles, including ethylene. Hypobaric conditions have also been shown by Goszczynska and Ryszard (1988) to accelerate the outward diffusion of gases from the internal tissues of horticultural crops during storage. Removal of volatiles may also have detrimental effects. Bangerth (1984) reported that apples stored for protracted periods under 50–75 mm Hg did not develop normal aroma and flavour when they were subsequently ripened in atmospheric pressure. There is some indication that this may be due to the low partial pressure of O_2, since the same effect has been reported for apples in CA storage (Burg, 2004). Wang and Dilley (2000) proposed that hypobaric ventilation removes a scald-related volatile substance that otherwise accumulates and partitions into the epicuticular wax of fruit stored under air at atmospheric pressure. They provided evidence that α-farnesene and 6-methyl-5-hepten-2-one (MHO) accumulation in the epicuticular wax associated with hypobaric storage may be involved. Burg (2010) reported that:

> The high gaseous diffusion rate at a low pressure eliminates the commodity's surface to centre O_2 gradient created by respiratory O_2 consumption, causing different commodity types to have nearly identical low O_2 tolerances, near 0.1%.

Under hypobaric conditions, Burg (2004) reported that fermentation is not induced at O_2 levels as low as 0.06–0.15 kPa, while fermentation can occur under atmospheric pressure. Burton (1989) and Mapson and Burton (1962) reported that gaseous diffusion through the periderm of mature potato tubers was entirely or almost entirely through the lenticels. They reported that permeability of tubers ranged from about 0.7 to about 2.5 $mm^3/cm^2/h/kPa$ depending on maturity, time in storage and cultivar, and O_2 diffusion would occur with an O_2 deficit of about 0.4–0.5 kPa.

Respiration rate

Hypobaric storage has been shown to reduce respiration rate compared with storage under atmospheric pressure. This has been shown on many fruit and vegetables, including sections of oat leaves (Veierskov and Kirk, 1986), oranges (Min and Oogaki, 1986), tomatoes (Workman *et al.*, 1957), apples (Bubb, 1975b), asparagus spears (Li *et al.*, 2006) and cranberries (Lougheed *et al.*, 1978). Cytochrome oxidase, the final electron-transferring enzyme of the respiratory chain, has a great affinity to O_2 but cytochrome oxidase is still able to operate under low O_2 pressure, such as 0.01 atm, without its activity being altered (Mapson and Burton, 1962; Burton, 1989). However, the reduction of respiration rate, during hypobaric storage, was attributed to the malfunctioning of oxidases such as PPO or AA oxidase. Hence, under the hypobaric condition of 0.70 atm (532 mm Hg), it is expected that O_2 partial pressure is not sufficiently reduced to cause the respiration rate to decrease. In the tissues of higher plants, the effect of exposure to pure O_2 was to stop CO_2 output, possibly the result of the inactivation of the associated enzyme systems (Caldwell, 1965).

Chilling injury

Hypobaric storage can reduce susceptibility to chilling injury. Chen *et al.* (2013a) reported that MDA content in Chinese bayberry fruit (*Myrica rubra*) was related to chilling injury since it is considered to be an indicator of membrane lipid peroxidation caused by oxidative stress. Electrolyte leakages, as well as MDA content, are indicators of cell membrane damage. MDA content and electrolyte leakage are used to indicate lipid peroxidation of membrane lipids and membrane permeability, respectively, which increase during low temperature storage (Zhao *et al.*, 2006). MDA content of the Chinese bayberry fruit under normal atmospheric pressure conditions increased gradually during storage, while storage for 15 days under hypobaric pressures of 85 ± 5, 55 ± 5 and 15 ± 5 kPa all inhibited the accumulation of MDA. Similar results had been obtained by Li *et al.* (2006), who reported that hypobaric storage could reduce MDA accumulation in asparagus.

Chlorosis

Nilsen and Hodges (1983) exposed bean leaves (*Phaseolus vulgaris*) to ethylene by dipping them in 30 ppm Ethephon and storing them at 26°C, which resulted in them becoming chlorotic more rapidly (reaching peak levels within 6 h) than those not treated. However, when the Ethephon-treated leaves were stored under hypobaric conditions (200 millibars, with O_2 and CO_2 compositions set to approximate normal atmospheric partial pressures), chlorophyll loss was prevented.

Desiccation

Reduced pressure inside the store can result in rapid water loss from the crop, since the boiling point of water reduces from 100°C at atmospheric pressure to 0°C at 4.6 mm Hg. Therefore there is a clear tendency to desiccation under hypobaric storage and fruit and vegetables need to be retained at humidity as close as possible to saturation in order to limit weight loss. The particular hypobaric pressures used for fruit and vegetable storage can also affect their weight loss. Apelbaum *et al.* (1977b) tested the effect of hypobaric storage on mango fruit and observed that at pressure below 50 mm Hg, mangoes were severely desiccated. Cicale and Jamieson (1978, quoted by Burg 2004) reported that their best results were storage of avocados at 61 mm Hg, since lower storage pressure resulted in higher desiccation. Patterson and Melsted (1977) reported that cherries under hypobaric conditions resulted in some problems with desiccation, especially at 41 mm Hg. Bubb (1975b) found that apples under hypobaric storage (35–40 mm Hg and 70–80 mm Hg) had higher weight losses than the apples under atmospheric pressure or controlled atmosphere storage. Hughes *et al.* (1981) also reported increased desiccation of capsicums during hypobaric storage compared with those stored under atmospheric pressure. An *et al.* (2009) reported that curled lettuce had high moisture loss during storage under both 190 and 380 mm Hg compared with those under atmospheric pressure. Spalding and Reeder (1976) reported that average weight loss and shrivelling of limes were higher during storage under hypobaric conditions compared with those stored at atmospheric pressure. Conversely, Burg (2004) found that at 0–3°C spring onions, in general, lost less weight during hypobaric storage than during storage in air at atmospheric pressure. He also claimed that the VacuFresh system of hypobaric storage had a very slow removal of air and therefore there was no desiccation problem of the fruit and vegetables transported in these containers. In general, the weight loss of radishes during storage at 1°C under 56 mm Hg and near-saturation humidity was less than that which occurred during storage under atmospheric pressure (McKeown and Lougheed, 1981). Cicale and Jamieson (1978, quoted by Burg 2004) found that avocados lost 1.2% in weight at various pressures ranging from 61 to 203 mm Hg compared with 5.7% under atmospheric pressure during storage at 6°C for 35 days. Spalding and Reeder (1976) reported that humidity did not appear to be a factor in the storage life of avocados, since the acceptability of avocados stored under hypobaric storage at 80–85% RH and 98–100% RH was not significantly different. All successful hypobaric systems have some form of humidification of the air as it enters the chambers, but Burg (2004) commented that in many cases these may have been insufficient.

In order to minimize desiccation, McKeown and Lougheed (1981) reported that the highest weight loss of asparagus was while the pressure was decreasing and humidity was not near saturation and they sprayed asparagus spears with water before hypobaric storage to reduce desiccation. Laurin *et al.* (2006) reported that water can be sprayed on cucumbers to resolve the problem of insufficient relative humidity, which causes desiccation during hypobaric storage.

In his hypobaric patent application, Burg (1976) states:

> Though relative humidities of 80% are usefully permissible, the preferable relative humidity of the air in the storage chamber should be higher than approximately 90% for the storage of foodstuffs such as fruits and vegetables.

Burg and Kosson (1983) commented that reduction in air pressure surrounding plant tissue will reduce the cellular hydrostatic pressure. This reduction can lead to decreased cellular water potential, but cellular activity will only be slightly altered by this pressure reduction and changing this by 20 atm will only change the water activity by 3% or 4%. Burton (1989) described the vapour pressure gradient between the surface of fruit or vegetables and the air in which they are in contact. A proportion of the water in fresh fruit, vegetables and flowers will pass from cell to cell and eventually to the surface, where it evaporates using latent heat from the commodity. This principle is used in vacuum cooling of fruit and vegetables. Some fruit and vegetables are coated with a waxy cuticle or, in the case of some root crops, a periderm that restricts this water loss. The permeability of fruit and some vegetable tissue to gas exchange is also affected by intercellular spaces and cuticle composition as well as the presence of stomata, lenticels and hydrothodes. Laurin *et al.* (2006) commented that it is likely that desiccation of fruit and vegetables under hypobaric storage is related to an increase in transpiration rate enhanced by the properties and action of stomata, lenticels, cuticle and epidermal cells. In grapes the epidermis does not contain a significant number of functional stomata; therefore water loss occurs mostly through the cuticle, which in turn restricts water loss. Stomata have been shown to be affected by storage conditions. For example, after 96 h of storage, cucumbers under hypobaric conditions had significantly more open stomata than those under atmospheric pressure. They also commented that it is likely that desiccation under hypobaric conditions is due to an increase in transpiration rate enhanced by the properties and action of stomata as well as moisture being more volatile at reduced atmospheric pressure.

Fungal infection

Although the high humidity maintained in hypobaric stores is generally suitable for fungal growth and decay development, there are several reports that show hypobaric conditions can reduce decay. Many authors have pointed out that hypobaric conditions can retard or limit pathogen growth (Burg and Kosson, 1983; Goszczynska and Ryszard, 1988; Lougheed *et al.*, 1978; Chau and Alvarez, 1983). Inhibition of the germination and growth of fungal spore at low O_2 levels was demonstrated in the early studies of Brown (1922). Couey *et al.* (1966) showed that postharvest decay in strawberry fruit was reduced at O_2 levels of 0.5% or less and they demonstrated a direct effect of such low O_2 environments on growth of mycelium and sporulation. Less pathogenic breakdown was observed in cranberries under hypobaric conditions of 76 mm Hg than in fruit stored under atmospheric pressure. Similarly, Chau and Alvarez (1983) reported that papaya inoculated with *Colletotrichum gloeosporioides* developed less infection when stored under about 15 mm Hg than when stored under normal atmospheric pressure. Apelbaum and Barkai-Golan (1977) showed that the degree of inhibition of fungal growth in hypobaric stores increased with the reduction in pressure below 150 mm Hg. They also reported that hypobaric pressure had direct fungi static effects on spore germination and mycelium growth of various storage fungi. In a review, the control of postharvest diseases by maintaining O_2 partial pressure in the region of 0.1–0.25 ± 0.008 kPa was reported by Burg (2004). However, these low O_2 levels could damage the fruit. In contrast, studies by Bangerth (1974) with various fruit and vegetables, including tomatoes, peppers and cucumbers, found a high incidence of decay after storage at hypobaric pressure. In order to suppress decay, he recommended a combination of hypobaric storage with postharvest fungicidal treatments. Barkai Golan and Phillips (1991) reported that in *in vitro* studies, different fungi responded differently to reduced pressures. For example, storage under 100 mm Hg inhibited spore germination of *Penicillium digitatum* compared with 760 mm Hg, while inhibition of spore germination of *Botrytis cinerea* and *Alternaria alternata* occurred only at 50 mm Hg. However, 50 mm Hg had no effect on the germination of *Geotrichum candidum* spores, but reducing the pressure to 25 mm Hg totally prevented spore germination of *P. digitatum*, *B. cinerea* and *A. alternata* though it had almost no effect on the germination of *G. candidum*. Transfer of inhibited cultures from hypobaric to atmospheric pressure resulted in renewed growth, suggesting that there was no irreversible damage to the fungi (Alvarez, 1980; Alvarez and Nishijima, 1987). In contrast, the effectiveness of short hypobaric treatments against postharvest diseases was investigated by Romanazzi *et al.* (2001) who found that it reduced fungal infections in sweet cherries, strawberries and table grapes. Adams *et al.* (1976) investigated the effects of a range of pressures (760–122 mm Hg) on the growth of *Penicillium expansum* and *P. patulum* and their production of the toxin patulin. They demonstrated that the amount of sporulation decreased with reduction in pressure, but mycelial growth was similar for 456 and 357 mm Hg and patulin production was lower at the lower pressures.

Insects

Disinfestation of fruit and vegetables by exposure to hypobaric conditions during export has been described by Burg (2010), Chen *et al.* (2005), Davenport *et al.* (2006), Johnson and Zettler (2009), Mbata and Philips (2001) and Navarro *et al.* (2001, 2007). Effective control of insects has been observed at O_2 concentrations of less than 6.6 kPa, and especially at 0.15–0.30 kPa (Burg, 2004). Burg (2004) and Aharoni *et al.* (1986) gave the optimal condition for transporting many tropical fruits at 13°C and where the pressure in the container could be reduced to 15–20 mm Hg this would kill most insects infesting the fruit. They claimed that in these conditions 98% of fruit fly eggs and larvae were killed within 1 week and all of them by the 11th day, and all the green peach aphids on wrapped head lettuce in 2½ days at 2°C. Insect mortality under hypobaric storage is predominantly caused by low O_2 concentrations (Navarro and Calderon, 1979), though the low humidity that could be generated in a hypobaric system could also enhance its lethal effect on insects (Navarro, 1978; Jiao *et al.*, 2012). When insects were placed into a hypoxic environment for a sufficient duration, adenosine triphosphate production was reduced, resulting in increasing membrane phospholipid hydrolysis (Herreid, 1980). Cell and mitochondrial membranes then become permeable, causing cell damage or death (Mitcham *et al.*, 2006). A patent was taken out in the USA in 2005 by Timothy K. Essert and Manuel C. Lagunas-Solar of the University of California (PCT/US2004/013225).

Contamination

Hypobaric conditions have been shown to be effective in enabling different crops to be stored together without mutual contamination. For example, carrots exposed to ethylene can synthesize isocoumarin, which gives them a bitter flavour (Lafuente *et al.*, 1989). Keeping apples, cabbages and carrots together in a store at 2°C and 60 mm Hg resulted in no isocoumarin detected in the carrots despite the assumed presence of ethylene from the apples (McKeown and Lougheed, 1981). A sensory panel compared these carrots with those stored in O_2 levels below 2 kPa and bags of slaked lime to absorb CO_2 and found that those from the hypobaric storage were superior.

Effects on Selected Crops

Apple

Optimum hypobaric conditions appear to vary with cultivars. In Korea, Kim *et al.* (1969, quoted by Ryall and Pentzer, 1974) reported that optimum conditions for 'Summer Pearmain' were 200 mm Hg and for 'Jonathan' it was 100 mm Hg. Bubb and Langridge (1974) found no extension in the storage life of 'Cox's Orange Pippin' at 3.3°C under 380 mm Hg (0.5 atm). However, they found that storage of 'Tydeman's Late Orange' at 3.3°C had reduced respiration rates and ethylene production under 76 mm Hg compared with those stored under atmospheric pressure or under 380 mm Hg. Bubb (1975b) harvested 'Cox's Orange Pippin' apples in September and compared the following storage conditions in 3.3°C: air at atmospheric pressure; CA of 2 kPa O_2 + 0 kPa CO_2 at atmospheric pressure; and hypobaric conditions of 35–40 mm Hg or 70–80 mm Hg. They found that the onset of ethylene biosynthesis was delayed by 50 days in the CA stored fruit, by 70 days in the hypobaric stored fruit at 70–80 mm Hg and by 100 days under the 35–40 mm Hg, all compared with those stored in air in atmospheric pressure.

Respiration rate was about half the level in fruit that had been stored under hypobaric conditions and slightly more in fruit that had been stored under CA compared with those stored in air under atmospheric pressure. This effect continued when fruit was removed to 10°C in air for 3 weeks. Those under hypobaric storage had higher weight losses than the other fruit, but those that had been stored under 35–40 mm Hg were firmer that those in air or CA storage. In a subsequent experiment, weight loss was reduced to 0.3% per month by improving the humidification system. The apples from both hypobaric conditions were assessed as having poorer flavour and a 'denser texture' than those that had been stored in air or CA conditions.

Bubb (1975a) also compared the storage of 'Cox's Orange Pippin' apples at 3.3°C in air with CA storage under 2 kPa O_2 + 0 kPa CO_2 and 1 kPa O_2 + 0 kPa CO_2 at atmospheric pressure and under hypobaric storage in 25–30 mm Hg and 50–60 mm Hg. He found that respiration rate of those under hypobaric storage was some 50% higher than those stored under CA storage; ethylene production was lower in those stored under 25–30 mm Hg compared with those under CA storage in 1 kPa O_2 + 0 kPa CO_2, but ethylene production was inhibited in fruit under 50–60 mm Hg only until mid-December, after which the production rate rose sharply. The ethylene production rate of the fruit under 2 kPa O_2 + 0 kPa CO_2 was similar to the latter but the rise in ethylene production rate was much slower. Little physiological damage was observed on any fruit until the end of April, when all samples showed lenticel rotting, with fruit under 50–60 mm Hg being the worst; and slight core flush and breakdown, with 25–30 mm Hg being the worst for both disorders. Sound fruit from all treatments was transferred to 12°C at that time and the apples that had been stored under 25–30 mm Hg continued to show depression in ethylene production for 2–3 weeks.

Sharples and Langridge (1973) found that 'Cox's Orange Pippin' apples, stored at 3.3°C with airflow of 15 l/h, had more lenticel blotch pitting for those stored under 380 mm Hg compared with those at atmospheric pressure. However, they reported that there was only 20% breakdown in the fruit stored under 380 mm Hg compared with 50% in those under atmospheric pressure in apples from poor keeping-quality orchards and they commented that this effect may have been related to higher weight loss from the fruit under hypobaric conditions. Bangerth (1984) reported that apples stored under hypobaric pressure of 51 mm Hg never produced autocatalytic ethylene or developed a respiratory climacteric during 11 months of storage. Only a slight decrease in fruit firmness was measured during that time. When ethylene was continuously supplied into the hypobaric containers, a considerable response was observed at the beginning of the storage period, but later the effect of ethylene was only marginal. He also found that there was no diminished response to ethylene in storage under 51 mm Hg, whatever the storage temperature tested. There was a similar decrease in sensitivity to ethylene in terms of respiration rate, softening and volatile flavour substances after shelf-life evaluation with the fruits that had been stored under hypobaric pressures for 2½, 5, 7 or 10 months.

Storing apples under hypobaric conditions resulted in delayed softening, control of physiological disorders and reduced decay development, which resulted in extended shelf-life after removal from storage (Laugheed *et al.*, 1978; Dilley *et al.*, 1982). Wang and Dilley (2000) also found that hypobaric conditions could prevent the development of scald. They reported that apples of the cultivars 'Law Rome' and 'Granny Smith' that were placed under hypobaric conditions of 38 mm Hg within 1 month of harvest did not develop scald during storage at 1°C, but scald developed if there was a delay of 3 months in establishing hypobaric conditions, as it did on fruit stored under atmospheric pressure throughout.

Jiao *et al.* (2013) suggested that exposure to 10 mm Hg at 10°C and greater than 98% RH had potential as an alternative disinfestation treatment against codling moth in apples and 15 days of exposure of the cultivar 'Red Delicious' had no detrimental effect on fruit quality. They studied eggs, 2nd to 3rd instar larvae, 5th instar larvae and pupae and found that the 5th instar larvae were the most tolerant stage for codling moth exposed to the treatment.

Asparagus

At 0°C and 20, 40 or 80 mm Hg, Dilley (1990) found that spears could be kept in marketable condition for 4–6 weeks with better AA retention at 20 mm Hg that at the other hypobaric conditions. Li and Zhang (2006) reported an extension in postharvest life of green asparagus in storage under 112.5 ± 37.5mm Hg. Li *et al.* (2006) found that storage life of asparagus at room temperature was 6 days, in refrigerated storage it was 25 days and in refrigerated storage with hypobaric conditions it was up to 50 days. They also reported that the spears under hypobaric storage had a lower respiration rate, lower losses of chlorophyll, AA, TA and TSS, reduced MDA accumulation and improved sensory quality. McKeown and Lougheed (1981) found that in storage at 3°C and near-saturation humidity under 61 mm Hg and also under 2 kPa O_2 + 0 kPa CO_2, CA-stored asparagus spears remained firm and green for 42 days while those in air were senescent. However, spears from all three treatments showed a disorder that resembled chilling injury. They concluded that there appears to be limited potential for the storage of asparagus at 3°C. The storage life of fresh green asparagus under hypobaric pressure of 266–342 mm Hg was extended to 50 days compared with 25 days refrigerated storage and only 6 days under room temperature, both at atmospheric pressure. Furthermore, the respiration rate was lower and it prevented loss of chlorophyll, vitamin C and acidity, improved sensory qualities and delayed postharvest senescence when stored under hypobaric conditions (Li *et al.*, 2006). Li (2006) stored the asparagus cultivar UC800 at −2°C and 85 ± 5% RH under hyperbaric conditions, starting at 10 kPa then transferred to 20 kPa and finally to 35 kPa. This combination was compared with storage under atmospheric pressure or constant reduced pressure levels. This three-stage hypobaric storage slowed the reduction of de-greening, sugar, soluble protein, AA and TA, compared with the normal atmosphere and constant hypobaric levels. It also reduced their respiration rate and ethylene emission and increased the activities of superoxide dismutase and catalase.

Avocado

Burg (2004) summarized his work over many years on the effects of hypobaric storage on avocados. The cultivar 'Choquette' stored at 14.4°C under atmospheric pressure started to ripen in 8–9 days and the fruits were fully ripe in 14 days. Softening of those under 40–101 mm Hg began after 25 days and when they were transferred to 20°C under atmospheric pressure all fruit developed normal taste with no internal blackening or decay. He subsequently found that in storage at 12.8°C, hypobaric conditions of 101–152 mm Hg were better than at 40–81 mm Hg; and in later work he reported that 21 mm Hg was optimal at 10°C. With the cultivar 'Waldin', he reported that in storage at 10°C their postharvest life was improved as the pressure was lowered from 101–152 mm Hg down to 60–81 mm Hg, with the fruit remaining firm for 30 days at 61–81 mm Hg compared with 12–16 days at atmospheric pressure. He reported similar results for avocados in storage at 12°C, but all fruit ripened quicker.

Spalding and Reeder (1976) compared storage of 'Waldin' at 7.2°C and 98–100% RH for 25 days at atmospheric pressure in air with CA storage under 2 kPa O_2 and 10 kPa CO_2 or 2 kPa O_2 and 0 kPa CO_2 and two hypobaric storage conditions in 91 mm Hg, one with added CO_2 at 10 kPa. After storage, all the fruit was ripened at 21.1°C. They found that 92% of the fruits stored in the CA of 2 kPa O_2 and 10 kPa CO_2 were acceptable and all those in the hypobaric conditions of 91 mm Hg plus 10 kPa CO_2, while none of those in the other treatments were acceptable. The factors that affected acceptability were anthracnose disease (*Colletotrichum gloeosporioides*) and chilling injury, both of which were completely absent in fruit stored under 91 mm Hg plus 10 kPa

CO_2. They defined acceptable fruit as having good appearance, free of moderate or severe decay and chilling injury and having no off-flavours. They also found no stem-end rot (*Diplodia natalensis*) directly after storage; and after ripening at 21.1°C again no stem-end rot was detected, except at low levels on those that had been stored under 91 mm Hg and higher levels in those that had been stored under 2 kPa O_2 and 10 kPa CO_2:

> Black pitted areas developed in lenticels during softening of avocados stored at atmospheric pressure or hypobaric plus 10 kPa CO_2. However, pitting was slight and was not considered to be objectionable to the average consumer. Tissue from the infected areas contained *Pestalotia* spp. fungus.

From this they concluded that high CO_2 was necessary for the successful storage of avocados which is borne out by many other workers. However, Burdon *et al.* (2008) found that the inclusion of CO_2 at 5 kPa under CA storage retarded fruit ripening but stimulated rot expression and they concluded that it should not be used for CA storage of New Zealand-grown 'Hass'. With the cultivar 'Lula', Burg (2004) reported that in storage at 7.2°C the fruits began to soften within 21 days and all were soft after 41 days under atmospheric pressure, while under 81–122 mm Hg it took 88 days for them to begin to soften. In another experiment at 8°C and 40–81 mm Hg, the fruits remained firm for 75–100 days compared with 23–30 days under atmospheric pressure. When they were stored at 61 mm Hg for 102 days, they became eating-ripe within 3–4 days when they were transferred to 26.7°C at atmospheric pressure (Burg, 2004). Spalding and Reeder (1976) compared storage of 'Lula' at 10°C and 98–100% RH for 6 weeks at atmospheric pressure in air, CA storage under 2 kPa O_2 and 10 kPa CO_2 and hypobaric storage under 76 and 152 mm Hg. After storage, all the fruit was ripened at 21.1°C and they found that 70% of the fruits stored under CA were acceptable and none of those in the other treatments were acceptable, which was mainly due to chilling injury symptoms and decay due to anthracnose. 'Lula' avocados stored under 2 kPa O_2 with 10 kPa CO_2 under atmospheric pressure were acceptable after softening and also this CA mixture inhibited the development of decay and chilling injury, confirming the finding that CO_2 is essential under CA storage (Spalding and Reeder, 1975).

With the cultivar 'Booth 8', Burg (2004) reported that after storage at 4.4°C under 40, 61, 81 or 122 mm Hg for 30 days they ripened in 2–3 days when transferred to 20°C under atmospheric pressure but had a poor flavour. In another experiment, waxed 'Booth 8' were stored at 7.8–10°C where they began to ripen in 8–22 days under atmospheric pressure, while under 40–81 mm Hg they did not soften during 50 days of storage but ripened without skin darkening when they were transferred to 23.9°C under atmospheric pressure. After 64 days of storage at 7.8–10°C under 40–81 mm Hg, fruits were still firm but they did not ripen to an acceptable quality when transferred to 23.9°C under atmospheric pressure.

With the cultivar 'Hass' it was reported that in storage at 5°C under atmospheric pressure fruits softened within 30 days, while under 15 to 40 mm Hg they were still hard and almost half of those under 61 mm Hg had begun to ripen (Cicale and Jamieson, 1978, quoted by Burg, 2004). In storage under 40 mm Hg, fruits began to soften after 38–45 days but under 15–21 kPa they remained firm and when removed to ambient conditions they ripened in 5.4 days, which was the same period as freshly harvested fruit. In another trial at 6°C for 35 days, fruit under atmospheric pressure lost 1.2% in weight at pressures ranging from 61 to 203 mm Hg compared with 5.7% under atmospheric pressure. After 70 days of storage the weight losses were 1.7% for 81 and 101 mm Hg and 3% for 61 mm Hg. They commented that their best results were storage at 61mm Hg, since this retarded softening, and fruits ripened normally when transferred to 14°C under atmospheric pressure and lower storage pressures resulted in higher desiccation.

Spores of *Glomerella cingulata* (which causes anthracnose) germinate on the surface of avocado fruit in the field and form appressoria. The fungus then remains quiescent until antifungal dienes in the skin of fruit break down due to degradation by lipoxygenase activity. Breakdown of the dienes has been shown to be delayed by various treatments, including hypobaric storage (Prusky *et al.*, 1983, 1995). Previous studies on hypobaric storage of avocados suggested that atmospheres both low in O_2 and high in CO_2 are required for successful suppression of anthracnose development (Spalding and Reeder, 1976).

Apricot

Salunkhe and Wu (1973b) and Haard and Salunkhe (1975) found that storage life of apricots could be extended from 53 days in cold storage to 90 days in cold storage combined with reduced pressure of 102 mm Hg. They found that hypobaric storage delayed carotenoid production, but after storage carotenoid, sugar and acid levels were the same as for those that had been in cold storage at atmospheric pressure.

Bamboo shoots

Chen *et al.* (2013b) stored freshly harvested bamboo shoots (*Phyllostachys violascens*) at 2 ± 1°C under various hypobaric conditions (101, 75, 50 and 25 kPa) for 35 days. They found that under 50 kPa there were reduced accumulations of lignin and cellulose in the shoots' cells. They also found that it inhibited ethylene production, reduced the rate of accumulation of MDA and hydrogen peroxide, and maintained significantly higher activities of superoxide dismutase (SOD), catalase and ascorbate peroxidase, but restrained the activities of PAL and POD. They therefore concluded that the delay in flesh lignification was due to maintenance of higher antioxidant enzyme activities and reduced ethylene production.

Banana

In his review, Burg (2004) reported that at 13.3–14.4°C the banana varieties 'Valery' and 'Gros Michel' ripened in 10 days under atmospheric pressure; they remained green for 40–50 days at 152 mm Hg, but mould could develop. He also reported that 'Valery' at 13.3°C remained green for more than 105 days in storage under 6.4, 9.5 or 12.7 kPa and ripened normally when they were transferred to atmospheric pressure and exposed to exogenous ethylene. There were no deleterious effects on flavour or aroma. Burg (2004) stored bananas under 49, 61, 76, 101, 122 and 167 mm Hg for 30 days and found they lost 1.1–3.6% in weight, with the higher losses at the higher pressures because of their higher respiration rates. All the fruits were still green and there were no apparent differences between the different hypobaric conditions. Bangerth (1984) found that there was no diminished response to ethylene in storage under 51 mm Hg, whatever the storage temperature tested. Apelbaum *et al.* (1977a) stored the variety 'Dwarf Cavendish' at14°C at 81 and 253 mm Hg and atmospheric pressure and found that they began to turn yellow after 30 days under atmospheric pressure, 60 days under 253 mm Hg and were still dark green under 81 mm Hg. When they were subsequently transferred to atmospheric pressure at 20°C and exposed to exogenous ethylene, they all ripened to a good flavour, texture and aroma. Bangerth (1984) successfully stored 'Cavendish' at 14°C under 51 mm Hg for 12 weeks and found that when they were subsequently ripened in exogenous ethylene at 50 µl/l they were the same quality as freshly harvested fruit or those that had been stored under atmospheric pressure.

Hypobaric storage has also been shown to delay the speed of ripening of bananas that have been initiated to ripen. Quazi and Freebairn (1970) found that fruit that had been initiated to ripen by exposure to exogenous ethylene for 2–5 h longer than required to initiate ripening did not ripen during hypobaric storage, but began to ripen within 1–2 days after transfer to ambient atmospheric pressure and had good eating quality and texture. However, fruit that had been initiated to ripen by exposure to ethylene with exposure time of 16 h longer than required did ripen during hypobaric storage, again with good eating quality and texture. Liu (1976b) initiated 'Dwarf Cavendish' to ripen by exposure to ethylene at 10 µl/l at 21°C. They then stored them at 14°C for 28 days under hypobaric storage of 51 and 79 mm Hg or CA storage of 1 kPa O_2 + 99 kPa N_2. All fruits remained green and firm and continued to ripen normally after they had been removed to ripening temperature in atmospheric pressure. Quazi and Freebairn (1970) showed that high CO_2 and low O_2 delayed the increased production of ethylene associated with the initiation of ripening in bananas, but the application of exogenous ethylene was shown to reverse this effect. Wade (1974) showed that bananas could be ripened in atmospheres of reduced O_2, even as low as 1 kPa, but the peel failed to de-green, which resulted in ripe fruit that were still green. Similar effects were shown at O_2 levels as high as 15 kPa. Since the de-greening process in Cavendish bananas is entirely due to chlorophyll degradation (Seymour *et al.*, 1987b; Blackbourn *et al.*, 1990), the CA storage effect was presumably due to suppression of this process. Hesselman and Freebairn (1969) showed that ripening of bananas that had already been initiated to ripen by ethylene was slowed in low O_2 atmospheres.

Bean

Green bean pods (*Phaseolus vulgaris*) could be kept in good condition in storage for 30 days at reduced pressure compared with 10–13 days in cold storage alone (Haard and Salunkhe, 1975). Spalding (1980) reported that the cultivars 'McCaslan 42' pole beans and 'Sprite' bush beans stored better at 7°C under 76 and 152 mm Hg for 2 weeks than similar beans stored at 760 mm Hg. Burg (1975) found that beans stored at 7.2°C and 60 mm Hg were in excellent condition after 26 days compared with those stored at the same temperature at 760 mm Hg, which were shrivelled and in poor condition. Knee and Aggarwal (2000) found that water-soaked lesions appeared on green beans kept in vacuum containers at 380 mm Hg; however, this effect is commonly associated with water condensation (Thompson, 2015).

Hypobaric storage has been shown to be effective on dried beans. Berrios *et al.* (1999) reported that the combined effect of refrigeration and hypobaric storage demonstrated potential for maintaining the fresh quality of black beans (*P. vulgaris*) in storage for up to 2 years. Black beans stored at 4.5°C and 50–60% RH and hypobaric pressure of 125 mm Hg exhibited quality factors characteristic of fresh beans, such as shorter cooking time, smaller quantities of solids loss, lower leaching of electrolytes and lower percentage of hard-shell than beans stored at 23–25°C and 30–50% RH. At 4.5°C, beans stored under 125 mm Hg had a germination rate of 93% while the rate for those stored at atmospheric pressure was 72%. Beans stored in ambient conditions exhibited hard-to-cook, which is a problem in the processing industry and means that their cooking time is considerably extended

Beet

In general, the weight loss after storage at 1°C under 61 mm Hg and near-saturation humidity and CA storage under 2 kPa O_2 + 0 kPa CO_2 was less than that which occurred after air storage. The beets also retained a fresh appearance after holding under 61 mm Hg and sensory evaluation indicated that those held under 61 mm Hg were similar to those held in air, while beets held under 2 kPa O_2 + 0 kPa CO_2 had a lower rating and off-flavours (McKeown and Lougheed, 1981).

Blueberry

Al-Qurashi *et al.* (2005) found that the cultivar 'Rabbiteye' stored under 13.5 mm Hg lost less weight, were firm, developed less decay and did not show any shrivelling during storage at 4°C for 28 days, compared with those stored at atmospheric pressure as a control. Borecka and Pliszka (1985) observed that blueberries stored under 38 mm Hg tasted good, contained less acid and had lower total soluble solid than other treatments. Burg (2004) reported that in storage at 0–2°C those under atmospheric pressure spoiled within 4 weeks while those under hypobaric storage ranging from 81 to 203 mm Hg were limited to 6 weeks. This was reported to be because of mould growth. He also quoted Dilley *et al.* (1989), who reported similar results on reduction in mould on the cultivar 'Jersey' during 44 days of storage, with 87% decay for those under atmospheric pressure and only some 11% for those under 21 mm Hg.

Broccoli

Broccoli heads were kept for 4 days at 1.1°C and then at 0°C and 10 mm Hg or atmospheric pressure for 21 days. They were assessed for quality after a shelf-life at 10°C. The ones that had been stored under atmospheric pressure had 60–90% yellowing and those that been in hypobaric storage had 40% yellowing with no off odours or off-flavours (Burg, 2004).

Brussels sprouts

Burg (2004) recommended that storage of Brussels sprouts under 10 mm Hg should be tested, since it could prevent the growth of microorganisms that could result in postharvest diseases, although Ward (1975) found no improvement in storage under 76 mm Hg and high humidity.

Cabbage

Cabbages, like other leaf vegetables, become chlorotic when exposed to ethylene. It was reported that white cabbages were successfully stored with apples at 0°C and 60 mm Hg (McKeown and Lougheed, 1981), presumably by limiting the effect of ethylene produced by the apples or removing the ethylene from the container before it could have an effect. Onoda *et al.* (1989) found that cabbages retained better appearance and had lower weight loss when stored in cycles between 100 and 300 mm Hg (with no humidification) than those stored at a constant atmospheric pressure.

Capsicum

Kopec (1980) compared storage of capsicum at atmospheric pressure with storage under 77 mm Hg and found that hypobaric storage retarded ripening, but once ripening started it progressed at the same rate as in atmospheric pressure. However, capsicums are classified as non-climacteric (Bosland and Votava, 2000) and so it is difficult to interpret these results, since climacteric fruits do not ripen. Perhaps what is being referred to is change of colour from green. Hughes *et al.* (1981) experimented with capsicums in hypobaric storage at 8.6–9.0°C and 82–90% RH under 38, 76 or 150 mm Hg. They found that they had a significantly higher weight loss during storage than those stored at normal pressure (Table 11.2). This was undoubtedly due to the comparatively low humidity, which should have been as close as possible to saturation. However, all had similar levels of 'sound' fruit after 20 days within the range of 60–72%. Hypobaric stored capsicums did not have an increased subsequent storage life compared with those stored under atmospheric pressure at the same temperature (Table 11.2). Burg (1975) found that capsicums could be kept for 7 weeks at 7.2°C under hypobaric storage without loss of colour or 'crispness'. Burg (2004) reported that under storage at 7.2°C and 80 mm Hg capsicums were in excellent condition after 28 days, while those stored under atmospheric pressure began to deteriorate after 16 days and were in poor condition after 21 days. Even after 46 days, capsicums under hypobaric storage were considered to be marketable except for a trace of mould on the cut stem. It was concluded that decay was the limiting factor in hypobaric storage. Jamieson (1980, quoted by Burg, 2004) found that after 50 days storage 87%, 80%, 47% and 0% were 'saleable' from 80, 40, 15 and 760 mm Hg, respectively, in experiments in the Grumman Allied Industries laboratory. Bangerth (1973) stored the cultivar 'Neusiedler Ideal' at 10–12°C for 23 days and found that those under 75 mm Hg were firmer, greener and had

Table 11.2. The effects of hypobaric conditions during storage and after removal on the mean percentage weight loss of fruit as a percentage of their original weight and on the mean percentage of sound marketable fruit. Modified from Hughes *et al.* (1981).

	20 days hypobaric storage at 8.8°C		7 days subsequent shelf-life in 760 mm Hg pressure at 20°C	
Pressure	Weight loss %	Marketable %	Weight loss %	Marketable %
760 mm Hg	0.03	68	0.87	43
152 mm Hg	0.16	72	0.76	46
76 mm Hg	0.22	68	1.00	36
38 mm Hg	0.15	60	1.10	42

slightly higher AA content and lower ethylene production than those stored under atmospheric pressure.

Cauliflower

Cauliflowers were kept for 4 days at 1.1°C then 0°C and 10, 20 or 40 mm Hg or atmospheric pressure for 21 days. They were assessed for quality after a shelf-life at 10°C by Burg (2004). At the end of the shelf-life test the ones that had been stored under atmospheric pressure had leaves that were yellow and dry and easily abscised with minimal handling. Those that been in hypobaric storage had some yellowing but remained firmly attached when handled and had a superior appearance, especially those that had been stored at 10 or 20 mm Hg. However, Ward (1975) found no improvement in storage under 76 mm Hg and high humidity.

Cherry

The cultivar 'Bing' stored well for 93 days in experimental hypobaric chambers at 102 mm Hg at 0°C. This was up to 33 days longer than those stored at 0°C under atmospheric pressure. However, their pedicels stayed green for only 60 days and pedicel browning can affect marketability. Hypobaric conditions delayed chlorophyll and starch breakdown in fruit as well as carotenoid formation and the decrease in sugars and total acidity (Salunkhe and Wu, 1973b). Patterson and Melsted (1977) reported that cherries could be stored for 6–10 weeks (depending on their condition at harvest) at 41–203 mm Hg with good retention of colour and brightness and delayed disease development. They also found that they stored equally well under CA storage with high CO_2, but under hypobaric conditions there were some problems with desiccation, especially at 41 mm Hg. CA storage recommendations included: –1°C to –0.6°C with 20–25 kPa CO_2 and 0.5–2 kPa O_2 which helped to retain fruit firmness, green pedicels and bright fruit colour (Hardenburg *et al.*, 1990); –1.1°C with 20–25 kPa CO_2 and 10–20 kPa O_2 (SeaLand, 1991); 20–30 kPa CO_2 reduced decay (Haard and Salunkhe, 1975); and 0–5°C with 10–12 kPa CO_2 and 3–10 kPa O_2 (Kader, 1989).

The effectiveness of short hypobaric treatments against postharvest rots was investigated by Romanazzi *et al.* (2001), who found that the cultivar Ferrovia exposed to 0.50 atm for 4 h had the lowest incidence of *Botrytis cinerea*, *Monilinia laxa* and total rots. Romanazzi *et al.* (2003) also found that the combination of spraying with chitosan at 0.1%, 0.5% or 1.0% 7 days before harvest and storage under 0.50 atm for 4 h directly after harvest effectively controlled fungal decay of sweet cherries during 14 days of storage at 0 ± 1°C, followed by a 7-day shelf-life. Fungi associated with rots included brown rot (*M. laxa*), grey mould (*B. cinerea*), blue mould (*Penicillium expansum*), Alternaria rot (*Alternaria* sp.) and Rhizopus rot (*Rhizopus* sp.).

Cranberry

Storing cranberries (*Vaccinium macrocarpon*) under 80 mm Hg resulted in a lower respiration rate and ethylene production as well as an extended shelf-life compared with those stored under atmospheric pressure (Pelter, 1975 quoted by Al-Qurashi *et al.*, 2005). Lougheed *et al.* (1978) explained that ethylene and respiration rate of cranberries stored under 80 mm Hg decreased compared with fruit stored under atmospheric pressure.

Cucumber

Hypobaric storage of cucumbers in 76 mm Hg extended the storage life to 7 weeks, compared with 3–4 weeks in cold storage (Bangerth, 1974), and reduced respiration rate by 67–75%. Burg (2004) reported that storage at 7.2°C resulted in chilling injury and reducing the pressure to 100, 120 or 160 mm Hg reduced chilling injury. However, under 80 mm Hg there were no symptoms after 7 weeks but chilling injury occurred within 1–2 days when they were removed to ambient conditions. Cucumbers were placed at 71 mm Hg for 6 h in the dark at 20°C to simulate air flight transportation and placed subsequently in cold storage facilities at 101 mm Hg at 20°C and 70% RH for 7 days. Results showed that cucumbers exposed to the hypobaric conditions had significantly more open stomata, compared with those at atmospheric pressure, after 96 h of subsequent storage (Laurin *et al.*, 2006).

Currants

Bangerth (1973) stored blackcurrants, redcurrants and white currants (all *Ribes sativum*) at about 3°C either in atmospheric pressure or under 76 mm Hg for up to 38 days. Those under the hypobaric conditions retained their AA content better than those under atmospheric pressure and also had higher sugar content, less decay and a better taste. Percentages of spoiled berries after 38 days of storage were 21–37% for those under atmospheric pressure and only 0.5–5% for those under hypobaric conditions.

Grape

Burg (1975) found that hypobaric storage extended the postharvest life of grapes from 14 days in cold storage in air to as much as 60–90 days. Burg (2004) subsequently reported that the cultivar Red Emperor could be stored for 90 days under hypobaric pressures of 30, 65, 80 or 160 mm Hg at 1.7–4.4°C, compared with 21–56 days at −0.5°C to 0°C at atmospheric pressure. Hypochlorous acid vapour was included to control diseases during hypobaric storage. Exposure to an air flow that had been in contact with hypochlorous acid at 0.25%, 1.5% or 5.25% had no mould after 60 days at 1.7–4.4°C. The effectiveness of short hypobaric exposure against postharvest rots had also been investigated by Romanazzi *et al.* (2001). They found that exposure of bunches of the cultivar Italia to 190 mm Hg for 24 h significantly reduced the incidence of grey mould during subsequent storage. Grapes were also wounded and inoculated after hypobaric treatment and it was determined that this treatment decreased infection and diameter of lesions significantly compared with the untreated fruits. The combination of pre-storage treatment with essential oils and storage under 50 kPa (380 mm Hg) gave good control of grey mould in grapes (Servili *et al.*, 2017).

Grapefruit

Haard and Salunkhe (1975) mentioned two reports that suggested that hypobaric storage could extend the postharvest life of grapefruit compared with cold storage alone. In one report the increase was from 20 days at normal atmospheric pressure to 3–4 months under hypobaric conditions and in the other from 30–40 days at normal atmospheric pressure to 90–120 days under hypobaric storage. Hypobaric storage at 380 mm (the lower limit of the experimental equipment) and 4.5°C had no effect on the incidence of chilling injury (Grierson, 1971).

Kohlrabi

Bangerth (1974) reported better leaf retention in kohlrabi in storage at 2–4°C under 75 mm Hg compared with storage under atmospheric pressure at the same temperature.

Leek

Ward (1975) found no improvement in storage of leeks under 76 mm Hg and high humidity.

Lettuce

It was claimed that hypobaric storage increased the storage life of lettuce from 14 days in conventional cold stores up to 40–50 days (Haard and Salunkhe, 1975). However, Ward (1975) found no improvement in storage under 76 mm Hg and high humidity and Bangerth (1973) found no improvement in storage under 75 mm Hg. Burg (2004) reported an increase in the physiological disorder, pink rib, when they were exposed to 30 or 80 mm Hg at 0–1°C for 4 weeks. Conversely Jamieson (1980, quoted by Burg, 2004) found that storage at 0.5°C and 90–95% RH under 10, 20, 40 or 80 mm Hg resulted in the lettuce remaining in excellent condition after 63 days of storage. The lettuce that had been stored in the same conditions at atmospheric pressure had significant incidence of pink rib, black heart and decay after only 37 days storage. Burg (2004) also reported that the manufacturer of hypobaric storage containers, Grumman Allied Industries, investigated marginal and pink discoloration and found that during storage at 5 mm Hg over 35 days the disorder was 'eliminated'. However, it was 'progressively accentuated' at 10–40 mm Hg and 'then decreased in frequency as the pressure was increased from 80 to 160 mm Hg'. It was concluded that the best storage condition was 2°C and 5 mm Hg for 21 days, but with storage for 36 days there was a high incidence of russet spotting. An *et al.* (2009) studied packaging of curled lettuce in small rigid containers under different hypobaric conditions. In both 190 and 380 mm Hg the lettuce had high moisture loss without any observable benefit in keeping quality compared with atmospheric pressure storage.

Lime

Spalding and Reeder (1976) reported that limes coated with wax containing 0.1% of either of the fungicides thiabendazole or benomyl remained green and suitable for marketing after 3–4 weeks under hypobaric storage of 170 mm Hg at 21.1°C. They also found that hypobaric storage did not affect chilling injury, but Pantastico (1975) reported that storage in 7kPa O_2 reduced the symptoms of chilling injury compared with storage in air. Haard and Salunkhe (1975) stated that 'Tahiti' limes could be stored for 14–35 days in cold storage, but this was extended to 60–90 days when it was combined with hypobaric conditions. Spalding and Reeder (1974a) stored 'Tahiti' limes under 152 mm Hg pressure for 6 weeks at 10°C. They showed only small changes in colour, rind thickness, juice content, TSS, total acids and AA. Decay averaged 7.8% compared with 8.3% for those stored in air. Those stored at 228 mm Hg maintained acceptable green colour, but were a slightly lighter green than limes at 152 mm Hg. Those stored under 76 mm Hg maintained acceptable green colour, but had low juice content, thick rinds and a high incidence of decay. Limes from all treatments had acceptable flavour. In subsequent work Spalding and Reeder (1976) found that 'Tahiti' limes retained green colour, juice content and flavour acceptable for marketing and had a low incidence of decay during storage at a pressure of 170 mm Hg for up to 6 weeks at 10°C or 15.6°C with 98–100% RH. Fruits stored under atmospheric pressure turned yellow within 3 weeks. They concluded that hypobaric storage, but not CA storage, could be used to extend the storage life of limes, which may provide an advantage for hypobaric storage. They also reported higher weight loss and shrivelling of limes during storage under hypobaric conditions compared with those stored at atmospheric pressure.

Loquat

Hypobaric storage at 40–50 mm Hg reduced decay by 87% and also reduced fruit browning, flesh 'leatheriness', respiration rate and ethylene production of loquats compared with those stored under atmospheric pressure at 2–4°C for 49 days. Peroxidase and PPO activities increased and then reached the highest values after 14 days in air storage. Hypobaric storage reduced the increase in PPO and POD activity compared with those under atmospheric pressure storage and delayed the onset of their peak activity (Gao *et al.*, 2006).

Jujube

Considerable work has been reported from China on the hypobaric storage of jujube (*Zizyphus jujuba*). Chang (2001) stored the cultivars 'LiZao' and 'DongZao' under hypobaric conditions. They found that during storage the decrease of firmness and the rate of browning were both significantly (p = 0.01) inhibited in storage under 154 mm Hg compared with those not under hypobaric conditions. Weight loss under hypobaric storage was less than 2.5%. It was also reported that the fruits turned red slowly under 154, 385 and 616 mm Hg. Wang *et al.* (2007b) found that during hypobaric storage of the jujube cultivar 'DongZao' the fruits softened more slowly and had better retention of AA and other organic acids than those stored at normal pressure. Cao (2005) found that hypobaric storage significantly retained firmness and vitamin C content, reduced acetaldehyde and ethanol contents in pulp and respiratory rate, inhibited AA oxidase and alcohol dehydrogenase activities and slowed down the rate of ethylene production in fruit. Hypobaric storage significantly reduced 'number of climacteric peak', but did not postpone 'climacteric peak'. The work was on the cultivars 'Huanghua', 'Zhanhual', 'Zhanhua 2', 'Shandong Wudi' and 'Dagang' harvested at different stages of maturity. Cui (2008) reported that hypobaric storage, under 20.3, 50.7 or 101.3 kPa, decreased the respiration rate, delayed loss of AA, decreased the rate of superoxide anion production and prolonged the postharvest life of the cultivar 'Lizao'. There were no significant differences between 20.3 kPa and 50.7 kPa; therefore 50.7 kPa was recommended, to reduce the cost. Jin *et al.* (2006) compared storage in different hypobaric conditions and found that 55.7 kPa resulted in retention of the best quality.

Mango

Ilangantileke *et al.* (1989) tested storage of the cultivar 'Okrang' at 13 °C and 60–100 mm Hg and found that the fruit kept well for 4 weeks. Apelbaum *et al.* (1977b) observed that at pressures below 0.05 MPa (50 mm Hg) mangoes underwent desiccation and, even upon ripening, did not develop their natural red and orange colour. However, fruits submitted to pressures of 0.013 and 0.01 MPa (10 and 13 mm Hg) had an extended storage life of 25 and 35 days, respectively. Several Florida varieties of mango that were ripened in air at ambient temperatures showed less anthracnose (*Colletotrichum gloeosporioides*) and stem-end rot (*Diplodia natalensis*) when they had previously been stored under hypobaric conditions at 13°C. The reduction in decay coincided with a retardation in fruit ripening, permitting prolonged storage at 13°C (Spalding and Reeder, 1977). It is possible that this effect on anthracnose and stem-end rot may have been a reflection of the delay in ripening, since postharvest diseases of mangoes, particularly anthracnose, develop as the fruit ripens (Thompson, 2015).

Okra

Knee and Aggarwal (2000) found that okra showed less darkening of seeds in vacuum containers at 380 mm Hg than in containers without a vacuum.

Onion

McKeown and Lougheed (1981), Hardenburg *et al.* (1990) and Burg (2004) reported that bulb onions stored at 26°C and 58% RH under 61 mm Hg for 28 days lost 12.2% in weight. McKeown and Lougheed (1981) reported that there was potential for the use of hypobaric conditions of 61 mm Hg in 26°C and low humidity for the curing of onions.

Burg (2004) found that at 0–3°C spring onions could be stored for almost 3 weeks under 50–60 mm Hg, compared with less than 6 days under atmospheric pressure. Also, storage under 100–150 mm Hg had only a slight beneficial effect compared with atmospheric pressure. Storage at 1°C under 61 mm Hg and near-saturation humidity had little effect on the content of leaf chlorophyll of spring onions after 21 days. In general, their weight loss after hypobaric storage or CA storage under 2kPa O_2 + 0kPa CO_2 was less than that which occurred after air storage. The spring onions also retained a fresh appearance after hypobaric storage, but there was no difference among treatments in the sensory evaluation (McKeown and Lougheed, 1981). They also found better chlorophyll retention in storage at 1°C under 55–60 mm Hg than under atmospheric pressure or CA storage. However, Ward (1975) found no improvement in storage under 76 mm Hg and high humidity.

Orange

Min and Oogaki (1986) reported a reduction in the respiration rate of oranges stored under hypobaric conditions compared with those stored under atmospheric pressure. The respiration rate remained lower even after the fruits were transferred to atmospheric pressure. The orange cultivar 'Fukuhara' had lower rates of ethylene production and respiration rate after hypobaric storage compared with those under atmospheric pressure throughout. Oranges stored at 190 mm Hg and 98% RH decayed more rapidly after return to atmospheric pressure than those stored at 190 mm Hg and 75% RH or atmospheric pressure and 86% RH throughout (Min and Oogaki, 1986). However, the fruits stored at these humidity levels and low pressure would probably have been severely desiccated.

Papaya

Papaya stored at 10°C and 98% RH and a pressure of 20 mm Hg ripened more slowly and had less disease development than fruit at atmospheric pressure in refrigerated containers (Alvarez, 1980). Alvarez and Nishijima (1987) also reported that hypobaric storage appeared to suppress postharvest disease development in papaya fruit.

Parsley

It was reported that storage at 3°C under hypobaric conditions of 75 mm Hg extended the storage life of parsley from 5 weeks in atmospheric pressure to 8 weeks without appreciable losses in protein, AA or chlorophyll content (Bangerth, 1974).

Peach

Hypobaric storage prolonged the postharvest life of peaches from 7 days at atmospheric pressure to 27 days. It delayed chlorophyll and starch breakdown, carotenoid synthesis and the decrease in sugars and acidity (Salunkhe and Wu, 1973a). Storage at 11–12°C in an aerated hypobaric system at either 190 or 76 mm Hg retarded ripening, compared with storage at the same temperature under atmospheric pressure (Kondou *et al.*, 1983). In the cultivar 'Okubo' that had been stored under hypobaric conditions of 108 mm Hg, the ripening rate after storage was slower than that of fruit that had been stored in air at atmospheric pressure throughout. The physiological disorder known as mealy breakdown was reduced under hypobaric storage, but no effect was found on flesh browning or on abnormal peeling (Kajiura, 1975). S.W. Porritt and R.E. Woodruff (quoted by Pattie and Lougheed, 1974) suggested that 'LPS (low pressure storage) has not proven suitable for peaches'. They recommended that additional research was needed to assess further the value of hypobaric storage to extend storage life of peaches. Wang *et al.* (2015) stored 'Honey' peach under 101, 10–20, 40–50 or 70–80 kPa for 30 days at 0°C and at 85–90% RH followed by 4 days at 25°C and 80–85% RH. They found that storage at 10–20 kPa delayed decay rates, maintained overall quality and extended the shelf-life. They showed that it effectively delayed increases in both O_2 radical dot production rate and ozone content, enhanced activities of catalase and superoxide dismutase and increased contents of adenosine triphosphate and adenosine diphosphate, while reducing lipoxygenase activity and adenosine monophosphate content during the storage period as well as the following shelf-life.

Pear

Hypobaric conditions prolonged the storage life of pears by 1.5–4.5 months compared with cold storage alone. It delayed chlorophyll and starch breakdown, carotenoid formation and the decrease in sugars and acidity (Salunkhe and Wu, 1973a).

Pineapple

Staby (1976) stated that storage of pineapple under hypobaric condition could extend the storage life for up to 30–40 days.

Plum

Prune (*Prunus domestica*) cultivars that grown mainly for dried fruit production (prune is the French word for plum) were successfully stored under 40–101 mm Hg by Patterson and Melsted (1978).

Potato

Greening and glycoalkaloids synthesis, including solanine, can occur in stored potato tubers when they are exposed to strong light (Mori and Kozukue, 1995). Burg (2004) reported that greening in light was inhibited in potato tubers stored at 15°C and 126 mm Hg but there was no effect on their solanine content. He also reported similar results during storage at 2.5°C, 5°C, 10°C or 20°C and 76–95 mm Hg and under CA storage with 2–2.5 kPa O_2.

Radish

McKeown and Lougheed (1981) reported that during storage at 1°C under 55–60 mm Hg for 7–28 days the weight losses were similar to the levels reported for CA storage with 2 kPa O_2 and slaked lime to control CO_2. Off-flavours developed in the CA stored radishes but not in those under hypobaric storage. Bangerth (1974) found that in storage at 2.2–2.8°C and 95% RH there was decreased losses of protein, AA and chlorophyll losses under 75 mm Hg compared with storage in the same conditions under atmospheric pressure. In general, the weight loss during storage at 1°C under 61 mm Hg and near-saturation humidity and CA storage in 2 kPa O_2 + 0 kPa CO_2 was less than that which occurred during storage under atmospheric pressure. Radishes also retained a fresh appearance after storage under 61 mm Hg and their sensory evaluation indicated that they were similar to those held in air, while those held under 2 kPa O_2 + 0 kPa CO_2 atmosphere had a lower rating and off-flavours (McKeown and Lougheed, 1981). However, Ward (1975) found no improvement in storage under 76 mm Hg and high humidity.

Spinach

Bangerth (1973) found that storage of spinach at 3°C and 95% RH with 75 mm Hg retained the green colour, vitamin C and total protein content for almost 7 weeks.

Squash

Burg (2004) reported that storage of 'Yellow Crookneck' squash at 7.2°C and 65, 80 or 150 mm Hg had no beneficial effects compared with storage at atmospheric pressure. McKeown and Lougheed (1981) carried out limited trials with 'Acorn' squash at 10°C and at low humidity under 61 mm Hg for 42 days after curing, which 'showed some potential'.

Strawberry

It was reported by Haard and Salunkhe (1975) that the strawberry cultivars 'Tioga' and 'Florida 90' in cold storage could be stored for 21 days under hypobaric pressure, compared with 5–7 days at atmospheric pressure. Bubb (1975c) compared storage of the cultivar 'Cambridge Favourite' at 3.3°C under atmospheric pressure with storage under 25 mm Hg and 101 mm Hg and found better flavour retention under hypobaric storage. When he compared this with storage in O_2 partial pressures similar to those obtained under the hypobaric conditions, he did not observe similar flavour retention. There was some evidence of slight tainting of the flavour after hypobaric storage, but after subsequent shelf-life at 18.5°C he reported that the tainting of the flavour of the fruit became objectionable. After 10 days of storage none of the samples exceeded 2% weight loss, but there were high levels of rotting in all samples, though no consistent effect on fruit firmness. Knee and Aggarwal (2000) found a higher proportion of strawberry fruits were infected by fungi in vacuum containers at 380 mm Hg than in containers with no vacuum. An *et al.* (2009) also studied hypobaric packaging in rigid small containers during storage at 3°C with intermittent fresh-air flushing and vacuum treatment to avoid creating an anoxic environment. Bacterial growth was slightly reduced in fruit in hypobaric packaging, but there was an operational drawback with the 25.3 kPa conditions in that fresh-air flushing and repetitive vacuum application were required too frequently for the densely packed strawberry packages.

The effectiveness of short hypobaric treatments against postharvest rots was investigated by Romanazzi *et al.* (2001). On 'Pajaro', the greatest reductions of *B. cinerea* and *Rhizopus stolonifer* rot were observed on fruits treated for 4 h at 0.25 and 0.50 atm, respectively. Hashmi *et al.* (2013a) exposed strawberries to hypobaric pressures of 190, 380 and 570 mm Hg for 4 h at 20°C and subsequently stored them at 20°C or 5°C. They found that exposure to 380 mm Hg consistently delayed rot development at both 20°C and 5°C and did not affect their weight loss and firmness, but an initial increase in respiration rate was observed. They also found (Hashmi *et al.*, 2013b) that exposure of strawberries to 380 mm Hg for 4 h reduced rot incidence from natural infection during subsequent storage for 4 days at 20°C. The same treatment had the same effect when the fruit had been inoculated with *B. cinerea* or *R. stolonifer* spores. They found that activities of defence-related enzymes were increased after exposure to hypobaric conditions. PAL and chitinase peaked 12 h after exposure, while POD increased immediately after exposure. PPO activity remained unaffected during subsequent storage for 48 h at 20°C. However, they found that exposure to atmospheric pressure for 4 h did not influence rot development.

Sweetcorn

Haard and Salunkhe (1975) found that hypobaric storage of sweetcorn could increase storage life to 21 days from only 4–8 days in a conventional cold storage at atmospheric pressure. Burg (2004) stored the cultivar 'Wintergreen' at 1.7 ± 1°C at pressures over the range of 10–760 mm Hg and found that the respiration rate decreased with decreasing pressure. After 7 days, a sensory evaluation panel determined that those stored under 20 mm Hg had the highest scores for sweetness and flavour, while after 11 days those that had been stored under 50 mm Hg had the best flavour score.

Tomato

Storage of mature-green tomatoes under hypobaric conditions of 646, 471, 278 or 102 mm Hg at 12.7°C resulted in a reduction in fruit respiration, especially under 102 mm Hg

(Wu and Salunkhe, 1972). Wu *et al.* (1972) stored tomatoes for 100 days under 102 mm Hg and then transferred them to 646 mm Hg, both at 12.8°C and 90–95% RH, where they ripened normally based on their lycopene, chlorophyll, starch, sugar and β-carotene content. However, their aroma was inferior to those that had been stored under atmospheric pressure. Dong *et al.* (2013) demonstrated that mid-climacteric tomatoes harvested at the breaker or pink stages exposed to 76 mm Hg for 6 h showed transient increased sensitivity to 1-MCP. Subsequently they found that mid-climacteric fruit exposed to 1-MCP at 500 nl/l under 76 or 16 mm Hg for 1 h showed acute disturbance of ripening. They concluded that high efficacy of 1-MCP applied under hypobaric conditions was due to rapid ingress and accumulation of internal gaseous 1-MCP.

Turnip

Onoda *et al.* (1989) found that turnips retained better appearance and lower weight loss when stored in cycles of hypobaric pressures (between 100 and 300 mm Hg with no humidification) than those stored at a constant atmospheric pressure.

Watercress

Hypobaric storage of watercress at 3°C and 75 mm Hg for 12 days resulted in them retaining their AA content better than those stored under atmospheric pressure. There was also better retention of protein and chlorophyll content under hypobaric storage (Bangerth, 1974).

12

Recommended CA Conditions

Recommended storage conditions vary considerably even for the same species. The factors that can affect storage conditions are dealt with in earlier chapters, but the definition of exactly what is meant by maximum or optimum storage period will also vary. So recommendations may be dependent on the subjective judgement of the experimenter or store operator. They usually include the assumption that the crop is in excellent condition before being placed in storage and may also assume that pre-storage treatments such as fungicides and pre-cooling have been applied, that there has been no delay between harvesting the crop and placing it in storage and that the fruit or vegetable had been harvested at an appropriate stage of maturity. In this chapter, recommended conditions are given by species and sometimes by cultivar within a species. It was considered that it could be confusing if the author gave his subjective judgement and so it is for the reader to make their own judgement on the basis of sometimes conflicting recommendations, depending on their own experience and objectives.

Apple (*Malus domestica, M. pumila*)

There are over 7500 known cultivars of apple (Wikipedia, 2018) and each growing region varies in cultivars grown and their recommended storage conditions. CA storage recommendations for each region are constantly evolving with changes in cultivars, clones of cultivars and new technologies, which means that recommendations over 10 years old could be of limited value to current store managers. Examples of well done regional storage recommendations are those of the UK Department for Environment, Food and Rural Affairs (Defra, 2017; Defra AHDB, 2017), a Swiss publication (Gasser and Siegrist, 2011) and two French publications (Mathieu-Hurtiger *et al.*, 2010, 2014). The following recommendations will refer primarily to static CA storage. DCA storage may be included as well, where available.

The following cultivar discussion is restricted to the most popular cultivars worldwide. In 2015, in terms of worldwide production, the top ten cultivars (including clones) were 'Fuji', 'Delicious', 'Golden Delicious', 'Gala', 'Granny Smith', 'Idared', 'Jonagold', 'Cripps Pink' (Pink Lady®), 'Braeburn' and 'Elstar'. The next ten cultivars were 'McIntosh', 'Honeycrisp' (Honeycrunch®), 'Jonathan', 'Rome Beauty', 'Gloster', 'Empire', 'Orin', 'Reinette du Canada', 'Melrose' and 'Tsugaru' (modified from O'Rourke, 2017). Between 2015 and 2025, amongst other cultivars, only four are expected to increase their proportion of total production:

'Scifresh' (Jazz®), 'Pinova' (Piñata®, Evelina®), 'Ambrosia' and 'Nicoter' (Kanzi®) (O'Rourke, 2017). Clearly the CA storage conditions for many more cultivars have been published and reviewed in previous editions of this book (Thompson, 1998, 2010) and elsewhere. In this edition we focus on the most recent literature of the 24 most important cultivars in terms of production. Also individual 'Delicious' strains would be removed and covered in the 'Delicious' section.

Ambrosia

'Ambrosia' is regarded as a low ethylene producer (DeLong *et al.*, 2016). There is no consistent evidence indicating whether it is CO_2 sensitive or not. DeEll (2012) recommended two CA options:

- For 6–7 months storage, 0°C in 2.5 kPa O_2 + 2.0 kPa CO_2.
- For 6–8 months storage, 0–1°C in 1.7 kPa O_2 + 1.0 kPa CO_2.

Since some research indicates that 'Ambrosia' has a tendency to produce low-temperature disorders, such as soft scald and core flush, a more recent recommendation is a storage temperature of 3°C with 2.0–2.5 kPa O_2 and CO_2 (John DeLong, personal communication, 2018).

Braeburn

'Braeburn' is considered to be CO_2 intolerant (Kupferman, 2003). Thus, CO_2 should remain well below the O_2 level at all times and temperatures should be held slightly elevated during CA storage (> 0°C), especially in cooler regions. This is less critical in warmer growing regions (e.g. Brazil, South Africa). Since the application of 1-MCP significantly increases CO_2 intolerance in 'Braeburn' fruit, 1-MCP is not recommended.

The recommendation for UK-grown 'Braeburn', for storage until March, is 1.5–2.0°C with 1.2–2.0 kPa O_2 and < 1.0 kPa CO_2, with a delay of 3 weeks before establishing CA (Defra AHDB, 2017). The recommendation for Switzerland is 1.5°C with 1.5 kPa O_2 + 1.0 kPa CO_2 (Gasser and Siegrist, 2011). In France, the 'Braeburn' recommendation is to store at 0.5–1.0°C with a choice of three CA recommendations: (i) CA at 2–3 kPa O_2 + 1.5 kPa CO_2 (March); (ii) ULO at 1.5–1.8 kPa O_2 + 1 kPa CO_2 (April); or (iii) XLO (extreme low oxygen) at 1 kPa O_2 + ≤ 0.8 kPa CO_2 (April to May) (Mathieu-Hurtiger *et al.*, 2010, 2014). In South Tyrol, Italy, the recommendation is 1.3°C with 1.5 kPa O_2 and 1.0 kPa CO_2 with the O_2 pulldown delayed by 2 weeks (Zanella and Rossi, 2015). The recommendation for 'Braeburn' in Washington State is 0–1°C with 1.5 kPa O_2 and 0.5 kPa CO_2; for New Zealand 0.5°C with 3.0 kPa O_2 and 0.5–1.0 kPa CO_2; and for South Africa –0.5°C with 1.5 kPa O_2 and 1.5 kPa CO_2 (Kupferman, 2003). In Patagonia, Argentina, the recommendation is 0.5–1°C with 1.2–3.0 kPa O_2 and 0.5–1.2 kPa CO_2 (Benítez, 2001). In Australia, Little *et al.* (2000) recommended step-wise cooling to a final temperature of 1.0°C with 2.5 kPa O_2 and 0.5 kPa CO_2. In commercial practice in Washington State in 2017, the store temperature was 1.7°C rather than 0–1°C as recommended by Kupferman (2003) (Helgard Ackermann, personal communication). Storage of 'Braeburn' at 1°C in DCA, which was set at 0.3–0.4 kPa O_2 + 0.7–0.8 kPa CO_2, maintained firmness during some 7 months of storage followed by a shelf-life period of 7 days at 20°C. DCA storage reduced 'Braeburn' browning disorder (BBD), compared with fruit stored in 1.5 kPa O_2 + 1.0 kPa CO_2. The susceptibility to low temperature breakdown (soft scald, +7.6%) and external CO_2 injury (+3.6%) was significantly higher in DCA storage than storage in 1.5 kPa O_2 + 1.0 kPa CO_2 (Lafer, 2008). Both of these disorders can be controlled by higher storage temperature and Gasser *et al.* (2008) successfully stored 'Braeburn' at 3°C for 200 days in either DCA-CF or DCA-RQ without any physiological disorders, while fruit firmness of DCA-stored fruit was, in general, significantly higher than in ULO-stored control fruit. Zanella and Rossi (2015) also reported greater firmness retention in DCA-CF-stored 'Braeburn' fruit stored at 1.3°C, compared with the recommended CA (1.5 kPa O_2 + 1.0 kPa CO_2). Since 'Braeburn' is prone to greater storage disorders if treated with 1-MCP, DCA has become a popular storage technology for 'Braeburn'. For example, by 2011–2012, over 40% of the 'Braeburn' apples in South Tyrol, Italy, were stored in DCA-CF (Zanella and Stürz, 2013).

Cripps Pink (including clones)

'Cripps Pink' is also marketed as 'Pink Lady' or by clone name, e.g. Rosy Glow, Lady in Red. 'Cripps Pink' originated from a cross between 'Lady Williams' and 'Golden Delicious' made by J.E.L. Cripps at the Stoneville Horticultural Research Station, near Perth in Western Australia, in 1973. The aim of the cross was to combine the sweet, superficial-scald-free fruit of 'Golden Delicious' with the firm, long-storing fruit of 'Lady Williams' (Cripps *et al.*, 1993). Unfortunately, superficial scald can appear in 'Cripps Pink' if the fruit are harvested early and stored in air, rather than in CA (Cripps *et al.*, 1993; Kupferman, 2003). 'Cripps Pink' is a firm apple with a very low water permeance, regardless of harvest date (Maguire *et al.*, 1997), which is a factor that may affect its CA storage requirements. 'Cripps Pink' is considered to be CO_2 intolerant (Jobling *et al.*, 2004). Thus, CO_2 should remain well below the oxygen level at all times and temperatures should be held slightly elevated during CA storage (> 0°C), especially in cooler growing regions (James *et al.*, 2010). This is less critical in warmer growing regions, e.g. Australia. In Australia, Cripps *et al.* (1993) stated that the storage life can be 8–9 months or longer if held at 0–1°C in CA (O_2 and CO_2 not specified) and Little *et al.* (2000) recommended that Australian 'Cripps Pink' be stored at 0.5°C with 1.5 kPa O_2 + 1.0 kPa CO_2 and suggested that, where possible, ethylene should be scrubbed during loading and for the first 10 days. In Argentina, Benítez (2001) recommended 0°C with 1.2–2.5 kPa O_2 + 0.8–1.5 kPa CO_2. In France, Mathieu-Hurtiger *et al.* (2010, 2014) stated that 'Cripps Pink' is sensitive to internal browning, caused by late harvests, the storage temperature being too low, high CO_2 levels and superficial scald (as a result of early harvesting). They gave three recommendations for 'Cripps Pink' with increasing storage durations: (i) CA at 1.5–2.5°C with 2–3 kPa O_2 + < 1.5 kPa CO_2; (ii) ULO at 2.0–2.5°C with 1.5–1.8 kPa O_2 + < 1.0 kPa CO_2; or (iii) XLO at 2.0–2.5°C with 1.0 kPa O_2 + ≤ 0.8 kPa CO_2. In South Tyrol in Italy, stores for 'Cripps Pink' are filled at 4.5–5.0°C, the temperature is decreased over 2 weeks to 2.5°C, at a maximum rate of 1.5°C each week, then CA is established at 1.8 kPa O_2 + 1.3 kPa CO_2 (Zanella and Rossi, 2015). A significant quantity of 'Cripps Pink' is stored commercially in DCA-CF in South Tyrol, primarily to achieve some quality benefits, including control of superficial scald (Zanella and Stürz, 2013). Zanella and Rossi (2015) showed that both DCA-CF and 1-MCP improve firmness retention in 'Cripps Pink', compared with CA at 1.8 kPa O_2.

Delicious (including clones, also known as strains)

There are several strains of 'Delicious' (also known as 'Red Delicious'), such as Starking, Starkrimson, Starkspur, Ultrared, Oregon Spur, Early Red One, Topred and Classic. Some publications will only identify the 'Delicious' apple by the clone name, omitting the fact that they are strains of 'Delicious'. Kupferman (2003) described 'Delicious' as somewhat CO_2 tolerant and also tolerant of rapid CA. In South Tyrol, Italy, Zanella and Rossi (2015) recommended two CA options, both at 1.3°C: (i) 1.5 kPa O_2 + 1.3 kPa CO_2; or (ii) 1.0 kPa O_2 + 1.0 kPa CO_2. In Patagonia, Argentina, Benítez (2001) recommended –0.5°C to +0.5°C in 2.0 kPa O_2 + 1.5–2.0 kPa CO_2. DeEll (2012) recommended two CA options for Ontario, Canada: (i) for 7–9 months of storage, 0°C in 2.5 kPa O_2 + 2.5 kPa CO_2; or (ii) for 8–10 months, 0°C in 0.7–1.5 kPa O_2 + 1.0–1.5 kPa CO_2. Little *et al.* (2000) recommended that Australian-grown 'Delicious' (all strains) be stored at 0.5°C in 1.8 kPa O_2 + 2.5 kPa CO_2. In France, the 'Delicious' (labelled as 'Rouge Américaines') recommendation is to store at 0.5–1.0°C with a choice of three CA recommendations (Mathieu-Hurtiger *et al.*, 2010, 2014) with increasing storage durations: (i) CA at 2–3 kPa O_2 + 3.0 kPa CO_2; (ii) ULO at 1.5–1.8 kPa O_2 + < 2.0 kPa CO_2; or (iii) XLO at 0.8–1.0 kPa O_2 + 0.8 kPa CO_2.

DeLong *et al.* (2007) found that at 0°C for up to 8 months the HarvestWatch chlorophyll fluorescence system (DCA-CF) retained firmness and quality to a greater extent than 0.9–1.1 kPa O_2 + 1.2–1.3 kPa CO_2. 'Delicious' was one of the first cultivars to be stored commercially in DCA. 'Delicious' is prone to superficial scald and DCA controls superficial scald as well as maintaining firmness (DeLong *et al.*, 2004). As a result, by 2011–2012, over 40% of the 'Delicious' apples in South Tyrol, Italy, were stored in DCA-CF (Zanella and Stürz, 2013).

Elstar

'Elstar', like 'Golden Delicious' (Kupferman, 2003), one of its parents, can be characterized as CO_2 tolerant and also tolerant of rapid CA. Thus, CO_2 can remain above the oxygen level at all times. Kupferman (2003) reported that 'Elstar' in Nova Scotia, Canada, can be stored at 0–0.5°C in 2.5 kPa O_2 + 4.5 kPa CO_2 and Netherlands 'Elstar' can be held at 1.8°C in 1–1.2 kPa O_2 + 2.5 kPa CO_2. In Switzerland, there are two options, both at 0.5°C: (i) 2.0 kPa O_2 + 3 kPa CO_2; or (ii) 1.0 kPa O_2 + 3.0 kPa CO_2 (Gasser and Siegrist, 2011). In France, the 'Elstar' recommendation is to store at 0.5–2.0°C with a choice of two CA recommendations with increasing storage durations (Mathieu-Hurtiger *et al.*, 2010, 2014): (i) CA at 2–3 kPa O_2 + 2–3 kPa CO_2; or (ii) ULO at 1.5–1.8 kPa O_2 + 1.5 kPa CO_2, with insufficient data for an XLO recommendation.

'Elstar' grown in some parts of Europe is prone to having skin flecking on the fruit in storage that can be exacerbated by 1-MCP. The influence of DCA-CF storage with stepwise O_2 reduction, in comparison with storage in 1.4 kPa O_2, was examined by Hennecke *et al.* (2008). The results showed significant improvements in flesh firmness after removal from DCA storage and also post-storage for 3 weeks under cold storage condition compared with 1.4 kPa O_2 storage. The occurrence of skin spots was also significantly reduced by DCA storage. Veltman *et al.* (2003) used a DCA storage based on ethanol production to establish the lowest possible O_2 concentration under which the fruit could be stored. They found that fruit was firmer, tended to develop less skin spots and had a better colour retention after DCA storage, especially after shelf-life, compared with traditional CA storage. Schouten (1997) compared storage at 1–2°C in either 2.5 kPa CO_2 + 1.2 kPa O_2 or < 0.5 kPa CO_2 + 0.3–0.7 kPa O_2. The fruit stored in 0.5 kPa CO_2 + 0.3–0.7 kPa O_2 was shown to be of better quality both directly after storage and after a 10-day shelf-life period in air at 18°C. Gasser *et al.* (2008) successfully stored 'Elstar' at 3°C for 200 days in either DCA-CF or DCA-RQ without any physiological disorders, while fruit firmness of DCA-stored fruit was similar to ULO-stored control fruit. Köpcke (2009) examined the influence of different O_2 concentrations in CA storage on the skin spots occurrence on 'Elstar' apples with and without previous SmartFresh (1-MCP) treatments and concluded that higher O_2 levels and previous SmartFresh treatments increased the disorder, whereas almost no skin spots were detected on fruit stored under DCA-CF. Köpcke (2015) examined the possibility of increasing the store temperature (to save energy) from the recommended 2.0°C in ULO of 1.4 kPa O_2 + 2.6 kPa CO_2 to higher temperatures of either 3.5°C or ≤ 10°C in ULO or DCA-CF (with and without 1-MCP, always with 2.6 kPa CO_2). The results showed that increasing the store temperature from 2.0°C to 3.5°C maintained most, but not all, quality measurements when using DCA-CF or 1-MCP alone, but combining DCA-CF and 1-MCP showed no loss in quality.

Empire

Recent research shows that 'Empire' should be considered a chilling and CO_2-sensitive cultivar. Wang *et al.* (2000c) reported a CO_2-linked disorder of 'Empire' with symptoms resembling superficial scald. Symptoms were effectively controlled by conditioning fruit at 3°C (but not at 0°C) for 3–4 weeks at

1.5–3 kPa O_2 + 0 kPa CO_2 prior to storage in 1.5 kPa O_2 + 3 kPa CO_2. Reduction of the disorder was also achieved by storage at 3°C in air, but excessive ripening and associated loss of flesh firmness occurred during subsequent CA storage. Defra AHDB (2017) recommended that UK 'Empire' fruit be stored until February at 3–4°C in 2.0 kPa O_2 + < 1.0 kPa CO_2. DeEll (2012) recommended four CA options for Ontario (Canada) 'Empire' fruit, all at 2°C: (i) stored for 5–7 months, 2.5 kPa O_2 + 2.0 kPa CO_2; (ii) for > 7 months, use 1-MCP and store in 2.5 kPa O_2 + < 0.5 kPa CO_2 (for the first 6 weeks) and then 2.0 kPa CO_2; (iii) for > 7 months, use 1-MCP and DPA (to control CO_2 damage) and store in 2.5 kPa O_2 + 2.0 kPa CO_2; or (iv) for 6–8 months, store in 1.5 kPa O_2 + 1.0-1.5 kPa CO_2.

Fuji

'Fuji' is a low ethylene producer (Benítez, 2001). It is considered to be CO_2 intolerant (Kupferman, 2003). Thus, CO_2 should remain well below the oxygen level at all times and temperatures should be held slightly elevated during CA storage (> 0°C), especially in cooler growing regions. This is less critical in warmer growing regions (e.g. Brazil). Little *et al.* (2000) recommend that Australian 'Fuji' be stored at 0.5°C with 2.5 kPa O_2 + 2.0 kPa CO_2. The recommendation for Washington State 'Fuji' is 1°C with 2.0 kPa O_2 + 0.5 kPa CO_2 (Kupferman, 2003). The recommended CA condition for 'Fuji' apple in the South Tyrol, Italy, is 1.3°C with 1.5 kPa O_2 + 1.3 kPa CO_2 (Zanella and Rossi, 2015). In Ontario, Canada, the recommendation is 0–1°C with 2.5 kPa O_2 + 1.0–2.0 kPa CO_2 (DeEll, 2012). In Patagonia, Argentina, the recommendation is 0.5–1°C with 1.2–2.5 kPa O_2 + 0.8–1.3 kPa CO_2 (Benítez, 2001). In France, the 'Fuji' recommendation is to store at 0.5 –1.0°C with a choice of two CA recommendations with increasing storage durations: (i) CA at 2–3 kPa O_2 + 1.5 kPa CO_2; or (ii) ULO at 1.5–1.8 kPa O_2 + 1.0 kPa CO_2, with no XLO recommendation (Mathieu-Hurtiger *et al.*, 2010, 2014). A significant amount of 'Fuji' is stored commercially in DCA-CF in South Tyrol, Italy, primarily to achieve some quality benefits, including superficial scald control (Zanella and Stürz, 2013). 'Fuji' is noted for having good firmness retention, perhaps due to its low ethylene production, and this was confirmed by Zanella and Rossi (2015), who showed that both DCA-CF and 1-MCP did little to improve firmness retention in 'Fuji', compared with CA (1.5 kPa O_2).

Gala (including clones)

There are numerous 'Gala' clones including Mondial Gala, Must Gala, Royal Gala, Galaxy Gala and Ruby Gala that have arisen by bud mutation, some of which have been patented. Kupferman (2003) considered 'Gala' (like 'Golden Delicious') to be tolerant of low storage temperature, CO_2 and rapid CA. Based on his survey of various recommendations he concluded that 'Gala' can be stored as low as 1.0 kPa O_2 with CO_2 levels up to 2.5 kPa at 1°C and that the O_2 should be raised as the temperature is lowered below this point. Defra AHDB (2017) recommended that UK 'Gala' be stored at +0.5°C in various combinations of O_2 and CO_2, depending on storage duration: (i) storage until late December in 2 kPa O_2 + < 1.0 kPa CO_2; (ii) storage until mid-February in 1.2 kPa O_2 + < 1.0 kPa CO_2; or (iii) storage until late March in 1 kPa O_2 + 5 kPa CO_2. Gasser and Siegrist (2011) recommended that Swiss 'Gala' be stored in CA at +0.5°C in either 2 kPa O_2 + 2–3 kPa CO_2 or 1 kPa O_2 + 3 kPa CO_2. In France, the 'Gala' recommendation is to store at 0.5–1.0°C with a choice of three CA recommendations with increasing storage durations: (i) CA at 2–3 kPa O_2 + 2–3 kPa CO_2; (ii) ULO at 1.5–1.8 kPa O_2 + 1.5 kPa CO_2; or (iii) XLO at 0.8–1.0 kPa O_2 + ≤ 0.8 kPa CO_2 (Mathieu-Hurtiger *et al.*, 2010, 2014). In Australia, Little *et al.* (2000) recommend storage at +0.5°C in 1.8 kPa O_2 + 2.5 kPa CO_2. In Brazil, Weber *et al.* (2011) recommended CA storage conditions for 'Gala' of 1°C (rather than 0°C or 0.5°C) in 1.0 kPa O_2 (rather than 0.6 or 0.8 kPa O_2) and 2.0 kPa CO_2 (rather than 1.0 or 1.5 kPa CO_2) for up to 7 months. In Argentina, CA-stored 'Gala' are sensitive to epidermal spotting at low temperatures; therefore, the pulp temperature is maintained between +0.5°C and +0.75°C (Benítez, 2001). Even though Argentina 'Gala' have shown some sensitivity to CO_2, its presence is very important due to its fungistatic effect. Some combinations used successfully are as follows: 1.5 kPa O_2 + 0.5–1.0 kPa CO_2, or 1.2 kPa O_2 + 1.2 kPa CO_2, or 3 kPa O_2 + 1 kPa CO_2. Some authors propose levels of up to 2.5 kPa CO_2, provided that they work with higher levels of 1.5 kPa O_2.

'Gala' stores well in DCA storage, and regardless of region and growing season, 'Gala' has probably the lowest LOL of any apple cultivar, consistently being lower than 0.4 kPa O_2 and as low as 0.06 kPa, as measured by DCA-CF (e.g. Brackmann *et al.*, 2015; Thewes *et al.*, 2015; Weber *et al.*, 2015; Nock *et al.*, 2016). There are several disorders on 'Gala', such as core browning and stem-end browning, that are eliminated or reduced in DCA-CF (Nock *et al.*, 2016, 2018). Zanella and Rossi (2015), in their examination of postharvest firmness retention, found that 'Gala' firmness was only slightly increased in DCA-CF, in comparison with ULO – CA of 1.3°C in 1.0 kPa O_2 + 2.0 kPa CO_2. There has been extensive research on DCA-RQ on 'Royal Gala' and 'Galaxy Gala' in Brazil (Brackmann *et al.*, 2015; Weber *et al.*, 2015; Thewes *et al.*, 2017c, 2018). Weber *et al.* (2015) used RQs of 2, 4 and 6 on 'Royal Gala' and concluded that an RQ of 2 was the best. Both Brackmann *et al.* (2015) and Thewes *et al.* (2018) preferred an RQ of 1.5 for 'Galaxy Gala', possibly based on the statement in Brackmann *et al.* (2015) that all 'Gala' clones store best at RQ = 1.5. However, Thewes *et al.* (2017c) compared RQ = 1.3 and 1.5 on 'Galaxy Gala' and concluded that RQ = 1.3 was better that RQ = 1.5.

Gloster (Gloster 69)

Knee and Tsantili (1988) state that 'Gloster' apples are unusual because they do not accumulate ethylene during storage at 2 kPa O_2 (+ 0 kPa CO_2) at 1.5°C or 3.5°C with continuous ethylene removal. Ethylene production remained low for up to 200 days, and increased on transfer of fruit to 15°C. In spite of this evidence that fruit remained preclimacteric, some softening and production of soluble pectin and volatile esters occurred at 3.5°C. These processes were suppressed at 1.5°C, but chlorophyll, starch, malate and sucrose losses and increases in glucose and fructose occurred at both temperatures. Hansen (1977) recommended 0–1°C with 3 kPa CO_2 + 2.5 kPa O_2, Meheriuk (1993) 0.5–2°C in 1–3 kPa CO_2 + 1–3 kPa O_2 for 6–8 months and Van Schaik (1985) 1–2°C in air for 4 months, or in scrubbed CA storage at 1–2°C in 0–1 kPa CO_2 + 1–3 kPa O_2 for 7½ months with a shelf-life of 20–22 days. Konopacka and Płocharski (2002) reported a positive effect on firmness at 3°C in 1.5 kPa CO_2 + 1.5 kPa O_2 compared with 5 kPa CO_2 + 3 kPa O_2, although this difference decreased during the shelf-life for 14 days at 18°C. Sharples and Stow (1986) recommended 1.5–2°C with < 1 kPa CO_2 and 2 kPa O_2. Older general recommendations have been summarized by Herregods (1993, personal communication) and Meheriuk (1993) from a variety of countries as follows:

	Temperature °C	kPa CO_2	kPa O_2
Belgium	0.8	2	2
Canada (Nova Scotia)	0	4.5	2.5
Canada (Nova Scotia)	0	1.5	1.5
Denmark	0	2.5–3	3–3.5
France	1–2	1–1.5	1.5
Germany (Saxony)	2	1.7–1.9	1.3–1.5
Germany (Westphalia)	1–2	2	2
Holland	1	3	1.2
Slovenia	1	3	1
Slovenia	1	3	3
Spain	0.5	2	3
Switzerland	2–4	3–4	2–3

Köpcke (2015) examined the possibility of increasing the store temperature for 'Gloster' (to save energy) from the previously recommended 2.0°C in ULO of 1.4 kPa O_2 + 1.8 kPa CO_2 to higher temperatures of either 3.5°C or ≤ 10°C in ULO or DCA-CF (with and without 1-MCP, always with 2.6 kPa CO_2). The results showed that increasing the store temperature from 2.0°C to 3.5°C or ≤ 10°C maintained some quality measurements when using DCA-CF or 1-MCP alone, but combining DCA-CF and 1-MCP was more beneficial at the higher temperature. Watercore in 'Gloster' dissipated faster with increasing temperature and DCA-CF prevented the transformation of watercore to flesh browning in 1-MCP-treated apples.

Golden Delicious

Kupferman (2003) considered 'Golden Delicious' (like 'Gala') to be tolerant of low storage temperature, CO_2 and rapid CA. Based on his survey of various recommendations, he concluded that 'Golden Delicious' can be stored as low as 1.0 kPa O_2 with CO_2 levels up to 2.5 kPa at 1°C and the O_2 should be raised as the temperature is lowered below this point. Defra AHDB (2017) recommended that UK 'Golden Delicious' should be stored at 1.5–3.5°C in 3 kPa O_2 + 5 kPa CO_2. Gasser and Siegrist (2011) recommended that Swiss 'Golden Delicious' should be stored in CA at 2°C in either 2 kPa O_2 + 4 kPa CO_2 or 1 kPa O_2 + 3 kPa CO_2. In France, the 'Golden Delicious' recommendation is to store at 0.5–1.0°C with a choice of three CA recommendations with increasing storage durations: (i) CA at 2–3 kPa O_2 + 3–5 kPa CO_2; (ii) ULO at 1.5–1.8 kPa O_2 + 2–3 kPa CO_2; or (iii) XLO at 0.8-1.0 kPa O_2 + 0.8 -1.0 kPa CO_2 (Mathieu-Hurtiger *et al.*, 2010, 2014). In Australia, Little *et al.* (2000) recommended storage at +0.5°C in 1.8 kPa O_2 + 3.0 kPa CO_2. In Argentina, CA-stored 'Golden Delicious' are held at 0–0.5°C in either (i) conventional CA of 2–3 kPa O_2 + 3–5 kPa CO_2, or (ii) 1–1.5 kPa O_2 + < 3 kPa CO_2. At the INTA High Valley Experimental Station they have obtained good results with 1.5 kPa O_2 + 1.5–2.0 kPa CO_2 (Benítez, 2001). In Ontario, Canada, the recommendation is 0°C in either (i) conventional CA of 2.5 kPa O_2 + 2.5 kPa CO_2, or (ii) 1.5 kPa O_2 + 1.5 kPa CO_2 (DeEll, 2012).

Kittemann *et al.* (2015a) tested the possibility of storing 'Golden Delicious' for 7 months using (i) ULO at a high storage temperature, 5°C, in 1 kPa O_2 + 2.5 kPa CO_2 + 1-MCP, compared with (ii) ULO at 1°C, in 1 kPa O_2 + 2.5 kPa CO_2, or (iii) DCA at 1°C, ≅ 0.7 kPa O_2 + 1.5 kPa CO_2. In addition, 'reverse samples' of 1-MCP-treated apples were stored at 1°C (non 1-MCP-treated ULO room) and apples without 1-MCP treatment were stored at 5°C (1-MCP-treated room). The firmness and rot incidence of 'Golden Delicious' did differ between treatments but apples stored at 5°C showed a lower weight loss, compared with those at 1°C.

The value of 'Golden Delicious' in some markets is determined by its skin colour and this can be influenced by CA conditions. Little *et al.* (2000) stated that 8.0 kPa CO_2 can be used for the first 10 days to improve firmness and green colour after storage. Similarly, Mathieu-Hurtiger *et al.* (2014) suggested that in ULO, the CO_2 concentration can be lower than 2–3 kPa if a more yellow skin is desired. 'Golden Delicious' responds well to DCA storage conditions. For example, by 2011–2012, 'Golden Delicious' was c. 10% of the DCA-stored apples in the South Tyrol industry (Prange *et al.*, 2013).

Granny Smith

This variety is considered to be CO_2 intolerant (Kupferman, 2003). Thus, CO_2 should remain well below the oxygen level at all times and temperatures should be held slightly elevated during CA storage (> 0°C), especially in cooler growing regions. This is less critical in warmer growing regions, e.g. South Africa. Gasser and Siegrist (2011) recommended that Swiss 'Granny Smith' should be stored in CA at 0.5°C in 1 kPa O_2 + 2 kPa CO_2. In France, the 'Granny Smith' recommendation is to store at 0.5–1.0°C with a choice of three CA recommendations with increasing storage durations: (i) CA – 2–3 kPa O_2 + 1.5–2.0 kPa CO_2; (ii) ULO – 1.5–1.8 kPa O_2 + < 1.2 kPa CO_2; or (iii) XLO – 0.8–1.0 kPa O_2 + < 0.8 kPa CO_2 (Mathieu-Hurtiger *et al.*, 2010, 2014). In Australia, Little *et al.* (2000) recommended storage at 1.0°C in 1.8 kPa O_2 + 1.0 kPa CO_2. In Argentina, CA-stored 'Granny Smith' are held at 0–0.5°C in either (i) conventional CA at 2 kPa O_2 + 1.5–2.0 kPa

CO_2, or (ii) ULO at 0.8–1.2 kPa O_2 + 1 kPa CO_2. Negative temperatures increase the risk of CO_2 damage, especially when using the ULO conditions (Benítez, 2001).

'Granny Smith' is very susceptible to superficial scald. Exposure to initial low oxygen stress (ILOS, 0.5 kPa O_2 for 10 days at 1°C) followed immediately by 20 weeks (pre-optimally harvested) or 12 weeks (optimally harvested) of CA at –0.5°C in 1.5 kPa O_2 + 1.0 kPa CO_2 or 1.5 kPa O_2 + 3.0 kPa CO_2 significantly inhibited the development of superficial scald, but only after 8 weeks for pre-optimally and 2 weeks for optimally harvested 'Granny Smith' (Van der Merwe *et al.*, 2003). Zanella *et al.* (2005) confirmed that the development of superficial scald on 'Granny Smith' was completely controlled by means of dynamic CA (DCA-CF) on a commercial scale during 6 months of storage, even after a subsequent shelf-life period of 7 or 14 days at 20°C. Mditshwa *et al.* (2017) found that intermittent DCA-CF treatment, alternating with RA, provided complete superficial scald control for 'Granny Smith' but only for fruit stored for less than 70 days.

Honeycrisp (Honeycrunch® in EU)

'Honeycrisp' is a high-demand, high-value cultivar that will rapidly deteriorate after harvest if its storage requirements are not provided. It is intolerant of rapid CA, low storage temperature and CO_2. In North American regions there is a difference of opinion on whether 'Honeycrisp' should be stored in CA (Beaudry *et al.*, 2009, 2014; Watkins and Nock, 2012b; Nichols *et al.*, 2008). Prior to either RA or CA, 'Honeycrisp' has to be pre-conditioned in RA for a period of time at a high temperature (5–7 days at 10–20°C, depending on region) in order to avoid soft scald, internal breakdown disorders and external CO_2 damage (Prange *et al.*, 2003a; DeLong *et al.*, 2004; Beaudry *et al.*, 2009). 'Honeycrisp' is a firm apple with a very low water permeance (Schotsmans, 2006) and so a preconditioning treatment generally does not reduce quality, if limited to 1 week or less. Nova Scotia 'Honeycrisp' can be stored in CA at 3–5°C in 2.0–2.5 kPa O_2 + 1.0 -1.5 kPa CO_2 (Beaudry *et al.*, 2009; DeLong *et al.*, 2004). In Michigan, the current recommendation is 3.5°C in 3.0 kPa O_2 plus keeping CO_2 'low in the first month, in much the same way as for "Empire", and then allow the CO_2 levels to increase' (Beaudry *et al.*, 2014). In Washington State, USA, the current recommendation is 3–4°C in 2–4 kPa O_2 + 0.5–1.0 kPa CO_2 (Hanrahan and Blakey, 2017). The Washington State recommendation also states that for lots with bitter pit as a major concern, CA can be initiated during the preconditioning period to decrease the incidence of bitter pit in storage. Nova Scotia 'Honeycrisp' fruit have been successfully stored experimentally in DCA-CF without injury for 6 months (DeLong *et al.*, 2004) and 9 months (Nichols *et al.*, 2008).

Idared

The recommended CA storage temperature is colder in North America, c. 0–3°C, compared with 3–4°C in Europe. 'Idared' is somewhat tolerant to CO_2 with the recommended CO_2 equal to O_2 or slightly higher. In Nova Scotia, the recommendation is 0–3°C in 2.0 kPa O_2 + 0.5–1.5 kPa CO_2 and in Michigan the recommendation is 0°C in 1.5 kPa O_2 + < 3.0 kPa CO_2 (Kupferman, 2003). In Ontario, Canada, the recommendation is 0°C in 2.5 kPa O_2 + 2.5 kPa CO_2 or in 1.5 kPa O_2 + 1.5 kPa CO_2 (DeEll, 2012). In the UK, the recommendation is 3–4°C in 3.0 kPa O_2 + 5.0 kPa CO_2 or in 1.0 kPa O_2 + < 1.0 kPa CO_2. Further south, Gasser and Siegrist (2011) recommended that Swiss-grown 'Idared' be stored at 4°C in 2.0 kPa O_2 + 3.0 kPa CO_2, or in 1.0 kPa O_2 + 1.5 kPa CO_2. In France, the recommendation is 2–4°C in (i) CA at 2–3 kPa O_2 + 3–4 kPa CO_2; or (ii) ULO at 1.5–1.8 kPa O_2 + 1.8–2.2 kPa CO_2.

'Idared' has been examined under DCA conditions. Gasser *et al.* (2003), using a laboratory-based DCA-RQ system, measured ACP in 'Idared' and found that the amount of CO_2 production depended on the rate of

O_2 reduction, fruit maturity, storage duration and previous storage conditions. After 5–9 months of storage at 4°C, the ACP increased in regular atmosphere (RA) but did not change in low-O_2 CA (1.0 kPa O_2 + 1.5 kPa CO_2).

Gasser *et al.* (2005) found that the concentration at the ACP of 'Idared' was 0.37 kPa O_2 just after harvest, with a similar level 5 months later, but it decreased to 0.17 kPa O_2 after 8 months of storage. Fruits held in 4 kPa CO_2 and at 1°C showed the same or a slightly higher respiration rate compared with fruit held at 4°C. Levels of organic volatiles increased 2–4 times and ethanol up to 10 times at or below the anaerobic compensation point. Gasser *et al.* (2008) successfully stored 'Idared' at 3°C for 200 days in either DCA-CF or DCA-RQ without any physiological disorders, while fruit firmness of DCA-stored fruit was, in general, significantly higher than in ULO-stored control fruit.

Jonagold (including clones)

There are many 'Jonagold' clones that are sometimes marketed by clone name, e.g. Jonagored, Red Prince. The storage requirements of 'Jonagold' are very similar to 'Golden Delicious' and 'Gala'. 'Jonagold' can be considered to be tolerant of low storage temperature, CO_2 and rapid CA. In Nova Scotia, Canada, the CA recommendation is 0–0.5°C in 1.5 kPa O_2 + 1.5 kPa CO_2 and in The Netherlands it is 1.0°C in 1.0–1.2 kPa O_2 + 4.5 kPa CO_2 (Kupferman, 2003). Defra AHDB (2017) recommended that UK 'Jonagold' should be stored at 1.5–2.0°C in 1.2 kPa O_2 + 4.0 kPa CO_2. Gasser and Siegrist (2011) recommended that Swiss 'Jonagold' should be stored in CA at 2°C in either 2 kPa O_2 + 4 kPa CO_2 or 1 kPa O_2 + 3 kPa CO_2. In France, the 'Jonagold' recommendation is to store at 0.5–1.0°C with a choice of two CA recommendations with increasing storage durations: (i) CA at 2–3 kPa O_2 + 3–4 kPa CO_2; or (ii) ULO at 1.5–1.8 kPa O_2 + 1.5–2.0 kPa CO_2 (Mathieu-Hurtiger *et al.*, 2010). In Australia, Little *et al.* (2000) recommended storage at +0.5°C in 1.8 kPa O_2 + 2.5 kPa CO_2. In Argentina, CA-stored 'Jonagold' are held at 0.8°C in 1.5 kPa O_2 + 2.5 kPa CO_2 (Benítez, 2001). In Ontario, Canada, the recommendation is either (i) conventional CA with 0°C in 2.0 kPa O_2 + 2.5 kPa CO_2, or (ii) 0–0.5°C in 1.5 kPa O_2 + 1.5 kPa CO_2 (DeEll, 2012).

'Jonagold' apples stored in DCA-CF stored for up to 9 months were firmer and did not accumulate fermentative volatiles, compared with fruit stored in CA with 1.5 kPa O_2 (DeLong *et al.*, 2004).

Kitteman *et al.* (2015a) tested the possibility of storing 'Jonagold' for 7 months using (i) ULO at a high storage temperature, 5°C, 1 kPa O_2 + 2.5 kPa CO_2 + 1-MCP, compared with (ii) ULO at 1°C, 1 kPa O_2 + 2.5 kPa CO_2, or (iii) DCA at 1°C, ≅ 0.7 kPa O_2 + 1.5 kPa CO_2. In addition, 'reverse samples' of 1-MCP-treated apples were stored at 1°C (non 1-MCP-treated ULO room) and apples without 1-MCP treatment were stored at 5°C (1-MCP-treated room). Firmness had the highest values in 1-MCP-treated fruit, without any difference between the 1°C (reverse) and 5°C treatments. Rot incidence was highest in storage condition 1 (ULO + 1-MCP at 5°C) but weight loss was lowest in both condition 1 (ULO + 1-MCP at 5°C) and 3 (DCA at 1°C), compared with condition 2 (ULO at 1°C). Sensory panels rated taste as being the same in all three conditions but purchase preference was poorest in storage condition 2 (ULO at 1 °C). Sensory texture was rated highest in condition 1 (ULO + 1-MCP at 5°C), followed by condition 3 (DCA at 1 °C) and lowest in condition 2 (ULO at 1°C).

Jonathan

'Jonathan', a parent of 'Jonagold', is considered to be slightly less tolerant of low storage temperature, and CO_2, compared with 'Jonagold'. In Michigan, USA, the CA recommendation is 0°C in 1.5 kPa O_2 + <3.0 kPa CO_2 (Kupferman, 2003). Gasser and Siegrist (2011) recommended that Swiss 'Jonathan' should be stored at 3–4°C but provided no CA recommendation. In Australia, Little *et al.* (2000) recommended storage at +0.5°C in 1.8 kPa O_2 + 1.0 kPa CO_2. They also said that a CA store of 'Jonathan' fruit should never be sealed unless CA generation can start immediately and to ensure that CO_2 does not exceed 1.0 kPa initially. There have been no studies examining storage of 'Jonathan' under DCA conditions. Older general recommendations have been summarized by Herregods (1993, personal communication) and Meheriuk (1993) from a variety of countries as follows:

	Temperature °C	kPa CO_2	kPa O_2
Australia (South)	0	3	3
Australia (South)	0	1	3
Australia (Victoria)	2	1	1.5
China	2	2–3	7–10
Germany (Westphalia)	3–4	3	1–2
Israel	–0.5	5	1–1.5
Slovenia	3	3	3
Slovenia	3	3	1.5
Switzerland	3	3–4	2–3
USA (Michigan)	0	3	1.5

McIntosh

'McIntosh', discovered in Canada in 1811, is probably the second oldest cultivar, after 'Reinette du Canada', in the list of top 20 cultivars grown in the world. It is mostly cultivated in colder apple regions of Canada, USA and parts of Europe. 'McIntosh' is known to be sensitive to low storage temperatures and not very tolerant to O_2 below 1.0 kPa. 'Marshall McIntosh' is especially sensitive to low O_2 and may develop low-O_2 injury. The CA recommendation is 3–3.5°C in 1.5–2.5 kPa O_2 + 1.5–4.5 kPa CO_2 (non-'Marshall McIntosh' clones) and 2.5 kPa O_2 + 2.5 kPa CO_2 ('Marshall McIntosh') (Kupferman, 2003). In Ontario, Canada, the CA recommendation is 3°C in 2.5 kPa O_2 + gradual increase in CO_2 (2.5 kPa for first 6 weeks and then gradually increase to 4.5 kPa CO_2) (DeEll, 2012). Since 'McIntosh' is more sensitive to CO_2 if treated with 1-MCP, the first 6 weeks is 0.5 kPa CO_2 and it can be kept at 2.5 kPa for the first 6 weeks if both 1-MCP and DPA (reduces CO_2 damage) are applied (DeEll, 2012). Gasser and Siegrist (2011) recommended Swiss 'McIntosh' be stored at 3–4°C but provided no CA recommendation. 'McIntosh' stores well in DCA-CF (Prange *et al.*, 2003b; DeLong *et al.*, 2004). Older general recommendations are summarized by Herregods (1993, personal communication) and Meheriuk (1993) from a variety of countries as follows:

	Temperature °C	kPa CO_2	kPa O_2
Canada (British Columbia)	3	4.5	2.5
Canada (British Columbia)	3	5	2.5
Canada (British Columbia)	1.7	5	2.5
Canada (Nova Scotia)	3	4.5	2.5
Canada (Nova Scotia)	3	1.5	1.5
Canada (Ontario)	3	5	2.5–3
Canada (Ontario)	3	1	1.5
Canada (Quebec)	3	5	2.5
Germany (Westphalia)	3–4	3–5	2
USA (Massachusetts)	1–2	5	3
USA (Michigan)	3	3	1.5
USA (New York)	2	3–5	2–2.5
USA (New York)	2	3–5	4
USA (Pennsylvania)	1.1–1.7	0–4	3–4

Melrose

The parents of 'Melrose' are 'Jonathan' and 'Delicious' and, like its parents, it is low-temperature and somewhat CO_2 tolerant. Hansen (1977) recommended 0–1°C in 6 kPa CO_2 + 2.5 kPa O_2. Meheriuk (1993) recommended 0–3°C in 2–5 kPa CO_2 + 1.2–3 kPa O_2 for 5–7 months. At 1°C and 85–90% RH, weight losses and decay were negligible and there were no physiological disorders in storage in 7 kPa CO_2 with 7 kPa O_2, while bitter pit was observed in fruit stored in air (Jankovic and Drobnjak, 1994). Jankovic and Drobnjak (1992) showed that storage at 1°C and 85–90% RH in 7 kPa O_2 + < 1 kPa CO_2 for 168 days reduced respiration, delayed ripening and produced firmer, juicier fruit judged to have better taste at the end of storage than those stored in air. CA also reduced the loss of total sugars, total acids, ascorbic acid and starch during storage compared with storage in air. There have been no studies examining storage of 'Melrose' under DCA conditions. Older general recommendations have been summarized by Herregods (1993, personal communication) and Meheriuk (1993) from a variety of countries as follows:

	Temperature °C	kPa CO_2	kPa O_2
Belgium	2	2	2–2.2
France	0–3	3–5	2–3
Germany (Westphalia)	2–3	3	1–2
Slovenia	1	3	3
Slovenia	1	3	1.2

Mutsu ('Crispin')

The parents of 'Mutsu' are 'Golden Delicious' and 'Indo', which are also the parents of 'Orin'. 'Mutsu', like 'Golden Delicious', is low-temperature and CO_2 tolerant. Fidler (1970) recommended 1°C in 8 kPa CO_2 with data supplied only for stores not fitted with a scrubber. Hansen (1977) recommended 0–1°C in 6 kPa CO_2 + 2.5 kPa O_2 and Meheriuk (1993) recommended 0–2°C in 1–3 kPa CO_2 + 1–3 kPa O_2 for 6–8 months. Dilley *et al.* (1989) recommended 0°C or 3°C in 1.5 kPa O_2 + up to 3 kPa CO_2. In Ontario, Canada, the recommendation is 0°C with 2.5 kPa O_2 and 2.5 kPa CO_2 (DeEll, 2012). There have been no studies examining storage of 'Mutsu' under DCA conditions. Older general recommendation have been summarized by Herregods (1993, personal communication) and Meheriuk (1993) from a variety of countries as follows:

	Temperature °C	kPa CO_2	kPa O_2
Australia (Victoria)	1	1	1.5
Australia (Victoria)	0	3	3
Denmark	0–2	3–5	3
Germany (Westphalia)	1–2	3–5	1–2
Japan	0	1	2
Slovenia	1	3	1
USA (Michigan)	0	3	1.5
USA (Pennsylvania)	–0.5 to +0.5	0–2.5	1.3–1.5

Nicoter (Kanzi®)

'Nicoter' is a relatively new cultivar, appearing commercially in 2006. Its parents are 'Gala' and 'Braeburn' which are also the parents of 'Scifresh' ('Jazz'). Like 'Braeburn', 'Nicoter' does not seem to be tolerant of low storage temperature and is CO_2 intolerant. Defra AHDB (2017) recommended that UK 'Kanzi®' could be stored until late March at 3.0–3.5°C in 2.5 kPa O_2 + < 0.7 kPa CO_2 and sealing (initiation of CA) should be delayed for 21 days. Kittemann *et al.* (2015a) concluded that for 'Kanzi' apples in the Lake Constance region of Germany, storage at 3°C in 1.0 kPa O_2 + < 1.0 kPa CO_2 and a CA start immediately after room filling seemed to be suitable to reduce the occurrence of internal browning and to maintain fruit quality during long-term storage. 'Nicoter' has been stored experimentally in DCA-CF for many years, beginning in 2007 (Zanella and Stürz, 2015), and is stored commercially in DCA-CF in the South Tyrol, Italy (Prange *et al.*, 2013).

Orin

The parents of 'Orin' are 'Golden Delicious' and 'Indo', which are also the parents of 'Mutsu'. In the absence of publicly available information, an acceptable CA recommendation for 'Orin' is likely to be very similar to 'Mutsu'. There have been no studies examining storage of 'Orin' under DCA conditions.

Pinova ('Corail', 'Sonata', Piñata®)

In France, the 'Pinova' recommendation is to store at 0.5–1.0°C with a choice of two CA recommendations with increasing storage durations: (i) CA at 2–3 kPa O_2 + 2–2.5 kPa CO_2; or (ii) ULO at 1.5–1.8 kPa O_2 + 1.5 kPa CO_2 (Mathieu-Hurtiger *et al.*, 2010). Gasser and Siegrist (2011) recommended 2°C in 2 kPa O_2 + 4 kPa CO_2 or in 1 kPa O_2 + 3 kPa CO_2. Kitteman *et al.* (2015a) tested the possibility of storing 'Pinova' for 7 months using (i) ULO at a high storage temperature, 5°C, 1 kPa O_2 + 2.5 kPa CO_2 + 1-MCP, compared with (ii) ULO at 1°C, 1 kPa O_2 + 2.5 kPa CO_2, or (iii) DCA at 1°C, ≅ 0.7 kPa O_2 + 1.5 kPa CO_2. In addition, 'reverse samples' of 1-MCP-treated apples were stored at 1°C (non 1-MCP-treated ULO room) and apples without 1-MCP treatment were stored at 5°C (1-MCP-treated room). The firmness did not differ amongst the various conditions. Rot incidence was lowest in storage condition 1 (ULO at a high storage temperature, 5°C, + 1-MCP) and weight loss was highest in storage condition 2 (ULO at 1°C). Sensory panels rated taste and purchase preference as being the same in all three conditions, but sensory texture was poorest in storage condition 2 (ULO at 1°C), compared with the other two conditions. 'Pinova' has been stored experimentally in DCA-CF (Zanella and Stürz, 2015) and is stored commercially in DCA-CF in the South Tyrol, Italy (Prange *et al.*, 2013).

Reinette du Canada

'Reinette du Canada', which is known from the early 19th century, is probably the oldest cultivar listed in the top 20 apple cultivars grown worldwide. Gasser and Siegrist (2009) recommended 4°C in 2–3 kPa O_2 + 3 kPa CO_2. Guerra *et al.* (2010) found that 'Reinette du Canada' could be stored for 30 weeks at 1.5°C in 3.5 kPa O_2 + 3 kPa CO_2, compared with only 10 weeks in refrigerated air. There have been no studies examining storage of 'Reinette du Canada' under DCA conditions.

Rome Beauty ('Rome', 'Red Rome')

'Rome Beauty' comes from a seedling discovered in 1817 in Ohio, USA. It is considered to be CO_2 intolerant (Kupferman, 2003). Thus, CO_2 should remain well below the oxygen level at all times and temperatures should be held slightly elevated during CA storage (> 0°C), especially in cooler growing regions. He reported that the CA recommendation for Michigan, USA, was 0°C in 1.5 kPa O_2 + < 3 kPa CO_2 and in Italy 1–2°C in 2.0 kPa O_2 + 2.0 kPa CO_2. ASHRAE (1968) recommended −1.1°C to 0°C in 2–3 kPa CO_2 + 3 kPa O_2. Singh *et al.* (1972) used 1.5 kPa CO_2 + 2.5 kPa O_2 for 90 days and found that the amino and organic acid contents were considerably increased and the respiration rate was reduced both during and after storage compared with those stored in air. Sugars were not affected. There have been no studies examining storage of 'Rome Beauty' under DCA conditions.

Scifresh (Jazz®)

'Scifresh' is a relatively new cultivar, appearing commercially in 2004. Its parents are 'Gala' and 'Braeburn', which are also the parents of 'Nicoter' ('Kanzi'). Defra AHDB (2017) recommended, based on experience in New Zealand and France, that Jazz® apples in the UK could be stored until April at 0.5–0.8°C in 2 kPa O_2 + 2 kPa CO_2 (if manually controlled) or 1.5–1.8 kPa O_2 + 1–1.5 kPa CO_2 (if automatically controlled). In France, Mathieu-Hurtiger *et al.* (2014) made the following comments on 'Scifresh' ('Jazz®'). Storage temperatures should start with 3 weeks at 3°C before dropping the temperature to 1°C. The fruit is sensitive to soft scald during the temperature drop. Their only CA recommendation was ULO (1.5–1.8 kPa O_2 + < 1.5 kPa CO_2), since it stores very well without XLO. Gasser and Siegrist (2011) recommended 3–3.5°C in either (i) CA at 2.0 kPa O_2 + 2.5 kPa CO_2 or (ii) ULO at 1.0 kPa O_2 + 2.5 kPa CO_2. Siegrist and Cotter (2012) conducted further trials that showed that a stepwise decrease in temperatures from 3.5°C to 1.5°C increased soft scald damage. Cooling the apples at 3.5°C and then storing them at 3°C under CA avoided soft scald without any negative effects on fruit quality. Both CA (2.0 kPa O_2 + 2.5 kPa CO_2) and ULO (1.0 kPa O_2 + 2.5 kPa CO_2) gave good results for 8 months of storage, with ULO being recommended for longer storage. In general, this cultivar offers outstanding high tolerance to storage diseases. 'Scifresh' has been stored experimentally in DCA-CF (Zanella and Stürz, 2015).

Tsugaru

Hong *et al.* (1997) showed that at 0°C fruit firmness retention was better in 1 kPa O_2 + 3 kPa CO_2 storage than in air. Also CA reduced internal browning, cork spot and decay, mainly caused by *Botrytis cinerea* and *Penicillium expansum*. There have been no studies examining storage of 'Tsugaru' under DCA conditions.

Apricot (*Prunus armeniaca*)

Storage recommendations include: 2–3 kPa CO_2 + 2-3 kPa O_2 (SeaLand, 1991) and 2–3 kPa CO_2 + 2–3 kPa O_2 at 0–5°C (Kader, 1989; Bishop, 1996) but Kader stated that CA storage was not used commercially. It was reported that CA storage can lead to an increase in internal browning in some cultivars (Hardenburg *et al.*, 1990). Fruits of two cultivars were stored at 0–0.5°C in four different CA with 0, 2.5, 5 or 7.5 kPa CO_2 + 5.0 kPa O_2. Both the cultivars ICAPI 17 COL, but particularly ICAPI 30 COL, proved suitable for CA storage for up to 3 weeks. Only in the cultivar ICAPI 30 COL did development of some ripening parameters seem to be related to CO_2 concentration (Andrich and Fiorentini, 1986). Folchi *et al.* (1995) stored fruit of the cultivar 'Reale d'Imola' at 0°C or 6°C, either in air or in 0.3 kPa O_2 for up to 37 days. Fruit stored in 0.3 kPa O_2 had significantly increased ethanol and acetaldehyde content at 6°C, whereas the methanol content was not significantly affected by low O_2. Low O_2 reduced TSS when fruit was stored at 6°C and increased fruit firmness at both storage temperatures. Physiological disorders, such as flesh discoloration and browning, appeared after 5 days of CA storage at 6°C and after 12 days at 0°C. Folchi *et al.* (1994) stored the cultivar 'Reale d'Imola' in < 0.1 kPa CO_2 + 0.3 kPa O_2 and showed an increase in fruit aldehyde and ethanol content with increased storage time. Truter *et al.* (1994) showed that at –0.5°C, storage in 1.5 kPa CO_2 + 1.5 kPa O_2 or 5 kPa CO_2 + 2 kPa O_2 resulted in a loss of reduced mass compared with those stored in air. Ali Koyuncu *et al.* (2009) stored apricots at 0°C and 90 ± 5% RH and found that those stored in CA retained their quality better than those in MA packages or in air.

Fruit of the cultivars 'Rouge de Roussillon' and 'Canino' were picked half-ripe or mature and then subjected to pretreatment with high levels of CO_2 (10–30%) for 24, 48 or 72 h before storage in air and were compared with fruits stored directly in air or in 5 kPa O_2 + 5 kPa CO_2. Underripe fruit exposed to 20 kPa CO_2 for 24 or 48 h, or CA storage, remained firm for longer than those stored in air. CO_2 pretreatment also appeared to reduce the incidence of brown rot caused by *Monilinia* sp. TA decreased continuously during cold storage and after 24 days was the same for CO_2 pretreated and air-stored fruit, but remained high for the CA-stored fruit. Respiration rate of apricots stored in air showed a typical climacteric peak, which was delayed by the CO_2 pretreatments and was completely inhibited by CA storage. The best CO_2 pretreatment was exposure to 20 kPa CO_2 for 24–48 h. Higher concentrations or a longer duration of exposure resulted in fermentation. The benefits of CO_2 pretreatments were most noticeable at the beginning of cold storage but gradually declined, and, once the temperature was increased, they disappeared within about 24 h (Chambroy *et al.*, 1991). For storage of apricots for canning, Van der Merwe (1996) recommended –0.5°C with 1.5 kPa CO_2 + 1.5 kPa O_2 for up to 2 weeks, while Hardenburg *et al.* (1990) recommended 2.5–3 kPa CO_2 + 2–3 kPa O_2 at 0°C for the cultivar 'Blenheim (Royal)' which retained their flavour better than air-stored fruit when they were subsequently canned.

Apricot, Japanese (*Prunus mume*)

Also called mume. Mature-green fruits of the cultivar 'Ohshuku' were stored in 2–3 kPa O_2 + 3, 8, 13 or 18 kPa CO_2 at 20°C and 100% RH. Surface yellowing and flesh softening were delayed at the higher CO_2 levels. Pitting injury occurred in air-stored fruits, but no injury was observed in CA-stored fruits for up to 23 days in 3–13 kPa CO_2. The results indicated that CA-stored fruits had a shelf-life of 15, 15, 19 and 12 days using 3, 8, 13 and 18 kPa CO_2, respectively (Kaji *et al.*, 1991). Mature-green fruits of the cultivars 'Gojiro', 'Nankou', 'Hakuoukoume' and 'Shirakaga' were stored for 7 days at 25°C in various CA combinations. Fermentation occurred at O_2 concentrations of up to 2 kPa at 25°C. Fruit held at O_2 concentrations of up to 1 kPa developed browning injury and produced ethyl alcohol at a high rate. The percentage of fruit with water core injury was high in conditions of low O_2 and zero CO_2. Ethylene production rates were suppressed at high CO_2 concentrations or at O_2 levels of < 5 kPa. They found that storing fruit in approximately 10 kPa CO_2 + 3–4 kPa O_2 could help to retain optimum quality at 25°C, but that such CA storage conditions delayed the respiratory climacteric slightly and increasde ethyl alcohol production after the third day of storage (Koyakumaru *et al.*, 1995). The cultivar 'Gojiro' was stored for 3 days at 25°C in several different CA conditions. Compared with fruit in ambient air, O_2 uptake and ethylene production decreased when fruit was exposed to 19.8 kPa CO_2 + 21 kPa O_2; they decreased even more in 5 or 2 kPa O_2. However, at 2 kPa O_2 and 0 CO_2, the percentage of physiologically injured fruits increased considerably in or after storage. They suggested that removing ethylene and maintaining about 8 kPa CO_2 + at least 2 kPa O_2 are important to retain fruit quality (Koyakumaru *et al.*, 1994).

Artichoke, Globe (*Cynara scolymus*)

Typical storage conditions were given as 0–5°C in 2–3 kPa CO_2 + 2–3%O_2 (Bishop, 1996), 0.5–1.5°C in 0–2.5 kPa CO_2 + 5 kPa O_2 for 20–30 days (Monzini and Gorini, 1974), 3–5 kPa CO_2 + 2–3 kPa O_2 (SeaLand, 1991), 3 kPa CO_2 + 3 kPa O_2 in cold storage (unspecified) for 1 month, which reduced browning (Pantastico, 1975), and 0–5°C in 2–3 kPa CO_2 + 2–3 kPa O_2, but CA storage only had a slight effect on storage life (Saltveit, 1989; Kader, 1992). Kader (1985) had previously recommended 0–5°C in 3–5 kPa CO_2 + 2–3 kPa O_2 but indicated that it was not used commercially. Poma Treccarri and Anoni (1969) recommended 3 kPa CO_2 + 3 kPa O_2 for up to 1 month. They showed that it reduced browning discoloration of the bracts. Miccolis and Saltveit (1988) described storage of large, mature artichokes in humidified air at 7°C for 1 week, 2.5°C for 2 weeks, or 0°C for 3 weeks. All storage conditions resulted in significant quality loss, and there was little or no beneficial effect of 1, 2.5 or 5 kPa O_2 or 2.5 or 5 kPa CO_2 in storage at 0°C. Bracts and receptacles blackened in all atmospheres after 4 weeks of storage. The degree of blackening was insignificant in air, but became severe as the O_2 level decreased over the range from 5% to 1% and CO_2 increased over the range of 2.5–5%. CA storage of large artichokes was therefore not recommended. The cultivar 'Violeta' was stored for 15–28 days at 1°C and 90–95% RH in 1–6 kPa O_2 + 2–8 kPa CO_2. The best results, for physical, chemical and organoleptic properties, and a post-storage shelf-life of 3–4 days, were obtained with storage for 28 days in 2 kPa O_2 + 6 kPa CO_2 (Artes-Calero *et al.*, 1981). Gil *et al.* (2003) showed that storage in 5 kPa O_2 + 10 kPa CO_2 prolonged storage life of 'Blanca de Tudela' for long periods but storage in 5 kPa O_2 + 15 kPa CO_2 caused off-odours and CO_2 injuries in the inner bracts and receptacles. They found that respiration rates decreased down to 15–25 ml CO_2/kg/h in 5 kPa O_2 + 10 kPa CO_2 and 10–20 ml CO_2/kg/h in 5 kPa O_2 + 15 kPa CO_2 storage, while it was 30–40 ml CO_2/kg/h in air. Storage trials were carried out by Mencarelli (1987a) at 1 ± 0.5°C and 90–95% RH in a range of

5 and 10 kPa O_2 + 2–6 kPa CO_2 for up to 45 days. Those stored in 5 and 10 kPa O_2 + 0 kPa CO2 remained turgid with good organoleptic quality. The addition of 2 kPa CO_2, however, controlled the development of superficial mould, but 4 kPa CO_2 had adverse effects.

Artichoke, Jerusalem (*Helianthus tuberosus*)

Depending on cultivar, tubers can be stored for up to 12 months at 0–2°C and 90–95% RH; therefore there is little justification for using CA storage (Steinbauer, 1932). Tubers freeze at −2.2°C but exposure to −5°C causes little damage. Tubers are not sensitive to ethylene (Kays, 1997). They are a major source of inulin and storage of tubers in 22.5 kPa CO_2 + 20 kPa O_2 significantly retarded the rate of inulin degradation, apparently through an effect on enzyme activity (Denny *et al.*, 1944).

Asian Pear, Nashi (*Pyrus pyrifolia, P. ussuriensis* var. *sinensis*)

Zagory *et al.* (1989) reviewed work on several cultivars and found that storage experiments at 2°C in 1%, 2% or 3 kPa O_2 of the cultivars 'Early Gold' and 'Shinko' showed no clear benefit compared with storage in air. 'Early Gold' fruit appeared to be of reasonably good quality after up to 6 months of storage at 2°C. 'Shinko' became prone to internal browning after the third month, perhaps due to CO_2 injury. Low O_2 atmospheres reduced yellow colour development, compared with air. They also reported that the cultivar '20th Century' was shown to be sensitive to CO_2 concentrations above 1% when exposed for longer than 4 months or to 5 kPa CO_2 for more than 1 month. Kader (1989, 2003) recommended 0–5°C in 0–1 kPa CO_2 + 2–4 kPa O_2but claimed that it had limited commercial use. In other work, Meheriuk (1993) gave provisional recommendations of 2°C with 1–5 kPa CO_2 + 3 kPa O_2 for about 3–5 months and said that a longer storage period may result in internal browning. Meheriuk also stated that some cultivars are subject to CO_2 injury but indicated that this is when they are stored in concentrations of over 4%.

Asparagus (*Asparagus officinalis*)

Storage at 0°C in 12 kPa CO_2 or at 5°C or just above in 7 kPa CO_2 retarded decay and toughening (Hardenburg *et al.*, 1990). Lill and Corrigan (1996) showed that during storage at 20°C, spears stored in 5–10 kPa O_2 + 5–15 kPa CO_2 had an increased postharvest life from 2.6 days in air to about 4.5 days. Other recommendations include: 0°C and 95% RH in 9 kPa CO_2 + 5 kPa O_2 (Lawton, 1996); 5–10 kPa CO_2 + 20 kPa O_2 (SeaLand, 1991); 5°C with a maximum of 10 kPa CO_2 and a minimum of 10 kPa O_2 (Fellows, 1988); 1°C in 10 kPa CO_2 + 5–10 kPa O_2 (Monzini and Gorini, 1974); 0–5°C in 5–10 kPa CO_2 + 21 kPa O_2, which had a good effect but was of limited commercial use (Kader, 1985, 1992); 1–5°C with 10–14 kPa CO_2 + 21 kPa O_2, which had only a slight effect (Saltveit, 1989); and 0–3°C in 10–14 kPa CO_2 (Bishop, 1996). Lipton (1968) showed that levels of 10 kPa CO_2 could be injurious, causing pitting, at storage temperatures of 6.1°C, but at 1.7°C no CO_2 injury was detected. However, spears were injured at both temperatures in atmospheres containing 15 kPa CO_2. Lipton also showed that storage in 10 kPa CO_2 reduced rots due to *Phytopthora*, but there were no effects in atmospheres containing 5 kPa CO_2. Other work has also shown that fungal development during storage could be prevented by flushing with 30 kPa CO_2 for 24 h or by maintaining a CO_2 concentration of 5–10% (Andre *et al.*, 1980a). Hardenburg *et al.* (1990) indicated that brief (unspecified) exposure to 20 kPa CO_2 can reduce soft rot at the butt end of spears. McKenzie *et al.* (2004) found that storage in 5 kPa O_2 + 10 kPa CO_2 maintained the activity of several sugar-metabolizing enzymes at, or above, harvest levels compared with storage in air, where activities fell. CA storage also eliminated the increase in acid invertase activity found during storage in air. There were significant differences in the pool sizes of sugars in the different compartments (vacuole, cytoplasm, free space). CA storage also resulted in a lower concentration of glucose in the free space, but it appeared to engage the ethanolic fermentation pathway, though only transiently. CA storage may have permitted more controlled use of whichever vacuole sugar the tissue normally drew upon (fructose) and delayed an increase in plasma membrane permeability. Exposing asparagus to total N_2 for 8 h at room temperature resulted in reduced respiration rate and slowing the decrease in headspace O_2 in LDPE MA packaging. In the total N_2 MA packaging, spears lost < 12% fresh weight after 8 days at 10°C, showed less increase in the accumulation of fibre and lignin and reduced loss of chlorophyll, sugars and AA (Techavuthiporn and Boonyaritthongchai, 2016).

Aubergine, Eggplant (*Solanum melongena*)

Storage in 0 kPa CO_2 + 3–5 kPa O_2 was recommended by SeaLand (1991). High CO_2 in the storage atmosphere can cause surface scald browning, pitting and excessive decay and these symptoms are similar to those caused by chilling injury (Wardlaw, 1938). CO_2 concentrations of 5%, 8% or 12% during storage all resulted in CO_2 injury, characterized by external browning without tissue softening (Mencarelli *et al.*, 1989). Viraktamath *et al.* (1963, quoted by Pantastico, 1975) claimed that atmospheres containing CO_2 levels of 7% or higher were injurious.

Avocado (*Persea americana*)

Fellows (1988) gave a general recommendation of a maximum of 5 kPa CO_2 and a minimum of 3 kPa O_2. In South Africa, Van der Merwe (1996) recommended 5.5°C with 10 kPa CO_2 + 2 kPa O_2 for up to 4 weeks. In Australia, CSIRO (1998) recommended 2–5 kPa O_2 + 3–10 kPa CO_2 at 12°C. In other work, 3–10 kPa CO_2 + 2–5 kPa O_2 was recommended for both Californian and tropical avocados (SeaLand, 1991). A general recommendation by Lawton (1996) was 5°C and 85% RH in 9 kPa CO_2 + 2 kPa O_2. Typical storage conditions were given as 10–13°C in 3–10 kPa CO_2 + 2–5%O_2 by Bishop (1996). A general comment was that storage at 5–13°C in 3–10 kPa CO_2 + 2–5 kPa O_2 had a good effect and was of some commercial use (Kader, 1985, 1992, 1993) and that the benefits of reduced O_2 and increased CO_2 were good. He reported that skin browning can occur and off-flavour can develop if the CO_2 level is too high, and internal flesh browning and off-flavour can occur if O_2 levels are too low. In experiments described by Meir *et al.* (1993) at 5°C, increasing CO_2 levels over the range of 0.5%, 1%, 3% or 8% and O_2 levels of 3% or 21% slowed fruit softening, inhibited peel colour change and reduced the incidence of chilling injury. The most effective CO_2 concentration was 8% with either 3 or 21 kPa O_2. They found that fruit could be stored for 9 weeks in 3 kPa O_2 + 8 kPa CO_2 and after CA storage the fruit ripened normally and underwent typical peel colour changes. Some injury to the fruit peel was observed, probably attributable to low O_2 concentrations, with peel damage seen in 10% of fruit held in 3 kPa O_2 with either 3 or 8 kPa CO_2, but it was reported to be too slight to affect their marketability.

Continuous exposure to 20 kPa CO_2 for 7 days at the beginning of storage or for 1 day a week throughout the storage period at either 2°C or 5°C maintained fruit texture and delayed ripening compared with fruit stored throughout in air (Saucedo-Veloz *et al.*, 1991). Storage in total N_2 or total CO_2, however, caused irreversible injury to the fruit (Stahl and Cain, 1940). CA storage was reported to reduce chilling injury symptoms. There were fewer chilling injury symptoms in the cultivars ‘Booth 8’, ‘Lula’ and ‘Taylor’ in CA storage than in air (Haard and Salunkhe, 1975; Pantastico, 1975). Spalding and Reeder (1975) also showed that in storage at 0 kPa CO_2 + 2 kPa O_2 or 10 kPa CO_2 + 21 kPa O_2 fruit had less chilling injury and less anthracnose (*Colletotrichum gloeosporioides*) during storage at 7°C than fruit stored in air. Specific recommendations have been made on some cultivars as follows.

Anaheim

Bleinroth *et al.* (1977) recommended 10 kPa CO_2 + 6 kPa O_2 at 7°C, which gave a storage life of 38 days compared with only 12 days in air at the same temperature.

Fuerte

Both Overholser (1928) and Wardlaw (1938) showed that 'Fuerte' stored at 45°F (7.2°C) had a 2-month storage life in 3 or 4–5 kPa CO_2 + 3 or 4–5 kPa O_2. This was more than a month longer than they could be stored at the same temperature in air. Biale (1950) reported similar results. Pantastico (1975) showed that storage at 5–6.7°C had double or triple the storage life in 3–5 kPa CO_2 + 3–5 kPa O_2 than in air. Bleinroth *et al.* (1977) recommended 10 kPa CO_2 + 6 kPa O_2 at 7°C, which gave a storage life of 38 days compared with only 12 days in air at the same temperature. Pre-storage treatment with 3 kPa O_2 + 97% N_2 for 24 h at 17°C significantly reduced chilling injury symptoms during subsequent storage at 2°C for 3 weeks. Fruit softening was also delayed by this treatment. The treated fruit had a lower respiration rate and ethylene production not only during storage at 2°C but also when removed to 17°C (Pesis *et al.*, 1994).

Gwen

Lizana *et al.* (1993) described storage at 6°C and 90% RH in 5 kPa CO_2 + 2 kPa O_2, 5 kPa CO_2 + 5 kPa O_2, 10 kPa CO_2 + 2 kPa O_2, 5 kPa CO_2 + 2 kPa O_2, 10 kPa CO_2 + 5 kPa O_2, or 0.03 kPa CO_2 + 21 kPa O_2 (control) for 35 days followed by 5 days at 6°C in air and 5 days at 18°C to simulate shelf-life. After 35 days all the fruits in CA storage retained their firmness and were in excellent condition, regardless of treatment, while the control fruits were very soft.

Hass

Yearsley *et al.* (2003) showed that storage at 5°C in 1 kPa O_2 resulted in stress that showed itself in the biosynthesis of ethylene, but levels rapidly returned to trace levels when fruit was returned to non-stressed atmospheres. No ethanol was detected in fruit after storage for up to 96 h in CO_2 levels of up to 20%. Lizana and Figuero (1997) compared several CA storage conditions and found that 6°C with 5 kPa CO_2 + 2 kPa O_2 was optimum. Under these conditions the fruit could be stored for at least 35 days and then kept at the same temperature in air for a further 15 days and was still in good condition. Corrales-Garcia (1997) found that fruit stored at 2°C or 5°C for 30 days had higher chilling injury stored in air than fruit in CA storage in 5 kPa CO_2 + 5 kPa O_2 or 15 kPa CO_2 + 2 kPa O_2. Storage in 15 kPa CO_2 + 2 kPa O_2 was especially effective in reducing chilling injury. Intermittent exposure to 20 kPa CO_2 increased storage life at 12°C and reduced chilling injury during storage at 4°C compared with those stored in air at the same temperatures (Marcellin and Chaves, 1983). Burdon *et al.* (2008) concluded that fruit quality was better after CA storage than after storage in air, and that DCA storage was better than 'standard' CA storage. The effect of DCA storage was to reduce the incidence of rots when ripe. Inclusion of CO_2 at 5% in CA retarded fruit ripening but stimulated rot expression and they concluded that it should not be used for CA storage of New Zealand-grown 'Hass'. DCA storage prolonged fruit storage life but when removed from storage the fruits ripened more rapidly and more uniformly with fewer rots than those from 'conventional' CA storage (Burdon *et al.*, 2009). They recommended setting the DCA O_2 level as soon as possible after harvest. Fruits were stored for 6 weeks at 5°C in different CA conditions and then ripened in air at 20°C by Burdon *et al.* (2008). Those that had been stored in < 3 kPa O_2 + 0.5 kPa CO_2 ripened in 4.6 days, compared with 7.2 days for those that had been stored in 5 kPa O_2 + 5 kPa CO_2 and 4.8 days for those that had been stored in air.

Lula

Stahl and Cain (1940) recommended 3 kPa CO_2 + 10 kPa O_2. Storage at 10°C in 3–5 kPa CO_2 + 3–5 kPa O_2 delayed fruit softening and in 9 kPa CO_2 + 1 kPa O_2 maintained acceptable eating quality and appearance for 60 days (Hardenburg *et al.*, 1990). Earlier work had also shown that 'Lula' could be stored for 60 days at 10°C in 9 kPa CO_2 + 1 kPa O_2 or for 40 days at 7.2°C in 10 kPa CO_2 + 1 kPa O_2 (Pantastico, 1975). Spalding and Reeder (1972) showed that 'Lula' could be stored for 8 weeks at 4–7°C and 98–100% RH in 10 kPa CO_2 + 2 kPa O_2.

Pinkerton

'Pinkerton' is very susceptible to grey pulp. Both CA and MA storage delayed, but did not prevent, this postharvest disorder (Kruger and Truter, 2003).

Banana (*Musa*)

There are very wide variations in recommendations for optimum CA conditions and 3 kPa O_2 + 5 kPa CO_2 at 14°C is used in some commercial shipments. In other cases, higher levels of CO_2 and lower levels of O_2 have shown beneficial effects, but these levels have also been reported to have detrimental effects. Recommended storage conditions for green (pre-climacteric) bananas include the following: 10–13.5 kPa O_2 for 'Latundan' (*Musa* AAB) (Castillo *et al.*, 1967 quoted by Pantastico, 1975); 15°C in 2 kPa O_2 + 8 kPa CO_2 (Smock *et al.*, 1967); 15°C and 5–8 kPa CO_2 + 3 kPa O_2 for 'Bungulan' (*Musa* AAA) in the Philippines (Calara, 1969; Pantastico, 1975); 5% or lower levels of CO_2 in combination with 2 kPa O_2 (Woodruff, 1969); 20°C in 3 kPa O_2 + 5 kPa CO_2 for 182 days for 'Williams' in Australia and the fruit still ripened normally when removed to air (McGlasson and Wills, 1972); 15°C in 2 kPa O_2 + 6–8 kPa CO_2 for 'Dwarf Cavendish' (Pantastico, 1975); 20°C in 1.5–2.5 kPa O_2 + about 7–10 kPa CO_2 (IIR, 1978); 20°C in 1.5–2.5 kPa O_2 + 7–10 kPa CO_2 for 'Williams' (Sandy Trout Food Preservation Laboratory, 1978); 5 kPa CO_2 + 4 kPa O_2 (Hardenburg *et al.*, 1990); O_2 as low as 2% at 15°C if the CO_2 level is around 8% for 'Lakatan' (*Musa* AA) in South-east Asia (Abdullah and Pantastico, 1990); 2–5 kPa CO_2 + 2–5 kPa O_2 (SeaLand, 1991); 12–16°C in 2–5 kPa CO_2 + 2–5 kPa O_2 (Kader, 1993); 12–16°C in 2–5 kPa CO_2 + 2–5% O_2 (Bishop, 1996); 11.5°C in 7 kPa CO_2 + 2 kPa O_2 in South Africa (Van der Merwe, 1996); 14–16°C and 95% RH in 2–5 kPa CO_2 + 2–5 kPa O_2 (Lawton, 1996); 14°C in 4 or 6 kPa O_2 + 4 or 6 kPa CO_2 for 'Robusta' from the Windward Islands (Ahmad *et al.*, 2001); 12°C in 3 kPa O_2 + 9 kPa CO_2 or 5 kPa O_2 + 5 kPa CO_2 for 'Prata' (*Musa* AAB) in Brazil for 30 days (Botrel *et al.*, 2004) and 12.5 ± 0.5°C and 98 ± 1.0% RH in 2 kPa O_2 + 4 kPa CO_2 or 3 kPa O_2 + 7 kPa CO_2 for 'Prata Ana' (*Musa* AAB) (Santos *et al.*, 2006c). They also showed that 'Prata Ana' at colour stage 2 (mature-green) stored in 2 kPa O_2 + 4 kPa CO_2 and 3 kPa O_2 + 7 kPa CO_2 did not develop to colour stage 7 even after 40 days, whereas those in air reached colour stage 7 in 24 days and those in 4 kPa O_2 + 10 kPa CO_2 in 32 days.

CO_2 above certain concentrations can be toxic to bananas. Symptoms of CO_2 toxicity are softening of green fruit and an undesirable texture and flavour. Parsons *et al.* (1964) observed that 'Cavendish' (*Musa* AAA) fruit stored in < 1 kPa O_2 failed to ripen normally and developed off-flavour and a dull yellow-to-brown skin with a 'flaky' grey pulp. Also Wilson (1976) showed that fermentation takes place when bananas are ripened at 15.5°C in an atmosphere containing 1 kPa O_2.

IIR (1978) reported that in experiments with 25 CA storage combinations of 0.5, 1.5, 4.5, 13.5 and 21 kPa O_2 combined with 0, 1, 2, 4 and 8 kPa CO_2, all at 20°C, the optimum conditions appeared to be 1.5–2.5 kPa O_2 + possibly 7–10 kPa CO_2. Under these conditions there was an extension of the green life of the fruit of about six times compared with fruit stored in air. They also reported the extension in the postharvest life of bananas at 20°C in CA storage at 0.5 kPa O_2 rather than at 1.5–2.5 kPa O_2. Abdul Rahman *et al.* (1997) indicated that fermentation occurred in 'Berangan' (*Musa* AAA) at 12°C in 0.5 kPa O_2 but not in 2 kPa O_2. They also showed that after storage for 6 weeks in both 2 kPa O_2 and 5 kPa O_2 fruits had a lower respiration rate at 25°C than those that had been stored for the same period in air.

Wills *et al.* (1990) stored 'Cavendish' in a total N_2 atmosphere at 20°C for 3 days soon after harvest. These fruit took about 27 days to ripen in subsequent storage in air, compared with non-treated fruit which ripened in about 19 days. Klieber *et al.* (2002) found that storage in total N_2 at 22°C did not extend storage life compared with storage in air, but resulted in brown discoloration. Parsons *et al.* (1964) found that bananas could be stored satisfactorily for several days at 15.6°C in an atmosphere of 99% N_2 + 1 kPa O_2. However, they stated that this treatment should only be applied to high-quality fruit without serious skin damage and with short periods of treatment, otherwise fruit failed to develop a full yellow colour when subsequently ripened, even though the flesh softened normally.

After treatment with sulfur dioxide at 2 or 8 μg/kg for 12 h, Williams *et al.* (2003) found that at 15°C bananas could be stored for 4 weeks in air and for 6 weeks in CA. CA conditions before the initiation of ripening had beneficial effects in delaying ripening, with no detrimental effects on subsequent ripening or eating quality of the fruit when ripened in air. Reduced O_2 appeared to be mainly responsible for delaying ripening (Ahmad and Thompson, 2006). The delaying effect of low O_2 on ripening was greater than that of high levels of CO_2. CA storage produced firmer bananas and extended their shelf-life. 'Robusta' stored at 4 or 6 kPa O_2 + 4 or 6 kPa CO_2 had an extended storage life by 12–16 days beyond that of fruit stored in air and was of good eating quality when ripened (Ahmad *et al.*, 2001).

CA has been shown to reduce postharvest disease. Sarananda and Wilson Wijeratnam (1997) found that fruits stored at 14°C in 1–5 kPa O_2 had lower levels of crown rot than those stored in air. This fungistatic effect continued even during subsequent ripening in air at 25°C. They also found that CO_2 levels of 5% and 10% actually increased rotting levels during storage at 14°C. In contrast, Botrel *et al.* (2004) showed that storage in 3 kPa O_2 + 9 kPa CO_2 improved firmness and reduced the incidence of anthracnose (*Colletotrichum musae*) better than storage in air or the other CA treatments tested.

CA storage can also affect the shelf-life of fruit after it has begun to ripen. Bananas that had been initiated to ripen by exposure to exogenous ethylene then immediately stored in 1 kPa O_2 at 14°C remained firm and green for 28 days but ripened almost immediately when transferred to air at 21°C (Liu, 1976a, b). Ahmad and Thompson (2006) showed that the marketable life of 'Giant Cavendish' could be extended 2.3–3.8 times, depending on the combination of O_2 + CO_2 used, compared with storage in air. However, they found that there were detrimental effects on fruit quality when 2 kPa O_2 was used and overall 4 kPa O_2 was most effective in extending storage life. CO_2, in the range tested (4–8%), appeared to have no positive or negative effects on marketable shelf-life or fruit quality. Klieber *et al.* (2003) also found that exposure of 'Williams' to O_2 below 1% at 22°C after ripening initiation induced serious skin injury that increased in severity with prolonged exposure from 6 h to 24 h and also did not extend shelf-life. Fruit of the cultivar 'Sucrier' that had been initiated to ripen with ethylene and then placed in PE bags (0.03 mm) at 20°C showed inhibited ripening and a fermentation flavour (Romphophak *et al.*, 2004). Storage of 'Williams', which had been initiated to ripen, in total N_2 at 22°C had a similar shelf-life to storage in air. However, areas of brown discoloration appeared on bananas placed in N_2 storage (Klieber *et al.*, 2002).

Basil (*Ocimum basilicum*)

In Thailand, Penchaiya *et al.* (2003) stored sweet basil at 13°C under different CA conditions and found that atmospheres containing 5 or 10 kPa CO_2 were detrimental compared with 0 kPa CO_2 and the optimum CA conditions of the ones they tested were 1.5 kPa O_2 + 0 kPa CO_2. However, Sirivatanapa (2006) found that in storage in Thai ambient conditions their shelf-life was 3–5 days, while CA storage at 0–5°C with

1–5% O_2 and 5–15% CO_2 extended their shelf life up to 25 days.

Bayberry, Chinese (*Myrica rubra*)

Qi *et al.* (2003) found that the optimum conditions for keeping the fruit fresh was 2 ± 1°C and 'the lower the O_2 concentration, the better the inhibition effects against microorganisms'. CA storage reduced the development of soft and mushy fruit by 67% after 22 days storage at 5°C compared with storage in air. This effect was attributed to a reduction in ethylene production.

Bean, Runner (*Phaseolus vulgaris*)

Also called green beans, French beans, kidney beans and snap beans. Guo *et al.* (2008) showed that the rate of respiration (peak rate of 109.2 mg CO_2/kg/h), weight loss, soluble solid and surface colour changes were lower during 12 days of storage at 0°C compared with 8°C or 25°C. Kader (1985) reported that storage at 0–5°C in 5–10 kPa CO_2 + 2–3 kPa O_2 had a fair effect on storage but was of limited commercial use. Saltveit (1989) recommended 5–10°C in 4–7 kPa CO_2 + 2–3 kPa O_2, which was said to have only a slight effect, but, for those destined for processing, 5–10°C in 20–30 kPa CO_2 + 8–10 kPa O_2 was recommended. Storage in 5–10 kPa CO_2 + 2–3 kPa O_2 retarded yellowing; also discoloration of the cut end of the beans could be prevented by exposure to 20–30 kPa CO_2 for 24 h (Hardenburg *et al.*, 1990). The same conditions were recommended by SeaLand (1991). High CO_2 levels could result in development of off-flavour, but storage at 45°F (7.2°C) in 5–10 kPa CO_2 + 2–3 kPa O_2 retarded yellowing (Anandaswamy and Iyengar, 1961). Snap beans may be held for up to 3 weeks in 8–18 kPa CO_2, depending on the temperature. CO_2 concentrations of 20% or greater always resulted in severe injury, as loss of tissue integrity followed by decay. At 1°C, 8 kPa CO_2 was the maximum level tolerated, while 18 kPa CO_2 caused injury at 4°C, but not at 8°C. The cultivars 'Strike' and 'Opus' snap beans were stored for up to 21 days at 1°C, 4°C or 8°C in 2 kPa O_2 + up to 40 kPa CO_2, then transferred to air at 20°C for up to 4 days (Costa *et al.*, 1994). Sanchez-Mata *et al.* (2003) found that at 8°C, 3 kPa O_2 + 3 kPa CO_2 was best in extending shelf-life and preserved the nutritive value compared with storage in air, 5 kPa O_2 + 3 kPa CO_2, or 1 kPa O_2 + 3 kPa CO_2.

Beet (*Beta vulgaris*)

Beets can be stored for many months using refrigerated storage at 2°C, but some varieties have a very short marketable life. For example, in Savoy beet, there was a decrease in the β-carotene, AA and chlorophyll contents which was faster in summer, when the shelf-life was 4 days, than in winter, when it was 6 days (Negi and Roy, 2000). SeaLand (1991) reported that unspecified CA storage conditions had only a slight to no effect on beet storage. Shipway (1968) showed that atmospheres containing over 5 kPa CO_2 could damage red beet. Monzini and Gorini (1974) recommended 0°C with 3 kPa CO_2 and 10 kPa O_2 for 1 month.

Blackberry (*Rubus* spp.)

Berries retained their flavour well for 2 days when they were cooled rapidly and stored in an atmosphere containing up to 40 kPa CO_2 (Hulme, 1971). Kader (1989) recommended 0–5°C in 10–15 kPa CO_2 + 5–10 kPa O_2 for optimum storage. Agar *et al.* (1994b) showed that the cultivar 'Thornefree' can be stored at 0–2°C in 20–30 kPa CO_2 + 2 kPa O_2 for 6 days and had up to 3 days subsequent shelf-life in air at 20°C. Perkins-Veazie and Collins (2002) found that storage of the cultivars 'Navaho' and 'Arapaho' at 2°C and 95% RH resulted in reduced decay at 15 kPa CO_2 + 10 kPa O_2 compared with those stored in air. However, there was some decrease in anthocyanins and some off-flavours were detected after 14 days for the CA-stored fruit. Storage in 20–40 kPa CO_2 could be used to maintain the quality of machine-harvested blackberries for processing during short-term storage at 20°C (Hardenburg *et al.*, 1990).There was a moderate reduction in vitamin C in blackberries during storage in high CO_2 concentrations (10–30 kPa CO_2), but reducing O_2 level had little effect (Agar *et al.*, 1997).

Blackcurrant (*Ribes nigrum*)

Stoll (1972) recommended 2–4°C in 40–50 kPa CO_2 + 5–6 kPa O_2. After storage at 18.3°C with 40 kPa CO_2 for 5 days, the fruit had a subsequent shelf-life of 2 days with 2–3% of the fruit being unmarketable due to rotting (Wilkinson, 1972). Skrzynski (1990) described experiments where fruit of the cultivar 'Roodknop' was held at 6–8°C for 24 h and then transferred to 2°C for storage for 4 weeks in one of the following: 20 kPa CO_2 + 3 kPa O_2; 20 kPa CO_2 + 3 kPa O_2 for 14 days, then 5 kPa CO_2 + 3 kPa O_2; 10 kPa CO_2 + 3 kPa O_2; 5 kPa CO_2 + 3 kPa O_2; or in air. The best retention of AA was in both the treatments with 20 kPa CO_2. In years with favourable weather preceding the harvest, storage in 20 kPa CO_2 completely controlled the occurrence of moulds caused mainly by *Botrytis*, *Mucor* and *Rhizopus* species. Agar *et al.* (1991) stored the cultivar 'Rosenthal' at 1°C in 10, 20 or 30 kPa CO_2, all with 2 kPa O_2, or a high CO_2 environment of 10, 20 or 30 kPa CO_2, all with > 15 kPa O_2. The optimum CO_2 concentration was found to be 20%. Ethanol accumulation was higher in CA storage than in a high CO_2 environment. Fruit could be stored for 3–4 weeks under CA storage or high CO_2, compared with 1 week for fruit stored at 1°C in air. For juice manufacture, Smith (1957) recommended storage at 2°C in 50 kPa CO_2 for 7 days, followed by a further 3 weeks at 2°C in 25 kPa CO_2. With longer storage periods there was an accumulation of alcohol and acetaldehyde but the juice quality was not affected.

CA storage can affect the nutrient content of blackcurrants. Harb *et al.* (2008a) found that storage of 'Titania' in air for up to 6 weeks did not significantly affect total terpene volatiles, which were similar to those in freshly harvested berries. They found that decreasing O_2 and increasing CO_2 levels retarded the capacity to synthesize terpenes for 3 weeks, but during storage for an additional 3 weeks it led to a partial recovery. However, storage in 18 kPa CO_2 + 2 kPa O_2 resulted, in most cases, in a lower biosynthesis of volatile constituents compared with storage in air for 6 weeks. Non-terpene compounds, mainly esters and alcohols, were also increased in fruit during storage in air compared with those in CA storage where there was an initial reduction, but they later recovered. There was a moderate reduction in vitamin C during storage in high CO_2 concentrations (10–30 kPa CO_2) in blackcurrants but reducing O_2 level had little effect (Agar *et al.*, 1997).

Blueberry, Bilberry, Whortleberry (*Vaccinium corymbosum*, *V. myrtillus*)

Kader (1989) recommended 0–5°C in 15–20 kPa CO_2 + 5–10 kPa O_2 and Ellis (1995) recommended 0.5°C and 90–95% RH in 10 kPa O_2 + 10 kPa CO_2 for 'medium term' storage. Storage for 7–14 days at 2°C in 15 kPa CO_2 delayed decay by 3 days after they had been returned to ambient temperatures, compared with storage in air at the same temperature (Ceponis and Cappellini, 1983). They also showed that storage in 2 kPa O_2 had no added effect over the CO_2 treatment. The optimum atmosphere for the storage of the cultivar 'Burlington' was 0°C with 15 kPa O_2 + 10 kPa CO_2 in which fruit maintained good quality for over 6 weeks (Forney *et al.*, 2003). After 6 weeks of storage at 0°C in 15 kPa O_2 + 25 kPa CO_2, the concentration of ethanol was about 18 times higher and ethyl acetate was about 25 times higher than when stored in 15 kPa O_2 + 0 or 15 kPa CO_2 (Forney *et al.*, 2003). Harb and Streif (2006) successfully stored the cultivar 'Bluecrop' for up to 6 weeks in 12 kPa CO_2 + 21 kPa O_2, which retained fruit firmness, had low decay and the absence of off-flavour. 'Duke' fruits could be kept in air at 0–1°C in acceptable condition for up to 3 weeks. However, for storage for up to 6 weeks, up to 12 kPa CO_2 + 18–21 kPa O_2 was found to give the best results. Storage in 6–12 kPa CO_2 maintained their firmness, while those stored in > 12 kPa CO_2 rapidly softened at both 2 kPa O_2 and 18 kPa O_2 and had poorer flavour, firmness and acidity content. Also with increasing CO_2 levels the respiration rate increased in all CO_2 levels to higher respiration rates than in fruit stored in air. Chiabrando and Peano (2006) found that storage at 1°C in 3 kPa O_2 + 11 kPa CO_2 for 60 days maintained fruit firmness, total soluble solids and TA better than storage in air at 3°C and 90–95% RH. DeEll (2002) reported that storage in15–20 kPa CO_2 + 5–10 kPa O_2 reduced the respiration rate and softening, but exposure to < 2 kPa O_2 and/or > 25 kPa CO_2 could cause off-flavours and brown discoloration, depending on the cultivar, duration of exposure and temperature. Zheng *et al.* (2003) showed that the antioxidant levels in 'Duke' increased in 60–100 kPa O_2 as compared with 40 kPa O_2 or air during 35 days of storage. O_2 levels of between 60% and 100% also resulted in an increase in total phenolics and total anthocyanins.

Schotsmans *et al.* (2007) found that fungal development was reduced by CA storage. After 28 days at 1.5°C in 2.5 kPa O_2 + 15 kPa CO_2, the cultivars 'Centurion' and 'Maru' had only half as much blemished fruit compared with those stored in air. Incidence of fungal diseases, mainly caused by *Botrytis cinerea*, could be efficiently controlled by CO_2 levels over 6% (Harb and Streif, 2004b). DeEll (2002) reported that 15–20 kPa CO_2 + 5–10 kPa O_2 reduced the growth of *B. cinerea* and other decay-causing organisms during transport and storage. Zheng *et al.* (2003) showed that fruit stored in ≥ 60 kPa O_2 had significantly less decay. Song *et al.* (2003) showed that the percentage marketabity of the cultivar 'Highbush' was 4–7% greater than in the controls after treatment with 200 ppb O_3 for 2 or 4 days in combination with 10 kPa CO_2 + 15 kPa O_2. Rodriguez and Zoffoli (2016) stored 'Highbush' blueberries at 0°C in atmospheres of 5/1 kPa, 7/15 kPa or 15/5 kPa CO_2/O_2 for up to 40 days. Partial pressures of CO_2 higher than 8 kPa or O_2 partial pressures lower than 2 kPa resulted in increased fruit softening, but this varied between the cultivars they tested. They also found that CO_2 partial pressure higher than 8 kPa or O_2 partial pressure lower than 2 kPa induced fruit softening in the cultivar 'Brigitta'. Paniagua *et al.* (2014) compared storage of blueberries at 0°C and 4°C in air, 10 kPa CO_2 + 2.5 kPa O_2 or 10 kPa CO_2 +20 kPa O_2. Atmospheres with 10 kPa CO_2 reduced decay incidence, particularly when combined with 2.5 kPa O_2, though the low O_2 fruit tended to soften more quickly.

After 28 days, Schotsmans *et al.* (2007) showed that storage at 1.5°C in 2.5 kPa O_2 + 15 kPa CO_2 resulted in only half as much blemished fruit compared with storage in air, and fungal development was minimized in one of the two cultivars tested by the CA storage. Cantín *et al.* (2012) tested various levels of CO_2 combined with 3 kPa O_2 on disease control of blueberries. They found that 24 kPa CO_2 increased fruit softening and resulted in off-flavours They also combined CA storage with SO_2 fumigation and found that SO_2 fumigation followed by storage in 3 kPa O_2 combined with either 6 or 12 kPa CO_2 reduced decay, extended market life and maintained nutritional value.

Breadfruit (*Artocarpus altilis, A. communis*)

The optimum conditions for CA storage were given by Sankat and Maharaj (2007) as 5 kPa O_2 + 5 kPa CO_2 at 16°C.

Broccoli, Sprouting (*Brassica oleracea* var. *italica*)

A considerable amount of research has been published on CA and MA storage. Klieber and Wills (1991) reported that at 0°C and 100% RH broccoli could be stored for about 8 weeks and the storage life could be further extended in < 1 kPa O_2, but storage in 6 kPa CO_2 resulted in injury after 4–5 weeks. Other recommendations include: 5–10 kPa CO_2 + 1–2 kPa O_2 (SeaLand, 1991; Jacobsson *et al.*, 2004b); 10 kPa CO_2 + 1 kPa O_2 (McDonald,

1985; Deschene *et al.*, 1991); 0°C in 2–3 kPa O_2 + 4–6 kPa CO_2 (Ballantyne *et al.*, 1988); 0–5°C in 5–10 kPa CO_2 + 1–2 kPa O_2 (Kader, 1985, 1992; Saltveit, 1989), the latter claiming that it had a 'high level of effect'; 0–5°C in 5–10 kPa CO_2 + 1–2%O_2 (Bishop, 1996); a maximum of 15 kPa CO_2 and a minimum of 1 kPa O_2 (Fellows, 1988); 1.8–10 kPa CO_2 + 2 kPa O_2 for 3–4 weeks (Gorini, 1988); 0°C or 5°C in 0.5 kPa O_2 + 10 kPa CO_2 or 10°C in 1 kPa O_2 + 10 kPa CO_2 for 'Marathon' florets (Izumi *et al.*, 1996a); 15°C in 2 kPa O_2 + 10 kPa CO_2 (Saijo, 1990); and 3 kPa O_2 + 2 kPa CO_2 + 95% N_2 at 2°C (Yang *et al.*, 2004). The respiration rate of the cultivar 'Emperor' decreased as the O_2 concentration of the atmosphere decreased even down to 0% over a 24 h period at 10°C or 20°C (Praeger and Weichmann, 2001). Kubo *et al.* (1989a, b) reported that 60 kPa CO_2 + 20 kPa O_2 reduced the respiration rate. However, Makhlouf *et al.* (1989b) found that after 6 weeks at 1°C in 10 kPa CO_2 or more, the rate of respiration increased at the same time as the development of undesirable odours and physiological injury.

Generally CA storage reduced the rate of loss of chlorophyll and other phytochemicals. The cultivar 'Stolto' was stored for 6 weeks at 1°C in 0 kPa CO_2 + 20 kPa O_2, 10 kPa CO_2 + 20 kPa O_2, 6 kPa CO_2 + 2.5 kPa O_2, 10 kPa CO_2 + 2.5 kPa O_2 or 15 kPa CO_2 + 2.5 kPa O_2. Chlorophyll retention was better in CA than in air, mainly due to increased CO_2 concentration (Makhlouf *et al.*, 1989a). Yang *et al.* (2004) also showed that losses of chlorophyll and AA were reduced during CA compared with storage in air. The chlorophyll degradation was effectively delayed by storage at 0–1°C in 5 kPa O_2 + 10 kPa CO_2 or in atmospheres with < 1.5 kPa O_2. Total carotenoid content remained almost constant during 8 weeks of storage in air or 0–5 kPa CO_2 + 0.8–3.0 kPa O_2, but increased in storage in 10 kPa CO_2 + 3 kPa O_2 (Yang and Henze, 1988). Storage in 5 and 10 kPa CO_2 + 3 kPa O_2 were, by visual assessment, the most effective in reducing chlorophyll loss. Most CA combinations had no effect on flavour, but the samples stored at 10 kPa CO_2 + 3 kPa O_2 had an undesirable flavour after 8 weeks. CO_2 and O_2 levels had little effect on broccoli firmness (Yang and Henze, 1987). Colour and texture were not significantly influenced by CA compared with storage at the same temperature in air (Berrang *et al.*, 1990). Storage for 13 weeks in 15 kPa O_2 + 6 kPa CO_2 resulted in a modest retention of chlorophyll and a 30% reduction of trim loss as compared with air. Storage of the cultivar 'Green Valiant' in 8 kPa CO_2 + 10 kPa O_2, 6 kPa CO_2 + 2 kPa O_2 or 8 kPa CO_2 + 1 kPa O_2 reduced chlorophyll loss by 13%, 9% and 32%, respectively, and reduced trim loss by 40%, 45% and 41%, respectively, and the respiration rate as compared with air (McDonald, 1985). When freshly cut heads of the cultivars 'Commander' and 'Green Duke' were stored in air at 23°C or 10°C, the florets rapidly senesced. Chlorophyll levels declined by 80–90% within 4 days at 23°C and within 10 days at 10°C. Storage at 5°C or 10°C in 5 kPa CO_2 + 3 kPa O_2 at approximately 80% RH strongly inhibited loss of chlorophyll (Deschene *et al.*, 1991). The effect of storage of the cultivars 'Marathon', 'Montop' and 'Lord' in either 2 kPa O_2 + 6 kPa CO_2 or 0.5–1 kPa O_2 + 10 kPa CO_2 on antioxidant activity was similar to that for storage in air, but antioxidant activity and the vitamin C content increased in stored broccoli compared with fresh broccoli florets (Wold *et al.*, 2006). Li *et al.* (2016c) compared five CA conditions on storage of broccoli at 10°C. The combinations were: 70 kPa O_2 + 30kPa CO_2; 60 kPa O_2 + 40 kPa CO_2; 50 kPa O_2 + 50 kPa CO_2; 40 kPa O_2 + 60 kPa CO_2; 30 kPa O_2 + 70 kPa CO_2 at 10°C. They found that the 50 kPa O_2 + 50 kPa CO_2combination reduced the metabolism of the broccoli, particularly the respiration rate, and delayed senescence, though ATP levels were relatively higher. Xu *et al.* (2006) investigated the effects of CA storage on changes in the phytochemical content. Glucoraphanin content and induction of quinone reductase (QR) activity (both have been shown to have beneficial effects on human health) in broccoli was affected by temperature and CA conditions. They compared storage in CA with air and showed that there was an increase in the glucoraphanin content and the QR activity over the first 5 days of storage at 5°C, while CA conditions with reduced O_2 concentrations (1 kPa O_2, 1 kPa O_2 + 10 kPa

CO_2) led to a steady decrease of glucoraphanin content and QR activity during 20 days of storage at 5°C. The highest content of glucoraphanin and QR activity was found in broccoli stored under 21 kPa O_2+10 kPa CO_2 at 5°C, which maintained their visual quality, glucoraphanin content and QR activity for 20 days. Eason *et al.* (2007) reported that storage of broccoli for 96 h after harvest at 20°C in 10 kPa CO_2 + 5 kPa O_2 resulted in the loss of less water and sugars, had lower protease activity and no significant loss of colour compared with storage in air. They examined differential gene expression in broccoli tissues in response to postharvest CA storage and found a number of novel genes without previously assigned functions with up-regulated and down-regulated expression.

CA storage generally reduced postharvest losses due to disease but results have been mixed. Among the atmospheres tested by Makhlouf *et al.* (1989b), 6 kPa CO_2 + 2.5 kPa O_2 was the best for storage for 3 weeks since it delayed the development of soft rot and mould and there was no physiological injury. Compared with storage in air, broccoli stored in atmospheres containing 5–10 kPa CO_2 + 3 kPa O_2 had a small reduction in decay, while storage in 0.8 and 1.5 kPa O_2 resulted in more rapid decay (Yang and Henze, 1987). Storage in 11 kPa O_2 + 10 kPa CO_2 at 4°C significantly reduced the growth of microorganisms and extended the length of time they were subjectively considered acceptable for consumption (Berrang *et al.*, 1990). The effects of CA storage on the survival and growth of *Aeromonas hydrophila* was examined. Two lots of each were inoculated with *A. hydrophila* 1653 or K144. A third lot served as a non-inoculated control. Following inoculation they were stored at 4°C or 15°C in a CA storage system previously shown to extend their storage life, or in ambient air. Without exception, CA storage lengthened the time that they were considered to be acceptable for consumption. However, CA storage did not significantly affect populations of *A. hydrophila* or *Listeria monocytogenes* (Berrang *et al.*, 1989).

CA has been evaluated on minimally processed broccoli florets. 'Marathon' florets were held for 4 days at 10°C in containers with an air flow (20.5 kPa O_2 and < 0.5 kPa CO_2), a restricted air flow (down to 17.2 kPa O_2 and up to 3.7 kPa CO_2), no flow (down to 1.3 kPa O_2 and up to 30 kPa CO_2) and N_2 flow (< 0.01 kPa O_2 and < 0.25 kPa CO_2). Sensory analysis of cooked broccoli indicated a preference for the freshly harvested and air-stored broccoli, with samples stored in no-flow conditions showing the opposite results (Hansen *et al.*, 1993). 'Green Valiant' heads and florets were stored at 4°C. Minimally processing broccoli heads into florets increased the rate of respiration throughout storage at 4°C in air, in response to wounding stress. Ethylene production was also stimulated after 10 days. Atmospheres for optimal preservation of the florets were evaluated using continuous streams of the following defined atmospheres: 0 kPa CO_2 + 20% O_2 (air control); 6 kPa CO_2 + 1% O_2; 6 kPa CO_2 + 2% O_2; 6 kPa CO_2 + 3% O_2; 3 kPa CO_2 + 2% O_2; and 9 kPa CO_2 + 2% O_2. The atmosphere consisting of 6 kPa CO_2 + 2 kPa O_2 resulted in extended storage of broccoli florets from 5 weeks in air to 7 weeks. Prolonged chlorophyll retention and reduced development of mould and offensive odours and better water retention were especially noticeable when the florets were returned from CA at 4°C to air at 20°C. It was concluded that minimal processing had little influence on optimal storage atmosphere, suggesting that recommendations for intact produce are useful guidelines for MA packaging of minimally processed vegetables (Bastrash *et al.*, 1993). Cefola *et al.* (2016) reported that CO_2 levels of 10–15% could be damaging during storage of fresh-cut broccoli and the optimum conditions were 5% O_2 and ≤ 5% CO_2.

Brussels Sprouts (*Brassica oleracea* var. *gemmifera*)

High-quality sprouts could be maintained for 10 weeks in storage at 1.5–2.0°C in air, which could be extended to 12 weeks in 5 kPa CO_2 (Peters and Seidel, 1987). Beneficial effects have been reported with storage in atmospheres containing 2.5, 5 and 7 kPa

CO_2 (Pantastico, 1975). SeaLand (1991) recommended 5–7 kPa CO_2 + 1–2 kPa O_2. Typical storage conditions were given as 0–5°C in 5–7 kPa CO_2 + 1–2% O_2 by Bishop (1996). Storage in 5–7.5 kPa CO_2 + 2.5–5 kPa O_2 helped to maintain quality at 5°C or 10°C, but internal discoloration could occur at 0°C with O_2 levels below 1% (Hardenburg *et al.*, 1990). Kader (1985) and Saltveit (1989) recommended 0–5°C + 5–7 kPa CO_2 and 1–2 kPa O_2 which had a good effect on storage but was of no commercial use. Kader (1989) subsequently recommended 0–5°C in 5–10 kPa CO_2 + 1–2 kPa O_2 which he reported to have excellent potential benefit, but limited commercial use. Other recommendations include 0–1°C in 6 kPa CO_2 + 15 kPa O_2 or 6 kPa CO_2 + 3 kPa O_2 for up to 4 months (Damen, 1985) and 7 kPa O_2 + 8 kPa CO_2 for 80 days (Niedzielski, 1984). 'Lunette', 'Rampart' and 'Valiant' in storage in 0.5, 1, 2 or 4 kPa O_2 had lower respiration rates relative to those in air, but rates were similar among the four low O_2 levels. Ethylene production was low at 2.5°C and 5°C in all atmospheres, but at 7.5°C it was 20–170% higher in air than in low O_2. Ethylene production virtually stopped during exposure to 1 kPa O_2 + 10 kPa CO_2, 2 kPa O_2 + 10 kPa CO_2 or 20 kPa O_2 + 10 kPa CO_2, but increased markedly when removed to air storage. Since low O_2 levels retarded yellowing and 10 kPa CO_2 retarded decay development, the combination of low O_2 with high CO_2 effectively extended the storage life of sprouts at 5°C and 7.5°C. The beneficial effect of CA storage was still evident after return of the samples to air. The sprouts retained good appearance for 4 weeks at 2.5°C, whether stored in CA or in air. Storage in 0.5 kPa O_2 occasionally induced a reddish tan discoloration of the heart leaves and frequently an extremely bitter flavour in the non-green portion of the sprouts (Lipton and Mackey, 1987). The cultivar 'Rampart' was stored either still on the stems or loose at 2–3°C and < 75% RH, or 1°C and < 95% RH on the stem in air or in 6 kPa CO_2 + 3 kPa O_2. Loose sprouts became severely discoloured at the stalk end after only 1 week in air and after 2 weeks in CA. Those on the stem remained fresh in CA for 9 weeks at 2–3°C and for 16 weeks at 1°C (Pelleboer, 1983).

Butter Bean, Lima Bean (*Phaseolus lunatus*)

Storage of fresh beans in CO_2 concentrations of 25–35% inhibited fungal and bacterial growth without adversely affecting their quality (Brooks and McColloch, 1938).

Cabbage (*Brassica oleracea* var. *capitata*)

Cabbage is perhaps the most common vegetable to be stored commercially in CA storage conditions. Stoll (1972) recommended 0°C in 3 kPa CO_2 + 3 kPa O_2 for red and savoy cabbages and 0–3 kPa CO_2 + 3 kPa O_2 for white cabbage. Danish cultivars were successfully stored for 5 months at 0°C in 2.5–5 kPa CO_2 + 5 kPa O_2 (Isenberg and Sayles, 1969). Hardenburg *et al.* (1990) showed that the optimum conditions for CA storage were 2.5–5 kPa CO_2 + 2.5-5 kPa O_2 at 0°C, while SeaLand (1991) recommended 5–7 kPa CO_2 + 3–5 kPa O_2 for green, red or savoy. Kader (1985) recommended 0–5°C in 5–7 kPa CO_2 + 3–5 kPa O_2 or 0–5°C in 3–6 kPa CO_2 + 2–3 kPa O_2 (Kader, 1992), which was claimed to have had a good effect and was of some commercial use for long-term storage of certain cultivars. Saltveit (1989) recommended 0–5°C in 3–6 kPa CO_2 + 2–3 kPa O_2, which had a high level of effect. Typical storage conditions were given as 0°C in 5 kPa CO_2 + 3% O_2 by Bishop (1996) for white cabbage. After storage of several white cabbage cultivars at 0–1°C for 39 weeks, the percentage recovery of marketable cabbage after trimming was 92% in 5 kPa CO_2 + 3 kPa O_2 and only 70% for those that had been stored in air (Geeson, 1984). Cabbage stored for 159 days in air had a 39% total mass loss and in 3–4 kPa CO_2 + 2–3 kPa O_2 for 265 days the total mass loss was only 17%. Cabbages in CA storage showed better retention of green colour, fresh appearance and texture than those in air (Gariepy *et al.*, 1985). The cultivars 'Lennox' and 'Bartolo' were stored in air, 3 kPa O_2 + 5 kPa CO_2 or 2.5 kPa O_2 + 3 kPa CO_2. Disease incidence was lower with both the CA storage conditions and there were no trimming losses. CA also helped to retain green colour in 'Lennox' (Prange and Lidster, 1991). Storage at 0.5–1.5°C and 60–75% RH in 3 kPa O_2 + 4.5–5 kPa CO_2 lengthened the period for which cabbage could be stored by at least 2 months and improved quality compared with storage at 2–7°C in 60–75% RH in air (Zanon and Schragl, 1988).

Storage of the cultivar 'Tip Top' at 5°C in 5 kPa CO_2 + 5 kPa O_2 gave the slowest rate of deterioration and at 2.5°C a CO_2 concentration > 2.5% was detrimental. The cultivar 'Treasure Island' had a longer storage life than 'Tip Top' in all CO_2/O_2 combinations. At 2.5°C in 5 kPa CO_2 + 20.5 kPa O_2, 97% of 'Treasure Island' was saleable after 120 days (Schouten, 1985). Gariepy *et al.* (1984) showed that after storage at 3.5–5 kPa CO_2 + 1.5–3 kPa O_2 for 198 days there was a total mass loss of 14%, compared with 40% for storage in air, and better retention of colour, fresher appearance and firmer texture. The cultivar 'Winter Green' was stored at 1.3°C in 5–6 kPa CO_2 + 2–3 kPa O_2 + 92% N_2 and traces of other gases for 32 weeks compared with storage in air at 0.3°C. The average trimming losses were < 10% for the CA-stored cabbage while those in air exceeded 30%; and CA-stored cabbages retained their colour, flavour and texture better (Raghavan *et al.*, 1984). Huang *et al.* (2002) stored the cultivar 'Chu-chiu' at 0–1°C in 2–6 kPa O_2 (mostly at 3%) + 3–5 kPa CO_2 (mostly at 5%) and found that the storage life was doubled to 4 months compared with air storage. The factors that terminated storage were high weight loss and trim loss, inferior colour and freshness, loss of flavour and rooting at the cut end. Two cultivars of savoy cabbage, 'Owasa' and 'Wirosa', were stored at 0–1°C in 4 kPa CO_2 + 3 kPa O_2 + 93% N_2, which retained their quality and slowed down the degradation of vitamin C and chlorophyll pigments compared with air storage (Krala *et al.*, 2007). The need for fresh air ventilation at regular intervals to maintain ethylene concentrations at low levels was emphasized by Meinl *et al.* (1988).

AA concentration increased during storage at 0°C in air. This increase was reduced in an atmosphere of 1 kPa O_2 which was shown also to delay the yellowing of the outer laminae and maintained higher chlorophyll content (Wang and Ji, 1989). Berard (1985) found that storage at 1°C in 2.5 kPa O_2 + 5 kPa CO_2 considerably delayed de-greening, and eliminated abscission and loss of dormancy during the first 122 days of storage, compared with those stored in air.

There is strong evidence that CA storage can reduce some diseases. Storage at –0.5–0°C in 5–8 kPa CO_2 prevented the spread of *Botrytis cinerea* and total storage losses were lower in CA storage than in air (Nuske and Muller, 1984). Pendergrass and Isenberg (1974) also reported less disease, also mainly caused by *B. cinerea*, and better head colour was observed with storage in 5 kPa CO_2 + 2.5 kPa O_2 + 92.5% N_2 compared with storage in air at 1°C and 75%, 85% or 100% RH. Berard (1985) described experiments in which 25 cultivars were placed at 1°C and 92% RH. She found that those in 2.5 kPa O_2 + 5 kPa CO_2 usually had reduced or zero grey speck disease and reduced incidence and severity of vein streaking compared with those stored in air for up to 213 days, but not in every case. Black midrib and necrotic spot were both absent at harvest but in comparison with storage in air those stored in 2.5 kPa O_2 + 5 kPa CO_2 had increased incidence of black midrib and it also favoured the development of inner head symptoms on susceptible cultivars. In CA the incidence of necrotic spot in the core of the heads of cultivar 'Quick Green Storage' was increased, which was particularly evident in a season when senescence of cabbage was most rapid. Even though both disorders were initiated in the parenchyma cells, black midrib and necrotic spot had a distinct histological evolution and affected different cultivars under similar conditions of growth and storage (Berard *et al.*, 1986).

Cactus Pear, Prickly Pear, Tuna, Opuntia (*Opuntia ficus indica, O. robusta*)

Testoni and Eccher Zerbini (1993) recommended 5°C in 2 or 5 kPa CO_2 + 2 kPa O_2 which reduced the incidence of chilling injury and rot development. Cantwell (1995) cited the beneficial effects of lining boxes of fruit with PE film, especially with paper or other absorbent material to absorb condensation.

Besides the fruit, the flattened stems of the prickly pear called *cladodes nopal* and *nopalitos* are eaten in Mexico. Guevara *et al.* (2003) tested the effects of passive and semi-active MA packaging on the postharvest life and quality of *cladodes* stored at 5°C. In semi-active MA packaging, they injected elevated partial pressures of CO_2 (20, 40 or 80 kPa) in the packages immediately after sealing. Passive MA packaging (where no CO_2 was added) had an atmosphere of up to 8.9 kPa O_2 and 7 kPa CO_2 after 35 days of storage at 5°C. Semi-active atmospheres with initial CO_2 pressures of 40 or 80 kPa increased the losses in texture, weight, chlorophyll content, dietary fibre content and colour. Passive MA packaging and semi-active MA packaging with 20 kPa CO_2 significantly decreased the losses in the above-mentioned parameters and also decreased the microbial counts (total aerobic mesophiles, mould and yeasts), but slightly increased the total anaerobic mesophiles counts. The microorganisms identified were *Pseudomonas, Leuconostoc, Micrococcus, Bacillus, Ruminicoccus, Absidia, Cladosporium, Penicillium* and *Pichia*. Therefore, fresh prickly pear cactus stems could be stored for up to 32 days in MA packaging with ≤ 20 kPa CO_2 without significant losses in quality or any significant increase in microbial counts.

Capsicum, Sweet Pepper, Bell Pepper (*Capsicum annum* var. *grossum*)

Recommendations include: 8–12°C in 0 kPa CO_2 + 2 or 3–5 kPa O_2, which had a slight to fair effect but was of limited commercial use (Kader, 1985, 1992; Saltveit, 1989); 0–3 kPa CO_2 + 3–5 kPa O_2 (SeaLand, 1991); and 8°C and > 97% RH in 2 kPa CO_2 + 4 kPa O_2 (Otma, 1989). Storage with O_2 levels of 3–5% was shown to retard respiration, but high CO_2 could reduce loss of green colour and also result in calyx discoloration (Hardenburg *et al.*, 1990). Storage life of 'California Wonder' at 8.9°C could be extended from 22 days in air to 38 days in 2–8 kPa CO_2 + 4–8 kPa O_2 (Pantastico, 1975). Storage of the cultivar 'Jupiter' for 5 days at 20°C in 1.5 kPa O_2 + 98.5% N_2 resulted in post-storage respiratory rate suppression for about 55 h after transfer to air (Rahman *et al.*, 1995). 'California Wonder' stored in low O_2 had lower internal ethylene contents than those in air and storage in 1 kPa O_2 resulted in significantly lower internal CO_2 than storage in 3, 5, 7 or 21 kPa O_2. Colour retention was greater in storage in low O_2 atmospheres than in air (Luo and Mikitzel, 1996). Rahman *et al.* (1993b) found that there was a residual effect on respiration rate of storage for 24 h at 20°C in 1.5, 5 or 10 kPa O_2. The residual effect lasted for 24 h when they were transferred to air, with 1.5 kPa O_2 having the most effect. Extending the storage period in 1.5 kPa O_2 to 72 h extended the residual effect from 24 h to 48 h.

Polderdijk *et al.* (1993) studied the interaction of CA and humidity. 'Mazurka' fruits were stored for 15 days at 8°C in atmospheres containing 3 kPa CO_2 + 3 kPa O_2 or air at 85%, 90%, 95% or 100%. Fruits were then stored for 7 days in air at 20°C and 70% RH and the incidence of an unspecified decay during this period increased relative to the humidity increased during storage. Storage in 3 kPa CO_2 + 3 kPa O_2 reduced the incidence of post-storage decay compared with storage in air. Luo and Mikitzel (1996) also reported that CA could reduce decay. After 2 weeks of storage at 10°C, 'California Wonder' had 33% decay in those stored in air but only 9% of those stored in 1 kPa O_2 and this reduction in decay continued throughout the 4 weeks of storage. Storage in 3 and 5 kPa O_2 atmospheres slightly reduced decay for a short time, while the incidence of decay in fruits stored in 7 kPa O_2 was not significantly different to those stored in air. Conesa *et al.* (2007)

evaluated several atmospheres on the storage of fresh-cut peppers and found that an atmosphere containing 50–80 kPa O_2 + 20 kPa CO_2 prevented fermentation and inhibited growth of spoilage microorganisms. Those exposed to 0, 0.5, 1, 3 or 9 kPa O_2 (all with 0 kPa CO_2), and to 0 kPa O_2 + 20 kPa CO_2, had lower respiration rates than those in the range 20–100 kPa O_2 in both 0 or 20 kPa CO_2. High O_2 had little effect on respiration rate at 14°C and no effect at 2°C and 7°C in stimulating both CO_2 production and O_2 consumption compared with normal air, but 20 kPa CO_2 in the range 20–100 kPa O_2 increased the respiration rate, probably because physiological injury occurred at 14°C.

Carambola, Star Fruit (*Averrhoa carambola*)

Storage at 7°C and 85–95% RH with either 2.2 kPa O_2 + 8.2 kPa CO_2 or 4.2 kPa O_2 + 8 kPa CO_2 resulted in low losses of about 1.2% during 1 month and fruit retained a bright yellow colour with good retention of firmness, TSS and acidity compared with fruit stored in air (Renel and Thompson, 1994). Fresh-cut carambola were dipped in AA, citric acid or Ca-EDTA and then stored in various concentrations of O_2 from 0.4 to 20.3kPa. They found that the combination of 1% AA and storage in 0.4 kPa O_2 reduced browning and 'loss of visual quality' for up to 12 days, which was 3 days longer than those stored in 0.4 kPa O_2 without AA treatment (Teixeira *et al.*, 2008).

Carrot (*Daucus carota* subsp. *sativus*)

CA storage was not recommended by Hardenburg *et al.* (1990), since carrots stored in 5–10 kPa CO_2 + 2.5-6 kPa O_2 had more mould growth and rotting than those stored in air. Increased decay was also reported by Pantastico (1975) in atmospheres of 6 kPa CO_2 + 3 kPa O_2 compared with storage in air. Ethylene or high levels of CO_2 in the storage atmosphere could give the roots a bitter flavour (Fidler, 1963). However, storage in 1–2 kPa O_2 at 2°C for 6 months was previously reported to have been successful (Platenius *et al.*, 1934) and Fellows (1988) recommended CA storage in a maximum of 4 kPa CO_2 and a minimum of 3 kPa O_2. Storage atmospheres containing 1, 2.5, 5 or 10 kPa O_2 inhibited both sprouting and rooting during storage at 0°C, but again was reported to result in increased mould infection. Atmospheres containing 21 or 40 kPa O_2 reduced mould infection, but increased sprouting and rooting (Abdel-Rahman and Isenberg, 1974). Ayhan *et al.* (2008) found that storing minimally processed carrots in 80 kPa O_2+ 10 kPa CO_2 retained their quality better than in 5% O_2 + 0 kPa CO_2. Izumi *et al.* (1996b) stored carrot slices, sticks and shreds in air or in 0.5 kPa O_2+ 10 kPa CO_2 at 0°C, 5°C and 10°C. CA storage was beneficial in reducing decay, weight loss and white discoloration on shreds and microbial growth on sticks. White discoloration and microbial growth occurred only at 0°C and 5°C.

Cassava, Tapioca, Manioc, Yuca (*Manihot esculenta*)

These can deteriorate within a day or so of harvesting, due to a physiological disorder called vascular streaking (Thompson and Arango, 1977). The cultivar 'Valencia' was harvested from 12-month-old plants, coated with paraffin wax and stored at 25°C for 3 days in either 54–56% RH or 95–98% RH, with O_2 levels of either 21% (air) or 1% by Aracena *et al.* (1993). They found that storage in 54–56% RH in air resulted in 46% vascular streaking, while storage in 1 kPa O_2 had reduced vascular streaking down to 15%. However, at high humidity, irrespective of the O_2 level, vascular streaking incidence was reduced to only 1.4%. They concluded that the occurrence of vascular streaking was primarily related to water stress in the tissue, while O_2 was secondarily involved.

Cauliflower (*Brassica oleracea* var. *botrytis*)

Respiration rate was decreased in 3 kPa O_2 compared to storage in air (Romo Parada *et al.*, 1989). Recommended storage conditions include: 0°C with 10 kPa CO_2 + 10 kPa O_2 for 5 weeks (Wardlaw, 1938); 0°C in 10 kPa CO_2 + 11 kPa O_2, also for 5 weeks (Smith, 1952); 0°C in 0–3 kPa CO_2 + 2–3 kPa O_2 (Stoll, 1972); 5–6 kPa CO_2 + 3 kPa O_2 (Tataru and Dobreanu, 1978); 0°C in 5–10 kPa CO_2 + 5 kPa O_2 for 50–70 days (Monzini and Gorini, 1974); 2–5 kPa CO_2 + 2–5 kPa O_2 (SeaLand, 1991); a maximum of 5 kPa CO_2 and a minimum of 2 kPa O_2 (Fellows, 1988); 0–5°C in 2–5 kPa CO_2 + 2–5 kPa O_2 which had a fair effect but was of no commercial use (Kader, 1985, 1992); and storage at 0°C in 3 kPa O_2 + 5 kPa CO_2 which reduced weight loss and discoloration but did not affect free sugar or glucosinolate profiles compared with those stored in air at the same temperature for up to 56 days (Hodges *et al.*, 2006). Adamicki (1989) described successful storage of autumn cauliflowers at 1°C in 2.5 kPa CO_2 + 1 kPa O_2 for 71–75 days or in the same atmospheres but at 5°C for 45 days. Romo Parada *et al.* (1989) showed that curds stored at 1°C and 100% RH in 3 kPa O_2 + either 2.5 or 5 kPa CO_2 were still acceptable after 7 weeks of storage, while 3 kPa O_2 + 10 kPa CO_2 caused softening, yellowing and increased leakage. Curds of the cultivar 'Primura' were stored successfully for 6–7 days at 0–1°C and 90–95% RH in circulating air and for 20–25 days in 4–5 kPa CO_2 + 16–27 kPa O_2 (Saray, 1988). Work in Holland showed very little benefit of CA storage, but storage at 0–1°C and > 95% RH with 5 kPa CO_2 + 3 kPa O_2 gave a better external appearance but had no effect on curd quality (Mertens and Tranggono, 1989). Both summer and autumn crops stored at 1°C or 5°C in 2.5 kPa CO_2 + 3 kPa O_2 or 5 kPa CO_2 + 3 kPa O_2 had better leaf colour, curd colour, firmness and market value than in air. Curds of the autumn crop stored in 2 kPa CO_2 + 3 kPa O_2 had a superior composition to those stored in other CA at both storage temperatures (Adamicki and Elkner, 1985). Kaynaş *et al.* (1994) successfully stored the cultivar 'Iglo' in 3 kPa CO_2 + 3 kPa O_2 for a maximum of 6 weeks with a shelf-life of 3 days at 20°C, which was double their storage life in air.

In other work, CA did not extend storage life and CO_2 levels of 5% or more or 2 kPa O_2 or less injured the curds (Hardenburg *et al.*, 1990). After storage at either 4.4°C or 10°C with levels of 5 kPa CO_2 in the storage atmosphere, some injury was evident after the curds were cooked (Ryall and Lipton, 1972). Mertens and Tranggono (1989) concluded that, in storage at 0–1°C and < 95% RH for 4 or 6 weeks with subsequent shelf-life studies at 10°C and 85% RH, storage in 5 kPa CO_2 + 3 kPa O_2 had a very small effect, if any, on the respiration rate of the curds. Work on the cultivar 'Pale Leaf 75' by Tomkins and Sutherland (1989) with curds stored at 1°C for up to 47 days in air, or in 5 kPa CO_2 + 2 kPa O_2 or 0 kPa CO_2 + 2 kPa O_2, showed that curds stored in 2 kPa O_2 alone suffered severe, irreversible injury and were discarded after 27 days of storage. After 47 days storage in 5 kPa CO_2 + 2 kPa O_2 plus the 4-day marketing period, curd quality was acceptable owing to the reduction in incidence of soft rot and black spotting noted in air-stored curds. In air, curds were unsaleable after only 27 days of storage plus the 4-day marketing period. Storage at 4°C in 18 kPa O_2 + 3 kPa CO_2 for 21 days had no significant effect on the growth of microorganisms compared with storage in air at the same temperature (Berrang *et al.*, 1990). Storage for 8 days at 13°C in air or 15 kPa CO_2 + 21 kPa O_2 + 64% N_2 accelerated the deterioration of microsomal membranes during storage and caused an early loss in lipid phosphate (Voisine *et al.*, 1993). Menniti and Casalini (2000) stored cauliflowers at 0°C and found that concentrations of 10, 15 or 20 kPa CO_2 in the atmosphere delayed leaf yellowing and rot caused by *Alternaria brassicicola*, but caused injury inside the stem and developed off-flavours and odours after cooking.

Celeriac, Turnip-rooted Celery (*Apium graveolens* var. *rapaceum*)

CA storage was not recommended, because 5–7 kPa CO_2 + low O_2 increased decay during 5 months of storage (Hardenburg *et al.*, 1990). Pelleboer (1984) also reported that storage in air at 0–1°C gave better results than CA storage. SeaLand (1991) reported that CA storage had a slight to no effect and Saltveit (1989) recommended 0–5°C in 2–3 kPa CO_2 + 2–4 kPa O_2, which had only a slight effect.

Celery (*Apium graveolens* var. *dulce*)

Recommended storage conditions include: 2–5 kPa CO_2 + 2–4 kPa O_2 (SeaLand, 1991); 0°C in 5 kPa CO_2 + 3 kPa O_2, which reduced decay and loss of green colour (Hardenburg *et al.*, 1990); 0–5°C in 0 kPa CO_2 + 2–4 kPa O_2 or 0–5°C in 0–5 kPa CO_2 + 1–4 kPa O_2, which had a fair effect (Kader, 1985, 1992); 0–5°C in 3–5 kPa CO_2 + 1–4 kPa O_2, which had only a slight effect (Saltveit, 1989); and 3–4 kPa CO_2 + 2–3 kPa O_2, which retained better texture and crispness than storage in air (Gariepy *et al.*, 1985). Storage at 0°C, 4°C or 10°C for 7 days in 25 kPa CO_2 resulted in browning at the base of the petioles, reduced flavour and a tendency for petioles to break away more easily (Wardlaw, 1938). Storage in 1–4 kPa O_2 was shown to preserve the green colour only slightly and although 2.5 kPa CO_2 may be injurious, levels of 9% during 1 month's storage caused no damage (Pantastico, 1975). Reyes (1989) reviewed work on CA storage and concluded that at 0–3°C in 1–4 kPa CO_2 + 1–17.7 kPa O_2, storage could be prolonged for 7 weeks; and he specifically referred to his recent work which showed that at 0–1°C in 2.5–7.5 kPa CO_2 + 1.5 kPa O_2 market quality could be maintained for 11 weeks. Total weight loss of < 10% over a 10-week period was reported by storing celery in 1 kPa O_2 + 2 or 4 kPa CO_2 at 0°C. Significant increases in marketable celery resulted when ethylene was scrubbed from some atmospheres. It was suggested that improved visual colour, appearance and flavour and increased marketable yield justified the use of 4 kPa CO_2 in storage (Smith and Reyes, 1988). The cultivar 'Utah' stored at 0–1°C in 1.5 kPa O_2 had better marketable quality after 11 weeks than when stored in air. Marketable level was improved by using 2.5–7.5 kPa CO_2 in the storage atmosphere, but not by 2–4 kPa CO_2 (Reyes and Smith, 1987).

CA storage was shown to affect disease development. At 1°C, the growth *in vitro* of *Sclerotinia sclerotiorum* on celery extract agar was most suppressed in an atmosphere containing 7.5 kPa CO_2 + 1.5 kPa O_2, but only slightly suppressed in 4 kPa CO_2 + 1.5 kPa O_2 or in 1.5 kPa O_2 compared with storage in air. Watery soft rot caused by *S. sclerotiorum* was severe on celery stored in air for 2 weeks at 8°C. A comparable level of severity took 10 weeks to develop at 1°C. At 8°C, suppression of this disease was greatest in atmospheres of 7.5–30 kPa CO_2 + 1.5 kPa O_2, but only slightly reduced in 4–16 kPa CO_2 + 1.5 kPa O_2 or in 1.5–6 kPa O_2 alone (Reyes, 1988). A combination of 1 or 2 kPa O_2 + 2 or 4 kPa CO_2 prevented black stem development during storage (Smith and Reyes, 1988). Decay was most severe on celery stored in 21 kPa O_2 compared with CA storage; *Botrytis cinerea* and *S. sclerotiorum* were isolated most frequently from decayed celery (Reyes and Smith, 1987). After storage at 1°C, celery grown in Ontario, Canada, developed a black discoloration of the stalks that appeared outwardly in a striped pattern along the vascular strands. In cross-section the vascular strands were discoloured and appeared blackened. A CA storage atmosphere of 3 kPa O_2 + 2 kPa CO_2 almost completely eliminated the disorder. Ethylene and pre-storage treatment with sodium hypochlorite had little or no influence on the occurrence of the disorder but there was some indication of difference in cultivar susceptibility (Walsh *et al.*, 1985). Gómez and Artés (2004a) stored celery stalks at 4°C for 35 days in atmospheres containing air or 5 kPa O_2 + 5, 15 or 25 kPa CO_2. All the CA combinations reduced the respiration rates, which were some 70% of that found in air. Growth of leaves was negatively correlated with CO_2 concentration: the higher the CO_2 level, the lower was the leaf growth. CA also decreased the development of pithiness and improved sensory quality, avoiding the cut butt end from browning and keeping the green colour. Furthermore, decay development, which for air affected 10% of stalks, was inhibited. No off-odours or off-flavours were detected in any treatment. However, under 5 kPa O_2 + 25 kPa CO_2 a slight browning of the internal petioles was observed. After 5 weeks under 5 kPa O_2 + 15 kPa CO_2, no decay developed and stalks showed the best quality. Storing fresh-cut celery, which had been inoculated with *Listeria monocytogenes*, in 95 kPa O_2 + 5 kPa N_2 proved most beneficial compared with storage in air, MA packaging and the other atmospheres tested in maximizing both safety and quality retention during 1 week of storage at 7°C (González-Buesa *et al.*, 2014).

Cherimoya (*Annona cherimola*)

Ludders (2002) observed that the main obstacles to the successful marketing of cherimoyas were their rapid perishability and susceptibility to chilling injury. It was suggested that this could be overcome using CA storage. Storage recommendations include: 8°C in 10 kPa CO_2 + 2 kPa O_2 (Hatton and Spalding, 1990); 10°C, with a range of 8–15°C, in 5 kPa CO_2 + 2–5 kPa O_2 (Kader, 1993); and 8.5°C and 90% RH in 10 kPa CO_2 + 2 kPa O_2 for 22 days (De la Plaza *et al.*, 1979). Fruit of the cultivar 'Fino de Jete' was stored at 9°C in air, 3 kPa O_2 + 0 kPa CO_2, 3 kPa O_2 + 3 kPa CO_2 or 3 kPa O_2 + 6 kPa CO_2. Low O_2 resulted in the greatest reductions in respiration rate, sugars content and acidity, whereas high CO_2 resulted in the greatest reductions in ethylene production and softening rate. CO_2 at 3% and 6% delayed the softening by 5 and 14 days beyond 3 kPa O_2 + 0 kPa CO_2 and air storage, respectively. This allowed sufficient accumulation of sugars and acids to reach an acceptable quality. The results suggest that 3 kPa O_2 + 3 kPa CO_2 and 3 kPa O_2 + 6 kPa CO_2 atmospheres can extend storage life by 2 weeks longer than storage in air (Alique and Oliveira, 1994). The cultivar 'Concha Lisa' was harvested in Chile 240 days after pollination. Respiration rates during storage at 10°C in air showed a typical climacteric pattern with a peak after some 15 days. The climacteric was delayed by storage in 15% or 10 kPa O_2 and fruit kept in 5 kPa O_2 did not show a detectable climacteric rise and did not produce ethylene. All fruit ripened normally after being transferred to air storage at 20°C. However, the time needed to reach an edible condition was affected by O_2 level, with 11 days in 5 kPa O_2, 6 days in 10 kPa O_2 and 3 days in 20 kPa O_2 (Palma *et al.*, 1993). Berger *et al.* (1993) showed that waxed (unspecified) fruit of the cultivar 'Bronceada' could be stored at 10°C, 90% RH in 0 kPa CO_2 + 5 kPa O_2 for 3 weeks without visible change. Fruit of 'Fino de Jete' was stored for 4 weeks at 10–12°C in chambers supplied with a continuous flow of CO_2. The CO_2 treatment prolonged the storage life of cherimoyas by at least 3 weeks compared with those stored in air (Martinez-Cayuela *et al.*, 1986). Escribano *et al.* (1997) found that pre-treatment of the cultivar 'Fino de Jete' for 3 days at 6°C in 20 kPa CO_2 + 20 kPa O_2 maintained fruit firmness and colour compared with those not treated. They also reported that there were some interactions between cultivar and CA storage treatment. Not all reports on CA storage have been positive. De la Plaza (1980) and Moreno and De la Plaza (1983) showed that fruit of the cultivars 'Fino de Jete' and 'Campa' stored in 10 kPa CO_2 + 2 kPa O_2 had a higher respiration rate and deteriorated more quickly than fruit stored at the same temperature in air.

Cherry, Sour (*Prunus cerasus*)

SeaLand (1991) recommended 10–12 kPa CO_2 + 3–10 kPa O_2 and Ionescu *et al.* (1978) recommended 0°C in 5 kPa CO_2 + 3 kPa O_2 but for up to 20 days, which resulted in about 7% loss. English Morello cherries were stored at 2°C in 25 kPa CO_2 + 10 kPa O_2, 15 kPa CO_2 + 10 kPa O_2, 5 kPa CO_2 + 10 kPa O_2 or air for 20 days by Wang and Vestrheim (2002). They found that for the cultivar 'Crisana (Paddy)', decay was greatly reduced in 25 kPa CO_2 + 10 kPa O_2, which also gave the best colour and retention of TA and TSS.

Cherry, Sweet (*Prunus avium*)

Storage recommendations include: 20–25 kPa CO_2 or 0.5–2 kPa O_2, which helped to retain firmness, green stems and bright fruit colour (Hardenburg *et al.*, 1990); 0–5°C and 95% RH in 10–15 kPa CO_2 + 3–10 kPa O_2 (Lawton,1996); 20–25 kPa CO_2 + 10–20 kPa O_2 (SeaLand, 1991); 0–5°C in 10–15 kPa CO_2

+ 3–10 kPa O_2 (Bishop, 1996); 0–5°C in 10–12 kPa CO_2 + 3–10 kPa O_2 (Kader, 1985); 0–5°C in 10–15 kPa CO_2 + 3–10 kPa O_2 (Kader, 1989); 0°C in 5 kPa CO_2 + 3 kPa O_2 for 30 days with 9% losses for cultivars 'Hedelfingen' and 'Germersdorf' (Ionescu *et al.*, 1978); 1°C + 95% RH in 10 kPa CO_2 + 2 kPa O_2 21 days plus 2 days at 20°C to simulate shelf-life (Luchsinger *et al.*, 2005); 0 ± 0.5°C in 20 or 25 kPa CO_2 + 5 kPa O_2 for the cultivar '0900 Ziraat' for up to 60 days (Akbudak *et al.*, 2008); and 1°C and 95% RH in 2 kPa CO_2 + 5 kPa O_2 recommended for 'Sweetheart' for up to 6 weeks, which retained the anthocyanin content and the PPO activity was at its lowest (Remón *et al.*, 2003). At 0°C, Wang and Long (2014) found that respiration rate of cherries declined logarithmically from about 10 kPa to about 1 kPa but respiration rate from 21 kPa to about 10 kPa was affected very little. They also showed that storage atmospheres of 1.8–14.4 kPa O_2 + 5.7–12.9 kPa CO_2, generated by commercial MA packaging, delayed fruit softening. Stow *et al.* (2004) found that there were no consistent effects of the 16 combinations of 0, 5, 10 and 20 kPa CO_2 with 1, 2, 4 and 21 kPa O_2 and, overall, CA storage was not superior to air storage for the cultivars 'Stella', 'German Late', 'Colney', 'Pointed Black' and 'Lapins'. It was concluded that the maximum storage life could be obtained if the fruit had been cooled to 1°C within 36 h of harvest and thereafter maintained at 0°C in air. Rotting was the major cause of loss during storage at 0°C and especially during subsequent shelf-life at 10°C or 20°C. Shellie *et al.* (2001) had previously shown similar results in experiments with the cultivar 'Bing' stored for 14 days at 1°C in 6 kPa O_2 + 17 kPa CO_2 + 82% N_2. They had similar market quality as cherries stored in air. In spite of these findings, there is considerable evidence that both CA and MA storage can be beneficial.

For the cultivars 'Napoleon', 'Stella' and 'Karabodur', Eris *et al.* (1994) found that the optimum conditions were at 0°C and 90–95% RH in 5 kPa CO_2 + 5 kPa O_2. The conditions they tested were 0 kPa CO_2 + 21 kPa O_2, 5 kPa CO_2 + 5 kPa O_2, 10 kPa CO_2 + 3 kPa O_2, 20 kPa CO_2 + 2 kPa O_2 and 0 kPa CO_2 + 2 kPa O_2. Chen *et al.* (1981) compared a range of CA storage treatments on the cultivar 'Bing' at −1.1°C for 35 days and found that 0.03 kPa CO_2 + 0.5-2 kPa O_2 maintained the greenness of the stems, brighter fruit colour and higher acidity than other treatments. Storage in 10 kPa CO_2 had similar effects, with the exception of maintaining the stem greenness. Folchi *et al.* (1994) stored the cultivar 'Nero 1' in < 0.1 kPa CO_2 + 0.3 kPa O_2 and showed an increase in aldehyde and ethanol content of fruit with increased storage time. In a comparison between atmospheres of 4 and 20 kPa O_2 + 5 or 12 kPa CO_2, the best results for up to 12 days for the cultivar 'Burlat' were 12 kPa CO_2, independently of O_2 concentration. In these conditions there was a higher acidity level. lower anthocyanin content and lower levels of peroxidase and polyphenoloxidase activities (Remón *et al.*, 2004). The cultivars 'Star', 'Kordia' and 'Regina' were stored at 1°C with 90–93% RH for up to 3 weeks in 10 kPa CO_2 + 10 kPa O_2, 15 kPa CO_2 + 10 kPa O_2 or 20 kPa CO_2 + 10 kPa O_2 followed by storage in air for 1 week at 20°C with 60% RH. CA storage decreased the respiration rate, compared with air, and improved stem colour retention and condition, but there was little difference between CA and air storage on sugar and acid levels (Gasser *et al.*, 2004). 'Lapins' cherries stored in 5 kPa O_2 + 10 kPa CO_2 were firmer and had higher vitamin C and TA than those in MA packaging or CA with higher O_2 levels, but TSS contents were not significantly affected by CA (Tian *et al.*, 2004). Storage at 1°C with 5 kPa O_2 + 10 kPa CO_2 inhibited the enzymatic activities of PPO and peroxidase, reduced MDA content, effectively prevented flesh browning, decreased fruit decay and extended storage life more than storage in air, MA or 70 kPa O_2 + 0 kPa CO_2. Storage in 70 kPa O_2 + 0 kPa CO_2 was more effective at inhibiting ethanol production in flesh and reducing decay than other treatments, but showed increased fruit browning after 40 days of storage. 'Sweetheart' cherries were stored for 6 weeks at 1°C and 95% RH with 2 kPa CO_2 + 5 kPa O_2 while maintaining an excellent quality throughout their storage and shelf-life of 3 days at 20°C. Under these conditions they had the highest acceptability and appearance score, anthocyanin content remained unchanged and PPO activity was at its lowest level (Remón *et al.*, 2003).

CA has been shown to reduce disease and Haard and Salunkhe (1975) reported that storage with CO_2 levels of up to 30% reduced decay. Tian *et al.* (2004) also reported that storage of the cultivar 'Lapins' at 1°C in 70 kPa O_2 + 0 kPa CO_2 was effective in reducing decay, but stimulated fruit browning after 40 days. Storage in 2 kPa O_2 + 5 kPa CO_2, 5 kPa O_2 + 10 kPa CO_2 and 5 kPa O_2 + 15 kPa CO_2, with N_2 the balance, inhibited grey mould (*Botrytis cinerea*) development after 18 days at 1°C plus 6 days at ambient (Wermund and Lazar, 2003). Serradilla *et al.* (2013) compared storage of cultivar 'Ambrunés' in air, 3 kPa O_2 + 10 kPa CO_2, 5 kPa O_2 + 10 kPa CO_2 and 8 kPa O_2 + 10 kPa CO_2. They found that both 5 kPa O_2 + 10 kPa CO_2 and 8 kPa O_2 + 10 kPa CO_2 effectively controlled the growth of mesophilic aerobic bacteria, psychrotrophs, *Pseudomonas* spp., yeasts and the fungi *Aureo basidium* spp., *Penicillium* spp., *Leuconostoc* spp., and *Rahnella* spp.

CA has been used for controlling insects for phytosanitary purposes. The treatment times required to completely kill specific insects by O_2 levels at or below 1 kPa suggest that low O_2 atmospheres are potentially useful as postharvest quarantine treatments for some fruit. Fruit of the cultivar 'Bing' was treated with 0.25 kPa or 0.02 kPa O_2 (balance N_2) at 0°C, 5°C or 10°C to study the effects of these insecticidal low-O_2 atmospheres on fruit postharvest physiology and quality attributes. Development of alcoholic off-flavour, associated with ethanol accumulation, was the most common and important detrimental effect that limited fruit tolerance to low O_2 (Ke and Kader, 1992b).

Chestnut, Chinese (*Castanea mollissima*)

Respiration rate was reduced to a steady low level at 0–5°C but was not further reduced by decreasing O_2 or increasing CO_2 in the atmosphere (Wang *et al.*, 2000b). The decay rate when exposed to 40 kPa CO_2 for 20 days was only 1% after subsequent cold storage for 120 days, but an off-flavour was detected when treated for more than 20 days. Also there were no adverse effects of any CO_2 concentrations tested if the treatment duration was not more than 10 days. The presence of an off-flavour was irreversible when the CO_2 concentration was > 50 kPa and treatment duration was longer than 20 days (Liang *et al.*, 2004). Storage of the cultivar 'Dahongpao' at 0°C in 0–5 kPa O_2 for up to 120 days showed that off-flavours were produced by exposure to concentrations below 2 kPa O_2 for 20 days or more. Storage in 0–2 kPa O_2 could stimulate the decomposition of starch, while in 3–5 kPa O_2 the decomposition rate could be slowed. Those exposed to 1–3 kPa O_2 had higher total sugars compared with other treatments, but the mean decay rate in 0–2 kPa O_2 was 10% and with those stored in air it was 7%. The decay level in 3–4 kPa O_2 atmosphere for 20 days or 5 kPa O_2 for 25 days was only 1% at the end of 120 days of cold storage. In general, 3 kPa O_2 for 20 days was the best (Wang *et al.*, 2004a).

Chestnut, Sweet (*Castanea sativa*)

Also called Spanish chestnut, Portuguese chestnut and European chestnut. The cultivars 'Catot' and 'Platella' were stored at 1°C in 20 kPa CO_2 + 2 kPa O_2 and their freshness, taste and flavour were maintained and after 105 days they looked as fresh and bright as those freshly harvested (Mignani and Vercesi, 2003). This treatment controlled fungal infections except for those caused by *Aspergillus niger*, but storage in 2.5 kPa CO_2 + 1.5 kPa O_2 was less effective. Rouves and Prunet (2002) found that the best storage conditions of those that they tested for the cultivars 'Marigoule' and 'Bouche de Betizac' were –1°C in 2 kPa O_2 + 5 kPa CO_2. There was 'no water loss', mould development was much slower and taste remained unaltered but some germination occurred in 'Marigoule'. 'Comballe' and 'Marron de Goujonac' did not store well in any of the conditions tested.

Chicory, Endive, Belgian Endive, Escarole, Radicchio, Witloof Chicory (*Cichorium* spp.)

The cultivated forms of *C. intybus* var. *foliosum* are grown for their leaves, which are used in salads, or *C. intybus* var. *sativum* for their roots. The latter is used mainly as a coffee substitute and only *C. Intybus* var. *foliosum* and *C. endivia* are dealt with here. Storage of witloof at 0°C in 4–5 kPa CO_2 + 3–4 kPa O_2 delayed greening of the tips in light and delayed opening of the heads (Hardenburg *et al.*, 1990). Saltveit (1989) also recommended 4–5 kPa CO_2 and 3–4 kPa O_2 for witloof, but at 0–5°C, and it was reported to have had only a slight effect. Storage at 5°C in 10 kPa O_2 + 10 kPa CO_2 prevented red discoloration, leaf edge discoloration and other negative quality aspects, but in storage at 1°C there was increased red discoloration (Vanstreels *et al.*, 2002). In a comparison of 5, 10, 15 and 20 kPa CO_2 in storage at 0°C, Bertolini *et al.* (2003) found that 10 kPa CO_2 was the most effective in suppressing *Botrytis cinerea* in red chicory. Later Bertolini *et al.* (2005) stored radicchio 'Rosso di Chioggia' at 0°C for up to 150 days. They artificially inoculated some heads with *B. cinerea* and found that lesions caused by *B. cinerea* decreased with increasing concentrations of CO_2 over the range of 5–20 kPa for up to 60 days. Subsequently only 10 and 15 kPa CO_2 were effective, while after 120 days all the concentrations had low efficacy. In naturally infected heads, 5 and 10 kPa CO_2 was effective in preventing *B. cinerea* even after 150 days storage and it also reduced the fungus spreading to adjacent heads. Heads stored in 15 kPa CO_2 for 150 days showed phytotoxic effects and increased their vulnerability to rots. Monzini and Gorini (1974) recommended 0°C in 1–5 kPa CO_2 + 1–5 kPa O_2 for 45–50 days. Van de Velde and Hendrickx (2001) investigated storage of cut Belgian endive in atmospheres ranging from 2 to 18 kPa CO_2 + 2–18 kPa O_2 and found that the optimum concentration was 10 kPa CO_2 + 10 kPa O_2 + 80 kPa N_2. Wardlaw (1938) reported that storage in 25 kPa CO_2 could cause the central leaves to turn brown.

Chillies (*Capsicum annum, C. frutescens*)

Also called chilli peppers, hot peppers, cherry peppers. Storage at 0 kPa CO_2 + 3–5 kPa O_2 was recommended by SeaLand (1991). Kader (1985, 1992) recommended 8–12°C in 0 kPa CO_2 + 3–5 kPa O_2 which he observed had a fair effect but was of no commercial use, but 10–15 kPa CO_2 was beneficial at 5–8°C. Saltveit (1989) recommended 12°C in 0–5 kPa CO_2 + 3–5 kPa O_2 for the fresh market, which had only a slight effect, and 5–10°C in 10–20 kPa CO_2 + 3–5 kPa O_2 was recommended for processing. Storage of green chillies for 6 weeks at 10°C and 80–90% RH in 3 kPa O_2 + 5 kPa CO_2 resulted in less decay and wrinkling and they were firmer with higher TSS and vitamin C than those stored in air or other CA tested (Ullah Malik *et al.*, 2009).

Chinese Cabbage (*Brassica pekinensis*)

The recommended storage conditions in air were 0–1°C, with a maximum storage period of 6 weeks (Mertens, 1985). Hardenburg *et al.* (1990) recommended 0°C in 1 kPa O_2. Wang and Kramer (1989) also showed that storage in 1 kPa O_2 extended storage life and reduced the decline in AA, chlorophyll and sugars in the outer leaf laminae. Saltveit (1989) recommended 0–5°C in 0–5 kPa CO_2 + 1–2 kPa O_2, which had only a slight effect. Apeland (1985) stored the cultivars 'Tip Top' and 'Treasure Island' at 2.5°C or 5.0°C in 0.5, 2.5 or 5.0 kPa CO_2 + 1–20.5 kPa O_2. With 'Tip Top' the storage was best at 5°C in 5 kPa CO_2 + 5 kPa O_2. The results with 'Treasure Island' were inconclusive, because the control heads kept very well but with an average of only 68% saleable after 120 days. Pelleboer and Schouten (1984) reported that after 4 months of storage of the cultivars 'Chiko' and 'WR 60' at 2–3°C in 0.5 kPa CO_2 + 3 kPa O_2 or less, the average percentage of healthy heads was 72%, and after 5 months 60% or more were healthy. After subsequent storage at 15°C and 85% RH to simulate shelf-life, the corresponding percentages were 58% and 43%. Following storage in 6 kPa CO_2 + 3 kPa O_2 or 6 kPa CO_2 + 15 kPa O_2 there was a rapid fall in quality, confirming that 6 kPa CO_2 could be harmful.

Hermansen and Hoftun (2005) stored the cultivar 'Nerva', inoculated with *Phytophthora brassicae*, at 1.5°C in air, 0.5 kPa CO_2 + 1.5 kPa O_2 or 3.0 kPa CO_2 + 3.0 kPa CO_2 for 94–97 days. The infection caused by *P. brassicae* was significantly higher in both the CA treatments than in air, but *in vitro* studies gave the opposite results. Brown midribs, spots or streaks of brown tissue of the leaves and midribs and pepper spots are common physiological disorders. Chilling injury (brown midribs) found in 'Parkin' was highest in heads stored in air. CA storage reduced this disorder and the lowest percentage of brown midribs was found in 3 kPa CO_2 + 3 kPa O_2. No chilling injury was found. Adamicki (2003), working with the F_1 hybrid cultivars 'Asten', 'Bilko', 'Gold Rush', 'Maxim', 'Morillo', 'Parkin' and 'RS 6064', also reported that low concentrations of O_2 and CO_2 greatly reduced the incidence of physiological disorders, including brown ribs. However, storage at 5 kPa CO_2 + 3 kPa O_2 resulted in lower percentage of marketable heads and higher percentage of damaged leaves and heads compared with storage in air for 100–130 days at 0°C, 2°C or 5°C and 95–98% RH. CO_2 concentrations > 5 kPa resulted in greater losses of weight and trim, caused mainly by leaf rotting (Adamicki and Gajewski, 1999). They also found that 1.5–3.0 kPa O_2 + 2.5 kPa CO_2 was the best atmosphere for long-term storage. For most of the F_1 cultivars tested, a storage temperature 2°C was better than 0°C.

Citrus Hybrids (*Citrus* spp.)

Chase (1969) stored 'Temple' oranges (*C. sinensis* × *C. reticulata*) and 'Orlando' tangelos (*C. sinensis* × *C. paradisi*) at 1°C in atmospheres containing 5 kPa CO_2 + 10 kPa O_2. They found little benefit from CA storage compared with storage in air for 'Orlando' but when 'Temple' oranges were stored for 5 weeks in CA followed by 1 week at 21°C in air their flavour was superior to those stored in air throughout.

Cranberry (*Vaccinium* spp.)

There are two species: the American cranberry (*Vaccinium macrocarpon*) and the European cranberry (*V. oxycoccus*). Kader (1989) recommended 2–5°C in 0–5 kPa CO_2 + 1–2 kPa O_2, but Hardenburg *et al.* (1990) reported that CA storage was not successful in increasing storage life. In contrast, Gunes *et al.* (2002) found that in storage of the cultivars 'Pilgrim' and 'Stevens' at 3°C in atmospheres of 21 kPa O_2 + 15 or 30 kPa CO_2 there was decreased bruising, physiological breakdown and decay of berries compared with those stored in air. Respiration rate and weight loss were also decreased, but fruit softening increased compared with storage in air. There was also an increase in acetaldehyde, ethanol and ethyl acetate during storage in 21 kPa O_2 + 15 or 30 kPa CO_2 but the levels varied with cultivar and storage atmosphere, with the highest in 2 and 70 kPa O_2 and in 100 kPa N_2. Overall, 30 kPa CO_2 + 21 kPa O_2 appeared to be optimum. No sensory analysis was included to confirm whether accumulations of fermentation products at this atmosphere were acceptable for consumers. However, they also found that the storage atmosphere did not affect the content of total phenolics or flavonoids, but the total antioxidant activity of the fruit increased overall by about 45% in fruit stored in air. This increase did not occur during storage in 30 kPa CO_2 + 21 kPa O_2.

Cucumber (*Cucumis sativus*)

Over a 24 h period at 10°C or 20°C, the respiration rate of the cultivar 'Tyria' decreased as the O_2 concentration decreased down to 0.5 kPa, but at 0 kPa O_2, the respiration rate increased because of fermentation (Praeger and Weichmann, 2001). CA recommendations include: 14°C with 5 kPa CO_2 + 5 kPa O_2 (Stoll, 1972); 5–7 kPa CO_2 + 3–5 kPa O_2 (SeaLand, 1991); a maximum of 10 kPa CO_2 and a minimum of 3 kPa O_2 (Fellows, 1988); 8.3°C in 3–5 kPa O_2, which gave a slight extension in storage life (Pantastico, 1975); 5 kPa CO_2 + 5 kPa O_2 (Ryall and Lipton, 1972);

8–12°C in 0 kPa CO_2 + 3–5 kPa O_2, which had a fair effect but was of no commercial use (Kader, 1985, 1992); 12°C in 0 kPa CO_2 + 1–4 kPa O_2 for the fresh market and 4°C in 3–5 kPa CO_2 + 3–5 kPa O_2 for pickling (Saltveit, 1989); and 12.5°C in 5% O_2 + 5 kPa CO_2 (Schales, 1985). Mercer and Smittle (1992) stored cucumbers of the cultivar 'Gemini II' in 0, 5 or 10 kPa CO_2 + 5 or 20 kPa O_2 at 5°C or 6°C for 2, 4 or 6 days, or at 5°C for 5 days, or at 3°C for 10 days then 2–4 days at 25°C. High CO_2 and low O_2 delayed the onset of chilling injury symptoms, but did not prevent their development. Chilling injury symptoms increased with longer exposure to chilling temperatures and were associated with solubilization of cell wall polysaccharides. Storage in 5 kPa O_2 was shown to retard yellowing (Lutz and Hardenburg, 1968). Pre-treatment with O_2 has been shown to be beneficial. In storage at 5°C, pretreatment with 100 kPa O_2 for 48 h lowered the respiration rate compared with storage in either 5 kPa O_2 or air. It also delayed the appearance of fungal decay by 2 days, delayed the onset and reduced severity of chilling injury, halved weight loss and delayed the appearance of shrivelling by 4 days compared with control (Srilaong *et al.*, 2005). Reyes (1989) showed that the virulence of mucor rot (*Mucor mucedo*) and grey mould (*Botrytis cinerea*) were suppressed in 7.5 kPa CO_2 + 1.5 kPa O_2.

Durian (*Durio zibethinus*)

Durian is a climacteric fruit. It was reported by Pan (2008) that in South China it could be stored in air at ambient temperatures for 3 weeks. Tongdee *et al.* (1990) showed that storage of 85% mature 'Mon Tong' at 22°C in atmospheres with 5 or 7.5 kPa O_2 inhibited ripening, but fruit ripened normally when returned to air. They also found that in 2 kPa O_2 ripening was inhibited and the fruit failed to ripen when removed and stored in air. CO_2 levels of up to 20 kPa in air did not affect the speed of ripening or the quality of the ripe fruit. Fruit of the cultivars 'Chanee', 'Kan Yao' and 'Mon Tong' was harvested at three maturity stages and stored at 22°C. The respiration rate and ethylene production at harvest and the peak climacteric respiratory value were higher in fruit harvested at a more advanced stage. Storage in 10 kPa O_2 resulted in a significant reduction in respiration rate and ethylene production, but the onset or ripening and ripe fruit quality were not affected. Ripening was inhibited in fruit stored in 5–7.5 kPa O_2 but recovered when fresh air was subsequently introduced. O_2 at this level did not affect ripening in fruit harvested at an advanced stage of maturity. Fruit stored in 2 kPa O_2 failed to resume ripening when removed to air. Fruit stored in 10 or 20 kPa CO_2 was either not affected or showed a slight reduction in ethylene production. Thus high levels of CO_2 alone did not influence the onset of ripening or other quality attributes of ripe fruit. Atmospheres of 5, 10, 15 or 20 kPa CO_2 + 10 kPa O_2 had a greater effect on the condition of the aril than either high CO_2 or low O_2 alone. The aril remained hard in the less mature fruit stored in 10 kPa O_2 + 15 or 20 kPa CO_2. Kader (1993) recommended 10.5°C in 5–20 kPa CO_2 + 3–5 kPa O_2.

Feijoa (*Feijoa sellowiana*)

Aziz Al-Harthy *et al.* (2009) reported that fruits of the cultivar 'Opal Star' stored at 4°C in 2 kPa O_2 + 0 kPa CO_2, 2 kPa O_2 + 3 kPa CO_2, 5 kPa O_2 + 0 kPa CO_2 or 0 kPa O_2 + 3 kPa CO_2 for up to 10 weeks retained their green colour while those stored in air went yellow. CA also had some effects on reducing the rate of softening compared with storage in air, with 2 kPa O_2 + 0 kPa CO_2 and 5 kPa O_2 + 0 kPa CO_2 giving the best results.

Fig (*Ficus carica*)

Storage in high CO_2 atmospheres reduced mould growth without affecting the flavour of the fruit (Wardlaw, 1938). Hardenburg *et al.* (1990) also showed that storage with enriched CO_2 was a useful supplement to refrigeration. Storage at 0–5°C in 15 kPa CO_2 + 5 kPa O_2 was recommended by Kader

(1985), as was 0–5°C + 15–20 kPa CO_2 and 5–10 kPa O_2 by Kader (1989, 1992). SeaLand (1991) also recommended 15 kPa CO_2 + 5 kPa O_2. Tsantili *et al.* (2003) stored the cultivar 'Mavra Markopoulou' at −1°C in either air or 2 kPa O_2 + 98 kPa N_2 for 29 days. Those stored in air became soft during storage for longer than 8 days, but those in CA remained in much better condition. Colelli *et al.* (1991) showed that good quality of 'Mission' figs was maintained for up to 28 days when they were kept at 0°C, 2.2°C or 5°C in atmospheres enriched with 15 or 20 kPa CO_2. The benefits of exposure to high CO_2 levels were a reduction of the incidence of decay and the maintenance of a bright external appearance. Ethylene production was lower and softening was slower in figs stored at high CO_2 concentrations compared with those kept in air. Ethanol content of the fruit stored in 15 or 20 kPa CO_2 increased slightly during the first 3 weeks and moderately during the fourth week, while acetaldehyde concentration increased during the first week and then decreased. It was concluded that postharvest life can be extended by 2–3 weeks at 0–5°C in atmospheres enriched with 15–20 kPa CO_2 but off-flavours could be a problem by the fourth week of storage.

Garlic (*Allium sativum*)

Storage in 0 kPa CO_2 + 1–2 kPa O_2 was recommended (SeaLand, 1991) and Monzini and Gorini (1974) recommended 3°C in 5 kPa CO_2 + 3 kPa O_2 for 6 months. Cantwell *et al.* (2003) showed that storage of the cultivars 'California Late' and 'California Early' at 0–1°C in CO_2 atmospheres of 5 kPa, 10 kPa, 15 kPa and 20 kPa reduced sprout growth, decay and discoloration but CO_2 concentrations over 15 kPa could lead to injury after 4–6 months.

For fresh peeled garlic storage at 5°C and 10°C with either 5–15 kPa CO_2 or 1–3 kPa O_2 was effective in retarding discoloration and decay for 3 weeks (Cantwell *et al.*, 2003). They also showed that, when garlic shoots were stored at 0°C in ten combinations of 3–6.5 kPa O_2 + 0–12 kPa CO_2 , the optimum atmosphere was 3 kPa O_2 + 8 kPa CO_2, which maintained levels of chlorophyll, reducing and total sugars and freshness for 235 days. There was some spoilage and rotting, but rotting decreased with decreasing O_2 concentration, while high CO_2 reduced mould growth (Zhou *et al.*, 1992a). Zhou *et al.* (1992b) and Zhang and Zhang (2005) used CA storage of sprouted garlic in chambers flushed with N_2 from a carbon molecular sieve to reduce the O_2 content to 1–5 kPa. CO_2 was allowed to increase to 2–7% by product respiration. After 240–270 days, the quality of sprouted garlic remained high. After fumigating, the shoots were treated with a fungicide and packed in plastic bags with a silicon window and stored at 0–1°C with 95% RH. These treatments reduced the respiration rate and the loss of chlorophyll and inhibited cellulose production. After 280 days, 96% had good freshness, greenness and crispness.

Gooseberry (*Ribes uva-crispa, R. grossularia*)

Robinson *et al.* (1975) showed that with storage in low O_2 the respiration rate was reduced at all temperatures tested (Table 12.1). Prange (not dated) stated that they could be stored at 0–1°C for 3 weeks but storage duration could be extended to 6–8 weeks, using 1°C in 10–15 kPa CO_2 + 1.5 kPa O_2. Harb and Streif (2004a) stored the cultivar 'Achilles' at 1°C in air, 6 kPa CO_2 + 18 kPa O_2, 6 kPa CO_2 + 18 kPa O_2, 12 kPa CO_2 + 18 kPa O_2, 18 kPa CO_2 + 18 kPa O_2,12 kPa CO_2 + 2 kPa O_2, 18 kPa CO_2 + 2 kPa O_2 or 24 kPa CO_2 + 2 kPa O_2. Storage in air led to a reduction in firmness, darkening of fruit colour, mealy texture and lower acidity level compared with CA storage. Storage in 2 kPa O_2 + 18 kPa CO_2 resulted in off-flavours, but in 12 kPa CO_2 +

Table 12.1. Effects of temperature and reduced O_2 level on the respiration rate (CO_2 production in mg/kg/h) of Leveller gooseberries (Robinson *et al.*, 1975).

	0°C	5°C	10°C	15°C	20°C
Air	10	13	23	40	58
3 kPa O_2	7		16		26

2 kPa O_2 no off-flavour was detected. They recommended 12–15 kPa CO_2 + 18 kPa O_2 for up to 7 weeks. Certain cultivars of gooseberries have been held for as long as 3 months at 0°C in air (McKay and Van Eck, 2006), but for longer storage they suggested harvesting at the green-mature stage and placing them at –0.5°C to –0.9°C with 93% RH in 2.5–3.0 kPa O_2 + 20–25 kPa CO_2. They also found that gooseberries were sensitive to ethylene.

Grape (*Vitus vinifera*)

Storage recommendations were 1–3 kPa CO_2 + 3–5 kPa O_2 (SeaLand, 1991). Magomedov (1987) showed that different cultivars required different CA storage conditions. The cultivars 'Agadai' and 'Dol'chatyi' stored best in 3 kPa CO_2 + 5 kPa O_2 whereas for 'Muskat Derbentskii' 5 kPa CO_2 + 5 kPa O_2 or 3 kPa CO_2 + 2 kPa O_2 were more suitable. 'Muskat Derbentskii', 'Dol'chatyi' and 'Agadai' had a storage life of up to 7, 6 and 5 months, respectively, in CA storage. The best results for storage of the cultivar 'Moldova' were in either 8 kPa CO_2 + 2–3 kPa O_2 or 10 kPa CO_2 + 2–3 kPa O_2. In these conditions, 89%, 80% and 75% of first grade grapes were obtained after 5, 6½ and 7½ months of storage, respectively. In another trial with 'Muscat of Hamburg' and 'Italia', grapes stored in air or in CA for 3–7 months were assessed in relation to profitability. The best results were again obtained in storage of 8 kPa CO_2 + 2–3 kPa O_2 for 'Muscat of Hamburg' and of 5 kPa CO_2 + 2–3 kPa O_2 for 'Italia' (Khitron and Lyublinskaya, 1991). Turbin and Voloshin (1984) showed that for storage for less than 5 months, 8 kPa CO_2 + 3–5 kPa O_2 was most suitable for 'Muskat Gamburgskii' ('Hamburg Muscat'), 5–8 kPa CO_2 + 3–5 kPa O_2 for 'Italia' and 8 kPa CO_2 + 5–8 kPa O_2 for 'Galan'. 'Waltham Cross' and 'Barlinka' were stored in CA at –0.5°C for 4 weeks. The percentage of loose berries in 'Waltham Cross' was highest (> 5%) in 21 kPa O_2 + 5 kPa CO_2 than in lower O_2 levels (Laszlo, 1985). The cultivar 'Agiorgitiko' was stored at 23–27°C for 10 days either in 100 kPa CO_2 or in air. Fruits from both treatments were held at 0°C for 20 h before analysis (Dourtoglou *et al.*, 1994). 'Kyoho' stored well in 4 kPa O_2 + 30 kPa CO_2 but an alcoholic flavour and browning occurred after 45 days. Storage in 4 kPa O_2 + 9 kPa CO_2 or 80 kPa O_2 + 20 kPa N_2 retained good quality during 60 days of storage, without off-flavours (Deng *et al.*, 2006). Crisosto *et al.* (2003a) studied 5, 10, 15, 20 and 25 kPa CO_2 factorially combined with 3, 6 and 12 kPa O_2 on 'Redglobe'. Optimum conditions for late-harvested fruit (19% TSS) were 10 kPa CO_2 + 3, 6 or 12 kPa O_2 for up to 12 weeks of storage and for early-harvested fruit (16.5% TSS) it was 10 kPa CO_2 + 6 kPa O_2 for up to 4 weeks. Pedicel browning was accelerated in grapes exposed to 10 kPa CO_2 for early-harvested and above 15% for late-harvested fruit. However, the same authors (Crisosto *et al.*, 2003b) found that the combination of 15 kPa CO_2 with 3, 6 or 12 kPa O_2 was optimum for late-harvested 'Thompson Seedless' for up to 12 weeks and CA should not be used for commercially early-harvested grapes.

Sahin *et al.* (2015) compared storage of 'Red Globe' grapes at 0°C with 95% RH for 90 days in 21 kPa O_2 + 0.03 kPa CO_2, 2 kPa O_2 + 0.03 kPa CO_2, 2 kPa O_2 + 2.5 kPa CO_2, 2 kPa O_2 + 5 kPa CO_2 or 1 kPa O_2 + 2.5 kPa CO_2. Those stored in 2 kPa O_2 + 5 kPa CO_2 were the firmest and had the highest TSS content and lowest decay, while those that had been stored in 1 kPa O_2 + 2.5 kPa CO_2 had the lowest rachis browning.

CA has been shown to reduce disease development and may be able to be developed as an alternative to sulfur dioxide fumigation. Kader (1989, 1992) recommended 0–5°C in 1–3 kPa CO_2 + 2–5 kPa O_2 but reported that CA storage was incompatible with sulfur dioxide fumigation. However, Berry and Aked (1997) working on 'Thompson Seedless' stored at 0–1°C for up to 12 weeks showed that atmospheres containing 15–25 kPa CO_2 inhibited infection with *Botrytis cinerea* by between 95% and 100% without detrimentally affecting flavour. Mitcham *et al.* (1997) showed that high CO_2 levels in the storage atmosphere could be used for controlling insect pests of grapes. In trials where four cultivars were exposed to 0°C in 45 kPa CO_2 + 11.5 kPa O_2, complete control of *Platynota stultana*, *Tetranychus pacificus* and *Frankliniella occidentalis* was achieved without injury to the grapes. 'Flame Seedless' and 'Crimson Seedless' grapes were inoculated with *B. cinerea* and then exposed to 40 kPa CO_2 at 0°C for 48 h and stored in either in air or 12 kPa O_2 + 12 kPa CO_2. This combined treatment (CO_2 + CA) reduced the grey mould incidence compared with storage in air from 22% to 0.6% after 4 weeks and from 100% to 7.4% after 7 weeks of storage, compared with fruit not treated and stored in air at 0°C (Teles *et al.*, 2014).

Grapefruit, Pummelo (*Citrus* × *paradisi*)

Typical storage conditions were given as 10–15°C in 5–10 kPa CO_2 + 3–10% O_2 by Bishop (1996) although CA storage was not considered beneficial. Erkan and Pekmezci (2000) found that 1 kPa CO_2 + 3 kPa O_2 at 10°C to be optimum for 'Star Ruby' grown in Turkey. They were stored in these conditions for 125 days without losing much quality. SeaLand (1991) recommended 5–10 kPa CO_2 + 3–10 kPa O_2 for grapefruit from California, Arizona, Florida, Texas and Mexico. Storage at 10–15°C in 5–10 kPa CO_2 + 3–10 kPa O_2 had a fair effect on storage, but Kader (1985, 1992) reported that CA storage was not used commercially. Storage experiments have shown that there was some indication that fruit stored at 4.5°C in 10 kPa CO_2 for 3 weeks had less pitting (a symptom of chilling injury) than those stored in air. Also pre-treatment with 20–40 kPa CO_2 for 3 or 7 days at 21°C reduced physiological disorders on fruit stored at 4.5°C for up to 12 weeks (Hardenburg *et al.*, 1990). Hatton *et al.* (1975) and Hatton and Cubbedge (1982) also showed that exposing grapefruit before storage to 40 kPa CO_2 at 21°C reduced chilling injury symptoms during subsequent storage.

Guava (*Psidium guajava*)

Short-term exposure to 10, 20 or 30 kPa CO_2 had no effect on respiration rate, but ethylene biosynthesis was reduced by all three levels of CO_2 (Pal and Buescher, 1993). Teixeira *et al.* (2009) recommended storage of the cultivar 'Pedro Sato' at 12.2° in 5 kPa O_2 + 1 kPa CO_2 for up to 28 days, but they found that storage in 15 or 20 kPa CO_2 resulted in CO_2 injury. Mature-green guavas were exposed to air or 5 kPa CO_2 + 10 kPa O_2 for 24 h before storage in air at either 4°C or 10°C for 2 weeks. They were then transferred to 20–23°C for 3 days to simulate shelf-life. The colour of the CA-treated fruit developed more slowly than in those kept in air throughout and they were considered to be of better quality after storage and shelf-life and showed no chilling injury symptoms even after storage at 4°C for 3 weeks, while chilling injury occurred on those stored in air throughout (Bautista and Silva, 1997). Freshly harvested mature-green 'Lucknow-49' fruit was stored at 8°C with 85–90% RH in 5 kPa O_2 + 2.5 kPa CO_2 or 10 kPa O_2 + 5 kPa CO_2. CA-stored fruits could be kept in an unripe condition for 1 month, while those stored in air showed severe chilling injury symptoms, high weight loss and spoilage softening and colour change (Pal *et al.*, 2007). Subsequently Singh and Pal (2008b) investigated 2.5, 5, 8 and 10 kPa O_2 factorially combined with 2.5, 5 and 10 kPa CO_2 (balance N_2) at 8°C and 85–90% RH. They successfully stored the cultivars 'Lucknow-49', 'Allahabad Safeda' and 'Apple Colour' for 30 days in 5 kPa O_2 + 2.5 kPa CO_2, 5 kPa O_2 + 5 kPa CO_2 or 8 kPa O_2 +5 kPa CO_2. The fruits were then transferred to ambient conditions of 25–28°C and 60–70% RH and they all ripened successfully. Chilling injury and decay incidence were reduced during ripening of fruit that had been stored in CA compared with those that had been stored in air. However, larger amounts of ethanol and acetaldehyde accumulated in fruit held in atmospheres containing 2.5 kPa O_2. Teixeira *et al.* (2016) stored the cultivar 'Pedro Santo' at 12.2°C in 5 kPa O_2 combined with 1, 5, 10, 15 or 20 kPa CO_2. CO_2 injury occurred in fruit stored in both 15 and 20 kPa CO_2. They therefore recommended 5 kPa O_2 + 1 or 5 kPa CO_2 as giving the optimum conditions over 28 days storage.

Horseradish (*Armoracia rusticana*)

CA storage was reported to have had only a slight to no effect (SeaLand, 1991) and was not recommended by Saltveit (1989). Weichmann (1981) studied different levels of CO_2 of up to 7.5 kPa during 6 months of storage at 0–1°C and found no detrimental effects of CO_2 but no advantages either over storage in air. He did, however, find that those stored in 7.5 kPa CO_2 had a higher respiration rate and higher total sugar content than those stored in air. Gui *et al.* (2006) reported that exposure to high levels of CO_2 could result in some reduction in enzyme activity but activity returned to what it was when they were removed and stored in air.

Jujube (*Ziziphus jujuba*)

Han *et al.* (2006) found that the best CA storage condition for the cultivar 'Dong' was 1.5 ± 0.5°C and 95% RH in 5–6 kPa O_2 + 0–0.5 kPa CO_2 for 90 days.

Kiwifruit (*Actinidia chinensis*)

Also called Chinese gooseberry and yang tao. Storage recommendations include: 5 kPa CO_2 + 2 kPa O_2 (SeaLand, 1991); 0°C and 90% RH in 5 kPa CO_2 + 2 kPa O_2 (Lawton, 1996); 0–5°C in 5 kPa CO_2 + 2 kPa O_2 (Kader, 1985); 0°C in 3–5 kPa CO_2 + 1–2 kPa O_2 (Kader, 1989, 1992); 0.5–0.8°C and 95% RH with either 1 kPa O_2 + 0.8 kPa CO_2 or 2 kPa O_2 + 4–5 kPa CO_2 for 6 months (Brigati and Caccioni, 1995); and 0–5°C in 5–10 kPa CO_2 and 1–2 kPa O_2 (Bishop, 1996). Intermittent storage at 0°C of 1 week in air and 1 week in air with 10 kPa or 30 kPa CO_2 was shown to reduce fruit softening (Nicolas *et al.*, 1989). Storage in CA was shown to increase storage life by 30–40%, with optimum conditions of 3–5 kPa CO_2 + 2 kPa O_2, but levels of CO_2 above 10 kPa were found to be toxic to the fruit (Brigati and Caccioni, 1995). Tulin Oz and Eris (2009) found that storage of 'Hayward' at 0°C and 85–90% RH in 2 kPa O_2 + 5 kPa CO_2 retained their quality for 5 months. They recommended harvesting when the TSS content was 5.5–6.5%. Yildirim and Pekmezci (2009) recommended 0°C and 90–95% RH in 2 kPa O_2 + 5 kPa CO_2 for 4 months, which suppressed ethylene production and delayed softening. However, they found that there was a slight decrease in vitamin C content in both CA storage and air storage. Storage at 0°C and 90–95% RH in 3 kPa CO_2 + 1–1.5 kPa O_2 maintained fruit firmness during long-term storage, but 1 kPa CO_2 + 0.5 kPa O_2 gave the best storage conditions while maintaining an acceptably low level of the incidence of rotting (Brigati *et al.*, 1989). Özer *et al.* (1999) found that 'Hayward' could be stored for 6 months at 0 ± 0.5°C and 90–95% RH, especially in 5 kPa CO_2 + 5 kPa O_2 or 5 kPa CO_2 + 2 kPa O_2. However, subsequent shelf-life at 20 ± 3°C and 60 ± 5% RH should be limited to 15 days. Steffens *et al.* (2007b) investigated factorial combinations of 0.5, 1.0 and 1.5 kPa O_2 combined with 8, 12 and 16 kPa CO_2 on 'Bruno'. Sensory analysis showed that fruit stored well in 1.0 + 8 kPa CO_2 and 1.5 kPa O_2 + 8 kPa CO_2 without developing off-flavours (Steffens *et al.*, 2007b). 'Hayward' was stored over two seasons in 0–21 kPa O_2 + 0–5 kPa CO_2 at 0–10°C. CO_2 delayed softening but lowering O_2 to near 0 kPa did not inhibit softening completely at 0°C. At temperatures higher than 3°C the additional effect of CA storage was limited (Hertog *et al.*, 2004). Testoni and Eccher Zerbini (1993) showed that storage in CO_2 concentrations higher than 5 kPa resulted in irregular softening in the core.

Tonini and Tura (1997) showed that storage at –0.5°C in combination with 4.8 kPa CO_2 + 1.8 kPa O_2 reduced rots (*Botrytis cinerea* and *Phialophora* spp.) and softening compared with storage in air. The combination of either 1 kPa CO_2 + 1 kPa O_2 or 1.5 CO_2 + 1.5 kPa O_2 was even more effective in controlling rots caused by *Phialophora* spp. In contrast, in central and northern Italy, Tonini *et al.* (1999) found that CA storage at –0.8°C for 120–140 days favoured the spread of *B. cinerea*. They found that postponing the reduction of O_2 and the increase in CO_2 for 30–50 days avoided increasing Botrytis rots without any adverse effects on fruit firmness. Storage at 0.5–1°C with 92–95% RH in 4½–5 kPa CO_2 + 2–2½ kPa O_2 and ethylene at 0.03 ppm or less delayed flesh softening but strongly increased the incidence of Botrytis stem-end rot (Tonini *et al.*, 1989). Brigati and Caccioni (1995) also showed that the high level of CO_2 could lead to an increase in the incidence of *B. cinerea*. Li *et al.* (2017a) confirmed the benefit of CA storage in maintaining fruit firmness and that 5 kPa CO_2 was better than 2 kPa CO_2. The effect of 5 kPa CO_2 was maintained for a couple of weeks after transfer from CA to air storage. Whilst the softening of all tissue zones was slowed by CA, the outer pericarp appeared to be slowed to a greater extent than the inner pericarp. A consequence of the differential effect of CA on the two pericarp tissue zones was that after CA storage the inner pericarp had become relatively softer than the outer pericarp when compared with air-stored fruit at the same fruit firmness value.

Kiwifruit is very susceptible to ethylene even at very low concentrations and Brigati and Caccioni (1995) recommended that ethylene should be maintained at < 0.03 ppm in CA stores. Storage in 2–5 kPa O_2 + 0–4 kPa CO_2 reduced ethylene production (Wang *et al.*, 1994) and Tulin Oz and Eris (2009) found that the main effect of 2 kPa O_2 in storage was to reduce the rate of softening and of 5 kPa CO_2 to reduce ethylene production. Antunes and Sfakiotakis (2002) found that after 60, 120 or 180 days of storage at 0°C there was an initiation of ethylene production when fruits were transferred to 20°C, with no lag period from those stored in air or 2% O_2 + 5% CO_2. However, those removed from storage in 0.7 kPa O_2 + 0.7 kPa CO_2 or 1% O_2 + 1% CO_2 showed reduced capacity to produce ethylene, mainly due to low ACC oxidase activity rather than reduced ACC production or ACC synthase activity. Zhang (2002) successfully stored kiwifruit at 0 ± 0.5°C in 2–3 kPa O_2 + 4–5 kPa CO_2 with a potassium permanganate absorber for over 180 days. After storage the fruit looked fresh with good colour, smell and taste. The percentage of firm fruit was > 92%, and the retention of AA was > 80% compared with levels at the beginning of storage.

Storage of two species of hardy kiwifruit, 'Ananasnaya' (*A. arguta*) and 'Bingo' (*A. purpurea* × *A. arguta*), at 1°C and 85% RH in air or 1.5 kPa O_2 + 1.5 kPa CO_2 showed that storage in air was adequate for 4 weeks, but for 8 weeks CA was required (Latocha *et al.*, 2014).

Kohlrabi (*Brassica oleracea* var. *gongylodes*)

Escalona *et al.* (2007) found that storage in air at 0°C at high humidity resulted in yellowing of the stalks that later fell off, which affected their appearance and marketability. Whole or fresh-cut kohlrabi could be stored for 28 days at 5°C and 95% RH in 5 kPa O_2 +15 kPa CO_2 followed by 3 days at 15°C and 60–70% RH in air and still retain good commercial quality without detrimentally affecting their stalks. However, SeaLand (1991) had previously reported that CA storage had only a slight to no effect.

Lanzones, Langsat (*Lansium domesticum*)

Pantastico (1975) recommended 14.4°C in 0 kPa CO_2 + 3 kPa O_2 which gave a 16-day postharvest life compared with only 9 days at the same temperature in air. He also indicated that the skin of the fruit turned brown during retailing and if they were sealed in PE film bags the browning was aggravated, probably due to CO_2 accumulation.

Leek (*Allium ampeloprasum* var. *porrum*)

Optimum storage conditions were reported to be: 0°C in 5–10 kPa CO_2 + 1–3 kPa O_2 for up to 4–5 months (Kurki, 1979); 0°C in 15–25 kPa CO_2 for 4½ months (Monzini and Gorini, 1974); 3–5 kPa CO_2 + 1–2 kPa O_2 (SeaLand, 1991); 0–5°C in 3–5 kPa CO_2 + 1–2 kPa O_2, which had a good effect but was claimed to be of no commercial use (Kader, 1985, 1992); and 0–5°C in 5–10 kPa CO_2 + 1–6 kPa O_2, which had only a slight effect (Saltveit, 1989). Goffings and Herregods (1989) showed that with storage at 0°C and 94–96 kPa RH in 2 kPa CO_2 + 2 kPa O_2 + 5 kPa carbon monoxide, leeks could be stored for up to 8 weeks compared with 4 weeks at the same temperature in air. Under those conditions the total losses were 19% while those stored in the same conditions but without carbon monoxide had 28% losses and those in air had 37% losses. Lutz and Hardenburg (1968) reported that CO_2 levels of 15–20 kPa caused injury.

Lemon (*Citrus limon*)

Storage recommendations include: 5–10 kPa CO_2 + 5 kPa O_2 (SeaLand, 1991); 10–15°C in 0–5 kPa CO_2 + 5 kPa O_2 (Kader, 1985); 10–15°C in 0–10 kPa CO_2 + 5–10 kPa O_2 (Kader, 1989); and 10–15°C with 0–10 kPa CO_2 + 5–10 kPa O_2 (Bishop, 1996). The rate of colour change could be delayed with high CO_2 and low O_2 in the storage atmosphere but 10 kPa CO_2 could impair the flavour (Pantastico, 1975). Wild *et al.* (1977) reported that lemons may be stored in good condition for 6 months in 10 kPa O_2 + 0 kPa CO_2 combined with the continuous removal of ethylene.

Lettuce (*Lactuca sativa*)

Storage recommendations include: 2 kPa CO_2 + 3 kPa O_2 for up to 1 month (Hardenburg *et al.*, 1990); 0 kPa CO_2 + 2–5 kPa O_2 (SeaLand, 1991); 0°C with 98% RH in a maximum of 1 kPa CO_2 and a minimum of 2 kPa O_2 (Fellows, 1988); 1.5 kPa CO_2 + 3 kPa O_2 (Lawton, 1996); 0–5°C in 0 kPa CO_2 + 1–3 kPa O_2 (Bishop, 1996); and 3–5 kPa CO_2 + 15 kPa O_2 for 3 weeks (Tataru and Dobreanu, 1978). Storage in 1.5 kPa CO_2 + 3 kPa O_2 inhibited butt discoloration and pink rib at 0°C; the effect did not persist during 5 days of subsequent storage at 10°C (Hardenburg *et al.*, 1990) but CO_2 above 2.5 kPa or O_2 levels below 1 kPa could injure lettuce. Brown stain on the mid ribs of leaves can be caused by levels of CO_2 of 2 kPa or higher, especially if combined with low O_2 (Haard and Salunkhe, 1975). Adamicki (1989) described successful storage of lettuce at 1°C in 3 kPa CO_2 + 1 kPa O_2 for 21 days with less loss in AA than those stored in air. Kader (1985) recommended 0–5°C in 0 kPa CO_2 + 2–5 kPa O_2 or 0–5°C in 0 kPa CO_2 + 1–3 kPa O_2 (Kader, 1992), which had a good effect and was of some commercial use when carbon monoxide was added at the 2–3 kPa level. Saltveit (1989) recommended 0–5°C in 0 kPa CO_2 + 1–3 kPa O_2 for leaf, head and cut and shredded lettuce; the effect on the former was moderate and on the latter it had a high level of effect. Storage in 0 kPa CO_2 with 1–8 kPa O_2 gave an extension in storage life and hypobaric storage increased storage life from 14 days in conventional cold stores to 40–50 days (Haard and Salunkhe, 1975).

Lime (*Citrus aurantiifolia*)

CA storage recommendations were 0–10 kPa CO_2 + 5 kPa O_2 by SeaLand (1991). Pantastico (1975) recommended storage in 7 kPa O_2, which reduced the symptoms of chilling injury; however, he showed that CA storage of 'Tahiti' limes increased rind scald decay and reduced juice content. Storage at 10–15°C in 0–10 kPa CO_2 + 5 kPa O_2 was recommended by Kader (1986) or 10–15°C in 0–10 kPa CO_2 + 5 kPa O_2 (Kader, 1989, 1992). These storage conditions were shown to increase the postharvest life of limes but were said to be not used commercially (Kader, 1985). CA storage was not considered beneficial but typical storage conditions were given as 10–15°C in 0–10 kPa CO_2 + 5–10 kPa O_2 by Bishop (1996). Fruits were stored in air or in CA containing 3 kPa O_2 + 3 kPa CO_2, 5 kPa O_2 + 3 kPa CO_2, 3 kPa O_2 + 5 kPa CO_2 and 5 kPa O_2 + 5 kPa CO_2 by Sritananan *et al.* (2006). Those stored in air had a higher ethylene production and respiration rate than those in CA. All the CA storage conditions reduced the loss of chlorophyll and change in peel colour compared with fruits stored in air.

Litchi, Lychee (*Litchi chinensis*)

Kader (1993) recommended 3–5 kPa CO_2 + 5 kPa O_2 at 7°C, with a range of 5–12°C, and reported that the benefits of reduced O_2 were good and those of increased CO_2 were moderate. Vilasachandran *et al.* (1997) stored the cultivar 'Mauritius' at 5°C in air or in 5, 10 or 15 kPa CO_2 + 3 or 4 kPa O_2. After 22 days all fruits were removed to air at 20°C for 1 day. Those stored in 15 kPa CO_2 + 3 kPa O_2 or 10 kPa CO_2 + 3 kPa O_2 were lighter in colour and retained TSS better than the other treatments, but had the highest levels of off-flavours. Those in all the CA storage treatments had negligible levels of black spot and stem-end rot compared with the controls. On the basis of the above, they recommended 5 kPa CO_2 + 3 kPa O_2 or 5 kPa CO_2 + 4 kPa O_2. Mahajan and Goswami (2004) stored the cultivar 'Bombay' at 2°C and 92–95% RH for 56 days in 3.5 kPa O_2 + 3.5 kPa CO_2. Fruits retained their red colour, while those stored in air had begun to turn brown. Loss of acidity and AA content and the smallest increase of firmness and pericarp puncture strength of fruits were for those stored in CA compared with those stored in air. The sensory evaluation of aril colour and taste showed that the fruits held in CA were rated 'good' throughout the 56 days of storage. Ali *et al.* (2016) found that the cultivar 'Gola' stored for 35 days at 5 ± 1°C turned completely brown within 28 days in air, but in an atmosphere of 1 kPa O_2 + 5 kPa CO_2 browning was delayed, antioxidant activity maintained and the fruit had better organoleptic properties.

Mamey (*Mammea americana*)

Manzano-Mendez and Dris (2001) showed that storage of the cultivar 'Amarillo' at 15 ± 2°C in 1 kPa CO_2 + 5.6 kPa O_2 for 2 weeks retarded maturation.

Mandarin, Satsuma (*Citrus* spp.)

These include the satsuma mandarin, *C. unshiu*, and common mandarin, *C. reticulate.* CA had only a slight or no effect on storage (SeaLand, 1991). At 25°C in 60 kPa CO_2 + 20 kPa O_2 + 20 kPa N_2 the respiration rate was not affected, but in 80 kPa CO_2 + 20 kPa air and 90 kPa CO_2 + 10 kPa air, respiration rates were significantly reduced (Kubo *et al.*, 1989b). Ogaki *et al.* (1973) carried out experiments over a number of years and showed that the most suitable atmosphere for storing satsumas was 1 kPa CO_2 + 6–9 kPa O_2. A humidity of 85–90% RH produced the best-quality fruit and resulted in only 3% weight loss. Pre-storage treatment at 7-8°C and 80–85% RH was also recommended. In a two-season trial, satsumas were stored for 3 months at 3°C and 92% RH in 2.8–6.5 kPa O_2 + 1 kPa CO_2 + 93.5–97.2 kPa N_2 or in air. Weight losses were 1.2–1.5% in CA and 6.5–6.7% in air; and fruit colour, flavour, aroma and consistency were also better in CA. AA content was 5.9% higher in the flesh and 10.3% in the peel of CA-stored fruits compared with the controls (Dubodel and Tikhomirova, 1985). Satsumas were stored at 2–3°C and 90% RH in 3–6 kPa O_2 + 1 kPa CO_2 or in air. The total sugar content of fruit flesh decreased in both CA- and air-stored fruit by 10.7% and 13.0%, respectively, and losses in the peel were even greater. Reducing sugars decreased during storage, especially in the flesh of CA-stored fruits and in the peel of control fruits (Dubodel *et al.*, 1984). Yang and Lee (2003) compared storage in air with 5 kPa CO_2 + 3 kPa O_2, 3 kPa CO_2 + 1 kPa O_2 and 10.5 kPa CO_2 + 3.9 kPa O_2 and found that organic acids in fruit in all three CA conditions were higher than those in air until 60 days, but there were no differences between treatments after 120 days of storage. They found that 5 kPa CO_2 + 3 kPa O_2 maintained the best quality among the CA conditions after 120 days and gave the best retention of firmness. Yang (2001), using the same CA conditions, reported that the TSS of fruit in cold storage increased until 2 months and thereafter sharply decreased, whereas TSS of CA-stored fruit increased slowly throughout the storage period of 120 days. It was concluded that sensory evaluation clearly showed that CA could extend the marketing period of satsumas and retain good flavour. Storage of *C. unshiu* in China in containers with a D45 M2 1 silicone window of 20–25 cm^2/kg of fruit gave what they described as the optimum CO_2 concentration of < 3 kPa together with $O_2 < 10$ kPa (Hong *et al.*, 1983).

Mango (*Mangifera indica*)

CA storage recommendations include: 5 kPa CO_2 + 5 kPa O_2 (Pantastico, 1975; SeaLand, 1991); 10°C with 90% RH in 10 kPa CO_2 + 5 kPa O_2 (Lawton, 1996); 10–15°C in 5–10 kPa CO_2 + 3–5 kPa O_2 (Bishop, 1996); 10–15°C in 5 kPa CO_2 + 5 kPa O_2 (Kader, 1986); 5–10 kPa CO_2 + 3–5 kPa O_2 (Kader, 1989, 1992); and 13°C, with a range of 10–15°C, in 5–10 kPa CO_2 + 3–5 kPa O_2, or 5–7 kPa O_2 for south-east Asian varieties (Kader, 1993). Fuchs *et al.* (1978) described an experiment where storage at 14°C in 5 kPa CO_2 + 2 kPa O_2 kept the mangoes green and firm for 3 weeks. Upon removal they attained full colour in 5 days at 25°C, but 9% had rots. An additional week in storage resulted in 40% developing rots during ripening. Bleinroth *et al.* (1977) reported that fermentation occurred with alcohol production during storage at 8°C and > 10 kPa CO_2 for 3 weeks. After 38 days of storage of the cultivar 'Delta R2E2' at 13 ± 1°C harvested at the mature-green stage, ethanol, acetaldehyde and esters were significantly higher in fruit in 1.5 kPa O_2 + 6 kPa CO_2, 1.5 kPa O_2 + 8 kPa CO_2 or 2 kPa O_2 + 8 kPa CO_2 than those in air. Storage in 3 kPa O_2 + 6 kPa CO_2 appeared promising and resulted in no significant fermentation (Lalel and Singh, 2006). Kim *et al.* (2007) stored mature-green fruit at 10°C for 2 weeks in 3 kPa O_2 + 97 kPa N_2, 3 kPa O_2 + 10 kPa CO_2 + 87 kPa N_2 or in air and then ripened them at 25°C in air. They found that CA delayed TA and colour changes and the overall decline in polyphenolic concentration compared with fruit stored in air. Paull and Chen (2004) recommended 3–5 kPa O_2 and 5–10 kPa CO_2, at 7–9°C and 90% RH. For Florida varieties the optimum conditions were reported to be 3–5 kPa O_2 and 5–10 kPa CO_2 (Bender *et al.*, 2000).

Detrimental effects of CA storage have also been reported. Storage in 1 kPa O_2 resulted in the production of off-flavour and skin discoloration but storage at 12°C in 5 kPa CO_2 + 5 kPa O_2 was possible for 20 days (Hatton and Reeder, 1966). Deol and Bhullar (1972) mentioned that there was increased decay in mangoes stored in either CA storage or MA packaging compared with those stored non-wrapped in air. This is in contrast with the work of Wardlaw (1938), Thompson (1971) and Kane and Marcellin (1979), who all showed that either CA storage or MA packaging reduced postharvest decay of mangoes. However, this difference may be related to anthracnose (*Colletotrichum gloeosporioides*), since the fungus infects the fruit before harvest and begins to show symptoms as the fruit ripens. Since CA storage and MA packaging delay ripening, they would also be expected to delay the development of anthracnose symptoms. CA has been tested on ripe fruit. In a comparison of various conditions ranging from 1.6 to 20.7 kPa O_2 + 0.2 to 10.2 kPa CO_2 with the balance being N_2 in a flow-through system at 5°C, 15°C or 25°C. The optimum combination was found to be around 10 kPa CO_2 + 5 kPa O_2 for the suppression of the respiration rate of ripe 'Irwin' (Nakamura *et al.*, 2003).

CA storage conditions have been recommended for specific varieties as follows.

Alphonso

Niranjana *et al.* (2009) showed that fruits stored in air at 8°C showed chilling injury symptoms. However, treatment with a fungicide or hot water treatment of 55°C for 5 min followed by storage at 8°C in 5 kPa O_2 + 5 kPa CO_2 resulted in fruit with no chilling injury symptoms. CA-stored fruit also retained their antioxidant levels and fresh, hard, green appearance and they ripened normally when subsequently removed to ambient conditions of 24–29°C with 60–70% RH. Sudhakar Rao and Gopalakrishna Rao (2009) stored mature-green fruit at 13°C in 5 kPa O_2 + 5 kPa CO_2, 3 kPa O_2 + 5 kPa CO_2, 5 kPa O_2 + 3 kPa CO_2, or 3 kPa O_2 + 3 kPa CO_2. CA that contained 5 kPa O_2 significantly reduced the respiration and ethylene peaks, but 3 kPa O_2 + 5 kPa CO_2 resulted in abnormal respiration and ethylene production. Storage in 5 kPa O_2 + 5 kPa CO_2 extended storage life by 4 or 5 weeks. The fruit then ripened to good quality in ambient conditions of 25–32°C. These fruits had a bright yellow skin colour, high TSS and total carotenoid and sugar content, and were firm and acceptable organoleptically.

Amelie

Storage at 11–12°C in 5 kPa O_2 + 5 kPa CO_2 for 4 weeks was reported to reduce storage rots and give the best eating quality compared with storage in air (Medlicott and Jeger, 1987).

Banganapalli

Sudhakar Rao and Gopalakrishna Rao (2009) stored mature-green fruit at 13°C in 5 kPa O_2 + 5 kPa CO_2, 3 kPa O_2 + 5 kPa CO_2, 5 kPa O_2 + 3 kPa CO_2, or 3 kPa O_2 + 3 kPa CO_2. CA storage in 5 kPa O_2 significantly reduced the respiration and ethylene peaks during ripening, but CA containing 3 kPa O_2 + 3 kPa CO_2 resulted in abnormal respiration and ethylene production. In 5 kPa O_2 + 3 kPa CO_2 storage life was extended by 5 weeks, followed by an additional week to become fully ripe in ambient conditions of 25–32°C. Fruits stored in optimum 5 kPa O_2 + 3 kPa CO_2 ripened normally to a bright yellow skin colour, with high TSS, total carotenoid and sugar contents, and were firm with an acceptable organoleptic quality.

Carabao

Storage in 5 kPa CO_2 + 5 kPa O_2 for 35–40 days was recommended by Mendoza (1978).

Carlotta

Storage recommendations include 10–11°C in 2 kPa O_2 + 1 or 5 kPa CO_2 for 6 weeks (Medlicott and Jeger, 1987) and 8°C in 10 kPa CO_2 + 6 kPa O_2 for 35 days (Bleinroth *et al.*, 1977).

Delta R2E2

Bleinroth *et al.* (1977) reported that fermentation of fruit occurred with alcohol production during storage at 8°C and > 10 kPa CO_2 for 3 weeks. After 38 days of storage at 13 ± 1°C harvested at the mature-green stage, ethanol, acetaldehyde and esters were significantly higher in fruit in 1.5 kPa O_2 + 6 kPa CO_2, 1.5 kPa O_2 + 8 kPa CO_2 or 2 kPa O_2 + 8 kPa CO_2 than those in air.

Jasmin

Storage in 10–11°C in 2 kPa O_2 + 1 or 5 kPa CO_2 for 6 weeks (Medlicott and Jeger, 1987) and 8°C in 10 kPa CO_2 + 6 kPa O_2 for 35 days were recommended (Bleinroth *et al.*, 1977).

Julie

Storage at 11–12°C in 5 kPa O_2 + 5 kPa CO_2 for 4 weeks was said to reduce storage rots and give the best eating quality compared with those stored in air (Medlicott and Jeger, 1987).

Haden

Storage at 10–11°C in 2 kPa O_2 + 1 or 5 kPa CO_2 for 6 weeks (Medlicott and Jeger, 1987) and 8°C in 10 kPa CO_2 + 6 kPa O_2 for 30 days (Bleinroth *et al.*, 1977) was recommended. Sive and Resnizky (1985) found that ripening was delayed at 13–14°C but lower temperatures resulted in chilling injury. They reported that in CA fruit could be kept in good condition for 6–8 weeks at 13–14°C.

Keitt

Storage at 13°C in 5 kPa CO_2 + 5 kPa O_2 was recommended, after which fruit ripened normally in air (Hatton and Reeder, 1966). Sive and Resnizky (1985) found that ripening was delayed at 13–14°C but lower temperatures resulted in chilling injury. They reported that in CA they could be kept in good conditions for 8–10 weeks at 13–14 °C. 'Keitt' mangoes were stored at 7°C in 10 kPa CO_2 + 6 kPa O_2 + 84 kPa N_2, 8 kPa CO_2 + 6 kPa

O_2 + 86 kPa N_2, 5 kPa CO_2 + 5 kPa O_2 + 90 kPa N_2 or air at 7°C, 10°C, 13°C or 21–24°C. Of those that were tested the optimum conditions were 7°C under 10 kPa CO_2 + 6 kPa O_2 for up to 6 weeks. However, it was claimed that those stored at 13°C in air for up to 21 days were better and showed no chilling injury effect although there was no mention that chilling injury occurred at 7°C (Hailu, 2016). Yahia and Tiznado Hernández (1993) stored Keitt for 0–5 days at 20°C in a continuous flow of 0.2–0.3 kPa O_2 (balance N_2). Fruits were evaluated during storage and again after being held in air at 20°C for 5 days. There was no fruit injury or reduction in organoleptic quality due to the low O_2 atmosphere and fruits ripened normally. These results indicate that applying low O_2 atmospheres postharvest can be used to control insects in mangoes without adversely affecting fruit quality.

Kensington

Within the concentrations studied by McLauchlan *et al.* (1994), the optimum atmosphere appeared to be around 4 kPa CO_2 + 2–4 kPa O_2 but they suggested that further research was required below 2 kPa O_2 and above 10 kPa CO_2, as well as at lower storage temperatures to control softening. They also reported that colour development was linearly retarded by decreasing O_2 from 10 kPa to 2 kPa and increasing CO_2 from 0 kPa to 4 kPa; there was no further effect from 4–10 kPa CO_2. Fruit from storage in low O_2 atmospheres also had high TA after storage, but after 5 days in ambient conditions acid levels fell and they continued to develop typical external colour. In trials by O'Hare and Prasad (1993) over two seasons, over-ripeness was more common after 5 weeks at 13°C (and 10°C in the first season) and was associated with a decline in TA. They found that chilling injury occurred after 1 week at 5°C, but storage in 5 or 10 kPa CO_2 alleviated chilling injury symptoms. Storage in 5 kPa O_2 had no significant effect on chilling injury. Increased pulp ethanol concentration was associated with CO_2 injury in 10 kPa CO_2, but 3 kPa O_2 + 6 kPa CO_2 at 13°C proved to be beneficial for extending the storage life of 'Kensington Pride' by up to 6 weeks, with good fruit quality and maintaining a high concentration of the major volatile compounds responsible for the aroma of ripe mangoes (Singh and Zaharah, 2015). Fresh-cut slices from ripe 'Kensington' mango were prepared aseptically and stored at 3°C. Storage in 2.5 kPa O_2 was effective at controlling tissue darkening and the development of a 'glassy' appearance, while increasing CO_2 levels to 5–40 kPa had little positive effect on shelf-life, appearing to promote tissue softening. A combination of low oxygen and calcium allowed 'Kensington' slices to be held for at least 15 days at 3°C. Green-mature 'Kensington Pride' mangoes were stored at 13°C in 90 l chambers containing 2 kPa O_2 + 3 kPa, 6 kPa and 9 kPa CO_2 or air in one experiment and O_2 at 1 kPa, 2 kPa and 3 % factorily combined with CO_2 at 6 kPa and 8 % for 21 and 35 days in both experiments. The fruits were allowed to ripen at 21±1°C to eating soft stage. Increased concentrations of CO_2 increased the concentration of total fatty acids as well as palmitic acid, palmitoleic acid, stearic acid and linoleic acid and caused the ratio of palmitic to palmitoleic acid to be more than 1 and reduced the production of monoterpenes, sesquiterpenes and aromatics in the pulp of ripe fruit. An increased concentration of CO_2 also increased the production of total esters as well as major individual esters, especially ethyl butanoate. Storage in 2 kPa O_2 + 3 kPa CO_2 or 3 kPa O_2 + 6 kPa CO_2 gave the best results in extending the postharvest life while maintaining a high concentration of the major volatile compounds responsible for the aroma of ripe mangoes (Lalel and Singh, 2004).

Kent

Antonio Lizana and Ochagavia (1997) reported that storage at 12°C in 10 kPa CO_2 + 5 kPa O_2 increased postharvest life to 29 days, which was 8 days longer than when stored in air. Storage of mature unripe fruit for 21 days at 12°C showed that CO_2 levels of 50 kPa and 70 kPa resulted in high levels of ethanol production rates and symptoms of

CO_2 injury, while storage in 3 kPa O_2 appeared to have little effect on ethanol production (Bender *et al.*, 1994).

Maya

Sive and Resnizky (1985) found that at 13–14°C ripening was delayed and lower temperatures caused chilling injury, but in CA at 13–14°C the fruit could be kept in good conditions for 6–8 weeks.

Rad

Storage at 13°C in combinations of 4, 6 and 8 kPa CO_2 + 4, 6 and 8 kPa O_2 was studied. Ripening was delayed by CA storage for 2 weeks compared with storage in air, with 4 kPa CO_2 + 6 kPa O_2 giving the highest acceptability by taste panellists (Noomhorm and Tiasuwan, 1988).

San Quirino

Storage at 10–11°C in 2 kPa O_2 + 1 or 5 kPa CO_2 for 6 weeks (Medlicott and Jeger, 1987) or at 8°C in 10 kPa CO_2 + 6 kPa O_2 for 35 days was recommended (Bleinroth *et al.*, 1977).

Tommy Atkins

Antonio Lizana and Ochagavia (1997) reported that in storage at 12°C in 5 kPa CO_2 + 5 kPa O_2 there was a postharvest life of 31 days. Sive and Resnizky (1985) found that at 13–14°C ripening was delayed and lower temperatures caused chilling injury but in CA the fruit could be kept in good conditions for 6–8 weeks. Storage of mature unripe fruit for 21 days at 12°C showed that CO_2 levels of 50 kPa and 70 kPa resulted in high levels of ethanol production rates and symptoms of CO_2 injury, while the 3 kPa O_2 concentration seemed to have little effect on ethanol production (Bender *et al.*, 1994). CA storage can be used to control pest infestation. In storage trials to control fruit fly in mango fruits, it was found that fruits exposed to 50 kPa CO_2 + 2 kPa O_2 for 5 days or 70–80 kPa CO_2 + < 0.1 kPa O_2 for 4 days did not suffer adverse effects when they were subsequently ripened in air (Yahia *et al.*, 1989). CA in combination with other treatments has been shown to reduce disease. Immersing fruit in hot water (52°C for 5 min) plus benomyl (Benlate 50 WP 1 g/l) provided good control of stem-end rot on mangoes following inoculation with either *Dothiorella dominicana* or *Lasiodiplodia* [*Botryodiplodia*] *theobromae* during storage for 14 days at 25–30°C. In the same storage conditions prochloraz (as Sportak 45EC), DPXH6573 (40EC), RH3866, (25EC) and calcium chloride did not control stem-end rot (*D. dominicana*). During long-term storage at 13°C in 5 kPa O_2 + 2 kPa CO_2 for 26 days followed by air for 11 days at 20°C, benomyl at 52°C for 5 min followed by prochloraz at 25°C for 30 s provided effective control of stem-end rot and anthracnose. The addition of guar gum to hot benomyl improved control of stem-end rot in the combination treatment. Benomyl at 52°C for 5 min alone was ineffective. Other diseases, notably Alternaria rot (*A. alternata*) and dendritic spot (*D. dominicana*), emerged as problems during CA storage. *A. alternata* and dendritic spot were also controlled by benomyl at 52°C for 5 min followed by prochloraz. *Penicillium expansum*, *Botrytis cinerea*, *Stemphylium vesicarium* and *Mucor circinelloides* were reported as postharvest pathogens of 'Kensington Pride' mango (Johnson *et al.*, 1990a, b).

Mangosteen (*Garcinia mangosteen*)

Pakkasarn *et al.* (2003b) found that storage for 28 days at 13°C in 10 kPa CO_2 + 21 kPa O_2 was effective in reducing ethylene production and calyx chlorophyll loss, delaying peel colour and firmness changes and decreasing weight loss and respiration rate. Fruit in air could be stored for only 16 days. Pakkasarn *et al.* (2003a) also found that fruit in 2 kPa O_2 + 15 kPa CO_2 for 24 days developed a fermented flavour. Storage at 13°C in 2–6 kPa O_2 + 10–15 kPa CO_2 had reduced weight loss and retarded peel colour development, softening and subsequent hardening, calyx chlorophyll loss, respiration rate and ethylene production compared with those stored in air.

Medlar (*Mespilus germanica*)

Selcuk and Erkan (2015) stored medlars in air, MA packages and CA of 2 kPa O_2 + 5 kPa CO_2 or 3 kPa O_2 + 10 kPa CO_2 at 0°C for 60 days. Total phenolics concentrations, total flavonoids concentrations, antioxidant activity, AA retention and total condensed tannins retention were the highest in 2 kPa O_2 + 5 kPa CO_2.

Melon (*Cucumis melo*)

Few positive effects of CA storage have been shown. Kader (1985, 1992) recommended 3–7°C in 10–15 kPa CO_2 + 3–5 kPa O_2 for cantaloupes, which had a good effect on storage but was said to have limited commercial use. For honeydew, Kader (1985, 1992) recommended 10–12°C in 0 kPa CO_2 + 3–5 kPa O_2, which had a fair effect but was of no commercial use. Saltveit (1989) recommended 5–10°C in 10–20 kPa CO_2 + 3–5 kPa O_2, which had only a slight effect. Storage in 10–15 kPa CO_2 + 3–5 kPa O_2 was recommended for cantaloupe and 5–10 kPa CO_2 + 3–5 kPa O_2 for honeydew and casaba (SeaLand, 1991). Typical storage conditions were given as 2–7°C in 10–20 kPa CO_2 + 3–5 kPa O_2 by Bishop (1996) for cantaloupes. Christakou *et al.* (2005) stored galia in 30 kPa CO_2 + 70 kPa N_2 at 10°C, which helped to reduce the loss in quality. However, vitamin C decreased during storage for those stored in CA and those stored in air. Perez Zungia *et al.* (1983) compared storage of the cultivar 'Tendral' in 0 kPa CO_2 + 10 kPa O_2 with storage in air and found that the CA-stored fruit had better overall quality. 'Durango' hybrid fruit were stored for 20 days at 10 or 15°C and supplied daily with a 5 min flow of air containing 5.1 kPa CO_2. CO_2 treatment ressulted in a reduced TSS, firmness and acidity and increased the rate of colour change (Rodriguez and Manzano, 2000). CA had limited effects on disease. Mencarelli *et al.* (2003) carried out storage experiments on organically grown winter melons at 10°C and 95% RH for 63 days plus 5 days in air at 20°C. They concluded that CA storage in 2 kPa O_2 + 0 kPa CO_2, 2 kPa O_2 + 5 kPa CO_2 or 2 kPa O_2 + 10 kPa CO_2 had few beneficial effects compared with storage in air. Storage in 2 kPa O_2 + 10 kPa CO_2 resulted in some control of decay and the fruit stored at 2 kPa O_2 tasted better after storage and were sweeter and firmer than those stored in air. Martinez-Javega *et al.* (1983) stored the cultivar 'Tendral' in 12 kPa CO_2 + 10 kPa O_2 at temperatures within the range of 2–17°C and found no effects on levels of decay or chilling injury compared with fruits stored at the same temperatures in air.

Melon, Bitter (*Momordica charantia*)

Fruits of bitter melon were harvested at 'horticultural maturity' and stored for 2 weeks in various conditions. Fruit quality was similar after storage at 15°C in 21, 5 or 2.5 kPa O_2 + 0, 2.5, 5 or 10 kPa CO_2 or in air. Fruits stored for 3 weeks in 2.5 kPa O_2 + 2.5 or 5 kPa CO_2 showed greater retention of green colour and had less decay (unspecified) and splitting than those stored in air (Zong *et al.*, 1995). Bitter melons are reported to be subject to chilling injury by Dong *et al.* (2005). They immersed fruit in hot water at 42°C for 5 min, then packed them in commercial PE film bags and stored them at 4°C for 16 days with no indication of chilling injury.

Mushroom (*Agaricus bisporus*)

Recommended optimum storage conditions include: 5–10 kPa CO_2 (Lutz and Hardenburg, 1968; Ryall and Lipton, 1979); 10–15 kPa CO_2 + 21 kPa O_2 (SeaLand, 1991); and 0°C in 8 kPa O_2 + 10 kPa CO_2 (Zheng and Xi, 1994). Kader (1985, 1992) and Saltveit (1989) recommended 0–5°C in 10–15 kPa CO_2 + 21 kPa O_2, which had a fair or moderate effect but was of limited commercial use. Marecek and Machackova (2003) found that storage in 1 or 2 kPa O_2 + 0 or 4 kPa CO_2 was best in keeping the fruit's quality and firmness. Storage with CO_2 levels of 10–20 kPa inhibited mould growth and retarded cap and stalk development, O_2 levels of < 1 kPa were injurious (Pantastico, 1975). Anonymous (2003) confirmed that high CO_2 prevented head opening at 12°C, but also found that atmosphere with high CO_2 concentration resulted in more cap browning and that O_2 concentration did not have any effect on colour change. Zheng and Xi (1994) found that colour deterioration was inhibited by storage at 0°C and also that low O_2 and 10 kPa CO_2 inhibited cap opening and internal browning but caused a 'yellowing' of the cap surface. Tomkins (1966) reported that 10 kPa CO_2 could delay deterioration but levels of over 10 kPa could cause a 'pinkish' discoloration. Button mushrooms were stored in MA packages with 95–100 kPa CO_2 and then ventilated after 0, 12, 24 or 48 h by puncturing the film at four corners. CO_2 exposure for 12 h reduced browning and increased antioxidant activity, which in turn maintained flavour, quality and consumer acceptance. In addition, the MDA content was significantly lower, while catalase and POD activities were significantly increased by high CO_2 treatment (Lin *et al.*, 2017).

Mushroom, Cardoncello (*Pleurotus eryngii*)

Amodio *et al.* (2003) stored slices at 0°C either in air or in 3 kPa O_2 + 20 kPa CO_2 for 24 days and found that those in CA had a better appearance than those in air. None of the treatments showed visible presence of microbial growth, though there was a slight increase in mesophilic and psychrophilic bacteria and yeasts for slices kept in air, while there was no increase, or a slight decrease, for those in CA. Lovino *et al.* (2004) also used 3 kPa O_2 + 20 kPa CO_2 at 0°C but found a combination of CA plus microperforated plastic film to be the optimum. After 3 weeks and 48 h shelf-life, MA packaging in CA storage showed higher marketable quality than the CA alone.

Mushroom, Oyster (*Pleurotus ostreatus*)

CA storage was shown to have little effect on increasing storage life at either 1°C or 3.5°C, but at 8°C with a combination of 10 kPa CO_2 + 2 kPa O_2, 20 kPa CO_2 + 21 kPa O_2 or 30 kPa CO_2 + 21 kPa O_2 the oyster mushrooms had a reduced respiration rate and retained their quality for longer than those stored in air (Bohling and Hansen, 1989). Henze (1989) recommended storage at 1°C and 94% RH in 30 kPa CO_2 + 1 kPa O_2 for about 10 days and Pantastico (1975) recommended 5 kPa CO_2 + 1 kPa O_2 for 21 days.

Mushroom, Shiitake (*Lentinus edodes*)

Storage at an unspecified temperature with 40 kPa CO_2 + 1–2 kPa O_2 extended storage life four times longer than when stored in air (Minamide, 1981, quoted by Bautista, 1990). Storage at 0°C in 5 kPa, 10 kPa, 15 kPa or 20 kPa CO_2 + 1 kPa, 5 kPa and 10 kPa O_2 plus an air control was studied by Pujantoro *et al.* (1993). They found that 1 or 5 kPa O_2 levels resulted in poor storage. Respiration rates were suppressed by decreasing O_2 and increasing CO_2 concentrations at 5°C, 15°C, 20°C and 30°C. The RQ breakpoints were mainly controlled by O_2 concentration and were little affected by CO_2 concentration, suggesting that better keeping quality could be obtained from low O_2 atmosphere conditions above the RQ breakpoint at low temperature (Hu *et al.*, 2003).

Natsudaidai (*Citrus natsudaidai*)

Storage in 60 kPa CO_2 + 20 kPa O_2 + 20 kPa N_2 had little or no effect on fruit quality (Kubo *et al.*, 1989a), but Kajiura and Iwata (1972) recommended storage at 4°C in 7 kPa O_2 + 93 kPa N_2. They showed that fruit could be stored for 2 months at 3°C or 4°C or for 1 month at 20°C. At 3°C and 4°C, TSS and TA were little affected by the O_2 concentration, but the fall in AA was less marked at low O_2. However, O_2 levels below 3 kPa resulted in a fermented flavour but the eating quality was similar in fruit stored at all O_2 concentrations above 5 kPa. At 20°C there was more stem-end rot but *Penicillium* sp. development was reduced by low O_2. Storage at 3°C and 4°C and below 1.5 kPa O_2 resulted in the albedo becoming watery and yellow and the juice sacs turning from orange to yellow, while respiration rate fell. Kajiura (1972) showed that fruit in 4°C had less button browning in 5 kPa CO_2 and above. TSS was not affected by CO_2 concentration, but TA was reduced in 25 kPa CO_2. At 20°C, button browning was retarded above 3.5–5.0 kPa CO_2 but there were no differences in TSS content and TA between different CO_2 levels. An abnormal flavour and a sweet taste developed in storage above 13 and 5 kPa CO_2, respectively. Above 13 kPa CO_2 the peel turned red. In subsequent work, Kajiura (1973) stored fruit at 4°C and either 98–100% RH or 85–95% RH in air mixed with 0, 5, 10 or 20 kPa CO_2 for 50 days and found that high CO_2 retarded button browning at both humidities and the fruit developed granulation and loss of acidity at 85–95% RH. In 98–100% RH combined with high CO_2 there was increased water content of the peel and the ethanol content of the juice but there was abnormal flavour. At 85–95% RH no injury occurred and CO_2 was beneficial, its optimum level being much higher. In another trial, fruit was stored in 0 kPa CO_2 + 5 kPa O_2 or 5 kPa CO_2 + 7 kPa O_2. In 0 kPa CO_2 + 5 kPa O_2 there was little effect on button browning or acidity but granulation was retarded and an abnormal flavour and ethanol accumulation occurred. In 5 kPa CO_2 + 7 kPa O_2 the decrease in acidity was retarded and there was abnormal flavour development at high humidity but at low humidity it provided the best quality stored fruit.

Nectarine (*Prunus persica* var. *nectarina*)

Storage recommendations are similar to peach, including 5 kPa CO_2 + 2.5 kPa O_2 for 6 weeks (Hardenburg *et al.*, 1990). Storage at 0.5°C and 90–95% RH in 5 kPa CO_2 + 2 kPa O_2 was recommended for 14–28 days (SeaLand, 1991). Other storage recommendations were given as 0–5°C in 5 kPa CO_2 + 1–2 kPa O_2 (Kader, 1985) or 0–5°C in 3–5 kPa CO_2 + 1–2 kPa O_2 which had a good effect, but CA storage had limited commercial use. The storage life at –0.5°C in air was about 7 days, but fruit stored at the same temperature in atmospheres containing 1.5 kPa CO_2 + 1.5 kPa O_2 could be stored for 5–7 weeks (Van der Merwe, 1996). Folchi *et al.* (1994) stored the cultivar 'Independent' in < 0.1 kPa CO_2 + 0.3 kPa O_2 and the fruit showed an increase in aldehyde and ethanol content with increased storage time. Lurie (1992) reported that the cultivars 'Fantasia', 'Flavourtop' and 'Flamekist' stored well for up to 6 weeks at 0°C in an atmosphere containing 10 kPa O_2 + 10 kPa CO_2, and internal breakdown and reddening were almost completely absent. Although this CA prevented internal breakdown and reddening, after extended storage fruit did not develop the increased TSS content or extractable juice during post-storage ripening that occurred in non-stored fruit. Therefore, while preventing storage disorders, CA did not reduce the loss of ripening ability that occurred during storage. Later, Lurie *et al.* (2002) suggested that a certain level of ethylene production was essential for normal ripening of nectarines after cold storage. They linked this with prolonged storage where they did not ripen and developed a dry, woolly texture. They found that the disorder can be alleviated by storing the fruit in the presence of exogenous ethylene.

Anderson (1982) stored the cultivar 'Regal Grand' at 0°C in 5 kPa CO_2 + 1 kPa O_2 for up to 20 weeks with intermittent warming to 18–20°C every 2 days, which almost entirely eliminated the internal browning associated with storage in air at 0°C. Conditioning the fruit prior to storage at 20°C in 5 kPa CO_2 + 21 kPa O_2 for 2 days until they reached a firmness of 5.5 kg was necessary.

Okra (*Hibiscus esculentus*, *Albelmoschus esculentus*)

Storage at 7.2°C in 5–10 kPa CO_2 kept okra in good condition (Pantastico, 1975) but 0 kPa

CO_2 + 3–5 kPa O_2 was recommended by SeaLand (1991). Kader (1985) also recommended 8–10°C in 0 kPa CO_2 + 3-5 kPa O_2 which had a fair effect but was of no commercial use, though CO_2 at 5–10% was beneficial at 5–8°C. Saltveit (1989) recommended 7–12°C in 4–10 kPa CO_2 + 21 kPa O_2 but it had only a slight effect. Ogata *et al.* (1975) stored okra at 1°C in air or 3 kPa O_2 + 3, 10 or 20 kPa CO_2 or at 12°C in air or 3 kPa O_2 + 3 kPa CO_2. At 1°C there were no effects of any of the CA treatments. At 12°C the CA storage resulted in lower AA retention but it improved the keeping quality. At levels of 10–12 kPa CO_2 off-flavour may be produced (Anandaswamy, 1963).

Olive (*Olea europaea*)

Storage of fresh olives at 8–10°C in 5–10 kPa CO_2 + 2–5 kPa O_2 had a fair effect, but was not being used commercially (Kader, 1985). Kader (1989, 1992) recommended 5–10°C in 0–1 kPa CO_2 + 2–3 kPa O_2. In other work Kader (1989) found that green olives of the cultivar 'Manzanillo' were damaged when exposed to CO_2 levels of 5 kPa and above, but storage was extended in 2 kPa O_2. In air they found that in storage life was 8 weeks at 5°C, 6 weeks at 7.5°C and 4 weeks at 10°C, but in 2 kPa O_2 the storage life was extended to 12 weeks at 5°C and 9 weeks at 7.5°C. Garcia *et al.* (1993) stored the cultivar 'Picual' at 5°C in air, 3 kPa CO_2 + 20 kPa O_2, 3 kPa CO_2 + 5 kPa O_2 or 1 kPa CO_2 + 5 kPa O_2. They found that 5 kPa O_2 gave the best results in terms of retention of skin colour, firmness and acidity, but had higher incidence of postharvest losses compared with those stored in air. They found no added advantage of increased CO_2 levels. Agar *et al.* (1997) compared storage of the cultivar 'Manzanillo' in air or 2 kPa O_2 + 98 kPa N_2 at 0°C, 2.2°C and 5°C. They found that fruit firmness was not affected by the CA storage compared with air storage, but decay was reduced in the low O_2 compared with air at all three temperatures. They concluded that the olives could be stored for a maximum of 4 weeks at 2.2–5°C in 2 kPa O_2. Ramin and Modares (2009) found that at 7°C increasing the level of CO_2 in storage over the range of 1.6–6.4 kPa resulted in improved retention of firmness and colour and lowest respiration rate, with 2 kPa O_2 + 6.4 kPa CO_2 giving the best results.

Onion (*Allium cepa* var. *cepa*)

CA stores with a total capacity of 9300 t have recently been constructed in Japan for onions (Yamashita *et al.*, 2009). Robinson *et al.* (1975) reported that low O_2 levels in storage reduced the respiration rate of bulbs at all temperatures studied (Table 12.2). CA storage conditions gave variable success and were not generally recommended, but 5 kPa CO_2 + 3 kPa O_2 was shown to reduce sprouting and root growth (Hardenburg *et al.*, 1990). SeaLand (1991) recommended 0 kPa CO_2 with 1–2 kPa O_2. Typical storage conditions were given as 0°C and 65–75% RH with 5 kPa CO_2 and 3 kPa O_2 by Bishop (1996). Monzini and Gorini (1974) recommended 4°C in 5–10 kPa CO_2 + 3–5 kPa O_2 for 6 months. Fellows (1988) recommended a maximum of 10 kPa CO_2 and a minimum of 1 kPa O_2. Kader (1985) recommended a temperature range of 0–5°C and 75% RH in 1–2 kPa O_2 + 0 kPa CO_2. In later work Kader (1989) recommended 0–5°C in 0–5 kPa CO_2 + 1–2 kPa O_2. Saltveit (1989) recommended 0–5°C in 0–5 kPa CO_2 and 0–1 kPa O_2, which was claimed to have had only a slight effect. He also found that CA storage gave better results when onions were stored early in the season just after curing.

Smittle (1988) found that more than 99% of the bulbs of the cultivar 'Granex' were marketable after 7 months of storage at 1°C in 5 kPa CO_2 + 3 kPa O_2, though the weight loss was 9%. Bulbs stored at 1°C in 5 kPa

Table 12.2. Effects of temperature and reduced O_2 level on the respiration rate (CO_2 production in mg/kg/h) of Bedfordshire Champion onions; adapted from Robinson *et al.* (1975).

	Temperature °C				
Atmosphere	0	5	10	15	20
Air	3	5	7	7	8
3 kPa O_2	2		4		4

CO_2 + 3 kPa O_2 kept well when removed from storage, while bulbs from 10 kPa CO_2 + 3 and 5 kPa O_2 became unmarketable at a rate of about 15% per week due to internal breakdown during the first month out of store. Bulb quality, as measured by low pungency and high sugar, decreased slowly when onions were stored at 27°C or at 1°C or 5°C at 70–85% RH in CA. Quality decreased rapidly when the cultivar 'Granex' was stored in air at 1°C or 5°C. Adamicki (1989) also described successful storage of onions at 1°C in 5 kPa CO_2 + 3 kPa O_2. Adamicki and Kepka (1974) found that there were no changes in the colour, flavour or chemical composition of onions after 2 months of storage even in 15 kPa CO_2. Chawan and Pflug (1968) examined storage of the cultivars 'Dowing Yellow Globe' and 'Abundance' in several combinations of 5 and 10 kPa CO_2 + 1, 3 and 5 kPa O_2 at 1.1°C, 4.4°C and 10°C. The best combination was 10 kPa CO_2 + 3 kPa O_2 at 4.4°C and the next best was 5 kPa CO_2 + 5 kPa O_2 at 4.4°C. Internal spoilage of bulbs was observed in 10 kPa CO_2 + 3 kPa O_2 at 1.1°C but none at 4.4°C. Adamicki and Kepka (1974) also observed a high level of internal decay in bulbs stored in 10 kPa CO_2 + 5 kPa O_2 at 1°C, but none at 5°C in the same atmospheric composition. The number of internally decayed bulbs increased with length of storage. Their optimum results for long-term storage were in 5 kPa CO_2 + 3 kPa O_2 at either 1°C or 5°C. Stoll (1974) stated that there were indications that the early trimmed lots did not store as well in CA conditions as in conventionally refrigerated rooms, and he reported better storage life at 0°C rather than at 2°C or 4°C when the same 8 kPa CO_2 + 1.5 kPa O_2 atmosphere was used. Sitton *et al.* (1997) also found that storage in high CO_2 caused injury but neck rot (*Botrytis* sp.) was reduced in CO_2 levels of greater than 8.9 kPa. They found that bulbs stored in 0.5–0.7 kPa O_2 were firmer, of better quality and had less neck rot than those stored in O_2 levels of < 0.7 kPa.

Waelti *et al.* (1992) converted a refrigerated highway trailer to CA storage and they found that the cultivar 'Walla Sweet' stored well for 84 days in 2 kPa O_2 but subsequently the shelf-life was only 1 week. Yamashita *et al.* (2009) tested 1°C and 80% RH in 1 kPa O_2 + 1 kPa CO_2 for storage up to 196 days and found that sprouting and root growth were inhibited in CA storage and 98.2% were considered marketable, compared with 69.2% for those stored at –0.5°C and 80% RH in air after 196 days. The onion cultivar 'Hysam' was stored in air or 2 kPa O_2 + 2 kPa CO_2 or 2 kPa O_2 + 8 kPa CO_2 at 0.5 ± 0.5°C and 80 ± 3% RH for 9 weeks. Both the CA levels resulted in reduced pungency and flavour. This was by reducing both flavour precursors and enzyme activity (Uddin and MacTavish, 2003).

Storage of calçots (floral stems of the second-year onion *Allium cepa* L.) for 60 days at 1°C in 1 kPa O_2 + 2 kPa CO_2 resulted in higher consumer preference and higher antioxidant activity (DPPH radical dot assay), while those stored in air had higher antioxidant activity (FRAP assay), total phenolic content and total flavonoids (Zudaire *et al.*, 2017).

Physiological disorders

CO_2 injury or internal spoilage of onion bulbs is a physiological disorder that can be induced by elevating the CO_2 concentration around the bulbs. This effect was aggravated by low temperature (< 5°C). Chawan and Pflug (1968) observed internal spoilage of bulbs in storage at 1.1°C in < 10 kPa CO_2 + 3 kPa O_2, but there was no spoilage in the same CA at 4.4°C. It was stated that 'internal spoilage of the bulbs was due to an adverse combined effect of temperature and gas concentration'. Also Adamicki and Kepka (1974) reported that Böttcher had stated that for the cultivars 'Ogata' and 'Inoue' there was very strong internal spoilage at concentrations of 10 kPa CO_2 and above. They also found that after storage at 1°C (but not at 5°C) for more than 220 days there were 23–56% of internally spoiled bulbs when CO_2 concentrations were 10% or higher. Adamicki *et al.* (1977) found that internal breakdown was observed in 68.6% of the bulbs when they were stored at 1°C in sealed PE bags with a CO_2 concentration higher than 10 kPa; also they found similar disorder in bulbs stored at 5°C in sealed PE bags, and this was probably due to the very high concentration of CO_2 (28.6 kPa). Adamicki and Kepka (1974) indicated that internal spoilage of onions was due to the combined effects of high CO_2 concentrations, low temperature and a relatively long period of storage. Smittle (1989) found that the cultivar 'Granex' stored at 5°C in 10 kPa CO_2 + 3 kPa O_2 for 6 months had internal breakdown of tissue due to CO_2 toxicity.

Sprouting

CA storage is being increasingly used for onions since it can replace MH, which is used for sprout suppression. Previously Isenberg (1979) had concluded that CA storage was an alternative to the use of sprout suppression in onions, but stressed the need for further testing of the optimum condition of O_2 and CO_2 required for individual varieties. Adamicki and Kepka (1974) also quoted work where onions stored in an atmosphere with 5–15 kPa CO_2 at room temperature showed a decrease in the percentage of bulbs that sprouted. They also reported that bulbs of cultivar 'Wolska' stored for 163 or 226 days in 5 kPa CO_2 + 3 kPa O_2 and then transferred to 20°C sprouted about 10 days later than those transferred from air storage. There is evidence that CA storage can have residual effects after the bulbs have been removed from store. Bulbs of cultivar 'Wolska' stored for 163 or 226 days in 5 kPa CO_2 + 3 kPa O_2 and then transferred to 20°C sprouted about 10 days later than those transferred from air storage (Adamicki and Kepka, 1974). Yamashita *et al.* (2009) reported that storage at 1°C and 80% RH under 1 kPa O_2 + 1 kPa CO_2 for up to 196 days inhibited sprouting and root growth. In 10 kPa CO_2 + 3 kPa O_2 at 4.4°C, Chawan and Pflug (1968) found that bulbs showed no sprouting after 34 weeks, while bulbs stored in air had 10% sprouting for 'Dowing Yellow Globe' and 15% for 'Abundance'. In a detailed study of 15 different cultivars and various CA combinations, Gadalla (1997) found that all the cultivars stored in 10 kPa CO_2 for 6 months or more had internal spoilage (but after 3 months) with more spoilage when 10 kPa CO_2 was combined with 3 kPa or 5 kPa O_2. All CA combinations, 1, 3 or 5 kPa O_2 + 0 or 5 kPa CO_2, reduced sprouting but those combinations that included 5 kPa CO_2 were the most effective. Also the residual effects of CA storage were still effective after 2 weeks at 20°C. No external root growth was detected when bulbs were stored in 1 kPa O_2 compared with 100% for the bulbs in air. There was also a general trend to increased rooting with increased O_2 over the range of 1–5 kPa. Bulbs stored in 5 kPa CO_2 had less rooting than those stored in 0 kPa CO_2. Both these latter effects varied between cultivars. Most of the cultivars stored in 1 kPa O_2 + 5 kPa CO_2 and some cultivars stored in 3 kPa O_2 + 5 kPa CO_2 were considered marketable after 9 months of storage. He also found that CA storage was most effective when applied directly after curing, but a delay of 1 month was almost as good. It was therefore concluded that, with a suitable cultivar and early application of CA, it is technically possible to store onions at 0°C for 9 months without chemical sprout suppressants. Sitton *et al.* (1997) found bulbs stored in 0.5–0.7 kPa O_2 in cold storage sprouted more quickly when removed to 20°C, compared with those stored in O_2 levels of > 0.7 kPa.

Orange (*Citrus sinensis*)

CA storage was not considered beneficial but typical storage conditions were given as 5–10°C in 0–5 kPa CO_2 + 5–10 kPa O_2 by Bishop (1996). Storage at 5–10°C in 5 kPa CO_2 + 10 kPa O_2 had a fair effect on storage but was not being used commercially (Kader, 1985). Kader (1989, 1992) recommended 5–10°C in 0–5 kPa CO_2 + 5–10 kPa O_2. Recommended storage of the cultivar 'Valencia' grown in Florida was 1°C in 0 or 5 kPa CO_2 + 15 kPa O_2 for 12 weeks followed by 1 week in air at 21°C. In these conditions the oranges retained better flavour and had less skin pitting than those stored in air, but CO_2 levels of 2.5–5 kPa, especially when combined with 5 or 10 kPa O_2, adversely affected flavour retention (Hardenburg *et al.*, 1990). SeaLand (1991) recommended 5 kPa

Table 12.3. The effects of CA storage on the quality of oranges; modified from Chase (1969).

O_2 kPa	CO_2 kPa	Rotting (%)[a]	Rind breakdown (%)[a]	Flavour score[b]
Air		97	0	66
15	0	58	32	97
15	2½	28	13	95
15	5	62	0	93
10	0	67	49	74
10	2½	60	3	69
10	5	81	3	52
5	0	47	9	38
5	2½	36	2	20
5	5	79	0	5

[a]After 20 weeks storage at 1°C plus 1 week at 21°C.
[b]After 12 weeks storage at 1°C plus 1 week at 21°C.

CO_2 + 10 kPa O_2. Smoot (1969) showed that for 'Valencia' those stored in the combinations of 0 kPa CO_2 + 15 kPa O_2 and 0 kPa CO_2 + 10 kPa O_2 had less decay than those stored in air. However, ASHRAE (1968) showed that CA storage could have deleterious effects on fruit quality, particularly through increased rind injury and decay or on fruit flavour. Chase (1969) also showed that CA storage could affect fruit flavour, rind breakdown and rotting (Table 12.3).

Pak Choi (*Brassica campestris* L. subsp. *chinensis* var. *communis*)

Harbaum-Piayda *et al.* (2010) stored pak choi at 2°C and 99% RH in either air or 1.5–2.5 kPa O_2 + 5–6 kPa CO_2. The level of flavonoids increased more in CA than in air, but hydroxycinnamic acids were unaffected.

Papaya (*Carica papaya*)

At 18°C, storage in 10 kPa CO_2 reduced decay (Hardenburg *et al.*, 1990). Storage recommendations include: 10–15°C in 5–8 kPa CO_2 + 2–5 kPa O_2 (Bishop, 1996); 10 kPa CO_2 + 5 kPa O_2 for surface transport (SeaLand, 1991); 10–15°C in 10 kPa CO_2 + 5 kPa O_2 had a fair effect but was said to be 'not being used commercially' (Kader, 1985); 10–15°C in 5–10 kPa CO_2 + 3–5 kPa O_2 (Kader, 1989, 1992); and 12°C or 10–15°C in 5–8 kPa CO_2 + 2–5 kPa O_2 (Kader, 1993). Sankat and Maharaj (1989) found that fruit of the cultivar 'Known You Number 1' stored at 16°C were still in an acceptable condition after 17 days of storage in 5 kPa CO_2 + 1.5–2 kPa O_2 compared with only 13 days in air. The cultivar 'Tainung Number 1' stored in the same conditions as 'Known You Number 1' had a maximum life of 29 days in CA storage and was considered unacceptable after 17 days stored in air. In Hawaii, storage in 2 kPa O_2 + 98 kPa N_2 was shown to extend the storage life of fruit at 10°C compared with storage in air at the same temperature (Akamine and Goo, 1969). Akamine (1969) showed that storage of 'Solo' in 10 kPa CO_2 for 6 days at 18°C reduced decay compared with fruit stored in air, but fruit stored in air after storage in 10 kPa CO_2 decayed rapidly. Akamine and Goo (1969) showed that storage at 13°C in 1 kPa O_2 for 6 days or 1.5 kPa O_2 for 12 days ripened about 1 day later than those stored in air. Chen and Paull (1986) also showed that storage in 1.5–5 kPa O_2 delayed ripening of 'Kapaho Solo' compared with those stored in air and that there were no further benefits on storage by increasing the CO_2 level to either 2 kPa or 10 kPa. Hatton and Reeder (1968) showed that after storage in 13°C for 3 weeks in 5 kPa CO_2 + 1 kPa O_2 followed by ripening at 21°C, fruit were in a fair condition with little or no decay and were of good flavour. Cenci *et al.* (1997) showed that the cultivar 'Solo' stored at 10°C in 8 kPa CO_2 + 3 kPa O_2 could be kept for 36 days and still have an adequate shelf-life of 5 days at 25°C. Fonseca *et al.* (2006) compared storage in 3 kPa O_2 + 3 or 6 kPa CO_2 with storage in air and found that the postharvest diseases of anthracnose, chocolate spot, stem-end rot and black spot caused more losses on both 'Sunrise Solo' and 'Golden' stored in air than in a CA with 6 kPa CO_2. However, they also found that the advantages of the CA were not effective in fruit harvested in the Brazilian summer. A cultivar was developed at the Malaysian Agriculture Research and Development Institute called 'Paiola'. It was shown to have a shelf-life of an extra 2 weeks compared with 'Eksotika'. After 30 days in CA the fruit was shown to retain its quality and could be transported by sea-freight over long distances (Kwok, 2010). Also in Malaysia, Broughton *et al.* (1977) showed that scrubbing ethylene from the store containing the cultivar 'Solo Sunrise' had no effects on storage life but they recommended that 20°C with 5 kPa CO_2 with ethylene removal was the optimum storage condition for 7–14 days. They also found that chilling injury could occur at 15°C. In later work it was shown that ethylene application can accelerate ripening by 25–50% (Wills, 1990).

Techavuthiporn *et al.* (2009) found that shredded unripe papaya stored at 4°C in 5 kPa O_2 + 10 kPa CO_2 retained their quality for 5 days. Shredded unripe papaya is the major ingredient of 'Som-Tam' in Thailand, which is complicated to prepare and deteriorates quickly.

Storage at 20°C in 0 kPa CO_2 + < 0.4 kPa O_2 as a method of fruit fly control in papaya resulted in increased incidence of decay, abnormal ripening and the development of off-flavour after 5 days of exposure (Yahia *et al.*, 1989).

Passionfruit (*Passiflora edulis*)

Yonemoto *et al.* (2004) stored hybrid passionfruit (*P. edulis* forma *edulis* × *P. edulis* forma *flavicarpa*) in 20.9, 23.0 or 29.0 kPa O_2 at either 20°C or 15°C. The acidity of fruit kept at 20°C decreased from 2.9% to 2.4%, 2.1% and 2.0% after 10 days of storage at 20.9, 23.0 and 29.0 kPa O_2, respectively. There were no differences in acidity of fruit kept at 15°C after 5 days of storage, but acidity was significantly lower in fruit in 23.0 or 29.0 kPa O_2 after 10 days compared with 20.9 kPa.

Peach (*Prunus persica*)

CA storage recommendations include: 5 kPa CO_2 + 2 kPa O_2 (SeaLand, 1991); 0–5°C in 3–5 kPa CO_2 + 1–2 kPa O_2 (Bishop, 1996); 0–5°C in 5 kPa CO_2 + 1–2 kPa O_2, which had a good effect but was claimed to be of limited commercial use (Kader, 1985); 0–5°C in 3–5 kPa CO_2 + 1–2 kPa O_2 for both freestone and clingstone peaches (Kader, 1989, 1992); 5 kPa CO_2 + 1 kPa O_2, which maintained quality and retarded internal breakdown for about twice as long as those stored in air (Hardenburg *et al.*, 1990); and –0.5°C in 1.5 kPa CO_2 + 1.5 kPa O_2 for up to 4 weeks for canning (Van der Merwe, 1996). The storage life at –0.5°C in air was about 7 days. It was found that conditioning the fruits at 20°C in 5 kPa CO_2 + 21 kPa O_2 for 2 days prior to storage until they reach a firmness of 5.5 kg (as measured with a Magness and Taylor penetrometer) was necessary. Wade (1981) showed that storage of the cultivar 'J.H. Hale' at 1°C resulted in chilling injury symptoms (flesh discoloration and soft texture), but storage at the same temperature in atmospheres containing 20 kPa CO_2 resulted in only moderate levels even after 42 days. Bogdan *et al.* (1978) showed that the cultivars 'Elbarta' and 'Flacara' could be stored at 0°C for 3–4 weeks in air or about 6 weeks in 5 kPa CO_2 + 3 kPa O_2. Truter *et al.* (1994) showed that the cultivars 'Oom Sarel', 'Prof. Neethling' and 'Kakamas' could be successfully stored at –0.5°C in either 1.5 kPa CO_2 + 1.5 kPa O_2 or 5 kPa CO_2 + 2 kPa O_2 for 6 weeks with a mass loss of only 1%. This compared with mass losses of 20.7% and 12.6% for 'Oom Sarel' or 'Prof. Neethling', respectively, stored in air at the same temperature. CA storage, however, did not affect decay incidence but the fruit was of an acceptable quality even after 6 weeks of storage and was also considered to be still highly suitable for canning. Brecht *et al.* (1982) showed that there was a varietal CA storage interaction. They stored five cultivars at –1.1°C in 5 kPa CO_2 + 2 kPa O_2 and found that 'Loadel' and 'Carolyn' could be successfully stored for up to 4 weeks while 'Andross', 'Halford' and 'Klamt' could be stored for only comparatively short periods. Lurie *et al.* (2002) suggested that a certain level of ethylene production is essential for normal ripening after cold storage and linked this with prolonged storage where the fruit did not ripen and developed a dry, woolly texture.

CA storage plus intermittent warming to 18°C in air for 2 days every 3–4 weeks has shown promising results (Hardenburg *et al.*, 1990). Anderson (1982) stored the cultivar 'Rio Oso Gem' at 0°C in 5 kPa CO_2 + 1 kPa O_2 for up to 20 weeks with intermittent warming to 18–20°C every 2 days, which almost entirely eliminated the internal browning associated with storage in air at O_2.

Pear (*Pyrus communis*)

CA storage recommendations include: a maximum of 5 kPa CO_2 and a minimum of 2 kPa O_2 for the cultivar 'Bartlett' (Fellows, 1988); 0.5°C in 0.5–1.0 kPa CO_2 + 2–2.2 kPa O_2 (Koelet, 1992); 0–1 kPa CO_2 + 2–3 kPa O_2 (SeaLand, 1991); 0°C and 93% RH in 0.5 kPa CO_2 + 1.5 kPa O_2 (Lawton, 1996); 0–5°C in 0–1 kPa CO_2 + 2–3 kPa O_2 (Kader, 1985); 0–5°C in 0–3 kPa CO_2 + 1–3 kPa O_2 (Kader, 1992); -1 to -0.5 °C in <1 kPa CO_2 + 2 kPa O_2 for both (Conference) and (Concorde) (Johnson, 1994), −1°C to −0.5°C in < 1 kPa CO_2 + 2 kPa O_2 for 'Conference' (Sharples and Stow, 1986); 0.8–1 kPa CO_2 + 2–2.5 kPa O_2 or 0.1 kPa CO_2 or less + 1–1.5 kPa O_2 for 'Anjou' (Hardenburg *et al.*, 1990); 0.5–1°C in 5 kPa CO_2 + 5 kPa O_2 for 'Conference' and 'Doyenne du Comice' and 0.5–1°C in 6 kPa CO_2 + about 15 kPa O_2 for 'William's Bon Chretien' produced in England (Fidler and Mann, 1972); and 0°C in 2 kPa CO_2 + 2 kPa O_2 for 'Conference', 'Doyenne du Comice' and 'William's Bon Chretien' produced in Switzerland (Stoll, 1972). Exposure of 'Anjou' to 12 kPa CO_2 for 2 weeks immediately after harvest had beneficial effects on the retention of ripening capacity (Hardenburg *et al.*, 1990).

Table 12.4. Recommended CA storage conditions for selected pear cultivars by country and region (Herregods, 1993).

	°C	kPa CO_2	kPa O_2
Cultivar Comice			
Belgium	0 to −0.5	< 0.8	2 to 2.2
France	0	5	5
Germany (Westphalia)	0	3	2
Italy	−1 to 0.5	3–5	3–4
Spain	−0.5	3	3
Switzerland	0	5	3
USA (Oregon)	−1	0.1	0.5
Cultivar Conference			
Belgium	−0.5	< 0.8	2–2.2
Denmark	−0.5 to 0	0.5	2–3
England	−1 to 5	< 1	2
Germany (Saxony)	1	1.1–1.3	1.3–1.5
Germany (Westphalia)	−1 to 0	1.5 2	1.5
Holland	−1 to −0.5	Much < 1	3
Italy	−1 to −0.5	2–4	2–3
Italy	−1 to −1.5	1–1.5	6–7
Slovenia	0	3	3
Spain	−1	2	3.5
Switzerland	0	2	2
Cultivar Packham's Triumph			
Australia (South)	−1	3	2
Australia (Victoria)	0	0.5	1
Australia (Victoria)	0	4.5	2.5
Germany (Westphalia)	−1 to 0	3	2
New Zealand	−0.5	2	2
Slovenia	0	3	3
Switzerland	0	2	2

Richardson and Meheriuk (1989) reported that the cultivars 'Alexander Lucas' and 'Gellerts Butterbirne' were very sensitive to CO_2 and advised a maximum storage period of only 6 months and 4–5 months, respectively. Herregods (1993, personal communication) and Meheriuk (1993) compared recommendations for CA storage for selected cultivars from different countries (Table 12.4). Van der Merwe (1996) recommended specific conditions for storage of fruit produced in South Africa (Table 12.5). Recommended storage conditions for various pear cultivars were given by Richardson and Meheriuk (1989) (Table 12.6). Kupferman (2001b) also summarized storage conditions for pears (Table 12.7).

Table 12.5. Recommended CA storage conditions for pears at –0.5°C (Van der Merwe, 1996).

Cultivar	kPa O_2	kPa CO_2	Duration
Bon Chretien	1	0	4 months
Buerré Bosc	1.5	1.5	4 months
Forelle	1.5	0–1.5	7 months
Packham's Triumph	1.5	1.5	9 months

Chen and Varga (1997) showed that storage in 0.5 kPa O_2 + < 0.1 kPa CO_2 resulted in a high incidence of scald and black speck. After 6½ months of storage at –0.5°C the average percentage of 'Beurre d'Anjou' with symptoms of scald ranged from zero in fruit stored in 1.0 kPa O_2 + 0 kPa CO_2 or 1.5 kPa O_2 + 0.5 kPa CO_2 to 2% in fruit stored in 2.5 kPa O_2 + 0.8 kPa CO_2 and 100% in fruit stored in air. For the cultivar 'Bartlett', Zlatić *et al.* (2016) showed that storage under 0.8 kPa O_2 + < 0.5 kPa CO_2 at –1°C or +1°C maintained firmness and suppressed synthesis of aroma volatiles and synthesis of esters, including two character-impacting compounds: methyl and ethyl (2*E*,4*Z*)-deca-2,4-dienoate. Synthesis of hexyl acetate was suppressed under ULO storage regardless of temperature, while ethyl acetate synthesis was suppressed only by ULO at –1°C. The levels of most aroma volatiles were recovered after the following 10 days of shelf-life, though with significantly lower recovery for methyl and ethyl (2*E*,4*Z*)-deca-2,4-dienoate in fruit under ULO storage. Although synthesis of aroma volatiles was most suppressed under ULO at –1°C, butyl and hexyl

Table 12.6. Recommended CA storage conditions for pears (Richardson and Meheriuk, 1989).

Cultivar	Temperature °C	kPa O_2	kPa CO_2
Abate Fetel	–1 to 0	3–4	4–5
Alexander Lucas	–1 to 0	3	1
Anjou	–1 to 0	1–2½	0.03–2
Bartlett	–1 to 0	1–3	0–3
Buerré Bosc	–1 to 0	1–3	0.03–4
Buerré Hardy	–1 to 0	2–3	0–5
Clapps Favorite	0	2	0–1
Conference	–1 to 0.5	1½–3	0–3
Decana Comizio	0	3	4–5
Decana Inverno	0	3	5
Diels Butterbirne	–1 to 0	2	2
Doyenne du Comice	–1 to 1	2–3	0.8–5
Gellerts Butterbirne	–1 to 0	2	1
Kaiser Alexander	0 to 1	3	3–4
Kosui	–0.5 to 5	1–2	–
Nijiseiki	–1 to 1	3	≤ 1
Packham's Triumph	–1 to 1	1–3	0–5
Passe Crassane	–1 to 1	1.5–3	2–5
Tsu Li	–0.5 to 1	1–2	–
20th Century	–1 to 1	3	≤ 1
Williams Bon Chretien	–1 to 0	1–3	0–3
Ya Li	–0.5 to 0.5	4–5	Up to 5

Table 12.7. Optimum CA conditions for selected pear cultivars; modified from Kupferman (2001b).

Cultivar	Country	°C	kPa O_2	kPa CO_2	Storage life in months
Anjou	Washington State, USA	–0.5 to 0	1.5	0.3	9
Beurre Bosc	South Africa	–0.5	1.5	1.5	4
Conference	Netherlands	–1	2.5	0.7	7.5
Doyenne du Comice	Netherlands	–0.5	2.5	0.7	5
Doyenne du Comice	South Africa	–0.5	1.5	1.5	6
Doyenne du Comice	New Zealand	–0.5	2	< 1	3
Forelle	South Africa	–0.5	1.5	0	7
Josephine	South Africa	–0.5	1.5	1	8
Packham's Triumph	New Zealand	–0.5	2	< 1	5
Packham's Triumph	South Africa	–0.5	1.5	2.5	9
Rosemarie	South Africa	–0.5	1.5	1	5
Williams Bon Chretien	South Africa	0 to –0.5	1	0	4
Williams Bon Chretien	South Africa	0 to –0.5	1	0	4
Williams Bon Chretien	Washington State, USA	–0.5 to 1	1.5	0.5	4

acetate levels recovered better in fruit under ULO storage at −1°C than at +1°C. Acetaldehyde and nonanal were the principal aldehydes present, with levels that were higher in early-harvested fruit and in fruit stored at +1°C. The cultivars 'Swiss Bartlett' and 'Cold Snap', with or without 1-MCP treatment, were stored at 0°C in air or CA (18 kPa O_2 + 2 kPa CO_2 or 2.5 kPa O_2 + 2 kPa CO_2) for at least 167 days. There was high incidence of senescent scald and internal breakdown in pears that had not been treated with 1-MCP during storage in air, but symptoms were reduced by both CA and 1-MCP, resulting in minimal to negligible incidence in 1-MCP + 2.5 kPa O_2. 1-MCP and CA reduced the rates of ethylene production and delayed softening and yellowing in 'Swiss Bartlett', but had negligible to slight effects with 'Cold Snap'. In both cultivars, γ-aminobutyrate accumulated, though 1-MCP and CA slightly reduced the levels in 'Cold Snap' and 1-MCP increased levels in 'Swiss Bartlett' fruit. Ascorbate levels were rapidly depleted in 'Cold Snap', regardless of treatment; these levels were better maintained by 1-MCP in 'Swiss Bartlett' in all storage atmospheres. In both cultivars, glutathione concentrations and redox status fluctuated during storage, though these levels were generally higher in 1-MCP-treated fruit (Lum *et al.*, 2017).

Pea, Garden Pea, Mangetout, Snow Pea, Sugar Pea (*Pisum sativum*)

Pantastico (1975) reported that in storage at 0°C peas in their pods could be kept in good condition for 7–10 days, while in 5–7 kPa CO_2 + 5–10 kPa O_2 they could be kept for 20 days. At 0°C their quality was better after 20 days in 5–7 kPa CO_2 than in air (Tomkins, 1957). Storage in 5–7 kPa CO_2 at 0°C maintained eating quality for 20 days (Hardenburg *et al.*, 1990). Fellows (1988) recommended a maximum of 7 kPa CO_2 and a minimum of 5 kPa O_2. However, Suslow and Cantwell (1998) reported that there was no benefit in CA storage compared with storage in air at the same temperature. Monzini and Gorini (1974) recommended 0°C in 5 kPa CO_2 + 21 kPa O_2 for 20 days. Saltveit (1989) recommended 0–10°C in 2–3 kPa CO_2 + 2–3 kPa O_2 for sugar peas, which had only a slight effect. Pariasca *et al.* (2001) showed that storage at 5°C in 5–10 kPa O_2 + 5 kPa CO_2 were the best conditions among those that they tested, since changes in organic acid, free amino acid and sugar contents and pod sensory attributes were slight. The appearance of pods stored in CA was much better than for those stored in air. They also found that 2.5 kPa O_2 + 5 kPa CO_2 and 5 kPa O_2 + 10 kPa CO_2 had a detrimental effect on quality, since the peas developed slight off-flavours, but this effect was partially alleviated after ventilation.

Pepino (*Solanum muricatum*)

Storing mature fruit at 5°C in 5 kPa O_2 + 15 kPa CO_2 resulted in retention of colour and firmness compared with fruit stored in air (Prono-Widayat *et al.*, 2003). They also recommended a storage life of 3 weeks for ripe fruit stored in 5 kPa O_2 + 20 kPa CO_2, which retained their quality.

Persimmon (*Diospyros kaki*)

Also called sharon fruit and kaki. General storage recommendations include: 0–5°C in 5–8 kPa CO_2 and 3–5 kPa O_2 but reported to have only a fair effect on storage and not commercially useful (Kader, 1985, 1992); 5–8 kPa CO_2 + 3–5 kPa O_2 (SeaLand, 1991); and 0–5°C in 5–8 kPa CO_2 + 3–5 kPa O_2 (Bishop, 1996). Liamnimitr *et al.* (2018) determined the lower O_2 limit of aerobic respiration as 0.65 kPa, based on the respiratory quotient breakpoint. Next, the parameters of a Michaelis–Menten type respiration model were estimated, and they were used to predict the O_2 consumption rate at 1.5 kPa, which was set as the target O_2 partial pressure inside the package and is considered safe to avoid fermentation. The results revealed that the O_2 concentration at a steady state in the designed bulk MAP could be established at 1.3–1.6 kPa, which successfully corresponded with the target value. Furthermore, losses in flesh firmness and colour, and the external damage of fruits were reduced in the bulk MAP when compared to storage in the other two package treatments. Overall, the bulk MAP designed could be used for ca. 4 months storage of 'Fuyu' persimmon fruit. The incidence of skin browning was reduced by storing fruit in 2 kPa O_2 + 5 kPa CO_2 with an ethylene absorber, compared with storage in air (Lee *et al.*, 1993). After 2 months of storage of 'Kyoto' at –0.5°C followed by 5 days at 20°C, optimum flesh firmness was maintained in 0.5 kPa O_2 + 5 kPa CO_2. The highest rot incidence was observed in fruit stored at 2 kPa O_2 + 10 kPa CO_2. Levels of CO_2 of 10–15 kPa increased skin blackening (Brackmann *et al.*, 2004). Storage of 'Rojo Brillante' in 97 kPa N_2 + air for 30 days at 15°C, retained 'commercial' firmness and removed astringency (Arnal *et al.*, 2008). Fruit from trees of the cultivar 'Triumph' that were sprayed with GA at 50 mg/l 2 weeks prior to harvest was 10 times less sensitive to ethylene than fruit from non-treated trees when subsequently stored at 10°C in air or 1.5–2.0 kPa CO_2 + 3.0–3.5 kPa O_2 (Ben Arie *et al.*, 1989). CA recommendations for the cultivar 'Fuyu' include: 1°C and 90–100% RH in 8 kPa CO_2 + 3–5 kPa O_2 (Hulme, 1971); 0°C in 5–8 kPa CO_2 + 2–3 kPa O_2 for 5–6 months (Kitagawa and Glucina, 1984); and 0–2°C in 9 kPa CO_2 + 2 kPa O_2 for 5 months (Chung and Son, 1994). Neuwald *et al.* (2008) found that the best CA condition for 'Fuyu' was 0.5–1 kPa O_2 + 5–15 kPa CO_2 at –0.5°C for 3 months with a shelf-life at 20°C of 3 days. Fruit stored in 0.5–1 kPa O_2 showed less rot than fruit stored in > 10 kPa O_2 + 15 kPa CO_2 or in air. Also skin browning was reduced when 0.5–1 kPa O_2 was combined with CO_2 up to 10 kPa, but 0.5 kPa O_2 + 15 kPa CO_2 increased skin browning. Martins *et al.* (2004) studied the effect of delaying implementing CA for up to 3 days on subsequent storage of 'Fuyu'. Storage conditions were 0 ± 0.5°C and 95 ± 3% RH throughout, and CA was 8 kPa CO_2 + 2 kPa O_2. They concluded that conservation and quality of the fruit were not significantly affected by the different periods of refrigerated storage in air prior to storage in CA. Lee *et al.* (2003b) identified four different types of browning disorders classified by symptoms and the surface zone affected: (i) typical top flesh browning called 'chocolate symptom' on the top; (ii) pitted specks scattered on the surface; (iii) flesh blotch browning usually on the equatorial to bottom portion; and (iv) pitted blotch browning on the equatorial zone. They suggested that top flesh browning was primarily induced by low O_2 (not by high CO_2) and that the critical level of O_2 was in a very narrow range. For external development, however, highly elevated CO_2 seemed to be a complementary factor. Speculative primary inducing factors for other disorders were: low O_2 for flesh blotch browning, and high CO_2 for pitted specks and pitted blotch browning.

Pineapple (*Ananas comosus*)

Typical storage conditions were given as 8–13°C in 5–10 kPa CO_2 + 2–5 kPa O_2 by Bishop (1996) but CA storage was not considered beneficial. Dull *et al.* (1967) also considered that CA storage had no obvious effect on fruit quality or the maintenance of fruit appearance. However, Akamine and Goo (1971) found that the storage life of the cultivar 'Smooth Cayenne' was significantly extended in 2 kPa O_2 at 7.2°C and Yahia (1998) tentatively recommended 2–5% O_2 + 5–10 kPa CO_2. Kader (1989) recommended 5 kPa O_2 + 10 kPa CO_2 at 10–15°C. Storage at 10–15°C in 10 kPa CO_2 + 5 kPa O_2 had a fair effect but was not being used commercially (Kader, 1985). Kader (1989, 1992) recommended 8–13°C in 5–10 kPa CO_2 + 2–5 kPa O_2. Kader (1993) also recommended 10°C, with a range of 8–13°C, in 5–10 kPa CO_2 + 2–5 kPa O_2. SeaLand (1991) recommended 10 kPa CO_2 + 5 kPa O_2. Paull and Rohrbach (1985) found that storage at 3 kPa O_2 + 5 kPa CO_2 or 3 kPa O_2 + 0 kPa CO_2 did not suppress internal browning (sometimes referred to as black heart) in the cultivar 'Smooth Cayenne' stored at 8°C. If fruit was exposed to 3 kPa O_2 in the first week of storage at 22°C followed by 8°C, it was shown that the symptom could be effectively reduced. Dull *et al.* (1967) described an experiment where fruit was harvested at Stage 4 maturity (< 12% shell colour) and stored in atmospheres containing 21, 10, 5 or 2.5 kPa O_2 with the balance N_2. As the O_2 concentration decreased, so did the respiration rate. Where 5 or 10 kPa CO_2 was added to the atmosphere, the only noticeable effect was a slight further reduction

Table 12.8. Effects of CA storage on the frequency and level of internal browning score (where 0 = none and 5 = maximum) on 'Smooth Cayenne' pineapples stored at 8°C for 3 weeks and then 5 days at 20°C (Haruenkit and Thompson, 1996).

Composition (kPa)			Internal browning score
O_2	CO_2	N_2	
1.3	0	98.7	0.8
2.2	0	97.8	0.8
5.4	0	94.6	3.8
1.4	11.2	87.4	0.3
2.3	11.2	86.5	0.3
20.8	0	79.2	4.1
LSD (p = 0.05)			1.4

in respiration rate. For fruit harvested at full maturity and stored at 8°C in 2 kPa O_2 + 0, 2 or 10 kPa CO_2 or in air, all the fruits were still in good condition after 3 weeks. However, in subsequent storage at 22°C, to simulate marketing, some fruits in all the treatments developed internal translucency within 3 days and fruits stored in air and 2 kPa O_2 + 0 kPa CO_2 developed internal browning (black heart) (Haruenkit and Thompson, 1993). CA storage was shown to reduce but not eliminate internal browning (Table 12.8). For fresh-cut pineapple, Chonhenchob *et al.* (2007) found that an atmosphere of 6 kPa O_2 and 14 kPa CO_2 increased shelf-life by up to 7 days at 10°C.

Plantain (*Musa*)

SeaLand (1991) recommended 2–5 kPa CO_2 + 2–5 kPa O_2. Agoreyo *et al.* (2007) found that CA storage increases the postharvest life when compared with fruit stored in air and that ethylene production in CA was not detected. Maintaining high humidity around the fruit can help to keep it in the pre-climacteric stage so that where fruits were stored in moist coir dust or individual fingers were sealed in PE film they could remain green and pre-climacteric for over 20 days in Jamaican ambient conditions of about 28°C (Thompson *et al.*, 1972b). Such fruit ripened quickly and normally when removed from the plastic film. Since packing in moist coir proved as effective as MA packaging, it could be that the effect of MA on plantains is due entirely or at least partially to humidity rather than changes in O_2 and CO_2 inside the bags (Thompson *et al.*, 1974d).

Plum (*Prunus domestica*)

CA recommendations include: 0–5 kPa CO_2 + 2 kPa O_2 (SeaLand, 1991); 0–5°C in 0–5 kPa CO_2 + 1–2 kPa O_2 (Bishop, 1996); 0–5°C in 0–5 kPa CO_2 + 1–2 kPa O_2 reported to have a good effect but was not being used commercially (Kader, 1985); 0 kPa CO_2 + 1 kPa O_2 for 12 weeks for the cultivar 'Soldam' without any quality changes (Chung and Son, 1994); and 1°C in 2 kPa O_2 + 5 kPa CO_2 for up to 3 weeks for 'Opal' and 'Victoria', picked when not fully ripe (Roelofs and Breugem, 1994). Van der Merwe (1996) recommended –0.5°C in 5 kPa CO_2 + 3 kPa O_2 for fruit ripened to a firmness of 5.5 kg before storage for up to 7–8 weeks, depending on cultivar. This recommendation was shown to vary between cultivars and for 'Santa Rosa' and 'Songold' it was recommended that they should be partially ripened to a firmness of approximately 4.5 kg, storage at 0.5°C in 4 kPa O_2 + 5 kPa CO_2 for 7 or 14 days, which was sufficient to keep fruit in an excellent condition for an additional 4 weeks in air at 7.5°C. Internal breakdown was almost eliminated by this treatment and the fruit was of excellent eating quality after ripening (Truter and Combrink, 1992). Fruit of the cultivar 'Angeleno' was kept in air, 0.25% or 0.02 kPa O_2 at 0°C, 5°C or 10°C. Exposures to the low O_2 atmospheres inhibited ripening, including reduction in ethylene production rate, retardation of skin colour changes and flesh softening, but maintained TA and increased resistance to CO_2 diffusion. The most important detrimental effect of the low O_2 treatments was the development of an alcoholic off-flavour (Ke *et al.*, 1991a). Naichenko and Romanshchak (1984) recommended storage at –2°C for the cultivar 'Vengerka Obyknovennaya'. At that temperature fruits were kept in good condition for only 19 days in air but they could be stored for up to 125 days in 5 kPa CO_2 + 3 kPa O_2 + 92 kPa N_2. 'Victoria' plums at different stages of ripeness were stored in 3 or 21 kPa O_2 and 0, 4 or 7 kPa CO_2 at 0.3°C or 0.5°C for up to 4 weeks. Some fruits were harvested unripe, others when not yet fully ripe but of good colour. In 3 kPa O_2, fruit colour development was delayed, fruits remained firm and after 4 weeks they tasted better than those stored in 21 kPa O_2. In high CO_2, fruit colour development was also delayed and fruit flavour was not affected. Average wastage after 2, 3 and 4 weeks of storage was 9%, 20% and 32%, respectively (Roelofs, 1993a). With the cultivar 'Opal', the best storage conditions were 0°C in 1 or 2 kPa O_2 + 5 kPa CO_2 and with the cultivar 'Monsieur Hatif' they were 0°C and 1 or 2 kPa O_2 + 0 kPa CO_2. Both cultivars could be stored for up to 3 weeks without excessive spoilage due to rots and fruit cracking (De Wild and Roelofs, 1992). Folchi *et al.* (1994) stored the cultivar 'President' in < 0.1 kPa CO_2 + 0.3 kPa O_2 but observed an increase in aldehyde and ethanol content with increased storage time. Truter and Combrink (1997) stored the cultivars 'Laetitia', 'Casselman' and 'Songold' at 0°C for 8 weeks in various CA storage conditions followed by 7 days in air at 10°C. The best treatments of those investigated were 5 kPa CO_2 + 3 kPa O_2 but, for 'Songold', for a maximum of 7 weeks. Storage in 1 kPa O_2 at 0.5°C with intermittent warming was shown to have beneficial effects (Hardenburg *et al.*, 1990). In storage at 0.5°C physiological injury to the cultivar 'Monarch' was some 25% less in 2.5 kPa CO_2 + 5 kPa O_2 than in air (ASHRAE, 1968). Tonini *et al.* (1993) showed that there was a reduction in the incidence of internal breakdown in the cultivar 'Stanley' in atmospheres containing 20 kPa CO_2 compared with those stored in air, both at 0°C. Higher concentrations proved phytotoxic. For the cultivar 'Angelo', phytotoxicity was observed during storage at CO_2 concentrations higher than 2.5%. At 1°C in 12 kPa CO_2 + 2 kPa O_2 the cultivar 'Buhler Fruhzwetsche' had a good appearance, taste and firmness after 4 weeks of storage (Streif, 1989). They detected no injury in 'Buhler Fruhzwetsche during storage at CO_2 concentrations below 16 kPa.

Tahir and Olsson (2009) found that CA storage suppressed pathogenic decay but there was a difference in the response of different cultivars to CA storage. They recommended 0.5°C and 90% RH in 1 kPa O_2 + 2 kPa CO_2 for 'Opal', 'Vallor' and 'Victoria' and 1 kPa O_2 + 1 kPa CO_2 for 'Jubileum' and 'Vision'.

Good control of plum fruit moth *Cydia funebrana* and the fungus *Monilinia* spp. in fruit harvested in dry weather and at an optimum maturity stage could be achieved by storage for 1 month in air at 0°C and for 6 weeks in 2 kPa O_2 + 2.5 kPa CO_2. Atmospheres with 1–1.5 kPa O_2 could limit browning of the flesh, maintain a crisp and juicy texture and slow down rot development (Westercamp, 1995).

Pomegranate (*Punica granatum*)

Fruit of the cultivar 'Hicaz' was stored at 6°C, 8°C or 10°C and 85–90% RH in 1 kPa CO_2 + 3 kPa O_2, 3 kPa CO_2 + 3 kPa O_2, 6 kPa CO_2 + 3 kPa O_2 or in air. Those stored in air at 6°C were acceptable after 5 months of storage but those stored in 6 kPa CO_2 + 3 kPa O_2 retained good quality for 6 months. Fruit stored in air at 8°C or 10°C had a storage life of 50 days, whereas fruit stored at these temperatures in CA had a storage life of 130 days (Kupper *et al.*, 1994). CA storage has been shown to affect the development of a postharvest superficial browning disorder called 'husk scald'. The most effective control in the cultivar 'Wonderful' was by storing late-harvested fruit in 2 kPa O_2 at 2°C. However, this treatment resulted in accumulation of ethanol, which resulted in off-flavours, but ethanol and off-flavour dissipated when the fruit was transferred to air at 20°C (Ben Arie and Or, 1986). Kader *et al.* (1984) reported that pomegranate fruit did not respond to pre-storage ethylene treatment.

Potato (*Solanum tuberosum*)

The amount of O_2 and CO_2 in the atmosphere of a potato store can affect the sprouting, rotting, physiological disorders, respiration rate, sugar content and processing quality of tubers (Table 12.9). Fellows (1988) recommended a maximum of 10 kPa CO_2 and a minimum of 10 kPa O_2. Black heart developed in tubers held in 1 and 0.5 kPa O_2 at 15°C or 20°C (Lipton, 1967).

Table 12.9. Sugars (g/100 g dry weight) in tubers of three potato cultivars stored for 25 weeks under different CA at 5°C and 10°C and reconditioned for two weeks at 20°C (modified from Khanbari and Thompson, 1996).

	Gas content							
Cultivar	kPa CO_2	kPa O_2	Sucrose	Reducing sugars	Non-reducing sugars	Sucrose	Reducing sugars	Non-reducing sugars
Record	9.4	3.6	0.757	0.216	0.973	0.910	0.490	1.400
Record	6.4	3.6	0.761	0.348	1.109	1.385	1.138	2.523
Record	3.6	3.6	0.622	0.534	1.156	1.600	0.749	2.349
Record	0.4	3.6	0.789	0.510	1.299	0.652	0.523	1.175
Record	0.5	21.0	0.323	0.730	1.053	0.998	0.634	1.632
	Mean		0.650	0.488	1.138	1.109	0.707	1.816
Saturna	9.4	3.6	0.897	0.324	1.221	0.685	0.233	0.918
Saturna	6.4	3.6	0.327	0.612	0.993	0.643	0.240	0.883
Saturna	3.6	3.6	0.440	0.382	0.822	0.725	0.358	1.083
Saturna	0.4	3.6	0.291	0.220	0.511	0.803	0.117	0.920
Saturna	0.5	21.0	0.216	0.615	0.831	0.789	0.405	1.194
	Mean		0.434	0.473	0.907	0.729	0.271	1.000
Hermes	9.4	3.6	0.371	0.480	0.851	0.256	0.219	0.475
Hermes	6.4	3.6	0.215	0.332	0.547	1.364	0.472	1.836
Hermes	3.6	3.6	0.494	0.735	1.229	0.882	0.267	1.149
Hermes	0.4	3.6	0.287	0.428	0.715	0.585	0.303	0.888
Hermes	0.5	21.0	0.695	0.932	1.627	0.617	0.510	1.127
	Mean		0.412	0.682	1.094	0.741	0.354	1.095

Mechanical properties

Olsen *et al.* (2003) reported that it appeared that tubers prior to storage had quantitatively stronger tissue compared with tubers after storage and that storage at 7°C in 2 kPa O_2 + 10 kPa CO_2 altered tissue mechanical properties as well as carbohydrate content and had physiochemical characteristics of tubers stored at 3°C in air.

Flavour

Lipton (1967) found that 'White Rose' tubers stored at 0.5 kPa O_2 had a sour off-odour when raw, and this off-odour persisted when cooked. Aeration for 8 days following 0.5 kPa O_2 storage diminished this off-odour and off-flavour. The taste was influenced only slightly at 1 kPa O_2, and no off-odour or off-flavour was found in tubers stored in 5 kPa O_2.

Sugars and processing

If potatoes are to be processed into crisps or chips, it is important that reducing sugars and amino acids are at a level to permit the Maillard reaction during frying. This gives them an attractive golden colour and their characteristic flavour. If levels of reducing sugars are too high, they are too dark after frying and unacceptable to the processing industry (Schallenberger *et al.*, 1959; Roe *et al.*, 1990). Potatoes stored at low temperatures (around 4°C) have higher levels of sugars than those stored at higher temperatures of 7–10°C (Khanbari and Thompson, 1994), which can result in the production of dark-coloured crisps. At low storage temperatures, potato tubers accumulate reducing sugars, but the effects of CO_2 can influence their sugar levels (Table 12.10).

With levels of 6 kPa CO_2 + 2–15 kPa O_2 in storage for 178 days at 8°C and 10°C, fry colour was very dark (Reust *et al.*, 1984). Daniels-Lake (2012) reported that, although darkening of fry colour was related to ethylene, this effect interacted with CO_2. Mazza and Siemans (1990), studying storage in 1–3.2 kPa CO_2, found that darkening of crisps occurred after there was a rise in CO_2 levels in stores. At levels of 8–12 kPa CO_2 the fry colour was darker than for tubers stored in air (Schmitz, 1991). At O_2 levels of 2 kPa

Table 12.10. Effects of CO_2 levels in storage on reducing sugar content in potatoes.

CO_2 %	Effect	Reference
4	Higher reducing sugars than in air	Workman and Twomey, 1969
5	Prevents accumulation of reducing sugars but increased accumulation of sucrose	Denny and Thornton, 1941
6	Accumulation of sugars, especially sucrose	Reust *et al.*, 1984
5–20	Initial reduction in reducing sugars but later increase to a higher level than air	Burton, 1989

sugar levels were reduced or there was no accumulation at low temperature, but 5 kPa O_2 was much less effective (Workman and Twomey, 1969). The cultivar 'Bintje' were stored at 6°C in atmospheres containing 0, 3, 6 or 9 kPa CO_2 and 21, 18, 15 or 12 kPa O_2. During the early phase, the unfavourable effect of high CO_2 on fry colour increased to a maximum at 3 months. In the later phase, fry colour deteriorated in tubers stored in zero CO_2, and CO_2 enrichment had less effect (Schouten, 1994). Khanbari and Thompson (1994) stored 'Record' tubers at 4°C in 0.7–1.8 kPa CO_2 + 2.1–3.9 kPa O_2 and found that it resulted in a significantly lighter crisp colour, low sprout growth and fewer rotted tubers compared with in air. Storing tubers in anaerobic conditions of total N_2 prevented accumulation of sugars at low temperature but it had undesirable side effects on the tubers (Harkett, 1971). Gökmen *et al.* (2007) stored tubers of 'Agria' and 'Russet Burbank' at 9 ± 1°C and found that there was only a limited increase in the concentrations of sugars during 6 month of storage in air, but the concentrations of sugars increased in CA where the O_2 level was sufficiently low to increase the respiration rate.

Sprouting and disease

Schouten (1992) discussed the effects of CA storage on sprouting of stored potatoes. He found that sprout growth was stimulated in 3 kPa CO_2 at 6°C, whereas some inhibition of growth occurred in 6 kPa CO_2. Sprout growth was strongly inhibited in 1 kPa O_2 at 6°C, but pathological breakdown may develop at this O_2 level. Internal disorders were found at O_2 levels of < 1 kPa. Stimulation of sprout growth occurred at slightly higher O_2 levels if CA storage started from the beginning of the season. Stimulation was less if low O_2 conditions were applied from January onwards in Holland. It was concluded that CA storage at 6°C was not an alternative to chemical control of sprout growth for potatoes used for consumption. In the early phase of storage, Schouten (1994) showed that sprouting was stimulated by 3–6 kPa CO_2 and inhibited by 9 kPa CO_2. During the later phase, all CO_2 concentrations inhibited sprouting. Khanbari and Thompson (1994) showed that high CO_2 with low O_2 combinations during storage completely inhibited sprout growth, but caused the darkest colour when the potatoes were processed into crisps. However, after reconditioning, tubers gave the same level of sprouting and crisps as light as the other CA storage combination. Reconditioning of stored tubers can reduce their sugar content (Table 12.11) and improve their fry colour (Khanbari and Thompson, 2004). Schouten (1994) recommended reconditioning tubers at 15°C for 2–4 weeks and showed that the treatment improved crisp fry colour, but he also found that reconditioning stimulated sprouting. Tubers stored in 0.7–1.8 kPa CO_2 + 2.1–3.9 kPa O_2 showed a significantly higher weight loss and shrinkage after reconditioning, compared with tubers that had previously been stored in air (Khanbari and Thompson, 1994). They also slowed that tubers stored in high CO_2, especially at 10% or 15%, had earlier onset of rotting. In 0.7–1.6 kPa CO_2 + 2–2.4 kPa O_2 there was also an increase in tuber rotting.

Quince (*Cydonia oblonga*)

Quince is a pome fruit that shows a climacteric respiratory pattern and flesh browning

Table 12.11. Sugar content of tubers from three potato cultivars after being stored for 25 weeks in 9.4% CO_2 + 3.7% O_2 and another 20 weeks in air at 5°C and then reconditioned for two weeks at 20°C; modified from Khanbari and Thompson (1996).

		Sugars g/100 g dry weight			
Reconditioning	Cultivar	Fructose	Glucose	Sucrose	Grey level[a]
None	Record	0.778	0.847	0.520	130.3
None	Saturna	0.660	0.685	0.540	136.8
None	Hermes	0.980	1.120	0.564	122.9
Two weeks at 20°C	Record	0.628	0.563	0.897	132.7
Two weeks at 20°C	Saturna	0.381	0.343	0.489	151.8
Two weeks at 20°C	Hermes	1.030	1.127	1.133	123.6

[a]This is for the fry colour of potato crisps where the minimum acceptable grey level = 149 or above.

is the most limiting factor in storage. It can be stored at 0–2°C for up to 6 months in air and for up to 7 months in 2 kPa O_2 + 3 kPa CO_2 at 2°C (Gunes, 2008).

Radish (*Raphanus sativus*)

Low O_2 storage was shown to prolong postharvest life (Haard and Salunkhe, 1975). Lipton (1972) found that storage of topped radishes at 5°C or 10°C in 1 kPa O_2 reduced sprouting by about 50% compared with storage in air. Although storage in 0.5 kPa O_2 further reduced sprouting, it resulted in physiological injury. Storage at 0.6°C in 5 kPa O_2 or at 5°C or 10°C in 1–2 kPa O_2 was recommended by Pantastico (1975). Saltveit (1989) recommended 0–5°C in 2–3 kPa CO_2 + 1–2 kPa O_2 for topped radishes but indicated that CA storage had only a slight effect.

Rambutan (*Nephellium lappaceum*)

Kader (1993) recommended 10°C, with a range of 8–15°C, in 7–12 kPa CO_2 + 3–5 kPa O_2. Storage decay could be controlled by dipping the fruit in 100 ppm benomyl fungicide. Boonyaritthongchai and Kanlayanarat (2003b) reported that spintern and peel browning were the main causes of loss in quality and they tested atmospheres containing 1, 5, 10 and 20 kPa CO_2 on storage of the cultivar 'Rong-rien' at 13°C or 20°C and 90–95% RH. In air the fruits could be stored for only 6 days at 20°C and 10 days at 13°C. CO_2 toxicity was observed at 20°C, but at 13°C all the CO_2 concentrations used were able to prolong storage life by at least 2–8 days, with 10 kPa CO_2 being optimum, prolonging storage life to 18 days. This was due to reduced rates of browning, weight loss, respiration rate and ethylene production. Ratanachinakorn (2001) found that the storage life at 13°C was 15 days in 2 or 4 kPa O_2, 10 days in 6 or 21 kPa O_2, 5 days in 1 kPa O_2 and 10 days in 5 or 10 kPa CO_2. Fruit was injured in 20 and 40 kPa CO_2. During storage there was no change on TSS, acidity or vitamin C. At 13°C in 2 kPa O_2 + 10 kPa CO_2 fruit could be stored for 16 days with spoilage of < 10%; in 2 kPa O_2 + 5 kPa CO_2 spoilage was 23%. O'Hare *et al.* (1994) found that storage of cultivar 'R162' in 9–12 kPa CO_2 retarded colour loss and extended shelf-life by 4–5 days but storage in 3 kPa O_2 or 5 ppm ethylene did not significantly affect the rate of colour loss.

Raspberry (*Rubus idaeus*)

There is strong evidence that CA storage with high levels of CO_2 reduced rots. Kader (1989) and Bishop (1996) recommended 0–5°C in 15–20 kPa CO_2 + 5–10 kPa O_2 and Hardenburg *et al.* (1990) reported that storage in 20–25 kPa CO_2 retarded the development of rots. Haffner *et al.* (2002) also reported that storage in 10 kPa O_2 + 15 kPa CO_2 or 10 kPa O_2 + 31 kPa CO_2 suppressed rotting significantly compared with storage in air. The cultivars 'Glen Clova', 'Glen Moy', 'Glen Prosen' and 'Willamette' stored in 2 kPa O_2 + 10 kPa CO_2 had reduced fruit rot (*Botrytis cinerea*), delayed ripening, slower colour development, slower breakdown of acids (including AA), slower breakdown of TSS and firmer fruits compared with those stored in air (Callesen and Holm, 1989). The black raspberry cultivar 'Bristol' was stored at 18°C or 0°C in air or 20 kPa CO_2. Storage in 20 kPa CO_2 greatly improved postharvest quality at both temperatures by reducing grey mould development (Goulart *et al.*, 1992). Goulart *et al.* (1990) showed that when 'Bristol' black raspberry was stored at 5°C in 2.6, 5.4 or 8.3 kPa O_2 + 10.5 or 19.6 kPa CO_2 the mass loss was less in CA after 3 days than when stored in air. When fruits were removed after 3 days and held for 4 days in air at 1°C, deterioration was highest in those that had been stored in 2.6, 5.4 or 8.3 kPa O_2 + 10 kPa CO_2. Fruits removed after 7 days retained their quality for up to 12 days at 1°C and showed least deterioration, with the 15 kPa CO_2 treatment giving the best results. In another experiment by Goulart *et al.* (1990), fruits of the cultivar 'Heritage' red raspberry were stored for 3 or 5 days at 5°C in 5 kPa O_2 + 95 kPa N_2, 20 kPa CO_2, in air or in plastic bags, where the equilibrium atmosphere was 5 kPa O_2 + 18.1 kPa CO_2. The deterioration was greatest with fruits held in 5 kPa O_2, followed by those in air, and no off-flavour was detected in raspberries held in a high CO_2 atmosphere. Robbins and Fellman (1993) recommended 0–0.5°C and 90–95% RH using high CO_2 and reduced O_2. This was effective in reducing the incidence of rots and maintaining fruit quality. There was almost no reduction in vitamin C for raspberries stored in high CO_2 concentrations (10–30 kPa CO_2), and reducing O_2 level had little effect (Agar *et al.*, 1997). Forney *et al.* (2015) compared storage in air with storage in 12.5 kPa CO_2 + 7.5 kPa O_2, both at 1°C and 95% RH. They found that loss of fruit firmness was similar during storage in both CA and air but that CA storage strongly suppressed fruit decay of all the cultivars they tested, with the lag period extended to > 45 days, compared with 19–29 days for those stored in air. They calculated the lag period of decay and physiological breakdown to the number of days until the first sign of deterioration.

Redcurrant (*Ribes sativum*)

Storage of soft fruit at 20 kPa CO_2 was shown to slow their respiration rate, double their storage life, suppress fungi and reduce decomposition by 50% both at low temperature and during marketing (Hansen, 1986). He also reported that storage at 0°C required only one-third the concentration of CO_2 required at 7°C. Typical storage recommendations include: 0–5°C in 12–20 kPa CO_2 + 2–5 kPa O_2 (Bishop, 1996); 0.5°C and 90–95% RH in 2 kPa O_2 + 18 kPa CO_2 (Ellis, 1995); and 20 kPa CO_2 + 2 kPa O_2 for more than 20 weeks (Roelofs, 1993b). The cultivars 'Rotet', 'Rondom', 'Rovada', 'Roodneus' and 'Augustus' were stored for 8–25 weeks at 1°C in 0, 10, 20 or 30 kPa CO_2 + 2 kPa O_2 or higher (not controlled, 15–21 kPa) O_2. The earliest harvested berries were more susceptible to rots, but increased CO_2 concentration reduced rotting, while reduced O_2 levels only reduced fruit rotting at low CO_2 concentrations. Levels above 20 kPa CO_2 resulted in fruit discolouring in some cultivars after 13 weeks of storage and low O_2 concentrations further increased internal breakdown (Roelofs, 1993b). In storage of the cultivars 'Rovada' (late ripening) and 'Stanza' (mid-season ripening) at 0–1°C in air it was shown that they could be kept for 2–3 weeks, while in atmospheres containing 20 kPa CO_2 or more they could be kept for 8–10 weeks (Van Leeuwen and Van de Waart, 1991). The cultivar 'Rotet' was stored at 1°C in 10, 20 or 30 kPa CO_2 + 2 kPa O_2 or 10, 20 or 30 kPa CO_2 + > 15 kPa O_2. The optimum CO_2 concentration was found to be 20 kPa. At 1°C fruit could be stored for 8–10 weeks in high CO_2 atmospheres, compared with 1 month for those stored in air (Agar *et al.*, 1991). Six cultivars were stored at 1°C in 1, 2 or 21 kPa O_2 + 0 or 10 kPa CO_2, in all combinations, for 4½ months. Spoilage was least in 1 kPa O_2 + 10 kPa CO_2 (Roelofs, 1992). The cultivars 'Rotet', 'Rondom', 'Rovada', 'Augustus', 'Roodneus', 'Cassa' and 'Blanka' were stored in combinations of 0, 20 or 25 kPa CO_2 + 2 or 18–21 kPa O_2 for up to 24 weeks. After 24 weeks, fruit stored in 0 kPa CO_2 at 1°C or –0.5°C had about 95% rotting. The average incidence of fungal rots was lowest (about 5%) at –0.5°C in 20 or 25 kPa CO_2 and increased to about 35% at 1°C in 25 kPa CO_2 and to about 50% at 1°C in 20 kPa CO_2. Fruit quality, however, was adversely affected in 20 kPa CO_2 and high O_2, but this effect was in contrast to the findings subsequently reported by Roelofs (1994). The cultivar 'Rotet' was stored at 1°C for 10 weeks in air, 10 kPa CO_2 + 19 kPa O_2, 20 kPa CO_2 + 17 kPa O_2, 30 kPa CO_2 + 15 kPa O_2, 10 kPa CO_2 + 2 kPa O_2, 20 kPa CO_2 + 2 kPa O_2 or 30 kPa CO_2 + 2 kPa O_2 by Agar *et al.* (1994c). They found that the chlorophyll content was most stable in high CO_2 and low O_2 and for fruit stored in air the decrease in chlorophyll content coincided with a decrease in fruit quality. There was almost no reduction in AA during storage in 10–30 kPa CO_2; also reducing O_2 level had little effect (Agar *et al.*, 1997).

Rocket (*Diplotaxis tenuifolia*)

Rocket leaves were stored at 4°C in air or 5 kPa O_2 + 5 kPa CO_2, 5 kPa O_2 + 10 kPa CO_2 and air with 10 kPa CO_2 for up to 14 days. After 10 days, the sensory and microbiological quality of samples stored in air was not commercially acceptable but those in 5 kPa O_2 + 10 kPa CO_2 maintained visual quality and controlled aerobic mesophilic and psychrotropic microorganisms as well as coliforms. The total flavonoid content remained constant during storage or even increased at the end of the shelf-life in CA, but it was degraded in those in air. In addition, the total content of vitamin C was higher in those in CA than in those in air. A decrease in total antioxidant capacity occurred during storage, particularly in samples stored in air (Martínez-Sánchez *et al.*, 2006).

Roseapple (*Syzygium malaccesis*)

Also called pomerac. Basanta and Sankat (1995) stored pomeracs at 5°C for up to 30 days in 5 kPa CO_2 + 2 kPa O_2, 5 kPa CO_2 + 4 kPa O_2, 5 kPa CO_2 + 6 kPa O_2, 5 kPa CO_2 + 8 kPa O_2, 5 kPa CO_2 + 1 kPa O_2, 8 kPa CO_2 + 1 kPa O_2, 11 kPa CO_2 + 1 kPa O_2 or 14 kPa CO_2 + 1 kPa O_2. The optimum storage conditions were found to be 11 kPa CO_2 + 1 kPa O_2 or 14 kPa CO_2 + 1 kPa O_2, where flavour and appearance were maintained for 25 days. Previous work is quoted where storage at 5°C in air resulted in a postharvest life of 20 days.

Salsify (*Tragopogen porrifolius*)

Storage in 3 kPa CO_2 + 3 kPa O_2 at 0°C was recommended by Hardenburg *et al.* (1990).

Sapodilla (*Manilkara zapota*)

CA storage recommendations were 20°C in 5–10 kPa CO_2 combined with scrubbing of ethylene (Broughton and Wong, 1979), though the optimum temperature for storage in air conditions varied from as low as 0°C (Thompson, 1996). In other work, fruits stored for 28 days at 10°C failed to ripen properly, indicating chilling injury (Abdul-Karim *et al.*, 1993).

Soursop (*Annona muricata*)

It was shown that fruit in storage is damaged by exposure to low temperatures (Wardlaw, 1938). Snowdon (1990) recommended 10–15°C. No specific data on CA storage could be found, but in a comparison between 16, 22 or 28°C the longest storage period (15 days) occurred in 16°C with fruits packed in high-density PE film bags without an ethylene absorbent. This packaging treatment also resulted in the longest storage duration at the two higher temperatures, but the differences between treatments at those temperatures were much less marked (Guerra *et al.*, 1995).

Spinach (*Spinaca oleraceace*)

Storage recommendations include: 0–5°C in 10–20 kPa CO_2 + 21 kPa O_2 which had a fair effect but was of no commercial use (Kader, 1985); 0–5°C in 5–10 kPa CO_2 + 7–10 kPa O_2 which had only a slight effect (Saltveit, 1989); 10–20 kPa CO_2 + 21 kPa O_2 (SeaLand, 1991); and a maximum of 20 kPa CO_2 (Fellows, 1988). Wardlaw (1938) found that storage in high concentrations of CO_2 could damage the tips of leaves but Hardenburg *et al.* (1990) reported that storage in 10–40 kPa CO_2 + 10 kPa O_2 retarded yellowing.

Young, tender spinach leaves 8–10 cm long and leaves harvested at the traditional length of more than 15 cm were stored at 7.5°C in air or in 5 kPa O_2 + 10 kPa CO_2. CA maintained initial appearance for up to 12 days but carotenoids were in a lower concentration than those that had been stored in air. Chlorophyll levels in both maturities of leaves was highest after storage in 5 kPa O_2 + 20 kPa CO_2 but leaves showed rapid deterioration in visual quality even more than those stored in air (Martinez-Damian and Trejo, 2002). Storage in air at 2°C prevented a reduction in AA content and L-ascorbate peroxidase activity. AA content, L-ascorbate peroxidase activity and yellowing decreased rapidly in air at 25°C. CA storage at 10°C in 2 kPa O_2 + 3 or 10 kPa CO_2 controlled enzyme activity and contributed to the preservation of AA and L-ascorbate peroxidase activity (Mizukami *et al.*, 2003).

Spring onion (*Allium cepa*)

Also called scallion and green onion. Optimum recommended storage conditions by Hardenburg *et al.* (1990) were 5 kPa CO_2 + 1 kPa O_2 at 0°C for 6–8 weeks, while 10–20 kPa CO_2 + 2–4 kPa O_2 was recommended by SeaLand (1991). Kader (1985) recommended 0–5°C in 10–20 kPa CO_2 + 1–2 kPa O_2 for green onions, but it had only a fair effect and was of limited commercial use. Saltveit (1989) recommended 0–5°C in 0–5 kPa CO_2 + 2–3 kPa O_2 which also had only a slight effect.

Squash (*Cucurbita* spp.)

SeaLand (1991) recommended 5–10 kPa CO_2 + 3–5 kPa O_2 which had a slight to no effect and Hardenburg *et al.* (1990) reported that storage at 5°C in low O_2 was of little or no value. However, Wang and Kramer (1989) found that storage in 1 kPa O_2 reduced chilling injury symptoms at 2.5°C compared with storage in air. Symptoms occurred after 9 days of storage compared with 3 days of storage in air. The *C. pepo* cultivar 'Romanesco' was stored for 19 days at 5°C in 21 ± 1 kPa O_2 + 0, 2½, 5 or 10 kPa CO_2 and then held for 4 more days at 13°C in air. High CO_2 levels reduced the respiration rate and the development of chilling injury symptoms at all three harvest maturities (lengths 16 cm, 20 cm and 22 cm). At the end of the 23-day storage period, 82% of the 22 cm fruits held in 10 kPa CO_2 appeared saleable, but they had a slight off-flavour and were soft; 79% of those held in 5 kPa CO_2 were saleable, firm and free from off-flavour; samples from storage in 2.5 kPa CO_2 and in air were unacceptable because of decay and pitting. It is suggested that CO_2 concentrations around 5 kPa may be useful for storing zucchini at about 5°C, a temperature that normally causes chilling injury in zucchini (Mencarelli, 1987b). The extent of surface pitting of the *C. pepo* cultivar 'Ambassador' stored at 2.5°C was less in 1 kPa O_2 than when stored in air (Wang and Ji, 1989). At 0°C those stored in 1 kPa O_2 had reduced chilling injury compared with those stored in air (Wang and Kramer, 1989). Mencarelli *et al.* (1983) also showed that storage of zucchini at 2.5°C resulted in chilling injury, but the effect could be delayed by reducing the O_2 to 1–4%. In later work *C. pepo* fruits stored for 2 weeks in 1, 2 or 4 kPa O_2 had reduced respiration rates and ethylene production, particularly at 5°C and 10°C. Both rates increased during subsequent storage in air for 2 days at 10°C, but much more in those that had been held previously at 2.5°C or 5°C than in those from 10°C. About 75% and 55%, respectively, of the burst in respiration rate during storage in air in samples from 2.5°C and 5°C was due to exposure to low temperature; the remainder was attributed to the effect of low O_2 levels. For ethylene production, the corresponding values were about 95% and 70%. All fruits stored at 5°C for 2 weeks were virtually free of chilling injury, surface mould, decay and off-flavour and almost all the fruits were rated good to excellent in appearance. About three-quarters of the fruits were still in this category after two additional days at 10°C. At 5°C, low O_2 atmospheres had no effect. Storage at 2.5°C resulted in severe chilling injury, which was ameliorated by holding them in 1, 2 or 4 kPa O_2 instead of 8 or 21 kPa O_2. The best quality maintenance of 'Astra' zucchini was in storage at 6°C in air and 3°C in 5 kPa CO_2 + 3 kPa O_2 or 3 kPa CO_2 + 3 kPa O_2 but storage at 3°C caused chilling injury of young fruits (10–16 cm long) after 2 weeks (Gajewski and Roson, 2001). Subsequently Gajewski (2003) reported that best results were obtained if 'Astra' zucchini was kept at 6°C in either 5 kPa CO_2 + 3 kPa O_2 or 3 kPa CO_2 + 3 kPa O_2. Zheng *et al.* (2008) stored the zucchini cultivar 'Elite' at 5°C in 21, 100 or 60 kPa O_2. Squash in 100 kPa O_2 had the lowest respiration rate and those in 60 kPa O_2 had the lowest ethylene production. The transcript levels of antioxidative genes were relatively constant during storage in high O_2, but the expressions of alternative oxidase, O_2 radical scavenging enzyme activities and total phenol content were slightly increased in 100 and 60 kPa O_2. They interpreted these changes in relation to high O_2 increasing tolerance to chilling injury.

Strawberry (*Fragaria* spp.)

CA recommendations include: 15–20 kPa CO_2 + 10 kPa O_2 (SeaLand, 1991); 1°C, 90% RH, 20 kPa CO_2 and 17 kPa O_2 (Lawton, 1996); 0–5°C in 15–20 kPa CO_2 + 5–10 kPa O_2 (Bishop, 1996); a maximum of 20 kPa CO_2 and a minimum of 2 kPa O_2 (Fellows, 1988); 1.5 kPa CO_2 + 1.5 kPa O_2 for 3 weeks compared with only 4 days in air (Van der Merwe, 1996); and 0–5°C in 15–20 kPa CO_2 + 5–10 kPa O_2 (Kader, 1985, 1989). Woodward and Topping (1972) showed that high levels of CO_2 in the storage atmosphere at 1.7°C reduced rotting caused by infection with *Botrytis cinerea*, but levels as high as 20 kPa CO_2 were injurious. However, they showed that there was a temperature interaction with CO_2 levels. In CO_2 concentrations up to 20%, fruit stored at 3°C was in good condition after 10 days, but after 15–20 days there was a distinct loss of flavour, which appeared to be the main limiting factor to storage in the UK. Storage in 10–30 kPa CO_2 or 0.5–2 kPa O_2 slowed respiration rate and reduced disease levels, but 30 kPa CO_2 or < 2 kPa O_2 might cause off-flavours (Hardenburg *et al.*, 1990). Storage in 20 kPa CO_2 by Hansen (1986) slowed the respiration rate of 'Elsanta' more than that of 'Elvira' and doubled storage life of both cultivars compared with those stored in air.

Fruits of the cultivar 'Redcoat' were forced-air cooled and stored in 0, 12, 15 or 18 kPa CO_2 for 0, 2, 7 or 14 days. CO_2 increased firmness but fruits stored for more than 7 days softened rapidly in all CO_2 concentrations and fruits stored for 14 days were as soft as those not exposed to CO_2 at all. The amount of storage decay after 14 days was less in 12, 15 or 18 kPa CO_2 than in 0.5 kPa CO_2. Organoleptic evaluation identified fruit quality differences between treatments but the results were not consistent (Smith *et al.*, 1993). 'Redcoat' fruit was stored at several temperatures and for various intervals in CA containing atmospheres within the ranges of 0–18 kPa CO_2 + 15–21 kPa O_2. CA storage indicated that the addition of CO_2 to the storage environment enhanced fruit firmness. Fruits kept in 15 kPa CO_2 for 18 h were 48% firmer than non-treated samples were initially. Response to increasing CO_2 concentration was linear. There was no response to changing O_2 concentration. Maximum retention of firmness was at 0°C. In some instances, there was a moderate enhancement of firmness as time in storage increased. CO_2 reduced the quantity of fruit lost due to rotting. Fruits that were soft and bruised at harvest became drier and firmer in a CO_2-enriched environment (Smith, 1992). Fruits of various strawberry cultivars were stored by Smith *et al.* (1993) at 0°C for 42 h in 15 kPa CO_2 + 18 kPa O_2 to study the effect on firmness. Compared with initial samples and control samples stored in air, firmness was increased in 21 of the 25 cultivars evaluated. The CO_2 treatment had no effect on colour, TSS or pH. Fruits of 15 strawberry cultivars were picked at two different stages of maturity, pre-cooled or not, and stored for 5 or 7 days at 0°C or 20°C in 10 or 20 kPa CO_2. Increased atmospheric CO_2 was associated with excellent control of storage decay caused by *Botrytis* and *Penicillium* species and it also slowed fruit metabolism, thus preserving aroma and quality (Ertan *et al.*, 1990).

Fruits of the cultivar 'Chandler' were kept in air, 0.25 kPa O_2, 21 kPa O_2 + 50 kPa CO_2 or 0.25 kPa O_2 + 50 kPa CO_2 (balance N_2) at 5°C for 1–7 days to study the effects of CA on volatiles and fermentation enzymes. Acetaldehyde, ethanol, ethyl acetate and ethyl butyrate concentrations were greatly increased, while isopropyl acetate, propyl acetate and butyl acetate concentrations were reduced by the three CA storage atmospheres compared with those of control fruits stored in air. CA storage enhanced pyruvate decarboxylase and alcohol dehydrogenase activities, but slightly decreased alcohol acetyl transferase activity. The results indicate that the enhanced pyruvate decarboxylase and alcohol dehydrogenase activities cause ethanol accumulation, which in turn drives the biosynthesis of ethyl esters; the increased ethanol concentration also competes with other alcohols for carboxyl groups for esterification reactions; the reduced alcohol acetyl transferase activity and limited availability of carboxyl groups due to ethanol competition decrease production of other acetate esters (Ke *et al.*, 1994b). Almenar *et al.* (2006) compared storage in air, 3 kPa CO_2 + 18 kPa O_2, 6 kPa CO_2 + 15 kPa O_2, 10 kPa CO_2 + 11 kPa O_2, and 15 kPa CO_2 + 6 kPa O_2 on 'Reina de los Valles' wild strawberry fruit. They found that 10 kPa CO_2 + 11 kPa O_2 at 3°C prolonged the shelf-life by maintaining the quality parameters within acceptable values, through inhibition of *Botrytis cinerea*, without significantly modifying consumer acceptance. Three-quarter-coloured 'Chandler' fruits responded better to storage at 4°C in 5 kPa O_2 + 15 kPa CO_2 than fully red fruits, maintaining better appearance, firmness and colour over 2 weeks of storage, while achieving similar acidity and TSS contents with minimum decay development. CA was more effective than air storage in maintaining initial anthocyanin and TSS contents of three-quarter-coloured fruits. Although three-quarter-coloured fruits darkened and softened in storage at 10°C, the CA-stored fruits remained lighter coloured and as firm as the at-harvest values of fully red fruits. CA maintained better strawberry quality than air storage even at an above-optimum storage temperature of 10°C, but CA was more effective at the lower temperature of 4°C (Nunes *et al.*, 2002). Agar *et al.* (1997) found that AA plus dehydro ascorbic acid was reduced during storage in 10–30 kPa CO_2. Reducing the O_2 concentration in the storage atmosphere in the presence of high CO_2 had little effect on the vitamin C content. After CA storage of strawberries, 16 proteins increased and 13 decreased in abundance, and after air storage 11 proteins increased and 17 decreased while expression profile of 12 proteins was significantly changed by both CA and air (Li *et al.*, 2015).

Sliced strawberries (cultivars 'Pajaro' and 'G 3') were dipped in solutions of citric acid, AA and/or calcium chloride and were stored in air or in CA for 7 days at 2.5°C followed by 1 day in air at 20°C. Whole fruits were also stored in the same conditions. The respiration rate of slices was higher than that of whole fruit at both temperatures. CA slowed the respiration rate and ethylene production of sliced fruits. Firmness of slices was maintained by storage in 12 kPa CO_2 and in 0.5 kPa O_2, or by dipping in 1% calcium chloride and storing in air or CA storage (Rosen and Kader, 1989).

Swede (*Brassica napus* var. *napobrassica*)

Also called rutabaga. CA storage was reported to have a slight to no effect on keeping quality (SeaLand, 1991). Coating with paraffin wax was reported to reduce weight loss but if this was too thick the reduced O_2 supply could result in internal breakdown (Lutz and Hardenburg, 1968).

Sweetcorn, Baby Corn (*Zea mays* var. *saccharata*)

Recommendations include: 0–5°C in 10–20 kPa CO_2 + 2–4 kPa O_2 which had a good effect but was of no commercial use (Kader, 1985); 0-5°C in 5–10 kPa CO_2 + 2–4 kPa O_2 but it had only a slight effect (Saltveit, 1989); 10–20 kPa CO_2 + 2–4 kPa O_2 (SeaLand, 1991); a maximum of 20 kPa CO_2 (Fellows, 1988); and 80 kPa CO_2 + 20 kPa air at 15°C or 25°C, which reduced respiration rate (Inaba *et al.*, 1989). CA has been shown to preserve sugar content; for example, 5°C in 2 kPa O_2 + 15 kPa CO_2 for 14 days preserved the highest sugar level and reduced deterioration in visual quality (Riad and Brecht, 2003). Ryall and Lipton (1972) also reported that storage in 5–10 kPa CO_2 retarded sugar loss while 10 kPa caused injury. CA of 2 kPa O_2 + 10 or 20 kPa CO_2 reduced respiration rate and maintained higher sugar levels during 10 days at 1°C of fresh-cut sweetcorn (Riad and Brecht, 2001). Storage in high CO_2 retarded conversion of starch to sugar (Wardlaw, 1938) but cobs may be injured with more than 20 kPa CO_2 or < 2 kPa O_2, though in an atmosphere of 2 kPa O_2 the sucrose content remained higher than in air (Hardenburg *et al.*, 1990). Brash *et al.* (1992) stored cobs of the cultivar 'Honey 'n' Pearl' at 0°C in 2.5 kPa O_2 or levels of CO_2 in the range of 5–15% for 3–4 weeks followed by 3 days of shelf-life assessment at 15°C in simulated sea-freight export trials in New Zealand. Atmospheres of 2.5 kPa O_2 + 6–10 kPa CO_2 ensured that the cobs retained sweetness and husk leaves stayed greener for longer during shelf-life, compared with air storage. CA storage for 3 weeks also reduced insect numbers. The use of CA to control insect pests on sweetcorn harvested in New Zealand was studied by Carpenter (1993), who concluded that storage at 0–1°C alone was as effective as treatment with CA.

Sweet Potato (*Ipomoea batatas*)

SeaLand (1991) indicated that CA storage had a slight or no effect compared with storage in air. Typical storage conditions were given as 0–5°C in 5–10 kPa CO_2 + 2–4% O_2 by Bishop (1996). Storage in 2–3 kPa CO_2 + 7 kPa O_2 kept roots in better condition than those stored in air, but CO_2 levels above 10 kPa or O_2 levels below 7 kPa could give alcoholic or off-flavour to the roots (Pantastico, 1975; Hardenburg *et al.*, 1990). The cultivar 'Georgia Jet' was stored at 15.5°C in air at 85% RH or in a 7 kPa O_2 + 93 kPa N_2 atmosphere for 3 months. CA storage had no effect on moisture content, β-carotene or total lipids (Charoenpong and Peng, 1990). In storage for 7 days at 20°C in air, 1 kPa O_2 + 99 kPa N_2 or 100 kPa N_2, the total soluble solid content increased but weak off-odours were detected in roots stored in 100 kPa N_2. The intensity of off-odours increased as the concentrations of acetaldehyde and ethanol increased in roots during storage. Ethanol concentrations were higher than those of acetaldehyde, which remained low during storage in 1 kPa O_2 and in air, but increased greatly in roots stored in 100 kPa N_2 (Imahori *et al.*, 2007).

Sweetsop, Sugar Apple (*Annona squamosa*)

Storage of the cultivar 'Tsulin', harvested at the green-mature stage, at 20°C in 5 kPa CO_2 + 5 kPa O_2 had reduced ethylene production and delayed ripening compared with those stored in air (Tsay and Wu, 1989). Kader (1993) recommended 15°C, with a range of 12–20°C, in 5–10 kPa CO_2 + 3–5 kPa O_2.

Tomato (*Lycopersicon esculentum*)

CA recommendations include: 10–15°C in 3–5 kPa CO_2 + 3–5%O_2 (Bishop, 1996); a maximum of 2 kPa CO_2 and a minimum of 3 kPa O_2 (Fellows, 1988); 12°C in 2.5–5 kPa CO_2 + 2.5 kPa O_2 for 3 months (Monzini and Gorini, 1974); 12.8°C in 3 kPa O_2 + 5 kPa CO_2 (Parsons and Spalding, 1972); 12–20°C in 0 kPa CO_2 + 3–5 kPa O_2 for mature-green fruits and 8–12°C in 0 kPa CO_2 and 3–5 kPa O_2 for partially ripe fruit (Kader, 1985); 12–20°C in 2–3 kPa CO_2 + 3–5 kPa O_2 (Saltveit, 1989); 5 kPa CO_2 + 5 kPa O_2 at 12°C for fruits harvested at the yellow or 'tinted stage'

(Wardlaw, 1938); 0 kPa CO_2 + 3–5 kPa O_2 for mature-green tomatoes or 0–3 kPa CO_2 + 3–5 kPa O_2 for breaker to light pink stages (SeaLand, 1991); 13°C in 9.1 kPa CO_2 + 5.5 kPa O_2 for 60 days for the cultivar 'Criterium' harvested at the pink stage of maturity (Batu and Thompson, 1995); and 10°C with 90% RH in 8 kPa CO_2 + 5 kPa O_2 for up to 36 days for the cultivars 'Angela' and 'Kada' harvested at the mature-green stage (Vidigal *et al.*, 1979). In mature green and pink fruits, respiration rates and ethylene production were lower in CA than in air (Lougheed and Lee, 1989). At 12.8°C fruit colour and flavour were maintained in an acceptable condition for 6 weeks in 0 kPa CO_2 + 3 kPa O_2, while in other work 1–2 kPa CO_2 + 4–8 kPa O_2 at 12.8°C for breaker or pink fruit was recommended (Pantastico, 1975). Adamicki (1989) described successful storage of mature-green tomatoes at 12–13°C in 0 kPa CO_2 + 2 kPa O_2, 5 kPa CO_2 + 3 kPa O_2 or 5 kPa CO_2 + 5 kPa O_2 for 6–10 weeks. Parsons *et al.* (1974) harvested tomatoes at the mature-green stage and stored them at 12.8°C for 6 weeks and found that 0 kPa CO_2 + 3 kPa O_2 was better than those stored in air. Fruits in air were fully red in 6 weeks, those in 0 kPa CO_2 + 3 kPa O_2 or those in 5 kPa CO_2 + 3 kPa O_2 fruits were still pink, with the latter combination having an additional delaying effect on ripening. Dennis *et al.* (1979) stored fruits harvested at the mature green stage at 13°C and 93–95% RH in air, 5 kPa CO_2 + 3 kPa O_2 or 5 kPa CO_2 + 5 kPa O_2 for 6–10 weeks. Two greenhouse-grown cultivars ('Sonato' and 'Sonatine') and three field-grown ('Fortuna', 'Hundredfold' and 'Vico') were used. CA-stored fruit were found to ripen more evenly and to have a better flavour when removed from storage than air-stored fruit, with the 5 kPa CO_2 + 3 kPa O_2 treatment giving the best-quality fruit. Mature-green fruits of the cultivars 'Saba' and 'Saul' were stored by Anelli *et al.* (1989) for 14 days at 12°C in a room equipped with an ethylene scrubber, or 40 days at 12°C in 3 kPa O_2 and 3 kPa CO_2. At the end of storage the fruits had marketable percentages of 90% and 85% for ethylene scrubber and CA storage, respectively. *Phytophthora* decay appeared on some fruits by the end of CA storage. Tataru and Dobreanu (1978) showed that tomatoes kept best in 3 kPa CO_2 + 3 kPa O_2. Greenhouse-grown mature-green and pink tomatoes (cultivar 'Veegan') were stored in air or in 0.5, 3.0 or 5.0 kPa O_2 in N_2 or argon. Colour development was not affected by the background gases N_2 or argon except at 3 kPa O_2, where ripening was delayed by the argon mixture. All the levels of O_2 suppressed colour development in mature-green fruits compared with those stored in air, with maximum suppression in 0.5 kPa O_2. Most fruits held in 0.5 kPa O_2 rotted after removal to air before colour development was complete.

Green-mature tomatoes of the cultivar 'Ramy' were stored for 20 days at 4°C or 8°C in 3 kPa O_2 + 0 kPa CO_2 with or without ethylene removal, or in 1.5 kPa O_2 + 0 kPa CO_2 with ethylene removal using potassium permanganate. Fruit ripening was delayed most in fruits stored in 1.5 kPa O_2 + 0 kPa CO_2 with ethylene removal (Francile, 1992). The colour of fruit harvested at the pink stage did not change when they were stored in 6.4 kPa CO_2 + 5.5 kPa O_2 and 9.1 kPa CO_2 + 5.5 kPa O_2 even after 50 days and in some cases after 70 days of storage. The red colour development of the tomatoes exposed to < 6.4 kPa CO_2 increased, whereas red colour decreased with CO_2 levels above 9.1 kPa during storage (Batu, 1995). Storage in atmospheres containing > 4 kPa CO_2 or < 4 kPa O_2 gave uneven ripening (Pantastico, 1975). Li *et al.* (1973) showed that O_2 levels below 1 kPa caused physiological injury during storage, owing presumably to fermentation. The minimum O_2 level for prolonged storage at 12–13°C was about 2–4 kPa O_2, which delayed the climacteric peak. Salunkhe and Wu (1973a) showed that exposing 'green-wrap' fruits of the cultivar 'DX-54' to low O_2 atmospheres at 12.8°C inhibited ripening and increased storage life. Storage at 10 kPa O_2 + 90 kPa N_2 resulted in a storage life of 62 days, and in 3 kPa O_2 + 97 kPa N_2 it was 76 days. The maximum storage life was 87 days at 1 kPa O_2 and 99 kPa N_2. Low O_2 atmospheres inhibited fruit chlorophyll and starch degradation and also lycopene, β-carotene and sugar synthesis. Parsons and Spalding (1972) inoculated fruit with soft rot bacteria and held them for 6 days at 12.8°C. Decay lesions were smaller on fruits stored in 3 kPa O_2 + 5 kPa CO_2 than on those stored in air, but CA storage did not control decay. However, although at 12.8°C inoculated tomatoes kept better in CA storage than in air, at 7.2°C or 18.3°C they kept equally well in CA storage and air.

Moura and Finger (2003) harvested 'Agriset' at breaker maturity (< 10% red colour) and stored them in 2, 3, or 4 kPa O_2 or air for 14 days at 12°C then ripening in air at 20°C. They found that there were no differences in volatile content of the fruit except for *cis*-3-hexenal and 2 + 3-methylbuthanol. The other volatiles they measured were acetone, methanol, ethanol, 1-penten-3-one, hexanal, trans-2-hexenal, trans-2-heptenal, MHO, *cis*-3-hexenol, 2-isobutylthiazole, 1-nitro-2-phenylethane, geranylacetone, and β-ionone levels among the treatments. Tomatoes were harvested at the mature-green and pink stages of maturity and then stored for 60 days in air or 3.2 kPa CO_2 + 5.5 kPa O_2 or 6.4 kPa CO_2 + 5.5 kPa O_2 or 9.1 kPa CO_2 + 5.5 kPa O_2 at 15°C for mature-green and 13°C for pink. Fruits stored in air at 15°C had significantly better flavour and were sweeter and more acceptable than those in any of the CA conditions (Batu, 2003).

Turnip (*Brassica rapa* var. *rapa*)

CA storage of turnip was reported to have had only a slight to no effect (SeaLand, 1991).

Turnip-rooted Parsley (*Petroselinum crispum* subsp. *tuberosum*)

Storage in 11 kPa CO_2 + 10 kPa O_2 helped to retain the green colour of the leaves during storage (Hardenburg *et al.*, 1990), though optimum temperature for storage in air is about 0°C (Thompson, 1996).

Watermelon (*Citrullus lanatus*)

It was stated (SeaLand, 1991) that CA storage had only a slight or no effect. This was supported by Radulovic *et al.* (2007) who said, 'Watermelons are exclusively fresh consumed, and CA storage does not offer any benefits to watermelon quality.' Fonseca *et al.* (2004) found that 'Royal Sweet' had an increased respiration rate in storage at 7–9°C in < 14 kPa O_2 compared with air and they therefore recommended storage at 1–3°C in > 14 kPa O_2 atmospheres, but they found limited applicability of CA or MA packaging. However, Tamas (1992) showed that the storage life of 'Crimson Sweet' was extended from 14–16 days in ambient conditions to 42–49 days by cooling within 48 h of harvest and storage at 7–8°C and 85–90% RH in 2 kPa CO_2 + 7 kPa O_2. Changes in fruit appearance, flavour and aroma were minimal but improvement in the colour intensity and consistency of the flesh were noticeable.

Yam (*Dioscorea* spp.)

Little information is available in the literature on CA storage but one report indicated that it had only a slight to no effect (SeaLand, 1991). Fully mature *D. opposita* tubers can be stored for long periods of time at high O_2 tensions at 5°C, which was also shown to reduce browning of tubers (Imakawa, 1967).

13

Transport

Airfreight is used to transport some high-value fruits, but the transport time is so short that CA conditions are not necessary, though small insulated airfreight containers, designed by Envirotainer Worldwide, were produced that used dry ice for cooling, which would have also increased the CO_2 content in the container. However, large and increasing amounts of fresh fruit and vegetables are transported by sea. Some ships are fitted with refrigerated holds (break bulk) but most is commonly by refrigerated (reefer) containers (Fig. 13.1). Reefer containers were first introduced in the 1930s but it was only in the 1950s that large numbers of reefers were transported on ships. Both ships' holds and reefer containers may be fitted with CA facilities.

International transport of fruit and vegetables has had its critics, especially in terms of its use of energy. Schäfer *et al.* (2014) investigated the carbon footprint of asparagus from four sources to supply the German market. Their product carbon footprint was 1.75 kg CO_2e/kg (CO_2e = carbon dioxide equivalent) for the locally grown asparagus, 0.82–1.3 kg CO_2e/kg from Turkey or Greece by truck and 8 kg CO_2e/kg airfreighted from Peru. Michalský and Hooda (2015) compared apples, cherries, strawberries, garlic and peas airfreighted into the UK with those produced locally. They found that those imported from non-European countries contained, on average, embedded greenhouse gas emissions of 10.16 kg CO_2e/kg (production, airfreight and distribution within the UK) compared with emissions of 0.5 kg CO_2e/kg for those produced and supplied locally. Berners-Lee (2010) reported the carbon footprint of one apple produced locally and seasonal as 10 g CO_2e, and for shipped, cold-stored and inefficiently produced as 150 g CO_2e. For a banana he gave 80 g CO_2e imported and for oranges 90 g CO_2e /kg shipped 2000 miles by boat and 500 g CO_2e/kg by truck and 5.5 kg CO_2e/kg by airfreight.

Kruger and Truter (2003) stated: 'During the last years, the use of CA storage has considerably contributed towards improving the quality of South African avocados exported to Europe.' Ludders (2002) suggested that the main obstacles to the successful marketing of cherimoya could be overcome by using CA containers for sea-freight transport. Harman (1988) recommended the use of CA containers for the New Zealand fruit industry and considerable research has been carried out by Lallu *et al.* (1997, 2005) and Burdon *et al.* (2005, 2008, 2009) relevant to CA transport of kiwifruit and avocados. Dohring (1997) claimed that:

> ...avocados, stone fruit, pears, mangoes, asparagus and tangerine made up over 70% of [CA] container volumes in recent years. Lower value commodities e.g. lettuce, broccoli, bananas and apples, make up a

Fig. 13.1. Reefer containers plugged in at the port awaiting loading on to a ship.

> greater percentage of the overall global fruit and vegetable trade volumes but cannot absorb the added CA costs in most markets.

Bananas are shipped commercially from tropical and subtropical countries in the refrigerated holds of ships, some of which have CA equipment installed. For example, over 1000 t of bananas are shipped weekly from the West Indies in CA reefer ships. CA reefer containers are also used to a limited extent for bananas. Experimental reports published over several decades have indicated various gas combinations that have been suitable for extending the storage life of bananas. In 1997 there were problems with bananas ripening during transport on reefer ships from the West Indies to the UK. Tests were made using CA reefer containers but eventually CA reefer ships were used, which solved the problem, but added about 2 US cents/kg to the shipping costs. At the present time the bananas are transported across the Atlantic from Central America and the West Indies to Europe, North Africa and the Middle East on reefer vessels using MA packaging, or otherwise CA vessels and CA containers.

Hansen (1986) reported that in transport of fruit and vegetables the benefits of reduced O_2 were very good and those of increased CO_2 were also good, in that they both delayed ripening and high CO_2 also reduced chilling injury symptoms. CO_2 in cylinders was used for soft fruit during transport in 20 kPa CO_2, but he reported that dry ice could be used, which is cheaper.

He also reported that storage at 0°C required only one-third of the concentration of CO_2 required at 7°C for soft fruit. DeEll (2002) reported that 15–20 kPa CO_2 + 5–10 kPa O_2 reduced the growth of *Botrytis cinerea* and other decay-causing organisms during transport of blueberry. SeaLand (1991) recommended 10 kPa CO_2 + 5 kPa O_2 for surface transport of papaya. For transport of bananas, an atmosphere of 3 kPa O_2 + 5 kPa CO_2 was successfully used.

Sea Freight

The length of the postharvest life of many fruits is the limiting factor to sea freight. Rupavatharam *et al.* (2015) reported that, in New Zealand, feijoa growers harvest fruit at a stage of maturity that gives a storage potential of 4 weeks at 4°C but reliable sea freight from New Zealand to Asian, European and North American markets requires a postharvest life of some 6 weeks. Therefore they suggested the need to investigate the possibility of developing a more immature stage of harvest maturity to prolong the potential postharvest period. Singh and Zaharah (2015) found that mangoes could be stored for only 2–3 weeks at 12–13°C coupled with substantial losses in fruit quality. However, CA storage at 13°C in 3 kPa O_2 + 6 kPa CO_2 extended the postharvest life of the cultivars 'Kensington Pride' and 'R2E2' by up to 6 weeks while retaining their good quality. This gave potential for sea-freight export.

Gas environments in purpose-built CA reefer vessels

Purpose-built CA reefer vessels have been used for transport of banana cargoes for decades. The general configuration of reefer carriers is to have four holds (Holds 1 to 4) in sequence from stem to stern, each of which is also divided vertically into four (or possibly five) decks (Decks A to D or E). In a typical CA vessel, each hold is also segregated into two CA 'zones' so that Decks A and B constitute a single CA gas control zone and Decks C and D/E constitute the second CA zone within the same hold. A vessel with four holds would, therefore, have 16 CA gas control zones. When transporting produce such as bananas, it is standard practice to close-pack palletized fruit on each deck. The air temperatures for each deck and the gas concentrations in each CA gas zone are monitored and controlled independently.

Cooling

For maximum quality maintenance throughout the marketing period, it is crucial to achieve and maintain the optimum storage conditions as soon as possible. This is called the cold chain. In a study of the cold chain for blueberry exports, Paniagua *et al.* (2014) found that delays in cooling had a small effect on final product weight, whereas variation in storage temperature and atmosphere during simulated transport influenced both firmness and the incidence of rotting. Defraeye *et al.* (2016) investigated warm loading of citrus fruit into reefers for cooling during marine transport as an alternative to forced air pre-cooling as it could simplify logistics and be more economic. They found that it was theoretically possible to cool the produce in less than 5 days in a reefer container, but these cooling rates were not achieved in practice in their trials. They also commented that cooling performance of containers depended on the way in which the fruit was stowed and convectively cooled. They packed citrus fruits in ventilated boxes, pre-cooled them in the reefer container and compared a channelling configuration, which reduced airflow by-pass between pallets, with a horizontal configuration, which forced air horizontally through the pallets. The fruit was cooled within about 3 days to seven-eighths cooling time with standard ambient loading. The channelling configuration exhibited similar cooling behaviour but the fruit lost less moisture, had a longer shelf-life and was of better quality. The horizontal configuration performance was worse in all aspects. In the case of bananas, pre-climacteric fruit is loaded into containers or ships' holds at ambient temperatures. The cargoes are then subjected to rapid cooling with the objective to reduce the temperature of the return air circulating through the cargo to within 2.2°C (4°F) of the designated storage temperature (set point) within a few days of loading. The duration of this 'reduction period' provides a measure of the efficiency with which the fruit have been cooled. If the reduction period for any consignment of bananas is protracted, this suggests that the respiration rate of the fruit is super-optimal and, as a consequence, the green life of the fruit will probably be less than anticipated.

Pest control

Controlling pests and diseases on crops being transported internationally is vital in preventing their introduction into countries where the specific pest or disease may not exist. This has led to legislation governing quarantine and other control measures.

Methyl bromide

MB, a widely used agent in disinfestation procedures, was scheduled to be withdrawn in many countries in 2005 due to its ozone-depleting properties. It was reported by UNEP (1992) that MB was identified, under the authority of the World Meteorological Organization with the National Oceanic and Atmospheric Administration (NOAA) and National Aeronautics and Space Administration (NASA), as causing significant damage to the earth's protective ozone layer and is scheduled for global phasing out under the Montreal Protocol, an international treaty developed to protect the earth from the detrimental effects of ozone depletion. MB for pre-shipment and quarantine purposes was declared exempt from these restrictions. The concentration of MB required depends on the pests to be controlled as well as temperature. For example, Zoffoli and Latorre (2011) reported that for false red spider mite (*Brevipalpus chilensis*) a dose rate of 24 g/m^3 at greater than 26.5°C was required. Phillips *et al.* (2015) simulated shipping of rock melons from Australia to New Zealand by sea- and airfreight. No eggs survived MB fumigation, but they took 3–20 days to die, whereas phytosanitary inspections of rock melons occurred within 2–7 days of fumigation, which could lead to misinterpretation of effectiveness. Delays were not influenced by MB concentration, but were significantly lengthened by cooler storage temperatures.

Fumigation with ethyl formate combined with high CO_2 for 2½ h in export cartons in simulated cool down from 15°C to 10°C were found to control long-tailed mealybug (*Pseudococcus longispinus*) and citrus mealybug (*Planococcus citri*) without damaging the fruit. These treatments were reported to satisfy 'generally regarded as safe' criteria and were suitable for use in grapes and grapefruits produced under organic certification (Lima, 2015).

Irradiation

Follett (2014) reported that the USA, Australia and the International Plant Protection Convention had approved the generic radiation dose of 150 Gy for quarantine treatment of Tephritidae fruit flies. The USA has also approved the generic dose of 400 Gy for other insects, except pupae and adult Lepidoptera. A development was reported by Cetinkaya *et al.* (2016), who showed that DNA Comet Assay (single-cell gel electrophoresis assay) may be a practical quarantine control method for irradiated citrus fruit, since it has been possible to estimate the applied low doses as small as 100 Gy when it is combined with image analysis.

Hot air-assisted microwave disinfestation can be performed at farms or centralized packhouses, since the capital cost would be comparatively lower than vapour heat or ionizing radiation treatments. Gamage *et al.* (2015) investigated hot air (40°C) exposure of apples in combination with a pentagonal microwave system (giving an average temperature of 53.4 ± 1.3°C at the core of the fruit) that gave 100% mortality of the most tolerant stage of Queensland fruit fly (*Bactrocera tryoni*) and Jarvis's fruit fly (*B. jarvisi*) without any adverse impact on fruit quality. The development of small-scale

X-ray machines could provide farmers and packhouses with in-house treatment capability and accelerate adoption of the technology. Also a recent change in import regulations by the USA allows for treatment upon entry, which makes it possible for exporting countries to explore new markets prior to investing in treatment facilities at the origin (Bustos-Griffin *et al.*, 2015).

The effects of irradiation on fruit may be detrimental, depending on the dose, but it can also be beneficial. For example, Kang *et al.* (2012) found that X-ray irradiation at 200 and 400 Gy could reduce the respiration rate of grapes and extended their shelf-life. They found that there were no significant other physical or chemical effects of irradiation and the irradiated fruit had a better appearance than the non-irradiated fruit after 14 days of subsequent storage at 1.5 ± 0.5°C and 75 ± 5% RH.

Ozone

A company, Purfresh Transport, developed a reefer container with patented technology that has a slow ozone input throughout the transport period. They claimed that 'ozone molecules kill moulds, yeasts, and bacteria in the air and on surfaces by up to 99%, as well as to consume and regulate ethylene levels'. It was reported to have been successfully tested on avocados, bananas, berries, citrus fruit, cucumbers, grapes, kiwifruit, mangoes, melons, onions, peppers, pineapples, pome fruit, potatoes, stone fruit, tomatoes and 'tropicals' (Bordanaro, 2009). The company carried out a trial on transport of nectarines from California to Taiwan taking 14 days. They reported that the Purfresh container 'outperformed the CA treatment on net weight loss, fruit pressure, brix content, and microbial counts'.

Quarantine

In-transit pest control was described by Fallik *et al.* (2012). They carried out experiments whose results provided the basis for the future establishment of an export quarantine treatment protocol for capsicum fruit. This was based on in-transit storage at temperatures of 1.5°C for 21 days to eliminate Mediterranean fruit fly (*Ceratitis capitate*) eggs and larvae from capsicum fruit while maintaining the commercial quality of the fruit. However, in other work, it was reported that storage of capsicums at 5°C for 2 weeks resulted in chilling injury (González-Aguilar, 2002). Unahawutti *et al.* (2014) found that infestation of mangosteens with the quarantine pest oriental fruit fly (*Bactrocera dorsalis*) and carambola fruit fly (*B. carambolae*) could only occur if fruit had cracks or mechanical injury.

Monitoring

The manner in which the quality of a crop is monitored and maintained relative to predetermined criteria during export can vary, but the random sampling of boxes at different points along the supply chain has been used for decades. For example, the Windward Islands banana industry in the 1990s used a comprehensive method (Table 13.1) where boxes were tested as they were being loaded on the ship in the Windward Islands and then another sample was taken for testing on arrival in the UK. The data was sent back to the exporting countries immediately so that any problem could be dealt with before the next weekly shipment. All the boxes had an identification code to enable the grower and packer to be identified.

In the past two decades there has been an impetus to develop technologies that would allow the remote monitoring of not only the location but also the condition of merchandise during transit. One of the first systems developed was that which deployed radio frequency identification technology (RFID). Morris *et al.* (2003) described technologies that had been developed for monitoring refrigerated shipping containers that made it possible for actual storage conditions during transport to be monitored in real time, thus forewarning shippers and carriers of potential problems with a consignment before its arrival at its destination or trans-shipping point. At that time, Morris discussed the possibility of integrating the

Table 13.1. Levels of defects allowed for Class 1 fruit as used by the banana export industry in the Windward Islands in the 1990s (0.4 inch^2 = 2.58 cm^2).

Defect	Tolerance level per cluster
Superficial bruise	Up to 0.4 inch^2
Latex staining	Up to 0.4 inch^2
Maturity stain	Up to 0.4 inch^2
Pest damage	Up to 0.4 inch^2
Red rust damage	Up to 0.4 inch^2
Scars	Up to 0.4 inch^2
Sooty mould	Up to 0.4 inch^2
Flower thrips	Up to 25%
Speckling/pinspotting	Up to 50%
Crown trimming	0
Damaged pedicel/ neck injury	0
Dirty	0
Finger end rot	0
Fused fingers	0
Misshapen/ malformed fingers	0
Mutilated fingers	0
Overgrade	0
Peel burn	0
Residue	0
Ripe and turning	0
Scruffy	0
Short finger	0
Stale fruit	0
Undeflowered	0
Undergrade	0

potential to monitor the physiological status of a consignment with storage life prediction models. At the time of writing, technological advances now allow for the routine real-time monitoring of O_2 and CO_2 concentration, temperature and relative humidity within CA reefer vessels and containers. In the past few years container operators have begun to develop services that more nearly approach those envisaged by Morris. Systems are now in place whereby the atmospheric conditions within containers can be monitored remotely and relayed to the shippers. As yet the interphase with storage life predictive models has yet to be achieved. This goal will require more investment, research and a more comprehensive understanding of the variability of temperature and gas concentration gradients within the cargoes being transported.

As in the case of CA reefer containers, the gas concentrations within each CA zone on board a ship are usually monitored by sensors positioned in the machine space external to or at the periphery of the stow where the circulating air returns to be recooled. In order to monitor air temperature, temperature probes are located within the airstreams entering and leaving the decks. These sensors record supply air temperatures (SAT) (also referred to as delivery air temperature (DAT)) and return air temperatures (RAT). On some vessels each deck within the holds may also be fitted with United States Department of Agriculture temperature probes/sensors (USDA probes) attached to flexible cables. It is common for at least two USDA probes to be present on any one deck. This allows air temperatures to be monitored at various locations to the port and starboard side of each deck. Temperature and humidity data loggers are also used to record environmental conditions within the cargo. Often these are placed adjacent to the merchandise in cartons on the top of or in the upper rows of individual pallets. The information recorded by such data loggers provides a more accurate estimate of the air temperatures and humidity in the immediate vicinity of the produce. In comparison, the gas and temperature sensors located externally to the stows provide a more general impression of the status of the air passing to or from the decks and gas control zones.

CA Transport Technology

Reefers are increasingly of standard size and capacity. Champion (1986) reviewed the state of CA transport as it existed at that time and defined the difference between CA containers and MA containers. CA containers had some mechanism for measuring the changes in gases and adjusting them to a pre-set level. MA containers had the appropriate mixture injected into the sealed container at the beginning of the journey with no subsequent control, which means that the atmosphere would constantly change during transport due to respiration and leakage. The degree of control over the gases in CA containers is affected by how gas tight the container is; some early systems had a leakage rate of 5 m^3/h or more, but with improved technology systems can be below 1 m^3 (Garrett, 1992). Much of the air leakage is through the door and fitting plastic curtains inside the door can reduce the leakage, but they are difficult to fit and maintain in practice. Some reefer containers introduced in 1993 had a single door instead of the double doors, which are easier to make gas tight. According to Garrett (1992), the system used to generate the atmosphere in a container falls into three categories:

1. The gases that are required to control the atmosphere are carried with the container in either a liquid or solid form.
2. Membrane technology is used to generate the gases by separation.
3. The gases are generated in the container and recycled with pressure absorption and swing absorption technology.

The first method involves injecting N_2 into the container to reduce the level of O_2 with often some enhancement of CO_2 (Anonymous, 1987). It was claimed that such a system could carry cooled produce for 21 days compared with an earlier model, using N_2 injection only, which could be used only on journeys not exceeding 1 week. The gases were carried in the compressed liquid form in steel cylinders at the front of the container, with access from the outside. O_2 levels were maintained by injection of N_2 if the leakage into the container was greater than the utilization of O_2 through respiration by the stored crop. If the metabolism of the crop was high, the O_2 could be replenished by ventilation.

In containers that use membrane technology, the CO_2 is generated by the respiration of the crop and N_2 is injected to reduce the O_2 level. The N_2 is produced by passing the air through fine porous tubes, made from polysulphones or polyamides, at a pressure of about 5–6 bar. These will divert most of the O_2 through the tube walls, leaving mainly N_2, which is injected into the container (Sharples, 1989b). A CA reefer container, which has controls that can give a more precise control over the gaseous atmosphere, was introduced in 1993. The containers used ventilation to control O_2 levels and a patented molecular sieve to control CO_2 as described in Chapter 8. The levels of gas, temperature and humidity within the container are all controlled by a computer, which is an integral part of the container. It monitors the levels of O_2 from a paramagnetic analyser and the CO_2 from an infra-red gas analyser and adjusts the levels to those that have been pre-set in the computer (Fig. 13.2).

Champion (1986) mentions the 'Tectrol' (total environment control) gas sealing specifications that were developed and first used by the Transfresh Corporation of California in 1969. They began with the shipment of stone fruit and avocados from Chile to the USA in 1992. In this system, after loading, the reefer containers were flushed with the desired mixture of gases and then sealed. The reefer unit had N_2 tanks that were used to correct the deviation of O_2. Lime was included to lower the CO_2 level and magnesium sulfate to reduce ethylene. This system was only satisfactory when O_2 was the only gas to be controlled and the journey was short. The name Tectrol is also used for a system where natural or propane gas is burned with outside air that is introduced into CA stores to lower the O_2 and increase CO_2.

A system between MA and CA called Fresh-Air Exchange was developed that has mechanical ventilation to provide some control over O_2 and CO_2. Taeckens (2007) described the version made by Carrier called AutoFresh™:

> ... when the CO_2 level reached a preset point, the system activates, drawing in outside air to add O_2 and ventilate excess CO_2. Because Fresh-Air Exchange systems rely on natural atmosphere, the interior of the container being ventilated will generally be approaching a 78 kPa N_2 level, with O_2 and CO_2 in some combination making up the balance.

He thus maintained that the system was useful where the optimal combined CO_2 + O_2 levels were around 21 kPa and gave the example for strawberries at 6 kPa O_2 + 16 kPa CO_2. A German company, Cargofresh (2009),

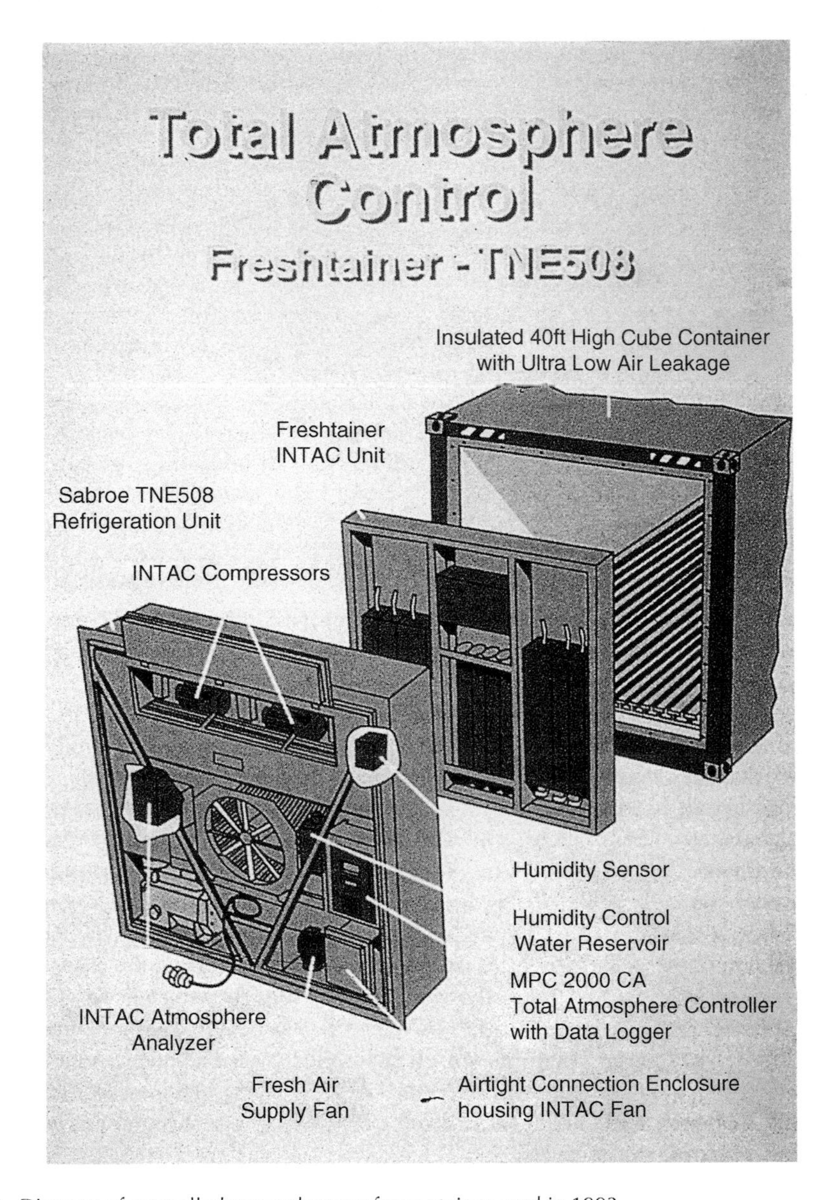

Fig. 13.2. Diagram of controlled atmosphere reefer container used in 1993.

reported that by connecting 'an external N_2 supply, an MA container can be "switched" to a fully fledged CA system. In this way, even with a central nitrogen supply, each individual container can be individually regulated.' They called this system 'CargoSwitch' and it was launched in February 2008.

There continue to be challenges in achieving the optimum CA conditions. During transport there can be considerable variation in temperature and gaseous atmosphere within a hold or container. This is due in part to stacking patterns not being adequate for air circulation. Respiratory heat at 'normal storage temperatures' in watts/tonne (W/t) was reported by Cambridge Refrigeration Technology (2001) as: avocado 183–465, asparagus 81–237, mango 133, banana 59–130, cauliflower 44, apple 9–18, grape 3.9–6.8, orange 18.9 and cherry 17–39. Roger Bancroft (2017, personal communication) stated:

> My experience with the out-turn quality from commercial CA transport is that the actual gas concentrations generated in containers is extremely variable so one container's CA is not equivalent to the next. In bulk carriers too although ships' data suggest gas concentrations and air temperatures are held within various parameters, in fact the temperature gradients within the cargo can differ within and between holds, so again the actual storage environment experienced by fruit at different locations can be very different. This will contribute to heterogeneous ripening behaviour [of climacteric fruit] after release from both CA and refrigeration.

Cronos CA containers

There are several patented systems available to control the gases in reefer containers. Cronos Containers supply inserts called the 'Cronos CA System', which can convert a standard reefer container to a CA container. The installation may be permanent or temporary and is self-contained, taking some 3 h the first time. If the equipment has already been installed in a container and subsequently dismantled, reinstallation can be achieved in about an hour. The complete units measure 2 m × 2 m × 0.2 m, which means that 50 of them can be fitted into a 40 ft (12 m) dry container for transport to where they will be used. This facilitates management of the system. It also means that they take up little of the cargo space (only 0.8 m^3) when fitted into the container. The unit operates alongside the container's refrigeration system and is capable of controlling, maintaining and recording the levels of O_2, CO_2 and humidity to the levels and tolerances pre-set into a programmable controller. Ethylene can also be removed from the container by scrubbing. It was claimed that this level of control was greater than any comparable CA storage system, increasing shipping range and enhancing the quality of fresh fruit, vegetables, flowers, fish, meat, poultry and similar products. The system is easily attached to the container floor and bulkhead and takes power from the existing reefer equipment, with minimal alteration to the reefer container. The design and manufacture are robust to allow operation in the harsh (marine) environments that will be encountered in typical use. The system fits most modern reefers and is easy to install, set up, use and maintain. A menu-driven programmable controller provides the interface to the operator, who simply has to pre-set the required percentages of each gas to levels appropriate for the produce in transit. The controls are located on the front external wall of the container, and once set up a display will indicate the measured levels of O_2, CO_2 and relative humidity. The system consists of a rectangular aluminium mainframe on to which the various sub-components are mounted (Fig. 13.3).

The compressor is located at the top left of the mainframe and is driven by an integral electric motor supplied from the control box. Air is extracted from the container and pressurized (up to 4 bar) before passing through the remainder of the system. A pressure relief valve is incorporated in the compressor along with inlet filters. A bleed supply of external air is ducted to the compressor and is taken via the manifold with a filter mounted externally to the container.

The compressed air is cooled prior to passing through the remainder of the system. A series of coils wound around the outside of the air cooler radiate heat back into the container. The compressed air then passes into this component, which removes the pressure pulses produced by the compressor and provides a stable air supply. The water trap passing into the CA storage system then removes moisture. Water from this component is ducted into the water reservoir and used to increase the humidity when required. Two activated alumina drier beds are used in this equipment, each located beneath one of the N_2 and CO_2 beds. Control valves are used to route the air through parts of the system as required by the conditioning process. Mesh filters are fitted to the outlet that vents N_2 and CO_2 back into the container and to the outlet that vents oxygen to the exterior of the container. N_2 and CO_2 beds are located above the dryer

Fig. 13.3. Cronos CA reefer container (a) exterior and (b) interior.

sieve beds and contain zeolite for the absorption of nitrogen and CO_2.

Ethylene can be removed from the container air if required. A single ethylene absorption bed is used which contains a mixture of activated alumina and Hisea material (a clay mineral-based system patented by BOC). Air from the container is routed to the bed that remains pressurized for several hours and then depressurized via the O_2 venting lines. The O_2 flow is then routed through the bed for 20 minutes in order to scrub it. After this the process is repeated.

To increase the relative humidity within the container, an atomized spray of water is injected as required into the main airflow through the reefer. The water supply is taken from a reservoir located at the base of the mainframe which is fed from the reefer defrost system and the water trap. Air from the instrument air buffer is used to form the spray by drawing up the water as required. A further stage of moisture removal is carried out using dryer beds that are charged with activated alumina. From here the air is routed either directly to the nitrogen and CO_2 beds, or via the ethylene bed, depending on the type of conditioning needed. The reefer process tends to decrease the humidity within the container, with water being discharged from the defrost equipment. When the CA storage system is used, this water is drained into a 5 l reservoir located within the mainframe and used to increase the humidity if required. Air from the instrument air buffer is used to draw the water into an atomizing injection system located in the main reefer airflow. The control valve is operated for a short period, and once the additional water spray has been mixed in with the air in the container the humidity level is measured and the valve operated again if required.

Four 10 l bottles of CO_2 are located in the mainframe. To increase the level of CO_2 within the container, the gas is vented via a regulator. CO_2 is retained in the molecular sieve along with the nitrogen. Normally this would be returned to the container, but if it is required to remove CO_2 then the flow of gas (which is mainly CO_2 at this point) is diverted for the last few seconds to be vented outside the container. This supply of CO_2 is intended to last for the duration of the longest trip. A check to ensure that the bottles are full is included as one of the pre-trip checks.

O_2 removal is accomplished using a molecular sieve, which is pressurized at up to 3 bar. N_2 and CO_2 are retained in the bed. The separated O_2 is taken from the sieve bed to the manifold at the bottom of the mainframe and then vented outside the container. It is possible to reduce the level of oxygen to around 4 kPa, though this does depend on a maximum air leakage into the container and in certain cases O_2

levels of 2 kPa or even 1.5 kPa can be maintained (Tim Bach, personal communication, 1997).

To add O_2, air from outside the container is allowed to enter by opening a control valve for a short period. Once the additional air has been mixed in with the air in the container, the oxygen level is measured and the valve operated again if required.

The N_2 and CO_2 retained in the main sieve beds are returned to the alumina dryer beds in order to recharge them and remove moisture from them. Finally the gas is returned to the main container via a valve located at the bottom of the mainframe. If the CO_2 is to be removed, a modified process occurs. The levels of O_2, CO_2 and humidity are continually monitored and the values passed back to the display. A supply of gas is taken from the general atmosphere within the container via a small pump located within the control box. The gas is passed on to an O_2 transducer and then a CO_2 transducer before being returned to the container. The location of these components is shown in Fig. 13.4. A transducer located in the airflow through the mainframe and connected into the control box measures humidity. The above sequence of operation is carried out according to instructions provided by the display. In particular, the measured values of O_2, CO_2 and humidity are compared in the display with those preset by the operator and the gas conditioning cycle is adjusted accordingly. Once set up, the process is automatic and no further intervention is required by the operator unless a fault occurs. Further detail on setting up the system is provided in the next section of the manual. In addition to providing the operator interface, the display has a communications socket that allows data to be transferred (e.g. to/from an external computer). The most frequent use of this facility will be the downloading of the data logged during normal operation of the system – the levels of O_2, CO_2 and relative humidity as periodically recorded. Data is output in a form that can be used in a spreadsheet. It is also possible to use the external computer to modify system settings and carry out other diagnostic operations.

Hypobaric Containers

Various other methods have been used for modifying the atmosphere during fruit and vegetable transport, including hypobaric containers (see also Chapter 11). A commercial hypobaric container was developed in the mid 1970s and used for fish and meat as well as fruit and vegetables. However, to prevent implosion it was very strongly constructed and therefore heavy and expensive. The Grumman Corporation in the USA constructed a hypobaric container that they called 'Dormavac'. It was operated at 2.2–2.8°C and a pressure of 15 mm Hg and they tested it in commercial situations, but were unable to make it profitable, resulting in losses of some $50 million (Grumman Park, undated). Alvarez (1980) described experiments where papaya fruits were subjected to sub-atmospheric pressure of 20 mm Hg at 10°C and 90–98% RH for 18–21 days during shipment in a hypobaric container from Hawaii to Los Angeles and New York. Both ripening and disease development were inhibited. Fruits ripened normally after removal from the hypobaric containers, but abnormal softening, unrelated to disease occurred in 4–45% of fruits of one packer. It was found that hypobaric stored fruits had 63% less peduncle infection, 55% less stem-end rot and 45% fewer fruit surface lesions than those stored in a refrigerated container at normal atmospheric pressure. A hypobaric system was described by Burg (2004) called VacuFresh^{sm}. It had a very slow removal of air and therefore it was claimed that there was no desiccation problem of the fruit and vegetables transported in these containers (Fig. 13.5).

MA Packaging

MA packaging has been developed for particular transport situations. For long-distance transport of bananas a system was developed and patented by the United Fruit Company called 'Banavac' (Badran, 1969).

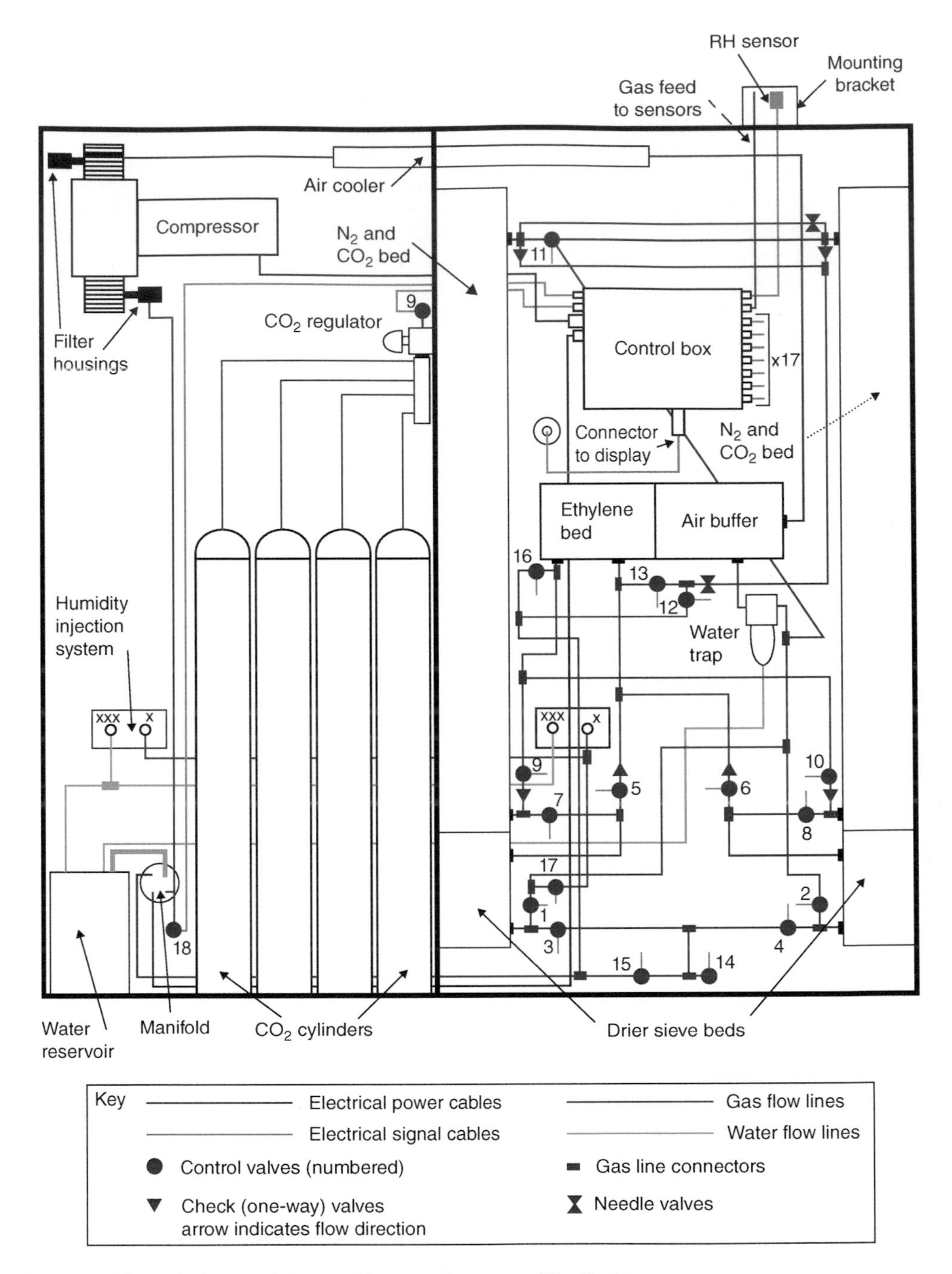

Fig. 13.4. Schematic layout of Cronos CA system (courtesy of Tim Bach).

The system uses PE film bags, 0.04 mm thick, in which the fruit is packed (typically 18.14 kg) and a partial vacuum is applied and the bags are sealed (Fig. 13.6). Typical gas contents that develop in the bags through fruit respiration are about 5 kPa CO_2 and 2 kPa O_2 during transport at 13–14°C (De Ruiter, 1991).

Fig. 13.5. Intermodal tank container from Welfit Oddy (Pty) Ltd, Port Elizabeth, South Africa.

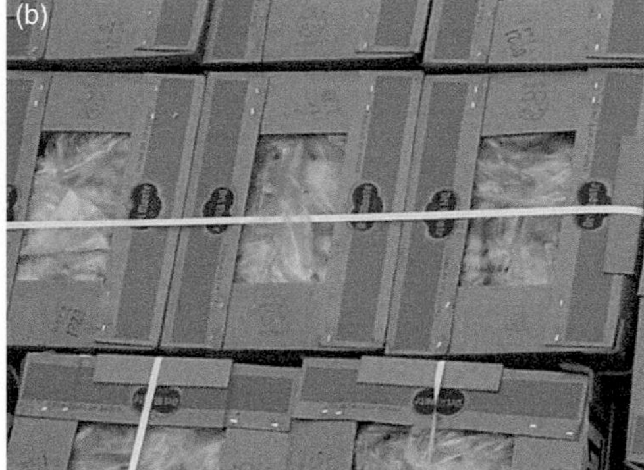

Fig. 13.6. Banavac MA packaging: (a) box with plastic bag before packaging; (b) packed boxes loaded on to a pallet and stowed in a ship for transport.

CA Transport Trials and Case Studies

Some sources of variability in MA and CA shipments

Despite improved understanding of physiology, and advances in handling practices and technology, those involved in the transportation of fresh produce are only too aware that there is always a risk that out-turn quality of a proportion of shipments will result in a biological and/or economic loss. Such outcomes may be the consequence of a single factor or the interaction of multiple factors. Companies and research establishments have worked on CA transport and the details of several experiments are described in the literature.

One of the recurrent problems that cause failure in CA-stored produce is the inefficient management of the required concentration of O_2 and CO_2. Whether or not CA stores function as required, occasionally the physiology status of the stored produce or a sub-component is such that neither refrigeration nor the gas conditions are able to prevent premature ripening, deterioration or senescence. Physiological stress and diseases may also play a role in negating the beneficial influence of MA or CA technologies. The following case studies seek to illustrate these points.

Mangoes

Maekawa (1990) reported that green-mature 'Irwin' mangoes were transported in a refrigerated truck at 12°C on a car ferry from Kume Island (Japan) to Tokyo Port and then from Tokyo to Tsukuba University by land. The total journey took 4 days. On arrival the fruits were stored in 5 kPa O_2 + 5 kPa CO_2 at 8°C, 10°C or 12°C. Fruit quality was retained at all temperatures for 1 month. No chilling injury was observed, even at the lowest temperature. Dos Prazeres and de Bode (2009) carried out shipment trials for 21 days at 10°C and 95% RH in 3 kPa O_2. They reported that all fruit arrived in good condition with 'no quality losses' and with a shelf-life of up to 10 days. They also did static trials of storage for 43 days and then an actual shipment taking 49 days and found that 'Kent' and 'Tommy Atkins' proved the best cultivars, since they arrived in good condition. All major aroma volatile compounds in ripe fruit decreased as the level of CO_2 was increased over the range of 3, 6 or 9 kPa at 2 kPa O_2 during 21 days storage of the cultivar 'Kensington Pride' at 13°C. Storage in 2 kPa O_2 + 6 kPa CO_2 at 13°C appeared to be promising for extending the shelf-life and maintaining fruit quality, while storage in 2 kPa O_2 + 2 kPa CO_2 appeared to be better at maintaining the aroma compounds of ripe fruit (Lalel *et al.*, 2005). Successful simulated commercial export of mangoes using the 'Transfresh' system of CA container technology was described in Ferrar (1988). Peacock (1988) described a CA transport experiment in Australia where mangoes were harvested from three commercial sites in Queensland when the TSS was 13–15%. Fruits were de-stalked in two of the three sources and washed, treated with 500 ppm benomyl at 52°C for 5 min, cooled and sprayed with prochloraz (200 ppm), dried and sorted and finally size graded and packed in waxed fibreboard cartons. After packing, fruits were transported by road to a pre-cooling room overnight (11°C). Pulp temperatures were 18–19°C the following morning. After 36 h of transport to Brisbane and overnight holding in a conventional cool room, the fruits were placed in a CA shipping container. At loading, the pulp temperature was 12°C. Fruits were stored in the container at 13°C with 5 kPa O_2 + 1 kPa CO_2 for 18 days. Ripening was significantly delayed in CA storage, though there were problems with CO_2 control, related to the efficiency of the scrubbing system. There was virtually no anthracnose disease but there were very high levels of stem-end rot. On removal from the CA container, fruits immediately began to ooze sap and the loss continued for at least 24 h.

Kiwifruit

Lallu *et al.* (1997) described experiments on CA transport of kiwifruit in containers where the atmosphere was controlled by either N_2 flushing and Purafil (potassium permanganate absorbed on to a clay mineral) or lime and Purafil and a third with no CA. N_2 flushing + Purafil maintained CO_2 levels of approximately 1 kPa and O_2 at 2–2.5 kPa while with lime + Purafil the CO_2 levels were 3–4 kPa and the O_2 levels increased steadily to 10 kPa. On arrival the ethylene levels in the three containers were less than 0.02 μl/l, indicating that the potassium permanganate was not necessary. However, they concluded that CA containers can result in benefits to fruit quality.

Blueberries

Storage in high concentrations of CO_2 is effective in delaying deterioration as well as reducing the susceptibility of fruit to fungal rots. The carriage instructions for a sample of eight containers of Chilean blueberries, analysed in 2013 (R.D. Bancroft, commercial unpublished data) and considered here, required refrigeration to −1°C and CA conditions of 8 kPa O_2 and 14 kPa CO_2. The CA units deployed had the capacity to regulate O_2 and CO_2 by the introduction of N_2 and CO_2, respectively. Analysis of the temperature logs for all the containers showed good compliance with the temperature set point. In six of the eight containers the concentration of O_2 reached the set point within 2–3 days of CA initiation. In one of the containers the concentration of O_2 was eventually reduced to a mean of 3.5 kPa, while in another container the average O_2 for the voyage was 11 kPa. As regards CO_2, the majority of the containers never reached consistent parity with the set point of 14 kPa. Certain CO_2 maxima were attained and then the concentration of this gas declined over time (Fig. 13.7).

These observations suggest that the gas atmosphere storage environments that became established in these containers were more variable than might be expected. A further complication was that each pallet stack was covered in a polyethylene cover (sometimes called a poly-foil). The covers were perforated with 12 mm holes at a frequency of 22 perforations/m^2. It seems likely, therefore, that the gas concentrations immediately adjacent to the blueberries deviated in some manner from those recorded by the CA units attached to the containers.

Analyses of the out-turn quality of the fruit transported in these containers did show some slight variation, with the proportion of uncompromised fruit ranging from 83% to 90%, but such trends were found to correlate with the variety and source of the fruit rather than the CA conditions.

In this study, although it transpired that the operation instructions cited 'Controlled Atmosphere' storage, in reality what was achieved was a 'primed' form of MA storage. Here the level of CO_2 was elevated at the beginning of the transit period and thereafter declined. This is understood to be 'normal' practice for the export of blueberry fruits

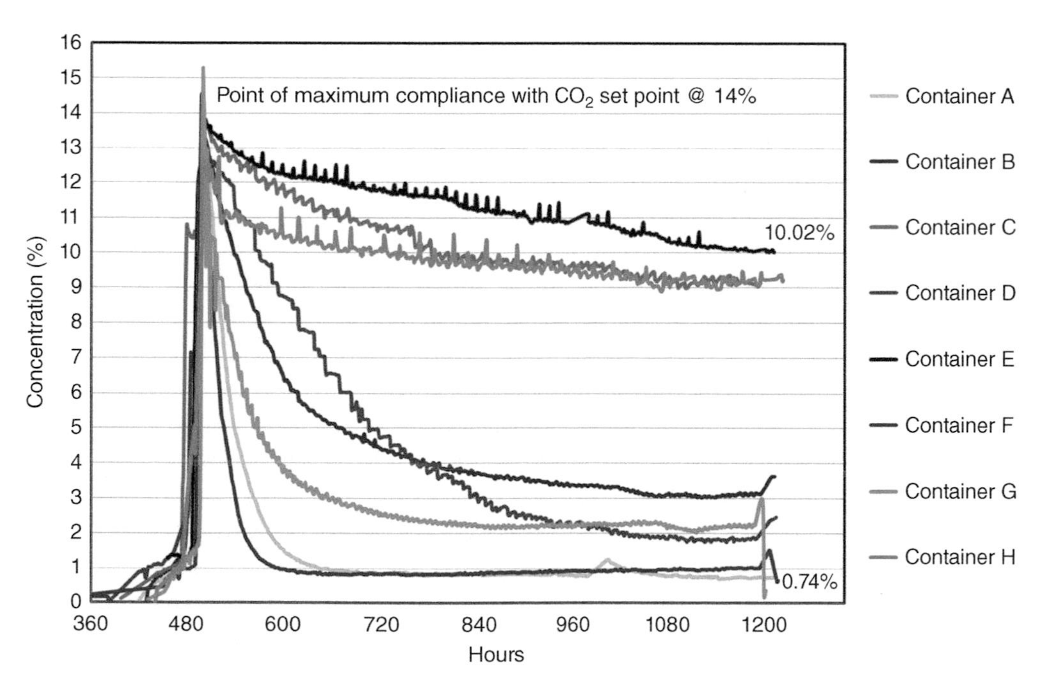

Fig. 13.7. The relative compliance of a series of eight CA containers to a CO_2 Set Point of 14%. Each container held palletized blueberries transported from Chile to the UK. The period of transit from gas initiation to de-vanning was approximately 32 days.

from Chile and it was advised that this system had worked well over the years. If so, then it is suggested that whatever the gas atmospheres registered in the CA control units, the partial pressures of O_2 and CO_2 within the pallets themselves must be adequate to permit the journey to Europe. From a commercial point of view it would be useful to know more precisely what gas levels really existed within the pallets and whether these could be achieved without aspiring to a 14 kPa CO_2 level at the beginning of the voyage and the maintenanco of 8 kPa O_2 throughout.

Avocados

Eksteen and Truter (1989) stored avocados in 'Freshtainer' 40 ft CA containers controlled by microprocessors. The set conditions were 7.4 ± 0.5°C for 8 days followed by 5.5 ± 0.5°C for 7 days, with 2 ± 0.5 kPa O_2 and a maximum of 10 ± 0.5 kPa CO_2. They reported that the containers were capable of very accurate control within the specified conditions. Also the fruit quality was better than that of the controls, which were 2 weeks storage at 5.5°C in air in insulated containers.

Bananas

The storage technologies deployed to transport bananas across the Atlantic to the European markets are well established. A small sample of 11 CA containers, investigated in 2016 (R.D. Bancroft, commercial unpublished data), suggested, however, that the degree to which the gas concentrations within individual containers conform to the standard set points can be very variable. The transit period for all containers discussed here was approximately 34 days. This involved a sea voyage of a nominal 9 days followed by road haulage across Europe to the UK. There were two transhipment periods. The set points for temperature, O_2 and CO_2 concentrations were 14°C, 3 kPa and 4 kPa, respectively. When powered, none of the containers failed to regulate their air temperatures as programmed. The out-turn quality of three of the containers, however, resulted in claims for loss of produce. Of the other eight CA containers, the control of CO_2 was satisfactory. Four of these containers also regulated O_2 to the set point, but the mean O_2 concentration of the remaining four containers varied from 11 kPa to 14 kPa.

Of the CA containers with poor out-turn results, one exhibited no gas control whatsoever throughout the entire transit period. As a consequence the endogenous levels of CO_2 began to rise during the last 6 days, due to the premature ripening of the banana fruit. The plots of CO_2 associated with the last two failed containers are reproduced in Figs 13.8 and 13.9. In one container (Fig. 13.8) the data indicates a discontinuity in CO_2 control following discharge from the vessel in Europe. This lasted for a period of about 10 days. Although CO_2 regulation was re-imposed during the later period of haulage, this was insufficient to prevent damage to the cargo.

In the second example (Fig. 13.9), despite appropriate cooling and CO_2 control during early transit, the concentration of CO_2 then rose consistently some 16 days after containerization. The data set then showed intermittent reductions in CO_2 that would have been associated with the opening of the container on a daily basis during the last 3 days of storage. Concurrent with the increase in CO_2 gas concentrations found in the air being recycled from within the container is a rise in return air temperature (RAT). This provided further evidence that the respiratory activity within the cargo had changed. The timing of this transition indicates not only that at least some of the bananas had begun to ripen but also that it is very likely that these fruit were already over-mature when they had been originally stowed. Once a sub-component of the cargo had begun to ripen, the process will have become autocatalytic and the remaining fruit will have also become susceptible to premature ripening. Under such conditions the CA control unit would no longer be able to influence the levels of CO_2 within the container. In this example, ultimately the inherent physiology of the fruit negated the potential of the CA unit to conserve the cargo.

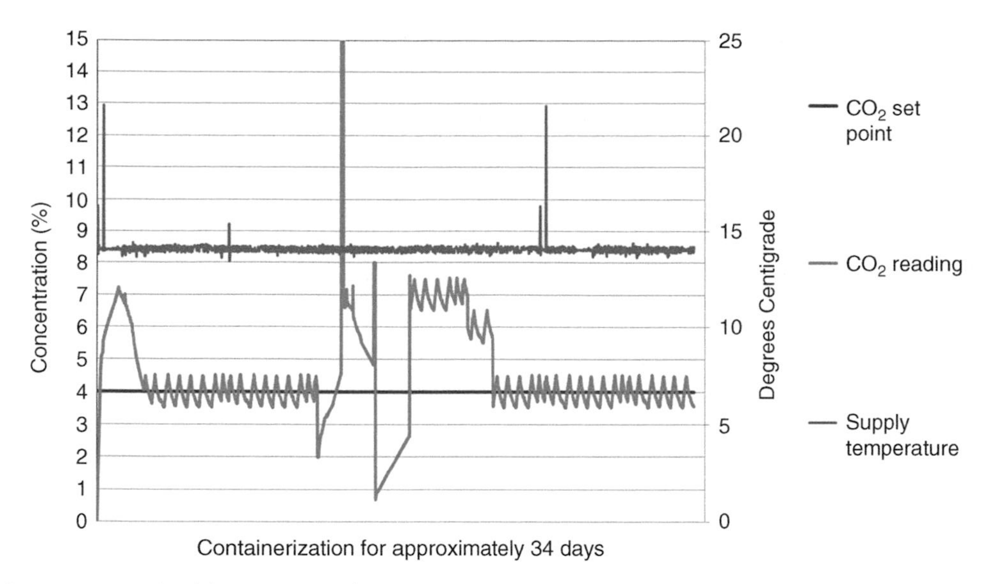

Fig. 13.8. Example of the progressive failure to control CO_2 concentration in a CA container transporting bananas across the Atlantic to Europe and finally by road haulage to the UK. This was attributed to periods of power loss or management oversight associated with transhipment.

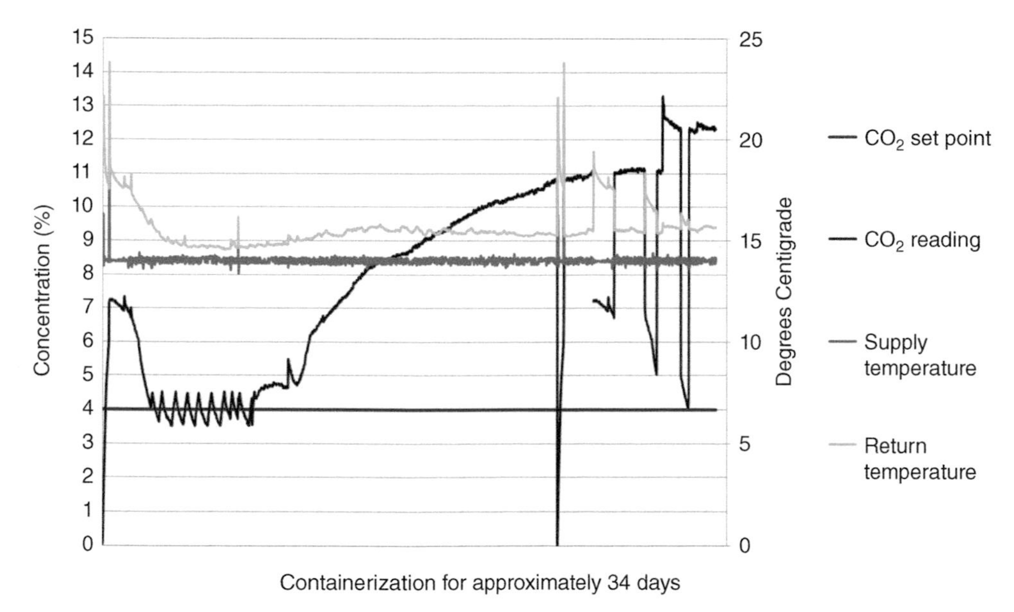

Fig. 13.9. Example of the progressive failure to control CO_2 concentration in a CA container transporting bananas across the Atlantic to Europe and finally by road haulage to the UK. The cause was the premature ripening of fruit.

Eksteen and Truter (1989) stored bananas in 'Freshtainer' 40 ft CA containers controlled by microprocessors. The set conditions for bananas were: 12.7 ± 0.5°C for 8 days followed by 13.5 ± 0.5°C for 11 days, with 2.0 ± 0.5 kPa O_2 and a maximum of 7 ± 0.5 kPa CO_2. The fruit quality was better than that of the controls, which were 2 weeks of storage at 13.5 ± 0.5°C in air in insulated containers.

Over-mature bananas

As seen in the previous case studies above, the efficiency with which the air temperature and gas concentrations may be managed within a storage environment does not preclude the loss in quality of a cargo. Figure 13.10 shows the O_2 concentrations established in six CA gas control zones aboard a CA reefer vessel transporting bananas from Central America to Europe in 2008. The set point for O_2 was 3%. Although the majority of the zones did not achieve this level of reduction during the initial phase of the voyage, experience suggests that O_2 concentrations between 3% and 8% would not necessarily have compromised the storage potential of the generality of the cargo. With regards to the management of CO_2, relative to a set point of 5% CO_2, following the imposition of CA conditions five of the six CA gas control zones equilibrated to approximately 6.8% CO_2 (Fig. 13.11). Although not ideal, in themselves such O_2 and CO_2 concentrations should have been sufficient to conserve the cargo during its passage to Europe. However, the elevated levels of CO_2 recovered from the different CA gas control zones indicated that the fruit was not quiescent and was respiring more than would have been expected. Within a few days it transpired that at least some of the bananas in all the CA holds had begun to ripen prematurely, so much so that the ship's crew terminated CA storage 9 days into the voyage and allowed ambient air into all the CA gas control zones.

On this particular vessel, Hold 2 transported bananas in Banavac packaging. With a temperature set point of 13.3°C, this will have generated an MA environment sufficient to conserve sound fruit for approximately 2 months. The evidence that the fruit committed to the CA holds was over-mature on loading was confirmed by an analysis of the air temperatures recorded by the USDA probes present on all the decks. Figure 13.11 depicts the mean air temperature registered by the USDA probes in the CA gas control zones in Holds 1, 3 and 4. This indicates that, on loading, the mean air temperature adjacent to the fruit was 21°C. Some 5 days later this had been reduced to 18°C before rising progressively from then on to some 23°C at discharge. A consistent reduction in air temperature to within 2.22°C of the set point did not occur in any of these

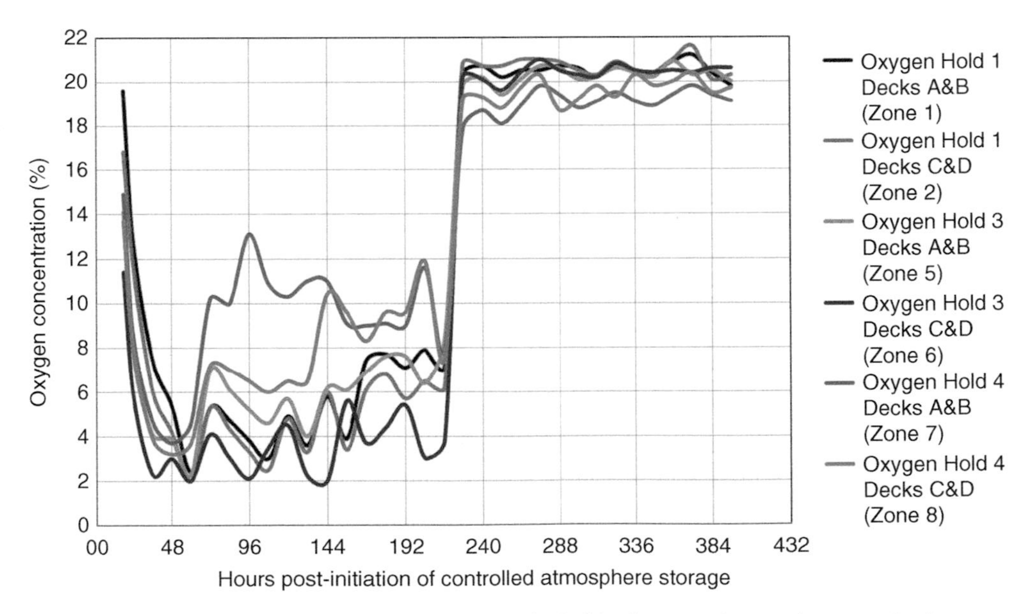

Fig. 13.10. Variation in O_2 concentrations recorded in the holds of a CA reefer vessel transporting bananas in 2008.The premature ripening of the cargo eventually resulted in the termination of active gas storage some 9 days into the voyage.

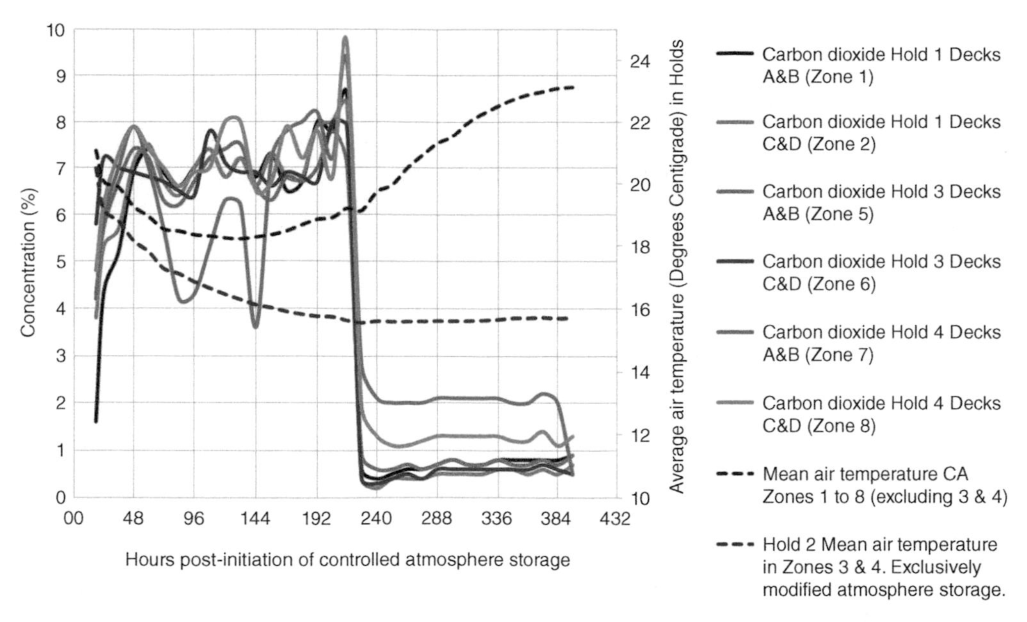

Fig. 13.11. Variation in CO_2 concentrations recorded in the holds of a CA reefer vessel transporting bananas in 2008.The mean air temperature recorded in the active CA holds (Gas Control Zones 1 to 8, excluding 3 and 4) indicated that the cargo had been over-mature on loading. In contrast, bananas held in modified atmosphere packaging in Hold 2 (Gas Control Zones 3 and 4) stored well and were discharged in good condition.

holds. In comparison, the air temperatures in Hold 2 show that the reduction period for the fruit held in MA packaging was 108 h. Thereafter the mean air temperature in Hold 2 was 15.8°C and a respectable 2.5°C above the set point for this cargo. There was no reported loss in quality in those fruit transported in Hold 2. There is no suggestion that the vessel failed to comply with the carriage instructions issued by the shipper and hence the loss in quality of the bananas recovered on discharge was due to their advance maturity on loading.

Mature bananas and latent diseases

Under normal circumstances, reefer vessels exporting bananas from ports in South or Central America and the Caribbean to discharge ports (disports) around the Mediterranean would expect to discharge their cargoes within 23–30 days. Polyethylene bags, often referred to as 'polypacks' (Stover and Simmonds, 1987) and Banavac plastic membranes are used to conserve bananas in MA conditions for many of these trade routes. Commercially, the green-life expectancy for sound, hard-green bananas in polypacks is some 28 days, while that in partially evacuated Banavac bags can be as long as 50–60 days.

Occasionally the routine transport of bananas to Europe and the Mediterranean is disrupted by mechanical failures of one sort or another. The degree to which the storage potential of the cargo may be foreshortened or jeopardized will often depend on whether the carrier can keep the merchandise adequately refrigerated and ventilated, the extent to which any transit period is prolonged, and the biological condition of the fruit.

In 2006 a cargo of bananas was being transported from Ecuador to the Middle East on a conventional reefer vessel. The carriage instructions indicated that the supply air temperature (SAT) within the holds was to be 13.3°C. Engine problems prolonged the voyage and discharge occurred 8–9 weeks after the first loading episode. The cargo consisted of several brands of bananas all conserved in Banavac bags. On account of the extended storage period, it was expected that at least some of the fruit would have ripened naturally in transit; however, analysis of a sample of cartons during discharge indicated that other factors were also responsible for the loss in quality of sub-components of the cargo (R.D. Bancroft, commercial unpublished data). The nature and proportion of defects observed varied according to brand and whether fruit was categorized as non-ripe or ripening. The premature ripening of many of the fruit was considered to have been caused by ethylene generated from damaged fruit and those chronically infected with anthracnose and/or crown rot. On discharge, the most expedient segregation of the cargo was into those cartons that contained non-ripened green bananas and those that had ripened or had begun to turn colour. Differences between brands immediately became apparent (Figs 13.12–13.15).

In Brand A, for example, of those fruit that remained green, 87% of the fruit sampled showed no external defects and, relative to a peel coloration scale of 1 (hard-green) to 7 (yellow with flecks of brown), registered a score of < 1.99 (Fig. 13.12). Where damage was observed in these fruit, 8% was attributable to wounds infected with fungi, principally anthracnose and the disease complex known as crown rot. Another 5% of the fruit showed signs of chilling injury. In contrast in Brand B, 82% of the consignment remained unblemished, with 68% registering a colour score of < 1.99 and the residual 14% having a peel colour in excess of 2 (Fig. 13.14). In Brand B, 15% of the fruit samples supported fungal infections and 3% had chilling injury. When the ripe or ripening fruit were analysed, 56% of the Brand A bananas were unblemished compared with 40% in Brand B (Figs 13.13 and 13.15). Damage caused by rough handling during loading accounted for 2–3% and that engendered during discharge adversely impacted 5% of fruit of Brand A and 10% of fruit of Brand B.

Of particular interest is the proportion of ripe or ripening bananas predated by anthracnose and crown rot. In Brand A, 12% of the samples showed signs of the development of latent anthracnose and 7% were clearly attacked by crown rot. A further 18% showed signs of infection by a mixture of

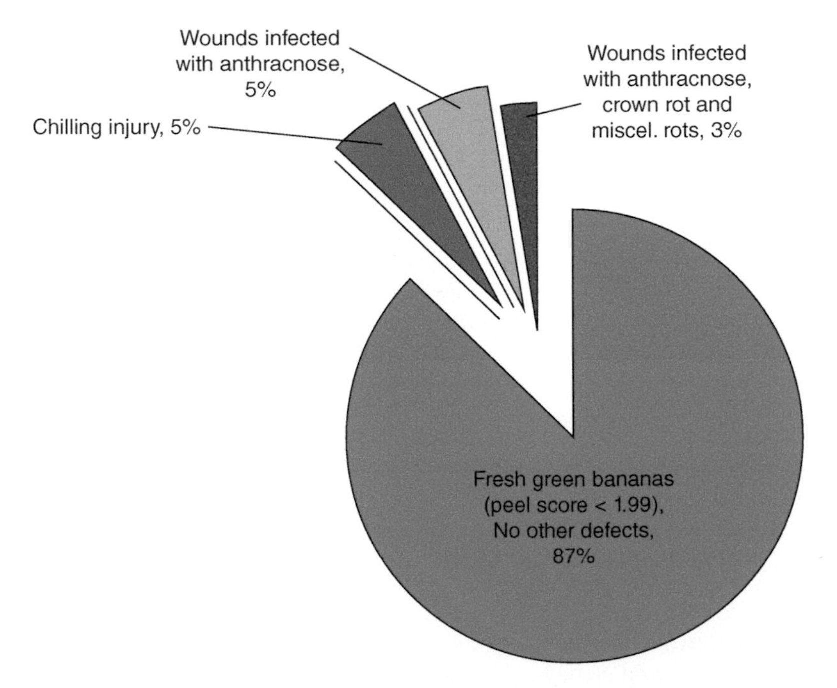

Fig. 13.12. Out-turn quality of non-ripened bananas of Brand A held in Banavac packaging and transported for 8–9 weeks in a conventional reefer vessel.

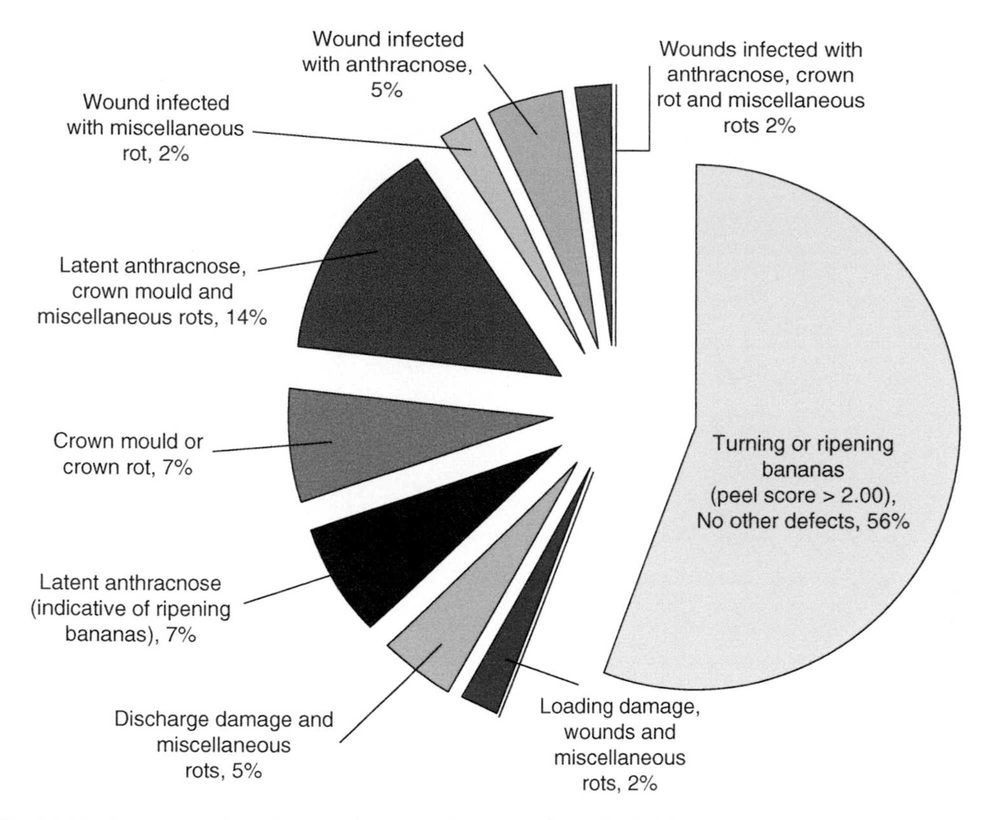

Fig. 13.13. Out-turn quality of ripe and ripening bananas of Brand A held in Banavac packaging and transported for 8–9 weeks in a conventional reefer vessel. Premature ripening of many of the fruit was considered to have been caused by ethylene gas generated from damaged fruit and those chronically infected with anthracnose and/or crown rot.

fungi, including anthracnose and crown rot. In Brand B, latent anthracnose occurring on its own was observed in 5% of fruit whereas crown rot was the principal disease in 25% of the bananas sampled. Mixed infections accounted for a further 15%. These observations suggested that latent anthracnose might have been more prevalent in Brand A while crown rot was found more often in Brand B. Both diseases will have arisen from latent infections. That is, although unobserved, these diseases will have been present in the fruit at the time of loading and will have become active during the voyage. The ripening gas ethylene is produced by damaged tissues and both crown rot (Fig. 13.16) and anthracnose pathogens (Fig. 13.17). The observations made at discharge indicate that, despite the retarding influence of the MA conditions created in the Banavac bags, damage to the cargo at loading augmented by the activity of anthracnose and crown rot is likely to have made a significant contribution to the premature ripening of the fruit.

Observations at out-turn also suggested that the storage potential of some of the Brand A cargo may have been compromised in Ecuador due to the extended period in which the fruit remained exposed to ambient temperature conditions during loading. This would have diminished the green-life potential of the fruit from the outset. Aware of the potential damage that could be caused by a protracted voyage, the SAT was reduced to 13°C some 38 days from the date of initial loading. In fact, on occasion the SAT was found to have dropped below 13°C. This is believed to have brought about the chilling injury observed in the cargo at discharge. It is unlikely that the merchandise would have

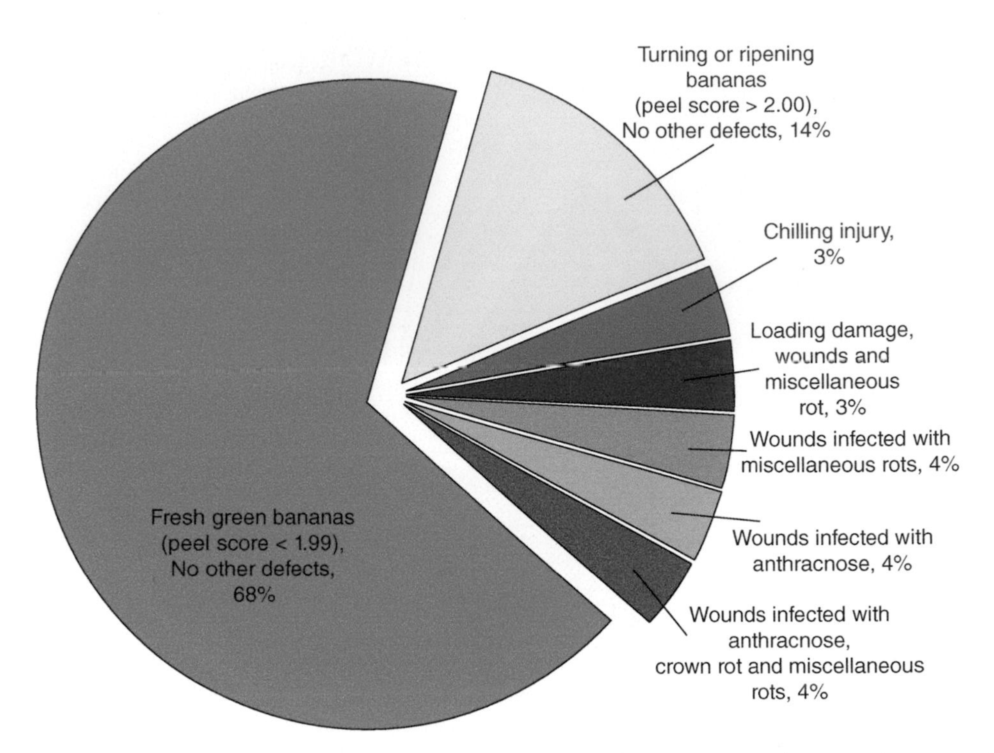

Fig. 13.14. Out-turn quality of non-ripened bananas of Brand B held in Banavac packaging and transported for 8–9 weeks in a conventional reefer vessel.

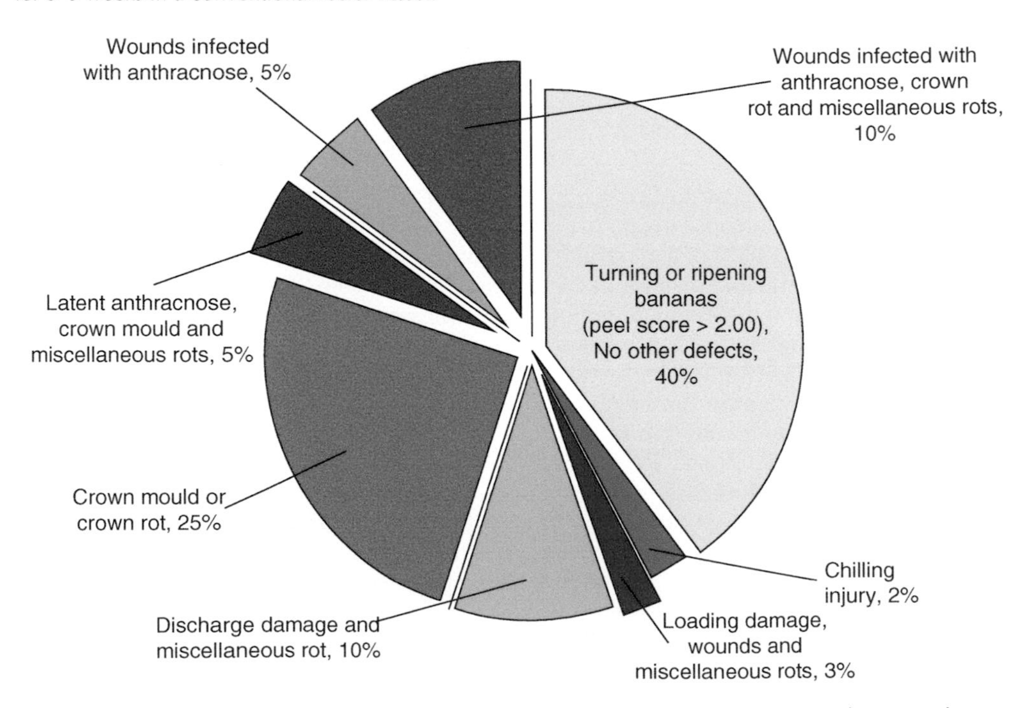

Fig. 13.15. Out-turn quality of ripe and ripening bananas of Brand B held in Banavac packaging and transported for 8–9 weeks in a conventional reefer vessel. Premature ripening of many of the fruit was considered to have been caused by ethylene gas generated from damaged fruit and those chronically infected with anthracnose and/or crown rot.

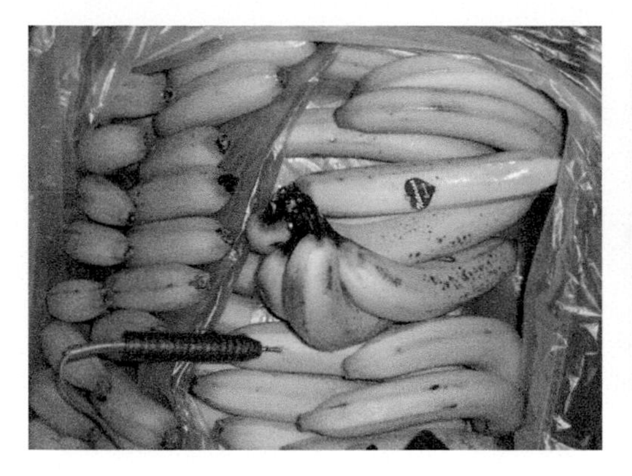

Fig. 13.16. Crown rot infecting the crown and pedicels of banana fruit. Ethylene released by the rot and damaged tissues will have triggered ripening in the surrounding fruit.

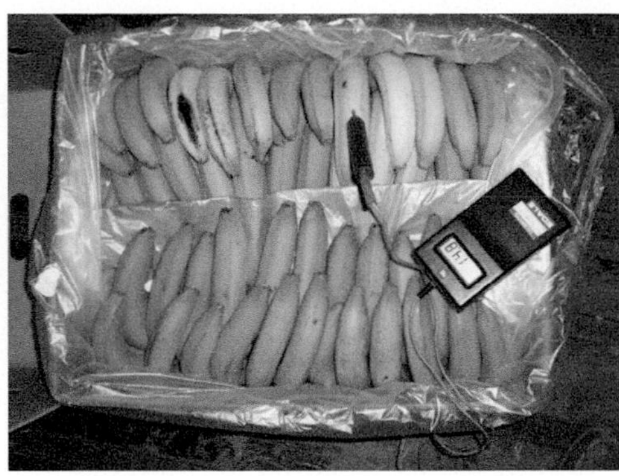

Fig. 13.17. An example of a wound infected with anthracnose generating ethylene, which has begun to trigger ripening in the surrounding hands of bananas.

expressed these defects if the temperature regime had remained unaltered.

It is apparent that the prolonged period of storage to which the banana cargo was exposed did engender premature ripening. However, the extent to which this occurred was also a reflection of the innate potential of the cargo. Of the losses identified at outturn, sampling suggests that a significant proportion was as a consequence of inherent infections and damage sustained by the merchandise prior to loading. The provision of MA packaging will have conserved healthy and robust fruit but would not have been able to prevent losses brought about by the advanced physiological age of some of the fruit, mechanical damage or the intervention of disease.

14

MA Packaging

Perhaps the first use of MA packaging in plastic film was for meat in 1979 by the retailer Marks and Spencer in the UK. Since then it has become increasingly used for a wide range of foods, including fresh fruit and vegetables. Bradshaw (2007) found that food packaging added more than 20% to the cost of buying fruit at a leading UK supermarket. A customer who buys fresh fruit worth £12.41 can expect to spend an additional £4.67 if they choose those that are pre-packed rather than loose. The same supermarket stated that they had reduced packaging of fresh fruit and vegetables by 33% since 2000. However, packaging can reduce losses and preserve quality in fruit and vegetables, especially MA packaging.

MA packaging can be defined as an alteration in the composition of gases in and around fresh produce by respiration and transpiration, when such commodities are sealed in plastic films (G.E. Hobson, personal communication). MA packaging, using various plastic films, has great potential in extending the postharvest life of fruit and vegetables, but it has never reached its full potential. It can be clearly demonstrated experimentally that it can have similar effects to CA storage on the postharvest life of crops and the mechanisms of the effects are doubtless the same. Its limited uptake by the industry probably reflects the limited control of the gases around and within the fruit and vegetables achieved by this technology. This can result in unpredictable effects on postharvest life and quality of the commodity packaged in this way. Some recent concerns about the safety of products in MA packaging may also militate against its use (Oliveira *et al.*, 2015).

Plastic films used in MA packaging for fruit and vegetables are permeable to gases. The degree of permeability depends on the chemical composition and thickness of the film as well as temperature. Other films, called barrier films, are designed to prevent the exchange of gases and are mainly used with non-respiring products. Developments in MA packaging were reviewed by Church (1994) and by Rai and Shashi (2007). These included low permeability films, high permeability films, O_2 scavenging technology, CO_2 scavengers and emitters, ethylene absorbers, ethanol vapour generators, tray-ready and bulk MA packaging systems, easy opening and resealing systems, leak detection, time–temperature indicators, gas indicators, combination treatments, and predictive/mathematical modelling. The concentration of the gases inside the film bag will also vary depending on mass, type, temperature and maturity of fruit or vegetable in the bag and the activity of microorganisms as well as the characteristics and thickness of the film used.

Given an appropriate gaseous atmosphere, MA packaging can be used to extend the postharvest life of fruit and vegetables in the same way as CA storage, but maintaining high humidity also seems to be involved. Plantains were shown to have a considerable extension in their storage life when packed in PE film, but keeping the fruit in a flow through system at high humidity had exactly the same effect. This effect of moisture content is more likely a reduction in stress within a climacteric fruit, which may be caused by a rapid rate of water loss in non-wrapped fruit. This in turn may result in increased ethylene production to internal threshold levels that can initiate ripening (Thompson *et al.*, 1974d). Porat *et al.* (2004) suggested that MA packaging reduces the development of rind disorders in citrus fruit in two ways. The first is by maintaining the fruit in high humidity; the second is maintaining elevated CO_2 and lowered O_2 levels. They found that storage of non-wrapped 'Shamouti' oranges in 95% RH reduced rind disorders to a similar level as those stored in macro-perforated plastic films. Johnson (1994a) reported that MA packaging during distribution and marketing may be at the expense of aroma and flavour.

Film Types

Since the 1970s, MA packages have become increasingly used for fruit and vegetables and new techniques have been developed, including micro-perforation, anti-fogging layers to improve product visibility and reduce rotting and equilibrium MA packaging. Companies have marketed films that they claimed could greatly increase the storage life of packed fruit and vegetables. These have been marketed with names such as Maxifresh®, Gelpack® and Xtend®.

Different films have differences in relation to O_2 and CO_2 permeability, which is a function of thickness, density, presence of additives and gradient concentration modification (Steffens *et al.*, 2007a). The manufacturing process for films has improved over the years and although the actual thicknesses of the film quoted may vary slightly, they are currently much more even and reliable. Many types of plastic have been used in the manufacture of films used in MA packaging and these are mainly by-products of the petrochemical industry.

Cellophane

Cellophane is a thin, transparent sheet made of regenerated cellulose from plant cells and has low gas permeability. It was invented by the Swiss chemist Jacques E. Brandenberger in 1900 and is completely biodegradable but releases carbon disulfide and other by-products of the manufacturing process. In the USA, 25 μ clear PP is sometimes referred to as cellophane. Cellophane is a registered trade mark of Innovia Films Ltd, UK. To make cellulose or cellophane film, cellulose is dissolved in a mixture of sodium hydroxide and carbon disulfide and then recast into sulfuric acid. The cellophane produced is very hydrophilic and therefore moisture sensitive; it has good mechanical properties, desirable strength, is a good water and O_2 barrier but is not heat sealable. Cellophane is often coated with nitrocellulose wax or PVDC (polyvinylidene chloride) to improve barrier properties and in such form it is used for packaging of baked goods, processed meat, cheese and candies. Cellulose acetate is biodegradable; it possesses relatively low gas and moisture barrier properties and has to be plasticized for film production. A plasticizer is a substance added to plastics to make them more pliable.

CPP

Cast film extrusion is used for laminating thin layers of plastic on top of paper or fibreboard by applying heat and pressure. Cast polypropylene (CPP) is one such film. Coextruded oriented polypropylene (COPP) has a good moisture vapour barrier and the gas barrier properties can be improved by coating it with PVDC. COPP can be perforated mechanically to improve gas exchange.

EVA

Ethylene-vinyl acetate copolymer has high flexibility in sheet form and has higher permeability to water vapour and gases than LDPE. It is mainly used as a component of the sealant layer in lids and base films. Ethylene-vinyl alcohol copolymer (EVOH) is expensive but moisture sensitive; it has a very high gas barrier and is sandwiched between the main formable and sealant layer to provide protection.

HDPE

High-density polyethylene is used as one of the layers in the lid material in coextruded forms; it has a higher softening point than LDPE and better gas barrier properties than linear LDPE (LLDPE) but poor clarity. HDPE has 75–90% crystalline structure, with an ordered linear arrangement of the molecules with little branching, and a molecular weight of 90,000 to 175,000 with a typical density range of 0.995–0.970 g/cm^3. It has greater tensile strength, stiffness and hardness than LDPE. HDPE film has the following specifications: 520–4000 O_2 cm^3/m^2/day, 3900–10,000 CO_2 cm^3/m^2/day, at 1 atm for 0.0254 mm thick at 22–25°C at various or unreported relative humidity and 4–10 water vapour g/m^2/day at 37.8°C and 90% RH (Schlimme and Rooney, 1994).

High-impact polystyrene

This is an opaque, thermoformable, moderately low gas-barrier material and is used as a component of a laminate or for coextrusion.

Ionomers

Ionomers are PE-based films. Al-Ati and Hotchkiss (2003) investigated altering PE ionomer films with permselectivities between 4–5 CO_2 and 0.8–1.3 O_2. The results on fresh-cut apples suggest that packaging films with CO_2:O_2 permselectivities lower than those commercially available (< 3) would further optimize O_2 and CO_2 concentration in MA packages, particularly of highly respiring and minimally processed produce. Al-Ati and Hotchkiss (2002) pressed sodium-neutralized ethylene–methacrylic ionomers into films at 120–160°C and found that the heat treatment may improve their potential for applications as selective barriers for MA packaging of fruit and vegetables.

LDPE

Low-density polyethylene is used for the ubiquitous 'polythene bag'. LDPE is produced by polymerization of the ethylene monomer which produces a branched-chain polymer with a molecular weight of 14,000–1,400,000 and a density ranging from 0.910 to 0.935 g/cm^3. LDPE film has the following specifications: 3900–13,000 O_2 cm^3/m^2/day, 7700–77,000 CO_2 cm^3/m^2/day, at 1 atm for 0.0254 mm thick at 22–25°C at various or unreported humidities and 6–23.2 water vapour g/m^2/day at 37.8°C and 90% RH (Schlimme and Rooney, 1994). It is an inert film with low permeability for water vapour but high gas permeability.

LLDPE

Linear low-density polyethylene is a variation of basic LDPE film but has better impact strength, tear resistance, higher tensile strength and elongation, greater resistance to environmental stress cracking and better puncture resistance. LLDPE combines the properties of LDPE film and HDPE film, giving a more crystalline structure than LDPE film but with a controlled number of branches, which makes it tougher and suitable for heat sealing. It is made from ethylene with butene, hexene or octene, with the latter two co-monomers giving enhanced impact resistance and tear strength. Permeability was given as 7000–9300 O_2 cm^3/m^2/day, at 1 atm for 0.0254 mm thick at 22–25°C at various or unreported humidities and

16–31 water vapour g/m^2/day at 37.8°C and 90% RH (Schlimme and Rooney, 1994).

MDPE

Medium-density polyethylene is used for shrink-film packaging, among other applications. It has a density in the range of 0.926 and 0.940 g/cm^3, 2600–8293 O_2 cm^3/m^2/day, 7700–38,750 CO_2 cm^3/m^2/day, at 1 atm for 0.0254 mm thick at 22–25°C at various or unreported humidities and 8–15 water vapour g/m^2/day at 37.8°C and 90% RH (Schlimme and Rooney, 1994).

OPP

Oriented polypropylene provides high moisture vapour and gas barrier, seven to ten times that of PE. OPP coated with PVDC in low gauge form provides a high barrier to moisture vapour.

PE

Polyethylene is a thermoplastic polymer consisting of long chains of the monomer ethylene. It is manufactured through polymerization of ethane and is classified mainly on its density and branching. Its mechanical properties depend on the extent and type of branching, the crystal structure and its molecular weight. It was first synthesized by the German chemist Hans von Pechmann in 1898 while heating diazomethane. However, it was not until 1933 that the first industrially practical PE was synthesized by Eric Fawcett, Reginald Gibson and Michael Perrin at the ICI works in the UK. Perrin developed a reproducible high-pressure synthesis for PE that became the basis for industrial LDPE in 1939. Subsequently catalysts have been developed to promote ethylene polymerization at milder temperatures and pressures. By the end of the 1950s both the Phillips catalyst (developed by 1951 by Robert Banks and John Hogan at Phillips Petroleum) and Ziegler type catalyst (developed by Karl Ziegler in 1953) were being used for HDPE production. A third type of catalyst system called metallocene was discovered in 1976 in Germany by Walter Kaminsky and Hansjörg Sinn. The Ziegler and metallocene catalysts are flexible at copolymerizing ethylene with other olefins and have become the basis for the wide range of polyethylene resins. Many different types of PE are currently available, including ultra-high molecular weight PE, high molecular weight PE, high-density PE and high-density cross-linked PE, but the ones used in MA packaging are: MDPE (medium-density PE), which is used in shrink film; LDPE and LLDPE, which have a high degree of toughness, flexibility and relative transparency; and VLDPE (very low-density PE) used in food packaging and stretch wrap. Polyethylene terephthalate (PET) can be used to manufacture thin oriented films with excellent thermal properties (–70V to +150°C), therefore it is used for pre-cooked meals and 'boil in the bag' but not for MA packaging of fresh fruit and vegetables.

PET or PETE

Polyethylene terephthalate is a saturated thermoplastic polyester resin made by condensing ethylene glycol and terephthalic acid and is commonly used as a synthetic fibre called polyester. In MA packaging this film is either laminated or extrusion-coated with polyethylene. It is used in various forms in MA packaging, such as as a low gauge oriented film of high clarity as lid material and in crystalline or amorphous form as preformed or thermoformed trays.

PO

Polyolefin is manufactured from olefins, especially ethylene and propylene, into a range of high-shrink multilayer high-speed machineable shrink-films with high tensile strength, high clarity and balanced heat shrinkage in all directions. For example, Cryovac is used for fruit, generally using foam trays.

PP

Polypropylene is chemically similar to PE and can be extruded or coextruded to provide a sealant layer. It was first polymerized in Spain by Karl Rehn in 1951 and Giulio Natta in 1954.

PS

Polystyrene is a clear brittle thermoplastic with a high tensile strength but a poor barrier to moisture vapour and gases. It can be blended with styrene-butadiene or styrene polybutadiene to achieve the required properties, but they can affect its clarity.

PVC

Polyvinyl chloride is the most widely used as the thermo-formable base for MA packages, since it provides good gas barrier and a moderate barrier to moisture vapour, which varies with thickness. The properties of PVC can range in form from soft and flexible to hard and rigid, either of which may be solid or cellular, depending on the type of plasticizer used. It was discovered in 1835 by the French chemist Henri Victor Regnault when it appeared as a white solid inside flasks of vinyl chloride that had been left exposed to sunlight. In the early 20th century Ivan Ostromislensky and Fritz Klatte of the German chemical company Griesheim-Elektron attempted to use PVC in commercial products, but difficulties in processing prevented development. In 1926, Waldo Semon of the B.F. Goodrich Company developed a method to plasticize PVC by blending it with various additives, which made it more flexible and easier to process, resulting in its widespread commercial use. Polyvinylidene chloride (PVDC) is a copolymer of vinylidene chloride and with vinyl chloride is used as a gas barrier coating on polyester and OPP for films used for lids and in film form as a sandwiched barrier layer. It has low permeability to water vapour and good barrier properties to O_2 and is heat sealable but it is not very strong.

Xtend®

These films are manufactured from polymers that, is claimed, give a 'correct balance' between O_2 and CO_2 while excess moisture is allowed to escape. Various films have been developed with a range of permeabilities. Xtend® films prevented water condensation inside the bags (Porat *et al.*, 2004) and are of two types: low water conductance, which give an internal humidity of 93%; and high water conductance, which give an internal humidity of 90% (Lichter *et al.*, 2005). Aharoni *et al.* (2008) used hydrophilic Xtend® films manufactured from various proprietary blends of polyamides with other polymeric and non-polymeric compounds with micro-perforations, which gave humidity levels inside the packs that prevented accumulation of condensed water. They showed that suitable combinations of O_2 and CO_2 were developed in the micro-perforated Xtend®.

Biodegradable MA Packaging

Discarded plastics, including plastic films used in MA packaging, represent a major source of environmental pollution, which is constantly increasing in spite of publicity and legislation to control their use and disposal. Most plastics are by-products of the petrochemical industry and can take hundreds of years to break down. Plastics made from plants break down relatively quickly and can even have beneficial effects for use in horticulture when composted. The Biodegradable Products Institute in New York, USA, defines biodegradable plastic as 'one in which degradation results from the action of naturally occurring micro-organisms' (http://www.bpiworld.org/). The use of biodegradable plastic films in MA packaging is still uncommon, with films manufactured from petrochemicals being used almost exclusively. The advantages of these petrochemical-based polymers include their high mechanical performance, ease of manufacture and good barrier properties but, overwhelmingly, their low cost.

Several biodegradable films have been developed and tested, including the following.

- PLA is a linear, semi-crystalline, aliphatic, biodegradable polyester that can be produced from lactic acid by the fermentation of renewable sources such as whey, corn, potatoes or molasses. It is used to make many products, including bottles and biodegradable medical devices as well as plastic films.
- Polybutylene succinate (PBS) is a biopolymer from poly-condensation of succinic acid and 1-4 butanediol, which has been used for biopolymer compounds. It is used for mulching crops, bags, bottles and medical devices as well as films
- Mater-Bi® is made from plant material including starches from cereals (particularly maize) and potatoes or from vegetable oils. It is a biodegradable film that has been used for many years as mulches, particularly for growing strawberries.

To test the rate of degradation of bags made from these three polymers in moist soil, Tengrang *et al.* (2015) packed 1 kg of rambutans in PLA, PBS, Mater-Bi® or LDPE bags and stored them at 13 ± 2°C and 78% RH. It was found that Mater-Bi® and PLA films took some 100 days to start degrading, while PBS film started to degrade in 16 days and its degradation increased to 31.25% in 161 days. Rambutans that had been stored in the PBS bags had the lowest weight loss and retained their yellowish/red skin colour and fresh green spinterns, which was superior to the other three packages where the fruit had higher weight losses and tended to develop a brown skin. They concluded that packaging made from PBS film could extend the postharvest life of rambutans for 15 days at 13 ± 2°C. Also they found that the PBS bags and the Mater-Bi® bags were easier to produce and use than the PLA bags. The highest levels of both water vapour and O_2 transmission rates were with Mater-Bi® film followed by PBS and PLA, respectively. Percentages of elongation of Mater-Bi® in machine direction and transverse were significantly higher than the other two films, whilst the tensile strength of Mater-Bi® was the lowest in both machine direction and transverse, followed by PBS and PLA, respectively.

Kantola and Helén (2001) compared the effects of Mater-Bi® with LDPE film. They stored tomatoes at 11 ± 1°C and 75–85% RH for 3 weeks in: perforated Mater-Bi® bags, Biopol®-coated paper board trays overwrapped with perforated Mater-Bi® bags, polylactide-coated paperboard trays overwrapped with perforated Mater-Bi® bags, perforated cellophane bags and perforated LDPE bags as a control. Package type had no significant influence on the tomatoes' sensory quality, but the tomatoes lost more weight in biodegradable packages than in LDPE packages. Giacalone and Chiabrando (2015) packed fresh cherries in plastic baskets, sealed with overwraps of either Mater-Bi® or PP film. They found that the quality of the cherries was similar after storage in both types of package, but the incidence of postharvest diseases was almost always lower in the fruit packed in Mater-Bi®, probably as a consequence of the higher CO_2 levels within the package. Koidea and Shib (2007) compared the effects of PLA packaging on the microbial and physico-chemical quality of green peppers. They compared the PLA with both perforated and non-perforated LDPE film and stored them for 7 days at 10°C. The levels of coliform bacteria were increased by less than 1 log CFU/g in the PLA film packaging, 2.3 log CFU/g in LDPE film non-perforated package and less than 1 log CFU/g in the perforated LDPE film package. The PLA film had higher water vapour permeability than the LDPE. They concluded that MA packaging of green peppers in PLA film can be used to maintain their quality and protect them from microbial and insect contamination. Mistriotis *et al.* (2016) developed a model simulating the gas diffusion through micro-perforations and the permeable membrane for MA packaging using PLA film. They reported that the relatively high water vapour permeability of PLA films compared with OPP films allowed the development of biodegradable-based packaging for the specific application.

Llana-Ruiz-Cabello *et al.* (2015) tested PLA containing 2%, 5% and 6.5% Proallium® (extract of *Allium* spp.) on MA packaging of minimally processed salads. Antimicrobial activity was observed, mainly in films containing 5% and 6.5% Proallium®. Packing salads in Proallium® film (6.5%) was effective in controlling bacteria for up to 5 days, and up to 7 days in controlling fungi. Mali and Grossmann (2003) made a film from 4.0g of yam starch plus 1.3 or 2.0g glycerol in 100g of 'filmogenic' solution. They tested the film on packaging strawberries stored at 4°C and 85% RH and compared it with PVC film packaging. Both films significantly reduced decay of the fruits compared with control, but PVC reduced softening and weight loss more than the yam starch film. Brandelero *et al.* (2016) prepared biodegradable films from cassava starch, polyvinyl alcohol and sodium alginate and studied the effects of adding 0.5% essential oils from copaiba (*Copaifera reticulata*) or lemongrass (*Cymbopogon citratus*). The 50μm biodegradable film they used had 11.43–8.11 MPa resistance and 11.3–13.22% elongation, with water vapour permeability of 0.5–4.04 × 10–12 g/s/Pa/m. They compared this film with PVC film on the storage of minimally processed lettuce at 6 ± 2°C. Lettuce stored in biodegradable films for 4 days remained similar to those freshly harvested, while those in PVC film lost quality after only 2 days, though they had up to eight times less mass loss than those stored in biodegradable films. The addition of essential oils to the films had no significant effects on the lettuce. They concluded that these biodegradable films were viable for the storage of minimally processed lettuce.

Peano *et al.* (2014) stored strawberries in two different single-layer films: a biodegradable film (from corn starch) and a 25 μm perforated PP film. These were stored at 2°C for 4 days or 2°C for 2 days followed by room temperature (20°C) for 2 days. The strawberries packed in biodegradable film and stored at 2°C retained their quality and maintained the optimum headspace composition during storage better than those in perforated PP film. Giuggioli *et al.* (2015) stored raspberries at 1°C for 2 days followed by 18°C for 2 days in the following films: non-commercial biodegradable and compostable film; non-perforated prototypes (Novamont, Italy); non-perforated commercial PP film (Trepack, Italy); and commercial PP macro-perforated film (6 mm holes; Trepack, Italy). After 2 days the fruit colour of those packed in biodegradable film was similar to that of those that had been freshly harvested for both the passive and active atmospheres. When the temperature was increased, it was only the biodegradable film that facilitated storage of fruit for up to 4 days. They had a headspace containing 24.4–25.9 kPa CO_2. Makino and Hirata (1997) used a film made from biodegradable laminate of a chitosan-cellulose and polycaprolactone for MA packaging for lettuce, broccoli, tomatoes, sweet corn and blueberries. They developed gas permeability coefficients that were designed to be close to the optimal composition needed for MA packaging. They concluded that this laminate was suitable as a packaging material for storage of fresh produce.

Biodegradable zein films are made from a hydrophobic protein produced from maize, but the films may swell and deform in prolonged contact with water, which seems to restrict their use in MA packaging of fruit and vegetables (Yoshino *et al.*, 2002). However, Rakotonirainy *et al.* (2008) reported that broccoli florets packaged in zein films (laminated or coated with tung oil) maintained their original firmness and colour during 6 days of storage. Films have been prepared by extrusion using acetylated and oxidized banana starches with LDPE (Torres *et al.*, 2008).

Film Permeability

Respiring fresh fruits and vegetables sealed in plastic films will cause the atmosphere to change, in particular O_2 levels to be depleted and CO_2 levels to be increased. The transmission of gases and water vapour though plastic films can vary with the type of material from which they are made, the use of additives in the plastic, temperature, humidity, the accumulation, concentration and gradient of the gas and the thickness of the material. Generally films are 4–6 times more permeable to CO_2 than to O_2. However, Barmore (1987) indicated that the relationship between CO_2 and O_2 permeability and that of water vapour is not so simple. Variation in transmission of water vapour can therefore be achieved to some extent, independently of transmission of CO_2 and O_2, using techniques such as producing multilayer films by coextrusion or applying adhesives between the layers. Achieving the exact O_2 and CO_2 levels in MA packaging can be difficult. Film permeability to gases is by active diffusion where the gas molecules dissolve in the film matrix and diffuse through in response to the concentration gradient (Kester and Fennema, 1986). Schlimme and Rooney (1994) showed that there was a range of permeabilities that could be obtained from films with basically the same specifications. A formula to describe film permeability was given by Crank (1975) as follows:

$$P = \frac{Jx}{A(p_1 - p_2)}$$

where J = volumetric rate of gas flow through the film at steady state, x = thickness of film, A = area of permeable surface, p_1 = gas partial pressure on side 1 of the film, and p_2 = gas partial pressure on side 2 of the film ($p_1 > p_2$).

Permselectivity is a medical term used to define the preferential permeation of certain ionic species through ion-exchange membranes. Al-Ati and Hotchkiss (2003) applied the term to MA packaging (see 'Ionomers', above).

Gas Flushing

The levels of CO_2 and O_2 can take some time to change inside the MA pack. In order to speed this process, the pack can be flushed with N_2 to reduce the O_2 rapidly, or the atmosphere can be flushed with an appropriate mixture of CO_2, O_2 and N_2. In other cases the pack can be connected to a vacuum pump to reduce the headspace so that the respiratory gases can change within the pack more quickly. Gas flushing is more important for non-respiring products such as meat or fish, but it can profitably be used with fresh fruit and vegetables. Zagory (1990) showed the effects of flushing fresh chilli peppers stored in plastic film with a mixture containing 10 kPa CO_2 + 1 kPa O_2 compared with no gas flushing (Fig. 14.1). In work described by Aharoni *et al.* (1973), yellowing and decay of leaves of the lettuce cultivar 'Hazera Yellow' were reduced when they were pre-packed in closed PE bags in which the O_2 concentration was reduced by flushing with N_2. Similar but less effective results were obtained when the lettuce was pre-packed in closed PE bags not flushed with N_2, or when open bags were placed in PE-lined cartons. Andre *et al.* (1980a) showed that fungal development during storage could be prevented in asparagus spears by packing them in PE bags with silicon elastomer windows and flushing with 30 kPa CO_2 for 24 h or by maintaining a CO_2 concentration of 5–10 kPa.

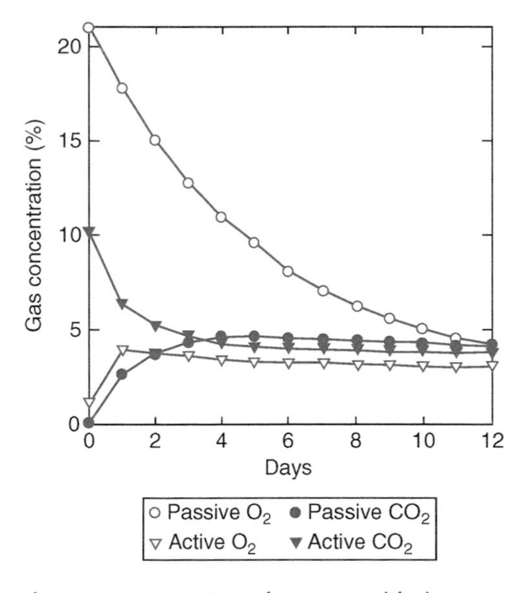

Fig. 14.1. A comparison of passive modified atmosphere packaging and active modified atmosphere packaging on the rate of change of CO_2 and O_2 in Anaheim chilli pepper fruit packed in Cryovac SSD-310 film. Reproduced from Zagory (1990) with permission.

Quantity of Product

The quantity of produce inside the sealed plastic film bag in relation to the size of the bag has been shown to affect the equilibrium gas content (Table 14.1), but the levels of CO_2 and O_2 do not always follow what would be predicted from permeability data and respiration load of the crop. Zagory (1990) also demonstrated the relationship between the weight of produce sealed in 20 cm × 30 cm Cryovac SSD-310 film bags and its O_2 and CO_2 content was linear (Fig. 14.2). However, this varied considerably with a four-fold variation in produce weight, which illustrates the importance of varying just one factor. Thompson *et al.* (1972b) also showed that the number of fruit packed in each plastic bag can alter the effect of MA packaging where plantains packed with six fruits in a bag ripened in 14.6 days compared with 18.5 days when fruits were packed individually during storage in Jamaican ambient of 26–34°C.

Table 14.1. Effects of the amount of asparagus spears sealed inside plastic bags on the equilibrium CO_2 and O_2 content at 20°C; modified from Lill and Corrigan (1996). Film used was W R Grace RD 106 polyolefin shrinkable multilayer film with anti-fog properties with 23,200 CO_2, 10,200 O_2 permeability at 20°C in ml/m³/atm/day.

Weight (g) of asparagus spears inside the bag	Equilibrium kPa CO_2	Equilibrium kPa O_2
100	2.5	9.7
150	3.2	6.2
200	3.5	4.1

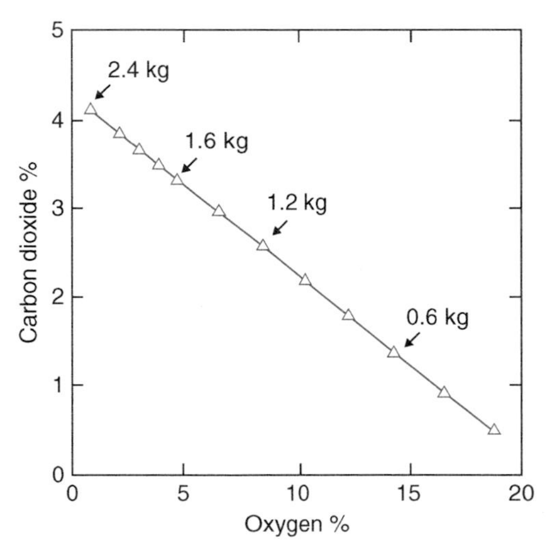

Fig. 14.2. Equilibrium gas concentration in a modified atmosphere package as a function of weight for Anaheim chilli pepper fruit packed in Cryovac SSD-310 film. Reproduced from Zagory (1990) with permission.

Perforation

The appropriate CO_2 and O_2 levels can be achieved for fruit or vegetables with low or medium respiration rates where they are sealed in films such as LDPE, PP, OPP and PVC. However, Rai and Shashi (2007) pointed out that for highly respiring produce such as mushroom, broccoli, asparagus and Brussels sprouts, packaging in these films can result in fermentation. Punching holes in the plastic can maintain a high humidity around the produce, but it may be less effective in delaying fruit ripening because it does not have the same effect on the CO_2 and O_2 content of the atmosphere inside the bag. The holes may be very small and in these cases they are commonly referred to as micro-perforations. Larger holes are sometimes referred to as macro-perforations, but there is no definition as to where one ends and the other begins.

Various studies have been conducted on the effects of perforations in the packing fruit and vegetables at different temperatures on the internal gas content. Perforated bags used for low-respiring fruit and vegetables can have both positive and negative effects. For example, Hardenburg (1955) found that onions in non-perforated bags tended to produce roots, but when the bags were perforated to address this problem the weight loss in the onions increased (Table 14.2). Pears of the cultivars 'Okusankichi' and 'Imamuraaki' had the highest levels of decay if placed in non-sealed 0.05 mm plastic bags and least in sealed bags with 5 pinholes, but bags with ten holes had more decay. Weight loss of fruit after 7 months of storage was 8–9% in non-sealed bags and < 1% in sealed bags. In sealed

Table 14.2. Effects of number and size of perforation in 3 lb (1.36 kg) 150 gauge PE film bags of 'Yellow Globe' onions on the relative humidity in the bags, rooting of the bulbs and weight loss after 14 days at 24°C (Hardenburg, 1955).

Number of perforations	Perforation size (mm)	% RH in bag	% bulbs rooted	% weight loss
0	–	98	71	0.5
36	1.6	88	59	0.7
40	3.2	84	40	1.4
8	6.4	–	24	1.8
16	6.4	54	17	2.5
32	6.4	51	4	2.5
Kraft paper with film window	–	54	0	3.4

bags, CO_2 concentration reached 1.9 kPa after 2 months of storage (Son *et al.*, 1983).

Adjustable Diffusion Leak

A simple method of CA storage was developed for strawberries (Food Investigation Board, 1920). The gaseous atmospheres containing reduced amounts of O_2 and moderate amounts of CO_2 were obtained by keeping the fruit in a closed vessel fitted with an adjustable diffusion leak. Marcellin (1973) described the use of PE bags with silicone rubber panels that allowed a certain amount of gas exchange and have been used for storing vegetables. Good atmosphere control within the bags was obtained for globe artichokes and asparagus at 0°C and green peppers at 12–13°C when the optimum size of the bag and the silicon gas-exchange panel was determined. Rukavishnikov *et al.* (1984) described CA storage in 150–300 μm PE film with windows of membranes selective for O_2 and CO_2 permeability made of polyvinyl trimethyl silane or silicon organic polymers. Apples and pears stored at 1–4°C under the covers had 93% and 94% sound fruit, respectively, after 6–7 months, whereas those stored non-wrapped at similar temperatures resulted in over 50% losses after 5 months. Hong *et al.* (1983) described the storage behaviour of *Citrus unshiu* fruits at 3–8°C in plastic films with or without a silicone window. The size of the silicone window was 20–25 cm²/kg fruit, giving < 3 kPa CO_2 + > 10 kPa O_2. After 110–115 days, 80.8–81.9% of fruits stored with the silicone window were healthy with good coloration of the peel and excellent flavour, while those without a silicone window had 59.4–76.8% with poor peel coloration and poor flavour. Those stored without wrappings for 90 days had 67% healthy fruit with poor quality and shrivelling of the peel, calyx browning and a high rate of moisture loss. The suitability of the silicone membrane system for CA storage of 'Winter Green' cabbage was studied by Gariepy *et al.* (1984) using small experimental chambers. Three different CA starting techniques were then evaluated and there was a control with the cabbages stored in air at the same temperature. The silicone membrane system maintained CA storage conditions of 3.5–5 kPa CO_2 + 1.5–3 kPa O_2, where 5% and 3%, respectively, were estimated theoretically. After 198 days of storage, total mass loss was 14% in CA storage compared with 40% in air. The three methods used to achieve the CA storage conditions did not have any significant effect on the storability of cabbage. Cabbage stored in CA showed better retention of colour, a fresher appearance and a firmer texture compared with those stored in air. Raghavan *et al.* (1982) used a silicone membrane system and found that carrots could be stored for up to 52 weeks and celery, swedes and cabbage for 16 weeks. Design calculations for selection of the membrane area were presented in the paper. Chinquapin fruit (*Castanea henryi*) packed in PE film bags, each with a silicone rubber window, and stored at 1°C had a longer storage life, a slower decline in vitamin C, sugars and starch, reduced desiccation and a lower occurrence of 'bad' fruits compared with those stored in gunny bags (Pan *et al.*, 2006).

Absorbents

Chemicals can be used with MA packaging to remove or absorb ethylene, O_2, CO_2 and water. Such systems are sometimes referred to as 'active packaging' and have been used for many years. They involve chemicals being placed within the package or they may be incorporated into the packaging material. One such product was marketed as 'Ageless' and used iron reactions to absorb O_2 from the atmosphere (Abe, 1990). AA-based sachets (which also generate CO_2) and cathecol-based sachets are also used as O_2 absorbers. The former are marketed as Ageless G® or Toppan C®, or Vitalon GMA® when combined with iron, and the latter as Tamotsu®. Mineral powders are incorporated into some films for fresh produce, particularly in Japan (Table 14.3).

Proprietary products such as Ethysorb® and Purafil® are available, which are made by impregnating an active alumina carrier (Al_2O_3) with a saturated solution of potassium permanganate and then drying it. The carrier is usually formed into small granules; the smaller the granules, the larger is the surface area and therefore the quicker are their absorbing characteristics. Any molecule of ethylene in the package atmosphere that comes into contact with the granule will be oxidized, so the larger surface area is an advantage. The oxidizing reaction is not reversible and the granules change colour from purple to brown, which indicates that they need replacing. Strop (1992) studied the effects of storing broccoli in PE film bags with and without Ethysorb®. She found that the ethylene content in the bags after 10 days at 0°C was 0.423 µl/l for those without Ethysorb® and 0.198 µl/l for those with Ethysorb®. However, Scott *et al.* (1971) showed that the inclusion of potassium permanganate in sealed packages reduced the mean level of ethylene from 395 µl/l to 1.5 µl/l and reduced brown heart from 68% to 36% in stored pears. Packing limes in sealed PE film bags inside cartons resulted in a weight loss of only 1.3% in 5 days, but all the fruits de-greened more rapidly than those that were packed without PE film, where the weight loss was 13.8% (Thompson *et al.*, 1974b). However, this de-greening effect could be countered by including an ethylene absorbent in the bags (Fig. 14.3). Where potassium permanganate was included in the bags containing bananas, the increase in storage life was 3–4 times compared with non-wrapped fruit and they could be stored for 6 weeks at 20°C or 28°C and 16 weeks at 13°C (Satyan *et al.*, 1992). The rate of ethylene removal from packages using potassium permanganate on an alumina carrier is affected by humidity (Lidster *et al.*, 1985). At the high humidity found in MA packages, the rate of ethylene removal of potassium permanganate was shown to be reduced. Kiwifruit of the cultivar 'Bruno' were stored at –1°C in sealed PE bags of varying thickness, with and without Ethysorb®. There was a slower rate of fruit softening and improved keeping quality during 6 months of storage of kiwifruit at –1°C in sealed PE bags 0.04–0.05 mm thick containing an ethylene absorbent. The average composition of the atmosphere in these bags was 3–4 kPa CO_2 + 15–16 kPa O_2 and ethylene at < 0.01 µl/l (Ben Arie and Sonego, 1985).

Moisture scavengers, which can modify package humidity, include silica gel, clays (e.g. montmorillonite), calcium oxide, calcium chloride and modified starch (Day, 2008). A moisture control system was developed to form a liner inside boxes of fruit

Table 14.3. Commercially available films in Japan (Abe, 1990).

Trade name	Manufacturer	Compound	Application
FH film	Thermo	Ohya stone/PE	Broccoli
BF film	BF distribution research	Coral sand/PE	Home
Shupack V	Asahi-kasei	Synthetic-zealite	–
Uniace	Idemitsu pet chem	Silicagel mineral	Sweetcorn
Nack fresh	Nippon unicar	Cristobalite	Broccoli and sweetcorn
Zeomic	Shinanen	Ag-zeolite/PE	–

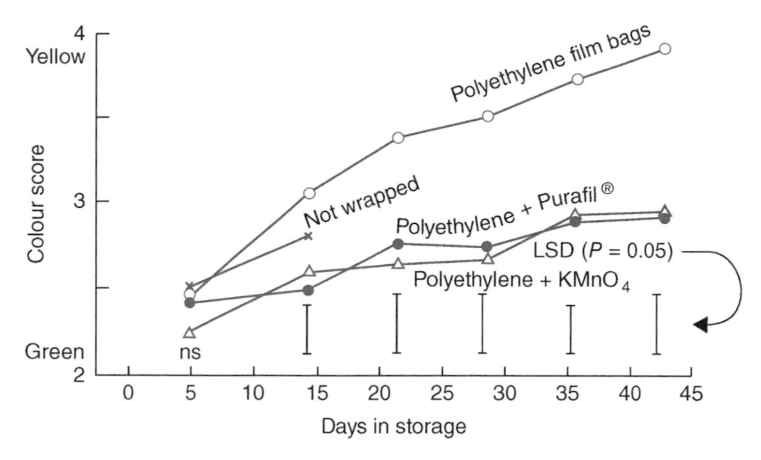

Fig. 14.3. Effects of modified atmosphere packaging in polyethylene film and an ethylene absorbent included within the pack on the rate of de-greening of limes stored in tropical ambient conditions of 31–34°C and 29–57% RH (Thompson *et al.*, 1974c).

to control humidity. Antifogging films allow the consumer to see the product clearly through the packaging films, which incorporate humidity absorbers, hydrophilic liners, or micro-perforations in the film (Ozdemir and Floros, 2004). However, moisture scavengers are primarily used in non-respiring foods, not fresh or minimally processed fruit or vegetables.

Modified interactive packaging (MIP) uses an LDPE impregnated with a concentration of minerals. The permeability of the substrate created by porous activated clay particles creates micro-cracks in the film when a container is formed. These micro-cracks allow for an interaction to take place between the produce, O_2, CO_2, water, other gases and the film. This allows for better control of the permeability of the film, but in tests the exchange rate did not exceed 3 parts O_2 to 1 part CO_2. Also high humidity is maintained inside the package with minimal condensation, which is claimed to reduce the risk of spoilage through fungal or bacterial infection. As excess condensation inside the punnet/liner/bag builds up, it was claimed that it was removed via 'osmotic diffusion through the film membrane and vaporised'. The major advantage is that produce moisture is not affected or removed, which would normally result in dehydration and early deterioration (Richard King, FreshTech, personal communication). One such film is marketed as 'Fresh 'n' Smart'. The film itself has a honeycomb structure with a number of tiny crevices, so that the PE molecular structure is broken down and yet retains its strength, toughness and elasticity. It is recommended that for optimum results, products should be packed in the bags dry and pre-cooled. Once sealed the bags should be kept in well-ventilated cartons or crates to allow air circulation. Fresh 'n' Smart uses resins in the manufacture of its film, which improves both the strength and elasticity. This means that the bags will be less prone to tearing and puncturing, resulting in a very durable and effective film. The resins are of approved food grade.

Adamicki (2001) cooled various vegetables to their optimum storage temperature and placed them either in MIP packages or into the boxes lined with PE film 35 μm thick without minerals. In the MIP packages a gas composition was maintained that resulted in maintenance of the quality and marketable value of the vegetable. The storage periods were 6–9 weeks for broccoli, 4–6 weeks for tomatoes, 4 weeks for sweet peppers and 4–5 weeks for 'Crisphead' lettuce in MIP packages. At 0°C the concentration of CO_2 inside the MIP increased up to 5.6–5.9 kPa in 2 days and then slowly decreased, reaching 3.5–3.7 kPa after 16 days, and stayed around this level to the end of the storage period. The concentration of O_2

inside the packages decreased from 21 to 3.5 kPa after 11 days and then remained at this level, with small fluctuation, throughout the period of storage. Subsequently Adamicki and Badełek (2006) reported that broccoli stored in MIP had better retention of green colour, firmness, quality and weight loss. Iceberg lettuce stored in MIP packages had less butt discoloration and rotting of leaves and improved marketable quality compared with those in boxes lined with PE film without minerals.

Ethylene-absorbing films

Various chemicals and formulations, as well as packing methods, have been used as inclusions within MA packaging of fresh fruit and vegetables in order to reduce their deterioration. These chemicals are mainly in the form of sachets that are sealed inside the plastic film bag along with the fruit or vegetables. Less successful are the chemicals that have been incorporated into the plastic film used for packaging. Joyce (1988) described an LDPE film extruded with finely divided crysburite ceramic and marketed by the Odja Shoji Co. Ltd, Japan, as BO Film, which was claimed to have ethylene-absorbing capacity. However, Chairat and Kader (1999) tested commercial products that could be incorporated into PE film and found that the ability of all the products they tested to remove ethylene was far lower than that of potassium permanganate. Zagory (1995) reviewed the subject and reported various materials that have been incorporated into plastic films to remove ethylene, including finely ground coral, a powdered zeolite in viscous oil or grease and oya stone. He also reported that an ethylene-absorbing PE film had been patented by Matsui (1989), which incorporated porous material derived from pumice, zeolite, active carbon, cristobalite or clinoptilolite with a small amount of metal oxide. These porous materials were dispersed and compacted into the PE by heat or pressure without melting the plastic. Plastic films that absorb ethylene are still available; for example, one is marketed by Active Pack International under the name of Bio Fresh®. Fibreboard is available for constructing boxes for fresh fruit and vegetables and one is marketed as PrimePro® (Chantler Packaging, Mississauga, Ontario). Gelpack® is a patented film made from PE blended with crushed volcanic rock during manufacture creating microscopic cavities in the film.

Activated carbon

Activated carbon, also called activated charcoal, can hold molecules such as ethylene and CO_2. When fresh air is passed through activated carbon the molecules are released. Incorporation of 30% activated carbon with 0.3% glucomannan in rice straw paper produced 77% ethylene-scavenging capacity (Sothornvit and Sampoompuang, 2012). This combination was proposed as a possible environmentally friendly scavenger that could be incorporated into MA packages.

Zeolites

Zeolite are a group of over 200 hydrated alumino-silicate minerals that are made up of interlinked tetrahedra of alumina (AlO_4) and silica (SiO_4). Zeolites have an open structure in which other molecules can be trapped and Abeles *et al.* (1992) reported that they can absorb 8% ethylene by weight.

Palladium

Palladium is not a very reactive metallic element but can absorb up to 900 times its own weight in hydrogen, which passes into the space between atoms. Catalytic converters, used in motor vehicles, may also contain Pd. In MA packaging, palladium chloride and light-activated titanium dioxide are metal catalysts that can be used to accelerate the oxidation reaction of potassium permanganate, thus increasing its absorption capacity by about sixfold (Martinez-Romero and Bailén 2007). Bailén *et al.* (2006) included granular activated carbon impregnated with palladium (ACPd) in MA packaging of tomatoes and compared them with MA packaging alone. ACPd changed the steady-state atmosphere from 4 kPa O_2 and 10 kPa CO_2 in MA packaging alone to 8 kPa O_2 and 7 kPa CO_2 in ACPd. The addition of ACPd also reduced ethylene accumulation inside packages and reduced the colour change, softening and weight loss and delayed decay. After storage, the tomatoes that had been stored with ACPd had higher taste-panel scores for sweetness, firmness, juiciness, colour, odour and flavour than tomatoes that had been stored in MA packaging alone. Terry *et al.* (2007) described a zeolite impregnated with palladium. At 20°C its ethylene absorption capacity was 4162 μl/g at approximately 100% RH and 45,600 μl/g under low humidity. They compared the ethylene absorption capacity of zeolite impregnated with palladium with potassium permanganate and found it outperformed it by approximately sixfold under high humidity and about 60-fold under low humidity. In a plug flow reactor containing ethylene at 200 μl/l using 0.1 g of zeolite impregnated with palladium, the ethylene level rapidly fell to zero (Fig. 14.4). Sachets containing zeolite impregnated with palladium are marketed by 'It's Fresh!' (Fig. 14.5). Cao *et al.* (2015) found that ethylene removal rate of acidified activated carbon powder impregnated with palladium chloride (10 mg/g) was increased with the addition of copper sulfate (30 mg/g) with a maximum removal of 21.77 ml/g at 25°C and 20.18 ml/g at 5°C. Its ethylene removal capacity could be regenerated to 81.6% by heating at 175°C for 20 min.

Oxygen scavengers

Sanjeev and Ramesh (2006) reviewed the use of low O_2 in packaged foods. Essentially O_2 scavengers are used in plant foods that have been thermally processed and therefore killed. However, they quote examples of minimally processed potatoes, apples, banana and peaches, where their minimal processing increased the oxidation/reduction potential and could accelerate the formation of such reactive O_2 species as peroxide ions and superoxide anions that could be detrimental. Excessive O_2 can cause the oxidation of phytochemicals, including vitamins,

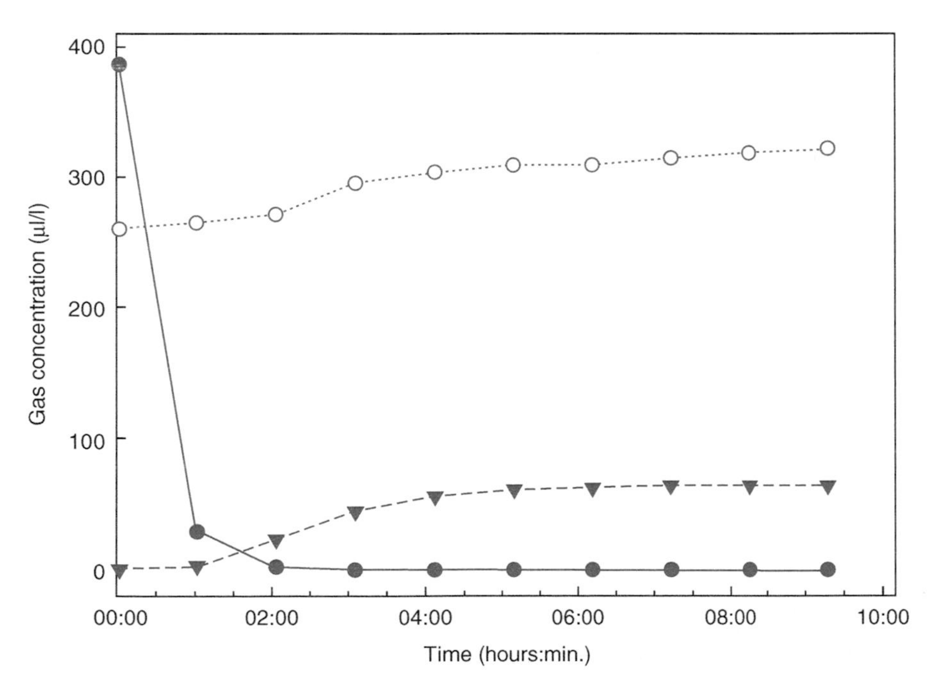

Fig. 14.4. Effect of palladium-promoted material on gaseous composition within sealed jars, initially containing 550 μl ethylene/l in air. CO_2 (○); ethylene (●); ethane (▼). Reproduced from Terry *et al.* (2007) with permission.

pigments, flavour compounds and lipids, as well as facilitating the growth of aerobic microorganisms. Most commercially available O_2 scavengers are used in processed food and moisture scavengers can also be used (Fig. 14.6). O_2 scavengers are reactive, based on iron oxidation to iron oxide, which can reduce the O_2 concentration in the headspace to less than 100 ppm in packages containing non-respiring foods.

Vapour Generation

Sulfur dioxide

Fumigation of fruit, especially fresh grapes, with SO_2 has been used commercially for many years to control postharvest diseases, mainly grey mould caused by *Botrytis cinerea*. SO_2 has primarily been used as a fumigant before storage or packaging of grapes, but it has also been used on other fruit; for example, in litchis to prevent discoloration (Ragnoi, 1989); and Rodriguez and Zoffoli (2016) found that 100-150 µl/l at 20°C for 30 min controlled decay of blueberries during subsequent storage at 0°C for up to 45 days. Packing SO_2 fumigated fruit into perforated LDPE bags (60 µm thick) was better than packaging the fruit in non-perforated LDPE bags, where the CO_2 steady state partial pressure that controlled decay was close to a level that injured the fruit. In non-perforated LDPE bags a steady-state content of 4.5–6.0 kPa CO_2 + 13–15.0 kPa O_2 developed, depending on cultivar.

Fig. 14.5. 'It's Fresh!' ethylene scavengers used in strawberry retail packs. Reproduced with permission from Simon Lee, It'sFresh!, Cranfield, UK.

Sodium metabisulfite has been used in pads that can be included in MA packaging. $Na_2S_2O_5$ reacts with water in the atmosphere and releases SO_2. Grey mould incidence of inoculated grapes was reduced from 91.7% to 1.1% when SO_2 pads were included in the MAP pack (Cayuela *et al.*, 2009). However, SO_2 pads have a limited life and Zutahy *et al.* (2008) reported that the SO_2 level had reduced to below 1 µl/l after 60–80 days at 0°C when SO_2 pads were included in packs of grapes. They described a dual-release SO_2 pad containing a quick-release plus a slow-release phase. In order to extend the life of the pads, they enclosed them in plastic laminate with macro-perforation and found that the optimum laminate had 32 perforations

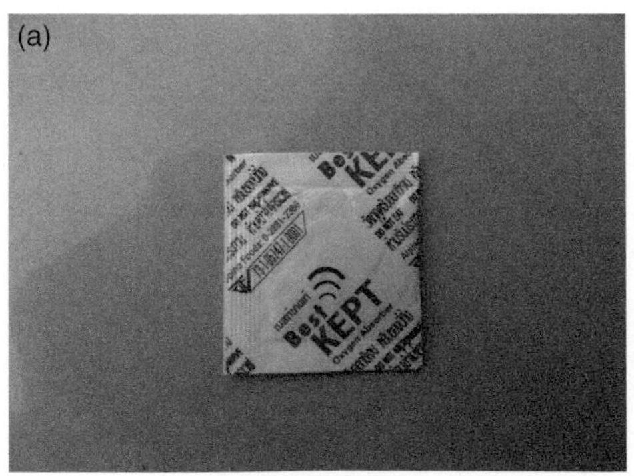

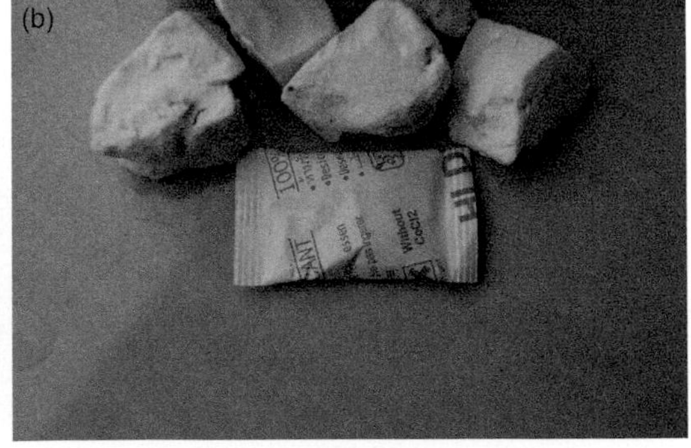

Fig. 14.6. (a) Sachet containing an oxygen scavenger used in processed food and (b) moisture scavenger from a sealed bag of freeze dried durian, both in food processing industry in Thailand.

of 6 mm diameter. This pad was tested in storage at 0°C for up to 116 days, and decay caused by *B. cinerea* and SO_2 damage were greater in the grapes with dual-release pads than the same pads in macro-perforated laminate. Ragnoi (1989), working with litchi, showed that fruits packed in sealed 37.5 μm PE film bags containing 2 kg of fruit and SO_2 pads could be kept in good condition at 2°C for up to 2 weeks, while fruits that were stored without packaging were discoloured and unmarketable.

Alcohol

Ethanol is in constant and common use for all kinds of cleaning purposes because of its biocidal properties. Chervin *et al.* (2003) reported preliminary results on the optimum ethanol dose for effective disease control in grapes in MA packages. Ethanol vapour was as effective as SO_2 pads in controlling grey mould. Consumer panels detected no significant difference in sensory perception between ethanol-treated and non-treated grapes. Stem browning was higher than with SO_2 treatments in fruit exposed to 5 ml ethanol/kg of fruit compared with a dose rate of 2 ml/kg. Karabulut *et al.* (2004) demonstrated that the addition of ethanol significantly improved the efficacy of a hot water treatment applied to grapes that had been inoculated with *B. cinerea*. A pre-storage dip of grapes in either 33% or 50% ethanol and then stored in MA packaging maintained their quality for up to 7 weeks with 3 days shelf-life at 20°C. The treatment was similar to or better than replacing the ethanol with SO_2 releasing pads (Lichter *et al.*, 2005). However, in *in vitro* studies on *B. cinerea*, ethanol fumigation with 100 μl/l for 3 or 6 min, or 200 μl/l for 6 min actually enhanced its growth (Lagopodi *et al.*, 2009), which may indicate that the fungus can utilize gaseous ethanol, at sub-toxic concentrations, for nutrition. Chu *et al.* (1999) inoculated cherries with spores of *B. cinerea* and then fumigated them with thymol, acetic acid or ethanol at 30 mg/l for 25 min before sealing them in MA packaging. After 10 weeks of storage, both thymol and acetic acid reduced grey mould from 36% to 0.5% and 6%, respectively, while those exposed to ethanol had similar levels to those not treated. Dong *et al.* (2016) fumigated fresh-cut burdock roots (*Arctium lappa*) with ethanol at 450–600 μl/l for 5 or 12 h and found that ethanol fumigation reduced their respiration rate, lowered the activity of PPO and PAL and reduced total phenolic content, thus prevented browning.

Hexenal

Trans-2-hexenal is released in many fruit and vegetables and is a major volatile compound in ripe tomatoes; it also contributes to the odour of other plants, including the smell of freshly cut green grass. Artes-Hernández *et al.* (2004) stored grapes at 0°C for 41 days in MA packaging sealed in micro-perforated PP film (35 μm thick) followed by 4 days in air also at 0°C and then 3 days at 15°C. They tested placing a filter paper soaked in hexenal (15 or 10 μl) in controlling postharvest diseases. The packs containing hexenal had the highest score for flavour and the lowest level of fungal disease (6%). Fumigation of grapes with 100 or 200 ml of *trans*-2-hexenal per 1.1 l container for 2 weeks at 2°C significantly lowered mould growth but did not affect fruit firmness or TSS content during subsequent storage at 2°C for up to 12 weeks (Archbold *et al.*, 1990). Guo *et al.* (2015) also showed that the incidence of natural or artificial infection by *B. cinerea* in tomatoes was significantly decreased after treatment with *trans*-2-hexenal. They also found that *trans*-2-hexenal increased the mRNA levels of genes encoding ethylene receptors, lipoxygenase and PAL, but did not induce the mRNA accumulation of the pathogenesis-related proteins, such as PR-1a or PR-5. The activities of peroxidases and PAL and the accumulation of phenolic compounds were significantly enhanced after *trans*-2-hexenal treatment. The inclusion of hexanal and *trans*-2-hexenal in MA packaging of apple slices gave a significant extension in their shelf-life and reduced the growth potential of naturally occurring bacteria (Corbo *et al.*, 2000). Exposure of pears to *trans*-2-hexenal vapour for at least

8 h at 20°C before storage at 1°C was required to control *Penicillium expansum* and reduce their patulin content. Treatment did not affect fruit quality and after 6 days at 20°C after storage *trans*-2-hexenal residue in treated fruits was less than the natural content in unripe fruits (Neri *et al.*, 2006).

Essential oils

Exposure to various essential oils has been successfully tested on insect control in stored grain (Lopez *et al.*, 2008), but also as supplement in MA packaging to control diseases. Fungal spore production, spore germination and germ tube length of *Colletotrichum coccodes, B. cinerea, C. herbarium* and *Rhizopus stolonifer* was inhibited *in vitro* with lemongrass oil. Also spraying or dipping fruit in essential oils could control postharvest diseases of several fruits (Tzortzakis, 2007). In *in vitro* studies of fumigation with the essential oils from *Illicium verum* and *Schizonepetate nuifolia*, both oils gave good control of *B. cinerea* and *C. gloeosporioides.* Shin *et al.* (2014) also found that fumigation with thymol (30 μg/ml) or linalool (120 μg/ml) significantly inhibited mycelial growth and conidia germination of *B. cinerea*. They also found that the occurrence rate of *B. cinerea* and other non-identified fungi was significantly reduced by fumigation of several grape cultivars with thymol. In this study, in addition to the reduced fungal infection, fumigation with thymol had no effect on the sugar content and firmness of the grapes. As was indicated above, Chu *et al.* (1999) found that cherries inoculated with spores of *B. cinerea* and then fumigated with thymol at 30 mg/l developed less grey mould in storage. Also the cherries that had been fumigated with thymol had lower TSS, higher TA and higher stem browning than other treated or non-treated cherries. Plotto *et al.* (2003) found that dipping tomatoes inoculated with *B. cinerea* and *Alternaria arborescens* in emulsions of oils of thyme and oregano at 5000 and 10,000 ppm reduced disease development. However, these treatments did not stop disease development in inoculated tomatoes and some vapours appeared to be phytotoxic after prolonged exposure. They concluded that essential oils may provide alternatives and supplements to conventional antimicrobial additives in foods. Arras and Usai (2001) in *in vitro* studies found that *Thymus capitatus* essential oil at 250 μl/l was toxic to *Penicillium digitatum, P. italicum, B. cinerea* and *Alternaria citri.* The fungitoxic activity of *T. capitatus* essential oil at 75, 150, and 250 ppm on healthy oranges, inoculated with *P. digitatum* by spraying and placed in 10 l desiccators, was weak at atmospheric pressure (3–10% inhibition at all three concentrations). In a vacuum (0.5 bar), conidial mortality on the exocarp was 90–97% at all three concentrations. These data proved not to be significantly different (p = 0.05) from treatments with thiabendazole at 2000 ppm. Mehra *et al.* (2013) tested biofumigation with essential oils (cinnamon oil, linalool, *p*-cymene, and peppermint leaf oil) and the plant oil-derived biofungicides Sporan (rosemary and wintergreen oils) and Sporatec (rosemary, clove, and thyme oils) on blueberries. The fruits were inoculated with conidial suspensions of *Alternaria alternata, B. cinerea, Colletotrichum acutatum*, or sterile deionized water as a control before fumigation and stored at 7°C for 1 week. Sporatec volatiles reduced disease incidence in most cases, whereas other treatments had no consistent effect on postharvest decay. Sensory analysis showed negative impacts on sourness, astringency, juiciness, bitterness and blueberry-like flavour. They concluded that the potential for postharvest biofumigation of blueberries under refrigeration appears limited. Valverde *et al.* (2005) showed that the addition of 0.5 μl of eugenol, thymol or menthol inside MA packages containing the cultivar 'Crimson Seedless' reduced weight loss and colour change, delayed rates of pedicel deterioration, retarded TSS:acid ratio development, maintained firmness and reduced decay. Also the total viable counts for mesophilic aerobics, yeasts and moulds were significantly reduced in the grapes packaged with eugenol, thymol or menthol. The authors concluded that the inclusion of one of these essential oils resulted in three additional weeks of storage compared with MA packaging only. The equilibrium atmosphere in PP bags in storage at 1°C (1.2–1.3 kPa CO_2 + 13–14 kPa O_2) was reached in 21 days. The inclusion of a sachet impregnated with either 0.5 ml thymol or 0.5 ml menthol inside the bags reduced yeasts, moulds and total aerobic mesophylic colonies and also improved the visual aspect of the rachis. The authors postulated that these essential oils could be an alternative to the use of SO_2 in controlling disease of grapes (Martinez-Romero *et al.*, 2005). Elgayyar *et al.* (2001) evaluated essential oils extracted from anise, angelica, basil, carrot, celery, cardamom, coriander, dill weed, fennel, oregano, parsley and rosemary against *Listeria monocytogenes*, *Staphylococcus aureus*, *Escherichia coli* O:157:H7, *Yersinia enterocolitica*, *Pseudomonas aeruginosa*, *Lactobacillus plantarum*, *Aspergillus niger*, *Geotrichum* and *Rhodotorula*. Oregano essential oil showed the greatest inhibition. Coriander and basil oils were also highly inhibitory to *E. coli* O:157:H7 and to the other bacteria and fungi tested. Anise oil had little inhibitory effect on bacteria but it was highly inhibitory to the fungi. There was no inhibition of either fungi or bacteria with carrot oil. The effects of sachets containing eugenol, carvacrol and *trans*-anethole extracted from herbs were tested in packages of fresh organic wild rocket (*Diplotaxis tenuifolia*), but they had no clear effect on the microbial load (Wieczyńska *el al.*, 2016).

Acetic acid

Acetic acid has been well established for its antifungal and antibacterial properties. For example, Sholberg and Gaunce (1995) found that fumigation with acetic acid gave good control of diseases caused by *B. cinerea*, *Penicillium expansum* and *P. italicum*. The levels of acetic acid they used were over the range of 2.0–5.4 mg/l in air at temperatures over the range of 2–20°C on citrus, apples, pears, tomatoes, grapes and kiwifruit. Increasing the humidity from 17% to 98% RH increased the effectiveness of acetic acid fumigation at both 5°C and 20°C. Acetic acid, at the concentrations used in their trials, had no apparent phytotoxic effects on the fruit. Sholberg *et al.* (2001) artificially or naturally contaminated apples with conidia of *P. expansum* then fumigated them with acetic acid vapour in a 1 m^3 gas-tight chamber at 10°C for times between 1 h and 24 h or dipped in thiabendazole at 450 µl/l. The fruits were then stored at 1°C for 3 months or more. Results showed that fumigation with acetic acid was as effective as thiabendazole in controlling blue mould decay. Perkins *et al.* (2017) found that red bayberries (*Morella rubra*) exposed for 1 h to ultrasonic fog treatment with acetic acid had 18% decay compared with 70% for non-treated fruit after 10 days at 5°C with no adverse effect on their organoleptic properties.

Acetic acid is the main organic compound in vinegar, but the use of vinegar with fresh fruit and vegetables is restricted by its effects on smell and taste. Krusong *et al.* (2015) carried out *in vitro* tests on the effects of baby corn fermented vinegar on cultures of *B. cinerea* and found complete inhibition of their growth. In order to counter the effects on smell and taste, they subsequently either sprayed or exposed strawberries to strawberry flavoured baby corn fermented vinegar vapour. Spraying extended shelf-life at 4°C to 7 days while the latter extended it to 11 days. A trained sensory panel found no significant effect on colour or texture between treated and non-treated, but they detected slight differences in flavour and taste directly after treatment, though not after longer storage. Lagopodi *et al.* (2009) showed that fumigation with acetic acid promoted, inhibited or had no influence on the growth of *B. cinerea in vitro*, with differences dependent on exposure time and dosage. Exposure to acetic acid at 4 µl/l for 6 min gave complete inhibition of the growth of *B. cinerea*.

Nitric oxide

Wills *et al.* (2000) showed that fumigation of strawberries with NO at 1–4000 µl/l immediately after harvest and then storage at

5°C or 20°C in air containing ethylene at 0.1 μl/l extended postharvest life at both temperatures. The most pronounced effect was obtained with NO in the concentration range 5–10 μl/l, which produced more than 50% extension in shelf-life. Hu *et al.* (2014b) inoculated mangoes with *Colletotricum gloeosporioides* (which causes anthracnose) then fumigated them with the NO donor (sodium nitroprusside, 0.1 mM) at 25°C for 5 min. This fumigation effectively suppressed lesion development. Lichanporn and Techavuthiporn (2013) fumigated longkongs for 3 h with NO vapour, which delayed pericarp browning. However, the fumigated fruit had lower levels of PAL, PPO and peroxidase than the control fruit. Huque *et al.* (2013) treated apple slices with NO (10 μl/l), which showed delayed development of surface browning during storage at 5°C. NO treatment also resulted in a lower level of total phenols, inhibition of PPO activity, reduced ion leakage and reduced rate of respiration, but had no significant effect on ethylene production or levels of lipid peroxide and hydrogen peroxide. The effectiveness of the treatment to inhibit development of surface browning may relate to their ability to minimize the level of phenols activity on the cut surface, possibly in conjunction with a reduced PPO activity. Liu *et al.* (2016b) applied 10 μM methyl jasmonate (MJ) and 1 mM sodium nitroprusside, a NO donor, postharvest to cucumbers and found a significant reduction in chilling injury during storage at 5°C. MJ and sodium nitroprusside reduced H_2O_2 in cucumber during storage, suggesting that both MJ and NO alleviated chilling injury by inhibiting H_2O_2 generation. The lower H_2O_2 induced by MJ and NO was coincident with enhanced catalase activity and gene expression, suggesting that catalase played an important role in control of H_2O_2 accumulation.

NO acts as an important signal molecule with diverse physiological functions in plants, including promoting cell death during pathogen attacks. Hu *et al.* (2014b) suggested that the resistance of NO-treated mangoes to anthracnose may be attributed to activation of defence responses as well as delay of ripening. NO may also be linked to cold tolerance. Dong *et al.* (2012) worked on the effects of a saccharide fraction prepared from yeast cell walls on chilling injury of cucumbers during storage at 4°C. They found that the yeast saccharide induced cold tolerance, which was linked with the induction of endogenous NO accumulation. Huque *et al.* (2013) also found that NO-treated apple slices had reduced ion leakage, which is associated with chilling injury.

There appear to be no reports of NO being included in MA packaging, but it seems to have potential and to be worthy of investigation; especially since it occurs naturally. The objective of using it in sachets would simply be to enhance levels already in the plant cells and therefore the effects that are already there.

Humidity

Thompson *et al.* (1974d) reported that the effect of MA packaging in extending the postharvest life of plantains was at least partly due to the maintenance of a high humidity and not simply to a reduction in O_2 or an increase in CO_2. The gas permeability of some plastics used for MA packaging is sensitive to environmental humidity. Yahia *et al.* (2005), extrapolating from work on *Opuntia*, conjectured that the humidity outside MA packages has a major effect on their gas exchange rates and hence on the levels of O_2 and CO_2 inside the package. Roberts (1990) showed that gas transmission of polyamides (nylons) can increase by about three times when the humidity is increased from 0% to 100% RH and with ethyl vinyl alcohol copolymers the increase can be as high as 100 times over the same range. Moisture given out by the produce can condense on the inside of the pack. This is especially a problem where there are large fluctuations in temperatures, because the humidity is high within the pack and easily reaches dew-point where the film surface is cooler than the pack air. Anti-fogging chemicals can be added during the manufacture of plastic films. These do not affect the quantity of moisture inside the packs, but they cause the moisture that has condensed on the inside of the pack to form sheets rather than discrete drops. These can eventually form puddles at the bottom of the pack.

Temperature

Temperature affects the respiration rate of the produce and the permeability of the film. Tano *et al.* (2007) studied temperature fluctuation between 3°C and 10°C over a 30-day period on packaged broccoli and found that it resulted in extensive browning, softening, weight loss increase, ethanol increase, and infection due to physiological damage and excessive condensation, compared with broccoli stored at constant temperature. At 3°C the atmosphere in the packages started at 3 kPa O_2 + 8 kPa CO_2 but when the temperature was raised to 10°C the CO_2 increased rapidly to a maximum of 15.5 kPa and O_2 decreased to < 1.5 kPa.

Chilling injury

There is strong evidence in the literature that, under certain conditions, MA packaging can reduce or eliminate the symptoms of chilling injury. Where this effect has been shown includes: carambola (Ali *et al.*, 2004); citrus (Porat *et al.*, 2004); melons (Kang and Park, 2000; Flores *et al.*, 2004); okra (Finger *et al.*, 2008); and papaya (Singh and Rao, 2005). The reason for this effect and its mode of action have not been clearly determined, but Ali *et al.* (2004) concluded that suppression of the enzyme activities in fruit in MA packaging appeared to contribute to increased tolerance to chilling injury. Martinez-Javega *et al.* (1983) reported that the reduction of ethylene production and water loss in MA packaging were necessary in preventing chilling injury symptoms in melons.

Shrink-wrapping

Besides reducing moisture losses and changing the O_2 and CO_2 levels, shrink-film wrapping can also protect the fruit from some damage, e.g. by scuffing during handling and transport, and possibly from some fungal infections. In Europe, cucumbers are commonly marketed in a shrink-wrapped plastic sleeve to help retain their texture and colour (Fig. 14.7). Thompson (1981) found that arracacha roots deteriorated quickly after harvest but could be stored for 7 days if packed in shrink film. Samsoondar *et al.* (2000) found that breadfruits that had been shrink-wrapped with PE film (60 gauge) had an extended storage life. Lemons stored in PVC shrink-wrap retained their quality, with reduced loss of weight and rotting compared with those stored non-wrapped or waxed (Neri *et al.*, 2004).

Fig. 14.7. Shrink-wrapped cucumber in a UK supermarket.

Vacuum Packing

Vacuum packaging uses a range of low- or non-permeable films (barrier films) or containers into which the fresh fruit or vegetable is placed and the air is sucked out. There are many different machines used for commercial application, including 'vertical form fill and seal', 'horizontal form and fill seal' and 'single chamber machines'. Banavac is a patented system that uses large PE film bags 0.04 mm thick in which typically 18.14 kg of green bananas are packed, then a vacuum is applied and the bags are sealed (see Fig. 13.6) (Badran, 1969). Nair and Tung (1988) reported that 'Pisang Mas' bananas stored at 17°C had an extension in their postharvest life of 4–6 weeks when they were kept in evacuated collapsed PE bags by applying a vacuum not exceeding 300 mm Hg. Knee and Aggarwal (2000) tested plastic containers, capable of being evacuated to 50%, on storage of celery, lettuces, grapes, green beans, melons and strawberries at 4°C and broccoli, okras and tomatoes at 8°C. O_2 levels fell below 5% after storage for 3 days for all types of produce in the vacuum containers. There were many negative effects on the produce and overall the vacuum containers showed little advantage over conventional containers.

Minimal Processing

Minimal processing is also called fresh cut and refers to fruit or vegetables that have been prepared and packaged to a form that can be directly eaten (Fig. 14.8). The damage inflicted in minimal processing of fruit and vegetables can reduce the postharvest life compared with those that remain intact, but MA packaging can be used to compensate for this to some degree. Bastrash *et al.* (1993) found that broccoli heads that had been cut into florets had an increased respiration rate in response to wounding stress throughout storage at 4°C in air and ethylene production was also stimulated after 10 days. They also found that the atmosphere for optimal preservation of broccoli florets was 6 kPa CO_2 + 2 kPa O_2 at 4°C, which delayed yellowing, reduced development of mould and offensive odours and resulted in better water retention during 7 weeks of storage. This compared with only 5 weeks in air. These effects were especially noticeable when the florets were returned from CA at 4°C to air at 20°C.

Fig. 14.8. Imported minimally processed mangoes in the UK.

Budu *et al.* (2001) found that the respiration rate of peeled pineapples was very close to that of intact fruits, but their respiration rate decreased in response to slight changes in O_2 and CO_2 concentrations. They concluded that MA packaging should have important commercial applications in reducing the respiration rate and increasing the shelf-life of minimally processed pineapple. For minimally processed pineapple slices Budu *et al.* (2007) found that an MA packaging giving 5 kPa O_2 + 15 kPa CO_2 appeared to be the most appropriate. Tancharoensukjit and Chantanawarangoon (2008) found that for fresh-cut pineapples storage in 5 or 10 kPa O_2 + 10 kPa CO_2 helped to maintain their lightness and visual quality and prolonged their storage life to 9 days. Antoniolli *et al.* (2007) sliced pineapples and dipped them in sodium hypochlorite (20 mg/l) for 30 s and then stored them at 5 ± 1°C in a flow-through system with different O_2 and CO_2 concentrations. None of the CA treatments seem to give an advantage, since the slices stored in air had little browning and were free of contamination that would affect the food safety at the end of the storage period. Antoniolli *et al.* (2006) did not detect ethylene production during the initial period of 12 h after minimal processing and storage at 5°C. They also found that the initial respiration rate of pineapple slices and chunks was double that observed in the peeled fruits. The respiration rate of slices and chunks was very similar during 14 days of storage. For fresh-cut pineapple Chonhenchob *et al.* (2007) found that an atmosphere of 6 kPa O_2 + 14 kPa CO_2, which was achieved at equilibrium in the headspace of MA packaging, increased the shelf-life by up to 7 days at 10°C .

Day (1996) also showed that the atmosphere inside the MA packaging can influence the rate of deterioration. In minimally processed fruit and vegetables, 70 and 80 kPa O_2 in MA packaging inhibited undesirable fermentation reactions, delayed browning caused by damage during processing and inhibited both aerobic and anaerobic microbial growth. Cut-surface browning and flesh softening were inhibited for fruits that had been stored in 100 or 1 kPa O_2 compared with those that had been stored in air. However, the slices from the 100 kPa O_2 and air contained a much lower content of fermentation products associated with off-flavours compared with the slices from apples from 1 kPa O_2. Lu and Toivonen (2000) sealed 'Spartan' apple slices in 40 μm LDPE film bags having a moderate O_2 transmission rate of about 2.28 cc/m^2/24 h at 23°C and they remained in good condition for up to 2 weeks at 1°C. Slices from apples that had been previously stored at 1°C in 1 kPa O_2 for up to 19 days developed more cut-surface browning and greater tissue solute leakage and enhanced accumulations of acetaldehyde, ethanol and ethyl acetate compared with slices cut from the fruit that had been stored in 100 kPa O_2 for up to 19 days.

Escalona *et al.* (2006b) concluded that 80 kPa O_2 must be used in MA packaging of fresh-cut butterhead lettuce in combination with 10–20 kPa CO_2 to reduce their respiration rate and avoid fermentation. Day (2003), working on high O_2 storage with iceberg lettuce packed in either 30 μm PVDC or 30 μm OPP and stored at 6–8°C for 7–10 days, found that the CO_2 content inside the PVDC film reached 30–40 kPa, which damaged the lettuce. Inside the OPP film the O_2 level did not exceed 25 kPa and the lettuce retained their good quality. In a study of minimally processed cabbages, Rinaldi *et al.* (2009) found that the CA for 10 days did not extend the storage life at 5°C or 10°C longer than for those stored in air. The CA they used in a flow-through system was combinations of 2–10 kPa O_2 + 3–10 kPa CO_2.

Equilibrium Modified Atmosphere Packaging

EMA packaging is a common term for the packaging technology used for minimally processed fruit and vegetables. The internal O_2 concentration of the packages can be predicted for the different steps of the simulated distribution chain by applying an integrated mathematical model and using highly permeable packaging films. For example, Del-Valle *et al.* (2009) developed a model to determine the EMA packaging for mandarin segments because of their high respiration rate. Mistriotis *et al.* (2016) developed a model simulating the combined gas diffusion through micro-perforations and the permeable polylactic acid (PLA) membrane for EMA packaging systems. By using this model, innovative biodegradable bio-based EMA packaging systems were designed specifically for selected fruits and vegetables using laser microperforation and the relatively high water vapour permeability of PLA films compared with oriented polypropylene (OPP) films.

Modelling

Day (1994) and Ben-Yehoshua *et al.* (1995) reviewed the concept of mathematical modelling of MA packaging for both whole and minimally processed fruit and vegetables. Evelo (1995) showed that the package volume did not affect the equilibrium gas concentrations inside an MA package but did affect the non-steady-state conditions. An MA packaging model was developed using a systems-oriented approach that allowed the selection of a respiration rate model according to the available data, with temperature dependence explicitly incorporated into the model. The model was suitable for assessing optimal MA packaging in realistic distribution chains. O_2 and CO_2 concentrations inside plastic film packages containing fruits of the mango cultivar 'Nam Dok Mai' were modelled by Boon-Long *et al.* (1994). Parameters for polyvinyl chloride (PVC), polyethylene (PE) and polypropylene (PP) films were included. A method of determining respiration rate from the time history of the gas concentrations inside the film package instead of from direct measurements was devised. It was claimed that satisfactory results were obtained in practical experiments carried out to verify the model.

Many other mathematical models have been published (e.g. Cameron *et al.*, 1989;

Lee *et al.*, 1991; Lopez-Briones *et al.*, 1993) which could help to predict the atmosphere around fresh produce sealed in plastic film bags. The latter suggested the following model:

$$\frac{1}{x}=\frac{\hat{A}m}{x_oS}\cdot\frac{1}{K}+\frac{1}{x_o}$$

where x = O_2 concentration in the pouches (%), x_o = initial O_2 concentration within the pouches (%), Â = proportionality between respiration rate and O_2 concentration including the effect of temperature (ml/g/day/atm), K = O_2 diffusion coefficient of the film (includes effects of temperature) (ml/m²/day/atm), S = surface area for gas exchange (m²), and m = weight of plant tissue (g).

Lee *et al.* (1991) obtained a set of differential equations representing the mathematical model for an MA system. In this case, the rate of reaction (r) is equal to:

$$r=V_mC_1/\left[K_M+C_1\left(1+C_2/K_1\right)\right]$$

where C_1 is the substrate concentration (which is O_2 concentration in the case of respiration rate), V_m and K_M are parameters of the classical Michaelis–Menten kinetics (V_m being the maximal rate of enzymatic reaction, and K_M the Michaelis constant), C_2 is the inhibitor concentration (CO_2 concentration in the case of respiration) and K_1 is the constant of equilibrium between the enzyme–substrate–inhibitor complex and free inhibitor.

Combining this equation with Fick's Law for O_2 and CO_2 permeation, Lee *et al.* (1991) estimated parameters of this model (V_m, K_M and K_1) from experimental data and then performed numerical calculation with the equations. CO_2 absorbents or a permeable window can be used to prevent atmospheres developing that could damage the crop. The concentration of the various respiratory gases, O_2, CO_2 and ethylene, within the crop is governed by various factors, including the gas exchange equation that is a modification of Fick's law:

$$-\delta s/\delta t=\left(C_{in}-C_{out}\right)\mathrm{DR}$$

where $-\delta s/\delta t$ = rate of gas transport out of the crop, C_{in} = the concentration of gas within the crop, C_{out} = the concentration of gas outside the crop, D = gaseous diffusion coefficient in air, and R = a constant specific to that particular crop.

The strawberry cultivars 'Pajaro' and 'Selva' were used to test a model describing gas transport through micro-perforated PP films and fruit respiration involved in MA packaging by Renault *et al.* (1994). They conducted some experiments with empty packs initially filled with either 100 kPa N_2 or 100 kPa O_2. Simulations agreed very well with experiments only if the cross-sectional area of the micro-perforations was replaced by areas of approximately half the actual areas in order to account for the resistance of air around the perforations. It was also possible to fit the model to gas concentration changes in packs filled with strawberries, although deviations were encountered due to contamination of strawberries by fungi. The model was used to quantify the consequences of the variability of pack properties of the number of micro-perforations per pack and cross-sectional area of these perforations on equilibrium gas concentrations and to define minimum homogeneity requirements for MA packaging. Del Nobile *et al.* (2007) used a simple mathematical model for designing plastic films for minimally processed prickly pear, banana and kiwifruit with a laminated PE/aluminium/PET film or a coextruded PO film. Despite the simplicity of the model, they reported that it satisfactorily described and predicted the respiration rate of the fruit during storage at 5°C.

Waghmare *et al.* (2013) confirmed that mathematical modelling can be used to predict respiration rate as a function of time and temperature in MA packaging and tested their model on MA packaging of fresh-cut coriander, cluster beans and beetroot during storage at 10°C, 20°C or 30°C. They concluded that the dependence of respiration rate on temperature and time was well described by Arrhenius and first-order decay models. The effect of impact stress on the expression of genes related to respiratory by-products and ethylene synthesis of the damaged cabbage was studied by Guillard *et al.* (2012). They used a mathematical model based on interval analysis rather than the probabilistic approach used to predict O_2 and CO_2 evolution in headspaces of MA packaging. They found that their approach needed a minimal amount of unverified assumption concerning uncertainties and also required only a few evaluations of the model. They successfully tested their approach on fresh chicory, mushrooms and blueberries in MA packaging.

Safety

This subject has been dealt with in detail in a review by Church and Parsons (1995). In a survey of 21 ready-to-eat fruits, 28 whole fresh vegetables, 15 sprout samples and 237 minimally processed salads in Spain, Usall *et al.* (2008) showed that the fresh-cut vegetables had the highest microorganism counts. These were associated with grated carrot, arugula and spinach, and the microorganisms were aerobic mesophilic microorganisms, yeast, moulds, lactic acid bacteria and Enterobacteriaceae. The lowest counts of these microorganisms were generally in fresh-cut endive and lettuce. Counts of psychrotrophic microorganisms were as high as those of mesophilic microorganisms. Microbiological counts for fresh-cut fruit were very low. Sprouts were highly contaminated with mesophilic psychrotrophic microorganisms and Enterobacteriaceae and showed a high incidence of *E. coli* (40% of the samples). Of the samples analysed, four were *Salmonella* positive and two harboured *L. monocytogenes*. None of the samples was positive for *E. coli* O157:H7, pathogenic *Yersinia enterocolitica* or thermo-tolerant *Campylobacter*. There is considerable legislation related to food sold for human consumption. In the UK, for example, the Food Safety Act 1990 states that 'it is an offence to sell or supply food for human consumption if it does not meet food safety requirements'. The use of MA packaging has health and safety implications. One factor that should be taken into account is that the gases in the atmosphere could possibly have a stimulating effect on microorganisms. Farber (1991) stated that 'while MA packaged foods have become increasingly more common in North America, research on the microbiological safety of these foods was still lacking'. The growth of aerobic microorganisms is generally optimum at about 21 kPa O_2 and falls off sharply with reduced O_2 levels, while with anaerobic microorganisms their optimum growth is generally at 0 kPa O_2 and falls as the O_2 level increases (Day, 1996). With many modified atmospheres containing increased levels of CO_2, the aerobic spoilage organisms, which usually warn consumers of spoilage, are inhibited, while the growth of pathogens may be allowed or even stimulated, which raises safety issues (Farber, 1991). Hotchkiss and Banco (1992) stated that extending the shelf-life of refrigerated foods might increase microbial risks in at least three ways in MA packaged produce:

- Increasing the time in which food remains edible increases the time in which even slow-growing pathogens can develop or produce toxin.
- Retarding the development of competing spoilage organisms.
- Packaging of respiring produce could alter the atmosphere so that pathogen growth is stimulated.

The ability of *E. coli* O157:H7 to grow on raw salad vegetables that have been subjected to processing and storage conditions simulating those routinely used in commercial practice has been demonstrated by Abdul Raouf *et al.* (1993). The influence of MA packaging, storage temperature and time on the survival and growth of *E. coli* O157:H7 inoculated on to shredded lettuce, sliced cucumber and shredded carrot was determined. Packaging in an atmosphere containing 3 kPa O_2 + 97 kPa N_2 had no apparent effect on populations of *E. coli* O157:H7, psychrotrophs or mesophiles. Populations of viable *E. coli* O157:H7 declined on vegetables stored for up to 14 days at 5°C and increased on vegetables stored at 12°C and 21°C. The most rapid increases in populations of *E. coli* O157:H7 occurred on lettuce and cucumbers stored at 21°C. These results suggest that an unknown factor or factors associated with carrots may inhibit the growth of

E. coli O157:H7. The reduction in acidity of vegetables was correlated with initial increases in populations of *E. coli* O157:H7 and naturally occurring microflora. Eventual decreases in *E. coli* O157:H7 in samples stored at 21°C were attributed to the toxic effect of accumulated acids. Changes in visual appearance of vegetables were not influenced substantially by growth of *E. coli* O157:H7. 'Golden Delicious', 'Red Delicious', 'McIntosh', 'Macoun' and 'Melrose' apples inoculated with *E. coli* O157:H7 promoted growth of the bacterium in bruised tissue independent of their stage of ripeness. Freshly picked (< 2 days after harvest) 'McIntosh' apples usually had no growth of *E. coli* O157:H7 for 2 days but growth occurred following 6 days of incubation in bruised McIntosh apple tissue. When apples were stored for 1 month at 4°C prior to inoculation with *E. coli* O157:H7, all the five cultivars supported its growth (Dingman, 2000).

Church and Parsons (1995) mentioned that there was a theoretical potential fatal toxigenesis through infections by *Clostridium botulinum* in the depleted O_2 atmospheres in MA packed fresh vegetables. It was claimed that this toxigenesis had not been demonstrated in vegetable products without some sensory indication (Zagory and Kader, 1988, quoted by Church and Parsons, 1995). Roy *et al.* (1995) showed that the optimum in-package O_2 concentration for suppressing cap opening of fresh mushrooms was 6 kPa and that lower O_2 concentrations in storage were not recommended because they could promote growth and toxin production by *C. botulinum*. In a review of hazards related to MA packaging of food, Betts (1996) indicated that vacuum packing of shredded lettuce had been implicated in a botulinum poisoning outbreak.

Jacxsens *et al.* (2002) reported that spoilage microorganisms can proliferate quickly on minimally processed bell peppers and lettuce, with yeasts being the main shelf-life limiting group. They reported that *Listeria monocytogenes* was found to multiply on cucumber slices, survived on minimally processed lettuce and decreased in number on bell peppers due to the combination of acidity and refrigeration. *Aeromonas caviae* multiplied on both cucumber slices and mixed lettuce but was also inhibited by the acidity of bell peppers. Storage temperature control was found to be of paramount importance in controlling microbial spoilage and safety of equilibrium MA packaged minimally processed vegetables.

Sziro *et al.* (2006) studied microbial safety of strawberry and raspberry fruits after storage in two permeable packaging systems, in combination with an ethylene-absorbing film, calculated to reach 3 kPa O_2 at equilibrium. The first was called high oxygen atmosphere (HOA) where the atmosphere started at 95 kPa O_2 in 5 kPa N_2m and the second was EMA starting from 3 kPa O_2 in 95 kPa N_2. *E. coli*, *Salmonella* spp. and *L. monocytogenes* were artificially inoculated on to packaged fruits and all three were able to survive in storage at 7°C. Raspberries showed an enhanced inactivation of *Salmonella* during storage in both types of packaging. Growth of *L. monocytogenes* was observed on the calyx of strawberries after 3 days. Generally, increasing the shelf-life of the fruits with EMA and HOA did not give an increased microbial risk.

Another safety issue is the possibility of the films being used in MA packaging being toxic. Schlimme and Rooney (1994) reviewed the possibility of constituents of the polymeric film used in MA packaging migrating to the food that they contain. They showed that it is unlikely to happen, because of their high molecular weight and insolubility in water. All films used can contain some non-polymerized constituents which could be transferred to the food and in the USA the Food and Drugs Administration and also the European Community have regulations related to these 'indirect additives'. The film manufacturer must therefore establish the toxicity and extraction behaviour of the constituents with specified food simulants.

There are other possible dangers in using high concentrations of CO_2 and O_2 in storage. Low O_2 and high CO_2 can have a direct lethal effect on human beings working in those atmospheres. Great care needs to be taken when using MA packaging containing 70 or 80 kPa O_2, or even higher levels, because of potential explosions.

Some MA Packaging Recommendations

Apple

In refrigerated conditions MA packaging was shown to extend the storage life of apples longer than those stored in air but shorter than those stored in CA. For example, in Germany, Hansen (1975) showed that the cultivar 'Jonagold' kept in cold stores at 0.5°C, 2.5°C or 3.5°C retained satisfactory organoleptic properties only until the end of January. However, when they were stored at 3.5°C in PE bags (with an equilibrium CO_2 content of 7 kPa) they retained their quality until April, but in CA storage (6 kPa CO_2 + 3 kPa O_2) at 2.5°C fruit quality remained satisfactory to the end of May. Under all conditions apples showed no rotting, CO_2 damage or physiological disorders. Hewett and Thompson (1988) also found that storage of the cultivar 'Golden Delicious' at 1°C for 49 days, then 7 days at ambient in PE bags 25 μm thick with 50–70 micro-perforations, made with a 1 mm diameter cold needle, lost less weight, were firmer and had better eating quality than fruit not stored in bags.

Generally, the acceptable time for which apples can remain in MA packages in ambient conditions varies with cultivar, but it is usually 2–4 weeks. An example of this was given by Geeson and Smith (1989), who showed that different apple cultivars required different film types and thicknesses for optimum effect. For 'Bramley's Seedling', 'Egremont Russet' and 'Spartan', they found that after CA storage, apples in 1 kg 30 μm LDPE pillow packs in ambient conditions showed slower flesh softening and skin yellowing and a longer shelf-life than those with no packaging. For 'Cox's Orange Pippin', 20 μm ethylvinyl acetate film was effective in retarding ripening without adversely affecting eating quality. Park *et al.* (2007) found that packaging in 25 μm MA film (film not specified) of the cultivar 'Fuji' for 24 weeks at 0°C was more effective than other film packaging treatments. TA and TSS were higher in MA packaging than control and decay was reduced from 6.8% in control to 2.4% in MA packaging.

Apricot

Ayhan *et al.* (2009) found that the cultivar 'Kabaasi' stored at 4°C in air could be kept for 7 days without loss of physical and sensory properties, but when they were packed in either bi-oriented polypropylene (BOPP) or cast polypropylene (CPP) films they could be successfully stored for 28 days. Ali Koyuncu *et al.* (2009) stored the cultivar 'Aprikoz' at 0°C and 90 ± 5% RH in 12, 16 or 20 μm stretch film or CA and found that those stored in CA retained their quality better than those in stretch film.

Arracacia

Scalon *et al.* (2002) found that *Arracacia xanthorrhiza* roots stored without packaging were unfit for marketing after 21 days. However, in ambient conditions in Bogotá (17–20°C and 68–70% RH), Thompson (1981) found that roots deteriorated much more quickly but could be stored for 7 days if wrapped in plastic cling film or plastic shrink film. Scalon *et al.* (2002) also found that roots packaged in PVC film had lower weight losses than those in cellophane film, but those in cellophane film had a better marketing appearance. Ramos *et al.* (2015) found that roots wrapped in PVC film and stored at 5°C or 10°C were commercially viable up to 60 days, while those stored non-wrapped at 5°C had a shelf-life of only 7 days, and at least 25% had chilling injury symptoms.

Artichoke

Mencarelli (1987a) sealed globe artichokes (immature flower heads) in films of differing permeability or perforated films for 35 days. Perforated heat-sealed films gave poor results, as did low-permeability films, but films sufficiently permeable to allow good gas exchange while maintaining high humidity gave good results, with high turgidity and good flavour and appearance. Storage in crates lined with perforated PE film was recommended by Lutz and Hardenburg (1968). Five storage trials carried out by Andre *et al.* (1980b) on the cultivar 'Violet de Provence' showed that the storage life in air was 1 week at room temperature and 3–4 weeks at 1°C. This was extended to 2 months by a combination of vacuum cooling, packing in PE bags and storage at 1°C. Gil-Izquierdo *et al.* (2004) found that the cultivar 'Blanca de Tudela' stored at 5°C had an equilibrium atmosphere of 14.4 kPa O_2 + 5.2 kPa CO_2 for those sealed in PVC bags and 7.7 kPa O_2 + 9.8 kPa CO_2 for those in LDPE bags. They showed that loss of water was the major cause of deterioration after 8 days of storage. The vitamin C content of the internal bracts (the edible part) decreased in MA storage, while for those stored without packaging it remained at levels similar to those at harvest. There was also a high phenolic content (896 mg/100 g fresh weight) in the internal bracts and this increased in those in LDPE and PVC but not in those stored without packaging.

Asparagus

Storage at 2°C in MA packaging retained sensory quality and AA levels (Villanueva *et al.*, 2005). In several trials cooling to 1°C within 6–8 h of harvest, followed by storage at 1°C in PE bags with silicon elastomer windows kept the spears in good condition for 20–35 days (Andre *et al.*, 1980a). Lill and Corrigan (1996) experimented with different MA packs and found that they all significantly extended the shelf-life of the spears by between 83% and 178%, depending on film type, compared with those not wrapped. The least permeable films (W.R. Grace RD 106 and Van Leer Packaging Ltd) gave the longer shelf-life extension; the former film had 2.5–3.5 kPa CO_2 + 4.1–9.7 kPa O_2 and the latter 3.6–4.6 kPa CO_2 + 2.0–4.5 kPa O_2 equilibrium gas content at 20°C.

Aubergine

MA packaging prolonged the shelf-life and flesh firmness compared with those stored non-wrapped (Arvanitoyannis *et al.*, 2005).

Avocado

Storage life was extended by 3–8 days at various temperatures by sealing individual fruits in PE film bags (Haard and Salunkhe, 1975). In earlier work it had been shown that fruit of the cultivar 'Fuerte' sealed individually in PE film bags 0.025 mm thick for 23 days at 14–17°C ripened normally on subsequent removal from the bags. The atmosphere inside the bags after 23 days was 8 kPa CO_2 + 5 kPa O_2 (Aharoni *et al.*, 1968). Thompson *et al.* (1971) showed that sealing various seedling varieties of West Indian avocados in PE film bags greatly reduced fruit softening during storage at various temperatures (Table 14.4).

Banana

There is strong evidence that MA packaging can extend the pre-climacteric life of banana, which has been shown in several varieties and at different temperatures. Packaging bananas in PE bags alone, gas flushed with 3 kPa O_2 + 5 kPa CO_2 or with a partial vacuum (400 mm Hg) resulted in shelf-life extension to 15, 24 and 32 days, respectively, compared with 12 days for those stored non-wrapped at 13 ± 1°C (Chauhan *et al.*, 2006). According to Shorter *et al.* (1987) the storage life was increased five times when bananas were stored in plastic film (where the equilibrium gas content was about 2 kPa O_2 + 5 kPa CO_2) with an ethylene scrubber, compared with fruit stored without wraps. Satyan *et al.* (1992) stored 'Williams' in PE tubes 0.1 mm thick at 13°C, 20°C and 28 C and found that the storage life was increased 2–3 times compared with fruit stored without

Table 14.4. Effects of storage temperature and packing in 125 gauge PE film bags compared with non-packed on the number of days to softening of avocados (modified from Thompson *et al.*, 1971).

	7°C	13°C	27°C
Not wrapped	32	19	8
Sealed film bags	38	27	11

wraps. Tiangco *et al.* (1987) also observed that green life of the variety 'Saba' (*Musa* BBB) held in MA at ambient temperature had a considerable extension of storage life compared with fruit stored without wraps, especially if combined with refrigeration. In Thailand, Tongdee (1988) found that the green life of 'Kluai Khai' (*Musa* AA) could be maintained for more than 45 days in PE bags at 13°C. This was longer by more than 20 days than fruit stored at 25°C in PE bags. In the Philippines, 'Latundan' (*Musa* AAB) could be stored in PE bags 0.08 mm thick at ambient temperature of 26–30°C for up to 13 days. It was also reported that 'Lakatan' (*Musa* AA) had a lag of 3 days before the beginning of colour change upon removal from the bags (Abdullah and Tirtosoekotjo, 1989; Abdullah and Pantastico, 1990). Marchal and Nolin (1990) found that storage in PE film reduced weight loss and respiration rate, which allowed the bananas to be stored for several weeks. Chamara *et al.* (2000) showed that packaging in LDPE bags 75 μm thick extended the green life of bananas up to 20 days at room temperature of about 25°C and 85% RH. Fruit ripened naturally within 4 days when removed from the packaging. Ali Azizan (1988, quoted by Abdullah and Pantastico, 1990) observed that the TSS and TA in 'Pisang Mas' changed more slowly during storage in MA at ambient temperature than those stored not wrapped.

Bowden (1993) showed that the effect on yellowing of bananas that had been initiated to ripen was related to the thickness of plastic and therefore the gas composition within the film. Film thickness also affected weight loss. For example, at 13°C Wei and Thompson (1993) showed that weight loss of 'Apple' bananas after 4 weeks storage in PE film was 1.5% in 200 gauge, 1.8% in 150 gauge and 2.1% in 100 gauge, while fruit stored without packaging was 12.2 %.

Potassium permanganate-soaked paper in sealed PE bags was the best packaging treatment for 'Harichhaal' (*Musa* AAA). Bavistin + PE bags (with 2% of the surface with perforations) were the most effective treatment in minimizing the microbial spoilage of fruit during storage (Kumar and Brahmachari, 2005). Clay bricks impregnated with potassium permanganate significantly ($p = 0.05$) lowered in-package ethylene and CO_2, retained higher O_2 content and resulted in minimum changes in firmness and TSS content (Chamara *et al.*, 2000). Ranasinghe *et al.* (2005) reported that potassium permanganate in combination with silica gel, as a desiccant, and soda-lime, as a CO_2 scrubber, further increased the storage life. They also recommended treatment with emulsions of cinnamon oils combined with MA packaging in 75 μm LDPE as a safe, cost-effective method for extending the storage life of 'Embul' (*Musa* AAB) up to 21 days at 14 ± 1°C + 90% RH and 14 days at 28 ± 2°C without affecting the organoleptic and physicochemical properties.

MA can also have negative effects. Wills (1990) mentioned that an unsuitable selection in packaging materials can still accelerate the ripening of bananas or enhance CO_2 injury when the ethylene accumulated over a certain period. Wei and Thompson (1993) found that on packaging the cultivar 'Apple' in PE film and storage at 13–14°C symptoms of CO_2 injury were observed when the CO_2 levels were between about 5% and 14%. These levels occurred with some fruit in 150 gauge PE bags and all the fruit in 200 gauge PE bags, but only after 3 weeks, and not at all in 100 gauge bags. Injury was characterized by darkening of the skin and softening of the outer pulp, while the inner pulp (core) remained hard and astringent. Often there would be a distinct irregular ring of dark brown tissue in the outer cross-section of the pulp. In some of the fingers the pulp developed a tough texture.

MA packaging of bananas in PE film bags is commonly used in international transport. With the 'Banavac' system patented by the United Fruit Company (see 'Vacuum packing', above), typical gas contents that develop during transport at 13–14°C in the bags through fruit respiration were about 5 kPa CO_2 + 2 kPa O_2 (De Ruiter, 1991). Besides retaining a desirable high humidity around the fruit, it has been shown that packing in non-perforated bags prolonged the pre-climacteric life of 'Mas' (*Musa* AA)

(Tan and Mohamed, 1990). Plastic bags are used commercially as liners for whole boxes and are mainly intact but some have macro-perforations. Those that were intact were often torn before being loaded into the ripening rooms, in order to ensure that ethylene penetrated to initiate ripening. With the use of forced-air ventilation in modern 'pressure' ripening rooms, this practice was found not to be necessary. Clusters of fruit are also packed individually in small PE bags at source, transported, ripened and sold in the same bags.

In Australia, bananas were transported internally packed in 13.6 kg commercial packs inside PE film bags. Scott *et al.* (1971) reported that in simulated transport experiments the fruit were kept in good condition at ambient temperatures during the 48 h required for the journey. Siriwardana *et al.* (2017) treated 'Embul' bananas (*Musa* AAB) with 1% aluminium sulfate, 1% aluminium sulfate + 0.4% basil oil or distilled water (control), then sealed them in LDPE bags and stored them at 12–14°C. After 14 days the bananas in all three treatments were in good condition and the atmosphere inside the bags was 3.1–3.7 kPa O_2 + 4.2–4.7 kPa CO_2. Aluminium sulfate + basil oil significantly controlled crown rot disease compared with the other two treatments and had no negative effects on consumer acceptability.

Basil

Patiño *et al.* (2018) stored basil leaves at 11°C and 85% RH in sealed LDPE bags and in open macro-perforated LDPE bags with and without previous exposition to 1-MCP at 0.3 cm^3/m^3 for 24 h, which they considered the optimum exposure. The MAP system they used gave a steady-state concentration inside the packaging of about 10.5 kPa O_2 + 4.2 kPa CO_2. The combined treatment of MAP and 1-MCP increased the storage life of the basil leaves up to 18 days, compared with 9 days for those stored in open bags without 1-MCP exposure, all at 11°C and 85% RH.

Beans

The quality of 15 cultivars of French beans stored in LDPE at 8–10°C was acceptable for up to 30 days, after which the colour changed to light brown in some of the cultivars. At that time the cultivar 'CH-913' had the highest percentage of beans (92.55%) that were still considered marketable (Attri and Swaroop, 2005).

Beet

Sugarbeet packed in PVC film maintained good appearance during 22 days storage at an ambient temperature of 15–26°C with a weight loss of 15.8% compared with 55% in the non-wrapped beets (Scalon *et al.*, 2000).

Black sapote

Black sapote (*Diospyros ebenaster*, *D. digyna*) is a climacteric fruit and the main factor limiting storage was reported to be softening. Fruits wrapped in plastic film had 40–50% longer storage life at 20°C than those not wrapped (Nerd and Mizrahi, 1993).

Blueberry

Hong *et al.* (2006) reported that packaging blueberries in 0.88 mm (*sic*) PE film significantly reduced the rates of weight loss, rotting and respiration rate during storage at 2°C. It also slowed the reduction in anthocyanin and inhibited enzyme activity compared with those stored non-wrapped. Tectrol MA pallet bags are used in reefer containers and were reported to have proven success in the storage of fresh blueberries for up to 65 days (DeEll, 2002).

Breadfruit

Storage of breadfruit at 12.5°C sealed in 150 gauge PE bags kept them in good condition for up to 13 days compared with 80% of the non-wrapped fruit being unmarketable after only 2 days (Thompson *et al.*, 1974a). At 28°C Maharaj and Sankat (1990) found that fruits could be stored for 5 days in sealed 100 gauge PE film bags, while at 12°C or 16°C in sealed 100 gauge PE film bags they could be stored for 14 days. Worrel and Carrington (1994) also showed that storing breadfruit sealed in plastic film maintained their green colour and fruit quality. They used 40 μm HDPE film or LDPE film and showed that fruit quality could be maintained for 2 weeks at 13°C. Breadfruits shrink-wrapped in 60 gauge PE film and stored at 16°C had reduced external skin browning and higher chlorophyll levels after 10 days of storage. This treatment also delayed fruit ripening as evidenced by changes in texture, TSS and starch content (Samsoondar *et al.*, 2000).

Broccoli

Serrano *et al.* (2006) reported that non-wrapped broccoli lost weight and turned yellow and the stems hardened; also there was a rapid decrease in total antioxidant activity, AA and total phenolics, giving a maximum storage period of only 5 days. These changes were delayed in MA packaging, especially in micro-perforated PP film and non-perforated PP film, with total antioxidant activity, AA and total phenolic compounds remaining almost unchanged during the whole storage period of 28 days. Broccoli stored at 5°C in 14 kPa O_2 + 10 kPa CO_2 or wrapped in PVC film for 3 weeks retained market quality significantly better than when left non-wrapped. Respiration rates of those in CA or in PVC film were reduced by 30–40% compared with the controls (Forney *et al.*, 1989). Cameron (2003) reported that low O_2 could retard yellowing at 5°C but could also produce highly undesirable flavours and odours when O_2 fell below 0.25 kPa. Rai *et al.* (2008) studied various MA packages in storage at 5°C and 75% RH and found that perforated PP film (2 holes, each of 0.3 mm diameter, with a film area of 0.1 m^2) could be used to retain chlorophyll and AA levels for 4 days. DeEll *et al.* (2006) reported that, overall, the use of sorbitol, a water absorbent, and sachets of potassium permanganate inside PD-961EZ bags at 0–1°C for 29 days reduced the amount of volatiles that are responsible for off-odours and off-flavours. This was reported to maintain their quality and marketability for longer. Acetaldehyde concentrations were higher in the bags with no sorbitol or potassium permanganate after 29 days, while ethanol was greater in both control bags and those with only potassium permanganate. After storage at 4°C or 10°C in LDPE film packages containing potassium permanganate with an equilibrium gas content of 5 kPa O_2 + 7 kPa CO_2, Jacobsson *et al.* (2004b) found that sensory properties were better than those in other films and similar to freshly harvested broccoli. The optimum equilibrium atmosphere was found to be 1–2 kPa O_2 and 5–10 kPa CO_2. Jacobsson *et al.* (2004a) measured the gas inside various packages and found that storage in OPP resulted in the highest CO_2 concentration of 6 kPa, while the lowest O_2 concentration of 9 kPa was found in the LDPE package. They also found that broccoli stored in PVC film deteriorated faster than when packaged in the other films and the influence of all the plastic packaging tested was greater at 10°C than at 4°C. Among the packages tested by Cefola *et al.* (2016), the combination of PP/PA in 5 kPa O_2 in 95 kPa N_2 containing CO_2-absorbing sachets reached an equilibrium condition very close to optimum for broccoli (5 kPa O_2 and ≤ 5 kPa CO_2). Fresh-cut broccoli stored under these conditions for 8 days showed negligible reduction of appearance and odour and were comparable to the freshly harvested samples.

Brussels sprouts

Storage in PVC film at 0°C for 42 days resulted in reduced browning of cut areas and reduced losses in weight and firmness compared with sprouts that were not wrapped. AA and total flavonoid contents remained almost constant in the PVC-wrapped sprouts, while radical scavenging activity increased (Vina *et al.*, 2007).

Cabbage

Peters *et al.* (1986) found that the best plastic covering for the cultivar 'Nagaoka King' stored in wooden boxes was PE sheets (160 cm × 60 cm × 0.02 mm) with 0.4% perforations, wrapped around the top and long sides of the boxes, with the slatted ends of the boxes left open. Omary *et al.* (1993) treated shredded white cabbage with citric acid and sodium erythorbate then inoculated it with *Listeria innocua* and packaged it in lots of 230 g. Four types of plastic film bags were used for packaging, with O_2 transmission rates of 5.6, 1500, 4000 and 6000 cc/m²/day and they were stored at 11°C. After 21 days, *L. innocua* populations increased in all packages but the increase was significantly less for cabbage packaged in film with the highest permeability.

Capsicum

Pala *et al.* (1994) studied storage in sealed LDPE film of various thicknesses at 8°C and 88–92% RH and found that non-wrapped peppers had a shelf-life of 10 days compared with 29 days for those sealed in 70 μm LDPE film (Table 14.5). Also this method of packaging gave good retention of sensory characteristics.

Nyanjage *et al.* (2005) showed that temperature was the major factor in determining the postharvest performance of the cultivar 'California Wonder'. For packaging in open trays, non-perforated or perforated PE bags did not significantly affect colour retention during storage at 4°C, 6.5°C or 17°C, but there was a significantly higher incidence of disease. Perforated PE packaging produced the best overall results.

In other work, the marketable value of peppers stored for 4 weeks at 8°C was highest for those in PE bags with 0.1% perforations and in non-perforated PE bags compared with non-packaged (Kosson and Stepowska, 2005). After 2 weeks of storage in non-perforated PE bags the atmosphere was about 5 kPa O_2 + 5 kPa CO_2 and with 0.0001% perforation it was about 18 kPa O_2 + 5 kPa CO_2. Water condensation on the inner surface of PE bags and on the stored peppers occurred in bags with 0.0001%, 0.001%, 0.01% and 0% perforations but not in those with 0.1% perforations. Banaras *et al.* (2005) found that there were no significant differences in AA and fruit firmness between the different perforated MA packaging of the cultivars 'Keystone', 'NuMex R Naky' and 'Santa Fe Grande'. However, perforated MA packaging was shown to reduce water loss, maintain turgidity of fruits and delay red colour and disease development at both 8°C and 20°C. MA packaging extended postharvest life for another 7 days at 8°C and for 10 days at 20°C compared with non-packaged fruit held at these temperatures. Postharvest water loss and turgidity were similar for fruit stored in packages

Table 14.5. Sensory score (1–9) of green capsicums during storage at 8°C and 88–92% RH, where 1–3.9 = non-acceptable, 4–6.9 = acceptable, 7–9 = good; modified from Pala *et al.* (1994).

		Sensory score			
	Storage time	Appearance	Colour	Texture	Flavour
Before storage	0 days	9	9	9	9
Non-wrapped control	15 days	4.3	4.6	2.7	4.2
20 μm LDPE	22 days	5.7	5.5	5.7	5.3
30 μm LDPE	20 days	5.8	6.0	6.0	4.5
50 μm LDPE	20 days	6.0	5.0	6.0	5.0
70 μm LDPE	29 days	7.8	7.8	7.6	7.8
100 μm LDPE	27 days	5.5	5.6	5.7	5.5

with and without 26 holes at 8°C and 20°C. Amjad *et al.* (2009) stored the cultivars 'Wonder King' and 'P-6' in various thicknesses of PE bags at 7°C, 14°C and 21°C. They found that the optimum thickness for 'Wonder King' was 21 μm at 7°C for minimizing weight loss, while 'P-6' fruit stored in 15 μm had the lowest weight loss and their maximum storage life was 20 days. AA levels and total phenolics and carotenoids also varied between cultivars, temperature and thickness of PE. Hughes *et al.* (1981) showed that capsicums sealed in various plastic films had a higher percentage of marketable fruit than those stored in air (see Table 11.2).

Carambola

Ali *et al.* (2004) harvested the cultivar 'B10' at the mature-green stage. They reported that MA packaging in LDPE film suppressed the incidence of chilling injury in storage at 10°C and also retarded softening and the development of fruit colour. Neves *et al.* (2004) stored the cultivar 'Golden Star' in various thicknesses of LDPE bags at 12 ± 0.5°C and 95 ± 3% RH for 45 days, then in ambient conditions of 22 ± 3°C and 72 ± 5% RH for a further 5 days. They found that fruits in LDPE bags 10μm thick best retained their firmness and TA and had the 'best colour standard', the lowest decay incidence and the highest acceptance scores by the taste panellists, but the lowest TSS.

Carrot

MA packaging is used commercially to extend the marketable life of immature carrots. Seljasen *et al.* (2003) found that carrots being gently handled and stored in perforated plastic bags at low temperature retained the most favourable taste in the Norwegian distribution chain. MA packaging did not result in an increase in off-flavours or an increase in 6-methoxymellein, which has been associated with off-flavours, or in a reduction in sugars. Storage in PE bags at 0°C was reported by Kumar *et al.* (1999) to slow the decrease in AA and reduce the levels of TSS and reduce the weight loss to 1% over an 8-day period compared with carrots stored in gunny bags in ambient conditions, which lost 16%.

Cassava

Thompson and Arango (1977) found that dipping the roots in a fungicide and packing them in PE film bags directly after harvest reduced the physiological disorder vascular streaking during storage for 8 days at 22–24°C compared with roots stored non-wrapped. Oudit (1976) showed that freshly harvested roots could be kept in good condition for up to 4 weeks when stored in PE bags. Coating with paraffin wax is commonly applied to roots exported from Costa Rica. Young *et al.* (1971) and Thompson and Arango (1977) found that this treatment kept them in good condition for 1–2 months at room temperature in Bogotá in Colombia.

Cauliflower

Menjura Camacho and Villamizar (2004) studied ways of reducing wastage of cauliflowers in the central markets of Bogotá and found that the best treatment was packing them in LDPE with 0.17% perforations. Menniti and Casalini (2000) found that PVC film wrapped around each cauliflower reduced weight loss and retarded yellowing during storage for 5 days at 20°C. Curd colour, odour and compactness of the curd remained acceptable for up to 14 days at 6°C in non-perforated PE bags (Rahman *et al.*, 2008). The highest AA content (41 mg/100 g and 23 mg/100 g) and β-carotene content (26 IU/100 g and 14 IU/100 g) were in non-perforated PE bags after 7 days and 14 days of storage, respectively, at 6°C. Kaynaş *et al.* (1994) stored the cultivar 'Iglo' in PE film bags 30 μm thick, PE film bags with one 5 mm diameter hole per kilogram head, 15 μm PVC film bags or not wrapped, at 1°C and

90–95% RH for various times. They found that they could be stored for a maximum of 6 weeks in PVC film bags with a shelf-life of 3 days at 20°C, which was double the storage life of those stored non-wrapped. The permeability of the PVC film used in the studies was given as 200 g of water vapour/m^2/day and 12,000 ml O_2/m^2/day/atm.

Celery

The equilibrium atmospheres within OPP and LDPE bags containing celery were reached after 10 days and were 8–9 kPa O_2 + 7 kPa CO_2 and 8 kPa O_2 + 5 kPa CO_2, respectively (Gómez and Artes, 2004b). The celery stored in OPP was rated to have an appearance most similar to that at harvest. Escalona *et al.* (2005) described an MA packaging for pallet loads by using a silicone membrane system for shipping celery to distant markets. The design of a system for 450 kg of celery per pallet was studied using an impermeable film with silicone membrane windows at 0.2 and 0.3 dm^2/kg, to reach an atmosphere of approximately 3–4 kPa O_2 + 7–8 kPa CO_2. Two kinds of silicone membrane windows were chosen, with permeability ranges of 4500–5250 ml O_2/dm^2/day and 10,125–14,000 ml CO_2/dm^2/day. The storage conditions for a trial were 21 days at 5°C and 95% RH followed by a shelf-life in air of 2 days at 15°C and 70% RH. As a control, partially unsealed packages were used. MA packaging resulted in slightly better quality and lower decay than those in the control. Rizzo and Muratore (2009) found that the most suitable MA packaging was polyolefin film (coextruded PE and PP) with an antifogging additive and storage at 4 ± 1°C and 90% RH for 35 days. They found tiny accumulations of condensate but it did not reduce shelf-life. Gómez and Artes (2004b) reported that decay developed in non-wrapped celery, but wrapping in either OPP or LDPE inhibited decay, decreased the development of pithiness and retained the sensory quality, reducing both the development of the butt end cut browning and chlorophyll degradation.

Cherimoya

Melo *et al.* (2002) found that the cultivar 'Fino of Jete' stored at 12 ± 1°C and 90–95% RH had a marketable life in air of 2 weeks, while fruits packed with zeolite film could be stored for 4 weeks.

Cherry

Bertolini (1972) stored the cultivar 'Durone Neo I' at 0°C for up to 20 days in air, 20 kPa CO_2 + 17 kPa O_2 or in 0.05 mm PE film bags. Cherries stored in the PE bags retained their freshness best as well as the colour of both the fruit and the stalks. The cultivars 'V-690618', 'V-690616' and 'Hedelfingen' were stored in polyolefin film bags with O_2 permeabilities of 3000, 7000 and 16,500 ml/m^2 (in 24 h at 23°C) for 6 weeks at 2 ± 1°C. The bags with the lowest O_2 permeabilities produced the best results and the cherries had reduced levels of decay. For the cultivar 'Hedelfingen', MA packaging reduced decay by 50% and significantly increased firmness (Skog *et al.*, 2003). Massignan *et al.* (2006) sealed the cultivar 'Ferrovia' in plastic bags and stored the cherries at 0.5°C in air for 30 days with 2 days of shelf-life at 15°C. They found that the cherries retained their quality throughout the storage and shelf-life period. The equilibrium atmosphere inside the bags was 10 kPa CO_2 + 10 kPa O_2. Kucukbasmac *et al.* (2008) packed the cultivar '0900 Ziraat' in either Xtend® CH-49 bags or PE bags in three sizes (500 g, 700 g and 1000 g) and stored them at 0°C for up to 21 days plus 3 days at 20°C to simulate shelf-life. The quality and firmness of the cherries from all three Xtend® CH-49 bags were maintained even after an additional 3 days shelf-life and were better than those stored in PE bags. Celikel *et al.* (2003) stored the cultivar Merton Bigarreau at 0°C in polystyrene trays (350 g/tray) over-wrapped with PVC stretch film or P-Plus PP film at different permeability, or left some not wrapped as a control. MA packaging doubled the storage life, compared with the control, to 8 weeks. The weight loss reached 22.5% for the control, 4.5% with PVC and 0.3–0.5% with PP after 8 weeks. Less permeable films, PP-90 and PE-120, also maintained the fruit quality and flavour better than high-permeable film. Alique *et al.* (2003) found that storage of the cultivar 'Navalinda' at 20°C in 15% micro-perforated films preserved fruit acidity and firmness while slowing the darkening of colour, loss of quality and decay and thus prolonging the shelf-life. However, they recommended that the levels of hypoxia reached at 20°C in micro-perforated films should restrict its use only to the distribution and marketing processes. Padilla-Zakour *et al.* (2004) stored the cultivars 'Hedelfingen' and 'Lapins' in the micro-perforated LDPE bags, called LifeSpan L204, which equilibrated at 4–5 kPa O_2 + 7–8 kPa CO_2, and LifeSpan 208, which equilibrated at 9–10 kPa O_2 + 8–9 kPa CO_2. After 4 weeks at 3°C with 90% RH, the fruit in the LDPE bags had green and healthy stems, better colour, better appearance and eating quality when compared with non-wrapped fruit. There was slightly better quality for LifeSpan 204, which had a lower O_2 permeability. Horvitz *et al.* (2004) stored the cultivar 'Sweetheart' at 0°C in LDPE bags for up to 42 days and found no significant changes in colour, firmness, TSS, pedicel dehydration and rotting. In PVC bags there was rapid pedicel deterioration, which affected approximately 50% of the fruit after 7 days of storage.

Sweet cherries were stored with the essential oils thymol, eugenol or menthol separately in trays sealed within PE bags. During 16 days storage at 1°C and 90% RH it was found that these treatments reduced weight loss, delayed colour changes and maintained fruit firmness compared with cherries packed without essential oils. Pedicels remained green in treated cherries while they became brown in those without essential oils. The equilibrium atmosphere inside the bags was 2–3 kPa CO_2 + 11–12 kPa O_2, which was reached after 9 days with or without the essential oils. The essential oils reduced moulds and yeasts and total aerobic mesophilic colonies (Serrano *et al.*, 2006).

Cherry, Suriname

Santos *et al.* (2006 a, b) harvested Suriname cherry (*Eugenia uniflora*) at three different maturities based on skin colour (pigment initiation, red-orange and predominant red) and stored them in different temperatures either not packaged or in PVC bags. Storage at 10°C or 14°C in PVC bags maintained TSS, TA, sugars and AA, and also resulted in a lower rate of increase in total carotenoids in fruits in the red-orange maturity stage for 8 days. Santos *et al.* (2006b) showed that fruit in PVC bags at 10°C or 14°C had a lower disease incidence and fruit shrinkage, giving a 4-day increase in postharvest life compared with those stored non-wrapped. Red-orange was the most suitable harvest maturity stage and 10°C the best temperature of those that were tested.

Chestnut, Japanese

It was reported by Lee *et al.* (1983) that the best storage results for the cultivar 'Okkwang' Japanese chestnut (*Castanea crenata*) were obtained when the cherries were sealed in 0.1 mm PE bags (30 cm × 40 cm) with 9–11 pin holes. The total weight loss was < 20% after 8 months of storage; the CO_2 concentration in the bags increased to 5–7 kPa and the O_2 concentration was reduced to 7–9 kPa. Panagou *et al.* (2006) showed that the atmosphere at equilibrium was 10.5 kPa O_2 + 10.9 kPa CO_2 for micro-perforated PET/PE (12 μm/40 μm) permeable film at 0°C and 8.3 kPa O_2 + 12 kPa CO_2 at 8°C. The total storage period was 110 days and sucrose content increased during the first 40 days in PET/PE micro-perforated and macro-perforated films, but changed only slightly thereafter, though starch and AA decreased throughout the storage period. In some less permeable film packages the O_2 concentration fell below 2 kPa, resulting in the development of off-odours. To create different O_2 and CO_2 compositions Homma *et al.* (2008) stored chestnuts in one, two or five layers of PE bags. The proportion of sound chestnuts decreased for all treatments after 2 months and starch breakdown was high in one or two layers of PE bag but was decreased when five layers of PE bag were used.

Chilli

Fresh chillies are often marketed in MA packaging. Fruits of the cultivar 'Nok-Kwang' were sealed in 0.025 mm PE film and stored at 18°C or 8°C. At both temperatures PE film reduced the weight loss below 1% and increased CO_2 concentration to 2–5 kPa (Eum and Lee, 2003). Amjad *et al.* (2009) found that the optimum thickness of PE film for maximum storage life was 15 µm at 7°C for 10 days.

Citrus

Porat *et al.* (2004) found that packaging citrus fruits in 'bag-in-box' Xtend® films effectively reduced the development of chilling injury as well as other types of rind disorders that are not related to chilling, such as rind breakdown, stem-end rind breakdown and shrivelling and collapse of the button tissue (ageing). In all cases, micro-perforated films (0.002% perforated area) that maintained 2–3 kPa CO_2 + 17–18 kPa O_2 inside the package were much more effective in reducing the development of rind disorders than macro-perforated films (0.06% perforated area), which maintained an equilibrium gas content of 0.2–0.4 kPa CO_2 + 19–20 kPa O_2. After 5 weeks at 6°C and 5 days of shelf-life in ambient conditions, chilling injury was reduced by 75% and rind disorders by 50% in 'Shamouti' orange and by 60% and 40%, respectively, in 'Minneola' tangerines for those in Xtend® compared with non-packaged fruit. Similarly, micro-perforated and macro-perforated Xtend® packages reduced the development of chilling injury after 6 weeks at 2°C and 5 days of shelf-life in 'Shamouti' oranges and 'Star Ruby' grapefruit. Choi *et al.* (2002) successfully stored the cultivar 'Tsunokaori of Tangor' ((*Citrus sinensis* × *C. unshiu*) × *C. unshiu*) [*C. sinensis* × *C. reticulata*], (*C. reticulata* × *C. sinensis* × *C. nobilis*)) in LDPE bags with or without calcium oxide and potassium permanganate in storage at 4 ± 1°C with 85 ± 5% RH. Calcium oxide decreased water loss of the fruit during storage but the sugar content did not change in any of the treatments. Storage of *C. unshiu* in China in containers with a D45 M2 1 silicone window of 20–25 cm²/kg of fruit gave the optimum concentration of < 3 kPa CO_2 + < 10 kPa O_2 (Hong *et al.*, 1983). Del-Valle *et al.* (2009) reported that the optimum equilibrium MA packaging for mandarin segments at 3°C was 19.8 kPa O_2 + 1.2kPa CO_2.

Cucumber

In Europe, cucumbers are commonly marketed in shrink-wrapped plastic sleeves that provide a modified atmosphere which helps to retain their texture and colour (see Fig. 14.7). 'Beit Alpha'-type cucumbers responded favourably to MA packaging with up to 8–9 kPa CO_2, showing reduced physiological disorders and decay, but CO_2 levels above 10 kPa were injurious. The combination of optimal atmosphere composition and humidity inhibited the cucumbers from toughening and preserved their tender texture and turgidity (Rodov *et al.*, 2003). Cucumbers stored in 100 kPa O_2 in plastic bags retained their fresh appearance but developed off-flavours; and at 5°C the level of O_2 in the bags had fallen to 50 kPa and CO_2 had risen to 20 kPa after 8 days (Srilaong *et al.*, 2005).

Endive

The quality of endive packed on polystyrene trays in LDPE bags or P-Plus PE coverings was commercially acceptable after 3 weeks storage at 5°C. Those covered with vinyl film (called Borden Vinilo) had the lowest microbial counts, with *Pseudomonas* and enterobacteria being the principal microbial contaminants (Venturini *et al.*, 2000). Charles *et al.* (2005) stored endive at 20°C in LDPE bags which gave a 3 kPa O_2 + 5 kPa CO_2 equilibrium atmosphere after

25 h storage with an O_2-absorbing packet, compared with 100 h without the O_2-absorbing packet. After 312 h of storage in both packages the total aerobic mesophile, yeast and mould population growth was reduced compared with those in macro-perforated OPP bags, which maintained gas composition close to that of air but also limited water loss.

Feijoa

Feijoas were stored at 6°C, 12°C or 17°C and 85% RH without packaging, or in PP bags with 0, 1, 2 or 3 perforations of 0.225 mm. Fruits stored in MA packaging systems deteriorated more slowly than those stored without packaging at 17°C: from 17 up to 25–31 days for the MA packaging at 12°C and 17°C, and 35–38 days for the MA packaging at 6°C. It was concluded that the most favourable storage was PP bags with one perforation at 6°C, which gave a steady-state atmosphere of 8.2 kPa O_2 + 5.8 kPa CO_2 (Castellanos *et al.*, 2016).

Fig

Fruits of the cultivar 'Masudohin' were sealed in 0.04 mm PE film bags; CO_2 was introduced into half of them and they all were stored for 10 days at 0°C followed by 2 days in ambient conditions. The bags with added CO_2 had an atmosphere of 12.3–14.0 kPa O_2 + 10.0–14.2 kPa CO_2 during the first 4 days, which progressively stabilized at 13.2–14.0 kPa O_2 + 1.1–1.2 kPa CO_2. Those with added CO_2 had better visual quality, a slower rate of softening and less incidence of decay than those without the CO_2 treatment (Park and Jung, 2000). Fruits of the cultivar 'Roxo de Valinhos' were packed in 50 µm PP film bags with or without flushing with 6.5 kPa O_2 + 20 kPa CO_2 and stored at 20 ± 2°C and 85 ± 5% RH for 7 days. Those in PP bags retained their appearance and had low weight loss of 1.7–2.5% compared with about 40% for non-wrapped fruit, but there were high levels of decay in all treatments (Souza and Ferraz, 2009). Villalobos *et al.* (2018) stored the cultivars 'Cuello Dama Negro' and 'San Antonio' in 100 µm micro-perforated films with 16, 5 or 3 holes or 9 mm macro-perforated film with 5 holes, all at 0°C for 14 days followed by 2 days at 20°C. 'San Antonio' fruit in the 100 µm micro-perforated films with 3 holes had delayed changes in the volatile profile, compared with the other treatments, without negative influence on their flavour.

Grape

The cultivars 'Kyoho' and 'Campbell Early' packed in 0.03 mm PE film bags were stored for 60 days at 0°C and 90% RH by Yang *et al.* (2007). They found that packaging in PE film maintained fruit firmness, TSS, TA and skin colour, and reduced respiration rate, decay and weight loss compared with those stored non-wrapped. In contrast, grapes in antifogging and perforated film had increased decay and abscission. Yamashita *et al.* (2000) stored the cultivar 'Italia' at 1°C and 85–90% RH followed by 25°C and 80–90% RH in three different film bags, all of which had high gas permeability. Fruit in all three films stored well but Cryovac PD-955 gave the longest storage life of 63 days, compared with 11–21 days for fruit stored without packaging. Artes-Hernández *et al.* (2003) found that an equilibrium atmosphere of about 15 kPa O_2 + 10 kPa CO_2 gave the best results and for the cultivar 'Autumn Seedless' this atmosphere was provided by micro-perforated PP film 35 µm thick.

Several workers have studied the use of MA packaging in combinations with other treatments as an alternative to sulfur dioxide treatment to control *Botrytis cinerea* in grapes. Artes-Hernández *et al.* (2007) used ozone (O_3) at 0.1 µl/l and found that the sensory quality of the fruit was preserved with MA packaging giving an atmosphere of 13–16 kPa O_2 + 8–11 kPa CO_2 and in CA storage in 5 kPa O_2 + 15 kPa CO_2. Although O_3 did not completely inhibit fungal development, its application increased antioxidant compounds. A pre-storage dip of the cultivar 'Superior' in either 33% or 50% ethanol then storage in sealed Xtend® or PE film bags maintained grape quality for up to 7 weeks with 3 days shelf-life at 20°C. The treatment was similar to or better than storage with SO_2-releasing pads (Lichter *et al.*, 2005). Artes-Hernández *et al.* (2003) stored the cultivar 'Napoleon' for 41 days in MA packaging or air followed by 4 days in air, all at 0°C, then 3 days at 15°C. The best results were by placing a soaked filter paper with 15 or 10 µl hexenal combined with sealing in micro-perforated PP of 35 µm thickness. MA packaging improved the visual appearance and reduced weight loss to 0.6% compared with non-packaged grapes, which showed a 3.7% weight loss. Hexenal-treated grapes had the highest score for flavour and the lowest level of fungal disease (6%). Valverde *et al.* (2005) showed that the addition of 0.5 µl of eugenol, thymol or menthol inside MA packages containing the cultivar 'Crimson Seedless' reduced weight loss and colour change, delayed rates of pedicel deterioration, retarded °brix:acid ratio development, maintained firmness and reduced decay. Also the total viable counts for mesophilic aerobics, yeasts and moulds were significantly reduced in the grapes packaged with eugenol, thymol or menthol. The authors concluded that the inclusion of one of these essential oils resulted in an additional 3 weeks of storage compared with MA packaging only. The equilibrium atmosphere in PP bags in storage at 1°C was reached in 21 days with 1.2–1.3 kPa CO_2 and 13–14 kPa O_2. The inclusion of a sachet impregnated with either 0.5 ml thymol or 0.5 ml menthol inside the bag reduced yeasts, moulds and total aerobic mesophylic colonies and also improved the visual aspect of the rachis. The authors postulated that these essential oils could be an alternative to the use of SO_2 in controlling disease in table grapes (Martinez-Romero *et al.*, 2005).

Ben Arie (1996) described a method of MA packaging for sea-freight transport. This involved whole pallet loads of half a tonne of fruit in cartons being wrapped in PE film. The advantage of this compared with individual cartons being lined with PE film was that any condensation remained on the outside of the box and pre-cooling was more rapid, since the boxes were pre-cooled before the plastic was applied.

Kiwifruit

'Hayward' kiwifruit stored in plastic trays and over-wrapped with 16 µm stretch film at both 10°C and 20°C ripened more slowly than non-wrapped fruit but to a similar flavour (Kitsiou and Sfakiotakis, 2003). Storage of 'Hayward' for 6 months at 0°C and 95% RH in sealed 50 cm × 70 cm PE bags with an ethylene absorbent resulted in firmer fruit, higher TA and AA and lower weight loss than when stored non-wrapped. The equilibrium atmosphere in the bags was 6–8 kPa O_2 + 7–9 kPa CO_2 after 6 months of storage (Pekmezci *et al.*, 2004).

Kohlrabi

Storage in antimist OPP bags 20 µm thick reduced weight loss and development of bacterial soft and black rots and extended the storage life to 60 days at 0°C plus 3 days at 12°C (Escalona *et al.*, 2007). The equilibrium atmosphere inside the bags was 4.5–5.5 kPa O_2 + 11–12 kPa CO_2.

Lanzones (langsat)

Pantastico (1975) indicated that the skin of the fruit turns brown during retailing and if they are sealed in PE film bags the browning was aggravated, probably due to CO_2 accumulation.

Lemon

Lemons stored at 25°C and 50–60% RH for up to 21 days in 100 gauge PVC 20% shrink-wrap retained their firmness and TSS, with reduced loss of weight and rotting compared with those stored non-wrapped or waxed (Neri *et al.*, 2004).

Lemongrass

Lemongrass leaves were hot-water treated at 55°C for 5 min, then hydrocooled at 3°C for 5 min and placed in PE bags, which were then flushed with various levels of O_2 + CO_2. They were then sealed and stored at 5°C and 90 ± 5% RH for 3 weeks. They found that *E. coli* was inhibited in 1 kPa O_2 + 10 kPa CO_2, coliforms and faecal coliforms were inhibited in 5 kPa O_2 and *Salmonella* at all gas mixtures for up to 14 days. Fungi were not controlled in any of the atmospheres tested (Samosornsuk *et al.*, 2009).

Lettuce

Cameron (2003) found that at 5°C browning was retarded when O_2 was below 1 kPa, while fermentation occurred below 0.3–0.5 kPa O_2; therefore MA packaging that gave an atmosphere between 0.5 and 1 kPa O_2 was recommended. Escalona *et al.* (2006b) concluded that 80 kPa O_2 should be used in MA packaging of fresh-cut butterhead lettuce in combination with 10–20 kPa CO_2 to reduce their respiration rate and avoid fermentation.

Lime

Storage at 6°C led to chilling injury and at 10°C to 'ageing' (Ladaniya, 2004). Packing limes in sealed PE film bags inside cartons in ambient conditions in Khartoum in the Sudan resulted in a weight loss of only 1.3% in 5 days, but all the fruits de-greened more rapidly than those that were packed in cartons without the PE liner, where the weight loss was 13.8% (Thompson *et al.*, 1974c). This de-greening effect could be countered by including potassium permanganate on vermiculite in the bags. The potential of 30 μm HDPE bags with 1 × 40, 2 × 40 or 3 × 40 micro-perforations was studied by Ramin and Khoshbakhat (2008). The greenest and firmest fruits were found in micro-perforated PE bags at a storage temperature of 10°C. At 20°C, fruits kept in micro-perforated bags were consistently greener than fruits that were stored non-packaged, but decay was high. Decay was highest in fruits with no package and lowest in PE with micro-perforations. Jadhao *et al.* (2007) found that freshness was retained and flavour and acceptability scores were the highest for the cultivar 'Kagzi' stored at 8 ± 1°C for up to 90 days in 25 μm PE bags with 0.5% area as punched holes, with minimum ageing and no chilling injury. Those stored without packing were spoiled completely within 20 days. The fruits stored in 100 gauge ventilated PE bags maintained maximum juice content, TA and AA and minimum TSS and spoilage for 30 days.

Litchi

Ragnoi (1989), working with the cultivar 'Hong Huai' from Thailand, showed that fruits packed in sealed 150 gauge PE film bags containing 2 kg of fruit and sulfur dioxide SO_2 pads could be kept in good condition at 2°C for up to 2 weeks, while fruits that were stored without packaging rapidly became discoloured and unmarketable. Anthocyanin levels in the pericarp are important, since they give the fruit an attractive appearance. Somboonkaew and Terry (2009) studied storage in a selection of films and found that anthocyanin concentrations in the cultivar 'Mauritius' wrapped in PropaFresh™ film (Table 14.6) were significantly higher than for other plastic films after 9 days of storage (see Fig. 14.4).

In storage at 14°C the cultivar 'McLean's Red' in PP film had 11.3% decay due mainly to *Alternaria alternata* and *Cladosporium* spp. However, fruits that had been dipped

Table 14.6. The water and gas permeability of the films used by Somboonkaew and Terry (2009). (More information on the films is available at www.innoviafilms.com.)

Packaging films	Water (g/m²/24 h)	O_2 (cc/m²/24 h/bar)
PropaFresh™	5	1600
NatureFlex™	360	3
Cellophane™	370	3

for 2 min at 15°C in *Bacillus subtilis* and stored at 14°C in PP film had no decay or pericarp browning and retained their colour and quality. The equilibrium atmosphere inside the PE bags was about 14 kPa O_2 + 5 kPa CO_2. Fruits stored in LDPE bags with *B. subtilis* had higher levels of decay and pericarp browning than those in PP bags. Higher yeast populations were observed in LDPE or LDPE with *B. subtilis* during storage at 2°C or 14°C and *Candida*, *Cryptococcus* and *Zygosaccharomyces* spp. were the predominant yeasts. The equilibrium atmosphere inside the LDPE bags was about 3 kPa O_2 + 10 kPa CO_2 (Sivakumar *et al.*, 2007).

Longan

Zhang and Quantick (1997) stored the cultivar 'Shixia' in PE film 0.03 mm thick for 7 days at room temperature followed by 35 days at 4°C. Atmospheres of 1, 3, 10 or 21 kPa O_2 were established in the bags and the former two were said to be effective in delaying peel browning and retaining TSS and AA content of the fruit, though taste panels detected a slight off-flavour in fruit stored in 1 kPa O_2. Seubrach *et al.* (2006) stored fruits of the cultivar 'Daw' in polystyrene boxes covered with 15 μm PVC film or LLDPE film of 10, 15 and 20 μm thickness, all at 4°C and 90–95 % RH. All the MA conditions extended the shelf-life to 20 days compared with 16 days for non-wrapped fruit. Fruit had slightly lower weight loss in the LLDPE films compared with those in the PVC film. Peel colour gradually deteriorated in all packages but less so for those in the PVC film, which also had the best appearance and better overall customer acceptance.

Loquat

Loquat cultivar 'Algerie' fruits were stored in five types of micro-perforated PP films for up to 6 weeks at 2°C then 4 days non-wrapped at 20°C. The equilibrium atmosphere changed from 1.2 to 8.5 kPa CO_2 and 19.5 to 13 kPa O_2 as film permeability decreased. Peel softening and colour development were decreased, as were changes in sugars and organic acids, especially after the shelf-life period, compared with the controls. They concluded that the most suitable atmosphere was about 2–4 kPa CO_2 + 16–18 kPa O_2, achieved in PA-80 and PA-60 films (Ding *et al.*, 2006). Similar results were reported by Amorós *et al.* (2008), who reported that postharvest life at 2°C without wrappings was 2 weeks with a 4-day shelf-life at 20°C, but in the same environment the postharvest life of those in PA-80 or PA-60 films was 6 weeks plus a 4-day shelf-life.

Mango

Mangoes stored in PE film bags at 21°C had almost twice the storage life of those stored without wraps (Thompson, 1971). Mature fruits of the cultivar 'Keitt' were stored at 20°C in air or in MA jars supplied with humidified CO_2 at 210 ml/min for 2 h before being sealed, resulting in 0.03–0.26 kPa O_2 + 72–82 kPa CO_2 for up to 4 days, or in jars ventilated with 2 kPa O_2 + 50 kPa CO_2 + 48 kPa N_2 at a continuous rate of 210 ml/min for up to 5 days. Both treatments delayed fruit ripening. Fruits showed no signs of internal or external injuries or off-flavour, either immediately after removal from storage or after transfer to air. There were no significant differences between air-stored and CA-stored fruits in all sensory attributes evaluated, but the overall acceptability of fruits stored in MA for over 72 h was lower than that of fruits stored in air (Yahia and Vazquez Moreno, 1993). The cultivar 'Keitt' stored in CA in 0.3 kPa O_2 or MA packaging in three types of LDPE films at 20°C had slower weight losses and remained firmer and with good appearance with a significant delay in ripening compared with controls. However, a few fruits packed in two of the three films developed a fermented flavour after 10 days of storage (Gonzalez-Aguilar *et al.*, 1994).

Melon

Martinez-Javega *et al.* (1983) reported that sealing fruits individually in PE film bags

0.017 mm thick reduced chilling injury during storage at 7–8°C. Flores *et al.* (2004) also found that MA packaging reduced chilling injury in the cultivar 'Charentais', which is subject to chilling injury when stored at temperatures around 2°C. They also found that MA packaging extended the postharvest life of wild-type melons and conferred additional chilling resistance on ethylene-suppressed melons during storage at 2°C and shelf-life at 22°C. Asghari (2009) reported that immersing the cultivar 'Semsory' in hot water at 55°C for 3 min or 59°C for 2 min and then sealing them in 30 μm PE bags kept them in good condition at 2.5°C for 33 days. This was longer than either treatment alone.

The effects of packing in ceramic films with thicknesses of 20, 40 or 80 μm and storage at either 3°C or 10°C showed that loss in fresh weight was greatly reduced to below 0.3% after 36 days at 3°C for fruits in ceramic films. CO_2 accumulated up to 15 and 30 kPa at 3°C and 10°C, respectively, in films 80 μm thick. The concentration of acetaldehyde was over 4 times higher in film 80 μm thick compared with the others at both 3°C and 10°C. Fruits in film 40 μm thick retained their firmness, TSS, TA and visual quality better than any other treatment and their storage life at 3°C was 36 days, which was twice as long as in the other treatments. Chilling injury symptoms occurred at 3°C but fruits packed in films 40 μm thick had lower levels (Kang and Park, 2000). Sa *et al.* (2008) stored cantaloupe in X-tend bags for 14 days at 3 ± 2°C and then removed them from the bags and stored them for a further 8 days at 23 ± 2°C and 90 ± 2% RH. The fruits retained their quality in the bags. Pre-treatment with 1-MCP or inclusion of potassium permanganate on vermiculite within the bags gave no additional effects. MA packaging extended the postharvest life of charentais-type melons by delaying ripening, in spite of the high in-package ethylene concentration of 120 μl/l (Rodov *et al.*, 2003).

Mushroom

Because of the high respiration rates and the very limited positive effects on quality, Anonymous (2003) suggested that it did not seem beneficial to pack mushrooms in MA packages. However, storage in MA packaging at 10°C and 85% RH for 8 days delayed maturation and reduced weight loss compared with storage without packaging (Lopez-Briones *et al.*, 1993). PVC overwraps on consumer-sized punnets (about 400 g) greatly reduced weight loss, cap and stalk development and discoloration, especially when combined with refrigeration (Nichols, 1971). MA packaging in microporous film delayed mushroom development, especially when combined with storage at 2°C (Burton and Twyning, 1989). Mushrooms were stored at 4°C in containers where atmospheres of 5 kPa O_2 + 10 kPa CO_2 were maintained and they were subjected to a sequence of temperature fluctuations of 10°C for 12 days. Temperature fluctuations had a major impact on the composition of the atmospheres and on product quality. CO_2 concentrations increased rapidly, reaching a maximum of 16 kPa, while O_2 concentrations decreased to less than 1.5 kPa. The temperature-fluctuating regime resulted in extensive browning, softening, weight loss increase, ethanol increase and infection due to physiological damage and excessive condensation, compared with storage at a constant temperature (Tano *et al.*, 2007). Hu *et al.* (2003) found that there was a positive effect on postharvest life when shiitake were stored in LDPE or PVC film bags. Respiration rates were suppressed by decreasing O_2 and increasing CO_2 concentrations at 5°C, 15°C, 20°C and 30°C, but there was little difference at 0°C. Lin *et al.* (2017) packed button mushrooms in MA packages with 95–100 kPa CO_2. After 12 h they punctured the film. This CO_2 exposure reduced browning and MDA and increased antioxidant, catalase and POD activities during subsequent storage. Wei *et al.* (2017) reported decreased PPO activities, delayed textural changes and reduced respiration rates and weight loss when pine mushrooms were stored in PE bags. They also tested PVC bags with silicon windows and found that odour changes were delayed and the bags contained lower amounts of ammonia. Both the package types tested resulted in increased catalase activity and AA content.

Okra

Hardenburg *et al.* (1990) reported that storage in 5–10 kPa CO_2 lengthened shelf-life by about a week. Finger *et al.* (2008) stored the Brazilian cultivar 'Amarelinho' at 25°C, 10°C or 5°C wrapped in PVC over a polystyrene tray or left not wrapped. The development of chilling symptoms (surface pitting) occurred at 5°C but was delayed in those covered with PVC film. Those stored at 25°C lost weight rapidly and became wilted, especially those not wrapped, therefore the best treatment was at 10°C covered in PVC. Mota *et al.* (2006) also found that PVC overwraps were efficient in reducing weight loss and in retaining the vitamin C contents and reducing browning during storage.

Onion

Packing onion bulbs in plastic film is uncommon, because the high humidity inside the bags can cause rotting and root growth (see Table 14.2).

Papaya

Wills (1990) reported that MA packaging in plastic film extended the storage life of papaya. Singh and Rao (2005) found that packaging the cultivar 'Solo' in LDPE or Pebax-C bags and storing at 7°C or 13°C and 85–90% RH prevented the development of chilling injury symptoms during storage at 7°C or 13°C for 30 days, but fruits stored at 7°C failed to ripen normally and showed blotchy appearance, due to skin scald, and also had internal damage. LDPE and Pebax-C packed fruits showed higher retention of AA, total carotenoids and lycopene after 30 days storage at 13°C and 7 days non-wrapped at 20°C compared with storage non-wrapped throughout. Rohani and Zaipun (2007) reported that the storage life of the cultivar 'Eksotika' could be extended to 4 weeks at 10–12°C when nine to ten fruits were stored in 0.04 mm LDPE bags measuring 72 cm × 66 cm. After 24 h the CO_2 level inside the bags was about 4–5 kPa and O_2 about 2–3 kPa. After 4 weeks of storage the fruit ripened normally within 3–4 days in ambient temperature when removed from the bags.

Parsley

Desiccation was the most important cause of postharvest loss. Storage in folded unsealed PE bags at 1°C combined with precooling gave a maximum storage life of about 6 weeks. If desiccation occurred, placing the petioles in water enabled them to recover from a maximum of 10% loss (Almeida and Valente, 2005).

Passionfruit

Mohammed (1993) found that both yellow and purple passionfruit stored best at 10°C in perforated LDPE film bags 0.025 mm thick.

Pea

Pariasca *et al.* (2001) stored snow pea pods in polymethyl pentene polymeric films of 25 and 35 μm thickness at 5°C. They found that internal quality was maintained and that appearance, colour, chlorophyll, AA, sugar content and sensory scores were retained better than for those stored without packaging. The equilibrium atmosphere in both thicknesses of bags was around 5 kPa O_2 + 5 kPa CO_2.

Peach

Malakou and Nanos (2005) treated the cultivar 'Royal Glory' in hot water at 46°C containing 200 mM sodium chloride for 25 min, then sealed them in PE bags and stored them at 0°C for up to 2 weeks. This combination resulted in good quality fruit after 1 week and gave an equilibrium atmosphere of > 15 kPa O_2 + < 5 kPa CO_2 inside the bags. Ripening was slowed in the PE bags compared with those not in bags. However, the gas content inside the bags changed to < 3k Pa O_2 + > 13 kPa CO_2 within 10 h at room temperature, which was found to damage the fruit. It was therefore recommended that they should be removed from the bags directly after removal from cold storage. De Santana *et al.* (2009) in a comparison of storage of the cultivar 'Douradão' at 1 ± 1°C and 90 ± 5% RH in LDPE bags of 30, 50, 60 and 75 μm found that 50 and 60 μm were the most suitable and that after 28 days of storage fruit quality was better than those in 30 or 75 μm LDPE. Zoffoli *et al.* (1997) stored the cultivars 'Elegant Lady' and 'O'Henry' at 0°C in film bags of various permeabilities. They found that the atmospheres inside the packs varied between 10 and 25 kPa CO_2 and 1.5 and 10 kPa O_2. The rates of fruit softening and flesh browning were reduced in all the high-CO_2 packages, but there was some development of mealiness and off-flavours.

Pear

Geeson *et al.* (1990a) found that when unripe 'Doyenné du Comice' were held in MA packages at 20°C softening was faster than chlorophyll loss, so that the appearance of the fruit failed to match its internal condition. However, a 4-day extension in shelf-life was obtained by MA packaging of fruits in a part-ripe condition and, although the effects on rate of softening were less than those for unripe fruits, the yellower ground colour was more consistent with the eating qualities of the fruit. For the cultivar 'Conference', Geeson *et al.* (1990b) found that in MA packs the rate of flesh softening was only partially slowed; chlorophyll degradation was completely inhibited but resumed when packs were perforated. The equilibrated atmosphere within the packs after about 3 days was 5–9 kPa CO_2 + < 5 kPa O_2. Pears retarded by MA packaging failed to develop the normal sweet, aromatic flavour and succulent, juicy texture of 'eating-ripe' fruit, even when the packs were perforated after 4 days.

Persimmon

In Korea, fruits are stored for 6 months at 2–3°C with minimal rotting or colour change, and storage in the same conditions with ten fruits sealed in PE film bags 60 μm thick was even better (A.K. Thompson, unpublished results). Astringency of persimmons was removed by storing fruits in PE bags that had been evacuated or with total N_2 or CO_2 atmospheres, but the fruits stored in the CO_2 atmosphere were susceptible to flesh browning. Fruits stored under vacuum or an N_2 atmosphere maintained high quality and firmness for 2 weeks at 20°C and 3 months at 1°C. Flesh appearance and taste after 14 weeks of storage at 1°C was best in the N_2 atmosphere (Pesis *et al.*, 1986). Liamnimitr *et al.* (2018) designed a bulk MAP for storage of 'Fuyu'. Initially, the lower O_2 limit of aerobic respiration was determined to be 0.65 kPa, based on the RQ breakpoint. They then used a Michaelis–Menten-type respiration model to predict the O_2 consumption rate at 1.5 kPa, which was set as the target O_2 level inside the package in order to avoid fermentation. They compared this bulk MAP with an individual package (LDPE film bag 60 μm thick) and non-packed fruit. After 4 months of storage and subsequent transport of fruit from Japan to Thailand, the O_2 partial pressure at a steady state in bulk MAP was 1.3–1.6 kPa, which was close to the 1.5 kPa target. Softening and loss of colour of fruits were reduced in the bulk MAP compared with the other two treatments.

Pineapple

Storage in PE bags led to condensation and mould growth. No condensation was observed in cellophane bags, as they allowed moisture to pass through them, but there was some drying out of the crowns. However, in both types of bag off-flavours were detected due to CO_2 accumulation and, because of the fruit's spiny nature. bags were considered impractical (Paull and Chen, 2003). Black heart development was less than

10% in storage of the cultivar 'Mauritius' in PE film bags for 2 weeks at 10°C with an equilibrium atmosphere inside the bags of about 10 kPa O_2 + 7 kPa CO_2 (Hassan *et al.*, 1985). They also suggested that pineapple should be kept in PE film bags until consumption in order to avoid the development of black heart.

Minimally processed pineapples are commonly packed in MA and 5 kPa O_2 + 15 kPa CO_2 appeared to be the most appropriate atmosphere for storage (Budu *et al.*, 2007). Minimally processed pineapple stored at 5 or 10 kPa O_2 + 10 kPa CO_2 helped to maintain their lightness and visual quality and prolonged the storage life up to 9 days (Tancharoensukjit and Chantanawarangoon, 2008). Pineapples were sliced and dipped in a solution of sodium hypochlorite at 20 mg/l for 30 seconds, then stored at 5 ± 1°C in a flow-through system with different O_2 and CO_2 concentrations. None of the CA treatments appeared to be advantageous, since the slices had little browning and were free of any contamination that would affect the food safety at the end of the storage period (Antoniolli *et al.*, 2007).

Plantain

The effects of PE film wraps on the postharvest life of plantains may be related to moisture conservation around the fruit as well as the change in the CO_2 and O_2 content. This was shown by Thompson *et al.* (1972b) where fruits stored in moist coir had a longer storage life than fruits that had been stored in dry coir or non-wrapped (Fig. 14.9). There was an added effect when fruits were stored in non-perforated PE bags, which was presumably due to the effects of the changes in the CO_2 and O_2 levels (Table 14.7). Thus the positive effects of storage of pre-climacteric fruits in sealed plastic films may be, in certain cases, a combination of its effects on the CO_2 and O_2 content within the fruit and the maintenance of high moisture content.

Pomegranate

The cultivar 'Hicaznar' could be kept in cold storage for 2–3 months in 8 μm film, but when packed in 12 μm stretch film the fruits maintained their quality and had no fungal disorders for 3–4 months (Bayram

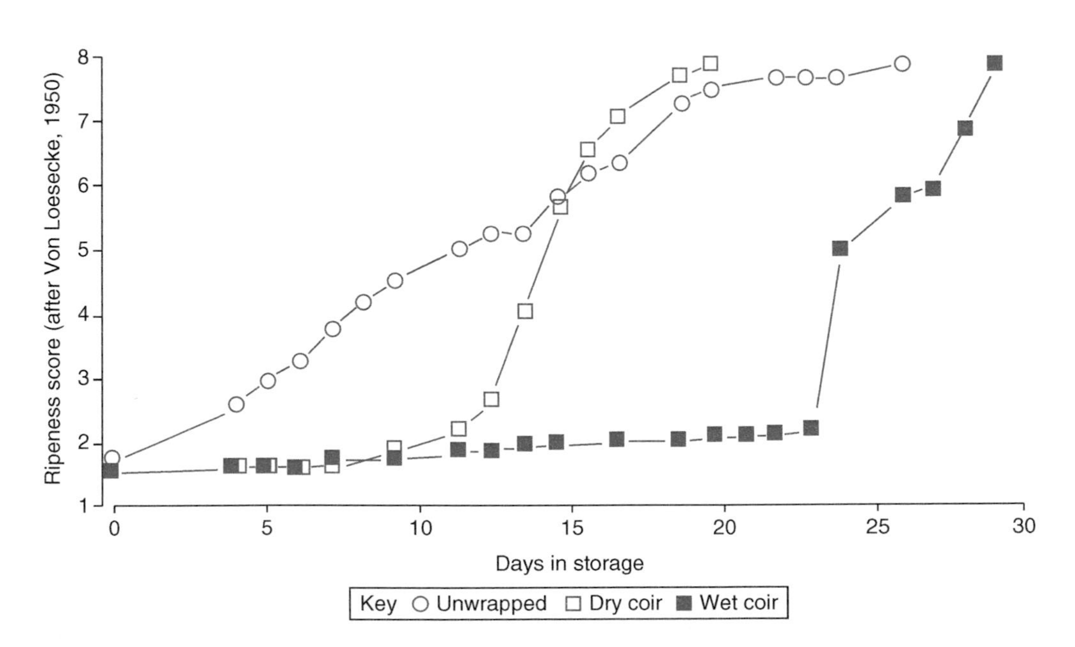

Fig. 14.9. Effects of storage in coir dust on ripening of plantains in Jamaican ambient conditions.

Table 14.7. Effects of wrapping and packing material on ripening and weight loss of plantains stored at tropical ambient conditions of 26–34°C and 52–87% RH; adapted from Thompson *et al.* (1972a).

Packing material	Days to ripeness	Weight loss at ripeness
Not wrapped	15.8	17.0%
Paper	18.9	17.9%
Moist coir fibre	27.2	(3.5%)[a]
Perforated PE	26.5	7.2%
PE	36.1	2.6%
LSD (p = 0.05)	7.28	2.81

[a]The fruit actually gained in weight.

et al., 2009). Porat *et al.* (2009) reported that packaging 4–5 kg in cartons lined with Xtend® film bags could maintain fruit quality for 3–4 months. They found that MA packaging reduced water loss, shrinkage, scald and decay.

Radish: Daikon

With Japanese radish, O_2 and CO_2 concentrations inside different micro-perforated film bags varied between 0.2 and 15 kPa O_2 and 5 and 16 kPa CO_2 at 15°C after 6 days of storage, depending on the degree of micro-perforation. Films that gave an equilibrium concentration of 9–13 kPa O_2 + 8–11 kPa CO_2 were suitable to maintain most of the qualitative parameters and avoiding the development of off-flavours during 6 days of storage (Saito and Rai, 2005).

Rambutan

In air the maximum shelf-life for the cultivars 'R162', 'Jit Lee' and 'R156' was given as 15 days at 7.5°C and 11–13 days at 10°C (O'Hare *et al.*, 1994). They also reported that storage in ethylene at 5 µl/l at 7.5°C or 10°C did not significantly affect the rate of colour loss. Fruit have been stored successfully at 10°C in PE film bags by Mendoza *et al.* (1972) but the storage life was given as about 10 days in perforated bags and 12 days in non-perforated bags. Fruits of the cultivar 'Rong-Rein' were stored in PE bags 40 µm thick at 13°C and 90–95% RH, giving an equilibrium atmosphere of 5–8 kPa O_2 + 7–10 kPa CO_2 (Boonyaritthongchai and Kanlayanarat, 2003a). They found that those in MA packaging had reduced pericarp browning, weight loss, respiration rate and ethylene production, giving a storage life of 16 days compared with fruits stored non-wrapped, which was about 12 days. Somboonkaew (2001) stored 'Rong-Rien' at 10°C for 22 days sealed in 0.036 mm LDPE bags where the air had been evacuated and then flushed with air, 5 kPa CO_2 + 5 kPa O_2, 10 kPa CO_2 + 5 kPa O_2 or 15 kPa CO_2 + 5 kPa O_2. She found that, irrespective of the initial level, the CO_2 stabilized at about 1 kPa after 3 days then rose to a peak of about 4 kPa after 2–3 weeks. O_2 levels stabilized at around 3 kPa after about 1 week. Little difference was detected in softening except for those in 15 kPa CO_2 + 5 kPa O_2, which were generally softer and less 'bright' after storage. On balance the fruit flushed with 5 kPa CO_2 + 5 kPa O_2 appeared best.

Raspberry

Siro *et al.* (2006) reported that raspberries had a shelf-life of 7 days in MA and 5 days in high O_2 atmospheres at 7°C. *Escherichia coli*, *Salmonella* spp. and *Listeria monocytogenes* were artificially inoculated on to packaged fruits and were able to survive at 7°C. Raspberries showed an enhanced inactivation of *Salmonella* spp. during storage. Generally, MA or high-O_2 atmospheres did not give an increased microbial risk. Jacxsens *et al.* (2003) found that storage of fruit in sealed film packages resulted in a prolonged shelf-life compared with macro-perforated packaging films. They used selective permeable films in order to obtain 3–5 kPa O_2 + 5–10 kPa CO_2 and limited ethylene accumulation. Storage in 95 kPa O_2 + 5 kPa N_2 in a semi-barrier film resulted in the accumulation of too-high CO_2 levels, resulting in fermentation and softening but an inhibition of mould development. Moor *et al.* (2009) stored the cultivar 'Polka' in Xtend®, LDPE or LDPE flushed with10 kPa O_2 + 15 kPa CO_2. Rotting was not suppressed in any of the bags, but after 4 days at 1.6°C those in LDPE bags that had not been flushed were of the best quality. However, after one more day at 6°C, to simulate shelf-life, the fruit in Xtend® bags were best.

Redcurrant

McKay and Van Eck (2006) described a successful packing system where fruit were placed in fibreboard boxes that were then stacked on pallets and covered with a large plastic bag, which fitted over the pallet. The bags were sealed at the bottom. In storage at –0.9°C to 0.5°C and 93% RH, the O_2 concentration was lowered to 2.5–3.0 kPa by N_2 injection and CO_2 was increased to 20–25 kPa during storage. They found that 93% RH was optimum and recommended pre-cooling to 0°C to reduce the chances of condensation; they also observed that the boxes could absorb some of the moisture, which helped to prevent condensation. They reported that the berries could lose their bright red colour and appear water-soaked if the O_2 level was too low.

Rhubarb

Unsealed PE film liners in crates reduced rhubarb weight loss (Lutz and Hardenburg, 1968) but there was no mention of other MA effects.

Roseapple

Worakeeratikul *et al.* (2007) reported storage where the fruits were dipped in whey protein concentrate and then successfully stored in PVC film at 5°C for 72 h.

Sapodilla

At ambient temperatures in Brazil, fruits stored in PVC film packaging had less weight loss and softened more slowly than those not wrapped. Decreases in acidity, TSS and total sugar contents were similar in wrapped and non-wrapped fruit (Miranda *et al.*, 2002). In India, Waskar *et al.* (1999) found that 'Kalipatti' could be stored for up to 9 days when packed in 100 gauge PE bags with 1.2% ventilation holes at room temperature of 31.7–36.9°C. They also found that in simple cool stores fruit in PE bags could be stored for 15 days at 20.2–25.6°C and 91–95% RH.

Soursop

In a comparison between 16°C, 22°C or 28°C the longest storage period (15 days) was at 16°C with fruits packed in HDPE bags without an ethylene absorbent. This packaging also resulted in the longest storage duration at 22°C and 28°C, but the differences between treatments at those temperatures were much less marked (Guerra *et al.*, 1995).

Spinach

Fresh-cut 'Hybrid 424' was packed in mono-oriented PP or LDPE bags. The type and permeability of film affected off-odour development, but did not influence visual sensory attributes or chlorophyll retention (Piagentini *et al.*, 2002). Díaz-Mula *et al.* (2017) stored baby spinach in MA packages. After 14 days the packages contained 0.3 kPa O_2 and 9.3 kPa CO_2 and off-odours were detected. They found 39 main compounds with olfactory activity, including alcohols associated with lipid peroxidation, sulfur compounds from amino acid degradation and alkanes from lipid autoxidation processes or carotenoid degradation.

Strawberry

A comparison of fruits in sealed plastic containers using impermeable PP film was made by Stewart (2003). Three internal atmospheres were injected as follows: air;

5 kPa O_2 + 5 kPa CO_2 + 90 kPa N_2; or 80 kPa O_2 + 20 kPa N_2. Overall the fruit stored in 80 kPa O_2 + 20 kPa N_2 proved most successful in terms of maintaining firmness, reduction in cell wall breakdown and improvement in appearance and taste.

Sweetcorn

Othieno *et al.* (1993) showed that the best MA packaging was non-perforated film, or film with the minimum perforation of one 0.4 mm hole per 6.5 cm^2. They found that cobs packed in PP film with no perforations showed a reduced rate in toughening, weight loss and reduced loss of sweetness, but they tended to develop off-odours. Manleitner *et al.* (2003) found that changes of carbohydrate composition of the cultivar 'Tasty Gold' were influenced by the permeability of the films during 20 days storage at 5°C. Sweetcorn in films with low permeability maintained carbohydrate content during storage, but had low sucrose after subsequent shelf-life of two days at 20°C. Losses of sucrose during the shelf-life phase were lowest in 20 μm LDPE film.

Taro

MA packaging in PE film bags has been shown to reduce postharvest losses. Both *Xanthosoma* and *Colocasia* stored in Trinidad ambient conditions of 27–32°C showed reduced weight losses (Table 14.8), though the recommended optimum temperature for storage in non-CA or non-MA conditions varied from 7°C to 13°C (Passam, 1982).

Tomato

At UK ambient temperature with about 90% RH, partially ripe tomatoes, packed in suitably permeable plastic film giving 4–6 kPa CO_2 + 4–6 kPa O_2, can be expected to have 7 days longer shelf-life than those stored without wrapping (Geeson, 1989). Under these conditions the eating quality was slowed but film packaging that results in higher CO_2 and lower O_2 levels may not only prevent ripening but also result in tainted fruits when they are ripened after removal from the packs. He also showed that film packages with lower water vapour transmission properties can encourage rotting. Batu (1995) showed that there were interactions between MA packaging and temperature in that MA packaging was more effective in delaying ripening at 13°C than at 20°C. MA packaging interacted also with harvest maturity: it was more effective in delaying ripening of fruits harvested at the mature-green stage than for those harvested at a more advanced stage of maturity. Batu and Thompson (1994) stored tomatoes at both the mature-green and the pink stage of maturity either non-wrapped or sealed in PE films of 20 μm, 30 μm or 50 μm thickness at either 13°C or 20°C for 60 days. All non-wrapped pink tomatoes were overripe and soft after 30 days at 13°C and after 10–13 days at 20°C. Green tomatoes sealed in 20 μm and 30 μm film reached their reddest colour after 40 days at 20°C and 30 days at 13°C. Fruits in 50 μm film still had not reached their maximum red colour even after 60 days at both 13°C and 20°C. All green tomatoes sealed in PE film were very firm even after 60 days of storage at 13°C and 20°C. Non-wrapped tomatoes remained

Table 14.8. Effects of packaging of the percentage weight loss of two species of taro (modified from Passam, 1982).

	Colocasia				Xanthosoma			
Storage time	Control	Dry coir	Moist coir	PE film	Control	Dry coir	Moist coir	PE film
2 weeks	11	7	0	1	8	7	0	1
4 weeks	27	21	3	5	15	19	4	3
6 weeks	35	30	7	9	28	28	21	6

Table 14.9. Effects of packaging material on the quality of yam (*Dioscorea trifida*) after 64 days storage at 20–29°C and 46–62% RH. Fungal score from 0 (= no surface fungal growth) to 5 (= tuber surface entirely covered with fungi). Necrotic tissue estimated by cutting the tuber into two, lengthways, measuring the area of necrosis and expressing it as % of the total cut surface.

Package type	% weight loss	Fungal score	% necrotic tissue
Paper bags	26.3	0.2	5
Sealed 0.03 mm thick PE bags with 0.15 % of the area as holes	15.7	0.2	7
Sealed 0.03 mm thick PE bags	5.4	0.4	4

acceptably firm for about 50 days at 13°C and 20 days at 20°C. Tano *et al.* (2007) stored mature-green tomatoes at 13°C in containers where an atmosphere of some 5 kPa O_2 + 5 kPa CO_2 was maintained and they were subjected to a sequence of temperature fluctuations of 10°C for 35 days. Temperature fluctuations had a major impact on the composition of the package atmospheres and on product quality. CO_2 concentrations increased rapidly, reaching a maximum of 11 kPa, while O_2 concentrations decreased to less than 1.5 kPa. The temperature-fluctuating regime resulted in extensive browning, softening, weight loss increase, ethanol increase and infection due to physiological damage and excessive condensation, compared with storage at constant temperature. Fagundes *et al.* (2015) showed that MA packaging in BOPP/LDPE bags decreased the ethylene production and respiration rates of cherry tomatoes and reduced weight loss and lycopene biosynthesis at 5°C. In a comparison of air and active MA packaging (5 kPa O_2 + 5 kPa CO_2) they found that the latter was preferable and extended their shelf-life to 25 days.

Wampee

Zhang *et al.* (2006) found that Wampee fruit (*Clausena lansium*) deteriorated quickly after harvest and that 2–4°C was optimum, which delayed the total sugar, acidity and AA losses. MA packaging in 0.025 mm PE film with a packing capacity of 5.4 kg/m had a positive additional extension to their storage life at 2–4°C.

Yam

Storing yams (*Dioscorea* spp.) in PE bags was shown to reduce weight loss, but had little effect on surface fungal infections and internal browning of tissue (Table 14.9).

References

AAFC. 2010, 2016. (Infohort online information system, Agriculture and Agri-Food Canada, Ottawa.) Available at: http://www4.agr.gc.ca/IH5_Reports/faces/cognosSubmitter.jsp accessed January 2010 (accessed January 2010, 14 March 2018).

Abadias, M., Usall, J., Anguera, M., Solsona, C. and Viñas, I. 2008. Microbiological quality of fresh, minimally-processed fruit and vegetables, and sprouts from retail establishments. *International Journal of Food Microbiology*, 31, 121–129.

Abbott, J. 2009. Horticulture Week. Available at: http://www.hortweek.com/news/921352/Interview-David-Johnson-agronomist-post-harvest-physiologist-East-Malling-Research/ (accessed October 2009).

Abdel-Rahman, M. and Isenberg, F.M.R. 1974. Effects of growth regulators and controlled atmosphere on stored carrots. *Journal of Agricultural Science*, 82, 245–249.

Abdul Rahman, A.S., Maning, N. and Dali, O. 1997. Respiratory metabolism and changes in chemical composition of banana fruit after storage in low oxygen atmosphere. *Seventh International Controlled Atmosphere Research Conference, July 13–18 1997*, University of California, Davis, California [abstract], 51.

Abdul-Karim, M.N.B., Nor, L.M. and Hassan, A. 1993. The storage of sapadilla *Manikara achras* L. at 10, 15, and 20°C. *Australian Centre for International Agricultural Research Proceedings* 50, 443.

Abdul Raouf, U.M., Beuchat, L.R. and Ammar, M.S. 1993. Survival and growth of *Escherichia coli* O157:H7 on salad vegetables. *Applied and Environmental Microbiology* 59, 1999–2006.

Abdullah, H. and Pantastico, E.B. 1990. *Bananas*. Association of Southeast Asian Nations-COFAF, Jakarta, Indonesia.

Abdullah, H. and Tirtosoekotjo, S. 1989. *Association of Southeast Asian Nations Horticulture Produce Data Sheets*. Association of Southeast Asian Nations Food Handling Bureau, Kuala Lumpur.

Abe, Y. 1990. Active packaging - a Japanese perspective. In: Day, B.P.F. (ed.) *International Conference On Modified Atmosphere Packaging Part 1*. Campden Food and Drinks Research Association, Chipping Campden, UK.

Abeles, F.B., Morgan, P.W. and Saltveit, M.E. 1992. *Ethylene in Plant Biology*. Academic Press, San Diego, California.

Adamicki, F. 1989. Przechowywanie warzyw w kontrolowanej atmosferze. *Biuletyn Warzywniczy Supplement*, I, 107–113.

Adamicki, F. 2001.The effect of modified interactive packaging (MIP) on post-harvest storage of some vegetables. *Vegetable Crops Research Bulletin* 54, 213–218.

Adamicki, F. 2003. Effect of controlled atmosphere and temperature on physiological disorders of stored Chinese cabbage (*Brassica rapa* L var. *pekinensis*). *Acta Horticulturae* 600, 297–301.

Adamicki, F. and Badełek, E. 2006. The studies on new technologies for storage prolongation and maintaining the high quality of vegetables. *Vegetable Crops Research Bulletin* 65, 63–72.

Adamicki, F. and Elkner, K. 1985. Effect of temperature and a controlled atmosphere on cauliflower storage and quality. *Biuletyn Warzywniczy* 28, 197–224.

Adamicki, F. and Gajewski, M. 1999. Effect of controlled atmosphere on the storage of Chinese cabbage (*Brassica rapa* L var. *pekinensis*). *Vegetable Crops Research Bulletin* 50, 61–70.

Adamicki, F. and Kepka, A.K. 1974. Storage of onions in controlled atmospheres. *Acta Horticulturae* 38, 53–73.

Adamicki, F., Dyki, B. and Malewski, W. 1977. Effects of carbon dioxide on the physiological disorders observed in onion bulbs during CA storage. *Qualitas Plantarum* 27, 239–248.

Adams, D.O. and Yang, S.F. 1979. Ethylene biosynthesis: Identification of 1-aminocyclopropane-1-carboxylic acid as an intermediate in the conversion of methionine to ethylene. *Proceedings of the National Academy of Sciences, USA* 76, 170–174.

Adams, K.B., Wu, M.T. and Salunkhe, D.K. 1976. Effects of subatmospheric pressure on the growth and patulin production of *Penicillium expansum* and *Penicillium patulum*. *LWT - Food Science and Technology*, 9, 153–155.

Addai, Z.R., Abdullah, A., Mutalib, S.A. and Musa, K.H. 2013. Effect of gum arabic on quality and antioxidant properties of papaya fruit during cold storage. *International Journal of Chemtech Research*, 5, 2854–2862.

Adesuyi, S.A. 1973. Advances in yam storage research in Nigeria. *Proceedings of the 3rd International Symposium on Tropical Root Crops*, IITA, Ibadan, Nigeria, pp. 428–433.

Agar, I.T., Streif, J. and Bangerth, F. 1991. Changes in some quality characteristics of red and black currants stored under CA and high CO_2 conditions. *Gartenbauwissenschaft* 56, 141–148.

Agar, I.T., Streif, J. and Bangerth, F. 1994a. Effect of high CO_2 and controlled atmosphere (CA) storage on the keepability of blackberry cv, 'Thornfree'. *Commissions C2,D1,D2/3 of the International Institute of Refrigeration International Symposium June 8–10*, Istanbul, pp. 271–280.

Agar, I.T., Bangerth, F. and Streif, J. 1994b. Effect of high CO_2 and controlled atmosphere concentrations on the ascorbic acid, dehydroascorbic acid and total vitamin C content of berry fruits. *Acta Horticulturae* 398, 93–100.

Agar, I.T., Cetiner, S., Garcia, J.M. and Streif, J. 1994c. A method of chlorophyll extraction from fruit peduncles: application to redcurrants. *Turkish Journal of Agriculture and Forestry* 18, 209–212.

Agar, I.T., Hess-Pierce, B., Sourour, M.M. and Kader, A.A. 1997. Identification of optimum preprocessing storage conditions to maintain quality of black ripe 'Mananillo' olives. *Seventh International Controlled Atmosphere Research Conference, July 13–18 1997*, University of California, Davis, California [Abstract], 118.

Agoreyo, B.O., Golden, K.D., Asemota, H.N. and Osagie, A.U. 2007. Postharvest biochemistry of the plantain (*Musa paradisiacal* L.). *Journal of Biological Sciences* 7, 136–144.

Aharoni, N., Rodov, V., Fallik, E., Porat, R., Pesis, E. and Lurie, S. 2008. Controlling humidity improves efficacy of modified atmosphere packaging of fruits and vegetables. *Acta Horticulturae* 804,121–128.

Aharoni, N., Yehoshua, S.B. and Ben-Yehoshua, S. 1973. Delaying deterioration of romaine lettuce by vacuum cooling and modified atmosphere produced in polyethylene packages. *Journal of the American Society for Horticultural Science* 98, 464–468.

Aharoni, Y. and Houck, L.G. 1980. Improvement of internal color of oranges stored in O_2-enriched atmospheres. *Scientia Horticulturae* 13, 331–338.

Aharoni, Y. and Houck, L.G. 1982. Change in rind, flesh, and juice color of blood oranges stored in air supplemented with ethylene or in O_2-enriched atmospheres. *Journal of Food Science* 47, 2091–2092.

Aharoni, Y., Nadel-Shiffman, M. and Zauberman, G. 1968. Effects of gradually decreasing temperatures and polyethylene wraps on the ripening and respiration of avocado fruits. *Israel Journal of Agricultural Research* 18, 77–82.

Aharoni, Y., Apelbaum, A.and Copel, A. 1986. Use of reduced atmospheric pressure for control of the green peach aphid on harvested head lettuce. *HortScience* 21, 469–470.

Ahmad, S. and Thompson, A.K. 2006. Effect of controlled atmosphere storage on ripening and quality of banana fruit. *Journal of Horticultural Science & Biotechnology* 81, 1021–1024.

Ahmad, S., Thompson, A.K., Asi, A.A., Mahmood Khan Chatha, G.A. and Shahid, M.A. 2001. Effect of reduced O_2 and increased CO_2 (controlled atmosphere storage) on the ripening and quality of ethylene treated banana fruit. *International Journal of Agriculture and Biology* 3, 486–490.

Ahmed, E.M. and Barmore, C.R. 1980. Avocado. In: Nagy, S. and Shaw, P.E. (eds) *Tropical and Subtropical Fruits: Composition, Properties and Uses*. AVI Publishing, Westport, Connecticut, 121–156.

Ahmed, J. and Ramaswamy, H.S. 2006. High pressure processing of fruits and vegetables. *Stewart Postharvest Review*, 1, 1–10.

Ahmed, Z.F.R. and Palta, J.P. 2016. Postharvest dip treatment with a natural lysophospholipid plus soy lecithin extended the shelf life of banana fruit. *Postharvest Biology and Technology*, 113, 58–65.

Ahrens, F.H. and Milne, D.L. 1993. Alternative packaging methods to replace SO_2 treatment for litchis exported by sea. *Yearbook South African Litchi Growers' Association*, 5, 29–30.

Akamine, E.K. 1969. Controlled atmosphere storage of papayas. *Proceedings. 5th Annual Meeting, Hawaii Papaya Industry Association, Sept 12–20*. Miscellaneous Publication 764. Hawaii Cooperative Extension Service, Hilo, Hawaii.

Akamine, E.K. and Goo, T. 1969. Effects of controlled atmosphere storage on fresh papayas *Carica papaya* L. variety Solo with reference to shelf-life extension of fumigated fruits. *Hawaii Agricultural Experiment Station, Honolulu, Bulletin* 144.

Akamine, E.K. and Goo, T. 1971. Controlled atmosphere storage of fresh pineapples (*Ananas comosus* [L.], Merr. 'Smooth Cayenne'). *Hawaii Agricultural Experment Station Research Bulletin* 152.

Akamine, E.K. and Goo, T. 1973. Respiration and ethylene production during ontogeny of fruit. *Journal of the American Society for Horticultural Science* 98, 381–383.

Akamine, E.K., Beaumont, J.H., Bowers, F.A.I., Hamilton, R.A., Nishida, T., Sherman G.D., Shoji, K. and Storey, W.B. 1957. *Passionfruit Cultivars in Hawaii*. University of Hawai'i Extension Circular, Manoa.

Akbas, M.Y. and Ölmez, H. 2007. Effectiveness of organic acid, ozonated water and chlorine dipping on microbial reduction and storage quality of fresh-cut iceberg lettuce. *Journal of the Science of Food and Agriculture* 87, 2609–2616.

Akbudak, B., Ozer, M.H. and Erturk, U. 2003. A research on controlled atmosphere (CA) storage of cv. 'Elstar' on rootstock of MM 106. *Acta Horticulturae* 599, 657–663.

Akbudak, B., Ozer, M.H. and Erturk, U. 2004. Effect of controlled atmosphere storage on quality parameters and storage period of apple cultivars 'Granny Smith' and 'Jonagold'. *Journal of Applied Horticulture (Lucknow)* 6, 48–54.

Akbudak, B., Tezcan, H. and Eris, A. 2008. Determination of controlled atmosphere storage conditions for '0900 Ziraat' sweet cherry fruit. *Acta Horticulturae* 795, 855–859.

Akhimienho, H.D. 1999. The control of vascular streaking in fresh cassava. MPhil thesis, Cranfield University.

Al Ubeed, H.M.S., Wills, R.B.H., Bowyer, M.C. and Golding, J.B. 2018. Comparison of hydrogen sulphide with 1-methylcyclopropene (1-MCP) to inhibit senescence of the leafy vegetable, pak choy. *Postharvest Biology and Technology* 137, 129–133.

Al-Ati, T. and Hotchkiss, J.H. 2002. Effect of the thermal treatment of ionomer films on permeability and permselectivity. *Journal of Applied Polymer Science* 86, 2811–2815.

Al-Ati, T. and Hotchkiss, J.H. 2003. The role of packaging film permselectivity in modified atmosphere packaging. *Journal of Agriculture and Food Chemistry* 51, 4133–4138.

Alegre Vilas, I., Viñas Almenar, I., Usalli Rodié, J., Abadiasi Sero, M.I., Altisent Rosell, R. and Anguera, M. 2013. Antagonistic effect of *Pseudomonas graminis* CPA-7 against foodborne pathogens in fresh-cut apples under simulated commercial conditions. *Food Microbiology* 33, 139–148.

Alegria, C., Pinheiro, J., Gonclaves, E.M., Fernandes, I., Moldao, M., and Abreu, M. 2009. Quality attributes of shredded carrot (*Daucuc carota* L. cv. Nantes) as affected by alternative decontamination processes to chlorine. *Innovative Food Science and Emerging Technology* 10, 61–69.

Ali, A., Maqbool, M., Ramachandran, S. and Alderson, P.G. 2010. Gum arabic as a novel edible coating for enhancing shelf life and improving postharvest quality of tomato (*Solanum lycopersicum* L.) fruit. *Postharvest Biology and Technology*, 58, 42–47.

Ali, A., Maqbool, M., Alderson, P.G. and Zahid, N. 2013. Effect of gum arabic as an edible coating on antioxidant capacity of tomato (*Solanum lycopersicum* L.) fruit during storage. *Postharvest Biology and Technology* 76, 119–124.

Ali, S., Khan, A.S., Malik, A.U. and Shahid, M. 2016. Effect of controlled atmosphere storage on pericarp browning, bioactive compounds and antioxidant enzymes of litchi fruits. *Food Chemistry* 206, 18–29.

Ali, Z.M., Hong, C.L., Marimuthu, M. and Lazan, H. 2004. Low temperature storage and modified atmosphere packaging of carambola fruit and their effects on ripening related texture changes, wall modification and chilling injury symptoms. *Postharvest Biology and Technology* 33, 181–192.

Ali Koyuncu, M., Dilmaçünal, T. and Özdemir, Ö. 2009. Modified and controlled atmosphere storage of apricots. *10th International Controlled & Modified Atmosphere Research Conference, 4–7 April 2009*, Turkey [Abstract], 10.

Ali Niazee, M.T., Richardson, D.G., Kosittrakun, M. and Mohammad, A.B. 1989. Non-insecticidal quarantine treatments for apple maggot control in harvested fruits. In: Fellman, J.K. (ed.) *Proceedings of the Fifth International Controlled Atmosphere Research Conference, Wenatchee, Washington, USA, 14–16 June 1989. Vol. 1: Pome fruit*. Washington State University, Pullman, Washington, pp. 193–205.

Alique, R. and De la Plaza, J.L. 1982. Dynamic controlled atmosphere for apple storage. *Proceeding of the Meeting IIR Commissions B2, C2 and D1*. Sofia, Bulgaria, 342–349.

Alique, R. and Oliveira, G.S. 1994. Changes in sugars and organic acids in Cherimoya *Annona cherimola* Mill. fruit under controlled atmosphere storage. *Journal of Agricultural and Food Chemistry* 42, 799–803.

Alique, R., Martinez, M.A. and Alonso, J. 2003. Influence of the modified atmosphere packaging on shelf-life and quality of Navalinda sweet cherry. *European Food Research and Technology* 217, 416–420.

Allen, F.W. and McKinnon, L.R. 1935. Storage of Yellow Newtown apples in chambers supplied with artificial atmospheres. *Proceedings of the American Society for Horticultural Science* 32, 146.

Allen, F.W. and Smock, R.M. 1938. CO_2 storage of apples, pears, plums and peaches. *Proceedings of the American Society for Horticultural Science* 35, 193–199.

Allwood, M.E. and Cutting, J.G.M. 1994. Progress report: gas treatment of 'Fuerte' avocados to reduce cold damage and increase storage life. *Yearbook South African Avocado Growers' Association* 17, 22–26.

Almeida, D.P.F. and Valente, C.S. 2005. Storage life and water loss of plain and curled leaf parsley. *Acta Horticulturae* 682, 1199–1202.

Almenar, E., Hernandez- Muñoz, P., Lagaron, J. M., Catala, R. and Gavara, R. 2006. Controlled atmosphere storage of wild strawberry (*Fragaria vesca* L.) fruit. *Journal of Agricultural and Food Chemistry* 54, 86–91.

Alonso, J.M., Hirayama, T., Roman, G., Nourizadeh, S, and Ecker, J.R. 1999. EIN2, a bifunctional transducer of ethylene and stress responses in Arabidopsis. *Science* 284, 2148–2152.

Al-Qurashi, A.D., Matta, F.B. and Garner, J.O. 2005. Effect of using low-pressure storage (LPS) on Rabbiteye blueberry 'Premier' fruits. *Journal of King Abdulaziz University: Meteorology, Environment and Arid Land Agriculture* 16, 3–14.

Altman, S.A. and Corey, K.A., 1987. Enhanced respiration of muskmelon fruits by pure oxygen and ethylene. *Scientific Horticulture* 31, 275–281.

Alvarez, A.M. 1980. Improved marketability of fresh papaya by shipment in hypobaric containers. *HortScience* 15, 517–518.

Alvarez, A.M. and Nishijima W.T. 1987. Postharvest diseases of papaya. *Plant Diseases* 71, 681–686.

Alves, R.E., Filgueiras, H.A.C., Almeida, A.S., Machado, F.L.C., Bastos, M.S.R. *et al.* 2005. Postharvest use of 1-MCP to extend storage life of melon in Brazil - current research status. *Acta Horticulturae* 682, 2233–2237.

Al-Zaemey, A.B.S., Magan, N. and Thompson, A.K. 1993. Studies on the effect of fruit-coating polymers and organic acids on growth of *Colletotrichium musae in vitro* and on postharvest control of anthracnose of bananas. *Mycological Research* 97, 1463–1468.

Amanatidou, A., Smid, E.J. and Gorris, L.G.M., 1999. Effect of elevated oxygen and carbon dioxide on the surface growth of vegetable-associated micro-organisms. *Journal of Applied Microbiology* 86, 429–438.

Amanatidou, A., Slump, R.A., Gorris, L.G.M. and Smid, E.J. 2000. High oxygen and high carbon dioxide modified atmospheres for shelf-life extension of minimally processed carrots. *Journal of Food Science*, 65, 61–66.

Amaro, A.L. and Almeida D.P.F. 2013. Lysophosphatidylethanolamine effects on horticultural commodities: a review. *Postharvest Biology and Technology* 78, 92–102

Amjad, M., Iqbal J., Rees, D., Iqbal, Q., Nawaz, A. and Ahmad, T. 2009. Effect of packaging material and different storage regimes on shelf life of green hot pepper fruits. *10th International Controlled & Modified Atmosphere Research Conference, 4–7 April 2009*, Turkey [Abstract].

Amodio, M.L., Colelli, G., de Cillis, F.M., Lovino, R. and Massignan, L. 2003. Controlled-atmosphere storage of fresh-cut 'Cardoncello' mushrooms (*Pleurotus eryngii*). *Acta Horticulturae* 599, 731–735.

Amodio, M.L., Rinaldi, R. and Colelli, G. 2005. Effects of controlled atmosphere and treatment with 1-methylcyclopropene (1-MCP) on ripening attributes of tomatoes. *Acta Horticulturae* 682, 737–742.

Amornputti, S., Ketsa, S. and van Doorn, W.G. 2016. 1-Methylcyclopropene (1-MCP) inhibits ethylene production of durian fruit which is correlated with a decrease in ACC oxidase activity in the peel. *Postharvest Biology and Technology* 114, 69–75.

Amorós, A., Pretel, M.T., Zapata, P.J., Botella, M.A., Romojaro, F. and Serrano, M. 2008. Use of modified atmosphere packaging with microperforated polypropylene films to maintain postharvest loquat fruit quality. *Food Science and Technology International* 14, 95–103.

An, D.S., Park, E. and Lee, D.S. 2009. Effect of hypobaric packaging on respiration and quality of strawberry and curled lettuce. *Postharvest Biology and Technology* 52, 78–83.

Anandaswamy, B. 1963. Pre-packaging of fresh produce. IV. Okra *Hibiscus esculentus*. *Food Science Mysore* 12, 332–335.

Anandaswamy, B. and Iyengar, N.V.R. 1961. Pre-packaging of fresh snap beans *Phaseolus vulgaris*. *Food Science* 10, 279.

Andersen, A.S. and Kirk, H.G. 1986. Influence of low pressure storage on stomatal opening and rooting of cuttings *Acta Horticulturae* 181, 305–311.

Anderson, R.E. 1982. Long term storage of peaches and nectarine intermittently warmed during controlled atmosphere storage. *Journal of the American Society for Horticultural Science* 107, 214–216.

Andre, P., Buret, M., Chambroy, Y., Dauple, P., Flanzy, C. and Pelisse, C. 1980a. Conservation trials of asparagus spears by means of vacuum pre-refrigeration, associated with controlled atmospheres and cold storage. *Revue Horticole* 205, 19–25.

Andre, P., Blanc, R., Buret, M., Chambroy, Y., Flanzy, C., Foury, C., Martin, F. and Pelisse, C. 1980b. Globe artichoke storage trials combining the use of vacuum pre-refrigeration, controlled atmospheres and cold. *Revue Horticole* 211, 33–40.

Andrich, G. and Fiorentini, R. 1986. Effects of controlled atmosphere on the storage of new apricot cultivars. *Journal of the Science of Food and Agriculture* 37, 1203–1208.

Andrich, G., Zinnai, A., Balzini, S., Silvestri, S. and Fiorentini, R. 1994. Fermentation rate of Golden Delicious apples controlled by environmental CO_2. *Commissions C2, D1, D2/3 of the International Institute of Refrigeration International Symposium June 8–10*,Istanbul, 233–242.

Anelli G., Mencarelli F. and Massantini, R. 1989. One time harvest and storage of tomato fruit: technical and economical evaluation. *Acta Horticulturae* 287, 411–415.

Anonymous. 1987. Freshtainer makes freshness mobile. *International Fruit World* 45, 225–231.

Anonymous. 1997. Squeezing the death out of food. *New Scientist* 2077, 28–32.

Anonymous. 2003. Quality changes and respiration behaviour of mushrooms under modified atmosphere conditions. *Champignonberichten* 205, 6–7.

Ansorena, M.R., Marcovich, N.E. and Roura, S.I. 2011. Impact of edible coatings and mild heat shocks on quality of minimally processed broccoli (*Brassica oleracea* L.) during refrigerated storage. *Postharvest Biology and Technology* 59, 53–63.

Antonio Lizana, L. and Ochagavia, A. 1997. Controlled atmosphere storage of mango fruits (*Mangifera indica* L.) cvs Tommy Atkins and Kent. *Acta Horticulturae* 455, 732–737.

Antoniolli, L R., Benedetti, B.C., Sigrist, J.M.M., Filho, S., Men de Sá, Alves, M and Ricardo, E. 2006. Metabolic activity of fresh-cut 'Pérola' pineapple as affected by cut shape and temperature. *Brazilian Journal of Plant Physiology* 18, 413–417.

Antoniolli, L.R., Benedetti, B.C., Sigrist, J.M.M., de Silveira, N.F.A. 2007. Quality evaluation of fresh-cut 'Perola' pineapple stored in controlled atmosphere. *Ciencia e Tecnologia de Alimentos* 27, 530–534.

Antunes, M.D.C. and Sfakiotakis, E.M. 2002. Ethylene biosynthesis and ripening behaviour of 'Hayward' kiwifruit subjected to some controlled atmospheres. *Postharvest Biology and Technology* 26, 167–179.

Antunes, M.D.C., Miguel, M.G., Metelo, S., Dandlen, S. and Cavaco, A. 2008. Effect of 1-Methylcyclopropene application prior to storage on fresh-cut kiwifruit quality. *Acta Horticulturae* 796, 173–178.

Anuradha, G. and Saleem, S. 1998. Certain biochemical changes and the rate of ethylene evolution in Ber fruits *Zizyphus mauritiana* Lamk. cultivar Umran as affected by the various post harvest treatments during storage. *International Journal of Tropical Agriculture* 16, 233–238.

Apeland, J. 1985. Storage of Chinese cabbage *Brassica campestris* L. *pekinensis* Lour Olsson in controlled atmospheres. *Acta Horticulturae* 157, 185–191.

Apelbaum, A. and Barkai-Golan, R. 1977. Spore germination and mycelia growth of postharvest pathogens under hypobaric pressure. *Phytopathology* 67, 400–403.

Apelbaum, A. Aharoni, Y. and Tempkin-Gorodeiski, N. 1977a. Effects of subatmospheric pressure on the ripening processes of banana fruits. *Tropical Agriculture* 54, 39–46.

Apelbaum, A., Zauberman G. and Fuchs Y. 1977b. Subatmospheric pressure storage of mango fruits. *Scientia Horticulturae* 7, 153–160.

Aracena, J.J., Sargent, S.A., Brecht, J.K. and Campbell, C.A. 1993. Environmental factors affecting vascular streaking, a postharvest physiological disorder of cassava root (*Manihot esculenta* Crantz). *Acta Horticulturae* 343, 297–299.

Archbold, D.D., Hamilton-Kemp, T.R., Clements A.M. and Collins, R.W. 1990. Fumigating 'Crimson Seedless' table grapes with (E)-2-hexenal reduces mold during long-term postharvest storage. *HortScience* 34, 705–707.

Arevalo-Galarza, L., Bautista-Reyes, B., Saucedo-Veloz, C. and Martinez-Damian, T. 2007. Cold storage and 1-methylcyclopropene (1-MCP) applications on sapodilla (*Manilkara zapota* (L.) P. Royen) fruits. *Agrociencia (Montecillo)* 41, 469–477.

Argenta, L.C., Fan, X. and Mattheis, J. 2000. Delaying establishment of controlled atmosphere or CO_2 exposure reduces 'Fuji' apple CO_2 injury without excessive fruit quality loss. *Postharvest Biology and Technology* 20, 221–229.

Argenta, L.C., Mattheis, J. and Fan, X. 2001. Delaying 'Fuji' apple ripening by 1-MCP treatment and management of storage temperature. *Revista Brasileira de Fruticultura* 23, 270–273.

Arnal, L., Besada, C., Navarro, P. and Salvador, A. 2008. Effect of controlled atmospheres on maintaining quality of persimmon fruit cv. 'Rojo Brillante'. *Journal of Food Science* 73, S26–S30.

Arnon, H., Zaitsev, Y., Porat, R. and Poverenov, E. 2014. Effects of carboxymethyl cellulose and chitosan bilayer edible coating on postharvest quality of citrus fruit. *Postharvest Biology and Technology* 87, 21–26.

Arpaia, M.L., Mitchell, F.G., Kader A.A. and Mayer, G. 1985. Effects of 2% O_2 and various concentrations of CO_2 with or without ethylene on the storage performance of kiwifruit. *Journal of the American Society for Horticultural Science* 110, 200–203.

Arras, G. and Usai, M. 2001. Fungitoxic activity of 12 essential oils against four postharvest citrus pathogens: chemical analysis of *Thymus capitatus* oil and its effect in subatmospheric pressure conditions. *Journal of Food Protection* 64, 1025–1029.

Artés Calero, F., Escriche, A., Guzman, G. and Marin, J.G. 1981. Violeta globe artichoke storage trials in a controlled atmosphere. *Congresso Internazionale di Studi sul Carciofo* 30, 1073–1085.

Artés-Hernandez, F., Artés, F., Villaescusa, R. and Tudela, J.A. 2003. Combined effect of modified atmosphere packaging and hexanal or hexenal fumigation during long-term cold storage of table grapes. *Acta Horticulturae* 600, 401–404.

Artés-Hernández, F., Aguayo, E. and Artés, F. 2004. Alternative atmosphere treatments for keeping quality of 'Autumn seedless' table grapes during long-term cold storage. *Postharvest Biology and Technology* 31, 59–67.

Artés-Hernandez, F., Tomás-Barberán, F.A. and Artez, F. 2006 Modified atmosphere packaging preserves quality of SO_2-free 'Superior Seedless' table grapes. *Postharvest Biology and Technology* 39, 15–24.

Artés-Hernandez, F., Aguayo, E., Artés, F. and Tomas-Barberan, F.A. 2007. Enriched ozone atmosphere enhances bioactive phenolics in seedless table grapes after prolonged shelf-life. *Journal of the Science of Food and Agriculture* 87, 824–831.

Artés-Hernández F., Robles P.A., Gómez P.A., Tomás-Callejas A. and Artés F. 2010. Low UV-C illumination for keeping overall quality of fresh-cut watermelon. *Postharvest Biology and Technology* 55, 114–120.

Arvanitoyannis, I.S., Khah, E.M., Christakou, E.C. and Bletsos, F.A. 2005. Effect of grafting and modified atmosphere packaging on eggplant quality parameters during storage. *International Journal of Food Science & Technology* 40, 311–322.

Asghari, M. 2009. Effect of hot water treatment and low density polyethylene bag package on quality attributes and storage life of muskmelon fruit, *10th International Controlled & Modified Atmosphere Research Conference, 4–7 April 2009*, Turkey [Abstract], 22.

ASHRAE. 1968. *Fruit and Vegetables*. Guide and Data Book. American Society of Heating, Refrigeration and Air-conditioned Engineering, New York.

Athanasopoulos, P., Katsaboxakis, K., Thanos, A., Manolopoulou, H., Labrinos, G. and Probonas, V. 1999. Quality characteristics of apples preserved under controlled atmosphere storage in a pilot plant scale. *Fruits (Paris)* 54, 79–86.

Attri, B.L. and Swaroop, K. 2005. Post harvest storage of French bean (*Phaseolus vulgaris* L.) at ambient and low temperature conditions. *Horticultural Journal* 18, 98–101.

Awad, M., de Oliveira, A.I. and Correa, D. de L. 1975. The effect of Ethephon, GA and partial vacuum on respiration in bananas (*Musa acuminata*). *Revista de Agricultura, Piracicaba, Brazil* 50, 109–113.

Awad, M.A.G., Jager, A., Roelofs, F.P., Scholtens, A. and de Jager, A. 1993. Superficial scald in Jonagold as affected by harvest date and storage conditions. *Acta Horticulturae* 326, 245–249.

Aydt, T.P., Weller, C.L. and Testin, R.F. 1991. Mechanical and barrier properties of edible corn and wheat protein films. *Transactions of the American Society of Agricultural Engineers* 34, 207–211.

Ayhan, Z., Estürk, O. and Tas, E. 2008. Effect of modified atmosphere packaging on the quality and shelf life of minimally processed carrots. *Turkish Journal of Agriculture* 32, 57–64.

Ayhan, Z., Eştürk, O. and Müftüoğlu, F. 2009. Effects of coating, modified atmosphere (MA) and plastic film on the physical and sensory properties of apricot. *10th International Controlled & Modified Atmosphere Research Conference, 4–7 April 2009*, Turkey [Abstract], 19.

Aziz Al-Harthy, A., East, A.R., Hewett, E.W. and Mawson, A.J. 2009. Controlled atmosphere storage of 'Opal Star' Feijoa. *10th International Controlled & Modified Atmosphere Research Conference, 4–7 April 2009*, Turkey [Abstract], 11.

Baba, T. and Ikeda F. 2003. Use of high pressure treatment to prolong the postharvest life of mume fruit (*Prunus mume*). *Acta Horticulturae* 628, 373–377.

Baba, T., Como, G., Ohtsubo, T. and Ikeda, F. 1999. Effects of high pressure treatment on mume fruit (*Prunus mume*). *Journal of the American Society for Horticultural Science* 124, 399–401.

Baba, T., Ito, S., Ikeda, F., and Manago, M. 2008. Effectiveness of high pressure treatment at low temperature to improve postharvest life of horticultural commodities. *Acta Horticulturae* 768, 417–422.

Back, E.A. and Cotton, R.T. 1925. The use of vacuum for insect control. *Journal of Agriculture Research* 31, 1035–1041.

Badran, A.M. 1969. Controlled atmosphere storage of green bananas. US Patent 17 June 3, 450, 542.

Bai, J., Alleyne, V., Hagenmaier, R.D., Mattheis, J.P. and Baldwin, E.A. 2003. Formulation of zein coatings for apples (*Malus domestica* Borkh). *Postharvest Biology and Technology* 28, 259–268.

Bailén, G., Guillén, F., Castillo, S., Serrano, M., Valero, D. and Martínez-Romero D. 2006. Use of activated carbon inside modified atmosphere packages to maintain tomato fruit quality during cold storage. *Journal of Agricultural and Food Chemistry* 54, 2229–2235.

Balasubramaniam, V.M., Farkas, D. and Turek, E.J. 2008. Preserving foods through high-pressure processing. *Food Technology* 62, 32–38.

Baldwin, E., Plotto, A. and Narciso, J. 2007. Effect of 1-MCP, harvest maturity, and storage temperature on tomato flavor compounds, color, and microbial stability. Available at: http://www.ars.usda.gov/research/publications/publications.htm?seq_no_115=207169 (accessed June 2009).

Ballantyne, A., Stark, R. and Selman, J.D. 1988. Modified atmosphere packaging of broccoli florets. *International Journal of Food Science and Technology* 23, 353–360.

Banaras, M., Bosland, P.W. and Lownds, N.K. 2005. Effects of harvest time and growth conditions on storage and post-storage quality of fresh peppers (*Capsicum annuum* L.). *Pakistan Journal of Botany* 37, 337–344.

Bancroft, R.D. 1989. The effect of surface coating on the development of postharvest fungal rots of pome fruit with special reference to 'Conference' pears. PhD thesis, University of Cambridge.

Bancroft, R.D. 1995. The use of a surface coating to ameliorate the rate of spread of post-harvest fungal diseases of top fruit. *International Biodeterioration & Biodegradation* 36, 385–405.

Bancroft, R.D. and Crentsil, D. 1995. Application of a low cost storage technique for fresh cassava (*Manihot esculenta*) roots in Ghana. In: Agbor-Egbe, T., Brauman, T.A., Griffon, D. and Treche, S. (eds) *Transformation alimentaire du Manioc (Cassava Food Processing)*. Insitut Française de Recherche Scientifique pour le Developpment en Cooperation, Paris, pp. 547–555.

Bancroft, R., Gray, A., Gallat, S., Crentsil, D. and Westby, A. 1998. The marketing system for fresh yam in Techiman, Ghana, and associated postharvest losses. *Tropical Agriculture* 75, 115–119.

Bancroft, R.D. Crentsil, D., Panni, J.Y., Aboagye-Nuamah, F. and Krampa L. 2005. The influence of postharvest conditioning and storage protocols on the incidence of rots in white yams (*Dioscorea rotundata* Poir.) in Ghana. *Acta Horticulturae* 682, 2183–2190.

Bandyopadhyay, C., Gholap, A.S. and Mamdapur, V.R. 1985. Characterisation of alkenyl resorcinol in mango Mangifers indica latex. *Journal of Agriculture and Food Chemistry* 33, 377–379.

Bangerth, F. 1973. The effect of hypobaric storage on quality, physiology and storage life of fruits, vegetables and cut flowers. *Gartenbauwissenschaft* 38, 479–508.

Bangerth, F. 1974. Hypobaric storage of vegetables. *Acta Horticulturae* 38, 23–32.

Bangerth, F. 1984. Changes in sensitivity for ethylene during storage of apple and banana fruits under hypobaric conditions. *Scientia Horticulturae* 24, 151–163.

Bangerth, F. and Streif, J. 1987. Fffect of aminoethoxyvinylglycine and low-pressure storage on the post-storage production of aroma volatiles by Golden Delicious apples. *Journal of the Science of Food and Agriculture* 41, 351–360.

Banks, N.H. 1984. Some effects of TAL-Prolong coating on ripening bananas. *Journal of Experimental Botany* 35, 127–137.

Barden, C.L. 1997. Low oxygen storage of 'Ginger Gold' apples. *Postharvest Horticulture Series* 16, 183–188. Department of Pomology, University of California.

Bare, C.O. 1948. The effect of prolonged exposure to high vacuum on stored tobacco insects. *Journal of Economic Entomology* 41, 109–110.

Barger, W.R. 1963. *Vacuum Precooling. A Comparison of Cooling of Different Vegetables*. Market Research Report 600. US Department of Agriculture, Washington, DC.

Barkai Golan, R. and Phillips, D.J. 1991. Postharvest heat treatment of fresh fruits and vegetables for decay control. *Plant Disease* 75, 1085–1089.

Barker, J. and Mapson, L.W. 1955. Studies in the respiratory and carbohydrate metabolism of plant tissues. VII. Experimental studies with potato tubers of an inhibition of the respiration and of a 'block' in the tricarboxylic acid cycle induced by 'oxygen poisoning'. *Proceedings of the Royal Society Biology* 143, 523–549.

Barmore, C.R. 1987. Packaging technology for fresh and minimally processed fruits and vegetables. *Journal of Food Quality* 10, 207–217.

Barmore, C.R. and Mitchell, E.F. 1977. Ethylene pre-ripening of mangoes prior to shipment. *The Citrus Industry* 58, 18–23.

Basanta, A.L. and Sankat, C.K. 1995. Storage of the pomerac Eugenia malaccensis. *Technologias de cosecha y postcosecha de frutas y hortalizas. Proceedings of a conference held in Guanajuato Mexico 20–24 February 1995*, 567–574.

Bastrash, S., Makhlouf, J., Castaigne, F. and Willemot, C. 1993. Optimal controlled atmosphere conditions for storage of broccoli florets. *Journal of Food Science* 58, 338–341, 360.

Batu, A. 1995. Controlled atmosphere storage of tomatoes. PhD thesis Silsoe College, Cranfield University.

Batu, A. 2003. Effect of long-term controlled atmosphere storage on the sensory quality of tomatoes. *Italian Journal of Food Science* 15, 569–577.

Batu, A. and Thompson, A.K. 1994. The effects of harvest maturity, temperature and thickness of modified atmosphere packaging films on the shelf life of tomatoes. *Commissions C2, D1, D2/3 of the International Institute of Refrigeration International Symposium June 8–10*, Istanbul, 305–316.

Batu, A. and Thompson, A.K. 1995. Effects of controlled atmosphere storage on the extension of postharvest qualities and storage life of tomatoes. *Workshop of the Belgium Institute for Automatic Control, Ostend June 1995*, 263–268.

Baumann, H. 1989. Adsorption of ethylene and CO_2 by activated carbon scrubbers. *Acta Horticulturae* 258, 125–129.

Bautista, O.K. 1990. *Postharvest Technology for Southeast Asian Perishable Crops*. Technology and Livelihood Resource Center, Philippines.

Bautista, P.B. and Silva, M.E. 1997. Effects of CA treatments on guava fruit quality. *Seventh International Controlled Atmosphere Research Conference, July 13–18 1997*, University of California, Davis, California [Abstract], 113.

Bayram, E., Dundar, O. and Ozkaya, O. 2009. The effect of different packing types on the cold storage Hicaznar pomegranate. *10th International Controlled & Modified Atmosphere Research Conference, 4–7 April 2009*, Turkey [Poster abstract], 62.

Beaudry, R.M. and Gran, C.D. 1993. Using a modified-atmosphere packaging approach to answer some post-harvest questions: factors influencing the lower O_2 limit. *Acta Horticulturae* 326, 203–212.

Beaudry, R., DeEll, J., Kupferman, E., Mattheis, J., Tong, C., Prange, R. and Watkins, C. 2009. Region-By-Region Storage Recommendations For Honeycrisp: Responses from apple storage researchers. Available at: postharvest.tfrec.wsu.edu/rep1020a (accessed March 2018).

Beaudry, R.M., Contreras, C. and Tran, D. 2014. Toward optimizing CA storage of Honeycrisp apples: minimizing pre-storage conditioning time and temperature. *New York Fruit Quarterly Fall 2014*, 9–14.

Been, B.O., Perkins, C. and Thompson, A.K. 1976. Yam curing for storage. *Acta Horticulturae* 62, 311–316.

Belay, Z.A., Caleb, O.J. and Opara, U.L. 2017. Impacts of low and super-atmospheric oxygen concentrations on quality attributes, phytonutrient content and volatile compounds of minimally processed pomegranate arils (cv. Wonderful). *Postharvest Biology and Technology* 124, 119–127.

Bellincontro, A., de Santis, D., Fardelli, A., Mencarelli, F. and Botondi, R. 2006. Postharvest ethylene and 1-MCP treatments both affect phenols, anthocyanins, and aromatic quality of Aleatico grapes and wine. *Australian Journal of Grape and Wine Research* 12, 141–149.

Ben, J.M. 2001. Some factors affecting flesh browning in 'Gala Must' apples. *Folia Horticulturae* 13, 61–67.

Ben Arie, R. 1996. Fresher via boat than airplane. *Peri News* 11, Spring 1996.

Ben Arie, R. and Or, E. 1986. The development and control of husk scald on 'Wonderful' pomegranate fruit during storage. *Journal of the American Society for Horticultural Science* 111, 395–399.

Ben Arie, R. and Sonego, L. 1985. Modified-atmosphere storage of kiwifruit *Actinidia chinensis* Planch with ethylene removal. *Scientia Horticulturae*,27, 263–273.

Ben Arie, R., Roisman, Y., Zuthi, Y. and Blumenfeld, A. 1989. Gibberellic acid reduces sensitivity of persimmon fruits to ethylene. *Advances in Agricultural Biotechnology* 26, 165–171.

Ben Arie, R., Levine, A., Sonego, L. and Zutkhi, Y. 1993. Differential effects of CO_2 at low and high O_2 on the storage quality of two apple cultivars. *Acta Horticulturae* 326, 165–174.

Ben Arie, R., Nerya, O., Zvilling, A., Gizis, A. and Sharabi-Nov, A. 2001. Extending the shelf-life of 'Triumph' persimmons after storage with 1-MCP. *Alon Hanotea* 55, 524–527.

Bender, R.J., Brecht, J.K. and Campbell, C.A. 1994. Responses of 'Kent' and 'Tommy Atkins' mangoes to reduced O_2 and elevated CO_2. *Proceedings of the Florida State Horticultural Society*, 107, 274–277.

Bender, R.J., Brecht, J.K., Sargent, S.A. and Huber, D.J. 2000. Low temperature controlled atmosphere storage for tree-ripe mangoes (*Mangifera indica* L.). *Acta Horticulturae* 509, 447–458.

Benítez, C.E. 2001. *Cosecha y poscosecha de Peras y Manzanas en los valles irrigados de la Patagonia*. Estación Experimental Agropecuaria Alto Valle, Instituto Nacional de Tecnologia Agropecuaria (INTA), Casilla de Correo 782, 8332 General Rocha, Rio Negro, Argentina.

Ben-Yehoshua, S. and Rodov, V. 2003. Transpiration and water stress. In: Bartz, J.A. and J.K. Brecht (eds.) *Postharvest Physiology and Pathology of Vegetables*. Marcel Dekker, New York, pp. 111–159.

Ben-Yehoshua, S., Kim, J.J. and Shapiro, B. 1989. Elicitation of reistance to the development of decay in sealed citrus fruit by curing. *Acta Horticulturae* 258, 623–630.

Ben-Yehoshua, S., Fang, DeQiu, Rodov, V., Fishman, S. and Fang, D.Q. 1995. New developments in modified atmosphere packaging, Part II. *Plasticulture* 107, 33–40.

Berard, J.E. 1821. Memoire sur la maturation des fruits. *Annales de Chimie et de Physique* 16, 152–183.

Berard, L.S. 1985. Effects of CA on several storage disorders of winter cabbage. Controlled atmospheres for storage and transport of perishable agricultural commodities. *4th National Controlled Atmosphere Research Conference, July 1985*, 150–159.

Berard, L.S., Vigier, B. Crete, R. and Chiang, M. 1985. Cultivar susceptibility and storage control of grey speck disease and vein streaking, two disorders of winter cabbage. *Canadian Journal of Plant Pathology* 7, 67–73.

Berard, L.S., Vigier, B. and Dubuc Lebreux, M.A. 1986. Effects of cultivar and controlled atmosphere storage on the incidence of black midrib and necrotic spot in winter cabbage. *Phytoprotection* 67, 63–73.

Berger, H., Galletti, Y.L., Marin, J., Fichet, T. and Lizana, L.A 1993. Efecto de atmosfera controlada y el encerado en la vida postcosecha de Cherimoya *Annona cherimola* Mill. cv. Bronceada. *Proceedings of the Interamerican Society for Tropical Horticulture* 37, 121–130.

Berners-Lee, M. 2010. *How Bad are Bananas? The Carbon Footprint of Everything*. Greystone Books, Vancouver.

Berrang, M.E., Brackett, R.E. and Beuchat, L.R. 1989. Growth of Aeromonas hydrophila on fresh vegetables stored under a controlled atmosphere. *Applied and Environmental Microbiology* 55, 2167–2171.

Berrang, M.E., Brackett, R.E. and Beuchat, L.R. 1990. Microbial, color and textural qualities of fresh asparagus, broccoli, and cauliflower stored under controlled atmosphere. *Journal of Food Protection* 53, 391–395.

Berrios J.D.J, Swanson B.G. and Cheong W.A. 1999. Physico-chemical characterization of stored black beans (*Phaseolus vulgaris* L.). *Food Research International* 32, 669–676.

Berry, G. and Aked, J. 1997. Controlled atmosphere alternatives to the post-harvest use of sulphur dioxide to inhibit the development of Botrytis cinerea in table grapes. *Seventh International Controlled Atmosphere Research Conference, July 13–18 1997*, University of California, Davis, California [Abstract], 100.

Bertolini, P. 1972. Preliminary studies on cold storage of cherries. *Universita di Bologna, Notiziatio CRIOF-Centro per la Protezione e Conservazione dei Prodotti Ortofrutticoli*, 3.11.

Bertolini, P., Pratella, G.C., Tonini, G. and Gallerani, G. 1991. Physiological disorders of 'Abbe Fetel' pears as affected by low-O_2 and regular controlled atmosphere storage. Technical-innovations-in-freezing-and-refrigeration-of-fruits-and-vegetables. Paper presented at a Conference held in Davis, California, 9–12 July, 1989, 61–66.

Bertolini, P., Baraldi, E., Mari, M., Trufelli, B. and Lazzarin, R. 2003. Effects of long term exposure to high-CO_2 during storage at 0 °C on biology and infectivity of *Botrytis cinerea* in red chicory. *Journal of Phytopathology* 151, 201–207.

Bertolini, P., Baraldi, E., Mari, M., Donati, I. and Lazzarin, R. 2005. High-CO_2 for the control of *Botrytis cinerea* rot during long term storage of red chicory. *Acta Horticulturae* 682, 2021–2027.

Besada, C., Arnal, L. and Salvador, A. 2008. Improving storability of persimmon cv. Rojo Brillante by combined use of preharvest and postharvest treatments. *Postharvest Biology and Technology* 50, 169–175.

Bessemans, N., Verboven, P., Verlinden, B.E. and B.M. Nicolaï, B.M. 2016. A novel type of dynamic controlled atmosphere storage based on the respiratory quotient (RQ-DCA). *Postharvest Biology and Technology* 115, 91–102.

Bessemans, N., Verboven, P., Verlinden, B.E.and Nicolaï, B.M. 2018. Model based leak correction of real-time RQ measurement for dynamic controlled atmosphere storage. *Postharvest Biology and Technology* 136, 31–41.

Betts, G.D. (ed.) 1996. *A code of practice for the manufacture of vacuum and modified atmosphere packaged chilled foods*. Guideline, 11. Campden and Chorleywood Food Research Association, Chipping Campden, UK.

Biale, J.B. 1946. Effect of oxygen concentration on respiration of the Fuerte avocado fruit. *American Journal of Botany* 33, 363–373.

Biale, J.B. 1950. Postharvest physiology and biochemistry of fruits. *Annual Review of Plant Physiology* 1, 183–206.

Biale, J.B. 1960. Respiration of fruits. In: Ruhland, W. (ed.) *Handbuch der Pflanzenphysiologie. Encyclopedia of Plant Physiology*. Springer-Verlag, Berlin, Vol. 12, 536–592.

Biale, J.B. and Barcus, D.E. 1970. Respsiratory patterns in tropical fruits of the Amazon basin. *Tropical Science* 12, 93–104.
Biale, J.B. and Young, R.E., 1947. Critical oxygen concentrations for the respiration of lemons. *American Journal of Botany* 34, 301–309.
Bishop, D.J. 1994. Application of new techniques to CA storage. *Commissions C2, D1, D2/3 of the International Institute of Refrigeration International Symposium, June 8–10* ,Istanbul, pp. 323–329.
Bishop, D.J. 1996. Controlled atmosphere storage. In: Dellino, C.J.V. (ed.) *Cold and Chilled Storage Technology*. Blackie, London.
Biton, E., Kobiler, I., Feygenberg, O., Yaari, M., Friedman, H. and Prusky, D. 2014. Control of alternaria black spot in persimmon fruit by a mixture of gibberellin and benzyl adenine, and its mode of action. *Postharvest Biology and Technology* 94, 82–88.
Blackbourn, H.D., Jeger, M.J., John, P. and Thompson, A.K. 1990. Inhibition of degreening in the peel of bananas ripened at tropical temperatures. III. Changes in plastid ultrastructure and chlorophyll protein complexes accompanying ripening in bananas and plantains. *Annals of Applied Biology* 117, 147–161.
Blank, H.G. 1973. The possibility of simple CO_2 absorption in CA storage. *Mitteilungen des Obstbauversuchsringes des Alten Landes* 28, 202–208.
Blankenship, S.M. and Sisler, E.C. 1991. Comparison of ethylene gassing methods for tomatoes. *Postharvest Biology and Technology* 1, 59–65.
Blazek, J. 2007. Apple cultivar 'Meteor'. *Nove Odrudy Ovoce*, 15–18.
Blednykh, A.A., Akimov, Yu. A., Zhebentyaeva, T.N. and Untilova, A.E. 1989. Market and flavour qualities of the fruit in sweet cherry following storage in a controlled atmosphere. *Byulleten' Gosudarstvennogo Nikitskogo Botanicheskogo Sada* 69, 98–102.
Bleinroth, E.W., Garcia, J.L.M. and Yokomizo, Y. 1977. Conservacao de quatro variedades de manga pelo frio e em atmosfera controlada. *Coletanea de Instituto de Tecnologia de Alimentos* 8, 217–243.
Blythman, J. 1996. *The Food We Eat*. Michael Joseph, London.
Bogdan, M., Ionescu, L., Panait, E. and Niculescu, F. 1978. Research on the technology of keeping peaches in cold storage and in modified atmosphere. *Lucrari Stiintifice, Institutul de Cerctari Pentru Valorifocarea Legumelor si Fructelor* 9, 53–60.
Bohling, H. and Hansen, H. 1977. Storage of white cabbage *Brassica oleracea* var. *capitata* in controlled atmospheres. *Acta Horticulturae* 62, 49.
Bohling, H. and Hansen, H. 1989. Studies on the metabolic activity of oyster mushrooms *Pleurotus ostreatus* Jacq. *Acta Horticulturae* 258, 573–577.
Boon-Long, P., Achariyaviriya, S. and Johnson, G.I. 1994. Mathematical modelling of modified atmosphere conditions. *Australian Centre for International Agricultural Research Proceedings* 58, 63–67.
Boonyaritthongchai, P. and Kanlayanarat, S. 2003a. Modified atmosphere and carbon dioxide shock treatment for prolonging storage life of 'Rong-rien' rambutan fruits. *Acta Horticulturae* 600, 823–828.
Boonyaritthongchai, P. and Kanlayanarat, S. 2003b. Controlled atmosphere storage to maintain quality of 'Rong-rien' rambutan fruits. *Acta Horticulturae* 600, 829–832.
Booth, R.H. 1975. *Cassava storage*. Series EE 16. The International Potato Centre, Lima, Peru.
Bordanaro, C. 2009. Available at: http://www.purfresh.com (accessed September 2009).
Borecka, H.W. and Pliszka, K. 1985. Quality of blueberry fruits (*Vaccinium corymbosum* L.) stored under LPS, CA and normal air storage. *Acta Horticulturae* 165, 241–246.
Bosland, P.W. and Votava, E.J. 2000. *Peppers, Vegetables and Spices*. CABI Publishing, New York and Oxford.
Both, V., Thewes, F.R., Brackmann, A., de Oliveira Anese, R., de Freitas Ferreira, D. and Wagner, R. 2017. Effects of dynamic controlled atmosphere by respiratory quotient on some quality parameters and volatile profile of 'Royal Gala' apple after long-term storage. *Food Chemistry* 215, 483–492.
Botrel, N., Fonseca, M.J. de O., Godoy, R.L.O. and Barboza, H.T.G. 2004. Storage of 'Prata Ana' bananas in controlled atmosphere. *Revista Brasileira de Armazenamento* 29, 125–129.
Bowden, A.P. 1993. Modified atmosphere packaging of Cavendish and Apple bananas. MSc thesis Cranfield University.
Bower, J.P., Cutting, J.G.M. and Truter, A.B. 1990. Container atmosphere, as influencing some physiological browning mechanisms in stored Fuerte avocados. *Acta Horticulturae* 269, 315–321.
Brackmann, A. 1989. Effect of different CA conditions and ethylene levels on the aroma production of apples. *Acta Horticulturae* 258, 207–214.
Brackmann, A. and Waclawovsky, A.J. 2000. Storage of apple (*Malus domestica* Borkh.) cv. Braeburn. *Ciencia Rural* 30, 229–234.

Brackmann, A., Streif, J. and Bangerth, F. 1993. Relationship between a reduced aroma production and lipid metabolism of apples after long-term controlled-atmosphere storage. *Journal of the American Society for Horticultural Science* 118, 243–247.

Brackmann, A., Streif, J. and Bangerth, F. 1995. Influence of CA and ULO storage conditions on quality parameters and ripening of preclimacteric and climacteric harvested apple fruits. II. Effect on ethylene, CO_2, aroma and fatty acid production. *Gartenbauwissenschaft* 60, 1–6, 23.

Brackmann, A., de Mello, A. M., de Freitas, S.T., Vizzotto, M. and Steffens, C.A. 2001. Storage of 'Royal Gala' apple under different temperatures and carbon dioxide and oxygen partial pressure. *Revista Brasileira de Fruticultura* 23, 532–536.

Brackmann, A., Benedetti, M., Steffens, C. A., and de Mello, A.M. 2002a. Effect of temperature and controlled atmosphere conditions in the storage of 'Fuji' apples with watercore incidence. *Revista Brasileira de Agrociencia* 8, 37–42.

Brackmann, A., Neuwald, D.A., Ribeiro, N.D. and de Freitas, S. T. 2002b. Conservation of three bean genotypes (*Phaseolus vulgaris* L.) of the group Carioca in cold storage and controlled atmosphere. *Ciencia Rural* 32, 911–915.

Brackmann, A., Steffens, C.A. and Waclawovsky, A.J. 2002c. Influence of harvest maturity and controlled atmosphere conditions on the quality of 'Braeburn' apple. *Pesquisa Agropecuaria Brasileira* 37, 295–301.

Brackmann, A., de Freitas, S.T., Giehl, R.F.H., de Mello, A.M., Benedetti, M., de Oliveira, V.R. and Guarienti, A.J.W. 2004. Controlled atmosphere conditions for 'Kyoto' persimmon storage. *Ciencia Rural* 34, 1607–1609.

Brackmann, A., Guarienti, A.J.W., Saquet, A.A., Giehl, R.F.H. and Sestari, I. 2005a. Controlled atmosphere storage conditions for 'Pink Lady' apples. *Ciencia Rural* 35, 504–509.

Brackmann, A., Pinto, J.A.V., Neuwald, D.A., Giehl, R.F.H. and Sestari, I. 2005b. Temperature and optimization conditions for controlled atmosphere storaging of Gala apple. *Revista Brasileira de Agrociencia* 11, 505–508.

Brackmann, A., Weber, A. and Both, V. 2015. CO_2 partial pressure for respiratory quotient and HarvestWatch™ dynamic controlled atmosphere for 'Galaxy' apples storage. *Acta Horticulturae* 1079, 435–440.

Bradshaw, B. 2007. Packaging adds 20 per cent to cost of store fruit. Available at: http://www.dailymail.co.uk/news/article-449276/Packaging-adds-20-cent-cost-store-fruit.html (accessed June 2009).

Bramlage, W.J., Bareford, P.H., Blanpied, G.D., Dewey, D.H., Taylor, S. *et al.* 1977. CO_2 treatments for 'McIntosh' apples before CA storage. *Journal of the American Society for Horticultural Science* 102, 658–662.

Brandelero, R.P.H., Brandelero, E.M., Almeida, F.M. and de Ciência, E. 2016. Biodegradable films of starch/PVOH/alginate in packaging systems for minimally processed lettuce (*Lactuca sativa* L.). *Agrotecnologia* 40, 510–521.

Brash, D.W., Corrigan, V.K. and Hurst, P.L. 1992. Controlled atmosphere storage of 'Honey 'n' Pearl"sweet corn. *Proceedings Annual Conference, Agronomy Society of New Zealand* 22, 35–40.

Brecht, J.K., Kader, A.A., Heintz, C.M. and Norona, R.C. 1982. Controlled atmosphere and ethylene effects on quality of California canning apricots and clingstone peaches. *Journal of Food Science* 47, 432–436.

Brednose, N. 1980. Effects of low pressure on storage life and subsequent keeping quality of cut roses. *Acta Horticulturae* 113, 73–79.

Brigati, S and Caccioni, D. 1995. Influence of harvest period, pre- and post-harvest treatments and storage techniques on the quality of kiwifruits. *Rivista di Frutticoltura e di Ortofloricoltura* 57, 41–3.

Brigati, S., Pratella, G.C. and Bassi, R. 1989. CA and low O_2 storage of kiwifruit: effects on ripening and disease. In: Fellman, J.K. (Ed.) *Proceedings of the Fifth International Controlled Atmosphere Conference, Wenatchee, Washsington, USA, 14–16 June. Vol. 2: Other commodities and storage recommendations.* Washington State University, Pullman, Washington, pp. 41–48.

Brizzolara, S., Santucci, C., Tenori, L., Hertog, M., Nicolai, B. *et al.* 2017. A metabolomics approach to elucidate apple fruit responses to static and dynamic controlled atmosphere storage. *Postharvest Biology and Technology* 127, 76–87.

Bron, I.U., Clemente, D.C., Vitti Kluge, R.A., de Arruda, M.C., Jacomino, A.P. and Lima, D.P.P. 2005. Influence of low temperature storage and 1-Methylcyclopropene on the conservation of fresh-cut watercress. *Brazilian Journal of Food Technology* 8, 121–126.

Brooks, C. and Cooley, J.S. 1917. Effect of temperature, aeration and humidity on Jonathan-spot and scald of apples in storage. *Journal of Agricultural Research* 12, 306–307.

Brooks, C., and McColloch, L.P. 1938. *Stickiness and spotting of shelled green lima beans.* Technical Bulletin 625. US Department of Agriculture, Washington, DC.

Brooks, H.J. 1964. Responses of pear seedlings to n-dimethyl-aminosuccinamic acid, a growth retardant. *Nature* 203, 1303.
Broughton, W.J. and Wong, H.C. 1979. Storage conditions and ripening of chiku fruits *Achras sapota*. *Scientia Horticulturae* 10, 377–385.
Broughton, W.J., Hashim, A.W., Shen, T.C. and Tan, I.K.P. 1977. Maturation of Malaysian fruits. I. Storage conditions and ripening of papaya *Carica papaya* L. cv. Sunrise Solo. *Malaysian Agricultural Research and Development Institute Research Bulletin* 5, 59–72.
Brown, W. 1922. On the germination and growth of fungi at various concentrations of oxygen and carbon dioxide. *Annals of Botany* 36, 257–283.
Browne, K.M., Geeson, J.D. and Dennis, C. 1984. The effects of harvest date and CO_2-enriched storage atmospheres on the storage and shelf life of strawberries. *Journal of Horticultural Science* 59, 197–204.
Bubb, M. 1975a. Hypobaric storage. *Annual Report of the East Malling Research Station, UK, for 1974*, pp. 77–78.
Bubb, M. 1975b. Hypobaric storage. *Annual Report of the East Malling Research Station, UK, for 1974*, p. 81.
Bubb, M. 1975c. Hypobaric storage. *Annual Report of the East Malling Research Station, UK, for 1974*, p. 83.
Bubb, M. and Langridge I.W. 1974. Low pressure storage. *Annual Report of the East Malling Research Station, UK, for 1973*, p. 104.
Budu, A.S., Joyce, D.C., Aked, J. and Thompson, A.K. 2001. Respiration of intact and minimally processed pineapple fruit. *Tropical Science* 41, 119–125.
Budu, A.S., Joyce, D.C. and Terry, L.A. 2007. Quality changes in sliced pineapple under controlled atmosphere storage. *Journal of Horticultural Science & Biotechnology* 82, 934–940.
Bufler, G. 2009. Exogenous ethylene inhibits sprout growth in onion bulbs. *Annals of Botany* 103, 23–28.
Bulens, I., Van de Poel, B., Hertog, M.L.A.T.M., De Proft, M.P., Geeraerd, A.H. and Nicolai, B.M. 2012. Influence of harvest time and 1-MCP application on postharvest ripening and ethylene biosynthesis of 'Jonagold' apple. *Postharvest Biology and Technology* 73, 11–19.
Burdon, J. 2009. Dynamic CA storage of avocados: Technology for managing exports? Available at: http://www.avocadosource.com/Journals/AUSNZ/AUSNZ_2009/BurdonJ2009.pdf (accessed July 2016).
Burdon, J. and Lallu, N. 2008. Dynamic controlled atmosphere storage of avocados: technology for managing exports? *New Zealand Avocado Growers' Association Annual Research Report* 8, 123–126.
Burdon, J., Lallu, N., Billing, D., Burmeister, D., Yearsley, C. *et al.* 2005. Carbon dioxide scrubbing systems alter the ripe fruit volatile profiles in controlled-atmosphere stored 'Hayward' kiwifruit. *Postharvest Biology and Technology* 35, 133–141.
Burdon, J., Lallu, N., Haynes, G., McDermott, K. and Billing, D. 2008. The effect of delays in establishment of a static or dynamic controlled atmosphere on the quality of 'Hass' avocado fruit. *Postharvest Biology and Technology* 49, 61–68.
Burdon, J., Lallu, N., Haynes, G., Pidakala, P., McDermott, K. and Billing, D. 2009. Dynamic controlled atmosphere storage of New Zealand grown 'Hass' avocado fruit. *10th International Controlled & Modified Atmosphere Research Conference, 4–7 April 2009*, Turkey [Abstract], 6.
Burdon, J., Lallu, N., Haynes, G., Pidakala, P., Billing, D. and McDermott, K. 2010. Dynamic controlled atmosphere storage of New Zealand-grown 'Hass' avocado fruit. *Acta Horticulturae* 876, 47–54.
Burfoot, D., Smith, D.L.O., Ellis, M.C.B. and Day, W. 1996. Modelling the distribution of isopropyl N-3-chlorophenyl carbamate [CIPC] in box potato stores. *Potato Research* 39, 241–251.
Burg, S.P. 1967. Method for storing fruit. US Patent3.333967 and US patent reissue Re. 28,995 (1976).
Burg, S.P. 1973. Hypobaric storage of cut flowers. *HortScience* 8, 202–205.
Burg, S.P. 1975. Hypobaric storage and transportation of fresh fruits and vegetables. In: Haard, N.F. and Salunkhe, D.K. (eds) *Postharvest Biology and Handling of Fruits and Vegetables*. AVI Publishing, Westpoint, Connecticut, pp. 172–188.
Burg, S.P. 1976. Low temperature hypobaric storage of metabolically active matter. US patents 3,958,028 and 4,061,483.
Burg, S.P. 1990. Theory and practice of hypobaric storage. In: Calderon, M. and Barkai-Golan, R. (eds) *Food Preservation by Modified Atmospheres*. CRC Press, Boca Raton, Florida, pp. 353–372.
Burg, S.P. 1993. Current status of hypobaric storage. *Acta Horticulturae* 326, 259–266.
Burg, S.P. 2004. *Postharvest Physiology and Hypobaric Storage of Fresh Produce*. CAB International, Wallingford, UK.
Burg, S.P. 2010. Experimental errors in hypobaric laboratory research. *Acta Horticulturae* 857, 45–62.
Burg, S.P. 2014. *Hypobaric Storage in Food Industry*, Academic Press, London.
Burg, S.P and Burg, E.A. 1965. Ethylene action and the ripening of fruits. *Science* 148, 1190–1196.
Burg, S.P and Burg, E.A. 1966a. Fruit storage at subatmospheric pressure. *Science* 153, 314–315.

Burg, S.P and Burg, E.A. 1966b. Relationship between ethylene production and ripening of bananas. *Botanical Gazette* 126, 200–204.

Burg, S.P. and Burg, E.A. 1967. Molecular requirements for the biological activity of ethylene. *Plant Physiology* 42, 144–152.

Burg, S.P. and Kosson, R. 1983. Metabolism, heat transfer and water loss under hypobaric conditions. In: Lieberman, M. (ed.) *Postharvest Physiology and Crop Preservation*. Plenum Corp., New York, pp. 399–424.

Burmeister, D.M. and Dilley, D.R. 1994. Correlation of bitter pit on Northern Spy apples with bitter pit-like symptoms induced by Mg^{2+} salt infiltration. *Postharvest Biology and Technology* 4, 301–308.

Burton, K.S. and Twyning, R.V. 1989. Extending mushroom storage life by combining modified atmosphere packaging and cooling. *Acta Horticulturae* 258, 565–571.

Burton, W.G. 1952. Studies on the dormancy and sprouting of potatoes. III. The effect upon sprouting of volatile metabolic products other than carbon dioxide. *New Phytologist* 51, 154–161.

Burton, W.G. 1974a. Some biophysical principles underlying the controlled atmosphere storage of plant material. *Annals of Applied Biology* 78, 149–168.

Burton, W.G. 1974b. The O_2 uptake, in air and in 5 % O_2, and the CO_2 out-put, of stored potato tubers. *Potato Research* 17, 113–137.

Burton, W.G. 1982. *Postharvest Physiology of Food Crops*. Longmans, London and New York.

Burton, W.G. 1989. *The Potato*. 3rd edn. Longmans, London and New York.

Bustos-Griffin, E., Hallman, G.J. and Griffin, R.L. 2015. Phytosanitary irradiation in ports of entry: a practical solution for less developed countries. *International Journal of Food Science & Technology* 50, 249–255.

Butchbaker, A.F., Nelson, D.C. and Shaw, R. 1967. Controlled atmosphere storage of potatoes. *Transactions of the American Society of Agricultural Engineers* 10, 534.

Butler, W., Cook, L. and Vayda, M.E. 1990. Hypoxic stress inhibits multiple aspects of the potato tuber wound responses. *Plant Physiology* 93, 264–270.

Butz, P., Serfert, Y., Fernandez, G.A., Dieterich, S., Lindaeur, R. *et al*. 2004. Influence of high pressure treatment at 25 and 80°C on folates in orange juice and model media. *Journal of Food Science* 69, 117–121.

Calara, E.S. 1969. The effects of varying CO_2 levels in the storage of 'Bungulan' Bananas. BSc thesis, University of the Philippines, Los Baños, Laguna.

Caldwell, J. 1965. Effects of high partial pressures of oxygen on fungi and bacteria. *Nature* 206, 321–323.

Callesen, O. and Holm, B.M. 1989. Storage results with red raspberry. *Acta Horticulturae* 262, 247–254.

Cambridge Refrigeration Technology. 2001. The biology of controlled atmosphere storage. Cambridge Refrigeration Technology, Publication.

Cameron, A.C. 2003. Modified-atmosphere packaging of perishable horticultural commodities can be risky business. *Acta Horticulturae* 600, 305–310.

Cameron, A.C., Boylan-Pett, W. and Lee, J. 1989. Design of modified atmosphere packaging systems: modelling O_2 concentrations within sealed packages of tomato fruits. *Journal of Food Science* 54, 1413–1421.

Campos, R.P., Kwiatkowski, A. and Clemente, E. 2011. Conservação pós-colheita de morangos recobertos com fécula de mandioca e quitosana. *Revista Ceres* 58 (5). Available at: http://dx.doi.org/10.1590/S0034-737X2011000500004 (accessed June 2017).

Cano-Salazar, J., López, M.L. and. Echeverría, G. 2013. Relationships between the instrumental and sensory characteristics of four peach and nectarine cultivars stored under air and CA atmospheres. *Postharvest Biology and Technology* 75, 58–67.

Cantín, C.M., Minas, I.M., Goulas, V., Jiménez, M., Manganaris, G.A., Michailides, T.G. and Crisosto, C.H. 2012. Sulfur dioxide fumigation alone or in combination with CO_2-enriched atmosphere extends the market life of highbush blueberry fruit. *Postharvest Biology and Technology* 67, 84–91.

Cantwell, M.I. 1995. Post-harvest management of fruits and vegetable stems. In: Barbara, G., Inglese, P and Pimienta-Barrios, E. (eds) *Agro-ecology, Cultivation and Uses of Cactus Pear.* FAO Plant Production and Protection Paper 132, pp. 120–136.

Cantwell, M.I., Reid, M.S., Carpenter, A., Nie X. and Kushwaha, L. 1995. Short-term and long-term high CO_2 treatments for insect disinfestation of flowers and leafy vegetables. Harvest and postharvest technologies for fresh fruits and vegetables. *Proceedings of the International Conference, Guanajuato, Mexico, 20–24 February*, 287–292.

Cantwell, M.I., Hong, G., Kang, J. and Nie, X. 2003. Controlled atmospheres retard sprout growth, affect compositional changes, and maintain visual quality attributes of garlic. *Acta Horticulturae* 600, 791–794.

Cao, J., Li, X., Wu, K., Jiang, W. and Qu, G. 2015. Preparation of a novel PdCl2–CuSO4–based ethylene scavenger supported by acidified activated carbon powder and its effects on quality and ethylene metabolism of broccoli during shelf-life. *Postharvest Biology and Technology* 99, 50–57.

Cao, S., Zheng, Y., Wang, K., Rui, H. and Tang, S. 2009a. Effect of 1-methylcyclopropene treatment on chilling injury, fatty acid and cell wall polysaccharide composition in loquat fruit. *Journal of Agricultural and Food Chemistry* 57, 8439–8443.

Cao, S., Zheng, Y., Wang, K., Rui, H. and Tang, S. 2009b. Effects of 1-methylcyclopropene on oxidative damage, phospholipases and chilling injury in loquat fruit. *Journal of the Science of Food and Agriculture* 89, 2214–2220.

Cao, Z.M. 2005. The study on hypobaric storage mechanism and technology in Dongzao jujube fruit. Master's thesis, Tianjin University of Science and Technology. Available at: http://www.dissertationtopic.net/doc/1089396 (accessed October 2013).

Cargofresh. 2009. Available at: www.cargofresh.com (accessed September 2009).

Carpenter, A. 1993. Controlled atmosphere disinfestation of fresh Supersweet sweet corn for export. *Proceedings of the Forty Sixth New Zealand Plant Protection Conference, 10–12 August*, 57–58.

Carrillo-Lopez, A., Ramirez-Bustamante, F., Valdez-Torres, J.B. and Rojas-Villegas, R. 2000. Ripening and quality changes in mango fruit as affected by coating with an edible film. *Journal of Food Quality* 23, 479–486.

Casals, C., Elmer, P.A.G., Viñas, I., Teixidó, N., Sisquella, M. and Usall, J. 2012. The combination of curing with either chitosan or Bacillus subtilis CPA-8 to control brown rot infections caused by Monilinia fructicola. *Postharvest Biology and Technology* 64, 126–132.

Castellanos, D.A., Polanía, W. and Herrera, A.O. 2016. Development of an equilibrium modified atmosphere packaging (EMA packaging) for feijoa fruits and modeling firmness and colour evolution. *Postharvest Biology and Technology* 120, 193–203.

Castro, E. de, Biasi, B., Mitcham, E., Tustin, S., Tanner, D. and Jobling, J. 2007. Carbon dioxide-induced flesh browning in Pink Lady apples. *Journal of the American Society for Horticultural Science* 132, 713–719.

Cayuela, J.A., Vázquez, A., Pérez, A.G. and García, J.M. 2009. Control of table grapes postharvest decay by ozone treatment and resveratrol induction. *Food Science and Technology International* 15, 495–502.

CBHPC. 1975. *Annotated Bibliography on Low Pressure (Hypobaric) Storage, 1968–1973.* 6326. Commonwealth Bureau of Horticulture and Plantation Crops, East Malling, UK.

Cefola, M., Amodio, M.L. and Colelli, G. 2016. Design of the correct modified atmosphere packaging for fresh-cut broccoli raab. *Acta Horticulturae* 1141, 117–122.

Celikel, F.G., Ozelkok, S. and Burak, M. 2003. A study on modified atmosphere storage of sweet cherry. *Acta Horticulturae* 628, 431–438.

Cen, H., Lu, R., Zhu, Q. and Mendoza, F., 2016. Non-destructive detection of chilling injury in cucumber fruit using hyperspectral imaging with feature selection and supervised classification. *Postharvest Biology and Technology* 111, 352–61.

Cenci, S.A., Soares, A.G., Bilbino, J.M.S. and Souza, M.L.M. 1997. Study of the storage of Sunrise Solo papaya fruits under controlled atmosphere. *Seventh International Controlled Atmosphere Research Conference, July 13–18 1997*, University of California, Davis [Abstract], 112.

Ceponis, M.J. and Cappellini, R.A. 1983. Control of postharvest decays of blueberries by carbon dioxide-enriched atmospheres. *Plant Disease* 67, 169–171.

Cetiner, A.I. 2009. Use of IPN (fire-resistant foam) sandwich panels in construction of CA storages. *International Controlled & Modified Atmosphere Research Conference, 4–7 April 2009*, Turkey [Poster abstract], 54.

Cetinkaya, N., Ercin, D., Özvatan, S. and Erel, Y. 2016. Quantification of applied dose in irradiated citrus fruits by DNA comet assay together with image analysis. *Food Chemistry* 192, 370–373.

Chairat, R. and Kader A. 1999. Evaluation of ethylene absorption capacity of 'Profresh'-based films. *Perishables Handling Quarterly* 97, 27.

Chamara, D., Illeperuma, K. and Galappatty, P.T. 2000. Effect of modified atmosphere and ethylene absorbers on extension of storage life of 'Kolikuttu' banana at ambient temperature. *Fruits (Paris)* 55, 381–388.

Chambroy, V., Souty, M., Reich, M., Breuils, L., Jacquemin, G. and Audergon, J.M. 1991. Effects of different CO_2 treatments on post harvest changes of apricot fruit. *Acta Horticulturae* 293, 675–684.

Champion, V. 1986. Atmosphere control – an air of the future. *Cargo System*, November issue, 28–33.

Chang, Y.P. 2001. Study on the physiological–biochemical changes and storage effects of jujube fruits under hypobaric (low pressure) condition. Master's thesis, Shanxi Agricultural University. Available at: http://www.dissertationtopic.net/doc/813061 (accessed October 2013).

Chapon, J.F. and Trillot, M. 1992. Pomme. L'entreposage longue duree en Italie du Nord. *Infos Paris* 78, 42–46.

Charles, F., Rugani, N. and Gontard, N. 2005. Influence of packaging conditions on natural microbial population growth of endive. *Journal of Food Protection* 68, 1020–1025.

Charm, S.E., Longmaid, H.E. and Carver, J. 1977. Simple system for extending refrigerated, nonfrozen preservation of biological-material using pressure. *Cryobiology* 14, 625–636.

Charoenpong, C. and Peng, A.C. 1990. Changes in beta-carotene and lipid composition of sweetpotatoes during storage. *Ohio Agricultural Research and Development Center, Special Circular* 121, 15–20.

Chase, W.G. 1969. Controlled atmosphere storage of Florida citrus. *Proceedings of the First International Citrus Symposium* 3, 1365–1373.

Chau, K.F. and Alvarez, A.M. 1983. Effects of LP storage on *Collectotrichum gloeosporioides* and postharvest infection of papaya. *HortScience* 18, 953–955.

Chauhan, O.P., Raju, P.S., Dasgupta, D.K. and Bawa, A.S. 2006. Modified atmosphere packaging of banana (cv. Pachbale) with ethylene, carbon di-oxide and moisture scrubbers and effect on its ripening behaviour. *American Journal of Food Technology* 1, 179–189.

Chávez-Franco, S.H., Saucedo-Veloz, C., Peña-Valdivia, C.B, Corrales, J.J.E. and Valle-Guadarrama, S. 2004. Aerobic–anaerobic metabolic transition in 'Hass' avocado fruits. *Food Science and Technology International* 10, 391–398.

Chávez-Murillo, C.E., Espinosa-Solís, V., Aparicio-Saguilán, A., Salgado-Delgado, R., Tirado-Gallegos, J.M. and Zamudio-Flores, P.B. 2015. Use of zein and ethylcellulose as biodegradable film on evaluation of post-harvest changes in tomato (*Lycopersicum esculentum*). *Journal of Microbiology, Biotechnology and Food Sciences* 4, 365–368.

Chawan, T. and Pflug, I.J. 1968. Controlled atmosphere storage of onion. *Michigan Agricultural Station Quarterly Bulleti*, 50, 449–475.

Cheah, L.H., Irving, D.E., Hunt, A.W. and Popay, A.J. 1994. Effect of high CO_2 and temperature on Botrytis storage rot and quality of kiwifruit. *Proceedings of the Forty Seventh New Zealand Plant Protection Conference, Waitangi Hotel, New Zealand, 9–11 August, 1994*, 299–303.

Cheema, M.U.A., Rees, D., Colgan, R.J., Taylor, M. and Westby, A. 2013. The effects of ethylene, 1-MCP and AVG on sprouting in sweet potato roots. *Postharvest Biology and Technology* 85, 89–93.

Chen, H., Yang, H., Gao, H., Long, J., Tao, F., Fang, X. and Jiang, Y. 2013a. Effect of hypobaric storage on quality, antioxidant enzyme and antioxidant capability of the Chinese bayberry fruits. *Chemistry Central Journal* 7, 4.

Chen, H., Ling, J., Wu, F., Zhang, L., Sun, Z. and Yang, H. 2013b. Effect of hypobaric storage on flesh lignification, active oxygen metabolism and related enzyme activities in bamboo shoots. *LWT - Food Science and Technology* 51, 190–195.

Chen, N. and Paull, R.E. 1986. Development and prevention of chilling injury in papaya fruits. *Journal of the American Society of Horticultural Science* 111, 639.

Chen, P.M. and Varga, D.M. 1997. Determination of optimum controlled atmosphere regimes for the control of physiological disorders of 'D'Anjou' pears after short-term, mid-term and long-term storage. *Seventh International Controlled Atmosphere Research Conference, July 13–18 1997*, University of California, Davis, California [Abstract], 9.

Chen, P.M., Mellenthin, W.M., Kelly, S.B. and Facteau, T.J. 1981. Effects of low oxygen and temperature on quality retention of 'Bing' cherries during prolonged storage. *Journal of the American Society for Horticultural Science* 105, 533–535.

Chen, Z.J., White, M.S., and Robinson, W.H. 2005. Low-pressure vacuum to control larvae of Hylotrupes bajulus (Coleoptera: Cerambycidae). In: Lee, C.Y. and Robinson, W.H. (eds) *Proceedings of the Fifth International Conference on Urban Pests*, Malaysia.

Chervin, C., Kreidl, S.L., Franz, P.R., Hamilton, A.J., Whitmore, S.R. *et al.* 1999. Evaluation of a non-chemical disinfestation treatment on quality of pome fruit and mortality of lepidopterous pests. *Australian Journal of Experimental Agriculture* 39, 335–344.

Chervin, C., Westercamp, P., El-Kereamy, A., Rache, P., Tournaire, A. *et al.* 2003. Ethanol vapours to complement or suppress sulfite fumigation of table grapes. *Acta Horticulturae* 628, 779–784.

Chiabrando, V. and Peano, G.G.C. 2006. Effect of storage methods on postharvest quality of highbush blueberry. *Italus Hortus* 13, 114–117.

Cho, S.S. and Ha, T.M. 1998. The effect of supplementary package materials for keeping freshness of fresh mushroom at ambient temperature. *RDA Journal of Industrial Crop Science* 40, 52–57.

Choi, Y.H., Ko, S.U, Kim, S.H., Kim, Y.H., Kang, S.K. and Lee, C.H. 2002. Influence of modified atmosphere packaging on fruit quality of 'Tsunokaori' tangor during cold storage. *Korean Journal of Horticultural Science & Technology* 20, 340–344.

Chonhenchob, V., Chantarasomboon, Y. and Singh, S.P. 2007. Quality changes of treated fresh-cut tropical fruits in rigid modified atmosphere packaging containers. *Packaging Technology and Science* 20, 27–37.

Chope, G.A., Terry, L.A. and White, P.J. 2007. The effect of 1-methylcyclopropene (1-MCP) on the physical and biochemical characteristics of onion cv. SS1 bulbs during storage. *Postharvest Biology and Technology* 44, 131–140.

Choudhury, J.K., 1939. Researches on plant respiration. V. On the respiration of some storage organs in different oxygen concentrations. *Proceedings of the Royal Society, London Series B* 127, 238–257.

Christakou, E.C., Arvanitoyannis, I.S., Khah, E.M. and Bletsos, F. 2005. Effect of grafting and modified atmosphere packaging (MAP) on melon quality parameters during storage. *Journal of Food, Agriculture & Environment* 3, 145–152.

Chu, C.L. 1992. Postharvest control of San Jose scale on apples by controlled atmosphere storage. *Postharvest Biology and Technology* 1, 361–369.

Chu, C.L., Liu, W.T., Zhou, T. and Tsao, R. 1999. Control of postharvest gray mold rot of modified atmosphere packaged sweet cherries by fumigation with thymol and acetic acid. *Canadian Journal of Plant Science* 79, 685–689.

Chung, D.S. and Son, Y.K. 1994. Studies on CA storage of persimmon *Diospyros kaki* T. and plum *Prunus salicina* L. *Rural Development Administration Journal of Agricultural Science, Farm Management, Agricultural Engineering, Sericulture, and Farm Products Utilization* 36, 692–698.

Church, I.J. and Parsons, A.L. 1995. Modified atmosphere packaging technology: a review. *Journal of the Science of Food and Agriculture* 67, 143–152.

Church, N. 1994. Developments in modified-atmosphere packaging and related technologies. *Trends in Food Science and Technology* 5, 345–352.

Ciccarese, A., Stellacci, A.M., Gentilesco, G. and Rubino, P. 2013. Effectiveness of pre- and post-veraison calcium applications to control decay and maintain table grape fruit quality during storage. *Postharvest Biology and Technology* 75, 135–141.

Clarke, B. 1996. Packhouse operations for fruit and vegetables. In: Thompson, A.K. *Posthavest Technology of Fruits and Vegetables*. Blackwell Science, Oxford, pp. 189–217.

Claypool, L.L. and Allen, F.W., 1951. The influence of temperature and oxygen level on the respiration and ripening of Wickson plums. *Hilgardia* 121, 29–160.

Clendennen, S.K. and May, G.D. 1997. Differential gene expression in ripening banana fruit. *Plant Physiology* 115, 463–469.

Coates, L., Cooke, A., Persley, D., Beattie, B., Wade, N. and Ridgeway, R. 1995. *Postharvest Diseases of Horticultural Produce, Volume 2: Tropical Fruit*. Queensland Department of Primary Industries, Brisbane.

Cogo, S.L.P., Chaves, F.C., Schirmer, M.A., Zambiazi, R.C., Nora, L., Silva, J.A. and Rombaldi, C.V. 2012. Low soil water content during growth contributes to preservation of green colour and bioactive compounds of cold-stored broccoli (*Brassica oleraceae* L.) florets. *Postharvest Biology and Technology* 60, 158–163.

Colelli, G. and Martelli, S. 1995. Beneficial effects on the application of CO_2-enriched atmospheres on fresh strawberries *Fragaria* X *ananassa* Duch. *Advances in Horticultural Science* 9, 55–60.

Colelli, G., Mitchell, F.G. and Kader, A.A. 1991. Extension of postharvest life of 'Mission' figs by CO_2 enriched atmospheres. *HortScience* 26, 1193–1195.

Colgan, R.J., Dover, C.J., Johnson, D.S. and Pearson, K. 1999. Delayed CA and oxygen at 1% or less to control superficial scald without CO_2 injury on Bramley's Seedling apples. *Postharvest Biology and Technology* 16, 223–231.

Collazo, C., Abadia, M., Colás-Medà, P., Iglesias, M.B., Granado-Serrano, A.B. *et al.* 2017. Effect of *Pseudomonas graminis* strain CPA-7 on the ability of *Listeria monocytogenes* and *Salmonella enterica* subsp. *enterica* to colonize Caco-2 cells after pre-incubation on fresh-cut pear. *International Journal of Food Microbiology* 262, 55–62.

Conesa, A., Verlinden, B.E., Artés-Hernández, F., Nicolaï, B. and Artés, F. 2007. Respiration rates of fresh-cut bell peppers under supertamospheric and low oxygen with or without high carbon dioxide. *Postharvest Biology and Technology* 45, 81–88.

Cooper, T., Retamales, J. and Streif, J. 1992. Occurrence of physiological disorders in nectarine and possibilities for their control. *Erwerbsobstbau* 34, 225–228.

Coquinot, J.P. and Richard, L. 1991. Methods of controlling scald in the apple Granny Smith without chemicals. *Neuvieme colloque sur les recherches fruitieres, 'La maitrise de la qualite des fruits frais', Avignon, 4–6 Decembre 1990*, 373–380.

Corbo, M.R., Lanciotti, R., Gardini, F., Sinigaglia, M. and Guerzoni, M.E. 2000. Effects of hexanal, trans-2-hexenal, and storage temperature on shelf life of fresh sliced apples. *Journal of Agricultural Food Chemistry* 48, 2401–2408.

Cordeiro, N., Sousa, L., Freitas, N. and Gouveia, M. 2013. Changes in the mesocarp of *Annona cherimola* Mill. 'Madeira' during postharvest ripening. *Postharvest Biology and Technology* 85, 179–184.

Corey, K.A., Bates, M.E., Adams, S.L. and MacElroy, R.D. 1996. Carbon dioxide exchange of lettuce plants under hypobaric conditions. *Advances in Space Research* 18, 301–308.

Corrales-Garcia, J. 1997. Physiological and biochemical responses of 'Hass' avocado fruits to cold-storage in controlled atmospheres. *Seventh International Controlled Atmosphere Research Conference, July 13–18 1997*, University of California, Davis, California [Abstract], 50.

Costa, M.A.C., Brecht, J.K., Sargent, S.A. and Huber, D.J. 1994. Tolerance of snap beans to elevated CO_2 levels. 107th Annual meeting of the Florida State Horticultural Society, Orlando, Florida, USA, 30 October–1 November 1994. *Proceedings of the Florida State Horticultural Society* 107, 271–273.

Couey, H.M. and Wells, J.M. 1970. Low oxygen and high carbon dioxide atmospheres to control postharvest decay of strawberries. *Phytopathology* 60, 47–49.

Couey, H.M., Follstad, M.N. and Uota, M. 1966. Low oxygen atmospheres for control of post-harvest decay of fresh strawberries. *Phytopathology* 56, 1339–1341.

Crank, J. 1975. *The Mathematics of Diffusion*, 2nd edn. Clarendon Press, Oxford.

Creech, D.L., Workman, M. and Harrison, M.D. 1973. The influence of storage factors on endogenous ethylene production by potato tubers. *American Potato Journal* 50, 145–150.

Cripps, J.E.L., Richards, L.A. and Mairata, A.M. 1993. 'Pink Lady' apple. *HortScience* 28, 1057.

Crisosto, C.H. 1997. Ethylene safety. *Central Valley Postharvest Newsletter Cooperative Extension University of California* 6, 4.

Crisosto, C.H., Garner, D. and Crisosto, G. 2003a. Developing optimum controlled atmosphere conditions for 'Redglobe' table grapes. *Acta Horticulturae*, 600, 803–808.

Crisosto, C.H., Garner, D. and Crisosto, G. 2003b. Developing optimal controlled atmosphere conditions for 'Thompson Seedless' table grapes. *Acta Horticulturae*, 600, 817–821.

CSIRO. 1998. *Fruit & Vegetables Storage and Transport Database V1*. CSIRO/Sydney Postharvest Laboratory.

Cui, Y. 2008. Effects of hypobaric conditions on physiological and biochemical changes of Lizao jujube. *Journal Anhui Agricultural Science* 36, 12900–12901.

Czynczyk, A. and Bielicki, P. 2002. Ten-year results of growing the apple cultivar 'Ligol' in Poland. *Sodininkyste ir Darzininkyste* 21, 12–21.

Dalrymple, D.G. 1967. The development of controlled atmosphere storage of fruit. Division of Marketing and Utilization Sciences, Federal Extension Service, US Department of Agriculture, Washington, DC.

Damen, P. 1985. Verlengen afzetperiode vollegrondsgroenten. *Groenten en Fruit* 40, 82–83.

Dang, K.T.H., Singh, Z. and Swinny, E.E. 2008. Edible coatings influence fruit ripening, quality and aroma biosynthesis in mango fruit. *Journal of Agriculture and Food Chemistry* 56, 1361–1370.

Daniels, J.A., Krishnamurthi, R. and Rizvi, S.S. 1985. A review of effects of CO_2 on microbial growth and food quality. *Journal of Food Protection* 48, 532–537.

Daniels-Lake, B.J. 2012. Effects of elevated CO_2 and trace ethylene present throughout the storage season on the processing colour of stored potatoes. *Potato Research* 55, 157.

Darko, J. 1984. An investigation of methods for evaluating the potential storage life of perishable food crops. PhD thesis, Cranfield Institute of Technology.

Davenport, T.L., Burg, S.P. and White, T.L. 2006. Optimal low pressure conditions for long-term storage of fresh commodities kill Caribbean fruit fly eggs and larvae. *HortTechnology* 16, 98–104.

Davies, D.H., Elson, C.M. and Hayes, E.R. 1988. N,O carboxymethyl chitosan, a new water soluble chitin derivative. *Fourth International Conference on Chitin and Chitosan, 22–24 August 1988*, Trondheim, Norway, 6.

Day, B.P.F. 1994. Modified atmosphere packaging and active packaging of fruits and vegetables. Minimal processing of foods, 14–15 Apr. 1994, Kirkkonummi, Finland. *VTT-Symposium* 142, 173–207.

Day, B.P.F. 1996. High O_2 modified atmosphere packaging for fresh prepared produce. *Postharvest News and Information* 7, 31N–34N.

Day, B.P.F. 2003. Novel MAP applications for fresh prepared produce. In: Ahvenainen, R. (ed.) *Novel Food Packaging Techniques*. Woodhead Publishing, Oxford.

Day, B.P.F. 2008. Active packaging of food. In: Kerry, J. and Butler, P. (eds) *Smart Packaging Technologies for Fast Moving Consumer Goods*. John Wiley, Hoboken, New Jersey, pp. 1–18.

Day, B.P.F., Bankier, W.J. and González, M.I., 1998. *Novel modified atmosphere packaging (MAP) for fresh prepared produce*. Research Summary Sheet 13. Campden and Chorleywood Food Research Association, Chipping Campden, UK.

De Freitas, S.T., do Amarante, C.V.T., Labavitch, J.M. and Mitcham, E.J. 2010. Cellular approach to understand bitter pit development in apple fruit. *Postharvest Biology and Technology* 57, 6–13.

De la Plaza, J.L. 1980. Controlled atmosphere storage of Cherimoya. *Proceedings of the International Congress on Refrigeration* 3, 701.

De la Plaza, J.L., Muñoz Delgado, L. and Inglesias, C. 1979. Controlled atmosphere storage of Cherimoya. *Bulletin Instiute International de Friod* 59, 1154.

De Martino, G., Mencarelli, F. and Golding, J.B. 2007. Preliminary investigation into the uneven ripening of banana (*Musa* sp.) peel. *New Zealand Journal of Crop and Horticultural Science* 35, 193–199.

De Reuck, K., Sivakumar, D. and Korsten, L. 2009. Integrated application of 1-methylcyclopropene and modified atmosphere packaging to improve quality retention of litchi cultivars during storage. *Postharvest Biology and Technology* 52, 71–77.

De Ruiter, M. 1991. Effect of gases on the ripening of bananas packed in banavac. MSc thesis, Cranfield Institute of Technology.

De Santana, L.R.R., Benedetti, B.C., Sigrist, J.M.M, Sato, H.H. and Sarantópoulos, C.I.G.L. 2009. Modified atmosphere packages and cold storage to maintain quality of 'Douradão' peaches. *10th International Controlled & Modified Atmosphere Research Conference, 4–7 April 2009*, Turkey [Abstract], 19.

De Wild, H. 2001. 1-MCP can make a big breakthrough for storage. 1-MCP kan voor grote doorbraak in bewaring zorgen. *Fruitteelt Den Haag* 91, 12–13.

De Wild, H. and Roelofs, F. 1992. Plums can be stored for 3 weeks. Pruimen zijn drie weken te bewaren. *Fruitteelt Den Haag* 82, 20–21.

DeEll, J.R. 2002. Modified atmospheres for berry crops. *Ohio State University Extension Newsletter* 6, no. 27.

DeEll, J.R. 2012. Controlled Atmosphere Storage Guidelines and Recommendations for Apples. Available at: http://www.omafra.gov.on.ca/english/crops/facts/12-045.htm (accessed December 2017).

DeEll, J.R. and Ehsani-Moghaddam, B. 2012. Delayed controlled atmosphere storage affects storage disorders of 'Empire' apples. *Postharvest Biology and Technology* 67, 167–171.

DeEll, J.R. and Lum, G.B. 2017. Effects of low oxygen and 1-methylcyclopropene on storage disorders in 'Empire' apples. *HortScience* 52, 1265–1270.

DeEll, J.R. and Murr, D.P. 2009. Fresh market quality program ca storage guidelines and recommendations for apples. Available at: http://www.omafra.gov.on.ca/english/crops/facts/03-073.htm (accessed January 2010).

DeEll, J.R. and Prange, R.K. 1998. Disorders in 'Cortland' apple fruit are induced by storage at 0°C in controlled atmosphere. HortScience 33, 121–122.

DeEll, J.R., Prange, R.K. and Murr, D.P. 1995. Chlorophyll fluorescence as a potential indicator of controlled-atmosphere disorders in 'Marshall' McIntosh apples. *HortScience* 30, 1084–1085.

DeEll, J.R., van Kooten, O., Prange, R.K. and Murr, D.P. 1999. Applications of chlorophyll fluorescence techniques in postharvest physiology. *Horticultural Reviews* 23, 69–107.

DeEll, J.R., Murr D.P., Wiley L. and Mueller, R. 2005. Interactions of 1-MCP and low oxygen CA storage on apple quality. *Acta Horticulturae* 682, 941–948.

DeEll, J.R., Toivonen, P.M.A., Cornut, F., Roger, C. and Vigneault, C. 2006. Addition of sorbitol with $KMnO_4$ improves broccoli quality retention in modified atmosphere packages. *Journal of Food Quality* 29, 65–75.

DeEll, J.R., Lum, G.B. and Ehsani-Moghaddam, B. 2016. Elevated carbon dioxide in storage rooms prior to establishment of controlled atmosphere affects apple fruit quality. *Postharvest Biology and Technology* 118, 11–16.

Defra. 2017. DEFRA recommendation for storage of apples. Department for Environment, Food and Rural Affairs, London.

Defra AHDB. 2017. Apple Best Practice Guide. Optimum storage conditions. Agriculture and Horticulture Development Board, Department for Environment, Food and Rural Affairs, London. Available at: http://apples.ahdb.org.uk/post-harvest-section9.asp (accessed December 2017).

Defraeye, T., Nicolai, B., Kirkman, W., Moore, S., van Niekerk, S., Verboven, P. and Cronjé, P. 2016. Integral performance evaluation of the fresh-produce cold chain: A case study for ambient loading of citrus in refrigerated containers. *Postharvest Biology and Technology* 112, 1–13.

Del Nobile, M.A., Licciardello, F., Scrocco, C., Muratore, G. and Zappa, M. 2007. Design of plastic packages for minimally processed fruits. *Journal of Food Engineering* 79, 217–224.

Delate, K.M. and Brecht, J.K. 1989. Quality of tropical sweetpotatoes exposed to controlled-atmosphere treatments for postharvest insect control. *Journal of the American Society for Horticultural Science* 114, 963–968.

Delate, K.M., Brecht, J.K. and Coffelt, J.A. 1990. Controlled atmosphere treatments for control of sweetpotato weevil *Coleoptera: Curculionidae*, in stored tropical sweetpotatoes. *Journal of Economic Entomology* 82, 461–465.

DeLong, J.M., Prange, R.K., Bishop, C., Harrison, P.A. and Ryan, D.A.J. 2003. The influence of 1-MCP on shelf-life quality of highbush blueberry. *HortScience* 38, 417–418.

DeLong, J.M., Prange, R.K., Leyte, J.C. and Harrison, P.A. 2004. A new technology that determines low-oxygen thresholds in controlled-atmosphere-stored apples. *HortTechnology* 14, 262–266.

DeLong, J.M., Prange, R.K. and Harrison, P.A. 2007. Chlorophyll fluorescence-based low-O_2 CA storage of organic 'Cortland' and 'Delicious' apples. *Acta Horticulturae* 737, 31–37.

DeLong, J.M., Harrison, P.A. and L. Harkness. 2016. Determination of optimal harvest boundaries for 'Ambrosia' apple fruit using a delta-absorbance meter. *Journal of Horticultural Science and Biotechnology* 91, 243–249.

Del-Valle, V., Hernandez-Muñoz, P., Catala, R. and Gavara, R. 2009. Optimization of an equilibrium modified atmosphere packaging (EMAP) for minimally processed mandarin segments. *Journal of Food Engineering* 91, 474–481.

Deng, W.M., Fan, L.H., Song, J., Mir, N., Verschoor, J. and Beaudry, R.M. 1997. MAP of apple fruit: effect of cultivar, storage duration, and carbon dioxide on the lower oxygen limit. *Postharvest Horticulture Series - Department of Pomology, University of California* 16, 156–161.

Deng, Y., Wu, Y. and Li, Y. 2005. Effects of high O_2 levels on post-harvest quality and shelf life of table grapes during long-term storage. *European Food Research and Technology* 221, 392–397.

Deng, Y., Wu, Y. and Li, Y. 2006. Physiological responses and quality attributes of 'Kyoho' grapes to controlled atmosphere storage. *Lebensmittelwissenschaft und Technologie* 39, 584–590.

Deng, Z., Jung, J., Simonsen, J. and Zhao, Y. 2017. Cellulose nanomaterials emulsion coatings for controlling physiological activity, modifying surface morphology, and enhancing storability of postharvest bananas (*Musa acuminate*). *Food Chemistry* 232, 359–368.

Dennis, C., Browne, K.M. and Adamicki, F. 1979. Controlled atmosphere storage of tomatoes. *Acta Horticulturae* 93, 75–83.

Denny, F.E. and Thornton, N.C. 1941. CO_2 prevents the rapid increase in the reducing sugar content of potato tubers stored at low temperatures. *Contributions of the Boyce Thompson Institute* 12, 79–84.

Denny, F.E., Thornton, N.C. and Schroeder, E.M. 1944. The effect of carbon dioxide upon the changes in the sugar content of certain vegetables in cold storage. *Contributions of the Boyce Thompson Institute* 13, 295–311.

Deol, I.S. and Bhullar, S.S. 1972. Effects of wrappers and growth regulators on the storage life of mango fruits. *Punjab Horticultural Journal* 12, 114.

Deschene, A., Paliyath, G., Lougheed, E.C., Dumbroff, E.B. and Thompson, J.E. 1991. Membrane deterioration during postharvest senescence of broccoli florets: modulation by temperature and controlled atmosphere storage. *Postharvest Biology and Technology* 1, 19–31.

Dessalegn, Y., Ayalew, A. and Woldetsadik, K. 2013. Integrating plant defence inducing chemical, inorganic salt and hot water treatments for the management of postharvest mango anthracnose. *Postharvest Biology and Technology* 85, 83–88.

Deuchande, T., Carvalho, S.M.P., Giné-Bordonaba, J., Vasconcelos, M.W. and Larrigaudière, C. 2017. Transcriptional and biochemical regulation of internal browning disorder in 'Rocha' pear as affected by O_2 and CO_2 concentrations. *Postharvest Biology and Technology* 132, 15–22.

DGCL. 2004. Dalian Refrigeration Company Limited, Dalian, China. Available at: http://en.daleng.cn/project/index.jsp?catid=148 (accessed March 2009).

Díaz-Mula, H.M., Serrano, M. and Valero, D. 2012. Alginate coatings preserve fruit quality and bioactive compounds during storage of sweet cherry fruit. *Food Bioprocess Technology* 5, 2990.

Díaz-Mula, H.M., Marin, A., Jordán, M.J. and Gil, M.I. 2017. Off-odor compounds responsible for quality loss of minimally processed baby spinach stored under MA of low O_2 and high CO_2 using GC–MS and olfactometry techniques. *Postharvest Biology and Technology* 129, 129–135.

Dick, E. and Marcellin, P. 1985. Effect of high temperatures on banana development after harvest. Prophylactic tests. *Fruits* 40, 781–784.

Digges, P. 1995. *NRI report describing the preliminary findings of the socio-economic assessment phase of the low-cost cassava fresh root storage technology*. 27 March to 12 April 1995. Natural Resources Institute, University of Greenwich, London.

Dijkink, B.H., Tomassen, M.M., Willemsen, J.H.A. and van Doorn, W.G. 2004. Humidity control during bell pepper storage, using a hollow fiber membrane contactor system. *Postharvest Biology and Technology* 32, 311–320.

Dilley, D.R. 1972. Hypobaric storage – a new concept for preservation of perishables. *Proceedings of the Michigan State Horticultural Society*, pp. 82–89.

Dilley, D.R. 1977. Application of the hypobaric system for storage and transportation of perishable agricultural commodities. *2nd Annual World's Fair for Technology Exchange, 7–11 February 1977*, pp. 135–149.

Dilley, D.R. 1990. Historical aspects and perspectives of controlled atmosphere storage. In: Calderon, M. and Barkai-Golan, R. (eds) *Food Preservation by Modified Atmospheres*. CRC Press, Boca Raton (Florida), Ann Arbor (Michigan), Boston (Massachusetts), 187–196.

Dilley, D.R. 2006. Development of controlled atmosphere storage technologies. *Stewart Postharvest Reviews* 6, 1–8.

Dilley, D.R. and Carpenter, W.J. 1975. Principles and application of hypobaric storage of cut flowers. *Acta Horticulturae* 41, 249–267.

Dilley, D.R., Irwin, P.L. and McKee, M.W. 1982. Low oxygen, hypobaric storage and ethylene scrubbing. In: Richardson, D.G. and Meheriuk, M. (eds.) *Controlled Atmosphere Storage and Transport of Perishable Agricultural Commodities*. Timber Press, Beaverton, Oregon, pp. 317–329.

Dilley, D.R., Lange, E. and Tomala, K. 1989. Optimizing parameters for controlled atmosphere storage of apples. In: Fellman, J.K. (ed.) *Proceedings of the Fifth International Controlled Atmosphere Research Conference, Wenatchee, Washington, USA, 14–16 June 1989. Vol. 1: Pome fruit.* Washington State University, Pullman, Washington, pp. 221–226.

Ding, C.-K., Wang, C.Y., Gross, K.C. and Smith, D.L. 2002. Jasmonate and salicylate induce the expression of pathogenesis-related-protein genes and increase resistance to chilling injury in tomato fruit. *Planta*, 214, 895–901.

Ding, Y., Sheng, J., Li, S., Nie, Y., Zhao, J. *et al*. 2015. The role of gibberellins in the mitigation of chilling injury in cherry tomato (*Solanum lycopersicum* L.) fruit. *Postharvest Biology and Technology* 101, 88–95.

Ding, Z., Tian, S., Wang, Y., Li, B., Chang, Z., Han, J. and Xu, Y. 2006. Physiological response of loquat fruit to different storage conditions and its storability. *Postharvest Biology and Technology* 41, 143–150.

Dingman, D.W. 2000. Growth of *Escherichia coli* O157:H7 in bruised apple (*Malus domestica*) tissue as influenced by cultivar, date of harvest, and source. *Applied and Environmental Microbiology* 66, 1077–1083.

Dirim, S.N., Esin, A. and Bayindirli, A. 2003. New protective polyethylene based film containing zeolites for the packaging of fruits and vegetables: film preparation. *Turkish Journal of Engineering and Environmental Science* 27, 1–9.

Doäÿan, A. and Erkan, M. 2014 A new storage technology: Palistore (Palliflex) storage system for horticultural crops. Available at: http://agris.fao.org/agris-search/search.do;jsessionid=330F06D27CA99DC3A39A0773E3858AF3?request_locale=ar&recordID=TR2016000060&query=&sourceQuery=&sortField=&sortOrder=&agrovocString=&advQuery=¢erString=&enableField (accessed 29 March 2017).

Dohring, S. 1997. Over sea and over land putting CA research and technology to work for international shipments of fresh produce. *Seventh International Controlled Atmosphere Research Conference, July 13–18 1997*, University of California, Davis, California [Abstract], 23.

Dolt, K.S., Karar, J., Mishra, M.K., Salim, J., Kumar, R., Grover, S.K. and Qadar Pasha, M.A. 2007. Transcriptional down regulation of sterol metabolism genes in murine liver exposed to acute hypobaric hypoxia. *Biochemical and Biophysical Research Communications* 354, 148–153.

Dong, H.Q., Jiang, Y.M., Wang, Y.H., Huang, J.B., Lin, L.C. and Ning, Z.X. 2005. Effects of hot water treatments on chilling tolerance of harvested bitter melon. *Transactions of the Chinese Society of Agricultural Engineering* 21, 186–188.

Dong, J., Yu, Q., Lu, L. and Xu, M. 2012. Effect of yeast saccharide treatment on nitric oxide accumulation and chilling injury in cucumber fruit during cold storage. *Postharvest Biology and Technology*, 68, 1–7.

Dong, T., Feng, Y., Shi, J., Cantwell, M.I., Guo, Y. and Wang, Q. 2016. Ethanol fumigation can effectively inhibit the browning of fresh-cut burdock (*Arctium lappa* L.). *Acta Horticulturae* 1141, 343–348.

Dong, X., Huber, D.J., Rao, J. and Lee, J.H. 2013. Rapid ingress of gaseous 1-MCP and acute suppression of ripening following short-term application to mid-climacteric tomato under hypobaria. *Postharvest Biology and Technology*, 86, 285–290.

Dori, S., Burdon, J.N., Lomaniec, E. and Pesis, E. 1995. Effect of anaerobiosis on aspects of avocado fruit ripening. *Acta Horticulturae* 379, 129–136.

Dos Prazeres, J.N. and de Bode, N. 2009. Successful shipments of tropical fruits (papaya, lime, mango and banana) under controlled atmosphere conditions for long distance. *10th International Controlled & Modified Atmosphere Research Conference, 4–7 April 2009*, Turkey [Abstract], 10.

dos Santos, I.D., Pizzutti, I.R., Dias, J.V., Fontana, M.E.Z., Brackmann, A., Anese, R.O., Thewes, F.R., Marques, L.N. and Cardoso, C.D. 2018. Patulin accumulation in apples under dynamic controlled atmosphere storage. *Food Chemistry* 255, 275–281.

Dourtoglou, V.G., Yannovits, N.G., Tychopoulos, V.G. and Vamvakias, M.M. 1994. Effect of storage under CO_2 atmosphere on the volatile, amino acid, and pigment constituents in red grape *Vitis vinifera* L. var. Agiogitiko. *Journal of Agricultural and Food Chemistry* 42, 338–344.

Drake, S.R. 1993. Short-term controlled atmosphere storage improved quality of several apple cultivars. *Journal of the American Society for Horticultural Science* 118, 486–489.

Drake, S.R. and Eisele, T.A. 1994. Influence of harvest date and controlled atmosphere storage delay on the color and quality of 'Delicious' apples stored in a purge-type controlled-atmosphere environment. *HortTechnology* 4, 260–263.

Drake, S.R. and Spayd, S.E. 1983. Influence of calcium treatment on 'Golden Delicious' apple quality. *Journal of Food Science* 48, 403–405.

Drake, S.R., Elfving, D.C., Drake, M.A., Eisele, T.A., Drake, S.L. and Visser, D.B. 2006. Effects of aminoethoxyvinylglycine, ethephon, and 1-methylcyclopropene on apple fruit quality at harvest and after storage. *HortTechnology* 16, 16–23.

Duan, J., Wu, R., Strik, B.C. and Zhao, Y. 2011. Effect of edible coatings on the quality of fresh blueberries (Duke and Elliott) under commercial storage conditions. *Postharvest Biology and Technology* 59, 71–79.

Dubodel, N.P. and Tikhomirova, N.T. 1985. Controlled atmosphere storage of mandarins. *Sadovodstvo* 6, 18.

Dubodel, N.P., Panyushkin, Yu.A., Burchuladze, A.Sh. and Buklyakova, N.N. 1984. Changes in sugars of mandarin fruits in controlled atmosphere storage. *Subtropicheskie Kul'tury* 1, 83–86.

Dull, G.G., Young, R.R. and Biale J.B. 1967. Respiratory patterns in fruit of pineapple, *Ananas comosus*, detached at different stages of development. *Plant Physiology* 20, 1059.

Eaks, I.L. 1956. Effects of modified atmospheres on cucumbers at chilling and non-chilling temperatures. *Proceedings of the American Society for Horticultural Science* 67, 473.

Eason, J.R., Ryan, D., Page, B., Watson, L. and Coupe, S.A. 2007. Harvested broccoli (Brassica oleracea) responds to high carbon dioxide and low oxygen atmosphere by inducing stress-response genes. *Postharvest Biology and Technology* 43, 358–365.

Eaves, C.A. 1934. Gas and cold storage as related fruit under Annapolis Valley conditions. *Annual Report Nova Scotia Fruit Growers Association* 71, 92–98.

Eaves, C.A. 1959. A dry scrubber for CA apple storages. *Transactions American Society Agricultural Engineers* 2, 127–128.

Eaves, C.A. 1963. Atmosphere generators for CA apple storages. *Annual Report Nova Scotia Fruit Growers Association* 100, 107–109.

Eaves, C.A., Forsyth, F.R. and Lockhart, C.L. 1969a. Influence of post-harvest anaerobiosis on fruit. *Proceedings of the XII International Congress of Refrigeration, Madrid, 1967*, 3, 307–313 [report 4.31].

Eaves, C.A., Forsyth, F.R. and Lockhart, C.L. 1969b. Recent developments in storage research at Kentville, Nova Scotia. *Canadian Institute of Food Technology* 2, 46–51.

Eggleston, V. and Tanner, D.J. 2005. Are carrots under pressure still alive? – The effect of high pressure processing on the respiration rate of carrots. *Acta Horticulturae* 687, 371–373.

Eksteen, G.J. and Truter, A.B. 1989. Transport simulation test with avocados and bananas in controlled atmosphere containers. *Yearbook of the South African Avocado Growers' Association* 12, 26–32.

Eksteen, G.J., van Bodegom, P. and van Bodegom, P. 1989. Current state of CA storage in Southern Africa. In: Fellman, J.K. (ed.) *Proceedings of the Fifth International Controlled Atmosphere Research Conference, Wenatchee, Washington, USA, 14–16 June 1989. Vol.1: Pome fruit.* Washington State University, Pullman, Washington, pp. 487–494.

Elgar, H.J., Burmeister, D.M. and Watkins, C.B. 1998. Storage and handling effects on a CO_2-related internal browning disorder of 'Braeburn' apples. *HortScience* 33, 719–722.

Elgayyar, M., Draughon, F.A., Golden, D.A. and Mount, J.R. 2001. Antimicrobial activity of essential oils from plants against selected pathogenic and saprophytic microorganisms. *Journal of Food Protection* 64, 1019–1024.

El-Ghaouth, A., Arul, J., Ponnampalam, R. and Boulet, M. 1991. Use of chitosan coating to reduce water loss and maintain quality of cucumber and cell pepper fruits. *Journal of Food Process Preservation* 15, 359.

El-Goorani, M.A. and Sommer, N.F. 1979. Suppression of postharvest plant pathogenic fungi by carbon monoxide. *Phytopathology* 69, 834–838.

Ella, L., Zion, A., Nehemia, A. and Amnon, L. 2003. Effect of the ethylene action inhibitor 1-methylcyclopropene on parsley leaf senescence and ethylene biosynthesis. *Postharvest Biology and Technology* 30, 67–74.

Ellis, G. 1995 Potential for all-year-round berries. *The Fruit Grower*, December, 17–18.

El-Sharkawy, I., Sherif, S., Qubbaj, T., Sullivan, A.J. and Jayasankar, S. 2016. Stimulated auxin levels enhance plum fruit ripening, but limit shelf-life characteristics. *Postharvest Biology and Technology* 112, 215–223.

El-Shiekh, A.F., Tong, C.B.S., Luby, J.J., Hoover, E.E. and Bedford, D.S. 2002. Storage potential of cold-hardy apple cultivars. *Journal of American Pomological Society* 56, 34–45.

Emmambux, M.N. and Stading, M. 2007. *In situ* tensile deformation of zein films with plasticizers and filler materials. *Food Hydrocolloids* 21, 1245–1255.

Enfors, S.O. and Molin, G. 1980. Effect of high concentrations of carbon dioxide on growth rate of *Pseudomonas fragi, Bacillus cereus* and *Streptococcus cremoris*. *Journal of Applied Bacteriology* 48, 409–416.

Eris, A., Turkben, C., Ozer, M.H., Henze, J. and Sass, P. 1994. A research on controlled atmosphere CA storage of peach cv. Hale Haven. *Acta Horticulturae* 368, 767–776.

Erkan, M. and Pekmezci, M. 2000. Investigations on controlled atmosphere (CA) storage of Star Ruby grapefruit grown in Antalya conditions. *Bahce* 28, 87–93.

Ertan, U., Ozelkok, S., Celikel, F. and Kepenek, K. 1990. The effects of pre-cooling and increased atmospheric concentrations of CO_2 on fruit quality and postharvest life of strawberries. *Bahce* 19, 59–76.

Ertan, U., Ozelkok, S., Kaynas, K. and Oz, F. 1992. Bazi onemli elma cesitlerinin normal ve kontrollu atmosferde depolanmalari uzerinde karsilastirmali arastirmalar – I. Akici sistem. *Bahce* 21, 77–90.

Escalona, V.H., Ortega, F., Artés-Hernandez, F., Aguayo, E. and Artés, F. 2005. Test of a respiration model for a celery plants modified atmosphere packaging system at commercial pallet scale. *Acta Horticulturae* 674, 531–536.

Escalona, V.H., Aguayo, E. and Artés, F. 2006a. Metabolic activity and quality changes of whole and fresh-cut kohlrabi (*Brassica oleracea* L. *gongylodes* group) stored under controlled atmospheres. *Postharvest Biology and Technology* 41, 181–190.

Escalona, V.H., Verlinden, B.E., Geysen, S. and Nicolai, B.N. 2006b. Changes in respiration of fresh-cut butterhead lettuce under controlled atmospheres using low and superatmospheric oxygen conditions with different carbon dioxide levels. *Postharvest Biology and Technology* 39, 48–55.

Escalona, V.H., Aguayo, E. and Artés, F. 2007. Extending the shelf-life of kohlrabi stems by modified atmosphere packaging. *Journal of Food Science* 72, S308–S313.

Escalona V.H., Aguayo E., Martínez-Hernández G.B. and Artés F. 2010. UV-C doses to reduce pathogen and spoilage bacterial growth *in vitro* and in baby spinach. *Postharvest Biology and Technology* 56, 223–231.

Escribano, M.I., Del Cura, B. Muñoz, M.T. and Merodio, C. 1997. High CO_2-low temperature interaction on ribulose 1,5-biphosphate carboxylase and polygalacturonase protein levels in cherimoya fruit. *Seventh International Controlled Atmosphere Research Conference, July 13–18 1997*, University of California, Davis, California [Abstract], 115.

Escribano, S., Lopez, A., Sivertsen, H., Biasi, W.V., Macnish, A.J. and Mitcham, E.J. 2016. Impact of 1-methylcyclopropene treatment on the sensory quality of 'Bartlett' pear fruit. *Postharvest Biology and Technology* 111, 305–313.

Escribano, S., Sugimoto, N., Macnish, A.J., Biasi, W.V. and Mitcham, E.J. 2017. Efficacy of liquid 1-methylcyclopropene to delay ripening of 'Bartlett' pears. *Postharvest Biology and Technology* 126, 57–66.

Eshel, D., Teper-Bamnolker, P., Vinokur, Y., Saad, I., Zutahy, Y. and Rodov, V. 2014. Fast curing: A method to improve postharvest quality of onions in hot climate harvest. *Postharvest Biology and Technology* 88, 34–39.

Estiarte, N., Crespo-Sempere, A., Marín, S., Sanchis, V. and Ramos, A.J. 2016. Effect of 1-methylcyclopropene on the development of black mold disease and its potential effect on alternariol and alternariol monomethyl ether biosynthesis on tomatoes infected with *Alternaria alternate*. *International Journal of Food Microbiology* 236, 74–82.

Eum, H.L. and Lee, S.K. 2003. Effects of methyl jasmonic acid on storage injury of 'Nok-Kwang' hot pepper fruits during modified atmosphere storage. *Journal of the Korean Society for Horticultural Science* 44, 297–301.

European Commission. 2005. 1-Methylcyclopropene draft review. Available at: http://ec.europa.eu/food/plant/protection/evaluation/newactive/1-methylcyclopropene_draft_review_report.pdf (accessed April 2009).

European Commission. 2006. Commission Regulation (EC) No. 401/2006 of 23 February 2006 laying down the methods of sampling and analysis for the official control of the levels of mycotoxins in foodstuffs. *Official Journal of the European Communities* (9 March 2006) L 70, pp. 12–34.

Evelo, R.G. 1995. Modelling modified atmosphere systems. *COST 94. The post-harvest treatment of fruit and vegetables: systems and operations for post-harvest quality. Proceedings of a Workshop, 14–15 September 1993, Milan, Italy*, 147–153.

Fabi, J.P., Cordenunsi, B.R., de Mattos Barreto, G.P., Mercadante, A.Z., Lajolo, F.M. and Oliveira do Nascimento, J.R. 2007. Papaya fruit ripening: response to ethylene and 1-methylcyclopropene (1-MCP). *Journal of Agricultural and Food Chemistry* 55, 6118–6123.

Fagundes, C., Moraes, K., Pérez-Gago, M.B., Palou, L., Maraschin, M. and Monteiro, A.R. 2015. Effect of active modified atmosphere and cold storage on the postharvest quality of cherry tomatoes. *Postharvest Biology and Technology* 109, 73–81.

Fallik, E., Perzelan, Y., Alkalai-Tuvia, S., Nemny-Lavy, E. and Nestel, D. 2012. Development of cold quarantine protocols to arrest the development of the Mediterranean fruit fly (*Ceratitis capitata*) in pepper (*Capsicum annuum* L.) fruit after harvest. *Postharvest Biology and Technology* 70, 7–12.

Fan, X., Blankenship, S.M. and Mattheis, J.P. 1999. 1-Methylcyclopropene inhibits apple ripening. *Journal of the American Society for Horticultural Science* 124, 690–695.
Fan, X., Argenta, L. and Mattheis, J.P. 2000. Inhibition of ethylene action by 1-methylcyclopropene prolongs storage life of apricots. *Postharvest Biology and Technology* 20, 135–142.
Farber, J.M. 1991. Microbiological aspects of modified-atmosphere packaging technology – a review. *Journal of Food Protection* 54, 58–70.
Farneti, B., Schouten, R.E. and Woltering, E.J. 2012. Low temperature-induced lycopene degradation in red ripe tomato evaluated by remittance spectroscopy. *Postharvest Biology and Technology* 73, 22–27
Fellman, J.K., Mattinson, D.S., Bostick, B.C., Mattheis, J.P. and Patterson, M.E. 1993. Ester biosynthesis in 'Rome' apples subjected to low-oxygen atmospheres. *Postharvest Biology and Technology* 3, 201–214.
Fellows, P.J. 1988. *Food Processing Technology*. Ellis Horwood, London, New York Toronto, Sydney, Tokyo and Singapore.
Feng, X., Apelbaum, A., Sisler, E.C. and Goren, R., 2000. Control of ethylene responses in avocado fruit with 1-methylcyclopropene. *Postharvest Biology and Technology* 20, 143–150.
Fernandes, P.A.R., Moreira, S.A., Fidalgo, L.G., Santos, M.D., Queirós, R.P., Delgadillo, I. and Saraiva, J.A. 2015. Food Preservation under pressure (hyperbaric storage) as a possible improvement/alternative to refrigeration. *Food Engineering Reviews* 7, 1–10.
Fernández-Trujillo, J.P., Serrano, J.M. and Martínez, J.A. 2009. Quality of red sweet pepper fruit treated with 1-MCP during a simulated post-harvest handling chain. *Food Science and Technology International* 15, 23–30.
Ferrar, P. 1988. Transport of fresh fruit and vegetables. *Australian Centre for International Agricultural Research Proceedings*, 23.
Ferrari, D.G and Di Matteo, M. 1996. High pressure stabilization of orange juice: evaluation of the effects of process conditions. *Italian Journal of Food Science* 2, 99–106.
Ferreira, S.A., Pitz, K.Y., Manshardt, R., Zee, F., Fitch, M.M. and Gonsalves, D. 2002. Virus coat protein transgenic papaya provides practical control of papaya ringspot virus in Hawai'i. *Plant Disease* 86, 101–105.
Ferrer, M.A., Gómez-Tena, M., Pedreño, M.A. and Barceló, A.R. 1996. Effects of ethrel on peroxidase of iceberg lettuce leaf tissue. *Postharvest Biology and Technology* 7, 301–307.
Ferruzzi, M.G., Failla, M.L. and Schwartz, S.J. 2001 Assessment of degradation and intestinal cell uptake of carotenoids and chlorophyll derivatives from spinach puree using an in vitro digestion and caco-2 human cell model. *Journal of Agricultural and Food Chemistry* 49, 2082–2089.
Feygenberg, O., Keinan, A., Kobiler, I., Falik, E., Pesis, E., Lers, A. and Prusky, D. 2014. Improved management of mango fruit though orchard and packinghouse treatments to reduce lenticel discoloration and prevent decay. *Postharvest Biology and Technology* 91, 128–133.
Fidalgo, L.G., Santos, M.D., Queirós, R.P., Inácio, R.S., Mota, M.J. *et al.* 2014. Hyperbaric storage at and above room temperature of a highly perishable food. *Food and Bioprocess Technology* 7, 2028–2037.
Fidler, J.C. 1963. Refrigerated storage of fruits and vegetables in the UK, the British Commonwealth, the United States of America and South Africa. *Ditton Laboratory Memoir*, 93.
Fidler, J.C. 1968. Low temperature injury of fruit and vegetables. *Recent Advances in Food Science* 4, 271–283.
Fidler, J.C. 1970. Recommended conditions for the storage of apples. *Report of the East Malling Research Station for 1969*, 189–190.
Fidler, J.C. and Mann, G. 1972. *Refrigerated Storage of Apples and Pears – a Practical Guide*. Commonwealth Agricultural Bureau, Farnham Royal, UK.
Fidler, J.C., Wilkinson, B.G., Edney, K.L. and Sharples R.O. 1973. *The Biology of Apple and Pear Storage*. Commonwealth Agricultural Bureaux Research Review, 3. Commonwealth Bureau of Horticultural and Plantation Crops, Farnham Royal, UK.
Finger, F.L., Della-Justina, M.E., Casali, V.W.D. and Puiatti, M. 2008. Temperature and modified atmosphere affect the quality of okra. *Scientia Agricola* 65, 360–364.
Fisher, D.V. 1939. Storage of delicious apples in artificial atmospheres. *Proceedings of the American Society for Horticultural Science* 37, 459–462.
Flores, F.B., Martinez-Madrid, M.C., Ben Amor, M., Pech, J.C., Latche, A. and Romojaro, F. 2004. Modified atmosphere packaging confers additional chilling tolerance on ethylene-inhibited cantaloupe Charentais melon fruit. *European Food Research and Technology* 219, 614–619.
Folch-Fortuny, A., Prats-Montalbán,, J.M., Cubero, S., Blasco, J., and Ferrer, A., 2016. VIS/NIR hyperspectral imaging and N-way PLS-DA models for detection of decay lesions in citrus fruits. *Chemometrics and Intelligent Laboratory Systems* 156, 241–248.
Folchi, A., Pratella, G.C., Bertolini, P., Cazzola, P.P. and Eccher Zerbini, P. 1994. Effects of oxygen stress on stone fruits. *COST 94. The post-harvest treatment of fruit and vegetables: controlled atmosphere storage of fruit and vegetables. Proceedings of a Workshop, 22–23 Apr. 1993, Milan, Italy*, 107–119.

Folchi, A., Pratella, G.C., Tian, S.P. and Bertolini, P. 1995. Effect of low O_2 stress in apricot at different temperatures. *Italian Journal of Food Science* 7, 245–253.

Follett, P. 2014. Phytosanitary irradiation for fresh horticultural commodities: generic treatments, current issues, and next steps. *Stewart Postharvest Review* 10, 1–7.

Fonberg-Broczek, M., Windyga, B., Szczawinski, J., Szczawinska, M., Pietrzak, D. and Prestamo, G. 2005. High pressure processing for food safety. *Acta Biochimica Polonica* 52, 721–724.

Fonseca, J.M., Rushing, J.W. and Testin, R.F. 2004. The anaerobic compensation point of fresh-cut watermelon and postprocess handling implications. *HortScience* 39, 562–564.

Fonseca, M.J. de O., Leal, N.R., Cenci, S.A., Cecon, P.R, and Smith, R.E.B. 2006. Postharvest controlled atmosphere storage of 'Sunrise Solo' and 'Golden' pawpaws. *Revista Brasileira de Armazenamento* 31, 154–161.

Food Investigation Board. 1919. *Food Investigation Board. Department of Scientific and Industrial Research Report for the Year, 1918.*

Food Investigation Board. 1920. *Food Investigation Board. Department of Scientific and Industrial Research Report for the Year, 1920*, 16–25.

Food Investigation Board. 1937. *Department of Scientific and Industrial Research. Annual Report for the Year 1936–37*, 185–186.

Food Investigation Board. 1958. *Food Investigation Board. Department of Scientific and Industrial Research Report for the Year 1957*, 35–36.

Forney, C.F., Rij, R.E. and Ross, S.R. 1989. Measurement of broccoli respiration rate in film wrapped packages. *HortScience* 24, 111–113.

Forney, C.F., Jordan, M.A. and Nicholas, K.U.K.G. 2003. Effect of CO_2 on physical, chemical, and quality changes in 'Burlington' blueberries. *Acta Horticulturae* 600, 587–593.

Forney, C.F., Jamieson, A.R., Munro Pennell, K.D., Jordan, M.A. and Fillmore, S.A.E. 2015. Relationships between fruit composition and storage life in air or controlled atmosphere of red raspberry. *Postharvest Biology and Technology* 110, 121–130.

Foukaraki, S.G., Cools, K. and Terry, L.A. 2016. Differential effect of ethylene supplementation and inhibition on abscisic acid metabolism of potato (*Solanum tuberosum* L.) tubers during storage. *Postharvest Biology and Technology* 112, 87–94.

Francile, A.S. 1992. Controlled atmosphere storage of tomato. *Rivista di Agricoltura Subtropicale e Tropicale* 86, 411–416.

Fraser, P.D., Truesdale, M.R., Bird, C.R., Schuch, W. and Bramley, P.M. 1994. Carotenoid biosynthesis during tomato development. *Plant Physiology* 105, 405–413.

Frenkel, C. 1975. Oxidative turnover of auxins in relation to the onset of ripening in Bartlett pear. *Plant Physiology* 55, 480–484.

Frenkel, C. and Garrison, S.A., 1976. Initiation of lycopene synthesis in the tomato mutant rin as influenced by oxygen and ethylene interactions. *HortScience* 11, 20–21.

Frenkel, C. and Patterson, M.E. 1974. Effect of CO_2 on ultrastructure of 'Bartlett pears'. *HortScience* 9, 338–340.

Fridovich, I., 1986. Biological effects of the superoxide radical. *Archives Biochemistry Biophysics* 247, 1–11.

Fu, L., Cao, J., Li, Q., Lin, L. and Jiang, W. 2007. Effect of 1-Methylcyclopropene on fruit quality and physiological disorders in Yali pear (*Pyrus bretschneideri* Rehd.) during storage. *Food Science and Technology International* 13, 49–54.

Fuchs, Y., Zauberman, G. and Yanko, U. 1978. Controlled atmosphere storage of mango. *Ministry of Agriculture, Institute for Technology and Storage of Agricultural Products: Scientific Activities 1974–1977*, 184.

Fukao, T., and Bailey-Serres, J., 2004. Plant responses to hypoxia – is survival a balancing act? *Trends in Plant Science* 9, 449–456.

Fulton, S.H. 1907. The cold storage of small fruits. *US Department of Agriculture, Bureau of Plant Industry, Bulletin*, 108, September 17.

Gadalla, S.O. 1997. Inhibition of sprouting of onions during storage and marketing. PhD thesis, Cranfield University, UK.

Gajewski, M. 2003. Sensory and physical changes during storage of zucchini squash (*Cucurbita pepo* var. *giromontina* Alef.). *Acta Horticulturae* 604, 613–617.

Gajewski, M. and Roson, W 2001. Effect of controlled atmosphere storage on the quality of zucchini squash (*Cucurbita pepo* var. *giromontina* Alef.). *Vegetable Crops Research Bulletin* 54, 207–211.

Galindo, F.G., Vaughan, D., Herppich, W., Smallwood, M., Sommarin, M., Gekas, V. and Sjoholm, I. 2004. Influence of cold acclimation on the mechanical strength of carrot (*Daucus carota* L.) tissue. *European Journal of Horticultural Science* 69, 229–234.

Gallat, S., Crentsil, D. and Bancroft, R.D. 1998. Development of a low cost cassava fresh root storage technology for the Ghanaian market. A paper presented at the Post-harvest Technology and Commodity Marketing Conference, International Trade Fair Centre, Accra, Ghana, 27–29 Nov. 1995. NRI Project Code F0021. In: Ferris, R.S.B. (ed.) *Postharvest Technology and Commodity Marketing. Proceedings of a Postharvest Conference, 2 November–1 December 1995, Accra, Ghana*. International Institute of Tropical Agriculture (IITA), Ibadan, Nigeria, pp. 77–84.

Gallerani, G., Pratella, G.C., Cazzola, P.P. and Eccher-Zerbini, P. 1994. Superficial scald control via low-O_2 treatment timed to peroxide threshold value. *COST 94. The post-harvest treatment of fruit and vegetables: controlled atmosphere storage of fruit and vegetables. Proceedings of a Workshop, 22–23 April 1993, Milan, Italy*, 51–60.

Galliard, T. 1975. Degradation of plant lipid by hydrolytic and oxidative enzymes. In: Galliard, T. and Mercer, E. (eds) *Recent Advances in Chemistry and Biochemistry of Plant Lipids*. Academic Press, New York, pp. 319–337.

Gamage, T.V., Sanguansri, P., Swiergon, P., Eelkema, M., Wyatt, P. *et al*. 2015. Continuous combined microwave and hot air treatment of apples for fruit fly (*Bactrocera tryoni* and *Bactrocera jarvisi*) disinfestation. *Innovative Food Science & Emerging Technologies* 29, 261–270

Gane, R. 1934. Production of ethylene by some ripening fruits. *Nature* 134, 1008.

Gao, H.Y., Chen, H.J., Chen, W.X., Yang, Y.T., Song, L.L., Jiang, Y.M. and Zheng, Y.H. 2006. Effect of hypobaric storage on physiological and quality attributes of loquat fruit at low temperature. *Acta Horticulturae* 712, 269–274.

Garcia, E. and Barrett, D.M. 2006. Assessing lycopene content in California processing tomatoes. *Journal of Food Processing and Preservation* 30, 56–70.

Garcia, J.M., Castellano, J.M., Morilla, A., Perdiguero, S. and Albi, M.A. 1993. CA-storage of Mill olives. *COST 94. The post-harvest treatment of fruit and vegetables: controlled atmosphere storage of fruit and vegetables. Proceedings of a Workshop, April 22–23, 1993, Milan, Italy*, 83–87.

Garcia-Martin, J.F., Olmo, M. and Garcia, J.M. 2018. *Pseudomonas graminis* strain CPA-7 differentially modulates the oxidative response in fresh-cut 'Golden Delicious' apples depending on the storage conditions. *Postharvest Biology and Technology* 138, 46–55.

Gariepy, Y., Raghavan, G.S.V. and Theriault, R. 1984. Use of the membrane system for long-term CA storage of cabbage. *Canadian Agricultural Engineering* 26, 105–109.

Gariepy, Y., Raghavan, G.S.V., Plasse, R., Theriault, R. and Phan, C.T. 1985. Long term storage of cabbage, celery, and leeks under controlled atmosphere. *Acta Horticulturae* 157, 193–201.

Gariepy, Y., Raghavan, G.S.V., Theriault, R. and Munroe, J.A. 1988. Design procedure for the silicone membrane system used for controlled atmosphere storage of leeks and celery. *Canadian Agricultural Engineering* 30, 231–236.

Garrett, M. 1992. Applications of controlled atmosphere containers. *BEHR'S Seminare Hamburg 16–17 November 1992, Munich, Germany*.

Gasser, F. and Siegrist, J.-P. 2009. Recommandations 2009–2010 aux entrepositaires de fruits et legumes. *Revue suisse Viticulture, Arboriculture, Horticulture* 41(5), 313–316.

Gasser, F. and Siegrist, J.-P. 2011. Recommandations 2011–2012 aux entrepositaires de fruits et legumes. *Revue suisse Viticulture, Arboriculture, Horticulture* 43, 316–319.

Gasser, F. and von Arx, K. 2015. Dynamic CA storage of organic apple cultivars. *Acta Horticulturae* 1071, 527–532.

Gasser, F., Dätwyler, D., Schneider, K., Naunheim, W. and Hoehn, E. 2003. Effects of decreasing oxygen levels in the storage atmosphere on the respiration of Idared apples. *Acta Horticulturae* 600, 189–192.

Gasser, F., Mattle, S. and Hohn, E. 2004. Cherry: storage trials 2003. *Obst und Weinbau* 140, 6–10.

Gasser, F., Dätwyler, D., Schneider, K., Naunheim, W. and Hoehn, E. 2005. Effects of decreasing oxygen levels in the storage atmosphere on the respiration and production of volatiles of 'Idared' apples. *Acta Horticulturae* 682, 1585–1592.

Gasser, F., Eppler, T., Naunheim, W., Gabioud, S. and Hoehn, E. 2008. Control of the critical oxygen level during dynamic CA storage of apples by monitoring respiration as well as chlorophyll fluorescence. *Acta Horticulturae* 796, 69–76.

Gasser, F., Eppler, T., Naunheim, W., Gabioud, S. and Bozzi Nising, A. 2010. Dynamic CA storage of apples: monitoring of the critical oxygen concentration and adjustment of optimum conditions during oxygen reduction. *Acta Horticulturae* 876, 39–46.

Gazit, S. and Blumenfeld, A. 1970. Response of mature avocado fruit to ethylene treatments before and after harvest. *Journal of the American Society of Horticultural Science* 95, 229–231.

Geeson, J.D. 1984. Improved long term storage of winter white cabbage and carrots. *Agricultural and Food Research Council, Fruit, Vegetable and Science* 19, 21.

Geeson, J.D. 1989. Modified atmosphere packaging of fruits and vegetables. *Acta Horticulturae* 258, 143–150.

Geeson, J.D. and Smith, S.M. 1989. Retardation of apple ripening during distribution by the use of modified atmospheres. *Acta Horticulturae* 258, 245–253.

Geeson, J.D., Genge, P.M., Sharples, R.O. and Smith, S.M. 1990a. Limitations to modified atmosphere packaging for extending the shelf-life of partly ripened Doyenné du Comice pears. *International Journal of Food Science & Technology* 26, 225–231.

Geeson, J.D., Genge, P.M., Smith, S.M. and Sharples, R.O. 1990b. The response of unripe Conference pears to modified atmosphere retail packaging. *International Journal of Food Science & Technology* 26, 215–224.

Giacalone, G. and Chiabrando, V. 2015. Modified atmosphere packaging of sweet cherries with different packaging systems: effect on organoleptic quality. *Acta Horticulturae* 1071, 87–95.

Gil, M.I., Conesa, M.A. and Artés, F. 2003. Effects of low-oxygen and high-carbon dioxide atmosphere on postharvest quality of artichokes. *Acta Horticulturae* 600, 385–388.

Gil-Izquierdo, A., Conesa, M., Ferreres, F. and Gil, M. 2004. Influence of modified atmosphere packaging on quality, vitamin C and phenolic content of artichokes (*Cynara scolymus* L.). *European Food Research and Technology* 215, 21–27.

Girard, B. and Lau, O.L. 1995. Effect of maturity and storage on quality and volatile production of 'Jonagold' apples. *Food Research International* 28, 465–471.

Giuggioli, N.R., Briano, R., Baudino, C. and Peano, C. 2015. Effects of packaging and storage conditions on quality and volatile compounds of raspberry fruits. *Journal of Food* 13, 512–521.

Goffings, G. and Herregods, M. 1989. Storage of leeks under controlled atmospheres. *Acta Horticulturae* 258, 481–484.

Goffings, G., Herregods, M. and Sass, P. 1994. The influence of the storage conditions on some quality parameters of Jonagold apples. *Acta Horticulturae* 368, 37–42.

Gökmen, V., Akbudak, B., Serpen, A., Acar, J., Turan, Z.M. and Eriş, A. 2007. Effects of controlled atmosphere storage and low-dose irradiation on potato tuber components affecting acrylamide and color formations upon frying. *European Food Research and Technology* 224, 743–748.

Golias, J. 1987. Methods of ethylene removal from vegetable storage chambers. Zpusoby odstraneni etylenu z atmosfery komory se skladovanou zeleninou. *Bulletin, Vyzkumny a Slechtitelsky Ustav Zelinarsky Olomouc* 31, 51–60.

Gomes, M.A., Ascheri, D.P.R. and de Campos, A.J. 2016. Characterization of edible films of *Swartzia burchelli* phosphated starches and development of coatings for post-harvest application to cherry tomatoes. *Semina: Ciências Agrárias (Londrina)* 37, 1897–1909.

Gómez, P.A. and Artés F. 2004a. Controlled atmospheres enhance postharvest green celery quality. *Postharvest Biology and Technology* 34, 203–209.

Gomez, P.A. and Artés, F. 2004b. Keeping quality of green celery as affected by modified atmosphere packaging. *European Journal of Horticultural Science* 69, 215–219.

González, D. 1998. Control of papaya ringspot virus in papaya: A case study. *Annual Review of Phytopathology* 36, 415–437.

González, M.A. and de Rivera, A.C. 1972. Storage of fresh yam (*Dioscorea alata* L.) under controlled conditions. *Journal of the Agricultural University of Puerto Rico* 56, 46–56.

González-Aguilar, G.A. 2002. Pepper. In: Gross, K.C., Wan, C.Y. and Saltveit, M. (eds) *The Commerical Storage of Fruits, Vegetables, and Florist and Nursery Stocks*. USDA Agricultural Handbook No. 66. USDA Agricultural Research Service (ARS). Available at: http://www.ba.ars.usda.gov/hb66/contents.html (accessed November 2012).

González-Aguilar, G., Vasquez, C., Felix, L., Baez, R., Siller, J. and Ait, O. 1994. Low O_2 treatment before storage in normal or modified atmosphere packaging of mango. Postharvest physiology, pathology and technologies for horticultural commodities: recent advances. *Proceedings of an International Symposium held at Agadir, Morocco, 16–21 January 1994*, 185–189.

González-Aguilar, G.A., Wang, C.Y., Buta, J.G. and Krizek, D.T. 2001. Use of UV-C irradiation to prevent decay and maintain postharvest quality of ripe 'Tommy Atkins' mangoes. *International Journal of Food Science and Technology* 36, 767–773.

González-Aguilar, G.A., Buta, J.G. and Wang, C.Y. 2003. Methyl jasmonate and modified atmosphere packaging (MAP) reduce decay and maintain postharvest quality of papaya 'Sunrise'. *Postharvest Biology and Technology* 28, 361–370.

González-Buesa, J., Page, N., Kaminski, C., Ryser, E.T., Beaudry, R. and Almenar, E. 2014. Effect of non-conventional atmospheres and bio-based packaging on the quality and safety of Listeria monocytogenes-inoculated fresh-cut celery (*Apium graveolens* L.) during storage. *Postharvest Biology and Technology* 93, 29–37.

González-Roncero, M.I. and Day, B.P.F., 1998. *The effect of elevated oxygen and carbon dioxide modified atmosphere on psychotrophic pathogens and spoilage microorganisms associated with fresh prepared produce*. Research Summary Sheet 98. Campden and Chorleywood Food Research Association, Chipping Campden, UK.

Goodenough, P.W. and Thomas, T.H. 1980. Comparative physiology of field grown tomatoes during ripening on the plant or retarded ripening in controlled atmosphere. *Annals of Applied Biology* 94, 445–455.

Goodenough, P.W. and Thomas, T.H. 1981. Biochemical changes in tomatoes stored in modified gas atmospheres. I Sugars and acids. *Annals of Applied Biology* 98, 507–516.

Gorini, F. 1988. Storage and postharvest treatment of brassicas. I. Broccoli. *Annali dell'Istituto Sperimentale per la Valorizzazione Tecnologica dei Prodotti Agricoli* Milano, 19, 279–294.

Goszczynska, D.M. and Ryszard M.R. 1988. Storage of cut flowers. *Horticultural Reviews* 10, 35–62.

Goto, M., Minamide, T. and Iwata, T. 1988. The change in chilling sensitivity in fruits of mume Japanese apricot, *Prunus mume* Sieb. et Zucc. depending on maturity at harvest and its relationship to phospholipid composition and membrane permeability. *Journal of the Japanese Society for Horticultural Science* 56, 479–485.

Goulart, B.L., Evensen, K.B., Hammer, P. and Braun, H.L. 1990. Maintaining raspberry shelf-life: Part 1. The influence of controlled atmospheric gases on raspberry postharvest longevity. *Pennsylvania Fruit News* 70, 12–15.

Goulart, B.L., Hammer, P.E., Evensen, K.B., Janisiewicz, B. and Takeda, F. 1992. Pyrrolnitrin, captan with benomyl, and high CO_2 enhance raspberry shelf-life at 0 or 18°C. *Journal of the American Society for Horticultural Science* 117, 265–270.

Goyette, B. 2010. Hyperbaric treatment to enhance quality attributes of fresh horticultural produce. PhD thesis, McGill University, Montreal, Canada.

Goyette, B., Charles, M.T., Vigneault, C. and Raghavan, G.S.V. 2007. Pressure treatment for increasing fruit and vegetable qualities. *Stewart Postharvest Review* 3, 5.1–5.6.

Goyette, B., Vigneault, C., Wang, N. and Raghavan, G.S.V. 2011. Conceptualization, design and evaluation of a hyperbaric respirometer. *Journal of Food Engineering* 105, 283–288.

Goyette, B., Vigneault, C., Charles, M.T. and Raghavan, G.S.V. 2012. Effect of hyperbaric treatments on the quality attributes of tomato. *Canadian Journal of Plant Science* 92, 541–551.

Gözlekçi, S., Erkan, M., Karaşahin, I. and Şahin, G. 2008. Effect of 1-methylcyclopropene (1-MCP) on fig (*Ficus carica* cv. Bardakci) storage. *Acta Horticulturae* 798, 325–330.

Graell, J. and Recasens, I. 1992. Effects of ethylene removal on 'Starking Delicious' apple quality in controlled atmosphere storage. *Postharvest Biology and Technology* 2, 101–108.

Grattidge, R. 1980. *Mango anthracnose control*. Australia Farm Note AGDEX 234/633 F18/Mar 80. Queensland Department of Primary Industries, Brisbane.

Grierson, D. 1993. Chairman's remarks. In: *Postharvest Biology and Handling of Fruit, Vegetables and Flowers*. Meeting of the Association of Applied Biologists, London, 8 December 1993.

Grierson, W. 1971. Chilling injury in tropical and subtropical fruits: IV. The role of packaging and waxing in minimizing chilling injury of grapefruit. *Proceedings of the Tropical Region, American Society for Horticultural Science* 15, 76–88.

Grozeff, G.G., Micieli, M.E., Gómez, F., Fernández, L., Guiamet, J.J., Chaves, A.R. and Bartoli, C.G. 2010. 1-Methyl cyclopropene extends postharvest life of spinach leaves. *Postharvest Biology and Technology* 55, 182–185.

Grumman Park 2012. *A broad look at the practical and technical aspects of hypobaric storage*. Dormavac Corp., Miami, Florida. Available at: http://www.grummanpark.org/content/dormavac (accessed 2 March 2018).

Guan, W., Fan, X. and Yan, R. 2012. Effects of UV-C treatment on inactivation of *Escherichia coli* O157:H7, microbial loads, and quality of button mushrooms. *Postharvest Biology and Technology* 64, 119–125.

Gudkovskii, V.A. 1975. Storage of apples in a controlled atmosphere. *Tematicheskii Sbornik Nauchnykh Rabot Instituta Plodovodstva i Vinogradarstva i Opytnykh Stantsii* 3, 256–268.

Guerra, M., Sanz, M.A. and Casquero, P.A. 2010. Influence of storage conditions on the sensory quality of a high acid apple. *International Journal of Food Science and Technology* 45, 2352–2357.

Guerra, N.B., Livera, A.V.S., da Rocha, J.A.M.R. and Oliveira, S.L. 1995. Storage of soursop Annona muricata, L. in polyethylene bags with ethylene absorbent. In: Kushwaha, L., Serwatowski, R. and Brook, R. (eds)

Technologias de cosecha y postcosecha de frutas y hortalizas / Harvest and Postharvest Technologies for Fresh Fruits and Vegetables. Proceedings of a conference held in Guanajuato, Mexico 20–24 February 1995. American Society of Agricultural Engineers, St Joseph, Missouri, pp. 617–622.

Guerreiro, A., Gago, C., Faleiro, M.L., Miguel, G. and Antunes, M.D. (2015). Alginate edible coatings enriched with citral for preservation of fresh and fresh-cut fruit. *Acta Horticulturae* 1091, 37–44.

Guerzoni, M.E., Gianotti, A., Corbo, M.R. and Sinigaglia, M. 1996. Shelf life modelling for fresh cut vegetables. *Postharvest Biology and Technology* 9, 195–207.

Guevara, J.C., Yahia, E.M., Brito de la Fuente, E. and Biserka, S.P. 2003.Effects of elevated concentrations of CO_2 in modified atmosphere packaging on the quality of prickly pear cactus stems (*Opuntia* spp.). *Postharvest Biology and Technology* 29, 167–176.

Gui, F.Q., Chen, F., Wu, J.H., Wang, Z.F., Liao, X.J. and Hu, X.S. 2006. Inactivation and structural change of horseradish peroxidase treated with supercritical carbon dioxide. *Food Chemistry* 97, 480–489.

Guillard, V., Guillaume, C. and Destercke, S. 2012. Parameter uncertainties and error propagation in modified atmosphere packaging modelling. *Postharvest Biology and Technology* 67, 154–166

Gull, D.D. 1981. *Ripening Tomatoes with Ethylene*. Vegetable Crops Fact Sheet VC 29. Vegetable Crops Department, University of Florida, Gainsville, Florida.

Gunes, G., Liu, R.H. and Watkins, C.B. 2002. Controlled-atmosphere effects on postharvest quality and antioxidant activity of cranberry fruits. *Journal of Agricultural and Food Chemistry* 50, 5932–5938.

Gunes, N.T. 2008. Ripening regulation during storage in quince (*Cydonia oblonga* Mill.) fruit. *Acta Horticulturae* 796, 191–196

Guo, L., Ma, Y., Sun, D.W. and Wang, P. 2008. Effects of controlled freezing-point storage at 0°C on quality of green bean as compared with cold and room-temperature storages. *Journal of Food Engineering* 86, 25–29.

Guo, M., Feng, J., Zhang, P., Jia, L. and Chen, K. 2015. Postharvest treatment with trans-2-hexenal induced resistance against *Botrytis cinerea* in tomato fruit. *Australasian Plant Pathology* 44, 121–128.

Guo, Y., Ma, S., Zhu, Y. and Zhao, G. 2007. Effects of 1-MCP treatment on postharvest physiology and storage quality of Pink Lady apple with different maturity. *Journal of Fruit Science* 24, 415–420.

Gwanpua, S.G., Verlinden, B.E.,. Hertog, M.L.A.T.M., Bulens, I., Van de Poel, B. *et al*. 2012. Kinetic modeling of firmness breakdown in 'Braeburn' apples stored under different controlled atmosphere conditions. *Postharvest Biology and Technology* 67, 68–74.

Gwanpua, S.G., Qian, Z. and East, A.R. 2018. Modelling ethylene regulated changes in 'Hass' avocado quality, *Postharvest Biology and Technology* 136, 12–22.

Haard, N.F. and Salunkhe, D.K. 1975. *Symposium: Postharvest Biology and Handling of Fruits and Vegetables*. AVI Publishing, Westpoint, Connecticut.

Haffner, K.E. 1993. Storage trials of 'Aroma' apples at the Agricultural University of Norway. *Acta Horticulturae* 326, 305–313.

Haffner, K., Rosenfeld, H.J., Skrede, G. and Wang, L. 2002. Quality of red raspberry *Rubus idaeus* L. cultivars after storage in controlled and normal atmospheres. *Postharvest Biology and Technology* 24, 279–289.

Hailu, Z. 2016. Effects of controlled atmosphere storage and temperature on quality attributes of mango. *Journal of Chemical Engineering & Process Technology* 7, 317.

Hamzaha, H.M., Osmana, A., Chin, P.T. Tan, C.P. and Ghazalia, F.M. 2013. Carrageenan as an alternative coating for papaya (*Carica papaya* L. cv. Eksotika). *Postharvest Biology and Technology* 75, 142–146.

Han, B., Wang, W.S. and Shi, Z.P. 2006. Effect of control atmosphere storage on physiological and biochemical change of Dong jujube. *Scientia Agricultura Sinica* 39, 2379–2383.

Han, J.H. and Floros, J.D. 2007. Active packaging. In: Tewari, G. and Juneja, V.K. (eds) *Advances in Thermal and Non-thermal Food Preservation*. Blackwell Professional, Ames, Iowa, pp. 167–183.

Han, T., Li, L.P. and Ge, X. 2000. Effect of exogenous salicylic acid on postharvest physiology of peach fruit. *Acta Horticulturae Sinica* 27, 367–368.

Han, Y., Fu, C., Kuang, J., Chen, J. and Lu, W. 2016. Two banana fruit ripening-related C2H2 zinc finger proteins are transcriptional repressors of ethylene biosynthetic genes. *Postharvest Biology and Technology*, 116, 8–15.

Hanrahan, I. and Blakey, R. 2017. Honeycrisp storage recommendations revisited. Available at: http://treefruit.wsu.edu/article/honeycrisp-storage-recommendations-revisited/ (accessed March 2018).

Hansen, H. 1975. Storage of Jonagold apples – preliminary results of storage trials. *Erwerbsobstbau* 17, 122–123.

Hansen, H. 1977. Storage trials with less common apple varieties. Lagerungsversuche mit neu im Anbau aufgenommenen Apfelsorten. *Obstbau Weinbau* 14, 223–226.

Hansen, H. 1986. Use of high CO_2 concentrations in the transport and storage of soft fruit. *Obstbau* 11, 268–271.

Hansen, H. and Rumpf, G. 1978. Quality and storability of the apple cultivar Undine. *Erwerbsobstbau* 20, 231–232.

Hansen, K., Poll, L., Olsen, C.E. and Lewis, M.J. 1992. The influence of O_2 concentration in storage atmospheres on the post-storage volatile ester production of 'Jonagold' apples. *Lebensmittel Wissenschaft and Technologie* 25, 457–461.

Hansen, M., Olsen, C.E., Poll, L. and Cantwell, M.I. 1993. Volatile constituents and sensory quality of cooked broccoli florets after aerobic and anaerobic storage. *Acta Horticulturae* 343, 105–111.

Harb, J. and Streif, J. 2004a. Quality and consumer acceptability of gooseberry fruits (*Ribes uvacrispa*) following CA and air storage. *Journal of Horticultural Science & Biotechnology* 79, 329–334.

Harb, J.Y. and Streif, J. 2004b. Controlled atmosphere storage of highbush blueberries cv. 'Duke'. *European Journal of Horticultural Science* 69, 66–72.

Harb, J. and Streif, J. 2006. The influence of different controlled atmosphere storage conditions on the storability and quality of blueberries cv. 'Bluecrop'. *Erwerbsobstbau*, 48, 115–120.

Harb, J., Bisharat, R. and Streif, J. 2008a. Changes in volatile constituents of blackcurrants (*Ribes nigrum* L. cv. 'Titania') following controlled atmosphere storage. *Postharvest Biology and Technology* 47, 271–279.

Harb, J., Streif, J. and Bangerth, K.F. 2008b. Aroma volatiles of apples as influenced by ripening and storage procedures. *Acta Horticulturae* 796, 93–103.

Harb, J., Streif, J., Bangerth, F. and Sass, P. 1994. Synthesis of aroma compounds by controlled atmosphere stored apples supplied with aroma precursors: alcohols, acids and esters. *Acta Horticulturae* 368, 142–149.

Harbaum-Piayda, B., Walter, B., Bengtsson, G.B., Hubbermann, E.M., Bilger, W. and Schwarz, K. 2010. Influence of pre-harvest UV-B irradiation and normal or controlled atmosphere storage on flavonoid and hydroxycinnamic acid contents of pak choi (*Brassica campestris* L. ssp. *chinensis* var. *communis*). *Postharvest Biology and Technology* 56, 202–208.

Hardenburg, R.E. 1955. Ventilation of packaged produce. Onions are typical of items requiring effective perforation of film bags. *Modern Packaging* 28, 140 199–200.

Hardenburg, R.E. and Anderson, R.E. 1962. *Chemical control of scald on apples grown in eastern United States*. Marketing Research Report 538. Agricultural Research Service, United States Department of Agriculture, Washington DC, pp. 51–54.

Hardenburg, R.E., Anderson, R.E. and Finney, E.E. Jr. 1977. Quality and condition of 'Delicious' apples after storage at 0 deg C and display at warmer temperatures. *Journal of the America Society for Horticultural Science* 102, 210–214.

Hardenburg, R.E., Watada, A.E. and Wang C.Y. 1990. *The commercial storage of fruits, vegetables and florist and nursery stocks*. Agriculture Handbook 66. Agricultural Research Service, United States Department of Agriculture, Washington DC.

Harkett, P.J. 1971. The effect of O_2 concentration on the sugar content of potato tubers stored at low temperature. *Potato Research* 14, 305–311.

Harman, J.E. 1988. Quality maintenance after harvest. *New Zealand Agricultural Science* 22, 46–48.

Harman, J.E. and McDonald, B. 1983. Controlled atmosphere storage of kiwifruit: effect on storage life and fruit quality. *Acta Horticulturae* 138, 195–201.

Harris, C.M. and Harvey, J.M. 1973. Quality and decay of California strawberries stored in CO_2 enriched atmospheres. *Plant Disease Reporter* 57, 44–46.

Hartmans, K.J., van Es, A. and Schouten, S. 1990. Influence of controlled atmosphere CA storage on respiration, sprout growth and sugar content of cv. Bintje during extended storage at 4 °C. *11th Triennial Conference of the European Association for Potato Research, Edinburgh, UK, 8–13 July, 1990*, 159–160.

Hartz, T.K., Johnson, P.R., Francis, D.M. and Miyao, E.M. 2005. Processing tomato yield and fruit quality improved with potassium fertigation. *Horticultural Science* 40, 1862–1867.

Haruenkit, R. and Thompson, A.K. 1993. Storage of fresh pineapples. *Australian Centre for International Agricultural Research Proceedings* 50, 422–426.

Haruenkit, R. and Thompson, A.K. 1996. Effect of O_2 and CO_2 levels on internal browning and composition of pineapples Smooth Cayenne. *Proceedings of the International Conference on Tropical Fruits, Kuala Lumpur, Malaysia 23–26 July 1996*, 343–350.

Hashemi, S.M.B., Khaneghah, A.M., Ghahfarrokhi, M.G. and Eş, I. 2017. Basil-seed gum containing Origanum vulgare subsp. viride essential oil as edible coating for fresh cut apricots. *Postharvest Biology and Technology* 125, 26–34.

Hashmi, M.S., East, A.R., Palmer, J.S. and Heyes, J.A. 2013a. Pre-storage hypobaric treatments delay fungal decay of strawberries. *Postharvest Biology and Technology* 77, 75–79.

Hashmi, M.S., East, A.R., Palmer, J.S. and Heyes, J.A. 2013b. Hypobaric treatment stimulates defence-related enzymes in strawberry. *Postharvest Biology and Technology* 85, 77–82.

Hassan, A., Atan, R. M. and Zain, Z.M. 1985. Effect of modified atmosphere on black heart development and ascorbic acid content in 'Mauritius' pineapple *Ananas comosus* cv. Mauritius during storage at low temperature. *Association of Southeast Asian Nations Food Journal* 1, 15–18.

Hatfield, S.G.S. 1975. Influence of post-storage temperature on the aroma production by apples after controlled-atmosphere storage. *Journal of Science Food and Agriculture* 26, 1611–1612.

Hatfield, S.G.S. and Patterson, B.D. 1974. Abnormal volatile production by apples during ripening after controlled atmosphere storage. *Facteurs and Regulation de la Maturation des Fruits*. Colleques Internationaux, CNRS, Paris, pp. 57–64.

Hatoum, D., Hertog, M.L.A.T.M., Geeraerd, A.H. and Nicolai B.M. 2016. Effect of browning related pre- and postharvest factors on the 'Braeburn' apple metabolome during CA storage. *Postharvest Biology and Technology* 111, 106–116.

Hatton, T.T. and Cubbedge, R.H. 1977. Effects of prestorage CO_2 treatment and delayed storage on stem end rind breakdown of 'Marsh' grapefruit. *HortScience* 12, 120–121.

Hatton, T.T. and Cubbedge, R.H. 1982. Conditioning Florida grapefruit to reduce chilling injury during low temperature storage. *Journal of the American Society for Horticultural Science* 107, 57.

Hatton, T.T. and Reeder W.F. 1965. Controlled atmosphere storage of Lula avocados-1965 tests. *Proceedings of the Caribbean Region of the American Society for Horticultural Science* 9, 152–159.

Hatton, T.T. and Reeder, W.F. 1966. Controlled atmosphere storage of 'Keitt' mangoes. *Proceedings of the Caribbean Region of the American Society for Hoticultural Science* 10, 114–119.

Hatton, T.T. and Reeder, W.F. 1968. Controlled atmosphere storage of papayas. *Proceedings of the Tropical Region of the American Society for Horticultural Science* 13, 251–256.

Hatton, T.T. and Spalding, D.H. 1990. Controlled atmosphere storage of some tropical fruits. In: Calderon, M. and Barkai-Golan, R. (eds) *Food Preservation by Modified Atmospheres*. CRC Press, Boca Raton (Florida), Ann Arbor (Michigan), Boston (Massachusetts), pp. 301–313.

Hatton, T.T., Reeder, W.F. and Campbell, C.W. 1965. *Ripening and storage of Florida mangoes*. Market Reseach Report 725, United States Department of Agriculture, Washington DC.

Hatton, T.T., Cubbedge, R.H. and Grierson, W. 1975. Effects of prestorage carbon dioxide treatments and delayed storage on chilling injury of 'Marsh' grapefruit. *Proceedings of the Florida State Society for Horticultural Science* 88, 335.

He, C.J. and Davies, F.T. 2012. Ethylene reduces plant gas exchange and growth of lettuce grown from seed to harvest under hypobaric and ambient total pressure. *Journal of Plant Science* 169, 369–378.

He, C.J., Davies, F.T., Lacey, R.E., Drew, M.C. and Brown, D.L. 2003. Effect of hypobaric conditions on ethylene evolution and growth of lettuce and wheat. *Journal of Plant Physiology* 160, 1341–1350.

Heimdal, H., Kuhn, B.F., Poll, L. and Larsen, L.M. 1995. Biochemical changes and sensory quality of shredded and MA-packaged iceberg lettuce. *Journal of Food Science* 60, 1265–1268.

Hellickson, M.L., Adre, N., Staples, J. and Butte, J. 1995. Computer controlled evaporator operation during fruit cool-down. *Technologias de cosecha y postcosecha de frutas y hortalizas / Harvest and postharvest technologies for fresh fruits and vegetables. Proceedings of a Conference held in Guanajuato, Mexico, 20–24 February*, 546–553.

Heltoft, P., Wold, A-B. and Molteberg, E.L. 2016. Effect of ventilation strategy on storage quality indicators of processing potatoes with different maturity levels at harvest. *Postharvest Biology and Technology* 117, 21–29.

Henderson, J.R. and Buescher, R.W. 1977. Effects of sulfur dioxide and controlled atmospheres on broken-end discoloration and processed quality attributes in snap beans. *Journal of the American Society for Horticultural Science* 102, 768–770.

Hendrickx, M., Ludikhuyze, L., Van den Broeck, I. and Weemaes, C. 1998. Effects of high pressure on enzymes related to food quality. *Trends in Food Science and Technology* 9, 197–203.

Hennecke, C., Köpcke, D., and Dierend, W. 2008. Dynamische Absenkung des Sauerstoffgehaltes bei der Lagerung von Äpfeln [Storage of apples in dynamic controlled atmosphere]. *Erwerbs-Obstbau* 50, 19–29.

Henz, G.P., Reifschneide, F.J.B. and dos Santos, F.F. 2005. Reação de genótipos de mandioquinha-salsa à podridão-mole das raízes causada por Pectobacterium chrysanthemi. *Pesquisa Agropecuária Brasileira* 40 no.1. doi: 10.1590/S0100-204X2005000100014

Henze, J. 1989. Storage and transport of Pleurotus mushrooms in atmospheres with high CO_2 concentrations. *Acta Horticulturae* 258, 579–584.

Hermansen, A. and Hoftun, H. 2005. Effect of storage in controlled atmosphere on post-harvest infections of *Phytophthora brassicae,* and chilling injury in Chinese cabbage (*Brassica rapa* L *pekinensis* (Lour) Hanelt). *Journal of the Science of Food and Agriculture* 85, 1365–1370.
Herreid, C.F. 1980. Hypoxia in invertebrates. *Comparative Biochemistry and Physiology* 67, 311–320.
Hertog, M.L.A.T.M., Nicholson, S.E. and Jeffery, P.B. 2004. The effect of modified atmospheres on the rate of firmness change of 'Hayward' kiwifruit. *Postharvest Biology and Technology* 31, 251–261
Hesselman, C.W. and Freebairn, H.T. 1969. Rate of ripening of initiated bananas as influenced by oxygen and ethylene. *Journal of the American Society for Horticultural Science* 94, 635–637.
Hewett, E.W. and Thompson, C.J. 1988. Modified atmosphere storage for reduction of bitter pit in some New Zealand apple cultivars. *New Zealand Journal of Experimental Agriculture* 16, 271–277.
Hewett, E.W. and Thompson, C.J. 1989. Modified atmosphere storage and bitter pit reduction in 'Cox's Orange Pippin' apples. *Scientia Horticulturae* 39, 117–129.
Hill, G.R. 1913. *Respiration of fruits and growing plant tissue in certain cases, with reference to ventilation and fruit storage.* Bulletin 330. Agricultural Experiment Station, Cornell University, Ithaca, New York.
Hill, S. 1997. Squeezing the death out of food. *New Scientist* 2077, 28–32.
Hite, B.H., Giddings, N.J. and Weakley, C.W. 1914. The effect of pressure on certain microorganisms encountered in the preservation of fruits and vegetables. *Bulletin 146, West Virginia University Experimental Station USA* 3–67.
HMSO. 1964. *Hygrometric Tables, Part III: Aspirated Psychrometer Readings Degrees Celsius.* Her Majesty's Stationery Office, London.
Ho, Q.T., Rogge, S., Verboven, P., Verlinden, B.E. and Nicolaï, B.M. 2016. Stochastic modelling for virtual engineering of controlled atmosphere storage of fruit. *Journal of Food Engineering* 176, 77–87.
Hodges, D.M., Munro, K.D., Forney, C.F. and McRae, K. 2006. Glucosinolate and free sugar content in cauliflower (*Brassica oleracea* var. *botrytis* cv. Freemont) during controlled-atmosphere storage. *Postharvest Biology and Technology* 40, 123–132.
Hoehn, E., Prange, R.K. and Vigneault, C. 2009. Storage technology and applications and storage temperature on the shelf-life and quality of pomegranate fruits cv. Ganesh. *Postharvest Biology and Technology* 22, 61–69.
Hofman, P.J., Joblin-Décor, M., Meiburg, G.F., MacNish, A.J. and Joyce, D.C. 2001. Ripening and quality responses of avocado, custard apple, mango and papaya fruit to 1-methylcyclopropene. *Australian Journal of Experimental Agriculture* 41, 567–572.
Holb, I.J., Balla, B., Vámos, A. and Gáll, J.M. 2012. Influence of preharvest calcium applications, fruit injury, and storage atmospheres on postharvest brown rot of apple. *Postharvest Biology and Technology* 67, 29–36.
Homma, T., Inoue, E., Matsuda, T. and Hara, H. 2008. Changes in fruit quality factors in Japanese chestnut (*Castanea crenata* Siebold & Zucc.) during long-term storage. *Horticultural Research (Japan)* 7, 591–598.
Hong, Y. Wang, C.Y., Gu, Y. and He, S.A. 2006. Effects of different packaging materials on the physiology and storability of blueberry fruits. *Journal of Fruit Science* 23, 631–634.
Hong, J.H., Hwang, S.K., Chung G.H. and Cowan, A.K. 2007. Influence of applied lysophosphatidylethanolamine on fruit quality in 'Thompson Seedless' table grapes. *Journal of Applied Horticulture* 9, 112–114.
Hong, Q.Z., Sheng, H.Y., Chen, Y.F. and Yang, S.J. 1983. Effects of CA storage with a silicone window on satsuma oranges. *Journal of Fujiart Agricultural College* 12, 53–60.
Hong, Y.P., Choi, J.H. and Lee, S.K. 1997. Optimum CA condition for four apple cultivars grown in Korea. *Postharvest Horticulture Series – Department of Pomology, University of California* 16, 241–245.
Horie, Y., Kimura, K., Ida, M., Yosida, Y. and Ohki, K. 1991. Jam preservation by pressure pasteurization. *Nippon Nogeiku Kaichi* 65, 975–980.
Horvitz, S., Yommi, A., Lopez Camelo, A. and Godoy, C. 2004. Effects of maturity stage and use of modified atmospheres on quality of sweet cherries cv. Sweetheart. *Revista de la Facultad de Ciencias Agrarias, Universidad Nacional de Cuyo* 36, 39–48.
Hotchkiss, J.H. and Banco, M.J. 1992. Influence of new packaging technologies on the growth of microorganisms in produce. *Journal of Food Protection* 55, 815–820.
Houck, L.G., Aharoni, Y. and Fouse, D.C. 1978. Colour changes in orange fruit stored in high concentrations of O_2 and in ethylene. *Proceedings of the Florida State Horticultural Society* 91, 136–139.
Hribar, J., Bitenc, F. and Bernot, D. 1977. Optimal harvesting date and storage conditions for the apple cultivar Stayman Red. *Jugoslovensko Vocarstvo* 10, 687–693.
Hribar, J., Plestenjak, A., Vidrih, R., Simcic, M. and Sass, P. 1994. Influence of CO_2 shock treatment and ULO storage on apple quality. *Acta Horticulturae* 368, 634–640.
HSE. 1991. *Confined Spaces.* Information Sheet 15. Health and Safety Executive UK.

Hu, H., Li, P., Wang, Y. and Gu, R. 2014a. Hydrogen-rich water delays postharvest ripening and senescence of kiwifruit. *Food Chemistry* 156, 100–109.

Hu, H., Zhao, S., Li, P. and Shen, W. 2018. Hydrogen gas prolongs the shelf life of kiwifruit by decreasing ethylene biosynthesis. *Postharvest Biology and Technology* 135, 123–130.

Hu, M., Yang, D., Huber, D.J., Jiang, Y., Li, M., Gao, Z. and Zhang, Z. 2014b. Reduction of postharvest anthracnose and enhancement of disease resistance in ripening mango fruit by nitric oxide treatment. *Postharvest Biology and Technology* 97, 115–122.

Hu, W., Sun, D.-W. and Blasco. J. 2017. Rapid monitoring 1-MCP-induced modulation of sugars accumulation in ripening 'Hayward' kiwifruit by Vis/NIR hyperspectral imaging. *Postharvest Biology and Technology* 125, 168–180.

Hu, W.Z., Tanaka, S., Uchino, T., Akimoto, K., Hamanaka, D. and Hori, Y. 2003. Atmospheric composition and respiration of fresh shiitake mushroom in modified atmosphere packages. *Journal of the Faculty of Agriculture, Kyushu University* 48, 209–218.

Huang, C.C., Huang, H.S. and Tsai, C.Y. 2002. A study on controlled atmosphere storage of cabbage with sealed plastic tent. *Journal of Agricultural Research of China* 51, 33–42.

Huang, X.M., Yuan, W.Q., Wang, C., Li, J.G., Huang, H.B., Shi, L. and Jinhua, Y. 2004. Linking cracking resistance and fruit desiccation rate to pericarp structure in litchi (*Litchi chinensis* Sonn.). *Journal of Horticultural Science & Biotechnology* 79, 897–905.

Huelin, F.E. and Tindale, G.B. 1947. The gas storage of Victorian apples. *Journal of Agriculture, Victorian Department of Agriculture* 45, 74–80.

Hughes, P.A., Thompson, A.K., Plumbley, R.A. and Seymour, G.B. 1981. Storage of capsicums *Capsicum annum* [L.], Sendt. under controlled atmosphere, modified atmosphere and hypobaric conditions. *Journal of Horticultural Science* 56, 261–265.

Hulme, A.C. 1956. CO_2 injury and the presence of succinic acid in apples. *Nature* 178, 218.

Hulme, A.C. 1970. *The Biochemistry of Fruits and their Products*. Vol. 1. Academic Press, London and New York.

Hulme, A.C. 1971. *The Biochemistry of Fruits and their Products*. Vol. 2. Academic Press, London and New York.

Hunsche, M., Brackmann, A. and Ernani, P.R. 2003. Effect of potassium fertilization on the postharvest quality of 'Fuji' apples. *Pesquisa Agropecuaria Brasileira* 38, 489–496.

Huque, R., Wills, R.B.H., Pristijono, P. and Golding, J.B. 2013. Effect of nitric oxide (NO) and associated control treatments on the metabolism of fresh-cut apple slices in relation to development of surface browning. *Postharvest Biology and Technology* 78, 16–23.

IIR. 1978. Banana CA storage. *Bulletin of the International Institute of Refrigeration* 18, 312.

Ilangantileke, S. and Salokhe, V. 1989. Low pressure atmosphere storage of Thai mango. In: Fellman, J.K. (ed.) *Proceedings of the Fifth International Controlled Atmosphere Research Conference, Wenatchee, Washington, USA, 14–16 June. Vol.* 2: *Other commodities and storage recommendations.* Washington State University, Pullman, Washington, pp. 103–117.

Ilangantileke, S.G., Turla, L. and Chen, R. 1989. *Pre-treatment and hypobaric storage for increased storage life of mango*. Paper 896036. Canadian/American Society of Agricultural Engineering, Quebec.

Imahori, Y., Kishioka, I., Uemura, K., Makita, E., Fujiwara, H. *et al.* 2007. Effects of short-term exposure to low oxygen atmospheres on phsyiological responses of sweetpotato roots. *Journal of the Japanese Society for Horticultural Science* 76, 258–265.

Imakawa, S. 1967. Studies on the browning of Chinese yam. *Memoir of the Faculty of Agriculture, Hokkaido University, Japan* 6, 181.

Inaba, A., Kiyasu, P. and Nakamura, R. 1989. Effects of high CO_2 plus low O_2 on respiration in several fruits and vegetables. *Scientific Reports of the Faculty of Agriculture, Okayama University* 73, 27–33.

Ionescu, L., Millim, K., Batovici, R., Panait, E. and Maraineanu, L. 1978. Resarch on the storage of sweet and sour cherries in cold stores with normal and controlled atmospheres. *Lucrari Stiintifice, Institutul de Cerctari Pentru Valorifocarea Legumelor si Fructelor* 9, 43–51.

Isenberg, F.M.R. 1979. Controlled atmosphere storage of vegetables. *Horticultural Review* 1, 337–394.

Isenberg, F.M.R., Thomas, T.H., Abed-Rahaman, M., Pendergrass, A., Carroll, J.C. and Howell, L. 1974. The role of natural growth regulators in rest, dormancy and regrowth of vegetables during winter storage. *Acta Horticulturae* 38, 95–125.

Isenberg, F.M.R. and Sayles, R.M. 1969. Modified atmosphere storage of Danish cabbage. *Proceedings of the American Society of Horticultural Science* 94, 447–449.

Ishigami, Y. and Goto, E. 2008. Plant growth under hypobaric conditions. *Journal of Science and High Technology in Agriculture* 20, 228–235.

Isidoro, N. and Almeida, D.P.F. 2006. Alpha-farnesene, conjugated trienols, and superficial scald in 'Rocha' pear as affected by 1-methylcyclopropene and diphenylamine. *Postharvest Biology and Technology* 42, 49–56.

Isshiki, M., Terai, H. and Suzuki, Y. 2005. Effect of 1-methylcyclopropene on quality of sudachis (*Citrus sudachi* hort. ex Shirai) during storage. *Food Preservation Science*, 31, 61–65.

Isolcell Italia. 2018. *ISOSTORE (DCA-CF). The dynamic controlled atmosphere by using fluorescence sensors.* Available at: http://storage.isolcell.com/en/pagina/1444726470-en/1479464957-en/1479466126-en (accessed 9 July 2018).

Itai, A. and Tanahashi, T. 2008. Inhibition of sucrose loss during cold storage in Japanese pear (*Pyrus pyrifolia* Nakai) by 1-MCP. *Postharvest Biology and Technology* 48, 355–363.

Ito, S., Kakiuchi, N., Izumi, Y. and Iba, Y. 1974. Studies on the controlled atmosphere storage of satsuma mandarin. *Bulletin of the Fruit Tree Research Station, B. Okitsu* 1, 39–58.

Itoo, S., Matsuo, T. Ibushi, Y. and Tamari, N. 1987. Seasonal changes in the levels of polyphenols in guava fruit and leaves and some of their properties. *Journal of the Japanese Society of Horticultural Science* 56, 107–113.

Iturralde-García, R.D., Borboa-Flores, J., Cinco-Moroyoqui, F.J., Riudavets, J., Toro-Sánchez, C.L.D. *et al.* 2016. Effect of controlled atmospheres on the insect *Callosobruchus maculatus* Fab. in stored chickpea. *Journal of Stored Products Research* 69, 78–85.

Izumi, H., Watada, A.E. and Douglas, W. 1996a. Optimum O_2 or CO_2 atmosphere for storing broccoli florets at various temperatures. *Journal of the American Society for Horticultural Science* 121, 127–131.

Izumi, H., Watada, A.E., Ko, N.P. and Douglas, W. 1996b. Controlled atmosphere storage of carrot slices, sticks and shreds. *Postharvest Biology and Technology* 9, 165–172.

Jacobsson, A., Nielsen, T. and Sjoholm, I. 2004a. Effects of type of packaging material on shelf-life of fresh broccoli by means of changes in weight, colour and texture. *European Food Research and Technology* 218, 157–163.

Jacobsson, A., Nielsen, T., Sjoholm, I. and Wendin, K. 2004b. Influence of packaging material and storage condition on the sensory quality of broccoli. *Food Quality and Preference* 15, 301–310.

Jacxsens, L., Devlieghere, F. and Debevere, J. 2002. Predictive modeling for packaging design: equilibrium modified atmosphere packages of fresh-cut vegetables subjected to a simulated distribution chain. *International Journal of Food Microbiology* 73, 331–341.

Jacxsens, L., Devliegherre, F., Van der Steen, C., Siro, I. and Debevere, J. 2003. Application of ethylene adsorbers in combination with high oxygen atmospheres for the storage of strawberries and raspberries. *Acta Horticulturae* 600, 311–318.

Jacxsens, L., Luning, P.A., van der Vorst, J.G.A.J., Devileghere, F., Leemans, R. and Uytendaele, M. 2010. Simulations modelling and risk assessment as tools to identify the impact of climate change on microbiological food safety – The case study of fresh produce supply chain. *Food Research International* 43, 1925–1935.

Jadhao, S.D., Borkar, P.A., Ingole, M.N., Murumkar, R.B. and Bakane, P.H. 2007. Storage of kagzi lime with different pretreatments under ambient condition. *Annals of Plant Physiology* 21, 30–37.

Jahn, O.L., Chace, W.G. and Cubbedge, R.H. 1969. Degreening of citrus fruits in response to varying levels of oxygen and ethylene. *Journal of the American Society for Horticultural Science* 94, 123–125.

James, H., Jobling, J., Tanner, D., Tustin, S. and Wilkinson, I. 2010. Climatic conditions during growth relate to risk of Pink Lady™ apples developing flesh browning during storage. *Acta Horticulturae* 857, 197–203.

Jameson, J. 1993. CA storage technology – recent developments and future potential. *COST 94. The postharvest treatment of fruit and vegetables: controlled atmosphere storage of fruit and vegetables. Proceedings of a Workshop, 22–23 Apr. 1993, Milan, Italy*, 1–12.

Jamieson, W. 1980. Use of hypobaric conditions for refrigerated storage of meats, fruits and vegetables. *Food Technology* 34, 64–71.

Janave, M.T. 2016. Biochemical changes induced due to Staphylococcal infection in spongy Alphonso mango (*Mangifera indica* L.) fruits. *Journal of Crop Science and Biotechnology* 10, 167–174.

Jankovic, M. and Drobnjak, S. 1992. The influence of the composition of cold room atmosphere on the changes of apple quality. Part 1. Changes in chemical composition and organoleptic properties. *Review of Research Work at the Faculty of Agriculture, Belgrade* 37, 135–138.

Jankovic, M. and Drobnjak, S. 1994. The influence of cold room atmosphere composition on apple quality changes. Part 2. Changes in firmness, mass loss and physiological injuries. *Review of Research Work at the Faculty of Agriculture, Belgrade* 39, 73–78.

Jansasithorn, R. and Kanlavanarat, S. 2006. Effect of 1-MCP on physiological changes in banana 'Khai'. *Acta Horticulturae* 712, 723–728.

Jay, J.M., Loessner, M.J., and Golden, D.V. 2005. *Modern Food Microbiology*, 3rd edn. Springer, New York.

Jeffery, D., Smith, C., Goodenough, P.W., Prosser, T. and Grierson, D. 1984. Ethylene independent and ethylene dependent biochemical changes in ripening tomatoes. *Plant Physiology* 74, 32.

Jiang, T. 2013. Effect of alginate coating on physicochemical and sensory qualities of button mushrooms (*Agaricus bisporus*) under a high oxygen modified atmosphere. *Postharvest Biology and Technology* 76, 91–97.

Jiang, T., Jahangir, M.M., Jiang, Z., Lu, X. and Ying, T. 2010. Influence of UV-C treatment on antioxidant capacity, antioxidant enzyme activity and texture of postharvest shiitake (*Lentinus edodes*) mushrooms during storage. *Postharvest Biology and Technology* 56, 209–215.

Jiang, W., Zhang, M., He, J. and Zhou, L. 2004. Regulation of 1-MCP-treated banana fruit quality by exogenous ethylene and temperature. *Food Science and Technology International* 10, 15–20.

Jiang, Y.M. 1999. Low temperature and controlled atmosphere storage of fruit of longan (*Dimocarpus longan* Lour.). *Tropical Science* 39, 98–101.

Jiang, Y.M. and Li, Y.B. 2001. Effects of chitosan coating on postharvest life and quality of longan fruit. *Food Chemistry* 73, 139–143.

Jiang, Y.M., Joyce, D.C. and MacNish, A.J. 1999. Extension of the shelf-life of banana fruit by 1-methylcyclopropene in combination with polyethylene bags. *Postharvest Biology and Technology* 16, 187–193.

Jiang, Y.M., Joyce, D.C. and Terry, L.A., 2001. 1-Methylcyclopropene treatment affects strawberry fruit decay. *Postharvest Biology and Technology* 23, 227–232.

Jiao, S., Johnson, J.A., Fellman, J.K., Mattinson, D.S., Tang, J., Davenport, T.L. and Wang, S. 2012. Evaluating the storage environment in hypobaric chambers used for disinfesting fresh fruits. *Biosystems Engineering* 111, 271–279.

Jiao, S., Johnson, J.A., Tang, J., Mattinson, D.S., Fellman, J.K., Davenport, T.L. and Wang, S. 2013. Tolerance of codling moth and apple quality associated with low pressure/low temperature treatments. *Postharvest Biology and Technology* 85, 136–140.

Jimenez-Cuesta, M., Cuquerella, J. and Martínez-Javega, J.M. 1981. Determination of a colour index for citrus fruit degreening. *Proceeding of the International Society of Citriculture* 2, 750–753.

Jin, A.X., Wang, Y.P. and Liang, L.S. 2006. Effects of atmospheric pressure on the respiration and softening of DongZao jujube fruit during hypobaric storage. *Journal of the Northwest Forestry University* 21, 143–146.

Joanny, P., Steinberg, J., Robach, P., Richalet, J.P., Gortan, C., Gardette, B. and Jammes, Y. 2001. Operation Everest III (Comex'97): the effect of simulated severe hypobaric hypoxia on lipid peroxidation and antioxidant defence systems in human blood at rest and after maximal exercise. *Resuscitation* 49, 307–314.

Jobling, J.J. and McGlasson, W.B. 1995. A comparison of ethylene production, maturity and controlled atmosphere storage life of Gala, Fuji and Lady Williams apples (*Malus domestica*, Borkh.). *Postharvest Biology and Technology* 6, 209–218.

Jobling, J.J., McGlasson, W.B., Miller, P. and Hourigan, J. 1993. Harvest maturity and quality of new apple cultivars. *Acta Horticulturae* 343, 53–55.

Jobling, J., Brown, G., Mitcham, E., Tanner, D., Tustin, S., Wilkinson, I. and Zanella, A. 2004. Flesh browning of Pink Lady™ apples: why do symptoms occur? Results from an international collaborative study. *Acta Horticulturae* 682, 851–858.

Jog, K.V. 2004. Cold storage industry in India. Available at: http://www.ninadjog.com/krishna/ColdStoragesIndia.pdf (accessed January 2010).

Johnson, D.S. 1994a. Prospects for increasing the flavour of 'Cox's Orange Pippin' apples stored under controlled atmosphere conditions. *COST 94. The post-harvest treatment of fruit and vegetables: quality criteria. Proceedings of a Workshop, 19–21 April 1994, Bled, Slovenia*, 39–44.

Johnson, D.S. 1994b. Storage conditions for apples and pears. *East Malling Research Association Review 1994–1995.*

Johnson, D.S. 2001. Storage regimes for Gala. *The Apple and Pear Research Council News* 27, 5–8.

Johnson, D.S. 2008. Factors affecting the efficacy of 1-MCP applied to retard apple ripening. *Acta Horticulturae* 796, 59–67.

Johnson, D.S. and Colgan, R.J. 2003. Low ethylene CA induces adverse effects on the quality of Queen Cox apples treated with AVG during fruit development. *Acta Horticulturae* 600, 441–448.

Johnson, D.S. and Ertan, U. 1983. Interaction of temperature and O_2 level on the respiration rate and storage quality of Idared apples. *Journal of Horticultural Science* 58, 527–533.

Johnson, D.S., Dover, C.J. and Pearson, K. 1993. Very low oxygen storage in relation to ethanol production and control of superficial scald in Bramley's Seedling apples. *Acta Horticulturae* 326, 175–182.

Johnson, G.I., Boag, T.S., Cooke, A.W., Izard, M., Panitz, M. and Sangchote, S. 1990a. Interaction of post harvest disease control treatments and gamma irradiation on mangoes. *Annals of Applied Biology* 116, 245–257.

Johnson, G.I., Sangchote, S. and Cooke, A.W. 1990b. Control of stem end rot *Dothiorella dominicana* and other postharvest diseases of mangoes cultivar Kensington Pride during short and long term storage. *Tropical Agriculture* 67, 183–187.

Johnson, J.A. and Zettler, J.L. 2009. Response of postharvest tree nut lepidopoteran pests to vacuum treatments. *Journal of Economic Entomology* 102, 2003–2010.

Johnson, P.R. and Ecker, J.R. 1998. The ethylene gas signal transduction pathway: a molecular perspective. *Annual Review of Genetics* 32, 227–254.

Jomori, M.L.L., Kluge, R.A. and Jacomino, A.P. 2003. Cold storage of 'Tahiti' lime treated with 1-methylcyclopropene. *Scientia Agricola* 60, 785–788.

Joyce, D.C. 1988. Evaluation of a ceramic-impregnated plastic film as a postharvest wrap. *HortScience* 23, 1088.

Kader, A.A. 1985. Modified atmosphere and low-pressure systems during transport and storage. In: Kader, A.A., Kasmire, R.F., Mitchell, F.G., Reid, M.S., Sommer, N.F. and Thompson, J.F. (eds). *Postharvest Technology of Horticultural Crops.* Division of Agriculture and Natural Resources, Cooperative Extension, University of California, Oakland, California, pp. 59–60.

Kader, A.A. 1986. Biochemical and physiological basis for effects on controlled and modified atmospheres on fruits and vegetables. *Food Technology* 40, 99–104.

Kader, A.A. 1988. Comparison between 'Semperfresh' and 'Nutri Save' coatings on 'Granny Smith' apples. *Perishables Handling, Postharvest Technology of Fresh Horticultural Crops* 63, 4–5.

Kader, A.A. 1989. A summary of CA requirements and recommendations for fruit other than pome fruits. In: Fellman, J.K. (ed.) *Proceedings of the Fifth International Controlled Atmosphere Conference, Wenatchee, Washington, USA, 14–16 June 1989. Vol. 2: Other commodities and storage recommendations.* Washington State University, Pullman, Washington, pp. 303–328.

Kader, A.A. 1992. *Postharvest Technology of Horticultural Crops,* 2nd edn. ANR Publications 3311. Division of Agriculture and Natural Resources, University of California, Oakland, California.

Kader, A.A. 1993. Modified and controlled atmosphere storage of tropical fruits. In *Postharvest handling of tropical fruits. Australian Centre for International Agricultural Research Proceedings* 50, 239–249.

Kader, A.A. 1997. A summary of CA requirements and recommendations for fruits other than pome fruits. *Seventh International Controlled Atmosphere Research Conference, July 13–18 1997,* University of California, Davis, California [Abstract], 49.

Kader, A.A. 2003. A summary of CA requirements and recommendations for fruits other than apples and pears. *Acta Horticulturae* 600, 737–740.

Kader, A.A. and Ben-Yehoshua, S. 2000. Effects of superatmospheric oxygen levels on postharvest physiology and quality of fresh fruits and vegetables. *Postharvest Biology and Technology* 20, 1–13.

Kader, A.A., Chastagner, G.A., Morris, L.L. and Ogawa, J.M. 1978. Effects of carbon monoxide on decay, physiological responses, ripening, and composition of tomato fruits. *Journal of the American Society for Horticultural Science* 103, 665–670.

Kader, A.A., Chordas, A. and Elyatem, S. 1984. Responses of pomegranates to ethylene treatment and storage temperature. *California Agriculture* 38, 14–15.

Kaewsuksaeng, S., Urano, Y., Aiamla-or, S., Shigyo, M. and Yamauchi, N. 2012. Effect of UV-B irradiation on chlorophyll-degrading enzyme activities and postharvest quality in stored lime (*Citrus latifolia* Tan.) fruit. *Postharvest Biology and Technology* 61, 124–130.

Kaji, H., Ikebe, T. and Osajima, Y. 1991. Effects of environmental gases on the shelf-life of Japanese apricot. *Journal of the Japanese Society for Food Science and Technology* 38, 797–803.

Kajiura, I. 1972. Effects of gas concentrations on fruit. V. Effects of CO_2 concentrations on natsudaidai fruit. *Journal of the Japanese Society for Horticultural Science* 41, 215–222.

Kajiura, I. 1973. The effects of gas concentrations on fruits. VII. A comparison of the effects of CO_2 at different relative humidities, and of low O_2 with and without CO_2 in the CA storage of natsudaidai. *Journal of the Japanese Society for Horticultural Science* 42, 49–55.

Kajiura, I. 1975. CA storage and hypobaric storage of white peach 'Okubo'. *Scientia Horticulturae* 3, 179–187.

Kajiura, I. and Iwata, M. 1972. Effects of gas concentrations on fruit. IV. Effects of O_2 concentration on natsudaidai fruits. *Journal of the Japanese Society for Horticultural Science* 41, 98–106.

Kamath, O.C., Kushad, M.M. and Barden, J.A. 1992. Postharvest quality of 'Virginia Gold' apple fruit. *Fruit Varieties Journal* 46, 87–89.

Kane, O. and Marcellin, P. 1979. Effects of controlled atmosphere on the storage of mango. *Fruits* 34, 123–129.

Kanellis, A.K., Solomos, T. and Mattoo, A.K. 1989a. Hydrolytic enzyme activities and protein pattern of avocado fruit ripened in air and in low oxygen, with and without ethylene. *Plant Physiology* 90, 259–266.

Kanellis, A.K., Solomos, T., Mehta, A.M. and Mattoo, A.K., 1989b. Decreased cellulase activity in avocado fruit subjected to 2.5 kPa O_2 correlates with lowered cellulase protein and gene transcripts levels. *Plant Cell Physiology* 30, 817–823.

Kanellis, A.K., Loulakakis, K.A., Hassan, M. and Roubelakis-Angelakis, K.A. 1993. Biochemical and molecular aspects of low oxygen action on fruit ripening. In: Pech, C.J., Latche, A., Balague, C. (eds) *Cellular and Molecular Aspects of the Plant Hormone Ethylene*. Kluwer Academic Publishers, Dordrecht, pp. 117–122.

Kang, F., Hallman, G.J., Wei, Y., Zhang, F. and Li, Z. 2012. Effect of X-ray irradiation on the physical and chemical quality of America red globe grape. *African Journal of Biotechnology* 11, 7966–7972.

Kang, H.M. and Park, K.W. 2000. Comparison of storability on film sources and storage temperature for Oriental melon in modified atmosphere storage. *Journal of the Korean Society for Horticultural Science* 41, 143–146.

Kantola, M. and Helén, H. 2001. Quality changes in organic tomatoes packaged in biodegradable plastic films. *Journal of Food Quality* 24, 167–176.

Karabulut, O.A., Gabler, F.M., Mansour, M. and Smilanick, J.L. 2004.Postharvest ethanol and hot water treatments of table grapes to control gray mold. *Postharvest Biology and Technology* 34, 169–177.

Karaoulanis, G. 1968. The effect of storage under controlled atmosphere conditions on the aldehyde and alcohol contents of oranges and grape. *Annual Report of the Ditton Laboratory, 1967–1968*. Agricultural Research Council, East Malling, UK.

Karbowiak, T., Debeaufort, F. and Voilley, A., 2007. Influence of thermal process on structure and functional properties of emulsion-based edible films. *Food Hydrocolloids* 21, 879–888.

Kashimura, Y., Hayama, H. and Ito, A. 2010. Infiltration of 1-methylcyclopropene under low pressure can reduce the treatment time required to maintain apple and Japanese pear quality during storage. *Postharvest Biology and Technology* 57, 14–18.

Kasim, R. and Kasim, M.U. 2012. UV-C treatment of fresh cut cress (*Lepidium sativum* L.) enhanced chlorophyll content and prevent leaf yellowing. *World Applied Science Journal* 17, 509–515.

Kawagoe, Y., Morishima, H., Seo, Y. and Imou, K. 1991. Development of a controlled atmosphere-storage system with a gas separation membrane part 1 – apparatus and its performance. *Journal of the Japanese Society of Agricultural Machinery* 53, 87–94.

Kay, D.E. 1987. *Crop and Product Digest No. 2 – Root Crops*. 2nd edn (revised by Gooding, E.G.B.). Tropical Development and Research Institute, London.

Kaynaş, K., Ozelkok, S. and Surmeli, N. 1994. Controlled atmosphere storage and modified atmosphere packaging of cauliflower. *Commissions C2,D1,D2/3 of the International Institute of Refrigeration International Symposium June 8–10 Istanbul Turkey*, 281–288.

Kaynaş, K., Sakaldaş, M., Kuzucu, F.C. and Uyar, E. 2009. The combined effects of 1-methylcyclopropane and modified atmosphere packaging on fruit quality of Fuyu persimmon fruit during storage, *10th International Controlled & Modified Atmosphere Research Conference, 4–7 April 2009 Turkey* [Abstract], 20.

Kays, S.J. 1997. *Postharvest Physiology of Perishable Plant Products*. Exon Press, Athens, Georgia.

Ke, D.Y. and Kader, A.A. 1989. Tolerance and responses of fresh fruits to O_2 levels at or below 1%. In: Fellman, J.K. (ed.) *Proceedings of the Fifth International Controlled Atmosphere Conference, Wenatchee, Washington, USA, 14–16 June, 1989. Vol. 2: Other commodities and storage recommendations*. Washington State University, Pullman, Washington, pp. 209–216.

Ke, D.Y. and Kader, A.A. 1992a. Potential of controlled atmospheres for postharvest insect disinfestation of fruits and vegetables. *Postharvest News and Information* 3, 31N–37N.

Ke, D.Y. and Kader, A.A 1992b. External and internal factors influence fruit tolerance to low O_2 atmospheres. *Journal of the American Society for Horticultural Science* 117, 913–918.

Ke, D.Y., Rodriguez Sinobas, L. and Kader, A.A. 1991a. Physiology and prediction of fruit tolerance to low O_2 atmospheres. *Journal of the American Society for Horticultural Science* 116, 253–260.

Ke, D.Y., Rodriguez-Sinobas, L. and Kader, A.A. 1991b. Physiological responses and quality attributes of peaches kept in low O_2 atmospheres. *Scientia-Horticulturae* 47, 295–303.

Ke, D.Y., El Wazir, F., Cole, B., Mateos, M. and Kader, A.A. 1994a. Tolerance of peach and nectarine fruits to insecticidal controlled atmospheres as influenced by cultivar, maturity, and size. *Postharvest Biology and Technology* 4, 135–146.

Ke, D.Y., Zhou, L. and Kader, A.A. 1994b. Mode of O_2 and CO_2 action on strawberry ester biosynthesis. *Journal of the American Society for Horticultural Science* 119, 971–975.

Keleg, F.M., Etman, A.A., Attia, M.M. and Nagy, N.M. 2003. Physical and chemical properties of Banzahir lime fruits during cold storage as affected by some postharvest treatments. *Alexandria Journal of Agricultural Research* 48, 77–84.

Kelly, M.O. and Saltveit Jr, M.E. 1988. Effect of endogenously synthesized and exogenously applied ethanol on tomato fruit ripening. *Plant Physiology* 88, 143–147.

Kende, H. 1993. Ethylene biosynthesis. *Annual Review of Plant Physiology and Plant Molecular Biology* 44, 283–307.

Kerbel, E., Ke, D. and Kader A.A. 1989. Tolerance of 'Fantasia' nectarine to low O_2 and high CO_2 atmospheres. In: Reid, D. (ed). In: Reid, D. (ed.) *Proceedings International Institute of Refrigeration Conference: Technical Innovations in Freezing and Refrigeration of Fruits and Vegetables, 9–12 July*, University of California, Davis, pp. 325–331.

Kerdchoechuen, O., Laohakunjit, N., Tussavil, P., Kaisangsri, N. and Matta, F.B. 2011. Effect of starch-based edible coatings on quality of minimally processed pummelo (*Citrus maxima* Merr.). *International Journal of Fruit Science* 11, 410–423.

Kesari, R., Trivedi, P.K. and Nath, P. 2007. Ethylene-induced ripening in banana evokes expression of defense and stress related genes in fruit tissue. *Postharvest Biology and Technology* 46, 136–143.

Kesari, R., Trivedi, P.K. and Nath, P. 2010. Gene expression of pathogenesis-related protein during banana ripening and after treatment with 1-MCP. *Postharvest Biology and Technology* 56, 64–70.

Kester, J.J. and Fennema, O.R. 1986. Edible films and coatings – a review. *Food Technology* 40, 46–57.

Khan, A.S. and Singh, Z. 2007. 1-MCP regulates ethylene biosynthesis and fruit softening during ripening of 'Tegan Blue' plum. *Postharvest Biology and Technology* 43, 298–306.

Khan, A.S., Singh, Z. and Swinny, E.E. 2009. Postharvest application of 1-Methylcyclopropene modulates fruit ripening, storage life and quality of 'Tegan Blue' Japanese plum kept in ambient and cold storage. *International Journal of Food Science & Technology* 44, 1272–1280.

Khanbari, O.S. and Thompson, A.K 1994. The effect of controlled atmosphere storage at 4°C on crisp colour and on sprout growth, rotting and weight loss of potato tubers. *Potato Research* 37, 291–300.

Khanbari, O.S. and Thompson, A.K. 1996. Effect of controlled atmosphere, temperature and cultivar on sprouting and processing quality of stored potatoes. *Potato Research* 39, 523–531.

Khanbari, O.S. and Thompson, A.K. 2004. The effect of controlled atmosphere storage at 8 °C on fry colour, sprout growth and rotting of potato tubers cv. Record. *University of Aden, Journal of Natural and Applied Sciences* 8, 331–340.

Khitron, Ya I. and Lyublinskaya, N.A 1991. Increasing the effectiveness of storing table grape. *Sadovodstvo i Vinogradarstvo* 7, 19–21.

Kidd, F. 1916. The controlling influence of CO_2: Part III the retarding effect of CO_2 on respiration. *Proceedings of the Royal Society, London*, 89B, 136–156.

Kidd, F. 1919. Laboratory experiments on the sprouting of potatoes in various gas mixtures nitrogen, O_2 and CO_2. *New Phytologist* 18, 248–252.

Kidd, F. and West, C. 1917a. The controlling influence of CO_2. IV. On the production of secondary dormancy in seeds of *Brassica alba* following a treatment with CO_2, and the relation of this phenomenon to the question of stimuli in growth processes. *Annals of Botany* 34, 439–446.

Kidd, F. and West, C. 1917b. The controlling influence of CO_2. IV. On the production of secondary dormancy in seeds of *Brassica alba* following treatment with CO_2, and the relation of this phenomenon to the question of stimuli in growth processes. *Annals of Botany* 31, 457–487.

Kidd, F. and West, C. 1923. Brown Heart – a functional disease of apples and pears. *Food Investigation Board Special Report* 12, 3–4.

Kidd, F. and West, C. 1925. The course of respiratory activity throughout the life of an apple. *Report of the Food Investigation Board London for 1924*, 27–34.

Kidd, F. and West, C. 1927a. A relation between the respiratory activity and the keeping quality of apples. *Report of the Food Investigation Board London for 1925, 1926*, 37–41.

Kidd, F. and West, C. 1927b. A relation between the concentration of O_2 and CO_2 in the atmosphere, rate of respiration, and the length of storage of apples *Report of the Food Investigation Board London for 1925, 1926*, 41–42.

Kidd, F. and West, C. 1930. The gas storage of fruit. II Optimum temperatures and atmospheres. *Journal of Pomology and Horticultural Science* 8, 67–77.

Kidd, F. and West, C. 1934. Injurious effects of pure O_2 upon apples and pears at low temperatures. *Report of the Food Investigation Board London for 1933*, 74–77.

Kidd, F. and West, C. 1935a. Gas storage of apples, *Report of the Food Investigation Board London. for 1934*, 103–109.

Kidd, F. and West, C. 1935b. The refrigerated gas storage of apples. *Department of Scientific and Industrial Research, Food Investigation Leaflet*, 6.

Kidd, F. and West, C. 1938. The action of CO_2 on the respiratory activity of apples. *Report of the Food Investigation Board London. for 1937*, 101–102.

Kidd, F. and West, C. 1939. The gas storage of Cox's Orange Pippin apples on a commercial scale. *Report of the Food Investigation Board London for 1938*, 153–156.

Kidd, F. and West, C. 1949. Resistance of the skin of the apple fruit to gaseous exchange. *Report of the Food Investigation Board London for 1939*, 64–68.

Kidmose, U., Hansen, S.L., Christensen, L.P., Edelenbos, M., Larsen, E. and Norbaek, R. 2004. Effects of genotype, root size, storage, and processing on bioactive compounds in organically grown carrots (*Daucus carota* L.). *Journal of Food Science* 69, S388–S394.

Kim, Y., Brecht, J.K. and Talcott, S.T. 2007. Antioxidant phytochemical and fruit quality changes in mango (*Mangifera indica* L.) following hot water immersion and controlled atmosphere storage. *Food Chemistry* 105, 1327–1334.

Kitagawa, H. and Glucina, P.G. 1984. *Persimmon Culture in New Zealand.* New Zealand Department of Scientific and Industrial Research (DISR), Information Series No. 159. Scientific Information Publishing Centre, Wellington.

Kitsiou, S. and Sfakiotakis, E. 2003. Modified atmosphere packaging of 'Hayward' kiwifruit: composition of the storage atmosphere and quality changes at 10 and 20 °C. *Acta Horticulturae* 610, 239–244.

Kittemann, D., McCormick, R. and Neuwald, D.A. 2015a. Effect of high temperature and 1-MCP application or dynamic controlled atmosphere on energy savings during apple storage. *European Journal of Horticultural Science* 80, 33–38.

Kittemann, D., Neuwald, D.A., and Streif, J. 2015b. Internal browning in 'Kanzi®' apples – Reasons and possibilities to reduce the disorder. *Acta Horticulture* 1079, 409–414.

Klaustermeyer, J.A. and Morris, L.L. 1975. The effects of ethylene and carbon monoxide on the induction of russet spotting on crisphead lettuce. *Plant Physiology* 56, (Supplement) 63.

Klein, J.D. and Lurie, S. 1992. Prestorage heating of apple fruit for enhanced postharvest quality: interaction of time and temperature. *HortScience* 27, 326–328.

Klieber, A. and Wills, R.B.H. 1991. Optimisation of storage conditions for 'Shogun' broccoli. *Scientia Horticulturae* 47, 201–208.

Klieber, A., Bagnato, N., Barrett, R. and Sedgley, M. 2002. Effect of post-ripening nitrogen atmosphere storage on banana shelf-life, visual appearance and aroma. *Postharvest Biology and Technology* 25, 15–24.

Klieber, A., Bagnato, N., Barrett, R. and Sedgley, M. 2003. Effect of post-ripening atmosphere treatments on banana. *Acta Horticulturae* 600, 51–54.

Kluge, K. and Meier, G. 1979. Flavour development of some apple cultivars during storage. *Gartenbau* 26, 278–279.

Knee, M. 1973. Effects of controlled atmosphere storage on respiratory metabolism of apple fruit tissue. *Journal of the Science of Food and Agriculture* 24, 289–298.

Knee, M. 1990. Ethylene effects on controlled atmosphere storage of horticultural crops. In: Calderon, M. and Barkai-Golan, R. (eds) *Food Preservation by Modified Atmospheres.* CRC Press, Boca Raton (Florida), Ann Arbor (Michigan), Boston (Massachusetts), pp. 225–235.

Knee, M. and Aggarwal, D. 2000. Evaluation of vacuum containers for consumer storage of fruits and vegetables. *Postharvest Biology and Technology* 19, 55–60.

Knee, M. and Bubb, M. 1975. Storage of Bramley's Seedling apples. II. Effects of source of fruit, picking date and storage conditions on the incidence of storage disorders. *Journal of Horticultura Science* 50, 121–128.

Knee, M. and Looney, N.E. 1990. Effect of orchard and postharvest application of daminozide on ethylene synthesis by apple fruit. *Journal of Plant Growth Regulation* 9, 175–179.

Knee, M. and Sharples R.O. 1979. Influence of CA storage on apples. *Quality in Stored and Processed Vegetables and Fruit Proceedings of a Symposium at Long Ashton Research Station, University of Bristol, 8–12 April 1979*, 341–352.

Knee, M. and Tsantili, E. 1988. Storage of 'Gloster 69' apples in the preclimacteric state for up to 200 days after harvest. *Physiologia Plantarum* 74, 499–503.

Kobiler, I., Akerman, M., Huberman, L. and Prusky, D. 2011. Integration of pre- and postharvest treatments for the control of black spot caused by Alternaria alternata in stored persimmon fruit. *Postharvest Biology and Technology* 59, 166–171.

Koch, K.E. 1996. Carbohydrate-modulated gene expression in plants. *Annual Review of Plant Physiology and Plant Molecular Biology* 47, 509–540.

Koelet, P.C. 1992. *Industrial Refrigeration*. MacMillan, London.

Koidea, S and Shib, J. 2007. Microbial and quality evaluation of green peppers stored in biodegradable film packaging. *Food Control* 18, 1121–1125.

Kollas, D.A. 1964. Preliminary investigation of the influence of controlled atmosphere storage on the organic acids of apples. *Nature* 204, 758–759.

Kondou, S., Oogaki, C. and Mim, K. 1983. Effects of low pressure storage on fruit quality. *Journal of the Japanese Society for Horticultural Science* 52, 180–188.

Konopacka, D. and Płocharski, W.J. 2002. Effect of picking maturity, storage technology and shelf-life on changes of apple firmness of 'Elstar', 'Jonagold' and 'Gloster' cultivars. *Journal of Fruit and Ornamental Plant Research* 10, 15–26.

Köpcke, D. 2009. Einfluss der Sauerstoff-Konzentration im Lager auf den Schalenfleckenbefall bei 'Elstar'. *Mitteilungen des Obstbauversuchsringes des Alten Landes* 64, 303–309.

Köpcke, D. 2015. 1-methylcyclopropene (1-MCP) and dynamic controlled atmosphere (DCA) applications under elevated storage temperatures: Effects on fruit quality of 'Elstar', 'Jonagold' and 'Gloster' apple (*Malus domestica* Borkh.). *European Journal of Horticultural Science* 80, 25–32.

Kopec, K. 1980. Zmeny plodov zeleninovej papriky pocas hypobarickeho uskladnovania. *Vedecke Prace Vyskumneho a Slachtitelskeho Ustavu Zeleniny a Specialnych Plodin v Hurbanove* 1, 34–41.

Korsten, I., De Villiers, E.E. and Lonsdale, J.H. 1993. Biological control of mango postharvest disease in the packhouse. *South African Mango Growers' Association Yearbook* 13, 117–121.

Kosson, R. and Stepowska, A. 2005. The effect of equilibrium modified atmosphere packaging on quality and storage ability of sweet pepper fruits. *Vegetable Crops Research Bulletin* 63, 139–148.

Kouassi, K.H.S, Bajj, M. and Jijakli, H. 2012. The control of postharvest blue and green moulds of citrus in relation with essential oil-wax formulations, adherence and viscosity. *Postharvest Biology and Technology* 73, 122–128.

Koyakumaru, T. 1997. Effects of temperature and ethylene removing agents on respiration of mature green mume Prunus mume Sieb. et Zucc. fruit held under air and controlled atmospheres. *Journal of the Japanese Society for Horticultural Science* 66, 409–418.

Koyakumaru, T., Adachi, K., Sakoda, K., Sakota, N. and Oda, Y. 1994. Physiology and quality changes of mature green mume *Prunus mume* Sieb. et Zucc. *Journal of the Japanese Society for Horticultural Science* 62, 877–887.

Koyakumaru, T., Sakoda, K., Ono, Y. and Sakota, N. 1995. Respiratory physiology of mature-green mume *Prunus mume* Sieb. et Zucc. *Journal of the Japanese Society for Horticultural Science* 64, 639–648.

Krala, L., Witkowska, M., Kunicka, A. and Kalemba, D. 2007. Quality of savoy cabbage stored under controlled atmosphere with the addition of essential oils. *Polish Journal of Food and Nutrition Sciences* 57, 45–50

Kramchote, S., Jirapong, C. and Wongs-Aree, C. 2008. Effects of 1-MCP and controlled atmosphere storage on fruit quality and volatile emission of 'Nam Dok Mai' mango. *Acta Horticulturae* 804, 485–492

Krishnamurthy, S. 1989. Effects of Tal-prolong on shelf life and quality attributes of mango. *Acta Horticulturae* 231, 675–678.

Kruger, F.J. and Truter, A.B. 2003. Relationship between preharvest quality determining factors and controlled atmosphere storage in South African export avocados. *Acta Horticulturae* 600, 109–113.

Krusong, W., Jindaprasert, A., Laosinwattana, C. and Teerarak, M. 2015. Baby corn fermented vinegar and its vapour control postharvest decay in strawberries. *New Zealand Journal of Crop and Horticultural Science* 43, 193–203.

Ku, V.V.V. and Wills, R.B.H. 1999. Effect of 1-methylcyclopropene on the storage life of broccoli. *Postharvest Biology and Technology* 17, 127–132.

Kuang, J.-F., Chen, L., Shan, W., Yang, S., Lu, W. and Chen, J. 2013. Molecular characterization of two banana ethylene signaling component MaEBFs during fruit ripening. *Postharvest Biology and Technology* 85, 94–101.

Kubo, Y., Inaba, A., Kiyasu, H. and Nakamura, R. 1989a. Effects of high CO_2 plus low O_2 on respiration in several fruits and vegetables. *Scientific Reports of the Faculty of Agriculture, Okayama University* 73, 27–33.

Kubo, Y., Inaba, A. and Nakamura, R. 1989b. Effects of high CO_2 on respiration in various horticultural crops. *Journal of the Japanese Society for Horticultural Science* 58, 731–736.

Kucukbasmac, F., Ozkaya, O., Agar, T. and Saks, Y. 2008. Effect of retail-size modified atmosphere packaging bags on postharvest storage and shelf-life quality of '0900 Ziraat' sweet cherry. *Acta Horticulturae* 795, 775–780.

Kumar, A. and Brahmachari, V.S. 2005. Effect of chemicals and packaging on ripening and storage behaviour of banana cv. Harichhaal (AAA) at ambient temperature. *Horticultural Journal* 18, 86–90.

Kumar, J., Mangal, J.L. and Tewatia, A.S. 1999. Effect of storage conditions and packing materials on shelf-life of carrot cv. Hisar Gairic. *Vegetable Science* 26, 196–197.

Kupferman, E. 1989. The early beginnings of controlled atmosphere storage. *Post Harvest Pomology Newsletter* 7, 3–4

Kupferman E. 2001a. *Storage scald of apples*. Tree Fruit Research and Extension Center, Washington State University. Available at: http://postharvest.tfrec.wsu.edu/EMK2000C.pdf. (accessed May 2009).

Kupferman, E. 2001b. *Controlled atmosphere storage of apples and pears*. Postharvest Information Network Article, Washington State University. Available at: http://postharvest.tfrec.wsu.edu/EMK2001D.pdf (accessed May 2009).

Kupferman, E. 2003. Controlled atmosphere storage of apples and pears. *Acta Horticulturae* 600, 729–735.

Kupferman, E. (undated) *Storage scald in apples and pears*. Tree Fruit Research and Extension Center, Washington State University. Available at: http://postharvest.tfrec.wsu.edu/EMK2000C.pdf (accessed May 2009).

Kupper, W., Pekmezci, M. and Henze, J. 1994. Studies on CA-storage of pomegranate *Punica granatum* L., cv. Hicaz. *Acta-Horticulturae* 398, 101–108.

Kurki, L. 1979. Leek quality changes during CA storage. *Acta Horticulturae* 93, 85–90.

Kwok, C.Y. 2010. Paiola. Available at: www.mafc.com.my (accessed January 2010).

Lachapelle, M., Bourgeois, G., DeEll, J., Stewart, K.A. and Séguin, P. 2013. Modeling the effect of preharvest weather conditions on the incidence of soft scald in 'Honeycrisp' apples. *Postharvest Biology and Technology* 85, 57–66.

Ladaniya, M.S. 2004. Response of 'Kagzi' acid lime to low temperature regimes during storage. *Journal of Food Science and Technology (Mysore)* 41, 284–288.

Lafer, G. 2001. Influence of harvest date on fruit quality and storability of 'Braeburn' apples. *Acta Horticulturae* 553, 269–270.

Lafer, G. 2005. Effects of 1-MCP treatments on fruit quality and storability of different pear varieties. *Acta Horticulturae* 682, 1227–1232.

Lafer, G. 2008. Storability and fruit quality of 'Braeburn' apples as affected by harvest date, 1-MCP treatment and different storage conditions. *Acta Horticulturae* 796, 179–184.

Lafer, G. 2009. Practical experience with new storage technologies in Austria – Dynamic CA (DCA) storage and SmartFresh™. *European Fruit Growers Magazine* 1(1), 24–27.

Lafer, G. 2011. Effect of different CA storage conditions on storability and fruit quality of organically grown 'Uta' pears. *Acta Horticulturae* 909, 757–760.

Lafuente, M.T., Cantwell, M., Yang, S.F. and Rubatzky, U. 1989. Isocoumarin content of carrots as influenced by ethylene concentration, storage temperature and stress conditions. *Acta Horticulturae* 258, 523–534.

Lafuente, M.T., Lopez-Galvez, G., Cantwell, M. and Yang, S.F., 1996. Factors influencing ethylene-induced isocoumar information and increased respiration in carrots. *Journal of the American Society for Horticultural Science* 121, 537–542.

Lagopodi, A.L., Cetiz, K., Koukounaras, A. and Sfakiotakis, E.M. 2009. Acetic acid, ethanol and steam effects on the growth of Botrytis cinerea in vitro and combination of steam and modified atmosphere packaging to control decay in kiwifruit. *Journal of Phytopathology* 157, 79–84.

Lalel, H.J.D. and Singh, Z. 2004. Biosynthesis of aroma volatile compounds and fatty acids in 'Kensington Pride' mangoes after storage in a controlled atmosphere at different oxygen and carbon dioxide concentrations. *Journal of Horticultural Science & Biotechnology* 79, 343–353.

Lalel, H.J.D. and Singh, Z. 2006. Controlled atmosphere storage of 'Delta R2E2' mango fruit affects production of aroma volatile compounds. *Journal of Horticultural Science & Biotechnology* 81, 449–457.

Lalel, H.J.D., Singh, Z. and Tan, S.C. 2005. Elevated levels of CO_2 in controlled atmosphere storage affects shelf-life, fruit quality and aroma volatiles of mango. *Journal of Horticultural Science & Biotechnology* 80, 551–556.

Lallu, N. and Burdon, J. 2007. Experiences with recent postharvest technologies in Kiwifruit. *Acta Horticulturae* 753, 733–740.

Lallu, N., Billing, D. and McDonald, B. 1997. Shipment of kiwifruit under CA conditions from New Zealand to Europe. *Seventh International Controlled Atmosphere Research Conference, July 13–18 1997*, University of California, Davis, California [Abstract], 28.

Lallu, N., Burdon, J., Billing, D., Burmeister, D., Yearsley, C., *et al.* 2005. Effect of carbon dioxide removal systems on volatile profiles and quality of 'Hayward' kiwifruit stored in controlled atmosphere rooms. *HortTechnology* 15, 253–260.

Landfald, R. 1988. Controlled-atmosphere storage of the apple cultivar Aroma. *Norsk Landbruksforskning* 2, 5–13.

Lange, E., Nowacki, J. and Saniewski, M. 1993. The effect of methyl jasmonate on the ethylene producing system in preclimacteric apples stored in low oxygen and high carbon dioxide atmospheres. *Journal of Fruit and Ornamental Plant Research* 1, 9–14.

Langridge, I.W. and Sharples, R.O. 1972. Storage under reduced atmospheric pressure. *Annual Report of the East Malling Research Station, UK, for 1971*, 76.

Larson, J., Vender, R. and Camuto, P. 1994. Cholestatic jaundice due to akee fruit poisoning. *American Journal of Gastroenterology* 89, 1577–1578.

Laszlo, J.C. 1985. The effect of controlled atmosphere on the quality of stored table grape. *Deciduous Fruit Grower* 35, 436–438.

Latocha, P., Krupa, T., Jankowski, P. and Radzanowska, J. 2014. Changes in postharvest physicochemical and sensory characteristics of hardy kiwifruit (*Actinidia arguta* and its hybrid) after cold storage under normal versus controlled atmosphere. *Postharvest Biology and Technology* 88, 21–33.

Lau, O.L. 1983. Effects of storage procedures and low O_2 and CO_2 atmospheres on storage quality of 'Spartan' apples. *Journal of the American Society for Horticultural Science* 108, 953–957.

Lau, O.L., 1989a. Storage of 'Spartan' and 'Delicious' apples in a low-ethylene, 1.5% O_2 plus 1.5% CO_2 atmosphere. *HortScience* 24, 478–480.

Lau, O.L., 1989b. Responses of British Columbia-grown apples to low-oxygen and low-ethylene controlled atmosphere storage. *Acta Horticulturae* 258, 107–114.

Lau, O.L. 1997. Influence of climate, harvest maturity, waxing, O_2 and CO_2 on browning disorders of 'Braeburn' apples. *Postharvest Horticulture Series – Department of Pomology, University of California* 16, 132–137.

Lau, O.L. and Yastremski, R. 1993. The use of 0.7% storage oxygen to attenuate scald symptoms in 'Delicious' apples: effect of apple strain and harvest maturity. *Acta Horticulturae* 326, 183–189.

Laugheed, E.C., Murr, D.P. and Berard, L. 1978. Low pressure storage for horticultural crops. *HortScience* 13, 21–27.

Laurin, É., Nunes, M.C.N., Émond, J.-P. and Brecht, J.K. 2006. Residual effect of low-pressure stress during simulated air transport on Beit Alpha-type cucumbers: stomata behaviour. *Postharvest Biology and Technology* 41, 121–127.

Laval Martin, D., Quennemet, J. and Moneger, R. 1975. Remarques sur l'evolution lipochromique et ultrastructurales des plastes durant la maturation du fruit de tomate 'cerise'. *Colleques Internationaux du CNRS Paris*, 374.

Lawless, J. 2014. *Encyclopaedia of Essential Oils: the complete guide to the use of aromatic oils in aromatherapy, herbalism, health and well being*. HarperCollins, London.

Lawton, A.R. 1996. *Cargo Care*. Cambridge Refrigeration Technology, Cambridge, UK.

Lee, B.Y., Kim, Y.B. and Han, P.J. 1983. Studies on controlled atmosphere storage of Korean chestnut, *Castanea crenata* var. *dulcis* Nakai. *Research-Reports, Office of Rural Development, S. Korea, Soil Fertilizer, Crop Protection, Mycology and Farm Products Utilization* 25, 170–181.

Lee, D.S., Hagger, P.E., Lee, J. and Yam, K.L. 1991. Model for fresh produce repiration in modified atmospheres based on the principles of enzyme kinetics. *Journal of Food Science* 56, 1580–1585.

Lee, H.D., Yun, H.S., Park, W.K. and Choi, J.U. 2003a. Estimation of conditions for controlled atmosphere (CA) storage of fresh products using modelling of respiration characteristics. *Acta Horticulturae* 600, 677–680.

Lee, H.D., Yun, H.S., Og Lee, W., Jeong, H. and Choe, S.Y. 2009. The effect of pressurized CA (Controlled Atmosphere) treatment on the storage qualities of peach. *10th International Controlled & Modified Atmosphere Research Conference, 4–7 April 2009 Turkey* [Abstract], 8.

Lee, S.K. and Kader, A.A. 2000. Preharvest and postharvest factors influencing vitamin C content of horticultural crops. *Postharvest Biology and Technology* 20, 207–220.

Lee, S.K., Shin, I.S. and Park, Y.M. 1993. Factors involved in skin browning of non astringent 'Fuju' persimmon. *Acta Horticulturae* 343, 300–303.

Lee, S.O., Park, I.-K., Choi, G., Lim, H.K., Jang, K.S. *et al.* 2007. Fumigant activity of essential oils and components of Illicium verum and Schizonepetate nuifolia against Botrytis cinerea and Colletotrichum gloeosporioides. *Journal of Microbiology and Biotechnology* 17, 1568–1572.

Lee, W.H., Kim, M.S., Lee, H., Delwiche, S.R., Bae, H. and Kim, D.Y. 2014. Hyperspectral near-infrared imaging for the detection of physical damages of pear. *Journal of Food Engineering* 130, 1–7.

Lee, Y.J., Lee, Y.M., Kwon, O.C., Cho, Y.S., Kim, T.C. and Park, Y.M. 2003b. Effects of low oxygen and high carbon dioxide concentrations on modified atmosphere-related disorder of 'Fuju' persimmon fruit. *Acta Horticulturae*, 601, 171–176.

Leiva-Valenzuela, G.A., Lu, R. and Aguilera, J.M., 2013. Prediction of firmness and soluble solids content of blueberries using hyperspectral reflectance imaging. *Journal of Food Engineering* 115, 91–98.

Leon, D.M., Cruz, J., Parkin, K.L. and Garcia, H.S. 1997. Effect of controlled atmospheres containing low O_2 and high CO_2 on chilling susceptibility of Manila mangoes. *Acta Horticulturae* 455, 635–642.

Lers, A., Jiang, W.B., Lomanies, E. and Aharoni, N. 1998. Gibberellic acid and CO_2 additive effect in retarding postharvest senescence of parsley. *Journal of Food Science* 63, 66–68.

Levin, A., Sonego, L., Zutkhi, Y., Ben Arie, R., and Pech, J.C. 1992. Effects of CO_2 on ethylene production by apples at low and high O_2 concentrations. In: Pech, J.C., Latché, A. and Balagué, C. (eds) *Cellular and Molecular Aspects of the Plant Hormone Ethylene. Current Plant Science and Biotechnology in Agriculture*, Vol. 16. Springer, Dordrecht, pp. 150–151.

Leyte, J.C. and Forney, C.F. 1999. Controlled atmosphere tents for storing fresh commodities in conventional refrigerated rooms. *HortTechnology* 9, 672–674.

Li, B.Q. and Tian, S.P. 2007. Effect of intracellular trehalose in Cryptococcus laurentii and exogenous lyoprotectants on its viability and biocontrol efficacy on *Penicillium expansum* in apple fruit. *Letters in Applied Microbiology* 44, 437–442.

Li, H.Y. and Yu, T. 2001. Effect of chitosan on incidence of brown rot, quality and physiological attributes of postharvest peach fruit. *Journal of the Science of Food and Agriculture* 81, 269–274.

Li ,H., Suo, J., Han, Y., Liang, C., Jin, M., Zhang, Z. and Rao, J. 2017a. The effect of 1-methylcyclopropene, methyl jasmonate and methyl salicylate on lignin accumulation and gene expression in postharvest 'Xuxiang' kiwifruit during cold storage. *Postharvest Biology and Technology* 124, 107–118.

Li, H., Billing, D., Pidakala, P. and Burdon, J. 2017b. Textural changes in 'Hayward' kiwifruit during and after storage in controlled atmospheres. *Scientia Horticulturae* 222, 40–45.

Li, J., Huang, W., Tian, X., Wang, C., Fan, S., and Zhao, C., 2016. Fast detection and visualization of early decay in citrus using Vis-NIR hyperspectral imaging. *Computers and Electronics in Agriculture* 127, 582–592.

Li, J., Lei, H., Song, H., Lai, T., Xu, X. and Shi, X. 2017c. 1-methylcyclopropene (1-MCP) suppressed postharvest blue mold of apple fruit by inhibiting the growth of *Penicillium expansum*. *Postharvest Biology and Technology* 125, 59–64.

Li, J.Y., Huang, W.N., Cai, L.X. and Hu, W.J. 2000. Effects of calcium treatment on physiological and biochemical changes in shiitake Lentinus edodes during the post harvest period. *Fujian Journal of Agricultural Sciences* 15, 43–47.

Li, T., Tan, D., Liu, Z., Jiang, Z., Wei, Y. *et al.* 2013. Exploring the apple genome reveals six ACC synthase genes expressed during fruit ripening. *Scientia Horticulturae* 157, 119–123.

Li, L., Luo, Z., Huang, X., Zhang, L., Zhao, P. *et al.* 2015. Label-free quantitative proteomics to investigate strawberry fruit proteome changes under controlled atmosphere and low temperature storage. *Journal of Proteomics* 120, 44–57.

Li, L., Lichter, A., Chalupowicz, D., Gamrasni, D., Goldberg, T., *et al.* 2016b. Effects of the ethylene-action inhibitor 1-methylcyclopropene on postharvest quality of non-climacteric fruit crops. *Postharvest Biology and Technology* 111, 322–329.

Li, L., Lv, F., Guo, Y. and Wang, Z. 2016c.Respiratory pathway metabolism and energy metabolism associated with senescence in postharvest Broccoli (*Brassica oleracea* L. var. italica) florets in response to O_2/CO_2 controlled atmospheres. *Postharvest Biology and Technology* 111, 330–336.

Li, P., Hu, H., Luo, S., Zhang, L. and Gao, J. 2017d. Shelf life extension of fresh lotus pods and seeds (*Nelumbo nucifera* Gaertn.) in response to treatments with 1-MCP and lacquer wax. *Postharvest Biology and Technology* 125, 140–149.

Li, T., Tan, D., Liu, Z., Jiang, Z., Wei, Y. *et al.* 2013. Exploring the apple genome reveals six ACC synthase genes expressed during fruit ripening. *Scientia Horticulturae* 157, 119–123.

Li, W.X. 2006. Studies on the technology and mechanism of green asparagus during three-stage hypobaric storage. PhD thesis, Jiangnan University [abstract]. Available at: http://www.dissertationtopic.net/doc/1615099 (accessed October 2013).

Li, W.X. and Zhang M. 2006. Effect of three-stage hypobaric storage on cell wall components, texture and cell structure of green asparagus. *Journal of Food Engineering* 77, 112–118.

Li, W., Zhang, M. and Yu, H. 2006. Study on hypobaric storage of green asparagus. *Journal of Food Engineering* 73, 225–230.

Li, W.X., Zhang, M. and Wang, S.J. 2008. Effect of three-stage hypobaric storage on membrane lipid peroxidation and activities of defense enzyme in green asparagus. *LWT – Food Science and Technology* 41, 2175–2181.

Li, X., Zhu, X., Wang, H., Lin, X. and Chen, W. 2018. Postharvest application of wax controls pineapple fruit ripening and improves fruit quality. *Postharvest Biology and Technology* 136, 99–110.

Li, Y., Wang, V.H., Mao, C.Y. and Duan, C.H. 1973. Effects of oxygen and carbon dioxide on after ripening of tomatoes. *Acta Botanica Sinica* 15, 93–102.

Liamnimitr, N., Thammawong, M., Techavuthiporn, C., Fahmy, K., Suzuki , T. and Nakano, K. 2018. Optimization of bulk modified atmosphere packaging for long-term storage of 'Fuyu' persimmon fruit. *Postharvest Biology and Technology* 135, 1–7.

Liang, L.S., Wang, G.X. and Sun, X.Z. 2004. Effect of postharvest high CO_2 shock treatment on the storage quality and performance of Chinese chestnut (*Castanea mollissima* Blume). *Scientia Silvae Sinicae* 40, 91–96.

Lichanporn, I. and Techavuthiporn, C. 2013. The effects of nitric oxide and nitrous oxide on enzymatic browning in longkong (*Aglaia dookkoo* Griff.). *Postharvest Biology and Technology* 86, 62–65.

Lichter, A., Zutahy, Y., Kaplunov, T., Aharoni, N. and Lurie, S. 2005. The effect of ethanol dip and modified atmosphere on prevention of Botrytis rot of table grapes. *HortTechnology* 15, 284–291.

Lidster, P.D., Lawrence, R.A., Blanpied, G.D. and McRae, K.B. 1985. Laboratory evaluation of potassium permanganate for ethylene removal from controlled atmosphere apple storages. *Transactions of the American Society of Agricultural Engineers* 28, 331–334.

Lill, R.E. and Corrigan, V.K. 1996. Asparagus responds to controlled atmospheres in warm conditions. *International Journal of Food Science and Technology* 31, 117–121.

Lima, C.P.F. de 2015. Fumigation of mealybugs in Australian table grapes and grapefruit. *Acta Horticulturae* 1091, 329–334.

Lima, L.C., Brackmann, A., Chitarra, M.I.F., Boas, E.V. de B.V. and Reis, J.M.R. 2002. Quality characteristics of 'Royal Gala' apple, stored under refrigeration and controlled atmosphere. *Ciencia e Agrotecnologia* 26, 354–361.

Lima, L.C., Hurr, B.M. and Huber, D.J. 2005. Deterioration of beit alpha and slicing cucumbers (*Cucumis sativus* L.) during storage in ethylene or air: responses to suppression of ethylene perception and parallels to natural senescence. *Postharvest Biology and Technology* 37, 265–276.

Lin, Q., Lu, Y., Zhang, J., Liu, W., Guan, W. and Wang, Z. 2017. Effects of high CO_2 in-package treatment on flavour, quality and antioxidant activity of button mushroom (*Agaricus bisporus*) during postharvest storage. *Postharvest Biology and Technology* 123, 112–118.

Lin, Y., Lin, H., Zhang, S., Chen, Y., Chen, M. and Lin, Y. 2014. The role of active oxygen metabolism in hydrogen peroxide-induced pericarp browning of harvested longan fruit. *Postharvest Biology and Technology,* 96, 42–48.

Liplap, P., Vigneault, C., Vijaya Raghavan, G.S. and Jenni, S. 2012. Storing avocado under hyperbaric pressure, Poster Board #035. *American Society for Horticultural Science, Annual Conference August 2012.*

Liplap, P., Charlebois, D., Charles, M.T., Toivonen, P., Vigneault, C. and Vijaya Raghavan, G.S. 2013a. Tomato shelf-life extension at room temperature by hyperbaric pressure treatment. *Postharvest Biology and Technology* 86, 45–52.

Liplap, P., Vigneault, C., Toivonen, P., Charles, M.T. and Vijaya Raghavan, G.S. 2013b. Effect of hyperbaric pressure and temperature on respiration rates and quality attributes of tomato. *Postharvest Biology and Technology* 86, 240–248.

Liplap, P., Vigneault, C., Toivonen, P., Boutin, J. and Vijaya Raghavan, G.S. 2013c. Effect of hyperbaric treatment on respiration rates and quality attributes of sweet corn. *International Journal of Postharvest Technology and Innovation* 3, 257–271.

Liplap, P., Boutin, J., LeBlanc, D.I., Vigneault, C., and Vijaya Raghavan, G.S. 2014a. Effect of hyperbaric pressure and temperature on respiration rates and quality attributes of Boston lettuce. *International Journal of Food Science and Technology* 49, 137–145.

Liplap, P., Toivonen, P., Vigneault, C., Boutin, J. and Vijaya Raghavan, G.S. 2014b. Effect of hyperbaric pressure treatment on the growth and physiology of bacteria that cause decay in fruit and vegetables. *Food Bioprocess Technology* 7, 2267–2280.

Liplap, P., Vigneault, C., Rennie, T.J. Boutin, J. and Vijaya Raghavan, G.S. 2014c. Method for determining the respiration rate of horticultural produce under hyperbaric treatment. *Food Bioprocess Technology* 7, 2461–2471.

Lipton, W.J. 1967. Some effects of low-oxygen atmospheres on potato tubers. *American Journal of Potato Research* 44, 292–299.

Lipton, W.J. 1968. *Market Quality of Asparagus – effects of maturity at harvest and of high CO_2 atmospheres during simulated transit.* USDA Marketing Research Report 817. US Department of Agriculture, Washington, DC.

Lipton, W.J. 1972. Market quality of radishes stored in low O_2 atmospheres. *Journal of the American Society for Horticultural Science* 97, 164.

Lipton, W.J. and Harris, C.M. 1974. Controlled atmosphere effects on the market quality of stored broccoli Brassica oleracea L Italica group. *Journal of the American Society for Horticultural Science* 99, 200–205.

Lipton, W.J. and Mackey, B.E. 1987. Physiological and quality responses of Brussels sprouts to storage in controlled atmospheres. *Journal of the American Society for Horticultural Science* 112, 491–496.

Little, C.R., Faragher, J.D. and Taylor, H.J. 1982. Effects of initial oxygen stress treatments in low oxygen modified atmosphere storage of Granny Smith apples. *Journal of the American Society for Horticultural Science* 107, 320–323.

Little, C.R., Holmes, R.J. and Faragher, J. (eds) 2000. *Storage Technology for Apples and Pears*. Institute for Horticulture Development Agriculture, Victoria, Knoxville, Australia.

Liu, F.W. 1976a. Banana response to low concentrations of ethylene. *Journal of the American Society for Horticultural Science* 101, 222–224.

Liu, F.W. 1976b. Storing ethylene pretreated bananas in controlled atmosphere and hypobaric air. *Journal of the American Society for Horticultural Science* 101, 198–201.

Liu, F.W. 1977. Varietal and maturity differences of apples in response to ethylene in controlled atmosphere storage. *Journal of American Society of Horticultural Science* 102, 93–95.

Liu, H., Jiang, W. Zhou, L., Wang, B. and Luo, Y. 2005. The effects of 1-methylcyclopropene on peach fruit (*Prunus persica* L. cv. Jiubao) ripening and disease resistance. *International Journal of Food Science & Technology* 40, 1–7.

Liu, R., Wang, Y. Qin, G. and Tian, S. 2016a. Molecular basis of 1-methylcyclopropene regulating organic acid metabolism in apple fruit during storage. *Postharvest Biology and Technology* 117, 57–63.

Liu, Y., Yang, X., Zhu, S. and Wang, Y. 2016b. Postharvest application of MeJA and NO reduced chilling injury in cucumber (*Cucumis sativus*) through inhibition of H_2O_2 accumulation. *Postharvest Biology and Technology* 119, 77–83.

Lizana, A. and Figuero, J. 1997. Effect of different CA on post harvest of Hass avocado. *Seventh International Controlled Atmosphere Research Conference, July 13–18 1997*, University of California, Davis, California [Abstract], 114.

Lizana, L.A., Fichet, T., Videla, G., Berger, H. and Galletti, Y.L. 1993. Almacenamiento de aguacates pultas cv. Gwen en atmosfera controlada. *Proceedings of the Interamerican Society for Tropical Horticulture* 37, 79–84.

Llana-Ruiz-Cabello, M., Pichardo, S., Baños, A., Núñez, C., Bermúdez, J.M. *et al.* 2015. Characterisation and evaluation of PLA films containing an extract of Allium spp. to be used in the packaging of ready-to-eat salads under controlled atmospheres. *LWT – Food Science and Technology* 64, 1354–1361

Lonsdale, J.H. 1992. In search of an effective postharvest treatment for the control of postharvest diseases of mangoes. *South African Mango Growers' Association Yearbook* 12, 32–36.

Lopez, M.L., Lavilla, T., Recasens, I., Riba, M. and Vendrell, M. 1998. Influence of different oxygen and carbon dioxide concentrations during storage on production of volatile compounds by Starking Delicious apples. *Journal of Agricultural and Food Chemistry* 46, 634–643.

Lopez, M.D., Jordan, M.J. and Pascual-Villalobos, M.J. 2008. Toxic compounds in essential oils of coriander, caraway and basil active against stored rice pests. *Journal of Stored Products Research, Manhattan* 44, 273–278.

Lopez-Briones, G., Varoquaux, P., Bareau, G. and Pascat, B. 1993. Modified atmosphere packaging of common mushroom. *International Journal of Food Science and Technology* 28, 57–68.

Lopez-Malo, A., Palou, E., Barbosa-Canovas, G.V., Welti-Chanes, J. and Swanson, B.G. 1998. Polyphenoloxidase activity and colour changes during storage of high hydrostatic pressure treated avocado purée. *Food Research International* 31, 549–556.

Louarn, S., Nawrocki, A., Thorup-Kristensen, K., Lund, O.S., Jensen, O.N., Collinge, D.B. and Jensen, B. 2013. Proteomic changes and endophytic micromycota during storage of organically and conventionally grown carrots. *Postharvest Biology and Technology* 76, 26–33.

Lougheed, E.C. and Lee, R. 1989. Ripening, CO_2 and C_2H_4 production, and quality of tomato fruits held in atmospheres containing nitrogen and argon. In: Fellman, J.K. (ed.) *Proceedings of the Fifth International Controlled Atmosphere Research Conference, Wenatchee, Washington, USA, 14–16 June 1989. Vol. 2: Other commodities and storage recommendations*. Washington State University, Pullman, Washington, pp. 141–150.

Lougheed, E.C, Franklin, E.W., Papple, D.J., Pattie, D.R., Malinowski, H.K. and Wenneker, A. 1974. *A Feasibility Study of Low-Pressure Storage*. Horticultural Science Department and School of Engineering, University of Guelph, Ontario, Canada.

Lougheed, E.C., Murr, D.P. and Berard, L. 1977. LPS – great expectations. In: Dewey, D.H. (ed.) *Proceedings of the 2nd National Controlled Atmosphere Research Conference, 5 to 7 April 1977, Michigan State University, East Lansing, Michigan USA*, pp. 38–44.

Lougheed, E.C., Murr, D.P. and Berard, L. 1978. Low pressure storage for horticultural crops. *HortScience* 13, 21–27.

Loulakakis, C.A., Hassan, M., Gerasopoulos, D. and Kanellis, A.K. 2006. Effects of low oxygen on *in vitro* translation products of poly (A)+ RNA, cellulase and alcohol dehydrogenase expression in preclimacteric and ripening-initiated avocado fruit. *Postharvest Biology and Technology* 39, 29–37.

Love, J.M. 1988. *Robert Smock and the diffusion of controlled atmosphere technology in the US apple industry, 1940–1960*. Cornell Agricultural Economics Staff Paper 88–20. Cornell University, Ithaca, New York.

Lovino, R., de Cillis, F. M. and Massignan, L. 2004. Improving the shelf-life of edible mushrooms by combined post-harvest techniques. *Italus Hortus* 11, 97–99.

Lu, C.W. and Toivonen, P.M.A. 2000. Effect of 1 and 100 kPa O_2 atmospheric pretreatment of whole 'Spartan' apples on subsequent quality and shelf-life of slices stored in modified atmosphere packages. *Postharvest Biology and Technology* 18, 99–107.

Lu, J., Charles, M.T., Vigneault, C., Goyette, B. and Raghavan, G.S.V. 2010. Effect of heat treatment uniformity on tomato ripening and chilling injury. *Postharvest Biology and Technology* 56, 155–162.

Luchsinger, L., Mardones, C. and Leshuk, J. 2005. Controlled atmosphere storage of 'Bing' sweet cherries. *Acta Horticulturae* 667, 535–537.

Ludders, P. 2002. Cherimoya (*Annona cherimola* Mill.) – Botany, cultivation, storage and uses of a tropical-subtropical fruit. *Erwerbsobstbau* 44, 122–126.

Ludikhuyze, L., Van Loey, A., Indrawati and Hendrickx, M. 2002. High pressure processing of fruit and vegetables. In: Jongen, W. (ed.) *Fruit and Vegetable Processing – Improving Quality*. Woodhead Publishing, Cambridge, UK, pp. 346–359.

Lulai, E.C. and Suttle, J.C. 2004. The involvement of ethylene in wound-induced suberization of potato tuber (*Solanum tuberosum* L.): a critical assessment. *Postharvest Biology and Technology* 34, 105–112.

Lum, G.B., Brikis, C.J., Deyman, K.L., Subedi, S., DeEll, J.R., Shelp, B.J. and Bozzo, G.G. 2016. Pre-storage conditioning ameliorates the negative impact of 1-methylcyclopropene on physiological injury and modifies the response of antioxidants and γ-aminobutyrate in 'Honeycrisp' apples exposed to controlled-atmosphere conditions. *Postharvest Biology and Technology* 116, 115–128.

Lum, G.B., DeEll, J.R., Hoover, G.J., Subedi, S., Shelp, J. and Bozzo, G.G. 2017. 1-Methylcylopropene and controlled atmosphere modulate oxidative stress metabolism and reduce senescence-related disorders in stored pear fruit. *Postharvest Biology and Technology* 129, 52–63

Luna, M.C., Tudela, J.A., Martínez-Sánchez, A., Allende, A., Marín, A. and Gil, M.I. 2012. Long-term deficit and excess of irrigation influences quality and browning related enzymes and phenolic metabolism of fresh-cut iceberg lettuce (*Lactuca sativa* L.) *Postharvest Biology and Technology* 73, 37–45.

Luo, Y. and Mikitzel, L.J. 1996. Extension of postharvest life of bell peppers with low O_2. *Journal of the Science of Food and Agriculture* 70, 115–119.

Luo, Z., 2007. Effect of 1-methylcyclopropene on ripening of postharvest persimmon (*Diospyros kaki* L.) fruit. *Food Science and Technology* 40, 285–291.

Luo, Z., Chen, C. and Xie, J. 2012. Effect of salicylic acid treatment on alleviating postharvest chilling injury of 'Qingnai' plum fruit. *Postharvest Biology and Technology* 62, 115–120.

Lurie, S. 1992. Controlled atmosphere storage to decrease physiological disorders in nectarines. *International Journal of Food Science & Technology* 27, 507–514.

Lurie, S., Pesis, E. and Ben-Arie, R., 1991. Darkening of sunscald on apples in storage is a non-enzymic and non-oxidative process. *Postharvest Biology and Technology* 1, 119–125.

Lurie, S., Zeidman, M., Zuthi, Y. and Ben Arie, R. 1992. Controlled atmosphere storage to decrease physiological disorders in peaches and nectarine. *Hassadeh* 72, 1118–1122.

Lurie, S., Lers, A., Zhou, H.W. and Dong, L. 2002. The role of ethylene in nectarine ripening following storage. *Acta Horticulturae* 592, 607–613.

Luscher, C., Schluter, O. and Knorr, D. 2005. High pressure–low temperature processing of foods: impact on cell membranes, texture, colour and visual appearance of potato tissue. *Innovative Food Science and Emerging Technologies* 6, 59–71.

Lutz, J.M. 1952. Influence of temperature and length of curing period on keeping quality of 'Puerto Rico' sweetpotatoes. *Proceedings of the American Society for Horticultural Science* 59, 421.

Lutz, J.M. and Hardenburg, R.E. 1968. *The Commercial Storage of Fruits, Vegetables and Florist and Nursery Stocks*. Agriculture Handbook 66. United States Department of Agriculture, Washington DC

Ma, G., Wang, R., Wang, C.-R., Kato, M., Yamawaki, K., Qin, F.F. and Xu, H.-L. 2009. Effect of 1-methylcyclopropene on expression of genes for ethylene biosynthesis enzymes and ethylene receptors in post-harvest broccoli. *Plant Growth Regulation* 57, 223–232.

Maekawa, T. 1990. On the mango CA storage and transportation from subtropical to temperate regions in Japan. *Acta Horticulturae* 269, 367–374.

MAFF ADAS. 1974. *Atmosphere Control in Fruit Stores*. Short Term Leaflet, 35. Ministry of Agriculture, Fisheries and Food, Agricultural Development and Advisory Service, London.

Magness, J.R. and Diehl, H.C. 1924. Physiological studies on apples in storage. *Journal of Agricultural Research* 27, 33–34.

Magomedov, M.G. 1987. Technology of grape storage in regulated gas atmosphere. *Vinodelie i Vinogradarstvo SSSR* 2, 17–19.

Maguire, K.M. and MacKay, B.R. 2003. A controlled atmosphere induced internal browning disorder of 'Pacific Rose' TM apples. *Acta Horticulturae* 600, 281–284.

Maguire, K., Banks, N. and Lang, S. 1997. Harvest and cultivar effects on water vapour permeance in apples. *CA '97. Proceedings of the 7th International Controlled Atmosphere Research Conference*, University of California, Davis, California, 2, 246–251.

Mahajan, P.V. and Goswami, T.K. 2004. Extended storage life of litchi fruit using controlled atmosphere and low temperature. *Journal of Food Processing and Preservation* 28, 388–403.

Maharaj, R. and Sankat, C.K. 1990. The shelf-life of breadfruit stored under ambient and refrigerated conditions. *Acta Horticulturae* 269, 411–424.

Mahfoudhi, N. and Hamdi, S. 2015. Use of almond gum and gum arabic as novel edible coating to delay postharvest ripening and to maintain sweet cherry (*Prunus avium*) quality during storage. *Journal of Food Processing and Preservation* 39, 1499–1508.

Makhlouf, J., Castaigne, F., Arul, J., Willemot, C. and Gosselin, A. 1989a. Long term storage of broccoli under controlled atmosphere. *HortScience* 24, 637–639.

Makhlouf, J., Willemot, C., Arul, J., Castaigne, F. and Emond, J.P. 1989b. Regulation of ethylene biosynthesis in broccoli flower buds in controlled atmospheres. *Journal of the American Society for Horticultural Science* 114, 955–958.

Makino, Y. and Hirata, T. 1997. Modified atmosphere packaging of fresh produce with a biodegradable laminate of chitosan-cellulose and polycaprolactone. *Postharvest Biology and Technology* 10, 247–254.

Malakou, A. and Nanos, G.D. 2005. A combination of hot water treatment and modified atmosphere packaging maintains quality of advanced maturity 'Caldesi 2000' nectarines and 'Royal Glory' peaches. *Postharvest Biology and Technology* 38, 106–114.

Mali, S. and Grossmann, M.V.E. 2003. Effects of yam starch films on storability and quality of fresh strawberries (*Fragaria ananassa*). *Journal of Agricultural and Food Chemistry* 51, 7005–7011.

Mandeno, J.L. and Padfield, C.A.S. 1953. Refrigerated gas storage of apples in New Zealand. I. Equipment and experimental procedure. *New Zealand Journal of Science and Technology* B34, 462–469.

Maneenuam, T., Ketsa, S. and van Doorn, W.G. 2007. High oxygen levels promote peel spotting in banana fruit. *Postharvest Biology and Technology* 43, 128–132.

Manganaris, G.A., Vicente, A.R. and Crisosto, C.H. 2008. Effect of pre-harvest and post-harvest conditions and treatments on plum fruit quality. *CAB Reviews: Perspectives in Agriculture, Veterinary Science, Nutrition and Natural Resources* 9, 1–10.

Manleitner, S., Lippert, F. and Noga, G. 2003. Post-harvest carbohydrate change of sweet corn depending on film wrapping material. *Acta Horticulturae* 600, 603–605.

Manzano-Mendez, J.E. and Dris, R. 2001. Effect of storage atmosphere and temperature on soluble solids in mamey amarillo (*Mammea americana* L.) fruits. *Acta Horticulturae* 553, 675–676.

Mapson, L.W. and Burton, W.G. 1962. The terminal oxidases of the potato tuber. *Biochemistry Journal* 82, 19–25.

Maqbool, M., Ali, A., Alderson, P.G., Zahid, N. and Siddiqui, Y. 2011. Effect of a novel edible composite coating based on gum arabic and chitosan on biochemical and physiological responses of banana fruits during cold storage. *Journal of Agricultural and Food Chemistry* 59, 5474–5482.

Maqbool, M., Ali, A., Alderson, P.G., Mohamed, M.T.M., Siddiqui, Y. and Zahid, N. 2012. Postharvest application of gum arabic and essential oils for controlling anthracnose and quality of banana and papaya during cold storage. *Postharvest Biology and Technology* 62, 71–76.

Marcellin, P. 1973. Controlled atmosphere storage of vegetables in polyethylene bags with silicone rubber windows. *Acta Horticulturae* 38, 33–45.

Marcellin, P. and Chaves, A. 1983. Effects of intermittent high CO_2 treatment on storage life of avocado fruits in relation to respiration and ethylene production. *Acta Horticulturae* 138, 155–163.

Marcellin, P. and LeTeinturier, J. 1966. Etude d'une installation de conservation de pommes en atmosphère controleé. *International Institution of Refrigeration Bulletin, Annex 1966–1*, 141–152.

Marcellin, P., Pouliquen, J. and Guclu, S. 1979. Refrigerated storage of Passe Crassane and Comice pears in an atmosphere periodically enriched in CO_2 preliminary tests. *Bulletin de l'Institut International du Froid* 59, 1152.

Marchal, J. and Nolin, J. 1990. Fruit quality. Pre- and post-harvest physiology. *Fruits*, Special Issue, 119–122.

Marecek, J. and Machackova, L. 2003. Development of compressive stress changes and modulus of elasticity changes of champignons stored at different variants of controlled atmosphere at temperature 8 °C. *Acta Universitatis Agriculturae, et Silviculturae, Mendelianae Brunensis* 51, 77–83.

Martin, F.W. 1974. Effects of type of wound, species and humidity on curing of yam *Dioscorea alata* L. tubers before storage. *Journal of the Agricultural University of Puerto Rico* 58, 211–221.

Martinez-Cayuela, M., Plata, M.C., Sanchez-de-Medina, L., Gil, A. and Faus, M.J. 1986. Changes in various enzyme activities during ripening of cherimoya in controlled atmosphere. *ARS Pharmaceutica* 27, 371–380.

Martinez-Damian, M.T. and de Trejo, M.C.. 2002. Changes in the quality of spinach stored in controlled atmospheres. *Revista Chapingo. Serie Horticultura* 8, 49–62.

Martínez-Esplá, A., Zapata, P.J., Castillo, S., Guillén, F., Martínez-Romero, D., Valero, D. and Serrano, M. 2014. Preharvest application of methyl jasmonate (MeJA) in two plum cultivars. 1. Improvement of fruit growth and quality attributes at harvest. *Postharvest Biology and Technology* 98, 98–105.

Martinez-Javega, J.M., Jimenez Cuesta, M. and Cuquerella, J. 1983. Conservacion frigoric del melon 'Tendral'. *Anales del Instituto Nacional de Investigaciones Agrarias Agricola* 23, 111–124.

Martinez-Romero, D. and Bailén, G. 2007. Tools to maintain postharvest fruit and vegetables quality through the inhibition of ethylene action: a review. *Critical Review of Food Nutrition* 47, 543–560.

Martinez-Romero, D., Castillo, S., Valverde, J. M., Guillen, F., Valero, D. and Serrano, M. 2005. The use of natural aromatic essential oils helps to maintain post-harvest quality of 'Crimson' table grapes. *Acta Horticulturae* 682, 1723–1729.

Martínez-Sánchez, A., Marín, A., Llorach, R., Ferreres, F. and Gil, M.I. 2006. Controlled atmosphere preserves quality and phytonutrients in wild rocket (*Diplotaxis tenuifolia*). *Postharvest Biology and Technology* 40, 26–33.

Martins, C.R., Girardi, C.L., Corrent, A.R., Schenato, P.G. and Rombaldi, C.V. 2004. Periods of cold preceding the storage in controlled atmosphere in the conservation of 'Fuyu' persimmon. *Ciencia e Agrotecnologia* 28, 815–822.

Massachusetts Agricultural Experiment Station, 1941. *Massachusetts Agricultural Experiment Station, Annual Report for 1940*, Bulletin 378.

Massignan, L., Lovino, R. and Traversi, D. 1999. Trattamenti post raccolta e frigoconservazione di uva da tavola 'biologica'. *Informatore Agrario Supplemento* 55, 46–48.

Massignan, L., Lovino, R., Cillis, F. M. de, Santomasi, F. 2006. Cold storage of cherries with a combination of modified and controlled atmosphere. *Rivista di Frutticoltura e di Ortofloricoltura* 68, 63–66.

Mateos, M., Ke, D., Cantwell, M. and Kader, A.A. 1993. Phenolic metabolism and ethanolic fermentation of intact and cut lettuce exposed to CO_2-enriched atmospheres. *Postharvest Biology and Technology* 3, 225–233.

Mathieu, D. 2006. *Handbook on Hyperbaric Medicine*. Springer, Berlin.

Mathieu-Hurtiger, V., Westercamp, P. and Coureau, C. 2010. *Conditions de conservation des principales variétés de pommes – 2010*. CTIFL and CEFEL Bulletin. Centre Techniques Interprofessionnel des Fruits et Légumes, Paris.

Mathieu-Hurtiger, V., Westercamp, P. and Coureau, C. 2014. *Conditions de conservation des variétés de pommes en ULO et Extrême ULO – 2014*. CTIFL and CEFEL Bulletin. Centre Techniques Interprofessionnel des Fruits et Légumes, Paris.

Mathooko, F.M., Tsunashima, Y., Kubo Y. and Inaba, A. 2004. Expression of a 1-aminocyclopropane-1-carboxylate (ACC) oxidase gene in peach (*Prunus persica* L.) fruit in response to treatment with carbon dioxide and 1- methylcyclopropene: possible role of ethylene. *African Journal of Biotechnology* 3, 497–502.

Matityahu, I., Marciano, P., Holland, D., Ben Arie, R. and Amir, R. 2016. Differential effects of regular and controlled atmosphere storage on the quality of three cultivars of pomegranate (*Punica granatum* L.). *Postharvest Biology and Technology* 115, 132–141.

Matsui, M. (1989) *Film for keeping freshness of vegetables and fruit*. US Patent No. 4847145.

Mattheis, J. 2007. *Harvest and postharvest practices for optimum quality*. Final Project Report. Project Number PR-04-433, Washington Tree Fruit Research Commission Research Reports. Available at: http://jenny.tfrec.wsu.edu/wtfrc/dbsearch.php (accessed November, 2017).

Mattheis, J.P. and Rudell, D. 2011. Responses of 'd'Anjou' pear (*Pyrus communis* L.) fruit to storage at low oxygen setpoints determined by monitoring fruit chlorophyll fluorescence. *Postharvest Biology and Technology* 60, 125–129.

Mattheis, J.P., Buchanan, D.A. and Fellman, J.K. 1991. Change in apple fruit volatiles after storage in atmospheres inducing anaerobic metabolism. *Journal of Agricultural and Food Chemistry* 39, 1602–1605.

Mattheis, J.P., Buchanan, D.A. and Fellman, J.K. 1998. Volatile compounds emitted by 'Gala' apples following dynamic atmosphere storage. *Journal of the American Society for Horticultural Science* 123, 426–432.

Mattus, G.E. 1963. Regular and automatic CA storage. *Virginia Fruit. June 1963*. Department of Horticulture, Virginia Tech College of Agriculture and Life Sciences, Blacksburg, Virginia.

May, B.K. and Fickak, A. 2003. The efficacy of chlorinated water treatments in minimizing yeast and mold growth in fresh and semi-dried tomatoes. *Drying Technology* 21, 1127–1135.

Mazza, G. and Siemens, A.J. 1990. CO_2 concentration in commercial potato storages and its effect on quality of tubers for processing. *American Potato Journal* 67, 121–132.

Mbata, G.N. and Philips, T.W. 2001. Effects of temperature and exposure time on mortality of stored-product insects exposed to low pressure. *Journal of Economic Entomology* 94, 1302–1307.

McCollum, T. and Maul, D. 2009. 1-MCP inhibits degreening, but stimulates respiration and ethylene biosynthesis in grapefruit. Available at: http://www.ars.usda.gov/research/publications/publications.htm?SEQ_NO_115=156365 (accessed January 2010).

McDonald, J.E. 1985. Storage of broccoli. *Annual Report, Research Station, Kentville, Nova Scotia*, 114.

McGarry, A. 1993. Mechanical properties of carrots. *Postharvest Biology and Handling of Fruit, Vegetables and Flowers. Meeting of the Association of Applied Biologists, London 8 December 1993.*

McGill, J.N., Nelson, A.I. and Steinberg, M.P. 1966. Effect of modified storage atmosphere on ascorbic acid and other quality characteristics of spinach. *Journal of Food Science* 31, 510.

McGlasson, W.B. and Wills, R.B.H. 1972. Effects of O_2 and CO_2 on respiration, storage life and organic acids of green bananas. *Australian Journal of Biological Sciences* 25, 35–42.

McGuire, R.G. 1993. Application of Candida guilliermondii in commercial citrus waxes for biocontrol of *Penicillium* on grapefruit. *Postharvest Handling of Tropical Fruit. Proceedings of an International Conference held in Chiang Mai, Thailand 19–23 July 1993. Australian Centre for International Agricultural Research Proceedings* 50, 464–468.

McKay, S. and Van Eck, A. 2006. Red currants and gooseberries: extended season and marketing flexibility with controlled atmosphere storage. *New York Fruit Quarterly* 14, 43–45.

McKenzie, M.J., Greer, L.A., Heyes, J.A. and Hurst, P.L. 2004. Sugar metabolism and compartmentation in asparagus and broccoli during controlled atmosphere storage. *Postharvest Biology and Technology*, 32, 45–56.

McKeown, A.W. and Lougheed, E.C. 1981. Low pressure storage of some vegetables. *Acta Horticulturae* 116, 83–96.

McLauchlan, R.L., Barker, L.R. and Johnson, G.I. 1994. Controlled atmospheres for Kensington mango storage: classical atmospheres. *Australian Centre for International Agricultural Research Proceedings* 58, 41–44.

McMillan, R.T. 1972. Enhancement of anthracnose control on mangos by combining copper with Nu Film 17. *Proceedings of the Florida State Horticultural Society* 85, 268–270.

McMillan, R.T. 1973. Control of anthracnose and powdery mildew of mango with systemic and non systemic fungicides. *Tropical Agriculture* 50, 245–248.

Mditshwa, A., Fawole, O.A., Vries, F., van der Merwe, K., Crouch, E. and Opara U.L. 2017. Impact of dynamic controlled atmospheres on reactive oxygen species, antioxidant capacity and phytochemical properties of apple peel (cv. Granny Smith). *Scientia Horticulturae* 216, 169–176.

Meberg, K.R., Gronnerod, K. and Nystedt, J. 1996. Controlled atmosphere (CA) storage of apple at the Norwegian Agricultural College. *Nordisk Jordbruksforskning* 78, 55.

Meberg, K.R., Haffner, K. and Rosenfeld, H.J. 2000. Storage and shelf-life of apples grown in Norway. I. Effects of controlled atmosphere storage on 'Aroma'. *Gartenbauwissenschaft* 65, 9–16.

Medlicott, A.P. and Jeger, M.J. 1987. The development and application of postharvest treatments to manipulate ripening of mangoes. In: Prinsley, R.T. and Tucker, G. (eds) *Mangoes – a Review*. Commonwealth Secretariat, London.

Meheriuk, M. 1989a. Storage chacteristics of Spartlett pear. *Acta Horticulturae* 258, 215–219.

Meheriuk, M. 1989b. CA storage of apples. In: Fellman, J.K. (ed.) *Proceedings of the Fifth International Controlled Atmosphere Conference, Wenatchee, Washington, USA, 14–16 June 1989. Vol. 2. Other commodities and storage recommendations*. Washington State University, Pullman, Washington, pp. 257–284.

Meheriuk, M. 1993. CA storage conditions for apples, pears and nashi. *Proceedings of the Sixth International Controlled Atmosphere Conference, June 15 to 17 1983, Cornell University, USA*, 819–839.

Mehra, L.K., MacLean, D.D., Shewfelt, R.L. Smith, K.C and Scherm, H. 2013. Effect of postharvest biofumigation on fungal decay, sensory quality, and antioxidant levels of blueberry fruit. *Postharvest Biology and Technology* 85, 109–115.

Mehyar, G.F. and Han, J.H. 2011. Active packaging for fresh-cut fruits and vegetables. In: Aaron Brody, L., Hong Zhuang and Jung H. Han (eds) *Modified Atmosphere Packaging for Fresh-Cut Fruits and Vegetables*. Blackwell Publishing, Oxford, pp. 267–284.

Meinl, G., Nuske, D. and Bleiss, W. 1988. Influence of ethylene on cabbage quality under long term storage conditions. *Gartenbau* 35, 265.

Meir, S., Akerman, M., Fuchs, Y. and Zauberman, G. 1993. Prolonged storage of Hass avocado fruits using controlled atmosphere. *Alon Hanotea* 47, 274–281.

Melo, M.R., de Castro, J.V., Carvalho, C.R.L. and Pommer, C.V. 2002. Cold storage of cherimoya packed with zeolit film. *Bragantia* 61, 71–76.

Mencarelli, F. 1987a. The storage of globe artichokes and possible industrial uses. *Informatore Agrario* 43, 79–81.

Mencarelli, F. 1987b Effect of high CO_2 atmospheres on stored zucchini squash. *Journal of the American Society for Horticultural Science* 112, 985–988.

Mencarelli, F., Lipton, W.J. and Peterson, S.J. 1983. Responses of 'zucchini' squash to storage in low O_2 atmospheres at chilling and non-chilling temperatures. *Journal of the American Society for Horticultural Science* 108, 884–890.

Mencarelli, F., Fontana, F. and Massantini, R. 1989. Postharvest practices to reduce chilling injury CI on eggplants. In: Fellman, J.K. (ed.) *Proceedings of the Fifth International Controlled Atmosphere Research Conference, Wenatchee, Washington, USA, 14–16 June 1989. Vol. 2: Other commodities and storage recommendations*. Washington State University, Pullman, Washington, pp. 49–55.

Mencarelli, F., Botondi, R., Kelderer, M. and Casera, C. 2003. Influence of low O_2 and high CO_2 storage on quality of organically grown winter melon and control of disorders of organically grown apples by ULO in commercial storage rooms. *Acta Horticulturae* 600, 71–76.

Mendoza, D.B. 1978. Postharvest handling of major fruits in the Philippines. *Aspects of Postharvest Horticulture in ASEAN*, 23–30.

Mendoza, D.B., Pantastico, E.B. and Javier, F.B. 1972. Storage and handling of rambutan (*Nephelium lappaceum* L.). *Philippines Agriculturist* 55, 322–332.

Menjura Camacho, S. and Villamizar, C.F. 2004. Handling, packing and storage of cauliflower (*Brassica oleracea*) for reducing plant waste in Central Markets of Bogotá, Colombia. *Proceedings of the Interamerican Society for Tropical Horticulture* 47, 68–72.

Menniti, A.M. and Casalini, L. 2000. Prevention of post-harvest diseases on cauliflower. *Colture Protette* 29, 67–71.

Mercer, M.D. and Smittle, D.A. 1992. Storage atmospheres influence chilling injury and chilling injury induced changes in cell wall polysaccharides of cucumber. *Journal of the American Society for Horticultural Science* 117, 930–933.

Mertens, H. 1985. Storage conditions important for Chinese cabbage. *Groenten en Fruit* 41, 62–63.

Mertens, H. and Tranggono 1989. Ethylene and respiratory metabolism of cauliflower *Brassica olereacea* L. convar. *botrytis* in controlled atmosphere storage. *Acta Horticulturae* 258, 493–501.

Miccolis, V. and Saltveit, M.E. Jr. 1988. Influence of temperature and controlled atmosphere on storage of 'Green Globe' artichoke buds. *HortScience* 23, 736–741.

Michalský, M. and Hooda, P.S. 2015. Greenhouse gas emissions of imported and locally produced fruit and vegetable commodities: a quantitative assessment. *Environmental Science & Policy* 48, 32–43.

Mignani, I. and Vercesi, A. 2003. Effects of postharvest treatments and storage conditions on chestnut quality. *Acta Horticulturae* 600, 781–785.

Miller, E.V. and Brooks, C. 1932. Effect of CO_2 content of storage atmosphere on carbohydrate transformation in certain fruits and vegetables. *Journal of Agricultural Research* 45, 449–459.

Miller, E.V. and Dowd, O.J. 1936. Effect of CO_2 on the carbohydrates and acidity of fruits and vegetables in storage. *Journal of Agricultural Research* 53, 1–7.

Mills, G., Earnshaw, R. and Patterson, M.F. 1998. Effects of high hydrostatic pressure on Clostridiuim sporogenes spores. *Letters Applied Microbiology* 26, 227–230.

Min, K. and Oogaki, C. 1986. Characteristics of respiration and ethylene production in fruits transferred from LP storage to ambient atmosphere. *Journal of the Japanese Society for Horticultural Science* 55, 339–347.

Miranda, M.R.A. de, da Silva, F.S., Alves, R.E., Filgueiras, H.A.C. and Araujo, N.C.C. 2002. Storage of two types of sapodilla under ambient conditions. *Revista Brasileira de Fruticultura* 24, 644–646.

Mistriotis, A., Briassoulis, D., Giannoulis, A. and D'Aquino, S. 2016. Design of biodegradable bio-based equilibrium modified atmosphere packaging (EMAP) for fresh fruits and vegetables by using micro-perforated poly-lactic acid (PLA) films. *Postharvest Biology and Technology* 111, 380–389.

Miszczak, A. and Szymczak, J.A. 2000. The influence of harvest date and storage conditions on taste and apples aroma. *Zeszyty Naukowe Instytutu Sadownictwa i Kwiaciarstwa w Skierniewicach* 8, 361–369.

Mitcham, E., Zhou, S. and Bikoba, V. 1997. Development of carbon dioxide treatment for Californian table grapes. *Seventh International Controlled Atmosphere Research Conference, July 13–18 1997*, University of California, Davis, California [Abstract], 65.

Mitcham, E.J., Martin, T., and Zhou, S. 2006. The mode of action of insecticidal controlled atmospheres. *Bulletin of Entomological Research* 96, 213–222.

Mitsuda, H., Kawai, F. and Yamamoto, A. 1972. Underwater and underground storage of cereal grains. *Food Technology* 26, 50–56.

Miyazaki, T. 1983. Effects of seal packaging and ethylene removal trom sealed bags on the shelf life of mature green Japanese apricot *Prunus mume* Sieb. Zucc. fruits. *Journal of the Japanese Society for Horticultural Science* 52, 85–92.

Mizobutsi, G.P., Borges, C.A.M. and de Siqueira, D.L. 2000. Conservacao pos colheita da lima acida 'Tahiti' Citrus latifolia Tanaka, tratada com acido giberelico e armazenada em tres temperaturas. *Revista Brasileira de Fruticultura* 22, (special issue) 42–47.

Mizukami, Y., Saito, T. and Shiga, T. 2003. Enzyme activities related to ascorbic acid in spinach leaves during storage. *Journal of the Japanese Society for Food Science and Technology* 50, 1–6.

Mohamed, S., Khin, M.M.K., Idris, A.Z., Yusof, S., Osman, A. and Subhadrabandhu, S. 1992. Effects of various surface treatments (palm oil, liquid paraffin, Semperfresh or starch surface coatings and LDPE wrappings) on the storage life of guava (*Psidium guajava* L.) at 10 °C. *Acta Horticulturae* 321, 786–794.

Mohammed, M. 1993. Storage of passionfruit in polymeric films. *Proceedings of the Interamerican Society for Tropical Horticulture* 37, 85–88.

Mohammed, M., Hajar Ahmad, S., Abu Bakar, R. and Lee Abdullah, T. 2011. Golden apple (*Spondias dulcis* Forst. syn. *Spondias cytherea* Sonn.) In: Yahia, E.M. (ed.) *Postharvest Biology and Technology of Tropical and Subtropical Fruits*. Woodhead Publishing, Oxford, Cambridge, Philadelphia, New Delhi, pp. 159–178.

Monzini, A. and Gorini, F.L. 1974. Controlled atmosphere in the storage of vegetables and flowers. *Annali dell'Istituto Sperimentale per la Valorizzazione Tecnologica dei Prodotti Agricoli* 5, 277–291.

Moor, U., Mölder, K., Tõnutare, T. and Põldma, P. 2009. Effect of active and passive MAP on postharvest quality of raspberry 'Polka'. *10th International Controlled & Modified Atmosphere Research Conference, 4–7 April 2009 Turkey* [Poster abstract], 65.

Morais, P.L.D., Miranda, M.R.A., Lima, L.C.O., Alves, J.D., Alves, R.E. and Silva, J.D. 2008.Cell wall biochemistry of sapodilla (*Manilkara zapota*) submitted to 1-methylcyclopropene. *Brazilian Journal of Plant Physiology* 20, 85–94.

Morales, H., Sanchis, V., Rovira, A., Ramos, A.J. and Marín, S. 2007. Patulin accumulation in apples during postharvest: effect of controlled atmosphere storage and fungicide treatments. *Food Control* 18, 1443–1448.

Moran, R.E. 2006. Maintaining fruit firmness of 'McIntosh' and 'Cortland' apples with aminoethoxyvinylglycine and 1-methylcyclopropene during storage. *HortTechnology* 16, 513–516.

Moreira, S.A., Fernandes, P.A.R., Duarte, R., Santos, D.I., Fidalgo, L.G. *et al.* 2015. A first study comparing preservation of a ready-to-eat soup under pressure (hyperbaric storage) at 25°C and 30°C with refrigeration. *Food Science & Nutrition* 3, 467–474.

Moreno, J. and De la Plaza, J.L. 1983. The respiratory intensity of cherimoya during refrigerated storage: a special case of climacteric fruit. *Acta Horticulturae* 138, 179.

Moretti, C.L., Araújo, A.L., Marouelli, W.A. and Silva, W.L.C. (undated) 1-methylcyclopropene delays tomato fruit ripening. Available at: http://www.scielo.br/scielo.php?script=sci_arttext&pid=S01020536 2002000400030 (accessed May 2009).

Mori, M. and Kozukue, N. 1995. The glyalkaloid contents of potato tubers: differences in tuber greening in various varieties and lines. *Report of the Kyushu Branch of the Crop Science Society of Japan* 61, 77–79.

Morris, L. and Kader, A.A. 1977. Physiological disorders of certain vegetables in relation to modified atmosphere. *Second National Controlled Atmosphere Research conference. Proceedings, Michigan State University Horticultural Report* 28, 266–267.

Morris, L., Yang, S.F. and Mansfield, D. 1981. Postharvest physiology studies. *Californian Fresh Market Tomato Advisory Board Annual Report 1980–1981*, 85–105.

Morris, S.C., Jobling, J.J., Tanner, D.J. and Forbes-Smith, M.R. 2003. Prediction of storage or shelf life for cool stored fresh produce transported by reefers. *Acta Horticulturae* 604, 305–311.

Mota, W.F. da, Finger, F.L., Cecon, P.R., Silva, D.J.H. da, Correa, P.C., Firme, L.P., and Neves, L.L.de M. 2006. Shelf-life of four cultivars of okra covered with PVC film at room temperature. *Horticultura Brasileira* 24, 255–258.

Moura, M.L. and Finger, F. 2003. L. Production of volatile compounds in tomato fruit (*Lycopersicon esculentum* Mill.) stored under controlled atmosphere. *Revista Brasileira de Armazenamento* 28, 25–28.

Moya-Leon, M.A., Vergara, M., Bravo, C., Pereira, M., and Moggia, C. 2007. Development of aroma compounds and sensory quality of 'Royal Gala' apples during storage. *Journal of Horticultural Science & Biotechnology* 82, 403–413.

Munera, S., Besada, C., Blasco, J., Cubero, S., Salvador, A., Talens, P, and Aleixos, N. 2017. Astringency assessment of persimmon by hyperspectral imaging. *Postharvest Biology and Technology* 125, 35–41.

Muñoz-Robredo, P., Rubio, P., Infante, R., Campos-Vargas, R., Manríquez, D., González-Agüero, M. and Defilippi, B.G. 2012. Ethylene biosynthesis in apricot: Identification of a ripening-related 1-aminocyclopropane-1-carboxylic acid synthase (ACS) gene. *Postharvest Biology and Technology* 63, 85–90.

Nagar, P.K. 1994. Physiological and biochemical studies during fruit ripening in litchi (*Litchi chinensis* Sonn.). *Postharvest Biology and Technology* 4, 225–234.

Nahor, H.B., Scheerlinck, N., Verboven, P., van Impe, J. and Nicolai, B. 2003. Combined discrete and continuous simulation of controlled atmosphere (CA) storage systems. *Communications in Agricultural and Applied Biological Sciences* 68, 17–21.

Nahor, H.B., Schotsmans, W., Scheerlinck, N. and Nicolaï, B.M. 2005. Applicability of existing gas exchange models for bulk storage of pome fruit: assessment and testing. *Postharvest Biology and Technology* 35, 15–24.

Naichenko, V.M. and Romanshchak, S.P. 1984. Growth regulators in fruit of the plum cultivar Vengerka Obyknovennaya during ripening and long term storage. *Fizioliya I Biokhimiya Kul'turnykhRastenii* 16, 143–148.

Naik, L., Sharma, R., Rajput, Y.S. and Manju, G. 2013. Application of high pressure processing technology for dairy food preservation-future perspective: a review. *Journal of Animal Production Advances* 3, 232–241.

Nair, H. and Tung, H.F. 1988. Postharvest physiology and Storage of Pisang Mas. *Proceedings of the UKM simposium Biologi Kebangsaan ketiga, Kuala Lumpur, Nov. 1988*, 22–24.

Nakamura, N., Sudhakar Rao, D.V., Shiina, T. and Nawa, Y. 2003. Effects of temperature and gas composition on respiratory behaviour of tree-ripe 'Irwin' mango. *Acta Horticulturae* 600, 425–429.

Nardin, K. and Sass, P. 1994. Scald control on apples without use of chemicals. *Acta Horticulturae* 368, 417–428.

Navarro, S. 1978. The effects of low oxygen tensions on three stored-product insect pests. *Phytoparasitica* 6, 51–58.

Navarro, S. and Calderon, M. 1979. Mode of action of low atmospheric pressures on *Ephestia cautella* (Wlk.) pupae. *Experientia* 35, 620–621.

Navarro, S., Donahaye, J.E., Dias, R., Azrieli, A., Rindner, M. *et al.* 2001. Application of vacuum in a transportable system for insect control. In: Donahaye, E.J., Navarro, S. and Leesch J.G. (eds) *Proceedings of the International Conference on Controlled Atmosphere and Fumigation in Stored Products, Fresno, CA. 29 Oct. to 3 Nov. 2000*. Executive Printing Services, Clovis, California, pp. 307–315.

Navarro, S., Finkelman, S., Donahaye, J. E., Isikber, A., Rindner, M. and Dias, R. 2007. Development of a methyl bromide alternative for the control of stored product insects using a vacuum technology. In: Donahaye, E.J., Navarro, S., Bell, C., Jayas, D, Noyes, R. and Phillips, T.W. (eds.) *Proceedings of the International Conference Controlled Atmosphere and Fumigation in Stored Products, Gold-Coast Australia. 8–13th August 2004*. FTIC Publishing, Israel, pp. 227–234.

Neale, M.A., Lindsay, R.T. and Messer, H.J.M. 1981. An experimental cold store for vegetables. *Journal of Agricultural Engineering Research* 26, 529–540.

Negi, P.S. and Roy, S.K. 2000. Storage performance of savoy beet (*Beta vulgaris* var. *bengalensis*) in different growing seasons. *Tropical Science* 40, 211–213.

Nerd, A. and Mizrahi, Y, 1993. Productivity and postharvest behaviour of black sapote in the Israeli Negeve desert. *Australian Centre for International Agricultural Research Proceedings* 50, 441 [Abstract].

Neri, D.M., Hernandez, F.A.D. and Guemes, V.N. 2004. Influence of wax and plastic covers on the quality and conservation time of fruits of Eureka lemon. *Revista Chapingo. Serie Ingenieria Agropecuaria* 7, 99–102.

Neri, F., Mari, M., Menniti, A.M. and Brigati, S. 2006. Activity of trans-2-hexenal against *Penicillium expansum* in 'Conference' pears. *Journal of Applied Microbiology* 100, 1186–1193.

Neuwald, D.A., Sestari, I., Giehl, R.F.H., Pinto, J.A.V., Sautter, C.K. and Brackmann, A. 2008. Maintaining the quality of the persimmon 'Fuyu' through storage in controlled atmosphere. *Revista Brasileira de Armazenamento* 33, 68–75.

Neuwald, D.A., Spuhler, M., Wünsche, J. and Kittemann, D. 2016. New apple storage technologies can reduce energy usage and improve storage life. *17th International. Conference on Organic Fruit-Growing, Stuttgart, Germany*. Available at: http://www.ecofruit.net/2016/29_Neuwald_184bis187.pdf (accessed January, 2018).

Neuwirth, G.R. 1988. Respiration and formation of volatile flavour substances in controlled atmosphere-stored apples after periods of ventilation at different times. *Archiv fur Gartenbau* 36, 417–422.

Neves, L.C., Bender, R.J., Rombaldi, C.V. and Vieites, R.L. 2004. Storage in passive modified atmosphere of 'Golden Star' starfruit (*Averrhoa carambola* L.). *Revista Brasileira de Fruticultura* 26, 13–16.

Nham, N.T., Willits, N., Zakharov, F. and Mitcham, E.J. 2017. A model to predict ripening capacity of 'Bartlett' pears (*Pyrus communis* L.) based on relative expression of genes associated with the ethylene pathway. *Postharvest Biology and Technology* 128, 138–143.

Nichols, D., Hebb, J., Harrison, P., Prange, R. and DeLong, J. 2008. *Atlantic Canada Honeycrisp: Harvest, Conditioning and Storage*. Agriculture and Agri-Food Canada Publication No. 10541E, 1–2. AAFC, Ottawa.

Nichols, R. 1971. *A Review of the Factors Affecting the Deterioration of Harvested Mushrooms*. Report 174. Glasshouse Crops Research Institute, Littlehampton, UK.

Nicolas, J., Rothan, C. and Duprat, F. 1989. Softening of kiwifruit in storage. Effects of intermittent high CO_2 treatments. *Acta Horticulturae* 258, 185–192.

Nicotra, F.P. and Treccani, C.P. 1972. Ripeness for harvest and for refrigeration in relation to controlled atmosphere storage in the apple cv. Morgenduft. *Rivista della Ortoflorofrutticoltura Italiana* 56, 207–218.

Niedzielski, Z. 1984. Selection of the optimum gas mixture for prolonging the storage of green vegetables. Brussels sprouts and spinach. *Industries Alimentaires et Agricoles* 101, 115–118.

Nigro F., Ippolito A., Lattanzio V., Venere D.D. and Salerno M. 2000. Effect of ultraviolet-C light on postharvest decay of strawberry. *Journal of Plant Pathology* 82, 29–37.

Nilsen, K.N. and Hodges, C.F. 1983. Hypobaric control of ethylene-induced leaf senescence in intact plants of *Phaseolus vulgaris* L. *Plant Physiology* 71, 96–101.

Nilsson, T. 2005. Effects of ethylene and 1-MCP on ripening and senescence of European seedless cucumbers. *Postharvest Biology and Technology* 36, 113–125.

Niranjana, P., Gopalakrishna, R.K.P., Sudhakar, R.D.V. and Madhusudhan, B. 2009. Effect of controlled atmosphere storage (CAS) on antioxidant enzymes and dpph-radical scavenging activity of mango (*Mangifera indica* L.) Cv. Alphonso. *African Journal of Food, Agriculture, Nutrition and Developemnt* 9, 779–792.

Nnodu, E.C. and Nwankiti, A.O. 1986. Chemical control of post-harvest deterioration of yam tubers. *Fitopatologia Brasileira* 11, 865–871.

No, H.K., Meyers, S.P., Prinyawiwatkul, W. and Xu, Z. 2007. Applications of chitosan for improvement of quality and shelf life of foods: a review. *Journal Food Science* 72, 87–100.

Nock, J.F., Al Shoffe, Y., Gunes, N., Zhang, Y., Wright, H., DeLong, J. and Watkins, C.B. 2016. Controlled atmosphere (CA) and dynamic CA-Chlorophyll Fluorescence (DCA-CF) storage of 'Gala'. *American Society for Horticultural Science Annual Conference, August 2016, Atlanta, Georgia, USA* [Abstract]. Available at: https://ashs.confex.com/ashs/2016/webprogram/Paper25121.html (accessed January 2018).

Nock, J.F., Doerflinger, F.C., Sutanto, G., Al Shoffe, Y., Gunes, N. *et al.* 2018. Managing stem-end flesh browning, a physiological disorder of 'Gala' apples. *Acta Horticulturae* (in press).

Nongtaodum, S. and Jangchud, A. 2009. Quality of fresh-cut mangoes (Fa-lun) during storage. *Kasetsart Journal* (*Natural Science*) 43, 282–289.

Noomhorm, A. and Tiasuwan, 1988. Effect of controlled atmosphere storage for mango. *N. Paper, American Society of Agricultural Engineers* 88, 6589.

North, C.J. and Cockburn, J.T., 1975. Ethyl alcohol levels in apples after deprivation of oxygen and the detection of alcohol vapour in controlled atmosphere storage using indicator tubes. *Journal of the Science of Food and Agriculture* 26, 1155–1161.

Nunes, M.C.N., Morais, A.M.M.B., Brecht, J.K. and Sargent, S.A. 2002. Fruit maturity and storage temperature influence response of strawberries to controlled atmospheres. *Journal of the American Society for Horticultural Science* 127, 836–842.

Nuske, D. and Muller, H. 1984. Erste Ergebnisse bei der industriemassigen Lagerung von Kopfkohl unter CA Lagerungsbedingungen. *Nachrichtenblatt fur den Pflanzenschutz in der DDR* 38, 185–187.

Nyanjage, M.O., Nyalala, S.P.O., Illa, A.O., Mugo, B.W., Limbe, A.E. and Vulimu, E.M. 2005. Extending post-harvest life of sweet pepper (*Capsicum annum* L. 'California Wonder') with modified atmosphere packaging and storage temperature. *Agricultura Tropica et Subtropica* 38, 28–32.

Ogaki, C., Manago, M., Ushiyama, K. and Tanaka, K. 1973. Studies on controlled atmosphere storage of satsumas. I. Gas concentration, relative humidity, wind velocity and pre-storage treatment. *Bulletin of the Kanagawa Horticultural Experiment Station* 21, 1–23.

Ogata, K., Yamauchi, N. and Minamide, T. 1975. Physiological and chemical studies on ascorbic acid in fruits and vegetables. 1. Changes in ascorbic acid content during maturation and storage of okra. *Journal of the Japanese Society for Horticultural Science* 44, 192–195.

O'Hare, T.J. and Prasad, A. 1993. The effect of temperature and CO_2 on chilling symptoms in mango. *Acta Horticulturae* 343, 244–250.

O'Hare, T.J., Prasad, A. and Cooke, A.W. 1994. Low temperature and controlled atmosphere storage of rambutan. *Postharvest Biology and Technology* 4, 147–157.

O'Hare, T.J., Wong, L.S., Prasad, A., Able, A.J. and King, G.J. 2000. Atmosphere modification extends the postharvest shelflife of fresh-cut leafy Asian brassicas. *Acta Horticulturae* 539, 103–107.

Oliveira, M., Abadias, M. Usall, J., Torres, R. and Teixidó, N. 2015. Application of modified atmosphere packaging as a safety approach to fresh-cut fruits and vegetables – a review. *Trends in Food Science & Technology* 46, 13–26.

Olsen, K. 1986. Views on CA Storage of Apples. *Post Harvest Pomology Newsletter* 4 (no. 2, July–August).

Olsen, N., Thornton, R.E., Baritelle, A. and Hyde, G. 2003. The influence of storage conditions on physical and physiological characteristics of Shepody potatoes. *Potato Research* 46, 95–103.

Olsson, S. 1995. Production equipment for commercial use. In: Ledweard, D.A., Johnston, D.E., Earnshaw, R.G. and Hasting, A.P.M. (eds) *High Pressure Processing of Foods*. Nottingham University Press, UK, p. 167.

Omary, M.B., Testin, R.F., Barefoot, S.F. and Rushing, J.W. 1993. Packaging effects on growth of *Listeria innocua* in shredded cabbage. *Journal of Food Science* 58, 623–626.

Onawunmi, G.O. 1989. Evaluation of the antimicrobial activity of citral. *Letters in Applied Microbiology* 9, 105–108.

Onoda, A., Koizumi, T., Yamamoto, K., Furruya, T., Yamakawa, H. and Ogawa, K. 1989. A study of varaible low pressure storage of cabbage and turnip. *Nippon Shokuhin Kogyo Gakkaishi* 36, 369–374.

Oriani, V.B., Molina, G., Chiumarelli, M., Pastore, G.M. and Hubinger, M.D. 2014. Properties of cassava starch based edible coating containing essential oils. *Journal of Food Science* 79, E189–E194.

O'Rourke, D. 2017. *World Apple Review*. Belrose, Inc., Pullman, Washington.

Ortiz, A. and Lara, I. 2008. Cell wall-modifying enzyme activities after storage of 1-MCP-treated peach fruit. *Acta Horticulturae* 796, 137–142.

Ortiz, A., Echeverría, G., Graell, J. and Lara, I. 2009. Overall quality of 'Rich Lady' peach fruit after air- or CA storage. The importance of volatile emission. *Food Science and Technology* 42, 1520–1529.

Ortiz, A., Graell, J. and Lara, I. 2011. Preharvest calcium applications inhibit some cell wall-modifying enzyme activities and delay cell wall disassembly at commercial harvest of 'Fuji Kiku-8' apples. *Postharvest Biology and Technology* 62, 161–167.

Othieno, J.K., Thompson, A.K. and Stroop, I.F. 1993. Modified atmosphere packaging of vegetables. *Postharvest Treatment of Fruit and Vegetables. COST 94 Workshop, September 14–15 1993, Leuven, Belgium*, 247–253.

Otma, E.C. 1989. Controlled atmosphere storage and film wrapping of red bell peppers *Capsicum annuum* L. *Acta Horticulturae* 258, 515–522.

Oudit, D.D. 1976. Polythene bags keep cassava tubers fresh for several weeks at ambient temperature. *Journal of the Agricultural Society of Trinidad and Tobago* 76, 297–298.

Overholser, E.L. 1928. Some limitations of gas storage of fruits. *Ice and Refrigeration* 74, 551–552.

Ozdemir, M. and Floros, J.D. 2004. Active food packaging technologies. *Critical Review of Food and Nutrition* 44, 185–193.

Özer, M.H., Eris, A. and Türk, R.S. 1999. A research on controlled atmosphere storage of kiwifruit. *Acta Horticulturae* 485, 293–300.

Özer, M.H., Erturk, U. and Akbudak, B. 2003. Physical and biochemical changes during controlled atmosphere (CA) storage of cv. Granny Smith. *Acta Horticulturae* 599, 673–679.

Özgen, M., Serçe, S., Akça, Y. and Hong, J.H. 2015. Lysophosphatidylethanolamine (LPE) improves fruit size, colour, quality and phytochemical contents of sweet cherry cv. '0900 Ziraat'. *Korean Journal of Horticultural Science & Technology* 33, 196–201

Ozturk, B., Kucuker, E., Karaman, S. and Ozkan, Y. 2012. The effects of cold storage and aminoethoxyvinylglycine (AVG) on bioactive compounds of plum fruit (*Prunus salicina* Lindell cv. 'Black Amber'). *Postharvest Biology and Technology* 73, 35–41.

Padilla-Zakour, O.I., Tandon, K.S. and Wargo, J.M. 2004. Quality of modified atmosphere packaged 'Hedelfingen' and 'Lapins' sweet cherries. *HortTechnology* 14, 331–337.

Paillart, M. 2013. Progress in physiology and techniques for storage and packaging of vegetables *Bioforsk Fokus* 8, 9.

Pakkasarn, S., Kanlayanarat, S. and Uthairantanakij, A. 2003a. Effect of controlled atmosphere on the storage life of mangosteen fruit (*Garcinia mangostana* L.). *Acta Horticulturae* 600, 759–762.

Pakkasarn, S., Kanlayanarat, S. and Uthairatanakij, A. 2003b. High carbon dioxide storage for improving the postharvest life of mangosteen fruit (*Garcinia mangosteen* L.). *Acta Horticulturae* 600, 813–816.

Pal, R.K. and Buescher, R.W. 1993. Respiration and ethylene evolution of certain fruits and vegetables in response to CO_2 in controlled atmosphere storage. *Journal of Food Science and Technology Mysore* 30, 29–32.

Pal, R.K., Singh, S.P., Singh, C.P. and Ram Asrey, 2007. Response of guava fruit (*Psidium guajava* L. cv. Lucknow-49) to controlled atmosphere storage. *Acta Horticulturae* 735, 547–554.

Pala, M., Damarli, E. and Alikasifoglu, K. 1994. A study of quality parameters in green pepper packaged in polymeric films. *Commissions C2, D1 ,D2/3 of the International Institute of Refrigeration International Symposium June 8–10 Istanbul Turkey*, 305–316.

Palma, T., Stanley, D.W., Aguilera, J.M. and Zoffoli, J.P. 1993. Respiratory behavior of cherimoya *Annona cherimola* Mill. under controlled atmospheres. *HortScience* 28, 647–649.

Pan, C.R., Lin, H.T. and Chen, J.Q. 2006. Study on low temperature and controlled atmosphere storage with silicone rubber window pouch of chinquapin. *Transactions of the Chinese Society of Agricultural Machinery* 37, 102–106.

Pan, Q.F., Chen, Y., Wang, Q., Yuan, F., Xing, S.H. *et al*. 2010. Effect of plant growth regulators on the biosynthesis of vinblastine, vindoline and catharanthine in *Catharanthus roseus*. *Plant Growth Regulation* 60, 133–141.

Pan, S.L., Huang, C.Y., Wang, H.B., Pang, X Q., Huang, X.M. and Zhang, Z.Q. 2007. Hydrogen peroxide induced chilling-resistance of postharvest banana fruit. *Journal of South China Agricultural University* 28, 34–37.

Pan, Y.G. 2008. The physiological changes of *Durio zibethinus* fruit after harvest and storage techniques. *South China Fruits* 4, 45–47.

Panagou, E.Z., Vekiari, S.A. and Mallidis, C. 2006. The effect of modified atmosphere packaging of chestnuts in suppressing fungal growth and related physicochemical changes during storage in retail packages at 0 and 8 degrees C. *Advances in Horticultural Science* 20, 82–89.

Paniagua, A.C., East, A.R. and Heyes, J.A. 2014. Interaction of temperature control deficiencies and atmosphere conditions during blueberry storage on quality outcomes. *Postharvest Biology and Technology* 95, 50–59.

Pantastico, Er.B. (ed.) 1975. *Postharvest Physiology, Handling and Utilization of Tropical and Sub-Tropical Fruits and Vegetables*. AVI Publishing, Westpoint, Connecticut.

Pariasca, J.A.T., Miyazaki, T., Hisaka, H., Nakagawa, H. and Sato, T. 2001. Effect of modified atmosphere packaging (MAP) and controlled atmosphere (CA) storage on the quality of snow pea pods (*Pisum sativum* L. var. *saccharatum*). *Postharvest Biology and Technology* 21, 213–223.

Park, H.J., Chinnan, M.S and Shewfelt, R.L. 1994. Edible corn-zein film coatings to extend storage life of tomatoes. *Journal of Food Processing and Preservation* 18, 317–331.

Park, H.J., Yoon, I.K. and Yang, Y.J. 2009. Quality changes in peaches 'Mibaekdo' and 'Hwangdo' from 1-MCP treatment. *10th International Controlled & Modified Atmosphere Research Conference, 4–7 April 2009 Turkey* [Poster abstract], 68.

Park, H.W., Cha, H.S., Kim, Y.H., Lee, S.A. and Yoon, J.Y. 2007. Change in the quality of 'Fuji' apples by using functional MA (modified atmosphere) film. *Korean Journal of Horticultural Science & Technology*, 25, 37–41.

Park, N.P., Choi, E.H. and Lee, O.H. 1970. Studies on pear storage. II. Effects of polyethylene film packaging and CO_2 shock on the storage of pears, cv. Changsyprang. *Korean Journal of Horticultural Science* 7, 21–25.

Park, Y.M. and Kwon, K.Y. 1999. Prevention of the incidence of skin blackening by postharvest curing procedures and related anatomical changes in 'Niitaka' pears. *Journal of the Korean Society for Horticultural Science* 40, 65–69.

Park, Y.S. and Jung, S.T. 2000. Effects of CO_2 treatments within polyethylene film bags on fruit quality of fig fruits during storage. *Journal of the Korean Society for Horticultural Science* 41, 618–622.

Parsons, C.S. and Spalding, D.H. 1972. Influence of a controlled atmosphere, temperature, and ripeness on bacterial soft rot of tomatoes. *Journal of the American Society for Horticultural Science* 97, 297–299.

Parsons, C.S., Gates, J.E. and Spalding D.H. 1964. Quality of some fruits and vegetables after holding in nitrogen atmospheres. *American Society for Horticultural Science* 84, 549–556.

Parsons, C.S., Anderson, R.E. and Penny, R.W. 1970. Storage of mature green tomatoes in controlled atmospheres. *Journal of the American Society for Horticultural Science* 95, 791–793.

Parsons, C.S., Anderson, R.E. and Penney, R.W. 1974. Storage of mature-green tomatoes in controlled atmospheres. *Journal of the American Society for Horticultural Science* 95, 791–794.

Pasentsis, K., Falara, V., Pateraki, I., Gerasopoulos, D. and Kanellis, A.K. 2007. Identification and expression profiling of low oxygen regulated genes from Citrus flavedo tissues using RT-PCR differential display. *Journal of Experimental Botany* 58, 2203–2216.

Passam, H.C. 1982. Experiments on the storage of eddoes and tannia (*Colocasia* and *Xanthosoma* spp.) under tropical ambient conditions. *Tropical Science* 24, 39–46.

Passam, H.C., Karapanos, I.C., Bebeli, P.J. and Savvas, D. 2007. A review of recent research on tomato nutrition, breeding and post-harvest technology with reference to fruit quality. *European Journal of Plant Science and Biotechnology* 1, 1–21.

Passaro-Carvalho, C.P., Navarro, P. and Salvador, A. 2012. Postharvest. In: *Citrus: Growing, Postharvest and Industrialization.* Lasallia Research and Science Series, Bogota, Colombia, pp. 223–285.

Patiño, L.S., Castellanos, D.A. and Herrera, A.O. 2018. Influence of 1-MCP and modified atmosphere packaging on the quality and preservation of fresh basil. *Postharvest Biology and Technology* 136, 57–65.

Paton, J.E. and Scriven, F.M. 1989. Use of NAA to inhibit sprouting in sweetpotatoes (*Ipomea batatas*). *Journal of the Science of Food and Agriculture* 48, 421–428.

Patterson, M.E. and Melsted, S.W. 1977. Sweet cherry handling and storage alternatives. *Proceedings of the 2nd National Controlled Atmosphere Research Conference. Michigan State University, East Lansing, Michigan*, p. 9.

Patterson, M.E. and Melsted, S.W. 1978. Improvement of prune quality and condition by hypobaric storage. *HortScience* 13, 351.

Pattie, D.R. and Lougheed, E.C. 1974. *Feasibility study of low pressure storage*. Department of Horticultural Science School of Engineering, University of Guelph, Ontario, Canada.

Paul, A.-L. and Ferl, R..J. 2006. The biology of low atmospheric pressure – implications for exploration mission design and advanced life support. *Gravitational and Space Biology* 19, 3–17.

Paull, R.E. and Chen , C.C. 2003. Postharvest physiology, handling and storage of pineapple. In: Bartholomew, D.P., Paull, R.E. and Rohrbach, K.G. (eds) *The Pineapple: Botany, Production and Uses*. CABI, Wallingford, UK, 253–279.

Paull, R.E. and Chen, C.C. 2004. Mango. In: Gross, K.C., Wang, C.Y. and Saltveit, M. (eds) *The Commercial Storage of Fruits, Vegetables and Florist and Nursery Stocks*. A draft version of the forthcoming revision to USDA, Agriculture Handbook 66 on the website of USDA, Agricultural Research Service. Available at: http://www.ba.ars.usda.gov/hb66/ (accessed 9 January 2009).

Paull, R.E. and Rohrbach, K.G. 1985. Symptom development of chilling injury in pineapple fruit. *Journal of the American Society for Horticultural Science* 110, 100–105.

Paull, R.E. Cavaletto, C.G. and Ragone, D. 2005. Breadfruit: potential new export crop for the Pacific. Available at: http://www.reeis.usda.gov/web/crisprojectpages/194131.html (accessed May 2009).

Paulo, F.J., Rosana, C.B., Gisela, P., de M.B., Adriana Z., M., Franco, M.L. and Roberto, O. do M.J. 2007. Papaya fruit ripening: response to ethylene and 1-Methylcyclopropene (1-MCP). *Journal of Agricultural and Food Chemistry* 55, 6118–6123.

Peacock, B.C. 1972. Role of ethylene in the initiation of fruit ripening. *Queensland Journal of Agriculture and Animal Science* 29, 137–145.

Peacock, B.C. 1988. Simulated commercial export of mangoes using controlled atmosphere container technology. *Australian Centre for International Agricultural Research Proceedings* 23, 40–44.

Peano, C., Girgenti, V. and Giuggioli, N.R. 2014. Effect of different packaging materials on postharvest quality of cv. Envie strawberry. *International Food Research Journal* 21, 1129–1134.

Pegoraro, C., dos Santos R.S., Krüge M. M., Tiecher A., da Mai L.C., Rombaldi C.V. and de Oliveira A.C. 2012. Effects of hypoxia storage on gene transcript accumulation during tomato fruit ripening. *Brazilian Journal of Plant Physiology* 24 no.2 (Campos dos Goytacazes Apr./June 2012). On-line version ISSN 1677–9452. Available at: http://dx.doi.org/10.1590/S1677-04202012000200007

Pekmezci, M., Erkan, M., Gubbuk, H., Karasahin, I. and Uzun, I. 2004. Modified atmosphere storage and ethylene absorbent enables prolonged storage of 'Hayward' kiwifruits. *Acta Horticulturae* 632, 337–341.

Pelayo, C., Ebeler, S.E. and Kader. A.A. 2003. Postharvest life and flavour quality of three strawberry cultivars kept at 5 °C in air or air + 20 kPa CO_2. *Postharvest Biology and Technology* 27, 171–183.

Pelleboer, H. 1983. A new method of storing Brussels sprouts shows promise. *Bedrijfsontwikkeling* 14, 828–831.

Pelleboer, H. 1984. A future for CA storage of open grown vegetables. *Groenten en Fruit* 39, 62–63.

Pelleboer, H. and Schouten, S.P. 1984. New method for storing Chinese cabbage is a success. *Groenten en Fruit* 40, 16–51.

Penchaiya, P., Niyomlao, W. and Kanlayanarat, S. 2003. Effect of controlled atmosphere storage on quality of sweet basil. *Agricultural Science Journal* 34, 4–6 (Suppl.), 127–129. Available at: http://www.phtnet.org/download/FullPaper/pdf/2ndSeminarKKU/af114.pdf (accessed 21 February 2018).

Pendergrass, A. and Isenberg, F.M.R. 1974. The effect of relative humidity on the quality of stored cabbage. *HortScience* 9, 226–227.

Peng, H., Yang, T. and Jurick, W.M. II. 2014. Calmodulin gene expression in response to mechanical wounding and *Botrytis cinerea* infection in tomato fruit. *Plants (Basel)* 3, 427–441.

Peppelenbos, W.H., Tijskens, L.M.M., Johan van't Leven, J., and Wilkinson, E.C. 1996. Modelling oxidative and fermentative carbon dioxide production of fruits and vegetables. *Postharvest Biology and Technology* 9, 283–295.

Perdones, A., Sánchez-González, L., Chiralt, A. and Vargas, M. 2012. Effect of chitosan-lemon essential oil coatings on storage-keeping quality of strawberry. *Postharvest Biology and Technology* 70, 32–41.

Pereira, A. de M., Faroni, L.R.D., de Sousa, A.H., Urruchi, W.I. and Roma, R.C.C. 2007. Immediate and latent effects of ozone fumigation on the quality of maize grains. *Revista Brasileira de Armazenamento* 32, 100–110.

Pereira, M.J., Amaro, A.L., Pintado, M. and Poças, M.F. 2017. Modeling the effect of oxygen pressure and temperature on respiration rate of ready-to-eat rocket leaves. A probabilistic study of the Michaelis–Menten model. *Postharvest Biology and Technology* 131, 1–9.

Perez Zungia, F.J., Muñoz Delgado, L. and Moreno, J. 1983. Conservacion frigoric de melon cv. 'Tendral Negro' en atmosferas normal y controlada. *Primer Congreso Nacional* II, 985–994.

Perkins, M.L., Yuan, Y. and Joyce, D.C. 2017. Ultrasonic fog application of organic acids delays postharvest decay in red bayberry. *Postharvest Biology and Technology* 133, 41–47.

Perkins-Veazie, P. and Collins, J.K. 2002. Quality of erect-type blackberry fruit after short intervals of controlled atmosphere storage. *Postharvest Biology and Technology* 25, 235–239.

Pesis, E. and Sass, P. 1994. Enhancement of fruit aroma and quality by acetaldehyde or anaerobic treatments before storage. *Acta Horticulturae* 368, 365–373.

Pesis, E., Levi, A., Sonego, L. and Ben Arie, R. 1986. The effect of different atmospheres in polyethylene bags on deastringency of persimmon fruits. *Alon Hanotea* 40, 1149–1156.

Pesis, E., Ampunpong, C., Shusiri, B., Hewett, E.W. and Pech, J.C. 1993a. High carbon dioxide treatment before storage as inducer or reducer of ethylene in apples. *Current Plant Science and Biotechnology in Agriculture* 16, 152–153.

Pesis, E., Marinansky, R., Zauberman, G. and Fuchs, Y. 1993b. Reduction of chilling injury symptoms of stored avocado fruit by prestorage treatment with high nitrogen atmosphere. *Acta Horticulturae* 343, 251–255.

Pesis, E., Marinansky, R., Zauberman, G. and Fuchs, Y. 1994. Prestorage low-oxygen atmosphere treatment reduces chilling injury symptoms in 'Fuerte' avocado fruit. *HortScience* 29, 1042–1046.

Peters, P. and Seidel, P. 1987. Gentle harvesting of Brussels sprouts and recently developed cold storage methods for preservation of quality. *Proceedings of a Conference held in Grossbeeren, German Democratic Republic, 15–18 June* 262, 301–309.

Peters, P., Jeglorz, J. and Kastner, B. 1986. Investigations over several years on conventional and cold storage of Chinese cabbage. *Gartenbau* 33, 298–301.

Phillips, C.B., Iline, I.I., Novoselov, M., McNeill, M.R., Richards, N.K., van Koten, C. and Stephenson, B.P. 2015. Methyl bromide fumigation and delayed mortality: safe trade of live pests? *Journal of Pest Science* 88, 121–134.

Phillips, W.R. and Armstrong, J.G. 1967. *Handbook on the Storage of Fruits and Vegetables for Farm and Commercial Use*. Publication 1260. Canada Department of Agriculture, Ottawa.

Piagentini, A.M., Guemes, D.R. and Pirovani, M.E. 2002. Sensory characteristics of fresh-cut spinach preserved by combined factors methodology. *Journal of Food Science* 67, 1544–1549.

Pina, A., Angel, D. and Nevarez, G. 2000. Effect of water stress and chemical spray treatments on postharvest quality in mango fruits cv. Haden, in Michoacan Mexico. *Acta Horticulturae* 509, 617–630.

Pinheiro, A.C.M., Boas, E.V. de B.V. and Mesquita, C.T. 2005. Action of 1-methylcyclopropene (1-MCP) on shelf-life of 'Apple' banana. *Revista Brasileira de Fruticultura* 27, 25–28.

Pinto, J.A.V., Brackmann, A., Steffens, C.A., Weber, A. and Eisermann, A.C. 2007. Temperature, low oxygen and 1-methylcyclopropene on the quality conservation of 'Fuyu' persimmon. *Ciencia Rural* 37, 1287–1294.

Pirov, T.T. 2001. Yield and keeping ability of onions under different irrigation regimes. *Kartofel' I Ovoshchi* 2, 42.

Platenius, H., Jamieson, F.S. and Thompson, H.C. 1934. Studies on cold storage of vegetables. *Cornell University Agricultural Experimental Station Bulletin* 602.

Plich, H. 1987. The rate of ethylene production and ACC concentration in apples cv. Spartan stored in low O_2 and high CO_2 concentrations in a controlled atmosphere. *Fruit Science Reports* 14, 45–56.

Plotto, A., Roberts, D.D. and Roberts, R.G. 2003. Evaluation of plant essential oils as natural postharvest disease control of tomato (*Lycopersicon esculentum*). *Acta Horticulturae* 628, 737–745.

Pocharski, W., Lange, E., Jelonek, W. and Rutkowski, K. 1995. Storability of new apple cultivars. *Materiay ogolnopolskiej konferencji naukowej Nauka Praktyce Ogrodniczej z okazji XXV-lecia Wydziau Ogrodniczego Akademii Rolniczej w Lublinie*, 93–97.

Polderdijk, J.J., Boerrigter, H.A.M., Wilkinson, E.C., Meijer, J.G. and Janssens, M.F.M. 1993. The effects of controlled atmosphere storage at varying levels of relative humidity on weight loss, softening and decay of red bell peppers. *Scientia Horticulturae* 55, 315–321.

Poldervaart, G. 2010a. KOB storage day 2010: DCA storage goes from strength to strength. *European Fruit Magazine* 2(8), 6–9.

Poldervaart, G. 2010b. DCA viable alternative for organic Topaz. *European Fruit Magazine* 2(8), 15.

Poma Treccarri, C. and Anoni, A. 1969. Controlled atmosphere packaging of polyethylene and defoliation of the stalks in the cold storage of artichokes. *Riv Octoflorofruttic Hal* 53, 203.

Pombo, M.A., Rosli, H.G., Martínez, G.A. and Civello, P.M. 2011. UV-C treatment affects the expression and activity of defense genes in strawberry fruit (*Fragaria* × *ananassa* Duch.). *Postharvest Biology and Technology* 59, 94–102.

Ponce-Valadeza, M., Fellman, S.M., Giovannonic, J., Gana Su-Sheng and Watkins, C.B. 2009. Differential fruit gene expression in two strawberry cultivars in response to elevated CO_2 during storage revealed by a heterologous fruit microarray approach. *Postharvest Biology and Technology* 51, 131–140.

Pongprasert, N., Sekozawa, Y., Sugaya, S. and Gemma, H. 2011. A novel postharvest UV-C treatment to reduce chilling injury (membrane damage, browning and chlorophyll degradation) in banana peel. *Scientia Horticulturae* 130, 73–77.

Porat, R., Weiss, B., Cohen, L., Daus, A. and Aharoni, N. 2004. Reduction of postharvest rind disorders in citrus fruit by modified atmosphere packaging. *Postharvest Biology and Technology* 33, 35–43.

Porat, R., Weiss, B., Fuchs, Y., Sandman, A., Ward, G., Kosto, I. and Agar, T. 2009. Modified atmosphere / modified humidity packaging for preserving pomegranate fruit during prolonged storage and transport. *Acta Horticulturae* 818, 299–304.

Poretta, S., Birzi, A., Ghizzoni, C. and Vicini, E. 1995. Effects of ultra-high hydrostatic pressure treatments on the quality of tomato juice. *Food Chemistry* 52, 35–41.

Povolny, P. 1995. Influence of exposure to light and cultivation system on resistance to Phoma foveata and Fusarium solani var. coeruleum in potato tubers – a pilot study. *Swedish Journal of Agricultural Research* 25, 47–50.

Praeger, U. and Weichmann, J. 2001. Influence of oxygen concentration on respiration and oxygen partial pressure in tissue of broccoli and cucumber. *Acta Horticulturae* 553, 679–681.

Prange, R.K (undated). Currant, gooseberry and elderberry. *The Commercial Storage of Fruits, Vegetables*. Available at: http://www.ba.ars.usda.gov/hb66/058currant.pdf (accessed 4 April 2009).

Prange, R.K. 1986. Chlorophyll fluorescence in vivo as an indicator of water stress in potato leaves. *American Potato Journal* 63, 325–333.

Prange, R.K. and Lidster, P.D. 1991. Controlled atmosphere and lighting effects on storage of winter cabbage. *Canadian Journal of Plant Science* 71, 263–268.

Prange, R.K., McRae, K.B., Midmore, D.J. and Deng, R. 1990. Reduction in potato growth at high temperature: role of photosynthesis and dark respiration. *American Potato Journal* 67, 357–369.

Prange, R.K., Delong, J.M., Leyte, J.C. and Harrison, P.A. 2002. Oxygen concentration affects chlorophyll fluorescence in chlorophyll-containing fruit. *Postharvest Biology and Technology* 24, 201–205.

Prange, R.K., DeLong, J. and Harrison, P. 2003a. Delayed cooling controls soft scald and other disorders in 'Honeycrisp' apple. *HortScience* 38(5), 862 (Abstract).

Prange, R.K., Delong, J.M., Harrison, P.A., Leyte, J.C. and McLean, S.D. 2003b. Oxygen concentration affects chlorophyll fluorescence in chlorophyll-containing fruit and vegetables. *Journal of the American Society for Horticultural Science* 128, 603–607.

Prange, R.K., Daniels-Lake, B.J., Jeong, J.-C. and Binns, M. 2005a. Effects of ethylene and 1-methylcyclopropene on potato tuber sprout control and fry color. *American Journal of Potato Research*, 82, 123–128.

Prange, R.K., Delong, J.M. and Harrison, P.A. 2005b. Quality management through respiration control: Is there a relationship between lowest acceptable respiration, chlorophyll fluorescence and cytoplasmic acidosis? *Acta Horticulturae* 682, 823–830.

Prange, R.K., DeLong, J.M. and Wright, A.H. 2010. Chlorophyll fluorescence: applications in postharvest horticulture. *Chronica Horticulturae* 50(1), 13–16.

Prange, R.K., DeLong, J.M. and Wright, A.H. 2011. Storage of pears using dynamic controlled-atmosphere (DCA), a non-chemical method. *Acta Horticulturae* 909, 707–717.

Prange, R.K., DeLong, J.M. and Wright, A.H. 2012. Improving our understanding of storage stress using chlorophyll fluorescence. *Acta Horticulturae* 945, 89–96.

Prange, R.K., Wright, A.H., DeLong, J.M. and Zanella A. 2013. History, current situation and future prospects for dynamic controlled atmosphere (DCA) storage of fruits and vegetables, using chlorophyll fluorescence. *Acta Horticulturae* 1012, 905–915.

Pratt, H.K. 1971. Melons. In: Hulme A.C. (ed.) *The Biochemistry of Fruits and their Products*. Vol. 2. Academic Press, London and New York, pp. 303–324.

Prencipe, S., Nari, L., Vittone, G., Gullino, M.L. and Spadaro, D. 2016. Effect of bacterial canker caused by *Pseudomonas syringae* pv. *actinidiae* on postharvest quality and rots of kiwifruit 'Hayward'. *Postharvest Biology and Technology* 113, 119–124.

Prono-Widayat, H., Huyskens-Keil, S. and Lu. 2003. Dynamic CA-storage for the quality assurance of pepino (*Solanum muricatum* Ait.). *Acta Horticulturae* 600, 409–412.

Prusky, D., Keen, N.T. and Eaks,I. 1983. Further evidence for the involvement of a preformed antifungal compound in the latency of *Colletotrichum gloeosporioides* on unripe avocado fruits. *Physiological Plant Pathology* 22, 189–198.

Prusky, D., Ohr, H.D., Grech, N., Campbell, S., Kobiler, I., Zauberman, G. and Fuchs, Y. 1995. Evaluation of antioxidant butylated hydroxyanisole and fungicide prochloraz for control of post-harvest anthracnose of avocado fruit during storage. *Plant Disease* 79, 797–800.

Pujantoro, L. Tohru, S. and Kenmoku, A. 1993. The changes in quality of fresh shiitake *Lentinus edodes* in storage under controlled atmosphere conditions. *Proceeding of ICAMPE '93 October 19–22 KOEX, Seoul, Korea, The Korean Society for Agricultural Machinery*, 423–432.

Pullano, G. 2017. Besseling makes inroads in North America cold storage. *Fruit Grower News*, July 26 2017.

Purseglove, J.W. 1968. *Tropical Crops, Dicotyledons*. Longmans, London.

Qi, S., Wang, C.S., Li, Z., Liu, Y., Hua, X. *et al*. 1989. Effects of two-dimensional dynamic controlled atmosphere storage on apple fruits. In: Fellman, J.K. (ed.) *Proceedings of the Fifth International Controlled Atmosphere Research Conference, Wenatchee, Washington, USA, 14–16 June 1989. Vol. 1: Pome fruit*. Washington State University, Pullman, Washington, pp. 295–305.

Qi, X.J., Liang, S.M., Zhou, L.Q. and Chai, X.Q. 2003. Study on effects of small controlled-atmosphere environment on keeping bayberry fruits fresh. *Acta Agriculturae Zhejiangensis* 15, 237–240.

Quazi, H.H. and Freebairn, H.T. 1970. The influence of ethylene, oxygen and carbon dioxide on the ripening of bananas. *Botanical Gazette* 131, 5–14.

Queirós, R.P., Santos, M.D., Fidalgo, L.G., Mota, M.J., Lopes, R.P., Inácio, R.S., Delgadillo, I. and Saraiva, J.A. 2014. Hyperbaric storage of melon juice at and above room temperature and comparison with storage at atmospheric pressure and refrigeration. *Food Chemistry* 147, 209–214.

Quimio, A.J. and Quimio, T.H. 1974. Postharvest control of Philippine mango anthracnose by benomyl. *Philippine Agriculturist* 58, 147–155.

Radulovic, M., Ban, D., Sladonja, B. and Lusetic-Bursic, V. 2007. Changes of quality parameters in watermelon during storage. *Acta Horticulturae* 731, 451–455.

Raffo, A., Kelderer, M., Paoletti, F. and Zanella, A. 2009. Impact of innovative controlled atmosphere storage technologies and postharvest treatments on volatile compound production in cv. Pinova apples. *Journal of Agricultural and Food Chemistry* 57, 915–923.

Raghavan, G.S.V., Tessier, S., Chayet, M., Norris, E.G. and Phan, C.T. 1982. Storage of vegetables in a membrane system. *Transactions of the American Society of Agricultural Engineers* 25, 433–436.

Raghavan, G.S.V., Gariepy, Y., Theriault, R., Phan, C.T. and Lanson, A. 1984. System for controlled atmosphere long term cabbage storage. *International Journal of Refrigeration* 7, 66–71.

Ragnoi, S. 1989. Development of the market for Thai lychee in selected European countries. MSc thesis, Silsoe College, Cranfield Institute of Technology.

Rahman, A.S.A., Huber, D. and Brecht, J.K. 1993a. Respiratory activity and mitochondrial oxidative capacity of bell pepper fruit following storage under low O_2 atmosphere. *Journal of the American Society for Horticultural Science* 118, 470–475.

Rahman, A.S.A., Huber, D. and Brecht, J.K. 1993b. Physiological basis of low O_2 induced residual respiratory effect in bell pepper fruit. *Acta Horticulturae* 343, 112–116.

Rahman, A.S.A., Huber, D.J. and Brecht, J.K. 1995. Low-O_2-induced poststorage suppression of bell pepper fruit respiration and mitochondrial oxidative activity. *Journal of the American Society for Horticultural Science* 120, 1045–1049.

Rahman, M.M., Zakaria, M., Ahmad, S., Hossain, M.M. and Saikat, M.M.H. 2008. Effect of temperature and wrapping materials on the shelf-life and quality of cauliflower. *International Journal of Sustainable Agricultural Technology* 4, 84–90.

Rai, D.R. and Shashi, P. 2007. Packaging requirements of highly respiring produce under modified atmosphere: a review. *Journal of Food Science and Technology (Mysore)* 44, 10–15.

Rai, D.R., Tyagi, S.K., Jha, S.N. and Mohan, S. 2008. Qualitative changes in the broccoli (*Brassica oleracea italica*) under modified atmosphere packaging in perforated polymeric film. *Journal of Food Science and Technology (Mysore)* 45, 247–250.

Rakotonirainy, A.M., Wang, Q. and Padua, I.G.W. 2008. Evaluation of zein films as modified atmosphere packaging for fresh broccoli. *Journal of Food Science* 66, 1108–1111.

Ramin, A.A. and Khoshbakhat, D. 2008. Effects of microperforated polyethylene bags and temperatures on the storage quality of acid lime fruits. *American Eurasian Journal of Agricultural and Environmental Science* 3, 590–594.

Ramin, A.A. and Modares, B. 2009. Improving postharvest quality and storage life of green olives using CO_2. *10th International Controlled & Modified Atmosphere Research Conference, 4–7 April 2009 Turkey* [Abstract], 39.

Ramm, A.A. 2008. Shelf-life extension of ripe non-astringent persimmon fruit using 1-MCP. *Asian Journal of Plant Sciences* 7, 218–222.

Ramos, P.A.S., Ribeiro, W.S., Araújo, F.F. and Finger, F.L. 2015. Extending shelf-life of arracacha roots by using PVC films. *Acta Horticulturae* 1071, 259–262.

Ranasinghe, L., Jayawardena, B. and Abeywickrama, K. 2005. An integrated strategy to control post-harvest decay of Embul banana by combining essential oils with modified atmosphere packaging. *International Journal of Food Science & Technology* 40, 97–103.

Ratanachinakorn, B. 2001. Annual Report 2000–2001, Horticultural Research Institute, Department of Agriculture. Ministry of Agriculture and Cooperative, Bangkok, Thailand, pp. 22–23.

Rees, D. 2014. The Fruit Grower, April 2014. www.actpub.co.uk (accessed June 2016).

Rees, D., Farrell, G. and Orchard, J. 2012. *Crop Post-Harvest: Science and Technology, Vol. 3: Perishables.* Wiley-Blackwell, Oxford.

Reichel, M. 1974. The behaviour of Golden Delicious during storage as influenced by different harvest dates. *Gartenbau* 21, 268–270.

Reichel, M., Meier, G. and Held, W.H. 1976. Results of experiments on the storage of apple cultivars in refrigerated or controlled atmosphere stores. *Gartenbau* 23, 306–308.

Remón, S., Marquina, P., Peiró, J.M. and Oria, R. 2003. Storage potential of sweetheart cherry in controlled atmospheres. *Acta Horticulturae* 600, 763–769.

Remón, S., Ferrer, A., Lopez-Buesa, P. and Oria, R. 2004. Atmosphere composition effects on Burlat cherry colour during cold storage. *Journal of the Science of Food and Agriculture* 84, 140–146.

Renault, P., Houal, L., Jacquemin, G. and Chambroy, Y. 1994. Gas exchange in modified atmosphere packaging. 2. Experimental results with strawberries. *International Journal of Food Science and Technology* 29, 379–394.

Renel, L. and Thompson, A.K. 1994. Carambola in controlled atmosphere. *Inter-American Institute for Co-operation on Agriculture, Tropical Fruits Newsletter* 11, 7.

Resnizky, D. and Sive, A. 1991. Storage of different varieties of apples and pears cv. Spadona in 'ultra-ultra' low O_2 conditions. *Alon Hanotea* 45, 861–871.

Retamales, J., Manríquez, D., Castillo, P. and Defilippi, B. 2003. Controlled atmosphere in Bing cherries from chile and problems caused by quarantine treatments for export to Japan. *Acta Horticulturae* 600, 149–153.

Reust, W., Schwarz, A. and Aerny, J. 1984. Essai de conservation des pommes de terre en atomsphere controlee. *Potato Research* 27, 75–87.

Reyes, A.A. 1988. Suppression of *Sclerotinia sclerotiorum* and watery soft rot of celery by controlled atmosphere storage. *Plant Disease* 72, 790–792.

Reyes, A.A. 1989. An overview of the effects of controlled atmosphere on celery diseases in storage. In: Fellman, J.K. (ed.) *Proceedings of the Fifth International Controlled Atmosphere Conference, Wenatchee, Washington, USA, 14–16 June 1989. Vol. 2 Other commodities and storage recommendations*. Washington State University, Pullman, Washington, pp. 57–60.

Reyes, A.A. and Smith, R.B. 1987. Effect of O_2, CO_2, and carbon monoxide on celery in storage. *HortScience* 22, 270–271.

Riad, G.S. and Brecht, J.K. 2001. Fresh-cut sweetcorn kernels. *Proceedings of the Florida State Horticultural Society* 114, 160–163.

Riad, G.S. and Brecht, J.K. 2003. Sweetcorn tolerance to reduced O_2 with or without elevated CO_2 and effects of controlled atmosphere storage on quality. *Proceedings of the Florida State Horticultural Society* 116, 390–393.

Richardson, D.G. and Meheriuk, M. (eds). 1982. *Controlled Atmospheres for Storage and Transport of Perishable Agricultural Commodities. Proceedings of the Third International Controlled Atmosphere Research Conference*. Timber Press, Beaverton, Oregon.

Richardson, D.G. and Meheriuk, M. 1989. CA recommendations for pears including Asian pears. In: Fellman, J.K. (ed.) *Proceedings of the Fifth International Controlled Atmosphere Conference, Wenatchee, Washington, USA, 14–16 June 1989. Vol. 2: Other commodities and storage recommendations*. Washington State University, Pullman, Washington, pp. 285–302.

Rickard, J.E. 1985. Physiological deterioration of cassava roots. *Journal of the Science of Food and Agriculture* 36, 167–176.

Rinaldi, M.M., Benedetti, B.C. and Moretti, C.L. 2009. Respiration rate, ethylene production and shelf life of minimally processed cabbage under controlled atmosphere. *10th International Controlled & Modified Atmosphere Research Conference, 4–7 April 2009 Turkey* [Abstract], 34.

Ritenour, M.A., Wardowski, W.F. and. Tucker, D.P. 2009. Effects of water and nutrients on the postharvest quality and shelf life of citrus. University of Florida, IFAS Extension solutions for your life publication #HS942. Available at: http://edis.ifas.ufl.edu/CH158 (accessed October 2009).

Riudavets, J., Alonso, M., Gabarra, R., Arnó, J., Jaques, J.A. and Palou, L. 2016. The effects of postharvest carbon dioxide and a cold storage treatment on *Tuta absoluta* mortality and tomato fruit quality. *Postharvest Biology and Technology* 120, 213–221.

Rizzo, V. and Muratore, G. 2009. Effects of packaging on shelf-life of fresh celery. *Journal of Food Engineering* 90, 124–128.

Rizzolo, A. Cambiaghi, P. Grassi, M. and Zerbini, P.E. 2005. Influence of 1-methylcyclopropene and storage atmosphere on changes in volatile compounds and fruit quality of Conference pears. *Journal of Agricultural and Food Chemistry* 53, 9781–9789.

Rizzolo, A., Vanoli, M., Grassi, M. and Eccher Zerbini, P. 2008. Gas exchange in 1-methylcyclopropene treated 'Abbè Fètel' pears during storage in different atmospheres. *Acta Horticulturae* 796, 143–146.

Robatscher, P., Eisenstecken, D., Sacco, F., Pohl, H., Berger, J., Zanella, A. and Oberhuber, M. 2012. Diphenylamine residues in apples caused by contamination in fruit storage facilities. *Journal of Agricultural and Food Chemistry* 60, 2205–2211.

Robbins, J.A. and Fellman, J.K. 1993. Postharvest physiology, storage and handling of red raspberry. *Postharvest News and Information* 4, 53N–59N.

Roberts, C.M. and Hoover, D.G. 1996. Sensitivity of Bacillus coagulans spores to combinations of high hydrostatic pressure, heat, acidity and nisin. *Journal of Applied Bacteriology* 81, 363–368.

Roberts, R. 1990. An overview of packaging materials for MAP. *International Conference on Modified Atmosphere Packaging Part 1*. Campden Food and Drinks Research Association, Chipping Campden, UK.

Robinson, J.E., Brown, K.M. and Burton, W.G. 1975. Storage characteristics of some vegetables and soft fruits. *Annals of Applied Biology* 81, 339–408.

Robinson, T.L., Watkins, C.B., Hoying, S.A., Nock, J.F. and Iungermann, K.I. 2006. Aminoethoxyvinylglycine and 1-methylcyclopropene effects on 'McIntosh' preharvest drop, fruit maturation and fruit quality after storage. *Acta Horticulturae* 727, 473–480.

Robitaille, H.A. and Badenhop, A.F. 1981. Mushroom response to postharvest hyperbaric storage. *Journal of Food Science* 46, 249–253.

Rodov, V., Horev, B., Vinokur, Y., Goldman, G. and Aharoni, N. 2003. Modified-atmosphere and modified-humidity packaging of whole and lightly processed cucurbit commodities: melons, cucumbers, squash. *Australian Postharvest Horticulture Conference, Brisbane, Australia, 1–3 October, 2003*, 167–168.

Rodriguez, J. and Zoffoli, I.P. 2016. Effect of sulfur dioxide and modified atmosphere packaging on blueberry postharvest quality. *Postharvest Biology and Technology* 117, 230–238.

Rodriguez, Z. and Manzano, J.E. 2000. Effect of storage temperature and a controlled atmosphere containing 5.1% CO on physicochemical attributes of the melon (Cucumis melo L.) hybrid Durango. *Proceedings of the Interamerican Society for Tropical Horticulture* 42, 386–390.

Roe, M.A., Faulks, R.M. and Belsten, J.L. 1990. Role of reducing sugars and amino acids in fry colour of chips from potato grown under different nitrogen regimes. *Journal of the Science of Food and Agriculture* 52, 207–214.

Roelofs, F. 1992. Supplying red currants until Christmas. (Rode bes aanvoeren tot kerst.) *Fruitteelt Den Haag* 82, 11–13.
Roelofs, F. 1993a. Choice of cultivar is partly determined by storage experiences. *Fruitteelt Den Haag* 83, 20–21.
Roelofs, F. 1993b. Research results of red currant storage trials 1992: CO_2 has the greatest influence on the storage result. *Fruitteelt Den Haag* 83, 22–23.
Roelofs, F. 1994. Experience with storage of red currants in 1993: unexpected quality problems come to the surface. *Fruitteelt Den Haag* 84, 16–17.
Roelofs, F. and Breugem, A. 1994. Storage of plums. Choose for flavour, choose for CA. *Fruitteelt Den Haag* 84, 12–13.
Rogiers, S.Y. and Knowles, N.R. 2000. Efficacy of low O_2 and high CO_2 atmospheres in maintaining the postharvest quality of saskatoon fruit (*Amelanchier alnifolia* Nutt.). *Canadian Journal of Plant Science* 80, 623–630.
Rohani, M.Y. and Zaipun, M.Z. 2007. MA storage and transportation of 'Eksotika' papaya. *Acta Horticulturae* 740, 303–311.
Rojas-Grau, M.A. and Martin-Belloso, O. 2008. Currect advances in quality maintenance of fresh-cut fruits. *Stewart Postharvest Review* 2, 6.
Romanazzi, G., Nigro, F., Ippolito, A. and Salerno, M. 2001. Effect of short hypobaric treatments on postharvest rots of sweet cherries, strawberries and table grapes. *Postharvest Biology and Technology* 22, 1–6.
Romanazzi, G., Nigro, F. and Ippolito, A. 2003. Short hypobaric treatments potentiate the effect of chitosan in reducing storage decay of sweet cherries. *Postharvest Biology and Technology* 7, 73–80.
Romanazzi, G., Mlikota Gabler, F., Smilanick, J.L., 2006. Preharvest chitosan and postharvest UV-C irradiation treatments suppress grey mold of table grapes. *Plant Disease* 90, 445–450.
Romanazzi, G., Nigro, F. and Ippolito, A. 2008. Effectiveness of a short hyperbaric treatment to control postharvest decay of sweet cherries and table grapes. *Postharvest Biology and Technology* 49, 440–442.
Romanazzi, G., Lichter, A., Mlikota Gabler, F. and Smilanick, J.L. 2012. Natural and safe alternatives to conventional methods to control postharvest grey mould of table grapes. *Postharvest Biology and Technology* 63, 141–147.
Romero, I., Sanchez-Ballesta, M.T., Maldonado, R., Escribano, M.I. and Merodio, C. 2008. Anthocyanin, antioxidant activity and stress-induced gene expression in high CO_2-treated table grapes stored at low temperature. *Journal of Plant Physiology* 165, 522–530.
Romo Parada, L., Willemot, C., Castaigne, F., Gosselin, C. and Arul, J. 1989. Effect of controlled atmospheres low O_2, high CO_2 on storage of cauliflower *Brassica oleracea* L., *Botrytis* group. *Journal of Food Science* 54, 122–124.
Romphophak, T., Siriphanich, J., Promdang, S. and Ueda, Y. 2004. Effect of modified atmosphere storage on the shelf-life of banana 'Sucrier'. *Journal of Horticultural Science & Biotechnology* 79, 659–663.
Rosen, J.C. and Kader, A.A. 1989. Postharvest physiology and quality maintenance of sliced pear and strawberry fruits. *Journal of Food Science* 54, 656–659.
Rosenfeld, H.J., Meberg, K.R., Haffner, K. and Sundell, H.A. 1999. MAP of highbush blueberries: sensory quality in relation to storage temperature, film type and initial high oxygen atmosphere. *Postharvest Biology and Technology* 16, 27–36.
Roser, B. and Colaco, C. 1993. A sweeter way to fresher food. *New Scientist* 15 May, 25–28.
Rothan, C., Duret, S., Chevalier, C. and Raymond, P., 1997. Suppression of ripening associated gene expression in tomato fruit subjected to a high CO_2 concentration. *Plant Physiology* 114, 255–263.
Rouves, M. and Prunet, J.P. 2002. New technology for chestnut storage: controlled atmosphere and its effects. *Infos Ctifl* 186, 33–35.
Roy, S., Anantheswaran, R.C. and Beelman, R.B. 1995. Fresh mushroom quality as affected by modified atmosphere packaging. *Journal of Food Science* 60, 334–340.
Rukavishnikov, A.M., Strel'tsov, B.N., Stakhovskii, A.M. and Vainshtein, I.I. 1984. Commercial fruit and vegetable storage under polymer covers with gas-selective membranes. *Khimiya v Sel'skom Khozyaistve* 22, 26–28.
Rupasinghe, H.P.V. Murr, D.P., Paliyath, G. and Skog, L. 2000. Inhibitory effect of 1-MCP on ripening and superficial scald development in 'McIntosh' and 'Delicious' apples. *Journal Horticultural Science & Biotechnology* 75, 271–276.
Rupavatharam, S., East, A.R. and Heyes, J.A. 2015. Opportunities to manipulate harvest maturity of New Zealand feijoa (*Acca sellowiana*) to enable sea freight export. *Acta Horticulturae* 1091, 223–229.
Rutkowski, K., Miszczak, A. and Plocharski, W. 2003. The influence of storage conditions and harvest date on quality of 'Elstar' apples. *Acta Horticulturae* 600, 809–812.

Ryall, A.L. 1963. Effects of modified atmospheres from liquefied gases on fresh produce. *Proceedings of the Seventeenth National Conference on Handling Perishable Agricultural Commodities (Purdue University), 11–14 March*, 17, 21–24.

Ryall, A.L. and Lipton, W.J. 1972. *Handling, Transportation and Storage of Fruits and Vegetables*. AVI Publishing, Westpoint, Connecticut.

Ryall, A.L. and Lipton, W.J. 1979. *Handling, Transportation and Storage of Fruits and Vegetables. Vol. 1. Vegetables and Melons*. 2nd edn. AVI Publishing, Westport, Connecticut.

Ryall, A.L. and Pentzer, W.T. 1974. *Handling, Transportation and Storage of Fruits and Vegetables* Vol. 2. AVI Publishing, Westpoint, Connecticut.

Sa, C.R.L., Silva, E. de O., Terao, D. and Oster, A.H. 2008. Effects of $KMnO_4$ and 1-MCP with modified atmosphere on postharvest conservation of Cantaloupe melon. *Revista Ciencia Agronomica* 39, 60–69.

Sabban-Amin, R., Feygenberg, O., Belausov, E. and Pesis, E. 2011. Low oxygen and 1-MCP pretreatment delay superficial scald development by reducing reactive oxygen species (ROS) accumulation in stored 'Granny Smith' apples. *Postharvest Biology and Technology* 62, 295–304.

Saberi, B., Golding, J.B., Marques, J.R, Pristijono, P., Chockchaisawasdee, S., Scarlett, C.J. and Stathopoulos, C.E. 2018. Application of biocomposite edible coatings based on pea starch and guar gum on quality, storability and shelf life of 'Valencia' oranges. *Postharvest Biology and Technology* 137, 9–20.

Sahin, G., Kurubas, M.S. and Erkan, M. 2015. Effects of modified atmosphere imposed with the palliflex system on postharvest fruit quality of 'Red Globe' table grapes. *Acta Horticulturae* 1071, 149–155.

Saijo, R. 1990. Post harvest quality maintenance of vegetables. *Tropical Agriculture Research Series* 23, 257–269.

Saito, M. and Rai, D.R. 2005. Qualitative changes in radish (*Raphanus* spp.) sprouts under modified atmosphere packaging in micro-perforated films. *Journal of Food Science and Technology (Mysore)* 42, 70–72.

Sale, A.J.H., Gould, G.W. and Hamilton, W.A. 1970. Inactivation of bacterial spores by hydrostatic pressure. *Journal of General Microbiology* 60, 323–334.

Saltveit, M.E. 1989. A summary of requirements and recommendations for the controlled and modified atmosphere storage of harvested vegetables. In: Fellman, J.K. (ed.) *Proceedings of the Fifth International Controlled Atmosphere Conference, Wenatchee, Washington, USA, 14–16 June 1989. Vol. 2: Other commodities and storage recommendations*. Washington State University, Pullman, Washington, pp. 329–352.

Saltveit, M.E. 2003. Is it possible to find an optimal controlled atmosphere? *Postharvest Biology and Technology* 27, 3–13.

Salunkhe, D.K. and Wu, M.T. 1973a. Effects of low oxygen atmosphere storage on ripening and associated biochemical changes of tomato fruits. *Journal of the American Society for Horticultural Science* 98, 12–14.

Salunkhe, D.K., and Wu M.T. 1973b. Effects of subatmospheric pressure storage on ripening and associated chemical changes of certain deciduous fruits. *Journal of the American Society for Horticultural Science* 98, 113–116.

Salunkhe, D.K. and Wu, M.T. 1974. Subatmospheric storage of fruits and vegetables. *Lebensmittel Wissenschaft und Technologie* 7, 261–267.

Salunkhe, D.K. and Wu, M.T. 1975. Subatmospheric storage of fruits and vegetables. In: Haard, N.F. and Salunkhe, D.K. (eds) *Postharvest Biology and Handling of Fruits and Vegetables*. AVI Publishing, Westpoint, Connecticut, pp. 153–171.

Samisch, R.M. 1937. Observations on the effect of gas storage upon Valencia oranges. *Proceedings of the American Society for Horticultural Science* 34, 103–106.

Samosornsuk, W., Bunsiri, A. and Samosornsuk, S. 2009. Effect of concentrations of oxygen and/or carbon dioxide on the microbial growth control in fresh-cut lemongrass. *10th International Controlled & Modified Atmosphere Research Conference, 4–7 April 2009 Turkey* [Abstract], 39.

Samsoondar, J., Maharaj, V. and Sankat, C.K. 2000. Inhibition of browning of the fresh breadfruit through shrink-wrapping. *Acta Horticulturae* 518, 131–136.

San Martín, M.F., Barbosa-Cánovas, G.V. and Swanson, B.G. 2002. Food processing by high hydrostatic pressure. *Critical Reviews in Food Science and Nutrition* 42, 627–645.

Sanchez, J.C.E. and Gontard, N. 2003. Active modified atmosphere packaging of fresh fruits and vegetables: modeling with tomatoes and oxygen absorber. *Journal of Food Science* 68, 1736–1743.

Sanchez-Mata, M.C., Camara, M. and Diez-Marques, C. 2003. Extending shelf-life and nutritive value of green beans (*Phaseolus vulgaris* L.), by controlled atmosphere storage: macronutrients. *Food Chemistry* 80, 309–315.

Sancho, F., Lambert, Y., Demazeau, G., Largeteau, A., Bouvier, J.-M. and Narbonne, J.-F. 1999. Effects of ultra-high hydrostatic pressure on hydrosoluble enzymes. *Journal of Food Engineering* 39, 247–253.

Sandy Trout Food Preservation Laboratory. 1978. Banana CA storage. *Bulletin of the International Institute of Refrigeration* 583, 12–16.

Sanjeev, K. and Ramesh, M.N. 2006. Low oxygen and inert gas processing of foods. *Critical Review of Food and Nutrition* 46, 423–451.

Sankat, C.K. and Maharaj, R. 1989. Controlled atmosphere storage of papayas. In: Fellman, J.K. (ed.) *Proceedings of the Fifth International Controlled Atmosphere Conference, Wenatchee, Washington, USA, 14–16 June 1989. Vol. 2 Other commodities and storage recommendations.* Washington State University, Pullman, Washington, pp. 161–170.

Sankat, C.K. and Maharaj, R. 2007. A review of postharvest storage technology of breadfruit. *Acta Horticulturae* 757, 183–191.

Santos, A.F. dos, Silva, S. de M. and Alves, R.E. 2006a. Storage of Suriname cherry under modified atmosphere and refrigeration: I – Postharvest chemical changes. *Revista Brasileira de Fruticultura* 28, 36–41.

Santos, A.F. dos, Silva, S. de M., Mendonca, R.M.N. and Filgueiras, H.A.C. 2006b. Storage of Suriname cherry under modified atmosphere and refrigeration: II – Quality and postharvest conservation. *Revista Brasileira de Fruticultura* 28, 42–45.

Santos, C.M.S., Boas, E.V.de B.V., Botrel, N. and Pinheiro, A.C.M. 2006c. Effect of controlled atmosphere on postharvest life and quality of 'Prata Ana' banana. *Ciencia e Agrotecnologia* 30, 317–322.

Sanz, C., Perez, A.G., Olias, R. and Olias, J.M. 1999. Quality of strawberries packed with perforated polypropylene. *Journal of Food Science* 64, 748–752.

Saquet, A.A., Streif, J. and Almeida, D.P.F. 2017. Responses of 'Rocha' pear to delayed controlled atmosphere storage depend on oxygen partial pressure. *Scientia Horticulturae* 222, 17–21.

Saraiva, J.A. 2014. Storage of foods under mild pressure (hyperbaric storage) at variable (uncontrolled) room: a possible new preservation concept and an alternative to refrigeration? *Journal of Food Processing & Technology* 5, 56 [Abstract].

Sarananda, K.H. and Wilson Wijeratnam, R.S. 1997. Changes in susceptibility to crown rot during maturation of Embul bananas and effect of low oxygen and high carbon dioxide on extent of crown rot. *Seventh International Controlled Atmosphere Research Conference, July 13–18 1997*, University of California, Davis, California [Abstract], 110.

Saray, T. 1988. Storage studies with Hungarian paprika *Capsicum annuum* L. var. *annuum* and cauliflower *Brassica cretica* convar. *botrytis* used for preservation. *Acta Horticulturae* 220, 503–509.

Satyan, S., Scott, K.J. and Graham, D. 1992. Storage of banana bunches in sealed polyethylene tubes. *Journal of Horticultural Science* 67, 283–287.

Saucedo-Veloz, C., Aceves Vega, E. and Mene Nevarez, G. 1991. Prolongacion del tiempo de frigoconservacion y comercialacion de frutos de aguacate 'Hass' mediante tratamientos con altas concentraciones de CO_2. *Proceedings of the Interamerican Society for Tropical Horticulture* 35, 297–303.

Sayyari, M., Babalar, M., Kalantari, S., Serrano, M. and Valero, D. 2009. Effect of salicylic acid treatment on reducing chilling injury in stored pomegranates. *Postharvest Biology and Technology* 53, 152–154.

Sayyari, M., Castillo, S., Valero, D., Díaz-Mula, H.M. and Serrano, M. 2012. Acetyl salicylic acid alleviates chilling injury and maintains nutritive and bioactive compounds and antioxidant activity during postharvest storage of pomegranates. *Postharvest Biology and Technology* 60, 136–142.

Scalon, S.P.Q., Scalon Filho, H., Sandre, T.A., da Silva, E.F. and Krewer, E.C.D. 2000. Quality evaluation and sugar beet postharvest conservation under modified atmosphere. *Brazilian Archives of Biology and Technology* 43, 181–184.

Scalon, S.P.Q., Vieira, M.C. and Zarate, N.A.H. 2002. Combinations of calcium, modified atmosphere and refrigeration in conservation postharvest of Peruvian carrot. *Acta Scientiarum* 24, 1461–1466.

Schäfer, F., Blanke, M. and Fels, J. 2014. Comparison of CO2e emissions associated with regional, heated and imported asparagus. *Proceedings of the 9th International Conference on Life Cycle Assessment in the Agri-Food Sector, San Francisco, California, USA, 8–10 October, 2014*, 1210–1214.

Schales, F.D. 1985. Harvesting, packaging, storage and shipping of greenhouse vegetables. In: Savage, A.J. (ed.) *Hydroponics Worldwide: State of the Art in Soilless Crop Production.* International Center for Special Studies, Honolulu, pp. 70–76.

Schallenberger, R.S., Smith, O. and Treadaway, R.H. 1959. Role of sugars in the browning reaction in potato chips. *Journal of Agricultural and Food Chemistry* 7, 274.

Schirra, M., D'hallewin, G., Inglese, P. and La Mantia, T. 1999. Epicuticular changes and storage potential of cactus pear [*Opuntia ficus-indica* Miller (L.)] fruit following gibberellic acid preharvest sprays and postharvest heat treatment. *Postharvest Biology and Technology* 17, 79–88.

Schlimme, D.V. and Rooney, M.L. 1994. Packaging of minimally processed fruits and vegetables. In: R.C. Wiley (ed.) *Minimally Processed Refrigerated Fruits and Vegetables*, Chapman and Hall, New York and London, pp. 135–182.

Schmitz, S.M. 1991. Investigation on alternative methods of sprout suppression in temperate potato stores. MSc thesis Silsoe College, Cranfield Institute of Technology, UK.

Schotsmans, W. 2006. Fruit moisture loss mechanism as it relates to postharvest disorders. *Nova Scotia Fruit Growers' Association, 2006 Annual Report*, 45–56.

Schotsmans, W., Molan, A. and MacKay, B. 2007. Controlled atmosphere storage of rabbiteye blueberries enhances postharvest quality aspects. *Postharvest Biology and Technology* 44, 277–285.

Schouten, R.E., Zhang, X, Verkerk, R., Verschoor, J.A., Otma, E.C., Tijskens, L.M.M. and van Kooten, O. 2009. Modelling the level of the major glucosinolates in broccoli as affected by controlled atmosphere and temperature. *Postharvest Biology and Technology* 53, 1–10.

Schouten, S.P. 1985. New light on the storage of Chinese cabbage. *Groenten en Fruit* 40, 60–61.

Schouten, S.P. 1992. Possibilities for controlled atmosphere storage of ware potatoes. *Aspects of Applied Biology* 33, 181–188.

Schouten, S.P. 1994. Increased CO_2 concentration in the store is disadvantageous for the quality of culinary potatoes. *Kartoffelbau* 45, 372–374.

Schouten, S.P. 1995. Dynamic control of the oxygen content during CA storage of fruits and vegetables. *Control Applications in Post-Harvest and Processing Technology: CAPPT '95, 1st IFAC/CIGR/EURAGENG/ISHS Workshop, Ostend, Belgium*, 163–168.

Schouten, S.P. 1997. Improvement of quality of Elstar apples by dynamic control of ULO conditions. *Seventh International Controlled Atmosphere Research Conference, July 13–18 1997*, University of California, Davis, California [Abstract], 7.

Schouten, S.P., Prange, R.K., Verschoor, J., Lammers, T.R. and Oosterhaven, J. 1997. Improvement of quality of Elstar apples by dynamic control of ULO conditions. *CA '97. Proceedings of the 7th International Controlled Atmosphere Research Conference*, University of California, Davis, California, 2, 71–78.

Schreiner, M., Huyskens-Keil, S., Krumbein, A., Prono-Widayat, H., Peters, P. and Lüdders, P. 2003. Interactions of pre- and post-harvest influences on fruit and vegetable quality as basic decision for chain management. *Acta Horticulturae* 604, 211–217.

Schulz, F.A. 1974. The occurrence of apple storage rots under controlled conditions. *Zeitschrift fur Pflanzenkrankheiten und Pflanzenschutz* 81, 550–558.

Scott, K.J. and Wills, R.B.H. 1973. Atmospheric pollutants destroyed in an ultra-violet scrubber. *Laboratory Practice* 22, 103–106.

Scott, K.J. and Wills, R.B.H. 1974. Reduction of brown heart in pears by absorption of ethylene from the storage atmosphere. *Australian Journal of Experimental Agriculture and Animal Husbandry* 14, 266–268.

Scott, K.J., Blake, J.R., Strachan, G., Tugwell, B.L. and McGlasson, W.B. 1971. Transport of bananas at ambient temperatures using polyethylene bags. *Tropical Agriculture Trinidad* 48, 245–253.

SeaLand. 1991. *Shipping Guide to Perishables*. SeaLand Services Inc., Iselim, New Jersey.

Segovia-Bravo, K.A., Guignon, B., Bermejo-Prada, A., Sanz, P.D. and Otero, L. 2012. Hyperbaric storage at room temperature for food preservation: a study in strawberry juice. *Innovative Food Science and Emerging Technologies* 15, 14–22.

Selcuk, N. and Erkan, M. 2015. The effects of modified and palliflex controlled atmosphere storage on postharvest quality and composition of 'Istanbul' medlar fruit. *Postharvest Biology and Technology* 99, 9–19.

Seljasen, R., Hoftun, H. and Bengtsson, G.B. 2001. Sensory quality of ethylene-exposed carrots (*Daucus carota* L., cv. 'Yukon') related to the contents of 6-methoxymellein, terpenes and sugars. *Journal of the Science of Food and Agriculture* 81, 54–61.

Seljasen, R., Hoftun, H. and Bengtsson, G.B. 2003. Critical factors for reduced sensory quality of fresh carrots in the distribution chain. *Acta Horticulturae* 604, 761–767.

Senevirathna, P.A.W.A.N.K. and Daundasekera, W.A.M. 2010. Effect of postharvest calcium chloride vacuum infiltration on the shelf life and quality of tomato (cv. 'Thilina'). *Ceylon Journal of Science* 39, 35–44.

Serradilla, M.J., del Carmen Villalobos, M., Hernández, A., Martín, A., Lozano, M. and de Guía Córdoba, M. 2013. Study of microbiological quality of controlled atmosphere packaged 'Ambrunés' sweet cherries and subsequent shelf-life. *International Journal of Food Microbiology* 166, 85–92.

Serrano, M., Martinez-Romero, D., Guillen, F., Castillo, S. and Valero, D. 2006. Maintenance of broccoli quality and functional properties during cold storage as affected by modified atmosphere packaging. *Postharvest Biology and Technology* 39, 61–68.

Servili, A., Feliziani, E. and Romanazzi, G. 2017. Exposure to volatiles of essential oils alone or under hypobaric treatment to control postharvest gray mold of table grapes. *Postharvest Biology and Technology* 133, 36–40.

Seubrach, P., Photchanachai, S., Srilaong, V. and Kanlayanarat, S. 2006. Effect of modified atmosphere by PVC and LLDPE film on quality of longan fruits (*Dimocarpus longan lour*) cv. Daw. *Acta Horticulturae* 712, 605–610.

Seymour, G.B., Thompson, A.K. and John, P. 1987a. Inhibition of degreening in the peel of bananas ripened at tropical temperatures. I. The effect of high temperature changes in the pulp and peel during ripening. *Annals of Applied Biology* 110, 145–151.

Seymour, G.B., John, P. and Thompson, A.K. 1987b. Inhibition of degreening in the peel of bananas ripened at tropical temperatures. 2. Role of ethylene, oxygen and carbon dioxide. *Annals of Applied Biology,* 110, 153–161.

Sfakiotakis, E., Niklis, N., Stavroulakis, G. and Vassiliadis, T. 1993. Efficacy of controlled atmosphere and ultra low O_2 - low ethylene storage on keeping quality and scald control of 'Starking Delicious' apples. *Acta Horticulturae* 326, 191–202.

Sharma, N. and Tripathi, A. 2006. Fungitoxicity of the essential oils of Citrus sinensis on post-harvest pathogens. *World Journal of Microbiology and Biotechnology* 22, 587–593.

Sharp, A.K. 1985. Temperature uniformity in a low-pressure freight container utilizing glycol-chilled walls. *International Journal of Refrigeration* 8, 37–42.

Sharples, R.O. 1967. A note on the effect of N dimethylaminosuccinamic acid on the maturity and storage quality of apples. *Annual Report of the East Malling Research Station for 1966*, 198–201.

Sharples, R.O. 1971.Storage under reduced atmospheric pressure. *Annual Report of the East Malling Research Station, UK, for 1970.*

Sharples, R.O. 1974. Hypobaric storage: apples and soft fruit. *Annual Report of the East Malling Research Station, UK, for 1973.*

Sharples, R.O. 1980. The influence of orchard nutrition on the storage quality of apples and pears grown in the United Kingdom. In: Atkinson, D., Jackson, J.E., Sharples, R.O. and Waller, W.M. (eds) *Mineral Nutrition of Fruit Trees*. Butterworths, London and Boston, pp. 17–28.

Sharples, R.O. 1986. Obituary Cyril West. *Journal of Horticultural Science* 61, 555.

Sharples, R.O. 1989a. Kidd, F. and West, C. In: Janick, J. (ed.) *Classical Papers in Horticultural Science*. Prentice Hall, New Jersey, pp. 213–219.

Sharples, R.O. 1989b. Storage of perishables. *Engineering Advances for Agriculture and Food.* In Cox S.W.R. Editor. Institution of Agricultural Engineers Jubilee Conference 1988, Cambridge, UK, Butterworths, 251–260.

Sharples, R.O. and Johnson, D.S. 1987. Influence of agronomic and climatic factors on the response of apple fruit to controlled atmosphere storage. *HortScience* 22, 763–766.

Sharples, R.O. and Langridge I.W. 1973. Reduced pressure storage. *Annual Report of the East Malling Research Station, UK, for 1972*, 108.

Sharples, R.O. and Stow, J.R. 1986. Recommended conditions for the storage of apples and pears. *Report of the East Malling Research Station for 1985*, 165–170.

Shellie, K.C. 1999. Muskmelon (*Cucumis melo* L.) fruit ripening and postharvest quality after a preharvest spray of aminoethoxyvinylglycine. *Postharvest Biology and Technology* 17, 55–62.

Shellie, K.C., Neven, L.G. and Drake, S.R. 2001. Assessing 'Bing' sweet cherry tolerance to a heated controlled atmosphere for insect pest control. *HortTechnology* 11, 308–311.

Sheng, K., Zheng, H., Shui, S., Yan, L. and Zheng, L. 2018. Comparison of postharvest UV-B and UV-C treatments on table grape: changes in phenolic compounds and their transcription of biosynthetic genes during storage. *Postharvest Biology and Technology* 138, 74–81.

Shi, J. and Le Maguer, M. 2000. Lycopene in tomatoes: chemical and physical properties affected by food processing. *Critical Reviews in Biotechnology* 20, 293–334.

Shibairo, S.I., Upadhyaya, M.K. and Toivonen, P.M.A. 1998. Influence of preharvest water stress on postharvest moisture loss of carrots (*Daucus carota* L.). *Journal of Horticultural Science & Biotechnology* 73, 347–352.

Shin, M.H., Kim, J.-H., Choi, H.W., Keum, Y.S. and Chun, S.C. 2014. Effect of thymol and linalool fumigation on postharvest diseases of table grapes. *Mycrobiology* 42, 262–268.

Shin, Y., Ryu, J.-A., Liu, R.H., Nock, J.F. and Watkins, C.B. 2008. Harvest maturity, storage temperature and relative humidity affect fruit quality, antioxidant contents and activity, and inhibition of cell proliferation of strawberry fruit. *Postharvest Biology and Technology* 49, 201–209.

Shinjiro, S., Kumi, C., Miho, N., Mami, O. and Reinosuke, A. 2002. Effect of 1-methylcyclopropene (1-MCP) on respiration, ethylene production and color change of pineapple fruit after harvest. *Food Preservation Science* 28, 235–241.

Shipway, M.R. 1968. *The Refrigerated Storage of Vegetables and Fruits*. Ministry of Agriculture Fisheries and Food, London.

Sholberg, P.L. and Gaunce, A.P. 1995. Fumigation of fruit with acetic acid to prevent postharvest decay. *HortScience* 30, 1271–1275.

Sholberg, P.L., Cliff, M. and Leigh Moyls, A. 2001. Fumigation with acetic acid vapor to control decay of stored apples. *Fruits* 56, 355 366.

Shorter, A.J., Scott, K.J. and Graham, D. 1987. Controlled atmosphere storage of bananas in bunches at ambient temperatures. *CSIRO Food Research Queensland* 47, 61–63.

Siegrist, J.-P. and Cotter, J.-Y. 2012. Entreposage frigorifique de pommes Jazz®, Scifresh[cov] en atmosphères controlées AC et ULO. *Revue suisse Viticulture, Arboriculture, Horticulture* 44, 106–111.

Simko, I., Jimenez-Berni, J.A. and Furbank R.T. 2015. Detection of decay in fresh-cut lettuce using hyperspectral imaging and chlorophyll fluorescence imaging. *Postharvest Biology and Technology* 106, 44–52.

Singh, A.K., Kashyap, M.M., Gupta, A.K. and Bhumbla, V.K. 1993. Vitamin-C during controlled atmosphere storage of tomatoes. *Journal of Research Punjab Agricultural University* 30, 199–203.

Singh, B., Littlefield, N.A. and Salunkhe, D.K. 1972. Accumulation of amino acids and organic acids in apple and pear fruits under controlled atmosphere storage conditions and certain associated changes in metabolic processes. *Indian Journal of Horticulture* 29, 245–251.

Singh, S.P. and Pal, R.K. 2008a. Response of climacteric-type guava (*Psidium guajava* L.) to postharvest treatment with 1-MCP. *Postharvest Biology and Technology* 47, 307–314.

Singh, S.P. and Pal, R.K. 2008b. Controlled atmosphere storage of guava (*Psidium guajava* L.) fruit. *Postharvest Biology and Technology* 47, 296–306.

Singh, S.P. and Rao, D.V.S. 2005. Effect of modified atmosphere packaging (MAP) on the alleviation of chilling injury and dietary antioxidants levels in 'Solo' papaya during low temperature storage. *European Journal of Horticultural Science*, 70, 246–252.

Singh, S.P., Saini, M.K., Sing, J., Pongener, A. and Sidhu, G.S. 2014. Preharvest application of AA promotes anthocyanins accumulation in pericarp of litchi fruit without adversely affecting postharvest quality. *Postharvest Biology and Technology* 96, 14–22.

Singh, Z. and Zaharah, S.S. 2015. Controlled atmosphere storage of mango fruit: challenges and thrusts and its implications in international mango trade. *Acta Horticulturae* 1066, 179–191.

Sirivatanapa, S. 2006. Packaging and transportation of fruits and vegetables for better marketing In: *Postharvest Management of Fruit and Vegetables in the Asia-Pacific Region. Reports of the APO seminar on Reduction of Postharvest Losses of Fruit and Vegetables held in India, 5–11 October 2004 and Marketing and Food Safety: Challenges in Postharvest Management of Agricultural/Horticultural Products in Islamic Republic of Iran, 23–28 July 2005*. Asian Productivity Organization, Tokoy, pp. 44–47.

Siriwardana, H., Abeywickrama, K., Kannangara, S., Jayawardena, B. and Attanayake, S. 2017. Basil oil plus aluminium sulfate and modified atmosphere packaging controls crown rot disease in Embul banana (*Musa acuminata*, AAB) during cold storage. *Scientia Horticulturae* 217, 84–91.

Siro, I., Devlieghere, F., Jacxsens, L., Uyttendaele, M. and Debevere, J. 2006. The microbial safety of strawberry and raspberry fruits packaged in high-oxygen and equilibrium-modified atmospheres compared to air storage. *International Journal of Food Science & Technology* 41, 93–103.

Sitton, J.W., Fellman, J.K. and Patterson, M.E. 1997. Effects of low-oxygen and high-carbon dioxide atmospheres on postharvest quality, storage and decay of 'Walla Walla' sweet onions. *Seventh International Controlled Atmosphere Research Conference, July 13–18 1997*, University of California, Davis, California [Abstract], 60.

Sivakumar, D. and Korsten, L. 2007. Influence of 1-MCP at simulated controlled atmosphere transport conditions and MAP storage on litchi quality. *South African Litchi Growers' Association Yearbook* 19, 39–41.

Sivakumar, D., Zeeman, K. and Korsten, L. 2007. Effect of a biocontrol agent (*Bacillus subtilis*) and modified atmosphere packaging on postharvest decay control and quality retention of litchi during storage. *Phytoparasitica* 35, 507–518.

Sivakumar, D., Arrebola, E. and Korsten, L. 2008. Postharvest decay control and quality retention in litchi (cv. McLean's Red) by combined application of modified atmosphere packaging and antimicrobial agents. *Crop Protection* 27, 1208–1214.

Sivakumar, D., Bill, M., Korsten, L. and Thompson, A.K. 2016. Integrated application of chitosan coating with different postharvest treatments in the control of postharvest decay and maintenance of overall fruit quality. In: Bautista-Banos, S., Jimenez-Aparicio, G. and Romanazzi, A. (eds) *Chitosan in the Preservation of Agricultural Commodities*. Academic Press, Oxford, UK, 127–154.

Sive, A. and Resnizky, D. 1979. Extension of the storage life of 'Red Rosa' plums by controlled atmosphere storage. *Bulletin de l'Institut International du Froid* 59, 1148.

Sive, A. and Resnizky, D. 1985. Experiments on the CA storage of a number of mango cultivars in 1984. *Alon Hanotea* 39, 845–855.

Skog, L.J., Schaefer, B.H. and Smith, P.G. 2003. On-farm modified atmosphere packaging of sweet cherries. *Acta Horticulturae* 628, 415–422.

Skrzynski, J. 1990. Black currant fruit storability in controlled atmospheres. I. Vitamin C content and control of mould development. *Folia Horticulturae* 2, 115–124.

Smith, G. 1991. Effects of ethephon on ripening and quality of fresh market pineapples. *Australian Journal of Experimental Agriculture* 31, 123–127.

Smith, R.B. 1992. Controlled atmosphere storage of 'Redcoat' strawberry fruit. *Journal of the American Society for Horticultural Science* 117, 260–264.

Smith, R.B. and Reyes, A.A. 1988. Controlled atmosphere storage of Ontario grown celery. *Journal of the American Society for Horticultural Science* 113, 390–394.

Smith, R.B., Skog, L.J., Maas, J.L. and Galletta, G.J. 1993. Enhancement and loss of firmness in strawberries stored in atmospheres enriched with CO_2. *Acta Horticulturae* 348, 328–333.

Smith, S.M. and Stow, J.R. 1984. The potential of a sucrose ester coating material for improving the storage and shelf-life qualities of 'Cox's orange pippin' apples. *Annals of Applied Biology* 104, 383–391.

Smith, S., Geeson, J. and Stow, J. 1987. Production of modified atmospheres in deciduous fruits by the use of films and coatings. *HortScience* 22, 772–776.

Smith, W.H. 1952. *The Commercial Storage of Vegetables*. Food Investigation Leaflet, 15. Department of Scientific and industrial Research, London.

Smith, W.H. 1957. Storage of black currents. *Nature* 179, 876.

Smith, W.H. 1965. Storage of mushrooms. *Ditton Laboratory Report 1965–66*, 13–14.

Smittle, D.A. 1988. Evaluation of storage methods for 'Granex' onions. *Journal of the American Society for Horticultural Science* 113, 877–880.

Smittle, D.A. 1989. Controlled atmosphere storage of Vidalia onions. In: Fellman, J.K. (ed.) *Proceedings of the Fifth International Controlled Atmosphere Conference, Wenatchee, Washington, USA, 14–16 June, 1989. Vol. 2: Other commodities and storage recommendations*. Washington State University, Pullman, Washington, pp. 171–177.

Smock, R.M. 1938. The possibilities of gas storage in the United States. *Refrigeration Engineering* 36, 366–368.

Smock, R.M. 1979. Controlled atmosphere storage of fruits. *Horticultural Reviews* 1, 301–336.

Smock, R.M. and Van-Doren, A. 1938. Preliminary studies on the gas storage of McIntosh and Northwestern Greening. *Ice and Refrigeration* 95, 127–128.

Smock, R.M. and Van-Doren, A. 1939. Studies with modified atmosphere storage of apples. *Refrigerating Engineering* 38, 163–166.

Smock, R.M. and Van-Doren, A. 1941. *Controlled-atmosphere Storage of Apples*. Cornell Agricultural Experiment Station Bulletin 762. Cornell University, Ithaca, New York.

Smock, R.M., Mendoza, D.B. and Abilay, R.M. 1967. Handling bananas. *Philippines Farms and Gardens* 4, 12–17.

Smoot, J.J. 1969. Decay of Florida citrus fruits stored in controlled atmospheres and in air. *Proceedings of the First International Citrus Symposium* 3, 1285–1293.

Snowdon, A.L. 1990. *A Colour Atlas of Postharvest Diseases and Disorders of Fruits and Vegetables. Volume 1: General introduction and fruits*. Wolfe Scientific, London.

Snowdon, A.L. 1992. *A Colour Atlas of Postharvest Diseases and Disorders of Fruits and Vegetables. Volume 2: Vegetables*. Wolfe Scientific, London.

Sobiczewski, P., Bryk, H. and Berczynski, S. 1999. Efficacy of antagonistic bacteria in protection of apples against *Botrytis cinerea* and *Penicillium expansum* under CA conditions. *Acta Horticulturae* 485, 351–356.

Solomos, T., Whitaker, B. and Lu, C., 1997. Deleterious effects of pure oxygen on 'Gala' and 'Granny Smith' apples. *HortScience* 32, 458.

Soltani, M., Alimardani, R., Mobli, H. and Mohtasebi, S.S. 2015. Modified atmosphere packaging: a progressive technology for shelf-life extension of fruits and vegetables. *Journal of Applied Packaging Research* 7, No. 3, Article 2.

Somboonkaew, N. 2001. Modified atmosphere packaging for rambutan (*Nephelium lappaceum* Linn. cv Rongrien). MSc thesis, Writtle College, University of Essex.

Somboonkaew, N. and Terry, L.A. 2009. Effect of packaging films on individual anthocyanins in pericarp of imported non-acid treated litchi. *10th International Controlled & Modified Atmosphere Research Conference, 4–7 April 2009 Turkey* [Abstract], 27.

Son, Y.K., Yoon, I.W., Han, P.J and Chung, D.S. 1983. Studies on storage of pears in sealed polyethylene bags. *Research Reports, Office of Rural Development, S. Korea, Soil Fertilizer, Crop Protection, Mycology and Farm Products Utilization* 25, 182–187.

Song, J., Fan, L., Forney, C.F., Jordan, M.A., Hildebrand, P.D., Kalt, W. and Ryan, D.A.J. 2003. Effect of ozone treatment and controlled atmosphere storage on quality and phytochemicals in highbush blueberries. *Acta Horticulturae* 600, 417–423.

Sothornvit, R. and Sampoompuang, C. 2012. Rice straw paper incorporated with activated carbon as an ethylene scavenger in a paper-making process. *International Journal of Food Science and Technology* 47, 511–517.

Souza, B.S. de, O'Hare, T.J., Durigan, J.F. and de Souza, P.S. 2006. Impact of atmosphere, organic acids, and calcium on quality of fresh-cut 'Kensington' mango. *Postharvest Biology and Technology* 42, 161–167.

Souza, F.C. and Ferraz, A.C.O. 2009. Variabilidade de índices de firmeza em figo utilizando ponteira cilíndrica e pratos planos. *Revista Brasileira de Fruticultura* 31, 257–261.

Soylu, E.M., Tok, F.M., Soylu, S., Kaya, A.D. and Evrendilek, G.A. 2005. Antifungal activities of the essential oils on post-harvest disease agent *Penicillium digitatum. Pakistan Journal of Biological Sciences* 8, 25–29.

Spalding, D.H. 1980. Low pressure hypobaric storage of several fruits and vegetables. *Proceedings of the Florida State Horticultural Society* 92, 201–203.

Spalding, D.H. 1982. Resistance of mango pathogens to fungicides used to control postharvest diseases. *Plant Disease* 66, 1185–1186.

Spalding, D.H. and Reeder, W.F. 1972. Quality of 'Booth 8' and 'Lula' avocados stored in a controlled atmosphere. *Proceedings of the Florida State Horticultural Society* 85, 337–341.

Spalding, D.H. and Reeder, W.F. 1974a. Current status of controlled atmosphere storage of four tropical fruits. *Proceedings of the Florida State Horticultural Society* 87, 334–339.

Spalding, D.H. and Reeder, W.F. 1974b. Quality of 'Tahiti' limes stored in a controlled atmosphere or under low pressure. *Proceedings of the Tropical Region American Society for Horticultural Science* 18, 128–135.

Spalding, D.H. and Reeder, W.F. 1975. Low-oxygen, high carbon dioxide controlled atmosphere storage for the control of anthracnose and chilling injury of avocados. *Phytopathology* 65, 458–460.

Spalding, D.H. and Reeder, W.F. 1976. Low pressure (hypobaric) storage of limes. *Journal of the American Society for Horticultural Science* 101, 367–370.

Spalding, D.H. and Reeder, W.F. 1977. Low pressure (hypobaric) storage of mangos. *Journal of the American Society for Horticultural Science* 102, 367–369.

Sposito, M.B., de Mourao Filho, F.A.A., Kluge, R.A. and Jacomino, A.P. 2000. Cold storage of Tahiti limes treated with gibberellic acid. *Revista Brasileira de Fruticultura* 22, 345–348.

Spotts, R.A., Sholberg, P.L., Randall, P., Serdani, M. and Chen, P.M. 2007. Effects of 1-MCP and hexanal on decay of d'Anjou pear fruit in long-term cold storage. *Postharvest Biology and Technology* 44, 101–106.

Srikul, S. and Turner, D. 1995. High N supply and soil water deficits change the rate of fruit growth of bananas (cv. Williams) and promote tendency to ripen. *Scientia Horticulturae* 62, 165–174.

Srilaong, V., Kanlayanarat, S. and Tatsumi, Y. 2005. Effects of high O_2 pretreatment and high O_2 map on quality of cucumber fruit. *Acta Horticulturae* 682, 1559–1564.

Sritananan, S., Uthairatanakij, A., Srilaong, V., Kanlayanarat, S. and Wongs-Aree, C. 2006. Efficacy of controlled atmosphere storage on physiological changes of lime fruit. *Acta Horticulturae* 712, 591–597.

Staby, G.L. 1976. Hypobaric storage – an overview. *Combined Proceedings of the International Plant Propagation Society* 26, 211–215.

Staby, G.L., Cunningham, M.S., Holstead, C.L., Kelly, J.W., Konjoian, P.S., Eisenbergi, B.A. and Dressier, B.S. 1984. Storage of rose and carnation flowers. *Journal of the American Society for Horticultural Science* 109, 193–197.

Staden, O.L. 1986. Post-harvest research on ornamentals in the Netherlands. *Acta Horticulturae* 181, 19–24.
Stahl, A.L. and Cain, J.C. 1937. Cold storage studies of Florida citrus fruit. III. The relation of storage atmosphere to the keeping quality of citrus fruit in cold storage. *Florida Agricultural Experiment Station Bulletin* 316, October.
Stahl, A.L. and Cain, J.C. 1940. Storage and preservation of micellaneous fruits and vegetables. *Florida Agricultural Experiment Station Annual Report*, 88.
Stange Jr, R.R. and McDonald, R.E. 1999. A simple and rapid method for determination of lignin in plant tissues – its usefulness in elicitor screening and comparison to the thioglycolic acid method. *Postharvest Biology and Technology* 15, 185–193.
Stange Jr, R.R., Midland, S.L., Holmes, G.J., Sims, J.J. and Mayer, R.T. 2001. Constituents from the periderm and outer cortex of *Ipomoea batatas* with antifungal activity against *Rhizopus stolonifer*. *Postharvest Biology and Technology* 23, 85–92.
Steffens, C.A., Brackmann, A. and Streck, N.A. 2007a. Permeability of polyethilene films and utilization in the storage of fruits. *Revista Brasileira de Armazenamento* 32, 93–99.
Steffens, C.A., Brackmann, A., Lopes, S.J., Pinto, J.A.V., Eisermann, A.C., Giehl, R.F.H. and Webber, A. 2007b. Internal breakdown and respiration of 'Bruno' kiwifruit in relation to storage conditions. *Ciencia Rural* 37, 1621–1626.
Steinbauer, C.E. 1932. Effects of temperature and humidity upon length of rest period of tubers of Jerusalem artichoke (*Helianthus tuberosus*). *Proceedings of the American Society for Horticultural Science* 29, 403–408.
Stenvers, N. 1977. Hypobaric storage of horticultural products. *Bedrijfsontwikkeling* 8, 175–177.
Stephens, B.E. and Tanner, D.J. 2005. The Harvest Watch system – measuring fruit's healthy glow. *Acta Horticulturae* 687, 363–364.
Stewart, D. 2003. Effect of high O_2 and N_2 atmospheres on strawberry quality. *Acta Horticulturae* 600, 567–570.
Stewart, J.K. and Uota, M. 1971. CO_2 injury and market quality of lettuce held under controlled atmosphere. *Journal of the American Society for Horticultural Science* 96, 27–31.
Stoddard, E.S. and Hummel C.E. 1957. Methods of improving food preservation in home refrigerators. *Refrigeration Engineering* 65, 33–38, 69, 71.
Stoll, K. 1972. Largerung von Früchten und Gemusen in kontrollierter Atmosphäre. *Mitt. Eidg. Forsch. Anst. Obst Wein Gartenbau Wädenswil Flugschrift* 77.
Stoll, K. 1974. Storage of vegetables in modified atmospheres. *Acta Horticulturae* 38, 13–23.
Stoll, K. 1976. Storage of the pear cultivar Louise Bonne. Lagerung der Birnensorte 'Gute Luise'. *Schweizerische Zeitschrift fur Obst und Weinbau* 112, 304–309.
Storex. 2018. Available at: www.storex.nl.
Stover, R.H. and Simmonds, N.W. 1987. *Bananas*, 3rd edn. Wiley, New York.
Stow, J.R. 1986. The effects of storage atmosphere and temperature on the keeping quality of Spartan apples. *Annals of Applied Biology* 109, 409–415.
Stow, J.R. 1989. Effects of O_2 cocentration on ethylene synthesis and action in stored apple fruits. *Acta Horticulturae* 258, 97–106.
Stow, J.R. 1995. The effects of storage atmosphere on the keeping quality of 'Idared' apples. *Journal of Horticultural Science* 70, 587–595.
Stow, J.R. 1996. Gala breaks through the storage barrier. *Grower* 126, 26–27.
Stow, J.R., Dover, C.J. and Genge, P.M. 2000. Control of ethylene biosynthesis and softening in 'Cox's Orange Pippin' apples during low-ethylene, low-oxygen storage. *Postharvest Biology and Technology* 18, 215–225.
Stow, J.R., Jameson, J. and Senner, K. 2004. Storage of cherries: the effects of rate of cooling, store atmosphere and store temperature on storage and shelf-life. *Journal of Horticultural Science & Biotechnology* 79, 941–946.
Streif, J. 1989. Storage behaviour of plum fruits. *Acta Horticulturae* 258, 177–184.
Streif, J. and Saquet, A.A. 2003. Internal flesh browning of 'Elstar' apples as influenced by pre- and post-harvest factors. *Acta Horticulturae* 599, 523–527.
Streif, J., Retamales, J., Cooper, T. and Sass, P. 1994. Preventing cold storage disorders in nectarine. *Acta Horticulturae* 368, 160–165.
Streif, J., Xuan, H., Saquet, A.A. and Rabus, C. 2001. CA-storage related disorders in 'Conference' pears. *Acta Horticulturae* 553, 635–638.
Streif, J., Saquet, A.A. and Xuan, H. 2003. CA-related disorders of apples and pears. *Acta Horticulturae* 600, 223–230.

Streif, J., Kittemann, D., Neuwald, D.A., McCormick, R. and Xuan, H. 2010. Pre- and post-harvest management of fruit quality, ripening and senescence. *Acta Horticulturae* 877, 55–68.

Strempfl, E., Mader, S. and Rumpolt, J. 1991. Trials of storage suitability of important apple cultivars in a controlled atmosphere. *Mitteilungen Klosterneuburg, Rebe und Wein, Obstbau und Fruchteverwertung* 41, 20–26.

Strop, I. 1992. Effects of plastic film wraps on the marketable life of asparagus and broccoli. MSc thesis, Silsoe College, Cranfield Institute of Technology.

Suchanek, M., Kordulska, M., Olejniczak, Z., Figiel, H. and Turek, K. 2017. Application of low-field MRI for quality assessment of 'Conference' pears stored under controlled atmosphere conditions. *Postharvest Biology and Technology* 124, 100–106.

Sudhakar Rao, D.V. and Gopalakrishna Rao, K.P. 2009. Controlled atmosphere storage of mango cultivars 'Alphonso and 'Banganapalli' to extend storage-life and maintain quality. *Journal of Horticultural Science & Biotechnology* 83, 351–359.

Sudto, T. and Uthairatanakij, A. 2007. Effects of 1-MCP on physico-chemical changes of ready-to-eat durian 'Mon-Thong'. *Acta Horticulturae*, 746, 329–334.

Suslow, T.V. and Cantwell, M. 1998. Peas-snow and snap pod peas. *Perishables Handling Quarterly, University of California Davis* 93, 15–16.

Sziro, I., Devlieghere, F., Jacxsens, L., Uyttendaele, M. and Debevere, J. 2006. The microbial safety of strawberry and raspberry fruits packaged in high oxygen and equilibrium modified atmospheres compared to air storage. *International Journal of Food Science and Technology* 41, 93–103.

Taeckens, J. 2007. Understanding Container Atmosphere Control Technologies. Available at: http://www.corp.carrier.com/static/ContainerFiles/Files/Knowledge_Center/Controlled_Atmosphere/ContainerAtmosphereControl.pdf (accessed May 2009).

Tahir, I.I. and Ericsson, N.A. 2003. Effect of postharvest heating and CA-storage on storability and quality of apple cv. 'Aroma'. *Acta Horticulturae* 600, 127–134.

Tahir, I. and Olsson, M. 2009. The fruit quality of five plum cultivars 'Prunus Domestica L.' Related to harvesting date and ultra low oxygen atmosphere storage. *10th International Controlled & Modified Atmosphere Research Conference, 4–7 April 2009 Turkey* [Poster abstract], 74.

Tamas, S. 1992. Cold storage of watermelons in a controlled atmosphere. *Elelmezesi Ipar* 46, 234–239, 242.

Tamer, C.E. and Çopur, O.U. 2010. Chitosan: an edible coating for fresh-cut fruits and vegetables. *Acta Horticulturae* 877, 619–624.

Tan, S.C. and Mohamed, A.A. 1990. The effect of CO_2 on phenolic compounds during the storage of 'Mas' banana in polybag. *Acta Horticulturae* 269, 389.

Tancharoensukjit, S. and Chantanawarangoon, S. 2008. Effect of controlled atmosphere on quality of fresh-cut pineapple cv. Phuket. *Proceedings of the 46th Kasetsart University Annual Conference, Kasetsart, Thailand, 29 January-1 February, 2008*, 73–80.

Tangwongchai, R., Ledward, D. A., and Ames, J. M. 2000. Effect of high-pressure treatment on the texture of cherry tomato. *Journal of Agricultural and Food Chemistry* 48, 1434–1441.

Tano, K., Oule, M.K., Doyon, G., Lencki, R.W. and Arul, J. 2007. Comparative evaluation of the effect of storage temperature fluctuation on modified atmosphere packages of selected fruit and vegetables. *Postharvest Biology and Technology* 46, 212–221.

Tariq, M.A. 1999. Effect of curing parameters on the quality and storage life of damaged citrus fruits. PhD thesis Cranfield University.

Tataru, D. and Dobreanu, M. 1978. Research on the storage of several vegetables in a controlled atmosphere. *Lucrari Stiintifice Institutul de Cercetari pentru Valorificarea Legumelor si Fructelor* 9, 13–20.

Tay, S.L. and Perera, C.O. 2004. Effect of 1-methylcyclopropene treatment and edible coatings on the quality of minimally processed lettuce. *Journal of Food Science* 69, 131–135.

Techavuthiporn, C. and Boonyaritthongchai, P. 2016. Effect of prestorage short-term Anoxia treatment and modified atmosphere packaging on the physical and chemical changes of green asparagus. *Postharvest Biology and Technology*, 117, 64–70.

Techavuthiporn, C., Kakaew, P., Puthmee, T. and Kanlayanarat, S. 2009. The effect of controlled atmosphere conditions on the quality decay of shredded unripe papaya. *10th International Controlled & Modified Atmosphere Research Conference, 4–7 April 2009 Turkey* [Poster abstract], 57.

Teerachaichayut, S. and Ho, H.T. 2017. Non-destructive prediction of total soluble solids, titratable acidity and maturity index of limes by near infrared hyperspectral imaging. *Postharvest Biology and Technology* 133, 20–25.

Teixeira, G.H.A., Durigan, J.F., Alves, R.E. and O'Hare, T.J. 2008. Response of minimally processed carambola to chemical treatments and low-oxygen atmospheres. *Postharvest Biology and Technology* 48, 415–421.

Teixeira, G., Durigan, J.F, Santos, L.O. and Ogassavara, F.O. 2009. High levels of carbon dioxide injures guava (*Psidium guajaba* L. cv. Pedro Sato) stored under controlled atmosphere. *10th International Controlled & Modified Atmosphere Research Conference, 4–7 April 2009 Turkey* [Abstract], 15.

Teixeira, G.H.A., Cunha Júnior, L.C., Ferraudo, A.S. and Durigan, J.F. 2016. Quality of guava (*Psidium guajava* L. cv. Pedro Sato) fruit stored in low-O_2 controlled atmospheres is negatively affected by increasing levels of CO_2. *Postharvest Biology and Technology* 111, 62–68.

Teles, C.S., Benedetti, B.C., Gubler, W.D. and Crisosto, C.H. 2014. Prestorage application of high carbon dioxide combined with controlled atmosphere storage as a dual approach to control *Botrytis cinerea* in organic 'Flame Seedless' and 'Crimson Seedless' table grapes. *Postharvest Biology and Technology* 89, 32–39.

Tengrang, S., Enmak, P., Leabwan, N., Mongkol Tunhaw, T., Pramart, K. *et al.* (2015) Packaging Technology Research and Development Project. Postharvest and Processing Research and Development Division, Department of Agriculture, Thailand.

Tennant, P., Fermin, G., Fitch, M.M., Manshardt, R.M., Slightom, J.L. and Gonsalves, D. 2001. Papaya ringspot virus resistance of transgenic Rainbow and SunUp is affected by gene dosage, plant development, and coat protein homology. *European Journal of Plant Pathology* 107, 645–653.

Terry, L.A., Ilkenhans, T., Poulston, S., Rowsell, L. and Smith, A.W.J. 2007. Development of new palladium-promoted ethylene scavenger. *Postharvest Biology and Technology* 45, 214–220.

Testoni, A. and Eccher Zerbini, P. 1993. Controlled atmosphere storage trials with kiwifruits, prickly pears and plums. *COST 94. The Post-harvest Treatment of Fruit and Vegetables: Controlled Atmosphere Storage of Fruit and Vegetables. Proceedings of a Workshop, April 22–23, 1993, Milan, Italy*, 131–136.

Testoni, A., Lovati, F., Nuzzi, M. and Pellegrino, S. 2002. First evaluations of harvesting date and storage technologies for Pink Lady[Reg] Cripps Pink apple grown in the Piedmont area. *Rivista di Frutticoltura e di Ortofloricoltura* 64, 67–73.

Thammawong, M., Orikasa, T., Umehara, H., Hewajulige, I.G.N., Kaneta, T. *et al.* 2014. Modeling of the respiration rate and gene expression patterns of cabbage in response to mechanical impact stress using a modified Weibull distribution. *Postharvest Biology and Technology* 96, 118–127.

Thammawong, M., Umehara, H., Nakamura, N., Ito, Y., Shiina, T. *et al.* 2015. Oscillations of respiration-relating gene expression in postharvest cabbage head under different controlled atmosphere (CA) storage regimes. *Acta Horticulturae* 1091, 303–310.

Tharanathan, R.N. 2003. Biodegradable films and composite coatings: past, present and future packaging films commonly used. *Trends in Food Science & Technology* 14, 71–78.

Thatcher, R.W. 1915. Enzymes of apples and their relation to the ripening process. *Journal of Agricultural Research* 5, 103–105.

Thewes, F.R., Both, V., Brackmann, A., Weber, A. and de Oliveira Anese, R. 2015. Dynamic controlled atmosphere and ultralow oxygen storage on 'Gala' mutants quality maintenance. *Food Chemistry* 188, 62–70.

Thewes, F.R., Brackmann, A., Both, V., Weber, A., de Oliveira Anese, R., dos Santos Ferrão, T. and Wagner, R. 2017a. The different impacts of dynamic controlled atmosphere and controlled atmosphere storage in the quality attributes of 'Fuji Suprema' apples. *Postharvest Biology and Technology* 130, 7–20.

Thewes, F.R., Brackmann, A., de Oliveira Anese, R., Bronzatto, E.S., Schultz, E.E. and Wagner, R. 2017b. Dynamic controlled atmosphere storage suppresses metabolism and enhances volatile concentrations of 'Galaxy' apple harvested at three maturity stages. *Postharvest Biology and Technology* 127, 1–13.

Thewes, F.R., Brackmann, A., de Oliveira Anese, R., Ludwig, V., Schultz, E.E., Ferreira dos Santos, L. and Wendt, L.M. 2017c. Effect of dynamic controlled atmosphere monitored by respiratory quotient and 1-methylcyclopropene on the metabolism and quality of 'Galaxy' apple harvested at three maturity stages. *Food Chemistry* 222, 84–93.

Thewes, F.R., Brackmann, A., de Oliveira Anese, R., Ludwig, V., Schultz, E.E. and Berghetti, M.R.P. 2018. 1-methylcyclopropene suppresses anaerobic metabolism in apples stored under dynamic controlled atmosphere monitored by respiratory quotient. *Scientia Horticulturae* 227, 288–295.

Thompson, A.K. 1971. The storage of mango fruit. *Tropical Agriculture* 48, 63–70.

Thompson, A.K. 1972a. Storage and transport of fruit and vegetables in the West Indies. In: *Proceedings of the Seminar/Workshop on Horticultural Development in the Caribbean, Maturin, Venezuela*, 170–176.

Thompson, A.K. 1972b. *Report on an assignment on secondment to the Jamaican Government as food storage advisor 1970–1972*. Report R278. Tropical Products Institute, London.

Thompson, A.K. 1998. *Controlled Atmosphere Storage of Fruits and Vegetables*. CAB International, Wallingford, UK.

Thompson, A.K. 1981. Reduction of losses during the marketing of arracacha. *Acta Horticulturae* 116, 55–60.

Thompson, A.K. 1985. *Postharvest losses of bananas, onions and potatoes in PDR Yemen*. Contract Services Report CO 485.Tropical Development and Research Institute, London.

Thompson, A.K. 1987. The development and adaptation of methods for the control of anthracnose. In: Prinsley, R.A. and Tucker, G. (eds) *Mangoes – A Review*. Commonwealth Secretariat, London, pp. 29–38.
Thompson, A.K. 1996. *Postharvest Technology of Fruit and Vegetables*. Blackwell Publishing, Oxford.
Thompson, A.K. 2003. *Fruit and Vegetables*. Blackwell Publishing, Oxford.
Thompson, A.K. 2010. *Controlled Atmosphere Storage of Fruits and Vegetables*, 2nd edn. CABI, Wallingford, UK.
Thompson, A.K. 2015. *Fruit and Vegetable Storage: Hypobaric, Hyperbaric and Controlled Atmosphere*. Springer, New York.
Thompson, A.K. and Arango, L.M 1977. Storage and marketing cassava in plastic films. *Proceedings of the Tropical Region of the American Horticultural Science* 21, 30–33.
Thompson, A.K. and Burden, O. J. 1995. Harvesting and fruit care. In Gowen S. (ed.) *Bananas and Plantains*. Chapman and Hall, London, pp. 389–412.
Thompson, A.K. and Seymour, G.B. 1984. *Inborja (CASCO) banana factory at Machala in Ecuador*. Tropical Development and Research Institute Report (unpublished).
Thompson, A.K., Mason, G.F. and Halkon, W.S. 1971. Storage of West Indian seedling avocado fruits. *Journal of Horticultural Science* 46, 83–88.
Thompson, A.K., Been, B.O. and Perkins, C.. 1972a Handling, storage and marketing of plantains. *Proceedings of the Tropical Region of the American Society of Horticultural Science* 16, 205–212.
Thompson, A.K., Booth, R.H. and Proctor, F.J. 1972b. Onion storage in the tropics. *Tropical Science* 14, 19–34.
Thompson, A.K., Been, B.O. and Perkins, C. 1974a. Storage of fresh breadfruit. *Tropical Agriculture Trinidad* 51, 407–415.
Thompson, A.K., Been, B.O. and Perkins, C. 1974b. Prolongation of the storage life of breadfruits. *Proceedings of the Caribbean Food Crops Society* 12, 120–126.
Thompson, A.K., Magzoub, Y. and Silvis, H. 1974c. Preliminary investigations into desiccation and degreening of limes for export. *Sudan Journal Food Science Technology* 6, 1–6.
Thompson, A.K., Been, B.O. and Perkins, C. 1974d. Effects of humidity on ripening of plantain bananas. *Experientia* 30, 35–36.
Thompson, A.K., Been, B.O. and Perkins, C. 1977. Fungicidal treatments of stored yams. *Tropical Agriculture* 54, 179–183.
Thompson, A.K. Ferris, R.S.B. and Al-Zaemey, A.B.S. 1992. Aspects of handling bananas and plantains. *Tropical Agriculture Association Newsletter* 12, 15–17.
Thornton, N.C. 1930. The use of carbon dioxide for prolonging the life of cut flowers, with special reference to roses. *American Journal of Botany* 17, 614–626.
Tian, S.P., Jiang, A.L., Xu, Y. and Wang, Y.S. 2004. Responses of physiology and quality of sweet cherry fruit to different atmospheres in storage. *Food Chemistry* 87, 43–49.
Tiangco, E.L., Agillon, A.B. and Lizada, M.C.C. 1987 Modified atmosphere storage of 'Saba' bananas. *Association of Southeast Asian Nations Food Journal* 3, 112–116.
Tindall, H.D. 1983. *Vegetables in the Tropics*. Macmillan Press, London.
Tirtosoekotjo, R.A. 1984. Ripening behaviour and physicochemical characteristics of 'Carabao' mango *Mangifera indica* treated with acetylene from calcium carbide. PhD thesis, University of the Philippines, Los Baños, Laguna.
Toivonen, P.M.A. and DeEll, J.R. 2001. Chlorophyll fluorescence, fermentation product accumulation, and quality of stored broccoli in modified atmosphere packages and subsequent air storage. *Postharvest Biology and Technology* 23, 61–69.
Tolle, W.E. 1969. *Hypobaric Storage of Mature Green Tomatoes*. Marketing Research Report 842. US Department of Agriculture, Washington, DC.
Tomás-Barberán, F.A. and Gil, M.I. (eds). 2008. *Improving the Health-promoting Properties of Fruit and Vegetable Products*. Woodhead, Cambridge, UK.
Tome, P.H.F., Santos, J.P., Cabral, L.C., Chandra, P.K. and Goncalves, R.A. 2000. Use of controlled atmosphere with CO_2 and N_2 for the preservation of technological qualities of beans (*Phaseolus vulgaris* L.) during storage. *Revista Brasileira de Armazenamento* 25, 16–22.
Tomkins, R.B. and Sutherland, J. 1989. Controlled atmospheres for seafreight of cauliflower. *Acta Horticulturae* 247, 385–389.
Tomkins, R.G. 1957. Peas kept for 20 days in gas store. *Grower* 48, 226–227.
Tomkins, R.G. 1965. Deep scald in Ellison's Orange apples. *Annual Report of the Ditton Laboratory 1964–1965*, 19.
Tomkins, R.G. 1966. The storage of mushrooms. *Mushroom Growers Association Bulletin* 202, 534, 537, 538, 541.
Tomkins, R.G. and Meigh, D.F. 1968. The concentration of ethylene found in controlled atmosphere stores. *Annual Report of the Ditton Laboratory 1967–68*, 33–36.

Tongdee, S.C. 1988. *Banana Postharvest Handling Improvements*. Report of the Thailand Institute of Science and Technology Research, Bangkok.

Tongdee, S.C., Suwanagul, A. and Neamprem, S. 1990. Durian fruit ripening and the effect of variety, maturity stage at harvest and atmospheric gases. *Acta Horticulturae* 269, 323–334.

Tonini, G. and Tura, E. 1997. New CA storage strategies for reducing rots (Botrytis cinerea and Phialophora spp.) and softening in kiwifruit. *Seventh International Controlled Atmosphere Research Conference, July 13–18 1997, University of California, Davis, California* [Abstract], 104.

Tonini, G., Brigati, S. and Caccioni, D. 1989. CA storage of kiwifruit: influence on rots and storability. In: Fellman, J.K. (ed.) *Proceedings of the Fifth International Controlled Atmosphere Conference, Wenatchee, Washington, USA, 14–16 June 1989. Vol. 2: Other commodities and storage recommendations*. Washington State University, Pullman, Washington, pp. 69–76.

Tonini, G., Caccioni, D. and Ceroni, G. 1993. CA storage of stone fruits: effects on diseases and disorders. *COST 94. The post-harvest treatment of fruit and vegetables: controlled atmosphere storage of fruit and vegetables. Proceedings of a Workshop, April 22–23, 1993, Milan, Italy*, 95–105.

Tonini, G., Barberini, K., Bassi, F. and Proni, R. 1999. Effects of new curing and controlled atmosphere storage technology on Botrytis rots and flesh firmness in kiwifruit. *Acta Horticulturae* 498, 285–291.

Tonutti, P., 2015. The technical evolution of CA storage protocols and the advancements in elucidating the fruit responses to low oxygen stress. *Acta Horticulturae* 1079, 53–60.

Torres, A.V., Zamudio-Flores, P.B., Salgado-Delgado, R. and Bello-Pérez, L.A. 2008. Biodegradation of low-density polyethylene-banana starch films. *Journal of Applied Polymer Science* 110, 3464–3472.

Tovar, B., Montalvo, E., Damián, B.M., García, H.S. and Mata, M. 2011. Application of vacuum and exogenous ethylene on Ataulfo mango ripening. *LWT – Food Science and Technology* 44, 2040–2046.

Tran, D.T., Verlinden, B.E., Hertog, M. and Nicolaï, B.M. 2015. Monitoring of extremely low oxygen control atmosphere storage of 'Greenstar' apples using chlorophyll fluorescence. *Scientia Horticulturae* 184, 18–22.

Trierweiler, B., Krieg, M. and Tauscher, B. 2004. Antioxidative capacity of different apple cultivars after long-time storage. *Journal of Applied Botany and Food Quality* 78, 117–119.

Truter, A.B. and Combrink, J.C. 1992. Controlled atmosphere storage of peaches, nectarine and plums. *Journal of the Southern African Society for Horticultural Sciences* 2, 10–13.

Truter, A.B. and Combrink, J.C. 1997. Controlled atmosphere storage of South African plums. *Seventh International Controlled Atmosphere Research Conference, July 13–18 1997, University of California, Davis, California* [Abstract], 47.

Truter, A.B., Eksteen, G.J. and Van der Westhuizen, A.J.M. 1982. Controlled-atmosphere storage of apples. *Deciduous Fruit Grower* 32, 226–237.

Truter, A.B., Combrick, J.C., Fourie, P.C. and Victor, S.J. 1994. Controlled atmosphere storage of prior to processing of some canning peach and apricot cultivars in South Africa. *Commissions C2,D1,D2/3 of the International Institute of Refrigeration International Symposium June 8–10 Istanbul Turkey*, 243–254.

Tsantili, E., Karaiskos, G. and Pontikis, C. 2003. Storage of fresh figs in low oxygen atmosphere. *Journal of Horticultural Science & Biotechnology* 78, 56–60.

Tsay, L.M. and Wu, M.C. 1989. Studies on the postharvest physiology of sugar apple. *Acta Horticulturae* 258, 287–294.

Tsiprush, R.Ya., Zhamba, A.I. and Bodyul, K.P. 1974. The effect of storage regime on apple quality in relation to growing conditions. *Sadovodstvo, Vinogradarstvo i Vinodelie Moldavii* 1, 52–54.

Tudela, J.A., Hernández, N., Pérez-Vicente, A. and Gil, M.I. 2017. Growing season climates affect quality of fresh-cut lettuce. *Postharvest Biology and Technology* 123, 60–68.

Tugwell, B. and Chvyl, L. 1995. Storage recommendations for new varieties. *Pome Fruit Australia*, May, 4–5.

Tulin Oz, A. and Eris, A. 2009. Effects of controlled atmosphere storage on differently harvested Hayward kiwifruits ethylene production. *10th International Controlled & Modified Atmosphere Research Conference, 4–7 April 2009 Turkey* [Poster abstract], 73.

Turbin, V.A. and Voloshin, I.A. 1984. Storage of table grape varieties in a controlled gaseous environment. *Vinodelie i Vinogradarstvo SSSR* 8, 31–32.

Tzortzakis, N.G. 2007. Maintaining postharvest quality of fresh produce with volatile compounds. *Innovative Food Science & Emerging Technologies* 8, 111–116.

Uddin, M.M. and MacTavish, H.S. 2003. Controlled atmosphere and regular storage-induced changes in S-alk(en)yl-l-cysteine sulfoxides and alliinase activity in onion bulbs (*Allium cepa* L. cv. Hysam). *Postharvest Biology and Technology* 28, 239–245.

Ullah Malik, A., Hameed, R., Imran, M. and Schouten, S. 2009. Effect of controlled atmosphere on storability and shelf-life and quality of green slender chilies (*Capsicum annuum* L.). *10th International Controlled & Modified Atmosphere Research Conference, 4–7 April 2009 Turkey* [Abstract], 9.

Unahawutti, U., Intarakumheng, R., Oonthonglang, P., Phankum, S. and Follett, P.A. 2014. Nonhost status of mangosteen to Bactrocera dorsalis and Bactrocera carambolae (Diptera: Tephritidae) in Thailand. *Journal of Economic Entomology* 107, 1355–1361.

UNEP, 1992. *Methyl bromide: its atmospheric science, technology and economics*. Synthesis report of the methyl bromide interim scientific assessment and methyl bromide interim technology and economic assessment. United Nations Environmental Programs, Montreal Protocol Assessment Supplement. US Government Printing Office, Washington, DC.

Urban, E. 1995. Postharvest storage of apples. Nachlagerungsverhalten von Apfelfruchten. *Erwerbsobstbau* 37, 145–151.

US Apple Association. 2016. *The Production and Utilization Analysis for the 2016 US Apple Association Apple Crop Outlook and Marketing Conference*, published by the US Apple Association. Available at: http://usapple.org/wp-content/uploads/2016/09/2016ProductionUtilizationAnalysis.pdf (accessed 14 March 2018).

US Federal Register. 2002. The United States Federal Register: July 26, 2002 Volume 67, Number 144 Pages 48796–48800. Available at: http://regulations.vlex.com/vid/pesticides-raw-commodities-methylcyclo propene-22883760 (accessed January 2010).

Usall, A.M, Anguera, J., Solsona, M. and Viñas I., C. 2008. Microbiological quality of fresh, minimally-processed fruit and vegetables, and sprouts from retail establishments. *International Journal of Food Microbiology* 31, 121–129.

Valero, D., Díaz-Mula, H.M., Zapata, P.J., Guillén, F., Martínez-Romero, D., Castillo, S. and Serrano, M. 2013. Effects of alginate edible coating on preserving fruit quality in four plum cultivars during postharvest storage. *Postharvest Biology and Technology* 77, 1–6.

Valverde, J.M., Guillen, F., Martinez-Romero, D., Castillo, S., Serrano, M. and Valero, D. 2005. Improvement of table grapes quality and safety by the combination of modified atmosphere packaging (MAP) and eugenol, menthol, or thymol. *Journal of Agricultural and Food Chemistry* 53, 7458–7464.

Van Amerongen CA Technology. 2018. Available at: https://www.van-amerongen.com/en/control-systems (accessed 26 March 2018).

Van der Merwe, J.A. 1996. Controlled and modified atmosphere storage. In: Combrink, J.G. (ed.) *Integrated Management of Post-harvest Quality*. South Africa INFRUiTEC ARC/LNR, pp. 104–112.

Van der Merwe, J.A., Combrick, J.C., Truter, A.B. and Calitz, F.J. 1997. Effect of initial low oxygen stress treatment and controlled atmosphere storage at increased carbon dioxide levels on the post-storage quality of South African-grown 'Granny Smith' and 'Topred'apples. *Seventh International Controlled Atmosphere Research Conference, July 13–18 1997, University of California, Davis, California* [Abstract], 8.

Van der Merwe, J.A., Combrink, J.C. and Calitz, F.J. 2003. Effect of controlled atmosphere storage after initial low oxygen stress treatment on superficial scald development on South African-grown Granny Smith and Topred apples. *Acta Horticulturae* 600, 261–265.

Van der Zwet, E. 2018. Pallet Fresh System. Available at: http://besseling-group.com/applications/pallet-fresh-system/ (accessed 24 March 2018).

Van Doren, A., Hoffman, M.B. and Smock, R.M. 1941. CO_2 treatment of strawberries and cherries in transit and storage. *Proceedings of the American Society for Horticultural Science* 38, 231–238.

Van Eeden, S.J. and Cutting, J.G.M. 1992. Ethylene and ACC levels in ripening 'Topred' apples after storage in air and controlled atmosphere. *Journal of the Southern African Society for Horticultural Sciences* 2, 7–9.

Van Eeden, S.J., Combrink, J.C., Vries, P.J. and Calitz, F.J. 1992. Effect of maturity, diphenylamine concentration and method of cold storage on the incidence of superficial scald in apples. *Deciduous Fruit Grower* 42, 25–28.

Van Leeuwen, G. and Van de Waart, A. 1991. Delaying red currants is worthwhile. *Fruitteelt Den Haag* 81, 14–15.

Van Schaik, A.C.R. 1985. Storage of Gloster and Karmijn. *Fruitteelt* 75, 1066–1067.

Van Schaik, A.C.R., Van de Geijn, F.G., Verschoor, J.A. and Veltman, R.H. 2015. A new interactive storage concept: dynamic control of respiration. *Acta Horticulturae* 1071, 245–251.

Van de Velde, M.D. and Hendrickx, M.E. 2001. Influence of storage atmosphere and temperature on quality evolution of cut Belgian Endives. *Journal of Food Science* 66, 1212–1218.

Vanoli, M., Rizzolo, A., Grassi, M. and Eccher Zerbini, P. 2007. Storage disorders and quality in 'Abbé Fétel' pears treated with 1-methylcyclopropene. *Novel approaches for the control of postharvest diseases and disorders. Proceedings of the International Congress, Bologna, Italy, 3–5 May*, 269–277.

Vanoli, M., Eccher Zerbini, P., Grassi, M. and Rizzolo, A. 2010a. Ethylene production and quality in 1-methylcyclopropene treated 'Abbé Fétel' pears after storage in dynamically controlled atmosphere. *Acta Horticulturae* 876, 31–38.

Vanoli, M., Grassi, M., Eccher Zerbini, P. and A. Rizzolo. 2010b. Fluorescence, conjugated trienes, α-Farnesene and storage disorders in 'Abbé Fétel' pears cooled with different speeds and treated with 1-MCP. *Acta Horticulturae* 858,191–198.

Vanstreels, E., Lammertyn, J., Verlinden, B.E., Gillis, N., Schenk, A. and Nicolai, B.M. 2002. Red discoloration of chicory under controlled atmosphere conditions. *Postharvest Biology and Technology* 26, 313–322.

Vargas, M., Pastor, C., Chiralt, A., McClements, D.J. and González-Martínez, C. 2008. Recent advances in edible coatings for fresh and minimally processed fruits. *Critical Reviews in Food Science and Nutrition* 48, 496–511.

Veierskov, B and Kirk, H. G. 1986. Senescence in oat leaf segments under hypobaric conditions. *Physiologia Plantarum* 66, 283–287.

Veltman, R.H. 2013. A method and apparatus for controlling the atmosphere in a space filled with agricultural or horticultural products. WIP/PCT. International publication date: August 29, 2013. Patent # WO2013/125944 A1.

Veltman, R.H., Sanders, M.G., Persijn, S.T., Peppelenbos, H.W. and Oosterhaven, J. 1999. Decreased ascorbic acid levels and brown care development in pears (*Pyrus communis* L. cv. Conference). *Physiology Plantarum* 107, 39–45.

Veltman, R.H., Verschoor, J.A. and Van Dugteren, J.H.R. 2003. Dynamic control system (DCS) for apples (*Malus domestica* Borkh. cv. 'Elstar'): optimal quality through storage based on product response. (Special issue: Optimal controlled atmosphere). *Postharvest Biology and Technology* 27, 79–86.

Venturini, M.E., Jaime, P., Oria, R. and Blanco, D. 2000. Efficacy of different plastic films for packing of endives from hydroponic culture: microbiological evaluation. *Alimentaria* 37, 87–94.

Verlent, I., Loey, A.V., Smout, C., Duvetter, T., Nguyen, B.L. and Hendrickx, M.E. 2004. Changes in purified tomato pectinmethylesterase activity during thermal and high pressure treatment. *Journal of the Science of Food and Agriculture* 84, 1839–1847.

Vermathen, M., , GaëlleDiserens, M., Baumgartner, D., Good, C., Gasser, F. and Vermathen, P. 2017. Metabolic profiling of apples from different production systems before and after controlled atmosphere (CA) storage studied by 1H high resolution-magic angle spinning (HR-MAS) NMR. *Food Chemistry* 233, 391–400.

Vidigal, J.C., Sigrist, J.M.M., Figueiredo, I.B. and Medina, J.C. 1979. Cold storage and controlled atmosphere storage of tomatoes. *Boletim do Instituto de Tecnologia de Alimentos Brasil* 16, 421–442.

Vigneault, C. and Raghavan, G.S.V. 1991. High pressure water scrubber for rapid O_2 pull-down in controlled atmosphere storage. *Canadian Agricultural Engineering* 33, 287–294.

Vigneault, C., Leblanc, D. I., Goyette, B. and Jenni, S. 2012. Invited review: engineering aspects of physical treatments to increase fruit and vegetable phytochemical content. *Canadian Journal of Plant Science* 92, 372–397.

Vilasachandran, T., Sargent, S.A. and Maul, F. 1997. Controlled atmosphere storage shows potential for maintaining postharvest quality of fresh litchi. *Seventh International Controlled Atmosphere Research Conference, July 13–18 1997*, University of California, Davis, California [Abstract], 54.

Villalobos, M.C., Serradilla, M.J., Martín, A., Aranda, E., López-Corrales, M. and Córdoba M.G. 2018. Influence of modified atmosphere packaging (MAP) on aroma quality of figs (*Ficus carica* L.). *Postharvest Biology and Technology* 136, 145–151.

Villanueva, M.J., Tenorio, M.D., Sagardoy, M., Redondo, A. and Saco, M.D. 2005. Physical, chemical, histological and microbiological changes in fresh green asparagus (*Asparagus officinalis* L.) stored in modified atmosphere packaging. *Food Chemistry* 91, 609–619.

Villatoro, C., Echeverria, G., Graell, J., Lopez, M. L. and Lara, I. 2008. Long-term storage of pink lady apples modifies volatile-involved enzyme activities: consequences on production of volatile esters. *Journal of Agricultural and Food Chemistry* 56, 9166–9174.

Villatoro, C., Lara, I., Graell, J., Echeverría. G. and López, M.L. 2009. Cold storage conditions affect the persistence of diphenylamine, folpet and imazalil residues in 'Pink Lady®' apples. *Food Science and Technology* 42, 557–562.

Vina, S.Z., Mugridge, A., Garcia, M.A., Ferreyra, R.M., Martino, M.N., Chaves, A.R. and Zaritzky, N.E. 2007. Effects of polyvinylchloride films and edible starch coatings on quality aspects of refrigerated Brussels sprouts. *Food Chemistry* 103, 701–709.

Viraktamath, C.S. *et al.* 1963. Pre-packaging studies on fresh produce. III. Brinjal eggplant *Solanum melongena*. *Food Science Mysore* 12, 326–331 [*Horticultural Abstracts* 1964].

Visai, C., Vanoli, M., Zini, M. and Bundini, R. 1994. Cold storage of Passa Crassana pears in normal and controlled atmosphere. *Commissions C2, D1, D2/3 of the International Institute of Refrigeration International Symposium, June 8–10, Istanbul Turkey*, 255–262.

Voisine, R., Hombourger, C., Willemot, C., Castaigne, D. and Makhlouf, J. 1993. Effect of high CO_2 storage and gamma irradiation on membrane deterioration in cauliflower florets. *Postharvest Biology and Technology* 2, 279–289.

Von Loesecke, H.W. 1950. *Bananas*, 2nd edn. Interscience Publishers, New York.

Wade, N.L. 1974. Effects of O_2 concentration and ethephon upon the respiration and ripening of banana fruits. *Journal of Experimental Botany* 25, 955–964.

Wade, N.L. 1979. Physiology of cold storage disorders of fruit and vegetables. In: Lyons, J.M., Graham, D. and Raison, J.K. (eds) *Low Temperature Stress in Crop Plants*. Academic Press, New York.

Wade, N.L. 1981. Effects of storage atmosphere, temperature and calcium on low temperature injury of peach fruit. *Scientia Horticulturae* 15, 145–154.

Waelti, H., Zhang, Q., Cavalieri, R.P. and Patterson, M.E. 1992. Small scale CA storage for fruits and vegetables. *American Society of Agricultural Engineers Meeting Presentation*, Paper no. 926568.

Waghmare, R.B., Mahajan, P.V. and Annapure, U.S. 2013. Modelling the effect of time and temperature on respiration rate of selected fresh-cut produce. *Postharvest Biology and Technology* 80, 25–30.

Walsh, J.R., Lougheed, E.C., Valk, M. and Knibbe, E.A. 1985. A disorder of stored celery. *Canadian Journal of Plant Science* 65, 465–469.

Wan Zaliha, W.S. and Singh, Z. 2013. Lysophosphatidylethanolamine improves fruit colour and accumulation of anthocyanin in 'Cripps Pink' apples. *Acta Horticulturae* 1012, 227–232.

Wang, B., Wang, J., Feng, X., Lin, L., Zhao, Y. and Jiang, W. 2007a. Effects of 1-MCP and exogenous ethylene on fruit ripening and antioxidants in stored mango. *Plant Growth Regulation* 57, 185–192.

Wang, C.S., Liang, L.S. and Wang, G.X. 2007b. Effects of low pressure on quality of DongZao jujube fruit in hypobaric storage. *Food Science* 28, 335–339.

Wang, C.Y. 1979. Effect of short-term high CO_2 treatment on the market quality of stored broccoli. *Journal of Food Science* 44, 1478–1482.

Wang, C.Y. 1983. Postharvest responses of Chinese cabbage to high CO_2 treatment or low O_2 storage. *Journal of the American Society for Horticultural Science* 108, 125–129.

Wang, C.Y. 1990. Physiological and biochemical effects of controlled atmosphere on fruit and vegetables. In: Calderon, M. and Barkai-Golan, R. (eds) *Food Preservation by Modified Atmospheres*. CRC Press, Boca Raton (Florida), Ann Arbor (Michigan), Boston (Massachusetts), pp. 197–223.

Wang, C.Y. and Ji, Z.L. 1988. Abscisic acid and 1-aminocyclopropane 1-carboxylic acid content of Chinese cabbage during low O_2 storage. *Journal of the American Society for Horticultural Science* 113, 881–883.

Wang, C.Y. and Ji, Z.L. 1989. Effect of low O_2 storage on chilling injury and polyamines in zucchini squash. *Scientia Horticulturae* 39, 1–7.

Wang, C.Y. and Kramer, G.F. 1989. Effect of low O_2 storage on polyamine levels and senescence in Chinese cabbage, zucchini squash and McIntosh apples. In: Fellman, J.K. (ed.) *Proceedings of the Fifth International Controlled Atmosphere Conference, Wenatchee, Washington, USA, 14–16 June, 1989. Vol. 2: Other commodities and storage recommendations*. Washington State University, Pullman, Washington, pp. 19–27.

Wang, G.X., Han, Y.S. and Yu, L. 1994. Studies on ethylene metabolism of kiwifruit after harvest. *Acta Agriculturae Universitatis Pekinensis* 20, 408–412.

Wang, G.X., Liang, L.S. and Sun, X.Z. 2004a. The effects of postharvest low oxygen treatment on the storage quality of chestnut. *Acta Horticulturae Sinica* 31, 173–177.

Wang, J., You, Y., Chen, W., Xu, Q., Wang, J., Liu, Y., Song, L. and Wu, J. 2015. Optimal hypobaric treatment delays ripening of honey peach fruit via increasing endogenous energy status and enhancing antioxidant defence systems during storage. *Postharvest Biology and Technology* 101, 1–9.

Wang, L. and Vestrheim, S. 2002. Controlled atmosphere storage of sour cherry (*Prunus cerasus* L.). *Acta Agriculturae Scandinavica* 52, 143–146.

Wang, Q., Geil, P. and Padua, G. 2004b. Role of hydrophilic and hydrophobic interactions in structure development of zein films. *Journal of Polymers & Environment* 12, 197–202.

Wang, S., Tang, J. and Younce, F. 2003. Temperature measurement. In: Heldman, D.R. (ed.) *Encyclopedia of Agricultural, Food, and Biological Engineering*. Marcel Dekker, New York, pp. 987–993.

Wang, X., Li, C., Tang, S. and Tang, W. 2000a. Mechanisms of chestnut rotting during storage. *HortScience* 35, 407.

Wang, Y. and Long, L.E. 2014.Respiration and quality responses of sweet cherry to different atmospheres during cold storage and shipping. *Postharvest Biology and Technology* 92, 62–69.

Wang, Y., Xie, X. and Song, J. 2016. Preharvest aminoethoxyvinylglycine spray efficacy in improving storability of 'Bartlett' pears is affected by application rate, timing, and fruit harvest maturity. *Postharvest Biology and Technology* 119, 69–76.

Wang, Y.P., Dai, G.F., Wu, J.A., Li, Z.W., Qin, G.Y. *et al.* 2000b. Studies on the respiration characters of chestnut during storage. *Journal of Fruit Science* 17, 282–285.

Wang, Z. and Dilley, D.R. 2000. Hypobaric storage removes scald-related volatiles during the low temperature induction of superficial scald of apples. *Postharvest Biology and Technology* 18, 191–199.

Wang, Z., Kosittrakun, M. and Dilley, D.R. 2000c. Temperature and atmosphere regimens to control a CO_2-linked disorder of 'Empire' apples. *Postharvest Biology and Technology* 18, 183–189.

Ward, C.M. 1975. Hypobaric storage. *Annual Report of the National Vegetable Research Station, Wellesbourne, UK for 1974*, p. 85.

Wardlaw, C.W. 1938. Tropical fruits and vegetables: an account of their storage and transport. *Low Temperature Research Station, Trinidad Memoir* 7, Reprinted from *Tropical Agriculture Trinidad* 14.

Wardlaw, C.W. and Leonard, E.R. 1938. The low temperature research station. *Tropical Agriculture* 15, 179–182.

Warren, O., Sargent, S.A., Huber, D.J., Brecht, J.K., Plotto, A. and Baldwin, E. 2009. Influence of postharvest aqueous 1-methylcyclopropene (1-MCP) on the aroma volatiles and shelf life of 'Arkin' carambola. Poster Board # 109. American Society for Horticultural Science, 28 July 28 2009, St Louis, Illinois. Available at: http://ashs.confex.com/ashs/2009/webprogram/Paper2624.html (accessed July 2009).

Washington, G. 2012. New technology brings new opportunities for NZ avocados. Available at: http://www.foodproductiondaily.com/Processing/New-technology-brings-new-opportunities-for-NZ-avocados (accessed June 2012).

Waskar, D.P., Nikam, S.K. and Garande, V.K. 1999. Effect of different packaging materials on storage behaviour of sapota under room temperature and cool chamber. *Indian Journal of Agricultural Research* 33, 240–244.

Wasternack, C. and Hause, B. 2013. Jasmonates-biosynthesis and role in stress responses and developmental processes. *Annals of Botany* 111, 1021–1058.

Watkins, C.B.and Miller, W. B. (2003) Implications of 1-methylcyclopropene registration for use on horticultural products. In: Vendrell, M., Klee, H., Pech, J.C. and Romojaro, F. (eds) *Biology and Technology of the Plant Hormone Ethylene III*. IOS Press, Amsterdam, Netherlands, pp. 385–390.

Watkins, C. and Nock, J. 2004. SmartFresh™ (1-MCP) – the good and bad as we head into the 2004 season! *New York Fruit Quarterly* 12, 2–7.

Watkins, C.B. and Nock, J.F. 2012a. Controlled-Atmosphere storage of 'Honeycrisp' apples. *HortScience* 47, 886–2012.

Watkins, C.B. and Nock, J.F. 2012b. Rapid 1-methylcyclopropene (1-MCP) treatment and delayed controlled atmosphere storage of apples. *Postharvest Biology and Technology* 69, 24–31.

Watkins, C. and Yahia, E. 2008. Apples, controlled atmosphere storage. Available at: file:///G:/ApplesControlledAtmosphereDr.%20Chris%20Watkins,%20Cornell%20University,%20Ithaca,%20New%20York,%20and%20Dr.%20Elhadi%20Yahia,%20Universidad%20Autonoma%20de%20Queretaro.pdf (accessed July 2017).

Watkins, C.B., Burmeister, D.M., Elgar, H.J. and Fu WenLiu 1997. A comparison of two carbon dioxide-related injuries of apple fruit. *Postharvest Horticulture Series – Department of Pomology University of California* 16, 119–124.

Weber, A., Brackmann, A., Anese, R.O., Both, V. and Pavanello, E.P. 2011. Royal Gala apple quality stored under ultralow oxygen concentration and low temperature conditions. *Pesquisa Agropecuária Brasileira* 46, 1597–1602.

Weber, A., Brackmann, A., Both, V., Pavanello, E.P., de Oliveira Anese, R. and Thewes, F.R. 2015. Respiratory quotient: innovative method for monitoring 'Royal Gala' apple storage in a dynamic controlled atmosphere. *Scientia Agricola* 72, 28–33.

Weber, J. 1988. The efficiency of the defence reaction against soft rot after wound healing of potato tubers. 1. Determination of inoculum densities that cause infection and the effect of environment. *Potato Research* 31, 3–10.

Wei, W., Lv, P., Xia, Q., Tan, F., and Cheng, J. 2017. Fresh-keeping effects of three types of modified atmosphere packaging of pine-mushrooms. *Postharvest Biology and Technology* 132, 62–70.

Wei, Y. and Thompson A.K. 1993. Modified atmosphere packaging of diploid bananas Musa AA. *COST '94. Post-harvest Treatment of Fruit and Vegetables. Proceedings of a Workshop, 14–15 September 1993, Leuven, Belgium*, 235–246.

Weichmann, J. 1973. Die Wirkung unterschiedlichen CO_2-Partialdruckes auf den Gasstoffwechsel von Mohren *Daucus carota* L. *Gartenbauwissenschaft* 38, 243–252.
Weichmann, J. 1981. CA storage of horseradish, *Armoracia rusticana* Ph. Gartn. B. Mey et Scherb. *Acta Horticulturae* 116, 171–181.
Wells, J.M. 1974. Growth of *Erwinia carotovora, E. atroseptica* and *Pseudomonas fluorescence* in low-oxygen and high-carbon dioxide atmospheres. *Phytopathology* 64, 1012–1015.
Wermund, U. and Lazar, E.E. 2003. Control of grey mould caused by the postharvest pathogen *Botrytis cinerea* on English sweet cherries 'Lapin' and 'Colney' by controlled atmosphere (CA) storage. *Acta Horticulturae* 599, 745–748.
Westercamp, P. 1995. Conservation de la prune President – influence de la date de recolte sur la conservation des fruits. *Infos Paris* 113, 34–37.
Wheatley, C.C. 1989. *Conservation of Cassava in Polyethylene Bags*. CIAT Study Guide 04SC-07.06. CIAT, Cali, Colombia.
Whitaker, B.D., Solomos, T. and Harrison, D.J., 1998. Synthesis and oxidation of α-farnesene during high and low O_2 storage of apple cultivars differing in scald susceptibility. *Acta Horticulturae* 464, 165–170.
Whiting, D.C. 2003. Potential of controlled atmosphere and air cold storage for postharvest disinfestation of New Zealand kiwifruit. *Acta Horticulturae* 600, 143–148.
Wieczyńska, J., Luca, A., Kidmose, U., Cavoski, I. and Edelenbos, M. 2016. The use of antimicrobial sachets in the packaging of organic wild rocket: Impact on microorganisms and sensory quality. *Postharvest Biology and Technology* 121, 126–134.
Wikipedia. 2018. Apple cultivars. Available at: https://en.wikipedia.org/wiki/List_of_apple_cultivars
Wild, B.L., McGlasson, W.B. and Lee, T.H. 1977. Long term storage of lemon fruit. *Food Technology in Australia* 29, 351–357.
Wild, B.L., Wilson, C.L. and Winley, E.L. 1997. Apple host defence reactions as affected by cycloheximide, phosphonate, and citrus green mould, *Penicillium digitatum*. In: Johnson, G.I., E. Highley, E. and Joyce, D.E. (eds) *Disease Resistance in Fruit. Proceedings of an International Workshop, Chiang Mai, Thailand, 18–21 May 1997*, pp. 155–160.
Wilkinson, B.G. 1972. Fruit storage. *East Malling Reseach Station Annual Report for 1971*, 69–88.
Wilkinson, B.G. and Sharples, R.O. 1973. Recommended storage conditions for the storage of apples and pears. *East Malling Reseach Station Annual Report*, 212.
Willaert, G.A., Dirinck, P.J., Pooter, H.L. and Schamp, N.N. 1983. Objective measurement of aroma quality of Golden Delicious apples as a function of controlled-atmosphere storage time. *Journal of Agricultural and Food Chemistry* 31, 809–813.
Williams, M.W. and Patterson, M.E. 1964. Non-volatile organic acids and core breakdown of 'Bartlett' pears. *Journal of Agricultural and Food Chemistry* 12, 89.
Williams, O.J., Raghavan, G.S.V., Golden, K.D. and Gariepy, Y. 2003. Postharvest storage of Giant Cavendish bananas using ethylene oxide and sulphur dioxide. *Journal of the Science of Food and Agriculture* 83, 180–186.
Wills, R.B.H. 1990. Postharvest technology of banana and papaya in ASEAN: an overview. *Association of Southeast Asian Nations Food Journal* 5, 47–50.
Wills, R.B.H. and Tirmazi, S.I.H. 1979. Effects of calcium and other minerals on the ripening of tomatoes. *Australian Journal of Plant Pathology* 6, 221–227.
Wills, R.B.H. and Tirmazi, S.I.H. 1981. Retardation of ripening of mangoes by postharvest application of calcium. *Tropical Agriculture Trinidad* 58, 137–141.
Wills, R.B.H. and Tirmazi, S.I.H. 1982. Inhibition of ripening of avocados with calcium. *Scientia Horticulturae* 16, 323–330.
Wills, R.B.H., Pitakserikul, S. and Scott, K.J. 1982. Effects of pre-storage in low O_2 or high CO_2 concentrations on delaying the ripening of bananas. *Australian Journal of Agricultural Research* 33, 1029–1036.
Wills, R.B.H., McGlasson, W.B., Graham, D., Lee, T.H. and Hall, E.G. 1989. *Postharvest*. BSP Professional Books, Oxford, London, Edinburgh, Boston, Melbourne.
Wills, R.B.H., Klieber, A., David, R. and Siridhata, M. 1990. Effect of brief pre-marketing holding of bananas in nitrogen on time to ripen. *Australian Journal of Experimental Agriculture* 30, 579–581.
Wills, R.B.H., Ku, V.W., Shohet, D. and Kim, G.H. 1999. Importance of low ethylene levels to delay senescence of non-climacteric fruit and vegetables. *Australian Journal of Experimental Agriculture* 39, 221–224.
Wills, R.B.H., Ku, V.V.V. and Leshem Y.Y. 2000. Fumigation with nitric oxide to extend the postharvest life of strawberries. *Postharvest Biology and Technology* 18, 75–79.

Wilson, L.G. 1976. Handling of postharvest tropical fruit. *Horticultural Science* 11, 120–121.
Win, N.K.K., Jitareerat, P., Kanlayanarat, S. and Sangchote, S. 2007. Effect of cinnamon extract, chitosan coating, hot water treatment and their combinations on crown rot disease and quality of banana fruit. *Postharvest Biology and Technology* 45, 333–340.
Win, T.O., Srilaong, V., Heyes, J., Kyu, K.L. and Kanlayanarat, S. 2006. Effects of different concentrations of 1-MCP on the yellowing of West Indian lime (*Citrus aurantifolia*, Swingle) fruit. *Postharvest Biology and Technology* 42, 23–30.
Withnall, M. 2008. Alternative 'dynamically controlled atmosphere' technique for control of long-term storage disorders. *The Fruit Grower (UK)* 20 May 2008, 19–20.
Wold, A.B., Lea, P., Jeksrud, W.K., Hansen, M., Rosenfeld, H.J., Baugerod, H., and Haffner, K. 2006. Antioxidant activity in broccoli cultivars (*Brassica oleracea* var. *italica*) as affected by storage conditions. *Acta Horticulturae* 706, 211–217.
Wolfe, G.C., Black, J.L. and Jordan, R.A. 1993. The dynamic control of storage atmospheres. *CA '93. Proceedings of the 6th International Controlled Atmosphere Research Conference, Cornell University, Ithaca, New York*, 323–331.
Wollin, A.S., Little, C.R. and Packer, J.S. 1985. Dynamic control of storage atmospheres. *Proceedings of the 4th National Controlled Atmosphere Research Conference, Raleigh, North Carolina*, 308–315.
Woltering, E.J., van Schaik, A.C.R. and Jongen, W.M.F. 1994. Physiology and biochemistry of controlled atmosphere storage: the role of ethylene. *COST 94. The post-harvest treatment of fruit and vegetables: controlled atmosphere storage of fruit and vegetables. Proceedings of a Workshop, 22–23 April, 1993, Milan, Italy*, 35–42.
Woodruff, R.E. 1969. Modified atmosphere storage of bananas. *Proceedings of the National Controlled Atmosphere Research Conference, Michigan State University, Horticultural Report* 9, 80–94.
Woodward, J.R. and Topping, A.J. 1972. The influence of controlled atmospheres on the respiration rates and storage behaviour of strawberry fruits. *Journal of Horticultural Science* 47, 547–553.
Worakeeratikul, W., Srilaong, V., Uthairatanakij, A. and Jitareerat, P. 2007. Effect of whey protein concentrate on quality and biochemical changes in fresh-cut rose apple. *Acta Horticulturae* 746, 435–441.
Workhwa, S., Suttiwijitpukdee, N. and Teerachaichayut, S. 2017. Assessment of pericarp hardening of mangosteen by near infrared hyperspectral imaging (in press).
Workman, M.N. and Twomey, J. 1969. The influence of storage atmosphere and temperature on the physiology and performance of Russet Burbank seed potatoes. *Journal of the American Society for Horticultural Science* 94, 260–263.
Workman, M., Pratt, H.K. and Morris, L.L. 1957. Studies on the physiology of tomato fruit. I. Respiration and ripening behaviour at 20°C as related to date of harvest. *Proceeding of the American Society for Horticultural Science* 69, 352–365.
Worrel, D.B. and Carrington, C.M.S. 1994. Post-harvest storage of breadfruit. *Inter-American Institute for Co-operation on Agriculture, Tropical Fruits Newsletter* 11, 5.
Worrell, D.B., Carrington, C.M.S. and Huber, D.J. 2002. The use of low temperature and coatings to maintain storage quality of breadfruit, *Artocarpus altilis* (Parks.) Fosb. *Postharvest Biology and Technology* 25, 33–40.
Wright, A.H., DeLong, J.M., Franklin, J.L., Lada, R.R. and Prange, R.K. 2008. A new minimum fluorescence parameter, as generated using pulse frequency modulation, compared with pulse amplitude modulation: F_α versus F_o. *Photosynthesis Research* 97, 205–214.
Wright, H., DeLong, J., Gunawardena, A. and Prange, R. 2009. Improving our understanding of the relationship between chlorophyll fluorescence-based F-alpha, oxygen, temperature and anaerobic volatiles. *10th International Controlled & Modified Atmosphere Research Conference, 4–7 April 2009 Turkey* [Abstract], 4.
Wright, H., DeLong, J., Harrison, P.A., Gunawardena, A.H.L.A.N. and Prange, R. 2010. The effect of temperature and other factors on chlorophyll a fluorescence and the lower oxygen limit in apples (*Malus domestica*). *Postharvest Biology and Technology* 55, 21–28.
Wright, A.H., DeLong, J.M., Gunawardena, A.H.L.A.N. and Prange, R.K. 2011. The interrelationship between the lower oxygen limit, chlorophyll fluorescence and the xanthophyll cycle in plants. *Photosynthesis Research* 107, 223–235.
Wright, A.H., DeLong, J.M., Gunawardena, A.H.L.A.N. and Prange, R.K. 2012. Dynamic controlled atmosphere (DCA): Does fluorescence reflect physiology in storage? *Postharvest Biology and Technology* 64, 89–30.
Wszelaki, A.L. and Mitcham, E.J. 1999. Elevated oxygen atmospheres as a decay control alternative on strawberry. *HortScience* 34, 514–515.

Wszelaki, A.L. and Mitcham, E.J. 2000. Effects of super-atmospheric oxygen on strawberry fruit quality and decay. *Postharvest Biology and Technology*, 20, 125–133.

WTO. 2008. *1-Methylcyclopropene; Amendment to an Exemption from the Requirement of a Tolerance.* In G/SPS/N/USA/1683/Add.1 dated 9 April 2008. Committee on Sanitary and Phytosanitary Measures, World Trade Organization. Available at: www.fedfin.gov.ae/uaeagricent/sps/files/GEN842.doc (accessed January 2010).

Wu, M.T. and Salunkhe, D.K. 1972. Subatmospheric pressure storage of fruits and vegetables. *Utah Science* 33, 29–31.

Wu, M.T., Jadhav, S.J. and Salunkhe, D.K. 1972. Effects of sub atmospheric pressure storage on ripening of tomato fruits. *Journal of Food Science* 37, 952–956.

Xie, D.S., Xie, C.Y. and Li, C.Y. 1990. Controlled atmosphere storage of Jinhong apples and its effects on nutrient composition. *Journal of Fruit Science* 7, 211–216.

Xu, C.-J., Guo, D.P., Yuan, J., Yuan, G.-F. and Wang, Q.-M. 2006. Changes in glucoraphanin content and quinone reductase activity in broccoli (*Brassica oleracea* var. italica) florets during cooling and controlled atmosphere storage. *Postharvest Biology and Technology* 42, 176–184.

Xu, F., Wang, H., Tang, Y., Dong, S., Qiao, X., Chen, X. and Zheng, Y. 2016. Effect of 1-methylcyclopropene on senescence and sugar metabolism in harvested broccoli florets. *Postharvest Biology and Technology* 116, 45–49.

Xu, X. and Tian, S. 2008. Salicylic acid alleviated pathogen-induced oxidative stress in harvested sweet cherry fruit. *Postharvest Biology and Technology* 49, 379–385.

Xu, X., Lei, H., Ma, X., Lai, T., Song, H., Shi, X. and Li, J. 2017. Antifungal activity of 1-methylcyclopropene (1-MCP) against anthracnose (*Colletotrichum gloeosporioides*) in postharvest mango fruit and its possible mechanisms of action. *International Journal of Food Microbiology* 241, 1–6.

Xue, Y.B., Yu, L. and Chou, S.T. 1991. The effect of using a carbon molecular sieve nitrogen generator to control superficial scald in apples. *Acta Horticulturae Sinica* 18, 217–220.

Yahia, E.M. 1989. CA storage effect on the volatile flavor components of apples. In: Fellman, J.K. (ed.) *Proceedings of the Fifth International Controlled Atmosphere Conference, Wenatchee, Washington, USA, 14–16 June, 1989. Vol. 1: Pome fruit.* Washington State University, Pullman, Washington, pp. 341–352.

Yahia, E.M. 1991. Production of some odor-active volatiles by 'McIntosh' apples following low-ethylene controlled-atmosphere storage. *HortScience* 26, 1183–1185.

Yahia, E.M. 1995. The current status and the potential use of modified and controlled atmospheres in Mexico. In: Kushwaha, L., Serwatowski, R. and Brook, R. (eds) *Technologias de cosecha y postcosecha de frutas y hortalizas. Proceedings of a conference held in Guanajuato, Mexico, 20–24 February 1995. Harvest and postharvest technologies for fresh fruits and vegetables*, 523–529.

Yahia, E.M. 1998. Modified and controlled atmospheres for tropical fruits. *Horticultural Reviews* 22, 123–183.

Yahia, E.M. 2011. Avocado (*Persea americana* Mill.). In: Yahia, E.M. (ed.) *Postharvest Biology and Technology of Tropical and Subtropical Fruits*, Vol. 2. Woodhead Publishing, Oxford, Cambridge, Philadelphia, New Delhi, pp. 125–185.

Yahia, E.M. and Kushwaha, L. 1995. Insecticidal atmospheres for tropical fruits. In: Kushwaha, L., Serwatowski, R. and Brook, R. (eds) *Technologias de cosecha y postcosecha de frutas y hortalizas. Proceedings of a conference held in Guanajuato, Mexico, 20–24 February 1995. Harvest and postharvest technologies for fresh fruits and vegetables*, 282–286.

Yahia, E.M. and Tiznado Hernández, M. 1993. Tolerance and responses of harvested mango to insecticidal low O_2 atmospheres. *HortScience* 28, 1031–1033.

Yahia, E.M. and Vazquez Moreno, L. 1993. Responses of mango to insecticidal O_2 and CO_2 atmospheres. *Lebensmittel Wissenschaft and Technologie* 26, 42–48.

Yahia, E.M., Medina, F. and Rivera, M. 1989. The tolerance of mango and papaya to atmospheres containing very high levels of CO_2 and/or very low levels of O_2 as a possible insect control treatment. In: Fellman, J.K. (ed.) *Proceedings of the Fifth International Controlled Atmosphere Conference, Wenatchee, Washington, USA, 14–16 June, 1989. Vol. 2: Other commodities and storage recommendations*. Washington State University, Pullman, Washington, pp. 77–89.

Yahia, E.M., Rivera, M. and Hernandez, O. 1992. Responses of papaya to short-term insecticidal O_2 atmosphere. *Journal of the American Society for Horticultural Science* 117, 96–99.

Yahia, E.M., Guevara, J.C., Tijskens, L.M.M. and Cedeño, L. 2005. The effect of relative humidity on modified atmosphere packaging gas exchange. *Acta Horticulturae* 674, 97–104.

Yamashita, F., Tonzar, A.C., Fernandes, J.G., Moriya, S. and Benassi, M. deT. 2000. Influence of different modified atmosphere packaging on overall acceptance of fine table grapes var. Italia stored under refrigeration. *Ciencia e Tecnologia de Alimentos* 20, 110–114.

Yamashita, K., Sasaki, W., Fujisaki, K. and Haga, H. 2009. Effects of controlled atmosphere storage on storage life and shelf life of the Japanese onion cultivar Super-Kitamomiji. *10th International Controlled & Modified Atmosphere Research Conference, 4–7 April 2009 Turkey* [Poster abstract], 53.

Yang, D.S., Balandrán-Quintana, R.R., Ruiz, C.F., Toledo, R.T. and Kays, S.J. 2009. Effect of hyperbaric, controlled atmosphere and UV treatments on peach volatiles. *Postharvest Biology and Technology* 51, 334–341.

Yang, H.S., Feng, G.P. and Li, Y.F. 2004. Contents and apparent activation energies of chlorophyll and ascorbic acid of broccoli under controlled atmosphere storage. *Transactions of the Chinese Society of Agricultural Engineering* 20, 172–175.

Yang, S.F. 1985. Biosynthesis and action of ethylene. *HortScience* 20, 41–45.

Yang, Y.C., Sun, D.W., Pu, H., Wang, N.N. and Zhu, Z. 2015. Rapid detection of anthocyanin content in lychee pericarp during storage using hyperspectral imaging coupled with model fusion. *Postharvest Biology and Technology* 103, 55–65.

Yang, Y.J. 2001. Postharvest quality of satsuma mandarin fruit affected by controlled atmosphere. *Korean Journal of Horticultural Science & Technology* 19, 145–148.

Yang, Y.J. and Henze, J. 1987. Influence of CA storage on external and internal quality characteristics of broccoli *Brassica oleracea* var. *italica*. I. Changes in external and sensory quality characteristics. *Gartenbauwissenschaft* 52, 223–226.

Yang, Y.J. and Henze, J. 1988. Influence of controlled atmosphere storage on external and internal quality features of broccoli *Brassica oleracea* var. *italica*. II. Changes in chlorophyll and carotenoid contents. *Gartenbauwissenschaft* 53, 41–43.

Yang, Y.J. and Lee, K.A. 2003. Postharvest quality of satsuma mandarin fruit affected by controlled atmosphere. *Acta Horticulturae* 600, 775–779.

Yang, Y.J., Hwang, Y.S. and Park, Y.M. 2007. Modified atmosphere packaging extends freshness of grapes 'Campbell Early' and 'Kyoho'. *Korean Journal of Horticultural Science & Technology* 25, 138–144.

Yao, H. and Tian, S. 2005. Effects of pre- and post-harvest application of salicylic acid or methyl jasmonate on inducing disease resistance of sweet cherry fruit in storage. *Postharvest Biology and Technology* 35, 253–262.

Yearsley, C.W., Banks, N.H., Ganesh, S. and Cleland, D.J. 1996. Determination of lower oxygen limits for apple fruit. *Postharvest Biology and Technology* 8, 95–109.

Yearsley, C.W., Lallu, N., Burmeister, D., Burdon, J. and Billing, D. 2003 Can dynamic controlled atmosphere storage be used for 'Hass' avocados? *New Zealand Avocado Growers Association Annual Reseach Report* 3, 86–92.

Yen, G.-C. and Lin, H.-T. 1996. Comparison of high pressure treatment and thermal pasteurisation effects on the quality and shelf life of guava purée. *International Journal of Food Science and Technology* 31, 205–213.

Yildirim, A.N. and Koyuncu, F. 2010. The effect of gibberellic acid applications on the cracking rate and fruit quality in the '0900 Ziraat' sweet cherry cultivar. *African Journal of Biotechnology* 9, 6307–6311.

Yildirim, I.K. and Pekmezci, M. 2009. Effect of controlled atmosphere (CA) storage on postharvest physiology of 'Hayward' kiwifruit. *10th International Controlled & Modified Atmosphere Research Conference, 4–7 April 2009 Turkey* [Poster abstract], 79.

Yonemoto, Y., Inoue, H. and Okuda, H. 2004. Effects of storage temperatures and oxygen supplementation on reducing titratable acid in 'Ruby Star' passionfruit (*Passiflora edulis* x *P. edulis* f. *flavicarpa*). *Japanese Journal of Tropical Agriculture* 48, 111–114.

Yordanov, D.G. and Angelova, G.V. 2010. High pressure processing for foods preserving. *Biotechnology & Biotechnological Equipment* 3, 1940–1945.

Yoshino, T., Isobe, S. and Maekawa, T. 2002. Influence of preparation conditions on the physical properties of Zein films. *Journal of the American Oil Chemists' Society* 79, 345–349.

Young, N., deBuckle, T.S., Castel Blanco, H., Rocha, D. and Velez, G. 1971. *Conservation of yuca fresca*. Report. Instituto Investigacion Tecnologia, Bogatá, Colombia.

Youssef, K., Ligorio, A., Nigro, F. and Ippolito, A. 2012. Activity of salts incorporated in wax in controlling postharvest diseases of citrus fruit. *Postharvest Biology and Technology* 65, 39–43

Yu, J. and Wang, Y. 2017. The combination of ethoxyquin, 1-methylcyclopropene and ethylene treatments controls superficial scald of 'd'Anjou' pears with recovery of ripening capacity after long-term controlled atmosphere storage. *Postharvest Biology and Technology* 127, 53–59.

Yu, L., Shao, X., Wei, Y., Xu, F. and Wang, H. 2017. Sucrose degradation is regulated by 1-methycyclopropene treatment and is related to chilling tolerance in two peach cultivars. *Postharvest Biology and Technology* 124, 25–34.

Yuen, C.M.C. 1993. Calcium and postharvest storage potential. Postharvest Handling of Tropical Fruit. *Australian Centre for International Agricultural Research Proceedings* 50, 218–227.

Zacarias, L. and Alférez, F. 2007. Regulation by carbon dioxide of wound-induced ethylene biosynthesis in the peel of citrus fruit. *Food Science and Technology International* 13, 497–504.

Zagory, D. 1990. Application of computers in the design of modified atmosphere packaging to fresh produce. *International Conference on Modified atmosphere Packaging Part 1.* Campden Food and Drinks Research Association, Chipping Campden, UK.

Zagory, D. 1995. Ethylene-removing packaging. In: Rooney, M.L. (ed.) *Active Food Packaging*. Blackie, Glasgow, and Springer, Dordrecht, pp. 38–54.

Zagory, D., Ke, D. and Kader, A.A. 1989. Long term storage of 'Early Gold' and 'Shinko' Asian pears in low oxygen atmospheres. In: Fellman, J.K. (ed.) *Proceedings of the Fifth International Controlled Atmosphere Conference, Wenatchee, Washington, USA, 14–16 June, 1989. Vol. 1: Pome fruit.* Washington State University, Pullman, Washington, pp. 353–357.

Zanella, A. and Rossi, O. 2015. Post-harvest retention of apple fruit firmness by 1-methylcyclopropene (1-MCP) treatment or dynamic CA storage with chlorophyll fluorescence (DCA-CF). *European Journal of Horticultural Science* 80, 11–17.

Zanella, A. and Stürz, S, I. 2013. Replacing DPA post-harvest treatment by strategical application of novel storage technologies controls scald in 1/10th of EU's apples producing area. *Acta Horticulturae* 1012, 419–426.

Zanella, A. and Stürz, S, I. 2015. Optimizing postharvest life of horticultural products by means of dynamic CA: Fruit physiology controls atmosphere composition during storage. *Acta Horticulturae* 1071, 59–68.

Zanella, A., Cazzanelli, P., Panarese, A., Coser, M., Cecchinel, M. and Rossi, O. 2005. Fruit fluorescence response to low oxygen stress: modern storage technologies compared to 1-MCP treatment of apple. *Acta Horticulturae* 682, 1535–1542.

Zanella, A., Cazzanelli, P. and Rossi, O. 2008. Dynamic controlled atmosphere (DCA) storage by the means of chlorophyll fluorescence response for firmness retention in apple. *Acta Horticulturae* 796, 77–82.

Zanon, K. and Schragl, J. 1988. Lagerungsversuche mit Weisskraut. *Gemuse* 24, 14–17.

Zapata, P.J., Martínez-Esplá, A., Guillén, F., Díaz-Mula, H.M., Martínez-Romero, D., Serrano, M. and Valero, D. 2014. Preharvest application of methyl jasmonate (MeJA) in two plum cultivars. 2. Improvement of fruit quality and antioxidant systems during postharvest storage. *Postharvest Biology and Technology* 98, 115–122.

Zapotoczny, P. and Markowski, M. 2014. Influence of hypobaric storage on the quality of greenhouse cucumbers. *Bulgarian Journal of Agricultural Science* 20, 1406–1412.

Zerbini, P.E. and Grassi, M. 2010. Chlorophyll fluorescence and gas exchanges in 'Abbé Fétel' and 'Conference' pears stored in atmosphere dynamically controlled with the aid of fluorescence sensors. *Acta Horticulturae* 857, 469–474.

Zerpa-Catanho, D., Esquivel, P., Mora-Newcomer, E., Sáenz, M.V., Herrera, R. and Jiménez, V.M. 2017. Transcription analysis of softening-related genes during postharvest of papaya fruit (*Carica papaya* L. 'Pococí' hybrid). *Postharvest Biology and Technology* 125, 42–51.

Zhang, B.-Y., Samapundo, S., Pothakos, V., Baenst, I. de, Sürengil, G., Noseda, B. and Devlieghere, F. 2013. Effect of atmospheres combining high oxygen and carbon dioxide levels on microbial spoilage and sensory quality of fresh-cut pineapple. *Postharvest Biology and Technology* 86, 73–84.

Zhang, D. and Quantick, P.C. 1997. Preliminary studies on effects of modified atmosphere packaging on postharvest storage of longan fruit. *Seventh International Controlled Atmosphere Research Conference, July 13–18 1997, University of California, Davis, California* [Abstract], 55.

Zhang, Y.L. 2002. Combined technology of kiwifruit storage and freshness-keeping with freshness-keeping reagent at low temperature and modified atmosphere. *Transactions of the Chinese Society of Agricultural Engineering* 18, 138–141.

Zhang, Y.L. and Zhang, R.G. 2005. Integrated preservation technologies of low temperature storage, controlled atmosphere packaging with silicon window bags and thiabendazole fumigation for garlic shoot storage. *Transactions of the Chinese Society of Agricultural Engineering* 21, 167–171.

Zhang, Z.H., Xu, J.H., Li, W.X. and Lin, Q.H. 2006. Proper storage temperature and MAP fresh-keeping technique of wampee. *Journal of Fujian Agriculture and Forestry University (Natural Science Edition)* 35, 593–597.

Zhao, H. and Murata, T. 1988. A study on the storage of muskmelon 'Earl's Favourite'. *Bulletin of the Faculty of Agriculture, Shizuoka University* 38, 713 [Abstract].

Zhao, Z., Jiang, W., Cao, J., Zhao, Y. and Gu, Y. 2006. Effect of cold-shock treatment on chilling injury in mango (*Mangifera indica* L. cv. Wacheng) fruit. *Journal of the Science of Food and Agriculture* 86, 2458–2462.

Zheng, Y.H. and Xi, Y.F. 1994. Preliminary study on colour fixation and controlled atmosphere storage of fresh mushrooms. *Journal of Zhejiang Agricultural University* 20, 165–168.

Zheng, Y. , Wang, C.Y., Wang, S.Y. and Zheng, W. 2003. Anthocyanins and antioxidant capacity. *Journal of Agricultural and Food Chemistry* 51, 7162–7169.

Zheng, Y., Fung, R.W.M., Wang, S.Y. and Wang, C.Y. 2008. Transcript levels of antioxidative genes and oxygen radical scavenging enzyme activities in chilled zucchini squash in response to superatmospheric oxygen. *Postharvest Biology and Technology* 47, 151–158.

Zhong, Q.P., Xia, W.S. and Jiang, Y.M. 2006. Effects of 1-methylcyclopropene treatments on ripening and quality of harvested sapodilla fruit. *Food Technology and Biotechnology* 44, 535–539.

Zhou, H.-W., Lurie, S., Lers, A., Khatchitski, A., Sonego, L. and Ben Arie, R. 2000. Delayed storage and controlled atmosphere storage of nectarines: two strategies to prevent woolliness. *Postharvest Biology and Technology* 18, 133–141

Zhou, L.L., Yu, L., Zhao, Y.M., Zhang, X. and Chen, Z.P. 1992a. The application of carbon molecular sieve generators in the storage of garlic sprouts. *Acta Agriculturae Universitatis Pekinensis* 18, 47–51.

Zhou, L.L., Yu, L. and Zhou, S.T. 1992b. The effect of garlic sprouts storage at different O_2 and CO_2 levels. *Acta Horticulturae Sinica* 19, 57–60.

Zhou, S.-T. and Yu, L. 1989. The application of carbon molecular sieve generator in CA storage of apple and tomato. In: Fellman, J.K. (ed.) *Proceedings of the Fifth International Controlled Atmosphere Conference, Wenatchee, Washington, USA, 14–16 June, 1989. Vol. 2: Other commodities and storage recommendations*. Washington State University, Pullman, Washington, pp. 241–248.

Zhu, X., Wang, Q., Cao, J. and Jiang, W. 2008. *Journal of Food Processing and Preservation* 32, 770–784.

Zlatić, E., Zadnik, V., Fellman, J., Demšar, L., Hribar, J., Čejić, Z. and Vidrih, R. 2016. Comparative analysis of aroma compounds in 'Bartlett' pear in relation to harvest date, storage conditions, and shelf-life. *Postharvest Biology and Technology* 117, 71–80.

Zoffoli, J.P. and Latorre, B.A. 2011. Table grape (*Vitus vinifera* L.). In: Yahia E.M. (ed.) *Postharvest Biology and Technology of Tropical and Subtropical Fruits*. Woodhead Publishing, Oxford, Cambridge, Philadelphia, New Delhi, pp. 179–212.

Zoffoli, J.P., Rodriguez, J. Aldunce, P. and Crisosto, C. 1997. Development of high concentration carbon dioxide modified atmosphere packaging systems to maintain peach quality. *Seventh International Controlled Atmosphere Research Conference, July 13–18 1997, University of California, Davis, California* [Abstract], 45.

Zoffoli, J.P., Latorre, B.A and Naranjo, P. 2009. Preharvest applications of growth regulators and their effect on postharvest quality of table grapes during cold storage. *Postharvest Biology and Technology* 51, 183–192.

Zong, R.J., Morris, L. and Cantwell, M. 1995. Postharvest physiology and quality of bitter melon (*Momordica charantia* L.). *Postharvest Biology and Technology* 6, 65–72.

Zudaire, L., Viñas, I., Abadias, M., Simó, J., Echeverria, G., Plaza, L. and Aguiló-Aguayo, I. 2017. Quality and bioaccessibility of total phenols and antioxidant activity of calçots (*Allium cepa* L.) stored under controlled atmosphere conditions. *Postharvest Biology and Technology* 129, 118–128.

Zutahy, Y., Lichter, A., Kaplunov, T. and Lurie, S. 2008. Extended storage of 'Red Globe' grapes in modified SO_2 generating pads. *Postharvest Biology and Technology* 50, 12–17.

Index

Note: bold page numbers indicate figures; italic page numbers indicate tables.